AF411401

Phytochemical Omics in Medicinal Plants

Phytochemical Omics in Medicinal Plants

Editor

Jen-Tsung Chen

MDPI • Basel • Beijing • Wuhan • Barcelona • Belgrade • Manchester • Tokyo • Cluj • Tianjin

Editor
Jen-Tsung Chen
Department of Life Sciences
National University of
Kaohsiung
Kaohsiung
Taiwan

Editorial Office
MDPI
St. Alban-Anlage 66
4052 Basel, Switzerland

This is a reprint of articles from the Special Issue published online in the open access journal *Biomolecules* (ISSN 2218-273X) (available at: www.mdpi.com/journal/biomolecules/special_issues/ omics_medicinal_plants).

For citation purposes, cite each article independently as indicated on the article page online and as indicated below:

LastName, A.A.; LastName, B.B.; LastName, C.C. Article Title. *Journal Name* **Year**, *Volume Number*, Page Range.

ISBN 978-3-0365-1294-5 (Hbk)
ISBN 978-3-0365-1293-8 (PDF)

Contents

About the Editor

Jen-Tsung Chen

Jen-Tsung Chen is currently a professor at the National University of Kaohsiung in Taiwan. He teaches cell biology, genomics, proteomics, medicinal plant biotechnology, and plant tissue culture in college. Dr. Chen's research interests include bioactive compounds, chromatography techniques, plant tissue culture, phytochemicals, and plant biotechnology. He serves as an editorial board member in *Plant Methods, Biomolecules, International Journal of Molecular Science*, and a guest editor in *Frontiers in Plant Science, Frontiers in Pharmacology, Journal of Fungi*.

Preface to "Phytochemical Omics in Medicinal Plants"

Since prehistoric times, medicinal plants have been used in traditional medical practices for curing diseases and benefiting human health. A huge array of phytochemicals have been identified in medicinal plants, including carotenoids, phenolic acids, flavonoids, and lignans, which have a wide range of biological activities, including antioxidant, antibacterial effect, anticancer, anti-inflammatory, neuroprotection, and so on. In general, naturally occurring compounds possess complicate bioactive effects and their efficacy is difficult clarify. Therefore, there is a demand for comprehensive approaches to the synthesis and biological function of phytochemicals. In the past decade, multi-omics has been an emerging field as a powerful tool to gain deep insights into complex biological systems. This book originated from the Special Issue entitled Phytochemical Omics in Medicinal Plants and it aims to integrate recent innovative approaches using high-throughput technologies (omics), such as genomics, transcriptomics, proteomics, and metabolomics, as well as bioinformatics and other related topics, including biotechnology attempts, to make significant progress in understanding the relationship between phytochemicals and human health. As the editor of this book, I greatly appreciate the invaluable contribution of all authors and reviewers as well as the efforts of Editors in *Biomolecules*: Natural and Bio-inspired Molecules.

Jen-Tsung Chen
Editor

Editorial

Phytochemical Omics in Medicinal Plants

Jen-Tsung Chen

Department of Life Sciences, National University of Kaohsiung, Kaohsiung 811, Taiwan; jentsung@nuk.edu.tw

Received: 16 June 2020; Accepted: 18 June 2020; Published: 21 June 2020

Medicinal plants are used to treat diseases and provide health benefits, and their applications are increasing around the world. A huge array of phytochemicals have been identified from medicinal plants, belonging to carotenoids, flavonoids, lignans, and phenolic acids, and so on, having a wide range of biological activities. In order to explore our knowledge of phytochemicals with the assistance of modern molecular tools and high-throughput technologies, this special issue collects recent innovative original research and reviews. As summarized below, 27 papers are published in this special issue, which can be divided into four subtopics, as described below.

1. Mechanistic Insights into Bioactivities

1.1. Anti-Inflammation

Inflammation undergoes physiological and pathological responses caused by diverse stimuli such as microbial infections, injury, and traumata. Zwirchmayr et al. identified anti-inflammatory compounds that exhibit nuclear factor kappa-light-chain-enhancer of activated B cells (NF-κB) inhibiting ability from the extract of masterwort (*Peucedanum ostruthium* (L.) Koch) through a holistic omics-based tool, namely Eliciting Nature's Activities (ELINA) [1]. They successfully isolated four furanocoumarins, one coumarin, and one chromone. According to the results presented, they suggest that ELINA is practicable and may effectively accelerate the process of natural product-based drug discovery in the future.

Inflammation is part of immunity, Yeh and Lin investigated the immunomodulatory potential of steam-distilled essential oils (SDEOs) from *Acorus gramineusand* and *Euodia ruticarpa* cultivated in Taiwan [2]. They found SDEOs to be rich in flavonoids, polyphenols, and saponins, and additionally, SDEOs, particularly from *Euodia ruticarpa*, possess immunomodulatory ability by shifting the balance of type 1 and type 2 T helper cells in primary splenocytes and inhibiting inflammation in peritoneal macrophages in vitro.

1.2. Anti-Oxidative Stress

Oxidative stress is one of the main processes in human disorders. Wang et al. contributed a review on bioactive compounds for treating oxidative stress-related human disorders involving the toxic reactive aldehyde 4-hydroxynonenal (4-HNE), an advanced lipid peroxidation end product [3]. A number of related disorders are listed, such as cardiovascular injury, eye damage, liver injury, neurotoxicity, neurological disorder and energy metabolism disorder in order to discuss their prevention and treatment by bioactive compounds from particular medicinal plants, and finally, possible strategies for future research and applications combating deleterious effects induced by 4-HNE are proposed.

1.3. Bioinformatics

In research related to traditional herbal medicine (THM), network pharmacology (NP) is a useful tool for exploring the potential mechanisms of therapeutic effects that act on multiple targets. Lee et al. constructed the networks of co-author and affiliation to analyze the trend of methodologies utilized by researchers [4]. Firstly, they found a dramatic increase in THM-NP studies in the last ten years.

Additionally, the Traditional Chinese Medicine Systems Pharmacology Database and Analysis Platform was the most frequently utilized, and the methodology for constructing a compound-target network had achieved the greatest progress.

Jeyasri et al. used a system pharmacology approach to develop a strategy for multi-target treatment based on traditional Ayurvedic medicine, *Bacopa monnieri*, for neurological diseases, focusing on Alzheimer's disease (AD) and Spinocerebellar ataxia (SCA) [5]. When compared with commercially available drugs, it was revealed that the constituents of *Bacopa monnieri* have asiatic acid and loliolide for treating AD and SCA. In addition, various potential actions from the bioactive compounds were predicted, particularly benefiting cognitive function.

The extracts of some *Garcinia* species have an array of bioactivities in the treatment of adipogenesis, obesity, cardiovascular diseases, diabetes, inflammation, and cancer. Chen et al. reviewed the molecular docking approach for searching candidate agents for the treatment of diabetes using a number of critical hypoglycemic targets [6]. Altogether, some benzopyrans and triterpenes that existed in *G. linii* were proposed to be the chief components for regulating blood glucose.

2. Treatment of Diseases

2.1. Cancers

Due to the high morbidity and mortality of oral cancer, there is a demand to develop drugs for inhibiting local invasion and metastasis of cancer cells. It has been reported that Withaferin A (WFA), a steroidal lactone isolated from *Withania somnifera*, could inhibit the migration of cancer cells at high cytotoxic concentrations. Interestingly, Yu et al. further prove that WFA in a relatively low concentration of 0.5 μM inhibited oxidative stress-mediated migration as well as invasion in oral cancer Ca9-22 cells, and they suggest that it has the potential to inhibit metastasis in oral cancer therapy [7]. Velmurugan et al. use a cell-proliferation inhibitory flavonoid, luteolin-7-O-glucoside, to further test its effect on metastasis of oral cancer [8]. On the basis of their results, they propose that luteolin-7-O-glucoside inhibits cell migration and invasion by regulating the expression of matrix metalloproteinase-2 and extracellular signal-regulated kinase pathway.

Non-small-cell lung cancer (NSCLC) accounts for approximately 85% of all lung cancer and is the major cause of death by cancer in the world. Oh et al. investigated the anti-cancer potential of licochalcone D (LCD), a flavonoid isolated from *Glycyrrhiza inflata*, using an epidermal growth factor receptor (EGFR) mutant NSCLC cell line [9]. They found that LCD inhibits the activity of EGFR and hepatocyte growth factor receptor, induces reactive oxygen species-dependent apoptotic cell death, and inhibits the proliferation of cancer cells.

Nasopharyngeal carcinoma (NPC) has a unique geographical and ethnic distribution. Its high prevalence in Asia accounts for over 80% of new cases globally. Liu et al. studied the anticancer activity of asiatic acid (AA), extracted from *Centella asiatica*, using two cisplatin-resistant NPC cell lines [10]. They demonstrated that AA significantly reduced the cell viability in both cell lines through p38-α mitogen-activated protein kinase (MAPK)-mediated phosphorylation and activation of bcl-2-like protein 4 (Bax) in vitro.

Inflammatory bowel disease (IBD) is chronic intestinal and colorectal inflammation, which is highly capable of developing into colorectal cancer (CRC). Zheng et al. evaluated bioactivities of a clerodane diterpene, 16-hydroxycleroda-3,13-dien-15,16-olide (HCD) using an IBD mouse model and two human CRC cell lines [11]. It was demonstrated that the inflammatory symptoms of IBD mice could be ameliorated by HCD treatments, and additionally, in vitro tests confirmed that HCD-induced apoptosis may involve both extrinsic and intrinsic pathways.

Khan et al. reviewed the recent advances in the use of active phytochemicals (APs) against cancers by searching relevant keywords in reliable academic databases, selecting twenty medicinal plants to discuss in detail [12]. It was revealed that APs were effective for treating a number of cancer cell lines, and altogether, the inhibitory effects were mainly exerted by damaging DNA and activating enzymes that cause apoptosis. In addition, the anticancer activity of some APs was confirmed using in vivo animal models.

There are more than 1300 *ent*-kaurane diterpenoids that have been identified, most of which belong to the genus *Isodon*, and many possess anticancer potential. Sarwar et al. provide a review of the plant sources, biological targets and mechanistic pathways of *ent*-kaurane diterpenoids [13]. They indicate that the anticancer activities of such compounds are chiefly contributed by the regulation of apoptosis, autophagy, cell cycle arrest, and metastasis. The key regulators in each process were discussed in detail; for example, the most common metastatic target proteins are matrix metalloproteinases (MMP-2 and MMP-9), vascular endothelial growth factor (VEGF), and VEGF receptor.

2.2. Neurological Diseases

Quercetin, a flavonoid with an array of pharmacological effects, is widely available in fruits and vegetables. It has been demonstrated that quercetin possesses a neuron-protective effect by alleviating oxidative stress and inflammation. Khan et al. further refine the effect and underlying mechanisms of quercetin on AD, with an emphasis on cognitive performance [14]. Finally, quercetin is proposed to offer potential as a lead compound for clinical application in such disease.

Another flavonoid, naringenin, also has neuroprotective ability. Nouri et al. refined the current knowledge of pharmacological targets, signaling pathways, molecular mechanisms, and the clinical perspective of such compounds [15]. Additionally, systems for delivering naringenin were discussed.

2.3. Liver Diseases

Liver failure frequently leads to hepatic encephalopathy (HE), which has a spectrum of neuropsychiatric abnormalities. Baek et al. studied the effect of *Rheum undulatum* and *Glycyrrhiza uralensis* extract mixture (RG) on HE based on their anti-inflammatory and antioxidant properties [16]. Firstly, they identified seven bioactive ingredients in RG that had been predicted to be effective in treating neurological diseases. Then, they used a CCl_4-induced HE mouse model to unravel the therapeutic mechanisms of RG. Based on their results, RG consistently relieved HE symptoms by preserving blood–brain barrier permeability, and thus offers considerable potential for treating chronic liver disease as well as HE.

2.4. Cardiovascular Diseases

Cardiovascular diseases (CVDs) are the leading cause of death in the world, and are mainly related to atherosclerosis. Toma et al. refined the recent knowledge regarding the effects and underlying molecular mechanisms of phenolic compounds on atherosclerosis, especially involving dyslipidemia, oxidative stress, and inflammation [17]. In this review, the authors suggest future research directions for alternative therapy of CVDs using phenolic compounds, and encourage the adequate consumption of foods containing natural phenolic compounds in order to prevent such diseases.

2.5. Fracture Healing

Green tea is a popular drink that benefits human health, including having positive effects that increase bone mineral density and bone volume, while also diminishing osteoporotic fractures. Based on previous reports, a bioactive compound of tea catechin, (-)-epigallocatechin-3-gallate (EGCG) could promote bone defect healing in the distal femur, and this may be partly through the effect of bone morphogenetic protein-2 (BMP-2). Lin et al. further studied the effect of EGCG on tibial fracture healing in rats, and they concluded that the local treatment of EGCG accelerated fracture healing at least partly via BMP-2 [18].

3. Profiling, Extraction and Identification

3.1. Profiling

Ultraviolet-B (UV-B) radiation could induce stress in plants, and consequently, promotes the production of secondary metabolites through elevated defense responses. It has been reported that of the two *Astragalus* varieties, *A. mongholicus* has the higher tolerance to UV-B. Liu et al. further profiled the metabolites, particularly the phenolic compounds of *A. mongholicus*, when subjected to UV-B radiation [19]. Overall, they found that UV-B radiation could induce a tissue-specific strong shift from carbon assimilation to carbon accumulation of phenolic metabolism in roots, and it was activated by the upregulation of some relevant genes.

Chinese sage (*Salvia miltiorrhiza*) has been used as medicine for thousands of years for the treatment of a number of disorders, particularly by benefiting blood circulation systems. In a previous report, the polysaccharide fraction (PSF) of a symbiont, *Trichoderma atroviride*, was found to have a positive effect on the accumulation of tanshinones in the hairy root culture. Peng et al. further profiled the proteomic responses of the symbiont relationship, and they proposed that the PSF-induced accumulation of tanshinones may contribute through defense-related signaling involving jasmonic acid, leading to leucine-rich repeat protein synthesis [20].

3.2. Extraction and Identification

Flos Chrysanthemi indici is a traditional herb that has a number of bioactivities, including the antibacterial, anti-inflammatory, and antioxidant properties. Jing et al. studied the effect of extraction method on the yield and bioactivities of essential oils [21]. They suggest that solvent-free microwave extraction and supercritical fluid extraction are both good methods with higher yield that could simultaneously retain the antimicrobial activity of essential oils.

Deep eutectic solvents (DESs) are homogeneous solutions of quaternary ammonium salts, which have advantages including easy preparation, ecological friendliness, and so on. Zeng et al. tested the performance of six DESs in combination with five macroporous resins on isolation and purification of flavonoids and 20-hydroxyecdysone from *Chenopodium quinoa* Willd by preparative high-performance liquid chromatography [22]. They found the best combination of DES (choline chloride/urea) and macroporous resin (D101). Subsequently, three flavonoids and 20-hydroxyecdysone were successfully purified, and their chemical structures were identified.

Luteolin is a flavonoid and possesses an array of biological effects on hypertension, inflammatory diseases, and cancer. Juszczak et al. reviewed the current chromatographic techniques for analyzing luteolin and some derivatives [23]. They discussed the separation conditions for determining such compounds, as well as the pros and cons of each method.

To solve the excess structural information in the isolation of metabolites from natural products by high-resolution mass spectrometry, Lee et al. applied the Global Natural Product Social Molecular Networking platform together with hierarchical clustering analysis on the selectively isolate of lignans from *Trachelospermum asiaticum* [24]. Using this combined tool, eventually, they efficiently identified five dibenzylbutylrolactone-type lignans that possess cytotoxic activities against cancer cells.

Curcuma species (Zingiberaceae) have long been used as traditional medicine, food flavor and cosmetics. Pintatum et al. identified bioactive compounds from the *Curcuma aromatica* rhizome, including curcumin, curcumenol, curdione, germacrone, and zederone, and found that curcumin has the highest anti-inflammatory ability [25]. Additionally, the crude extract of *C. aromatica* demonstrated the highest inhibitory response to NF-κB activity when compared to the other three *Curcuma* species.

4. Biotechnology

4.1. Gene Transfer

Black nightshade (*Solanum nigrum*) is regarded as an herb that possesses bioactive compounds, particularly those with anti-tumor activities. Chhon et al. used transgenic black nightshade that overexpressed *AtPAP1* (*Arabidopsis thaliana* production of anthocyanin pigment 1), to investigate the effect of transgene on anthocyanin biosynthesis [26]. They found that transgenic plants produced a higher content of anthocyanin when compared to the control, and thus suggested that it could be a suitable platform for further studying AtPAP1-induced anthocyanin accumulation in response to environmental stress.

4.2. Nanoparticles

Green synthesis of nanoparticles (NPs) is an emerging field that is regarded as "green chemistry" due to the use of biological reducing agents including plant sources. Abbasi et al. made zinc oxide NPs that conjugated with the leaf extract from *Geranium wallichianum* (GW), namely GW-ZnONPs [27]. They tested a number of bioactivities, as well as the biocompatibility of GW-ZnONPs, and finally confirmed that it has considerable potential in biological applications.

5. Conclusions and Perspectives

Phytochemical omics is clearly an interesting topic and has attracted intensive studies focusing on anticancer, anti-inflammation, immunoregulation, anti-oxidative stress, anti-diabetes, fracture healing, and treatments for neurological, liver, and cardiovascular diseases. Additionally, recent advances in profiling, extraction, identification, and biotechnology are included. The findings offered by the contributors to this special issue may greatly aid the progress of phytochemical research. In the future, I hope that with the rapid development of molecular tools and approaches, particularly in integrative multi-omics and genome editing technologies, scientists will be able to further unravel the efficacy of phytochemicals for benefiting human health.

Funding: This research received no external funding.

Conflicts of Interest: The author declares no conflict of interest.

References

1. Zwirchmayr, J.; Grienke, U.; Hummelbrunner, S.; Seigner, J.; de Martin, R.; Dirsch, V.M.; Rollinger, J.M. A Biochemometric Approach for the Identification of In Vitro Anti-Inflammatory Constituents in Masterwort. *Biomolecules* **2020**, *10*, 679. [CrossRef]
2. Yeh, T.-H.; Lin, J.-Y. *Acorus gramineus* and *Euodia ruticarpa* Steam Distilled Essential Oils Exert Anti-Inflammatory Effects Through Decreasing Th1/Th2 and Pro-/Anti-Inflammatory Cytokine Secretion Ratios In Vitro. *Biomolecules* **2020**, *10*, 338. [CrossRef]
3. Wang, F.-X.; Li, H.-Y.; Li, Y.-Q.; Kong, L.-D. Can Medicinal Plants and Bioactive Compounds Combat Lipid Peroxidation Product 4-HNE-Induced Deleterious Effects? *Biomolecules* **2020**, *10*, 146. [CrossRef] [PubMed]
4. Lee, W.-Y.; Lee, C.-Y.; Kim, Y.-S.; Kim, C.-E. The Methodological Trends of Traditional Herbal Medicine Employing Network Pharmacology. *Biomolecules* **2019**, *9*, 362. [CrossRef] [PubMed]
5. Jeyasri, R.; Muthuramalingam, P.; Suba, V.; Ramesh, M.; Chen, J.-T. *Bacopa monnieri* and Their Bioactive Compounds Inferred Multi-Target Treatment Strategy for Neurological Diseases: A Cheminformatics and System Pharmacology Approach. *Biomolecules* **2020**, *10*, 536. [CrossRef]
6. Chen, T.-H.; Tsai, M.-J.; Fu, Y.-S.; Weng, C.-F. The Exploration of Natural Compounds for Anti-Diabetes from Distinctive Species *Garcinia linii* with Comprehensive Review of the *Garcinia* Family. *Biomolecules* **2019**, *9*, 641. [CrossRef]

7. Yu, T.-J.; Tang, J.-Y.; Ou-Yang, F.; Wang, Y.-Y.; Yuan, S.-S.F.; Tseng, K.; Lin, L.-C.; Chang, H.-W. Low Concentration of Withaferin a Inhibits Oxidative Stress-Mediated Migration and Invasion in Oral Cancer Cells. *Biomolecules* **2020**, *10*, 777. [CrossRef] [PubMed]

8. Velmurugan, B.K.; Lin, J.-T.; Mahalakshmi, B.; Chuang, Y.-C.; Lin, C.-C.; Lo, Y.-S.; Hsieh, M.-J.; Chen, M.-K. Luteolin-7-O-Glucoside Inhibits Oral Cancer Cell Migration and Invasion by Regulating Matrix Metalloproteinase-2 Expression and Extracellular Signal-Regulated Kinase Pathway. *Biomolecules* **2020**, *10*, 502. [CrossRef]

9. Oh, H.-N.; Lee, M.-H.; Kim, E.; Kwak, A.-W.; Yoon, G.; Cho, S.-S.; Liu, K.; Chae, J.-I.; Shim, J.-H. Licochalcone D Induces ROS-Dependent Apoptosis in Gefitinib-Sensitive or Resistant Lung Cancer Cells by Targeting EGFR and MET. *Biomolecules* **2020**, *10*, 297. [CrossRef]

10. Liu, Y.-T.; Chuang, Y.-C.; Lo, Y.-S.; Lin, C.-C.; Hsi, Y.-T.; Hsieh, M.-J.; Chen, M.-K. Asiatic Acid, Extracted from *Centella asiatica* and Induces Apoptosis Pathway through the Phosphorylation p38 Mitogen-Activated Protein Kinase in Cisplatin-Resistant Nasopharyngeal Carcinoma Cells. *Biomolecules* **2020**, *10*, 184. [CrossRef]

11. Zheng, J.-H.; Lin, S.-R.; Tseng, F.-J.; Tsai, M.-J.; Lue, S.-I.; Chia, Y.-C.; Woon, M.; Fu, Y.-S.; Weng, C.-F. Clerodane Diterpene Ameliorates Inflammatory Bowel Disease and Potentiates Cell Apoptosis of Colorectal Cancer. *Biomolecules* **2019**, *9*, 762. [CrossRef] [PubMed]

12. Khan, T.; Ali, M.; Khan, A.; Nisar, P.; Jan, S.A.; Afridi, S.; Shinwari, Z.K. Anticancer Plants: A Review of the Active Phytochemicals, Applications in Animal Models, and Regulatory Aspects. *Biomolecules* **2020**, *10*, 47. [CrossRef]

13. Sarwar, M.S.; Xia, Y.-X.; Liang, Z.-M.; Tsang, S.W.; Zhang, H.-J. Mechanistic Pathways and Molecular Targets of Plant-Derived Anticancer *ent*-Kaurane Diterpenes. *Biomolecules* **2020**, *10*, 144. [CrossRef]

14. Khan, H.; Ullah, H.; Aschner, M.; Cheang, W.S.; Akkol, E.K. Neuroprotective Effects of Quercetin in Alzheimer's Disease. *Biomolecules* **2020**, *10*, 59. [CrossRef] [PubMed]

15. Nouri, Z.; Fakhri, S.; El-Senduny, F.F.; Sanadgol, N.; Abd-ElGhani, G.E.; Farzaei, M.H.; Chen, J.-T. On the Neuroprotective Effects of Naringenin: Pharmacological Targets, Signaling Pathways, Molecular Mechanisms, and Clinical Perspective. *Biomolecules* **2019**, *9*, 690. [CrossRef] [PubMed]

16. Baek, S.Y.; Lee, E.H.; Oh, T.W.; Do, H.J.; Kim, K.-Y.; Park, K.-I.; Kim, Y.W. Network Pharmacology-Based Approaches of *Rheum undulatum* Linne and *Glycyrriza uralensis* Fischer Imply Their Regulation of Liver Failure with Hepatic Encephalopathy in Mice. *Biomolecules* **2020**, *10*, 437. [CrossRef]

17. Toma, L.; Sanda, G.M.; Niculescu, L.S.; Deleanu, M.; Sima, A.V.; Stancu, C.S. Phenolic Compounds Exerting Lipid-Regulatory, Anti-Inflammatory and Epigenetic Effects as Complementary Treatments in Cardiovascular Diseases. *Biomolecules* **2020**, *10*, 641. [CrossRef]

18. Lin, S.-Y.; Kan, J.Y.; Lu, C.-C.; Huang, H.H.; Cheng, T.-L.; Huang, H.-T.; Ho, C.-J.; Lee, T.-C.; Chuang, S.-C.; Lin, Y.-S.; et al. Green Tea Catechin (-)-Epigallocatechin-3-Gallate (EGCG) Facilitates Fracture Healing. *Biomolecules* **2020**, *10*, 620. [CrossRef]

19. Liu, Y.; Liu, J.; Abozeid, A.; Wu, K.-X.; Guo, X.-R.; Mu, L.-Q.; Tang, Z.-H. UV-B Radiation Largely Promoted the Transformation of Primary Metabolites to Phenols in *Astragalus mongholicus* Seedlings. *Biomolecules* **2020**, *10*, 504. [CrossRef]

20. Peng, W.; Ming, Q.-L.; Zhai, X.; Zhang, Q.; Rahman, K.; Wu, S.-J.; Qin, L.-P.; Han, T. Polysaccharide Fraction Extracted from Endophytic Fungus *Trichoderma atroviride* D16 Has an Influence on the Proteomics Profile of the *Salvia miltiorrhiza* Hairy Roots. *Biomolecules* **2019**, *9*, 415. [CrossRef]

21. Jing, C.-L.; Huang, R.-H.; Su, Y.; Li, Y.-Q.; Zhang, C.-S. Variation in Chemical Composition and Biological Activities of *Flos Chrysanthemi indici* Essential Oil under Different Extraction Methods. *Biomolecules* **2019**, *9*, 518. [CrossRef] [PubMed]

22. Zeng, J.; Shang, X.; Zhang, P.; Wang, H.; Gu, Y.; Tan, J.-N. Combined Use of Deep Eutectic Solvents, Macroporous Resins, and Preparative Liquid Chromatography for the Isolation and Purification of Flavonoids and 20-Hydroxyecdysone from *Chenopodium quinoa* Willd. *Biomolecules* **2019**, *9*, 776. [CrossRef] [PubMed]

23. Juszczak, A.M.; Zovko-Končić, M.; Tomczyk, M. Recent Trends in the Application of Chromatographic Techniques in the Analysis of Luteolin and Its Derivatives. *Biomolecules* **2019**, *9*, 731. [CrossRef]

24. Lee, J.; Yang, H.S.; Jeong, H.; Kim, J.-H.; Yang, H. Targeted Isolation of Lignans from *Trachelospermum asiaticum* Using Molecular Networking and Hierarchical Clustering Analysis. *Biomolecules* **2020**, *10*, 378. [CrossRef] [PubMed]

25. Pintatum, A.; Maneerat, W.; Logie, E.; Tuenter, E.; Sakavitsi, M.E.; Pieters, L.; Berghe, W.V.; Sripisut, T.; Deachathai, S.; Laphookhieo, S. In Vitro Anti-Inflammatory, Anti-Oxidant, and Cytotoxic Activities of Four *Curcuma* Species and the Isolation of Compounds from *Curcuma aromatica* Rhizome. *Biomolecules* **2020**, *10*, 799. [CrossRef] [PubMed]
26. Chhon, S.; Jeon, J.; Kim, J.; Park, S.U. Accumulation of Anthocyanins through Overexpression of AtPAP1 in *Solanum nigrum* Lin. (Black Nightshade). *Biomolecules* **2020**, *10*, 277. [CrossRef]
27. Abbasi, B.A.; Iqbal, J.; Ahmad, R.; Zia, L.; Kanwal, S.; Mahmood, T.; Wang, C.; Chen, J.-T. Bioactivities of *Geranium wallichianum* Leaf Extracts Conjugated with Zinc Oxide Nanoparticles. *Biomolecules* **2020**, *10*, 38. [CrossRef]

 *biomolecules*

Article

In Vitro Anti-Inflammatory, Anti-Oxidant, and Cytotoxic Activities of Four *Curcuma* Species and the Isolation of Compounds from *Curcuma aromatica* Rhizome

Aknarin Pintatum [1], Wisanu Maneerat [1,2], Emilie Logie [3], Emmy Tuenter [4], Maria E. Sakavitsi [5], Luc Pieters [4], Wim Vanden Berghe [3,*], Tawanun Sripisut [6], Suwanna Deachathai [1] and Surat Laphookhieo [1,2,*]

1   Center of Chemical Innovation for Sustainability (CIS) and School of Science, Mae Fah Luang University, Chiang Rai 57100, Thailand; p.aknarin@gmail.com (A.P.); wisanu.man@mfu.ac.th (W.M.); suwanna.dea@mfu.ac.th (S.D.)
2   Medicinal Plants Innovation Center of Mae Fah Luang University, Chiang Rai 57100, Thailand
3   Lab Protein Chemistry, Proteomics & Epigenetic Signalling (PPES), Department Biomedical Sciences, University of Antwerp, 2610 Wilrijk, Belgium; emilie.logie@uantwerpen.be
4   Natural Products & Food Research and Analysis (NatuRA), Department of Pharmaceutical Sciences, University of Antwerp, 2610 Wilrijk, Belgium; emmy.tuenter@uantwerpen.be (E.T.); luc.pieters@uantwerpen.be (L.P.)
5   Department of Pharmacognosy and Natural Products Chemistry, Faculty of Pharmacy, National and Kapodistrian University of Athens, Zografou, 15771 Athens, Greece; msakavitsi@pharm.uoa.gr
6   School of Cosmetic Science, Mae Fah Luang University, Chiang Rai 57100, Thailand; tawanun.sri@mfu.ac.th
*   Correspondence: wimk.vandenberghe@gmail.com (W.V.B.); surat.lap@mfu.ac.th (S.L.); Tel.: +32-3265-2657 (W.V.B.); +66-5391-6782 (S.L.)

Received: 22 April 2020; Accepted: 19 May 2020; Published: 21 May 2020

Abstract: The genus *Curcuma* is part of the Zingiberaceae family, and many *Curcuma* species have been used as traditional medicine and cosmetics in Thailand. To find new cosmeceutical ingredients, the in vitro anti-inflammatory, anti-oxidant, and cytotoxic activities of four *Curcuma* species as well as the isolation of compounds from the most active crude extract (*C. aromatica*) were investigated. The crude extract of *C. aromatica* showed 2,2-diphenyl-1-picrylhydrazyl (DPPH) radical scavenging activity with an IC_{50} value of 102.3 µg/mL. The cytotoxicity effect of *C. aeruginosa*, *C. comosa*, *C. aromatica*, and *C. longa* extracts assessed with the 3-[4,5-dimethylthiazol-2-yl]-2,5-diphenyl tetrazolium bromide (MTT) assay at 200 µg/mL were 12.1 ± 2.9, 14.4 ± 4.1, 28.6 ± 4.1, and 46.9 ± 8.6, respectively. *C. aeruginosa* and *C. comosa* presented apoptosis cells (57.7 ± 3.1% and 32.6 ± 2.2%, respectively) using the CytoTox-ONE™ assay. Different crude extracts or phytochemicals purified from *C. aromatica* were evaluated for their anti-inflammatory properties. The crude extract of *C. aromatica* showed the highest potential to inhibit NF-κB activity, followed by *C. aeruginosa*, *C. comosa*, and *C. longa*, respectively. Among the various purified phytochemicals curcumin, germacrone, curdione, zederone, and curcumenol significantly inhibited NF-κB activation in tumor necrosis factor (TNF) stimulated HaCaT keratinocytes. Of all compounds, curcumin was the most potent anti-inflammatory.

Keywords: *Curcuma aromatica*; sesquiterpene; anti-inflammatory; luciferase assay; cytotoxicity

---

1. Introduction

The genus *Curcuma* is part of the family Zingiberaceae and over 120 species have been identified [1]. Many *Curcuma* species have been used as traditional medicine for the treatment of various diseases [2], or as ingredients for coloring in cosmetics as well as enhancing food flavors [3–6]. Previous phytochemical investigations of *Curcuma* species resulted in the isolation and identification of sesquiterpenoids and diarylheptanoids as major constituents and many of them showed promising pharmacological activities including anti-inflammatory activity, cytotoxicity against cancer cell lines, and antioxidant activities [5–9].

C. aromatica is widely used in Thai and Chinese traditional medicine for anti-tumor therapy [6], blood stasis [10], throat infections [3], to eliminate body waste, and to promote wound healing [11]. It showed various pharmacological activities such as antioxidant, anti-inflammatory, and anti-carcinogenic activities [12]. The rhizome extract of this plant is well-known as a rich source of sesquiterpenes [5,13]. *C. comosa* has been used in Thai traditional medicine for the alleviation of postpartum uterine pain [14]. This plant showed various biological properties such as antioxidant, anti-inflammatory, insecticidal [15], and inhibitory effects on cell proliferation [16]. Sesquiterpenoids [8] and diarylheptanoids [15] were isolated as major compounds from the rhizome of *C. comosa*. The rhizome of *C. aeruginosa* has been traditionally used for the treatment of asthma, cancer, fever, inflammation, and skin diseases [17]. Pharmacological activities such as antioxidant, anti-inflammatory, and cytotoxic activities have been reported for extracts of this species. [18]. The phytochemical profile of the rhizome of *C. aeruginosa* is characterized by the presence of diarylheptanoids, curcuminoids, and sesquiterpenoids [17,19,20]. *C. longa* is commonly known as turmeric and its rhizome is used as food and in traditional medicine for the treatment of inflammation, infections or tumors, as carminative, and as diuretic [21–23]. In this study, we compared in vitro anti-inflammatory and anti-oxidant activity, and cytotoxicity of four *Curcuma* species namely, *C. aromatica*, *C. comosa*, *C aeruginosa*, and *C. longa*. In addition, over a dozen compounds were isolated from *C. aromatica* rhizome and its phytochemical profile was compared to that of the other three *Curcuma* species by means of Ultra-Performance Liquid Chromatography–High Resolution Mass Spectrometry (UPLC-HRMS) analysis.

2. Materials and Methods

2.1. Plant Material

The rhizome of *C. aromatica* (N: 20.1924°, E: 99.4854°), *C. comosa* (N: 20.1922°, E: 99.4852°), and *C. longa* (N: 20.1927°, E: 99.4855°) were collected from Doi Tung, Chiang Rai Province, Thailand in May 2016, while the rhizome of *C. aeruginosa* was purchased from Mae-Ca-Chan local markets, Chiang Rai Province, Thailand in June 2016. Plant authentication was verified by Mr. Martin Van de Bult and voucher specimens (MFU-NPR0192, MFU-NPR0193, MFU-NPR0194, and MFU-NPR0195, respectively) were deposited at the Natural Products Research Laboratory of Mae Fah Luang University.

2.2. Chemicals

L-Ascorbic acid, 2,2′ -azino-bis(3-ethylbenzothiazoline-6-sulfonic acid) diammonium salt (ABTS), 2,2-diphenyl-1-picrylhydrazyl (DPPH), 3-[4,5-dimethylthiazol-2-yl]-2,5-diphenyl tetrazolium bromide (MTT), sodium dodecyl sulfate (SDS), and dimethyl sulfoxide (DMSO) were purchased from Sigma-Aldrich (St. Louis, MO, USA). All chemicals and solvents used in this study were of analytical grade.

2.3. Extraction

The rhizomes of the four *Curcuma* species were cleaned, chopped, and air-dried at room temperature for three days. The air-dried rhizomes (1 kg) of each plant were macerated in EtOAc (3 × 10 L) at room temperature. The extracts were filtered and evaporated under reduced pressure to obtain the EtOAc extracts of *C. aromatica* (21.67 g), *C. comosa* (24.49 g), *C. aeruginosa* (20.21 g), and *C. longa* (19.76 g). Additionally, dried powder (100 g) of each plant was extracted with 80% ethanol (3 × 500 mL) at room temperature. Removal of the solvent under reduced pressure yielded the crude ethanolic extracts of *C. aromatica* (2.2 g), *C. comosa* (2.5 g), *C. aeruginosa* (2.0 g), and *C. longa* (2.1 g).

2.4. Fractionation and Isolation

The EtOAc extract of *C. aromatica* was selected for fractionation and isolation, based on the fact that it showed the most promising biological activities. The EtOAc extract was subjected to quick column chromatography (QCC) over silica gel, eluting with a gradient system of *n*-hexane/EtOAc (100% hexanes to 100% EtOAc) to give 13 fractions (A-M). Fraction B (1.45 g) was further separated by CC over Sephadex LH-20 (100% MeOH) to give compound **1** (4.5 mg). Fraction C (2.26 g) was separated by CC (1:4 CH_2Cl_2/*n*-hexane) to give fraction CP21-B5 (443.3 mg), which was further purified by CC over Sephadex LH-20 (100% MeOH) to give compound **7** (15.4 mg). Fraction E (540.1 mg) was separate by CC (1:3 CH_2Cl_2/*n*-hexane) to give nine fractions (CP6-01 to CP6-09). Compound **4** (9.9 mg) was obtained from fraction CP6-06 (263.0 mg) by repeated CC over Sephadex LH-20 (1:4 CH_2Cl_2/MeOH), while compound **5** (7.0 mg) yielded from fraction CP6-08 (108.5 mg) by repeated CC (1.5:8.5 CH_2Cl_2/*n*-hexane). Fraction F (4.05 g) was fractionated by CC (1:19 EtOAc/*n*-hexane) to give fraction CP30-02 (75.1 mg), which was further purified by CC (1:99 acetone/*n*-hexane) to afford compound **6** (5.2 mg). Compound **2** (217.4 mg) was obtained from fraction G (654.7 mg) by CC (2:3 CH_2Cl_2/*n*-hexane). Fraction H (3.13 g) was submitted to CC (1:49 EtOAc/*n*-hexane) to give fraction CP32-A (1.12 g), which was further purified by RP-18 (7:3 MeOH/H_2O) to afford compounds **3** (79.3 mg) and **15** (55.8 mg). Fraction I (957.2 mg) was subjected to CC (1:1 CH_2Cl_2/*n*-hexane) to give fraction CP7-2 (198.2 mg), then purified by CC (15:1:34 CH_2Cl_2/EtOAc/*n*-hexane) to give compound **8** (9.6 mg). Fraction J (1.30 g) was subjected to CC over Sephadex LH-20 (100% MeOH), followed by CC (3:7 CH_2Cl_2/*n*-hexane) to afford compounds **12** (3.1 mg) and **13** (3.1 mg). Fraction K (2.77 g) was fractionated by CC (1:4 EtOAc/*n*-hexane) to give fraction CP35-BC (1.03 g), then repeated CC (1:49 acetone/*n*-hexane and 1:9 CH_2Cl_2/*n*-hexane) to afford compound **9** (6.8 mg). Fraction L (2.07 g) was subjected to CC (1:99 acetone/CH_2Cl_2) to give compound **11** (31.1 mg) and six fractions (CP17-02 to CP17-07). Compound **14** (31.1 mg) was obtained from fraction CP17-05 (215.7 mg) by CC (1:49 acetone/CH_2Cl_2). Compound **10** (5.1 mg) was obtained from fraction CP17-06 (1.56 g) by CC over Sephadex LH-20 (100% MeOH) followed by CC (1:1:3 acetone/EtOAc/*n*-hexane).

2.5. Characterization of Curcuma Extracts by UPLC-HRMS

Crude extracts of the four *Curcuma* species, prepared with 80% ethanol/20% water were analyzed by Ultra-Performance Liquid Chromatography–High Resolution Mass Spectrometry (UPLC-HRMS) together with 8 of the 15 purified compounds isolated from *C. aromatica*, in order to determine whether these compounds were present in *C. longa*, *C. comosa* and *C. aeruginosa* too. Liquid chromatography analysis was performed on an Acquity® UPLC System (Waters, Milford, MA, USA). Detection was carried out on an LTQ-Orbitrap® XL hybrid mass spectrometer equipped with an Electrospray Ionization (ESI) source (Thermo Scientific, Waltham, MA, USA) for accurate mass. Separation was achieved on an Acquity UPLC® Peptide BEH C18 column (2.1 × 100 mm, 1.7 μm, Waters corporation®, Wexford, Ireland) using a gradient containing water with 0.1% (*v/v*) formic acid (A) and acetonitrile (B). The gradient elution was performed as follows: 0–2 min eluent B 2%; 2–18 min eluent B 2–100%; 18–20 min eluent B 100%; 21–25 min column equilibration-eluent B 2%. A flow rate of 0.4 mL/min was employed for elution. The column was maintained at 40 °C, the samples at 7 °C, and the flow rate was set to 0.4 mL/min. The 80% ethanol extracts (10 μL at 300 μg/mL) were injected. All samples were analyzed in the full scan *m/z* range of 115–1000, in negative and positive mode at a resolving power of 30,000 and data-dependent MS/MS events were acquired. In both modes the data-dependent acquisition was simultaneously performed using a collision induced dissociation C-trap (CID) with normalized collision energy at 35 V and a mass resolution of 10,000. In negative mode capillary temperature was set to 350 °C and the source voltage was 2.7 kV. Tube lens and capillary voltage were respectively tuned at −100 V and −30 V. In positive mode capillary temperature was set to 350 °C and the source voltage was 3.50 kV. Tube lens and capillary voltage were respectively tuned at +120 V and +40 V. In both modes the arbitrary units were used for sheath gas, auxiliary gas, and sweep gas was nitrogen at (40, 10, 0 arbitrary units, respectively). The control of the system and the spectral interpretation was performed using the XcaliburTM (Version 2.2, Thermo Scientific, Waltham, MA, USA) software.

2.6. DPPH Radical-Scavenging Activity Assay

The antioxidant activity was determined by the DPPH radical scavenging assay as described previously, with slight modifications [24]. In brief, 100 μL of extracts and compounds at different concentrations were mixed with 100 μL of 60 μM DPPH methanol solution in a 96-well microplate. The solution was incubated at room temperature in darkness for 30 min, then absorbance was measured at 517 nm. Ascorbic acid was used as positive control. The DPPH radical scavenging activity was expressed as the concentration at 50% inhibition (IC_{50}), which was calculated by plotting percent inhibition against concentration of the sample.

2.7. ABTS Radical Cation Scavenging Assay

The ABTS radical cation scavenging activity of extracts and compounds was determined using the method described previously [24] with some modifications. The $ABTS^{·+}$ solution was prepared from the reaction of equal volumes of 7 mM of ABTS and 2.45 mM potassium persulfate in a dark place at room temperature for 16 h before use. Prior to the assay, the $ABTS^{·+}$ solution was adjusted to the absorbance of 0.70 ± 0.05 at 734 nm with EtOH. Twenty microliters of extracts and compounds at different concentrations were mixed with 180 μL of $ABTS^{·+}$ solution in a 96-well microplate and incubated at room temperature for 5 min. Next, the absorbance was measured at 734 nm. Ascorbic acid was used as positive control. The ABTS radical cation scavenging activity was expressed as the concentration at 50% inhibition (IC_{50}), which was calculated by plotting percent inhibition against concentration of the sample.

2.8. Cell Culture

HaCaT keratinocyte cells with a stable transfected NF-κB luciferase reporter gene cassette has previously been described [25]. Cells were cultured in Dulbecco's modified eagle's medium, supplemented with 10% fetal bovine serum, 2% of sodium bicarbonate (7.5% solution), 1% of sodium pyruvate (100 mM), and 1% of penicillin–streptomycin (10,000 units/mL). The cells were incubated in a humidified 37 °C, 5% CO_2 incubator.

2.9. MTT Assay

Adverse anti-proliferative or toxic effects of various extracts and purified phytochemicals compounds on HaCaT cells were evaluated by MTT colorimetric assay. Cells were seeded into 96-well plates at 2×10^4 cells/well and incubated under the abovementioned conditions for 24 h. The extracts or pure compounds at different concentrations were added for another 24 h, after which 10 μL of MTT reagent (5 mg/mL) was added to each well and incubated for 4 h. Cells were lysed with 90 μL 10 mM HCl solution containing 10% SDS and OD value was measured at 595 nm with the Envision Plate Reader (Perkin Elmer, USA). Withaferin A was used as positive control.

2.10. CytoTox-ONE™ Cytotoxicity Assay

Cell cytotoxicity was measured by determining membrane integrity of HaCaT cells following treatment with crude extracts or purified phytochemicals by means of the CytoTox-ONE™ Assay according to the manufacturer's instructions (Promega, WI, USA). In brief, cells were plated at 2×10^4 cells/well in 96-well plates and incubated under the above-mentioned conditions for 24 h. Extracts or pure compounds at different concentrations were added to the cells and left to incubate for 24 h at 37 °C and 5% CO_2. After incubation, the assay plates were transferred to 22 °C for 5 min, 100 μL of the CytoTox-ONE™ reagent was added to all wells and incubated at 22 °C for 10 min. After that, 50 μL of stop solution was added to all wells and plates were shaken at 500 rpm for 10 s. The fluorescence signal was measured with an excitation wavelength of 560 nm and an emission wavelength of 590 nm with the Tecan GENios Microplate Reader (Tecan Trading AG, Männedorf, Switzerland). Withaferin A was used as positive control. The triplicate wells without cells were used as negative control to determine background fluorescence. Vehicle control was triplicate cells with untreated cells, the same solvent used to deliver the test compounds. In addition, 2 µL of lysis solution was used as maximum LDH release control.

2.11. Luciferase Assay

NFκB-luciferase-dependent reporter assays were performed in HaCaT cells stably expressing p(NFκB)$_3$50-luc as previously described [25]. In brief, cells were plated at a density of 10^5 cells/well in 24-well plates and grown overnight. Cells were subsequently treated with a dose range of crude extracts or purified compounds for 2 h, followed by TNF stimulation (2 ng/mL) for 6 h. Finally, cells were lysed in 1 X lysis buffer (25 mM Tris-phosphate (pH 7.8), 2 mM DTT, 2 mM CDTA, 10% glycerol, and 1% Triton X-100) and 25 µL of lysate was assayed for luciferase activity by adding 50 µL of luciferase substrate (1 mM luciferin or luciferin salt, 3 mM ATP, and 15 mM MgSO$_4$ in 30 mM HEPES buffer, pH 7.8). After 10 s of mixing, bioluminescence was measured for 1 s using the Envision multilabel reader (Perkin Elmer, Waltham, MA, USA). Withaferin A was used as positive control.

2.12. Data Analysis

All analyses were performed in triplicate and data were expressed as means ± standard deviation (SD) from at least three independent biological experiments. The results were analyzed by one-way analysis of variance (ANOVA) with the Dunnett test, significant difference ($p < 0.05$) using IBM SPSS Statistics, version 23 (IBM Crop.).

3. Results and Discussion

3.1. Isolation of Compounds

The EtOAc extract of *C. aromatica* was fractionated by column chromatography to afford 15 known compounds (Figure 1). The compounds were identified as germacrone (**1**) [25], curdione (**2**) [26], dehydrocurdione (**3**) [25], furanodienone (**4**) [27], zederone (**5**) [28], curzerenone (**6**) [27], curzeone (**7**) [29], comosone II (**8**) [30], gweicurculactone (**9**) [31], curcumenol (**10**) [25], isoprocurcumenol (**11**) [32], zedoarondiol (**12**) [33], vanillin (**13**) [34], curcumin (**14**) [35], and β-sitosterol (**15**) [36] by comparison of their spectroscopic data with those reported in the literature. Sesquiterpenes **7** and **8** were isolated from the rhizome of *C. aromatica* for the first time, while all remaining sesquiterpenes were similar to previous reports [5,13].

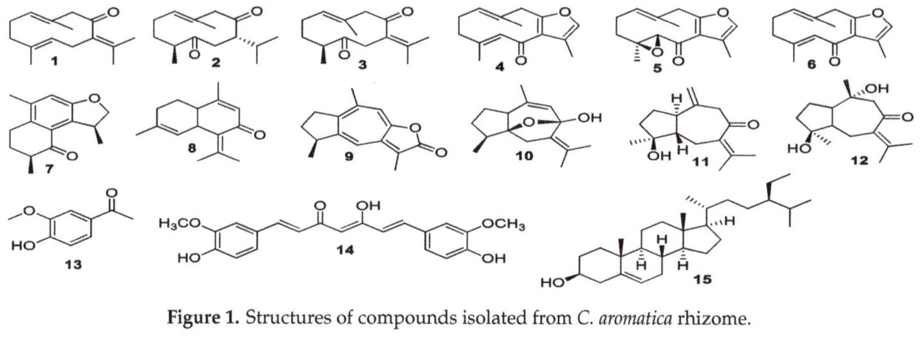

Figure 1. Structures of compounds isolated from *C. aromatica* rhizome.

3.2. Characterization of Curcuma Extracts by UPLC-HRMS

Eight of the purified compounds, germacrone (**1**), curdione (**2**), dehydrocurdione (**3**), zederone (**5**) curcumenol (**6**), zedoarondiol (**12**), curcumin (**14**), and β-sitosterol (**15**), were analyzed by UPLC-HRMS, together with the 80% EtOH extracts of *C. aromatica*, *C. longa*, *C. comosa*, and *C aeruginosa*. Except for compounds **12** and **15**, all compounds were detected in ESI$^+$ mode, while **5** and **13** could be detected in ESI$^+$ and ESI$^-$ mode. Table 1 shows the retention time and MS data obtained for the purified compounds. In addition, it is indicated whether these compounds could be detected in the crude extracts. Compounds **12** and **15** were not clearly detected in either of the detection modes, possibly due to poor ionization properties or their low abundance.

As expected, all six detected compounds were found in the crude extract of C. aromatica, since the compounds were purified from this Curcuma species as described in Sections 2.1 and 3.1 Also C. longa was found to contain these six compounds. Five out of six compounds could be identified in the 80% EtOH extracts of C. aeruginosa; only curcumin (**13**) was found to be absent in this Curcuma species. The C. comosa extract did not contain curcumin either, nor did it contain curcumenol. Our results about the phytochemical composition of different Curcuma species are in line with results reported by other groups [26,27].

Table 1. Chromatographic and spectral data, obtained with Ultra-Performance Liquid Chromatography–High Resolution Mass Spectrometry (UPLC-HRMS)analysis.

| Compound | Mol. Formula | RT (min) | ESI$^+$ | | | | | ESI$^-$ | | | | | Present in Extract | | | |
			Measured m/z	Ion	Calculated m/z	Δ (ppm)	MS fragments	Measured m/z	Ion	Calculated m/z	Δ (ppm)	MS fragments	*C. aromatica*	*C. aeruginosa*	*C. comosa*	*C. longa*
Germacrone (1)	$C_{15}H_{22}O$	13.8	219.1751	[M + H]$^+$	219.1749	0.91		n.d.					x	x	x	x
Curdione (2)	$C_{15}H_{24}O_2$	11.8	237.1858	[M + H]$^+$	237.1855	1.26		n.d.					x	x	x	x
Dehydrocurdione (3)	$C_{15}H_{22}O_2$	10.9	235.1703	[M + H]$^+$	235.1698	2.13		n.d.					x	x	x	x
Zederone (5)	$C_{15}H_{18}O_3$	11.2	247.1339	[M + H]$^+$	247.1334	2.02		245.1180	[M − H]$^-$	245.1178	0.82		x	x	x	x
Curcumenol (6)	$C_{15}H_{22}O_2$	11.1	235.1701	[M + H]$^+$	235.1698	1.28	217.1593; 199.1486; 189.1642; 177.1277	n.d.					x	x		x
Curcumin (13)	$C_{21}H_{20}O_6$	11.1	369.1345	[M + H]$^+$	369.1338	1.90	285.1129; 245.1814; 175.0756	367.1181	[M − H]$^-$	367.1182	−0.27	217.0504, 173.0608	x			x

3.3. Antioxidant Activity

The antioxidant radical scavenging activity of extracts were evaluated using DPPH and ABTS assays (Table 2), and purified compounds were tested in the DPPH assay as shown in Figure 2. Regarding antioxidant activity, the *C. aromatica* extract showed the most promising IC_{50} values (102.4 ± 1.9, 127.0 ± 1.9 µg/mL), followed by *C. longa* (134.9 ± 1.5, 170.8 ± 1.6 µg/mL), *C. comosa* (137.7 ± 5.2, 171.9 ± 1.9 µg/mL), and *C. aeruginosa* (187.4 ± 22.1, 217.9 ± 1.8 µg/mL). Ascorbic acid was used as positive control, with IC_{50} values of 1.80 ± 0.01 and 5.2 ± 0.8 for DPPH and ABTS assay, respectively. In addition, curcumin exhibited strong antioxidant activity with 68.9% ± 0.6% percent inhibition of at 25 µg/mL, whereas other compounds showed moderate activities, see Figure 2. Since curcumin was only detected in *C. aromatica* and *C. longa* and not in *C. comosa* and *C. aeruginosa*, the activity of the first two extracts may in part be attributed to the presence of curcumin. However, since *C. comosa* showed antioxidant activity similar to *C. longa*, and *C. aeruginosa* showed significant antioxidant activity too, curcumin cannot be the only active compound and other constituents might also contribute too to overall antioxidant activity.

Table 2. Antioxidant activities of EtOH extract from the rhizome of *C. aromatica*, *C. longa*, *C. comosa*, and *C. aeruginosa*.

Sample	Antioxidant (IC_{50}, µg/mL)	
	DPPH	**ABTS**
C. aromatica	102.4 ± 1.9	127.0 ± 1.9
C. longa	134.9 ± 1.5	170.8 ± 1.6
C. comosa	137.7 ± 5.2	171.9 ± 1.9
C. aeruginosa	187.4 ± 22.1	217.9 ± 1.8
Ascorbic acid	1.80 ± 0.01	5.2 ± 0.8

Note: Values are the mean ± SD, n = 3; DPPH: 2,2-diphenyl-1-picrylhydrazyl; ABTS: 2,2′-azino-bis(3-ethylbenzothiazoline-6-sulfonic acid) diammonium salt.

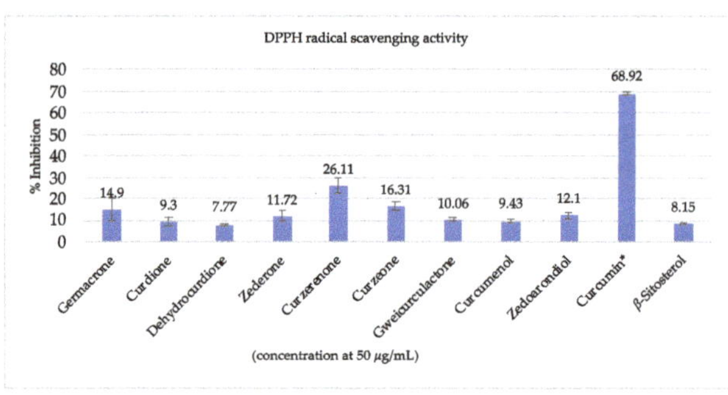

Figure 2. 2,2-Diphenyl-1-picrylhydrazyl (DPPH) radical scavenging activity of compounds isolated from *C. aromatica*, * = concentration of 25 µg/mL.

3.4. Cell Viability and Cytotoxicity

Cell viability and cytotoxicity of crude extracts and pure compounds were assessed by MTT assay and the CytoTox-ONE™ Homogeneous Membrane Integrity Assay using HaCaT keratinocyte cells, respectively. The MTT colorimetric assay estimates the number of viable cells based on the ability of mitochondrial enzymes to reduce the tetrazolium dye MTT to a purple colored formazan [37], whereas the CytoTox-ONE™ assay is a fluorometric-based method to detect loss of membrane integrity of dying cells. MTT results showed that exposure to 200 µg/mL of *C. aeruginosa*, *C. comosa*, *C. aromatica*, or *C. longa* extract inhibited the growth of cells, with relative percentages of cell viability being 12.1 ± 2.9, 14.4 ± 4.1, 28.6 ± 4.1, and 46.9 ± 8.6, respectively (Figure 3a). Interestingly, CytoTox-ONE™

showed a slightly different outcome with estimated cell death being lower compared to the MTT results. Treatment of HaCaT cells with 200 µg/mL concentrations of *C. aeruginosa* and *C. comosa* extract resulted in 57.7 ± 3.1% and 32.6 ± 2.2% cell death respectively, while no cytotoxicity could be observed with *C. aromatica* and *C. longa* treatments at the same concentration (Figure 4a). This suggests that all extracts mainly affect mitochondrial reduction capacity and cell proliferation, and that only *C. aeruginosa* and *C. comosa* extracts negatively impact membrane integrity at concentrations above 100 µg/mL [38–40]. In contrast, none of the purified phytochemicals inhibit cell viability (MTT) or cytotoxicity (CytoTox-One™) at concentrations 1–20 µM, whereas a reference cytotoxic anti-cancer compound withaferin A [28] dose dependently kills the HaCaT cells, as shown in Figures 3b and 4b.

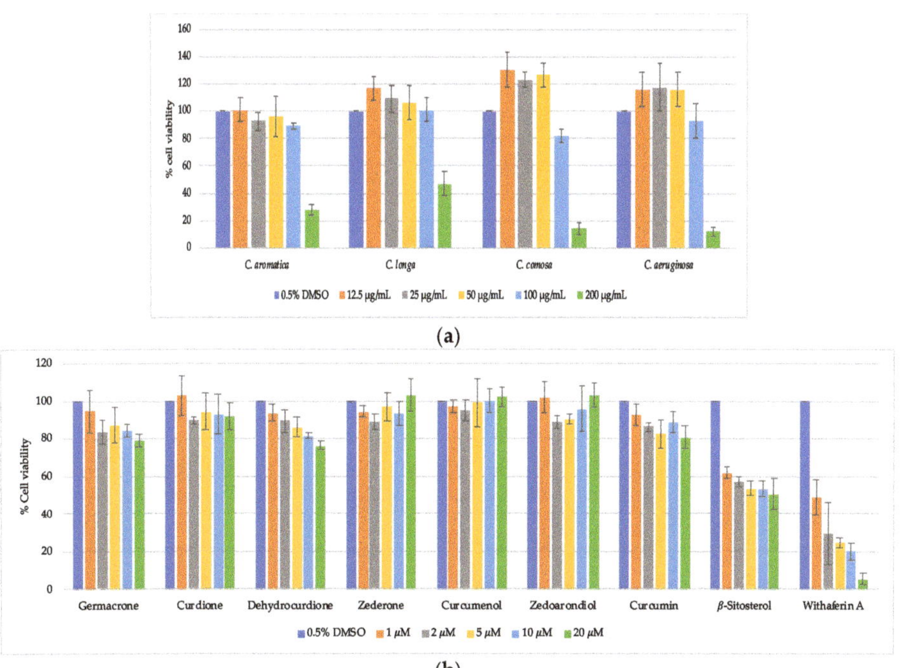

(a)

(b)

Figure 3. (a) Relative HaCaT viability by increasing concentrations of four *Curcuma* species. (b) Relative HaCaT viability (%) by increasing concentrations of pure compounds isolated from *C. aromatica* and the reference cytotoxic anti-cancer compound withaferin A in HaCaT cells.

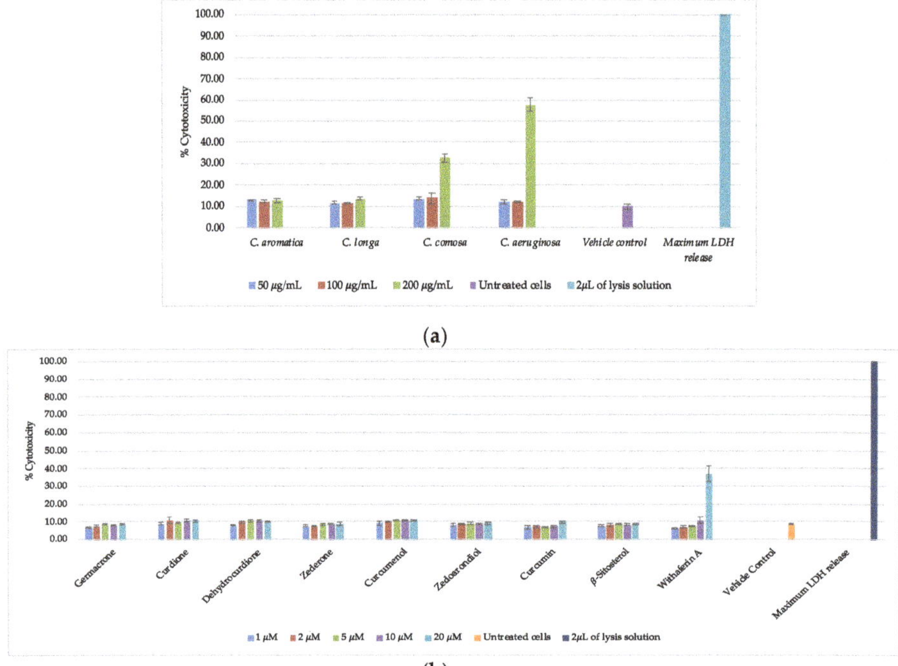

(**a**)

(**b**)

Figure 4. Disruption of membrane integrity measured by the release of lactate dehydrogenase (LDH) (CytoTox-ONE™). (**a**) Relative cytotoxicity (%) of four *Curcuma* species in HaCaT cells. (**b**) Relative cytotoxicity (%) of pure compounds isolated from *C. aromatica* and the reference cytotoxic anti-cancer compound withaferin A in HaCaT cells.

3.5. Anti-Inflammatory Activity

HaCaT NF-κB reporter gene cells were left untreated or pretreated for 2 h with various crude extracts or its purified phytochemicals, followed by 3 h combination treatment with the pro-inflammatory stimulus TNF. After 5 h treatment, cells were lysed and corresponding luciferase reporter gene activity was measured in lysates in presence of ATP/luciferin reagent (Promega, WI, USA) by measuring the total emitted bioluminescence (relative light units, RLU) during 30s (Envision multiplate reader, Perkin Elmer). As expected, and as shown in Figure 5a, the proinflammatory NF-κB activator TNF strongly increases luciferase gene expression in HaCaT NF-κB reporter cells, as compared to the control samples without TNF. Upon combination treatment of the different extracts with TNF, we observed dose dependent decrease of luciferase gene expression for all four extracts, suggesting anti-inflammatory effects on NF-κB activity. *C. aromatica* showed the strongest anti-inflammatory NF-κB effects, followed by *C. aeruginosa*, *C. comosa*, and *C. longa*, respectively.

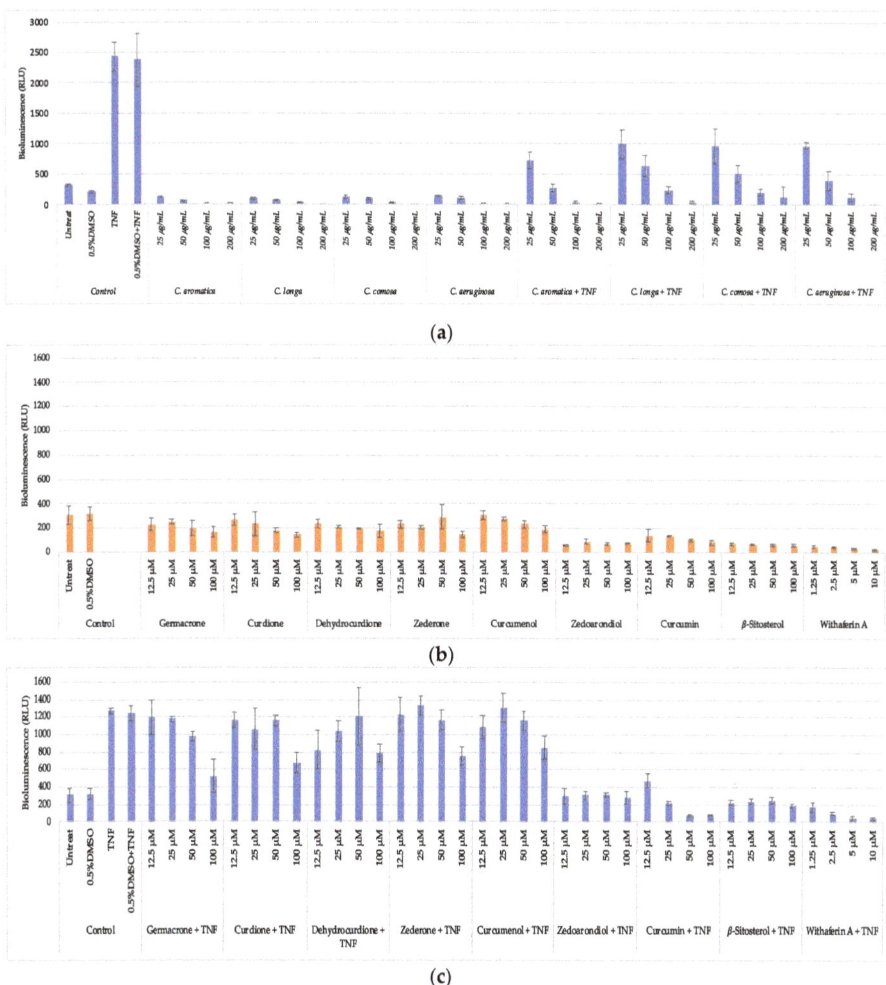

(a)

(b)

(c)

Figure 5. Anti-inflammatory effects of four *Curcuma* species and pure compounds isolated from *C. aromatica* measured in HaCaT NF-κB reporter gene cells. (**a**) Dose dependent effects of crude extracts of *Curcuma* species on basal and inflammation induced NF-κB reporter gene (luciferase relative light units) expression. (**b**) Dose dependent effect of pure compounds isolated from *C. aromatica* and the reference NF-κB inhibitor compound (withaferin A) on basal and inflammation induced NF-κB reporter gene (luciferase relative light units) expression. (**c**) Dose dependent effect of pure compounds isolated from *C. aromatica* and the reference NF-κB inhibitor compound (withaferin A) on basal and inflammation induced NF-κB reporter gene (luciferase relative light units) expression.

Next, stable phytochemicals isolated in sufficient quantities isolated from *C. aromatica* were further evaluated for their NF-κB inhibiting activity in TNF stimulated HaCaT keratinocytes, as compared to the reference inhibitor compound withaferin A [41]. As shown in Figure 5b, curcumin was found to be the most potent NF-κB inhibitor, although less potent the reference NF-κB inhibitor withaferin A, in line with previous research [11,41]. *C. aromatica*, which contains curcumin, indeed was the most potent NF-kB inhibiting extract. Thus, it's traditional use in the prevention and treatment of inflammatory diseases may be justified. However, the other three extracts, of which *C. longa* contains curcumin, whereas *C. comosa* and *C. aeruginosa* do not, show a comparable activity. This suggests that besides curcumin, additional constituents may be responsible for NF-κB inhibition in *C. comosa* and *C. aeruginosa* extracts. Indeed, germacrone, curdione, zederone, and curcumenol show moderate inhibition of NF-κB reporter gene expression in TNF stimulated HaCaT keratinocytes too. In addition, zedoarondiol and β-sitosterol show strong NF-κB inhibition, although they may be low abundant, since UPLC-HRMS analysis failed to detect significant amounts in the four extracts.

4. Conclusions

Sesquiterpenes are major bioactive constituents in the rhizome extract of *C. aromatica*. Of the four *Curcuma* species, *C. aromatica*, with its secondary metabolite curcumin, showed the highest antioxidant activity and most potent anti-inflammatory properties with the lowest toxicity. Besides curcumin, we purified additional anti-inflammatory bioactives in *C. aromatica*, *C. aeruginosa*, *C. comosa*, and *C. longa*, such as germacrone, curdione, zederone, curcumenol, zedoarondiol, and β-sitosterol present, which deserve further investigation.

In conclusion, our results suggest that the rhizome of *C. aromatica* holds promise to be developed as a safe cosmeceutical or functional skin care products for anti-aging and to reduce inflammatory skin irritation.

Author Contributions: Conceptualization, A.P. and S.L.; formal analysis, A.P., E.T., M.E.S., and E.L.; investigation, A.P., E.T., E.L., W.M., W.V.B., L.P., and S.L.; writing—original draft preparation, A.P.; writing—review and editing, S.L., E.T., M.E.S., E.L., W.M., T.S., S.D., L.P., and W.V.B.; supervision, S.L., W.V.B., and L.P.; funding acquisition, S.L. and W.M. All authors have read and agreed to the published version of the manuscript.

Funding: This research was funded by Thailand Research Fund through the Research and Researchers for Industries (RRi) PH.D. Program (PHD59I0050). Partial financial supports from Health Herb Center Co., Ltd., the Thailand Science Research and Innovation (DBG6280007), and Mae Fah Luang University are also acknowledged.

Acknowledgments: We would like to thank Mae Fah Luang University and University of Antwerp for providing access to their laboratory facilities.

Conflicts of Interest: The authors declare no conflicts of interest.

References

1. Mishra, S.; Verma, S.S.; Rai, V.; Awasthee, N.; Arya, J.S.; Maiti, K.K.; Gupta, S.S. Curcuma raktakanda induces apoptosis and suppresses migration in cancer cells: Role of reactive oxygen species. *Biomolecules* **2019**, *9*, 159. [CrossRef] [PubMed]
2. Srivilai, J.; Rabgay, K.; Knorana, N.; Waranuch, N.; Nuengchamnong, N.; Wisuitiprot, W.; Chuprajob, T.; Changtam, C.; Suksamrarn, A.; Chavasiri, W.; et al. Anti-androgenic curcumin analogues as steroid 5-alpha reductase inhibitors. *Med. Chem. Res.* **2017**, *26*, 1550–1556. [CrossRef]
3. Dong, S.; Luo, X.; Liu, Y.; Zhang, M.; Li, B.; Dai, W. Diarylheptanoids from the root of *Curcuma aromatica* and their antioxidative effects. *Phytochem. Lett.* **2018**, *27*, 148–153. [CrossRef]
4. Chan, E.W.C.; Lim, Y.Y.; Wong, L.F.; Lianto, F.S.; Wong, S.K.; Lim, K.K.; Joe, C.E.; Lim, T.Y. Antioxidant and tyrosinase inhibition properties of leaves and rhizomes of ginger species. *Food Chem.* **2008**, *109*, 477–483. [CrossRef]
5. Bamba, Y.; Yun, Y.S.; Kunugi, A.; Inoue, H. Compounds isolated from *Curcuma aromatica* Salisb. inhibit human P450 enzymes. *J. Nat. Med.* **2011**, *65*, 583–587. [CrossRef]
6. Liu, B.; Gao, Y.Q.; Wang, X.M.; Wang, Y.C.; Fu, L.Q. Germacrone inhibits the proliferation of glioma cells by promoting apoptosis and inducing cell cycle arrest. *Mol. Med. Rep.* **2014**, *10*, 1046–1050. [CrossRef]
7. Keeratinijakal, V.; Kongkiatpaiboon, S. Distribution of phytoestrogenic diarylheptanoids and sesquiterpenoids components in *Curcuma comosa* rhizomes and its related species. *Rev. Bras. Farm.* **2017**, *27*, 290–296. [CrossRef]
8. Qu, Y.; Xu, F.; Nakamura, S.; Matsuda, H.; Pongpiriyadacha, Y.; Wu, L.; Yoshikawa, M. Sesquiterpenes from *Curcuma comosa*. *J. Nat. Med.* **2009**, *63*, 102–104. [CrossRef]
9. Suksamrarn, A.; Ponglikitmongkol, M.; Wongkrajang, K.; Chindaduang, A.; Kittidanairak, S.; Jankam, A.; Yingyongnarongkul, B.; Kittipanumat, N.; Chokchaisiri, R.; Khetkam, P.; et al. Diarylheptanoids, new phytoestrogens from the rhizomes of *Curcuma comosa*: Isolation, chemical modification and estrogenic evaluation. *Bioorg. Med. Chem.* **2008**, *16*, 6891–6902. [CrossRef]

10. Dong, S.; Li, B.; Dai, W.; Wang, D.; Qin, Y.; Zhang, M. Sesqui- and diterpenoids from the radix of *Curcuma aromatica*. *J. Nat. Prod.* **2017**, *80*, 3093–3102. [CrossRef]

11. Ramsewak, R.S.; Dewitt, D.L.; Nair, M.G. Cytotoxicity, antioxidant and anti-inflammatory activities of Curcumins I-III from *Curcuma longa*. *J. Phytomed.* **2000**, *7*, 303–308. [CrossRef]

12. Booker, A.; Frommenwiler, D.; Johnston, D.; Umealajekwu, C.; Reich, E.; Heinrich, M. Chemical variability along the value chains of turmeric (*Curcuma longa*): A comparison of nuclear magnetic resonance spectroscopy and high performance thin layer chromatography. *J. Ethnopharmacol.* **2014**, *152*, 292–301. [CrossRef]

13. Agnihotri, V.K.; Thakur, S.; Pathania, V.; Chand, G. A new dihomosesquiterpene, termioic acid A, from *Curcuma aromatica*. *Chem. Nat. Compd.* **2014**, *50*, 665–668. [CrossRef]

14. Keeratinijakal, V.; Kladmook, M.; Laosatit, K. Identification and characterization of *Curcuma comosa* Roxb., phytoestrogens-producing plant, using AFLP markers and morphological characteristics. *J. Med. Plants Res.* **2010**, *4*, 2651–2657.

15. Jurgens, T.M.; Frazier, E.G.; Schaeffer, J.M.; Jones, T.E.; Zink, D.L.; Borris, R.P. Novel nematocidal agents from *Curcuma comosa*. *J. Nat. Prod.* **1994**, *57*, 230–235. [CrossRef]

16. Jariyawat, S.; Thammapratip, T.; Suksen, K.; Wanitchakool, P.; Nateewattana, J.; Chairoungdua, A.; Suksamrarn, A.; Piyachaturawat, P. Induction of apoptosis in murine leukemia by diarylheptanoids from *Curcuma comosa* Roxb. *Cell Biol. Toxicol.* **2011**, *27*, 413–423. [CrossRef]

17. Simoh, S.; Zainal, A. Chemical profiling of *Curcuma aeruginosa* Roxb. rhizome using different techniques of solvent extraction. *Asian Pac. J. Trop. Biomed.* **2015**, *5*, 412–417. [CrossRef]

18. Waras, N.; Nurul, K.; Muhamad, S.; Maria, B.; Ardyani, I.D.A.A.C. Phytochemical screening, antioxidant and cytotoxic activities in extracts of different rhizome parts from *Curcuma aeruginosa* Roxb. *Int. J. Res. Ayurveda Pharm.* **2015**, *6*, 634–637. [CrossRef]

19. Jose, S.; Thamas, T.D. Comparative phytochemical and anti-bacterial studies of two indigenous medicinal plants *Curcuma caesia* Roxb. and *Curcuma aeruginosa* Roxb. *Int. J. Green Pharm.* **2014**, *8*, 65–71.

20. Takano, I.; Yasuda, I.; Takeya, K.; Itokawa, H. Guaiane sesquiterpene lactones from *Curcuma aeruginosa*. *Phytochemistry* **1995**, *40*, 1197–1200. [CrossRef]

21. Li, S.; Yuan, W.; Deng, G.; Wang, P.; Yang, P.; Aggarwal, B.B. Chemical composition and product quality control of turmeric (*Curcuma longa* L.). *Pharm. Crop.* **2011**, *2*, 28–54. [CrossRef]

22. Priya, R.; Prathapha, A.; Raghu, K.G.; Menon, A.N. Chemical composition and in vitro antioxidative potential of essential oil isolated from *Curcuma longa* L. leaves. *Asian Pac. J. Trop. Biomed.* **2012**, S695–S699. [CrossRef]

23. Gounder, D.K.; Lingamallu, J. Comparison of chemical composition and antioxidant potential of volatile oil from fresh, fried and cured turmeric (*Curcuma longa*) rhizomes. *Ind. Crop. Prod.* **2012**, *38*, 124–131. [CrossRef]

24. Kanlayavattanakul, M.; Lourith, N. Sapodilla seed coat as a multifunctional ingredient for cosmetic applications. *Process. Biochem.* **2011**, *46*, 2215–2218. [CrossRef]

25. Firman, K.; Kinoshita, T.; Itai, A.; Sankawa, U. Terpenoids from *Curcuma heyneana*. *Phytochemistry* **1988**, *27*, 3887–3892. [CrossRef]

26. Harimaya, K.; Gao, J.F.; Ohkura, T.; Kawamata, T.; Iitaka, Y.; Guo, Y.T.; Inayama, S. A series of sesquiterpenes with a 7α-isopropyl side chain and related compounds isolated from *Curcuma wenyujin*. *Chem. Pharm. Bull.* **1991**, *39*, 834–853. [CrossRef]

27. Hikino, H.; Konno, C.; Agatsuma, K.; Takemoto, T.; Horibe, I.; Tori, K.; Ueyama, M.; Takeda, K. Sesquiterpenoids part XLVII structure configuration conformation and thermal rearrangement of furanodienone, isofuranodienone, curzerenone, epicuraerenone, and pyrocurzerenone, sesquiterpenoids of *Curcuma zedoaria*. *J. Chem. Soc.* **1975**, *5*, 401–524.

28. Shibuya, H.; Hamamoto, Y.; Cai, Y.; Kitagawa, I. A reinvestigation of the structure of zederone, a furanogermacrane-type sesquiterpene from zedoary. *Chem. Pharm. Bull.* **1987**, *35*, 924. [CrossRef]

29. Shiobara, Y.; Asakawa, Y.; Kodama, M.; Takemoto, T. Zedoarol, 13-hydroxygermacrone and curzeone, three sesquiterpenoids from *Curcuma zedoaria*. *Phytochemistry* **1986**, *6*, 1351–1353. [CrossRef]

30. Xu, F.; Nakamura, S.; Qu, Y.; Matsuda, H.; Pongpiriyadacha, Y.; Wu, L.; Yoshikawa, M. Structures of new sesquiterpenes form *Curcuma comosa*. *Chem. Pharm. Bull.* **2008**, *56*, 1710–1716. [CrossRef]

31. Hamdi, O.A.A.; Ye, L.J.; Kamarudin, M.N.A.; Hazni, H.; Paydar, M.; Looi, C.Y.; Shilpi, J.A.; Kadir, H.A.; Awang, K. Neuroprotective and antioxidant constituents from *Curcuma zedoaria* rhizomes. *Rec. Nat. Prod.* **2015**, *9*, 349–355.

32. Kuroyanagi, M.; Ueno, A.; Koyama, K.; Natori, S. Structures of sesquiterpenes of *Curcuma aromatica* Salisb. II. Studies on minor sesquiterpenes. *Chem. Pharm. Bull.* **1990**, *38*, 55–58. [CrossRef]

33. Kuroyanagi, M.; Ueno, A.; Ujiie, K.; Sato, S. Structures of sesquiterpenes from *Curcuma aromatica* Salisb. *Chem. Pharm. Bull.* **1987**, *35*, 53–59. [CrossRef]

34. Bandyopadhyay, B.; Banik, B.K. Bismuth nitrate-induced microwave-assisted expeditious synthesis of vanillin from curcumin. *Org. Med. Chem. Lett.* **2012**, *2*, 15. [CrossRef] [PubMed]

35. Payton, F.; Sandusky, P.; Alworth, W.L. NMR study of the solution structure of curcumin. *J. Nat. Prod.* **2007**, *70*, 143–146. [CrossRef]

36. Pierre, L.L.; Moses, M.N. Isolation and characterization of stigmasterol and β-sitosterol from *Odontonema strictum* (Acanthaceae). *J. Innov. Pharm. Biol. Sci.* **2015**, *2*, 88–95.

37. Chirumamilla, C.S.; Palagani, A.; Kamaraj, B.; Declerck, K.; Verbeek, M.W.C.; Oksana, R.; Bosscher, K.D.; Bougarne, N.; Ruttens, B.; Gevaert, K.; et al. Selective glucocorticoid receptor properties of GSK886 analogs with cysteine reactive warheads. *Front. Immunol.* **2017**, *8*, 1324. [CrossRef]

38. Sun, W.; Wang, S.; Zhao, W.; Wu, C.; Guo, S.; Hongwei, G.; Hongxun, T.; Lu, J.J.; Wang, Y.; Chen, X.; et al. Chemical constituents and biological research on plants in the genus *Curcuma*. *Food Sci. Nutr.* **2016**, *57*, 1451–1523.

39. Kannamangalam, U.; Varakumar, S.; Singhal, R.S. A comparative account of extraction of oleoresin from *Curcuma aromatica* Salisb by solvent and supercritical carbon dioxide: Characterization and bioactivities. *J. Food Sci. Technol.* **2019**, *116*, 108564.

40. Hassannia, B.; Logie, E.; Vandenabeele, P.; Berghe, T.V. Withaferin A: From ayurvedic folk medicine to preclinical anti-cancer drug. *Biochem. Pharmacol.* **2020**, *173*, 113602. [CrossRef]

41. Kaileh, M.; Berghe, W.V.; Heyerick, A.; Horion, J.; Piette, J.; Libert, C.; Keukeleire, D.D.; Essawi, T.; Haegeman, G. Withaferin A strongly elicits lkB kinase β hyperphosphorylation concomitant with potent inhibition of its kinase activity. *J. Biol. Chem.* **2006**, *282*, 4253–4264. [CrossRef] [PubMed]

 *biomolecules*

Article

Low Concentration of Withaferin a Inhibits Oxidative Stress-Mediated Migration and Invasion in Oral Cancer Cells

Tzu-Jung Yu [1], Jen-Yang Tang [2,3], Fu Ou-Yang [4], Yen-Yun Wang [5,6,7], Shyng-Shiou F. Yuan [5,7,8], Kevin Tseng [9], Li-Ching Lin [10,11,12,*] and Hsueh-Wei Chang [5,7,13,14,*]

[1] Graduate Institute of Medicine, College of Medicine, Kaohsiung Medical University, Kaohsiung 80708, Taiwan; u107500035@kmu.edu.tw
[2] Department of Radiation Oncology, Faculty of Medicine, College of Medicine, Kaohsiung Medical University, Kaohsiung 80708, Taiwan; reyata@kmu.edu.tw
[3] Department of Radiation Oncology, Kaohsiung Medical University Hospital, Kaohsiung 80708, Taiwan
[4] Division of Breast Surgery and Department of Surgery, Kaohsiung Medical University Hospital, Kaohsiung 80708, Taiwan; swfuon@kmu.edu.tw
[5] Center for Cancer Research, Kaohsiung Medical University, Kaohsiung 80708, Taiwan; wyy@kmu.edu.tw (Y.-Y.W.); yuanssf@kmu.edu.tw (S.-S.F.Y.)
[6] School of Dentistry, College of Dental Medicine, Kaohsiung Medical University, Kaohsiung 80708, Taiwan
[7] Cancer Center, Kaohsiung Medical University Hospital, Kaohsiung 80708, Taiwan
[8] Translational Research Center, Kaohsiung Medical University Hospital, Kaohsiung 80708, Taiwan
[9] Shanghai Jiao Tong University School of Medicine, Shanghai 200025, China; kevintseng192@sjtu.edu.cn
[10] Department of Radiation Oncology, Chi-Mei Foundation Medical Center, Tainan 71004, Taiwan
[11] School of Medicine, Taipei Medical University, Taipei 11031, Taiwan
[12] Chung Hwa University Medical Technology, Tainan 71703, Taiwan
[13] Department of Medical Research, Kaohsiung Medical University Hospital, Kaohsiung 80708, Taiwan
[14] Department of Biomedical Science and Environmental Biology, College of Life Sciences, Kaohsiung Medical University, Kaohsiung 80708, Taiwan
* Correspondence: 8508a6@mail.chimei.org.tw (L.-C.L.); changhw@kmu.edu.tw (H.-W.C.); Tel.: +886-6-281-2811 (ext. 53501) (L.-C.L.); +886-7-312-1101 (ext. 2691) (H.-W.C.)

Received: 28 February 2020; Accepted: 15 May 2020; Published: 17 May 2020

Abstract: Withaferin A (WFA) has been reported to inhibit cancer cell proliferation based on high cytotoxic concentrations. However, the low cytotoxic effect of WFA in regulating cancer cell migration is rarely investigated. The purpose of this study is to investigate the changes in migration and mechanisms of oral cancer Ca9-22 cells after low concentrations of WFA treatment. WFA under 0.5 µM at 24 h treatment shows no cytotoxicity to oral cancer Ca9-22 cells (~95% viability). Under this condition, WFA triggers reactive oxygen species (ROS) production and inhibits 2D (wound healing) and 3D cell migration (transwell) and Matrigel invasion. Mechanically, WFA inhibits matrix metalloproteinase (MMP)-2 and MMP-9 activities but induces mRNA expression for a group of antioxidant genes, such as nuclear factor, erythroid 2-like 2 (*NFE2L2*), heme oxygenase 1 (*HMOX1*), glutathione-disulfide reductase (*GSR*), and NAD(P)H quinone dehydrogenase 1 (*NQO1*)) in Ca9-22 cells. Moreover, WFA induces mild phosphorylation of the mitogen-activated protein kinase (MAPK) family, including extracellular signal-regulated kinases 1/2 (ERK1/2), c-Jun N-terminal kinase (JNK), and p38 expression. All WFA-induced changes were suppressed by the presence of ROS scavenger *N*-acetylcysteine (NAC). Therefore, these results suggest that low concentration of WFA retains potent ROS-mediated anti-migration and -invasion abilities for oral cancer cells.

Keywords: Withaferin A; oral cancer; oxidative stress; migration; invasion; matrix metalloproteinases; antioxidant signaling

---

1. Introduction

Oral cancer leads to high morbidity and mortality [1]. It invades local tissues [2] and reoccurs occasionally [3]. Local invasions are associated with metastasis, which is important to oral carcinogenesis [4]. Therefore, discovery of a drug that inhibits metastasis or local invasion is of great importance for oral cancer therapy.

Withaferin A (WFA), a triterpenoid derived from the root or leaf of the medicinal plant *Withania somnifera*, is reported to exhibit antiproliferative properties and can induce apoptosis in several types of cancers such as leukemia [5], cervical [6], pancreatic [7], breast [8], lung [9], colorectal [10], and oral [11,12] cancer cells. These anticarcinogenic effects for WFA were based on high cytotoxic concentrations.

These cytotoxic concentrations of WFA were reported to induce reactive oxygen species (ROS)-mediated apoptosis in oral [12] and colon [10] cancer cells. ROS may induce a number of reactions such as apoptosis [5–12], autophagy, and endoplasmic reticulum stress [13]; however, its effect on migration has rarely been reported.

Migration inhibitory effects of WFA against cancer cells had been reported recently [14,15]. For example, WFA exhibits G2/M cell cycle arrest, apoptosis, and antiproliferation, as well as migration inhibition in gastric cancer AGS cells [14]. However, its migration inhibitory effects were based on wound healing and invasive assays at >10 μM and >1 μM WFA, where the IC_{50} value for WFA in AGS cells was 0.75 μM [14]. WFA also showed antiproliferative effects against breast cancer cells (MDA-MB-241) and could exhibit migration inhibitory effect using the concentration of IC_{50} value for WFA (12 μM) [15]. These migration inhibitory effects of WFA against cancer cells were based on high cytotoxic concentrations. The migration modulating effect of low concentration of WFA with low or no cytotoxicity warrants for detailed investigation.

To date, the migration inhibitory effects of WFA against oral cancer cells had rarely been investigated. Since ROS is a vital factor for cell migration regulation [16], the migration inhibitory effects of low concentration of WFA, as well as the role of WFA-generated ROS in regulating oral cancer cell migration warrants detailed investigation. Accordingly, the aim of this study is to evaluate the migration regulation of low concentration WFA and explore the involvement of oxidative stress in the migration-modulating mechanisms in oral cancer cells.

2. Materials and Methods

2.1. Cell Culture and Reagents

Ca9-22 oral cancer cell line (Japanese Collection of Research Bioresources Cell Bank; JCRB) were incubated in Dulbecco's Modified Eagle Medium (DMEM)/Nutrient Mixture F-12 containing 10% bovine serum and penicillin/streptomycin (Gibco, Grand Island, NY, USA), as described previously [17]. WFA and the antioxidant *N*-acetylcysteine (NAC) [18,19] were purchased from Selleckchem.com (Houston, TX, USA) and Sigma-Aldrich (St. Louis, MO, USA).

2.2. Cell Viability

Cell viability was determined through mitochondrial enzyme activity detection using MTS assay (Promega Corporation, Madison, WI, USA) as described previously [20].

2.3. ROS Flow Cytometry

Cellular ROS content was detected by Accuri C6 flow cytometer (BD Biosciences; Franklin Lakes, NJ, USA) using ROS interacting dye 2′,7′-Dichlorodihydrofluorescein diacetate (DCFH-DA) (Sigma-Aldrich, St. Louis, MO, USA) [21] under the following conditions: 10 μM, 37 °C for 30 min.

2.4. Wound Healing Assay

Wound healing assay was used to detect 2D migration ability as described previously [22,23]. The non-migrated cell-free area for vehicle, NAC, WFA, and NAC + WFA (NAC pretreatment and WFA posttreatment) in oral cancer cells were measured using the free software "TScratch" (https://www.cse-lab.ethz.ch/software/).

2.5. Cellular 3D Migration and Invasion Assays

Three-dimensional migration ability was detected using 8 μm pore transwell chambers (Greiner Bio-One; Vilvoorde, Belgium). Three-dimensional invasion ability was detected using 0.5% Matrigel (BD Matrigel Basement Membrane Matrix, BD Biosciences, Bedford, MA, USA) topped transwell chambers. For these two assays, cells were plated under serum-free medium in the transwell top chambers, which were soaked in 10% FBS-containing medium with vehicle, NAC, WFA, and NAC + WFA for 21 h in the bottom chamber. Other detailed steps were described previously [23]. Finally, the 3D migration and invasion abilities were analyzed using Image J software.

2.6. Zymography for Matrix Metalloproteinase (MMP)-2 and MMP-9 Activities

Cell invasion ability were proportional to the MMP-2 and MMP-9 activities [24], which were detected using zymography analysis. Cells were seeded overnight, washed with 1X PBS, and treated with vehicle, NAC, WFA, and NAC + WFA in serum-free medium for 48 h. The conditioned medium used for gelatin zymography was described previously [23]. Gelatinase-based MMP-2 and MMP-9 activities were measured by the area of clear zone using Image J software.

2.7. Quantitative RT-PCR (qRT-PCR) for Antioxidant-Associated Genes

Total RNA, prepared by Trizol reagent (Invitrogen, Carlsbad, CA, USA), was reverse- transcribed to cDNA using the OmniScript RT kit (Qiagen, Valencia, CA, USA) as described previously [25]. qRT-PCR was performed by iQ SYBR Green Supermix (Bio-Rad Laboratories, Hercules, CA, USA) using a MyiQ real-time machine (Bio-Rad). Touch-down PCR program [26] was used for the antioxidant-associated genes [27], including nuclear factor erythroid 2-like 2 (*NFE2L2*), glutathione-disulfide reductase (*GSR*), glutamate-cysteine ligase catalytic subunit (*GCLC*), glutathione peroxidase 1 (*GPX1*), thioredoxin (*TXN*), catalase (*CAT*), superoxide dismutase 1 (*SOD1*), heme oxygenase 1 (*HMOX1*), NAD(P)H quinone dehydrogenase 1 (*NQO1*), and *GAPDH*. Their primer and PCR amplicon information are provided in Table 1. The comparative method (2–ΔΔCt) was used for analyzing relative mRNA expression (fold activation) [28].

Table 1. Primer information for antioxidant-associated genes *.

Genes	Forward Primers (5′→3′)	Reverse Primers (5′→3′)	Length
TXN	GAAGCAGATCGAGAGCAAGACTG	GCTCCAGAAAATTCACCCACCT	270 bp
GSR	GTTCTCCCAGGTCAAGGAGGTTAA	CCAGCAGCTATTGCAACTGGAGT	297 bp
CAT	ATGCAGGACAATCAGGGTGGT	CCTCAGTGAAGTTCTTGACCGCT	274 bp
SOD1	AGGGCATCATCAATTTCGAGC [29]	CCCAAGTCTCCAACATGCCTC	211 bp
HMOX1	CCTTCTTCACCTTCCCCAACAT	GGCAGAATCTTGCACTTTGTTGC	251 bp
NFE2L2	GATCTGCCAACTACTCCCAGGTT	CTGTAACTCAGGAATGGATAATAGCTCC	302 bp
NQO1	GAAGGACCCTGCGAACTTTCAGTA	GAAAGCACTGCCTTCTTACTCCG	258 bp
GCLC	ACAAGCACCCTCGCTTCAGTACC	CTGCAGGCTTGGAATGTCACCT	232 bp
GPX1	AACCAGTTTGGGCATCAGGAG	AGTTCCAGGCAACATCGTTGC	256 bp
GAPDH	CCTCAACTACATGGTTTACATGTTCC [30]	CAAATGAGCCCCAGCCTTCT [31]	220 bp

* Primers without reference were designed in this study.

2.8. Western Blotting for Mitogen-Activated Protein Kinase (MAPK) Expressions

Total protein (45 μg) was electrophoresed by 10% SDS-PAGE. After PVDF transferring and blocking, primary antibodies recognized extracellular-signal-regulated kinase 1/2 (ERK1/2), c-Jun

N-terminal kinase 1/2 (JNK 1/2), p38 (MAPK Family Antibody Sampler Kit; #9926, Cell Signaling Technology, Inc., Danvers, MA, USA), and their phosphorylated forms (Phospho-MAPK Family Antibody Sampler Kit; #9910, Cell Signaling Technology, Inc., Danvers, MA, USA) as well as GAPDH (#GTX627408; GeneTex International Corp.; Hsinchu, Taiwan) were used and other detailed steps were described previously [23]. The band intensity was analyzed using Image J software.

2.9. Statistical Analysis

Multiple comparisons were analyzed using the Tukey HSD test (JMP13; SAS Institute, Cary, NC, USA). Treatments without the same letter characters show a significant difference.

3. Results

3.1. Identification of the Optimal Concentrations of WFA for Oral Cancer Cell Migration Assay

In the MTS assay (Figure 1), oral cancer cells (Ca9-22) were treated with 0, 0.25, and 0.5 µM of WFA for 24 h with or without NAC pretreatment (2 mM, 1 h). Neither the WFA nor the NAC + WFA (NAC pretreatment and WFA posttreatment) affect the viability of Ca9-22 cells. This result suggests that WFA under 0.5 µM in the single treatment (WFA) or the combined treatment (NAC + WFA) both exhibited no cytotoxic to oral cancer cells (>95% viability). These concentrations were chosen for the following migration related experiments.

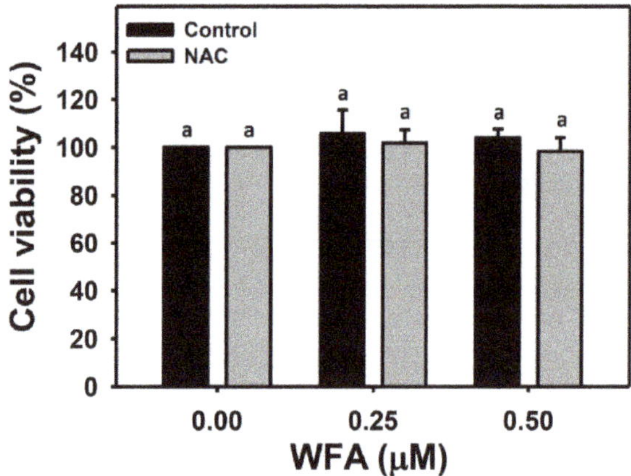

Figure 1. Viability of low concentration of Withaferin A (WFA) treatment in oral cancer cells. Oral cancer cells (Ca9-22) were pretreated with or without *N*-acetylcysteine (NAC) (2 mM, 1 h) and post-treated with different concentrations of WFA for 24 h. For multiple comparison, treatments with the same letter character show nonsignificant difference. Data, mean ± SD (n = 3).

3.2. ROS Generation of Oral Cancer Ca9-22 Cells at Low Concentrations of WFA

Figure 2A presented ROS patterns of Ca9-22 cells after NAC and/or WFA treatment. The ROS (+) (%) of Ca9-22 cells after low concentrations of WFA treatments were higher than those of the control, whereas this ROS generation was suppressed by NAC pretreatment (Figure 2B). Therefore, low concentrations of WFA triggered moderate ROS generation in oral cancer Ca9-22 cells.

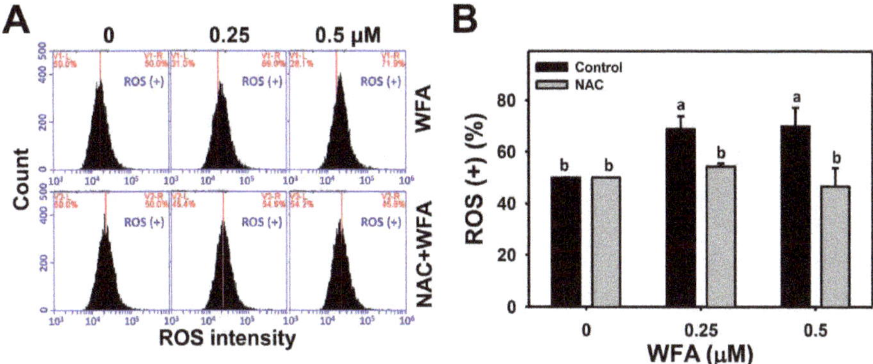

Figure 2. ROS generation effects of low concentrations of WFA in oral cancer cells. (**A**) ROS patterns of Ca9-22 cells after NAC and/or WFA treatments. Cells were pretreated with or without NAC (2 mM, 1 h) and post-treated with different concentrations of WFA for 24 h, i.e., NAC + WFA vs. WFA. ROS-positive population is marked as ROS (+). (**B**) Statistics of ROS change in Figure 2A. For multiple comparison, treatments without the same labels (a,b) indicate the significant difference. $p < 0.05\sim0.001$. Data, mean ± SD ($n = 3$).

3.3. 2D Migration of Oral Cancer Ca9-22 Cells at Low Concentrations of WFA

Figure 3A demonstrated the wound healing patterns of Ca9-22 cells after NAC and/or WFA treatments. Figure 3B showed that the cell-free area (%) of Ca9-22 cells after low concentrations of WFA treatments was greater than that of the untreated control over time. In contrast, this WFA-induced increase of cell-free area (%) was suppressed by NAC pretreatment. Therefore, low concentrations of WFA triggered 2D migration inhibition in Ca9-22 cells.

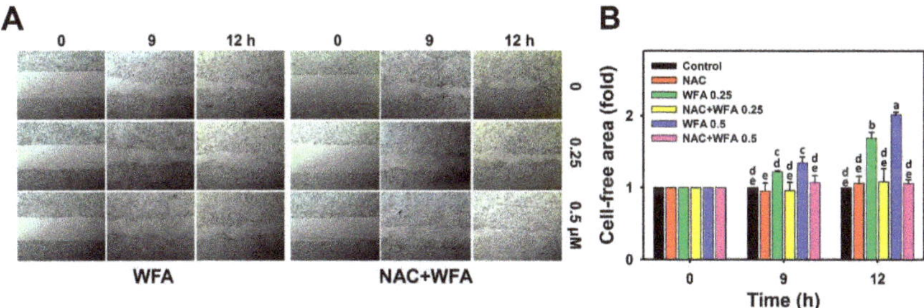

Figure 3. Two-dimensional anti-migration effects of low concentrations of WFA in oral cancer cells. (**A**) Two-dimensional migration (wound healing) images of Ca9-22 cells after NAC and/or WFA treatments. Cells were pretreated with or without NAC (2 mM, 1 h) and post-treated with different concentrations of WFA for 0, 9 and 12 h. (**B**) Statistics of 2D migration change in Figure 3A. For multiple comparison, treatments without the same labels (a–e) indicate the significant difference. $p < 0.05\sim0.0001$. Data, mean ± SD ($n = 3$).

3.4. 3D Migration and Invasion Changes in Oral Cancer Ca9-22 Cells at Low Concentrations of WFA

To further confirm the 2D migration inhibitory effect of WFA, the 3D migration and invasion assays of Ca9-22 cells were performed (Figure 4A,C, respectively). Figure 4B,D showed that low concentrations of WFA suppressed transwell migration and the Matrigel invasion abilities of Ca9-22 cells in a dose-response manner. In contrast, the WFA-induced 3D migration inhibition and invasion were suppressed by NAC pretreatment. Therefore, low concentrations of WFA triggers inhibitory 3D migration and invasion in Ca9-22 cells.

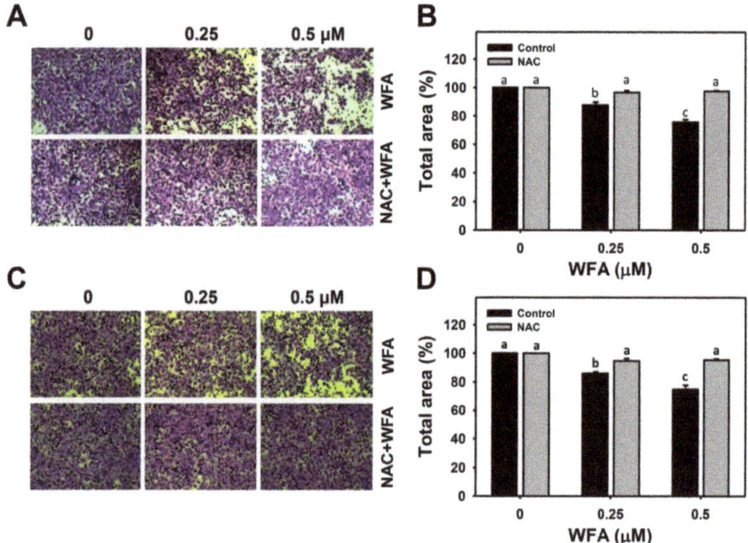

Figure 4. Three-dimensional anti-migration and -invasion effects of low concentrations of WFA in oral cancer cells. (**A,C**) 3D migration and invasion images of Ca9-22 cells after NAC and/or WFA treatments. Cells were pretreated with or without NAC (2 mM, 1 h) and post-treated with different concentrations of WFA for 21 h. (**B,D**) Statistics of 3D migration and invasion changes in Figure 4A,B. For multiple comparison, treatments without the same labels (a–c) indicate the significant difference. $p < 0.001\sim0.0001$ (**B**) and $p < 0.01\sim0.001$ (**D**). Data, mean $\pm$ SD ($n = 3$).

3.5. MMP-2 and MMP-9 Zymography of Oral Cancer Ca9-22 Cells at Low Concentrations of WFA

MMP-2 and MMP-9 activities were proportional to the cell invasion ability [32]. To detect MMP-2 and MMP-9 activities after low concentrations of WFA treatment, a zymography assay was performed. Figure 5 demonstrated the clear zone pattern of MMP-2 and MMP-9 in Ca9-22 cells after NAC and/or WFA treatment. It showed that the MMP-2 and MMP-9 activities of Ca9-22 cells were decreased after WFA treatment. In contrast, these WFA-induced inhibitions of MMP-2 and MMP-9 activities were suppressed by NAC pretreatment. Therefore, low concentrations of WFA triggers inhibition of MMP-2 and MMP-9 activities in Ca9-22 cells.

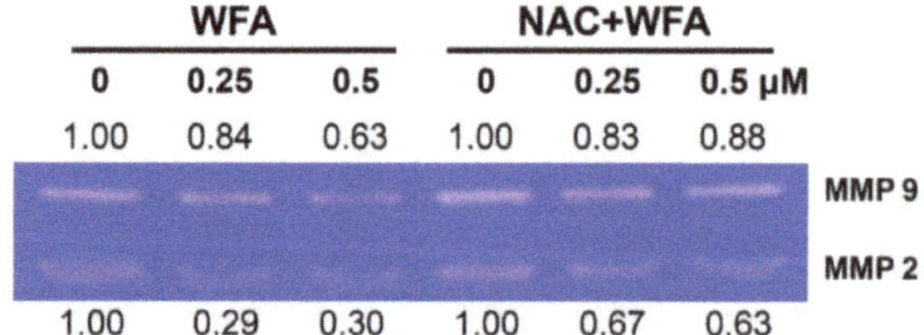

Figure 5. MMP-2 and MMP-9 activities of low concentrations of WFA in oral cancer cells. Zymography-detecting MMP-2 and MMP-9 activities in Ca9-22 cells after NAC and/or WFA treatments. Cells were pretreated with or without NAC (2 mM, 1 h) and post-treated with different concentrations of WFA for 48 h. Similar experiments were repeated 3 times.

3.6. Antioxidant Gene Expressions of Oral Cancer Ca9-22 Cells at Low Concentrations of WFA

Under oxidative stress, ROS may activate antioxidant pathways [33,34]. Since moderate ROS is induced by low concentrations of WFA, the mRNA expressions of antioxidant genes [27], including

NFE2L2, GSR, GCLC, GPX1, TXN, CAT, SOD1, HMOX1, and *NQO1*, were examined. Figure 6 showed that low concentrations of WFA significantly induced mRNA expressions of *NFE2L2, HMOX1, GSR*, and *NQO1* genes while expressions of other genes were not significantly affected. Therefore, low concentrations of WFA triggers some antioxidant signaling in Ca9-22 cells.

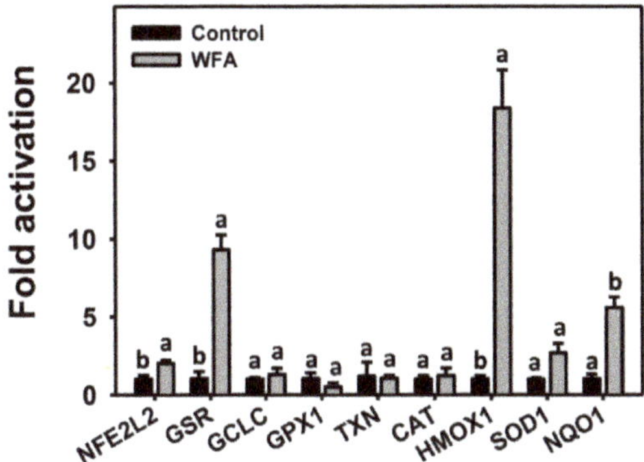

Figure 6. mRNA expressions of antioxidant genes of low concentrations of WFA in oral cancer cells. Cells were treated with or without 0.5 µM of WFA for 24 h. Treatments (control vs. WFA) without the same labels (a,b) indicate the significant difference. $p < 0.05\sim0.01$. Data, mean ± SD ($n = 2$).

3.7. Mitogen-Activated Protein Kinase (MAPK) Expressions of Oral Cancer Ca9-22 Cells at Low Concentrations of WFA

To further detect the potential upstream antioxidant signaling in oral cancer cells after low concentrations of WFA treatment, the activation of three members of MAPK, including ERK, JNK, and p38 MAPK was examined. Figure 7 showed that WFA induced phosphorylation of three MAPK members, i.e., p-ERK1/2, p-JNK1/2, and p-p38. In contrast, these WFA-induced MAPK phosphorylations were suppressed by NAC pretreatment. Therefore, low concentrations of WFA triggers MAPK phosphorylations in Ca9-22 cells.

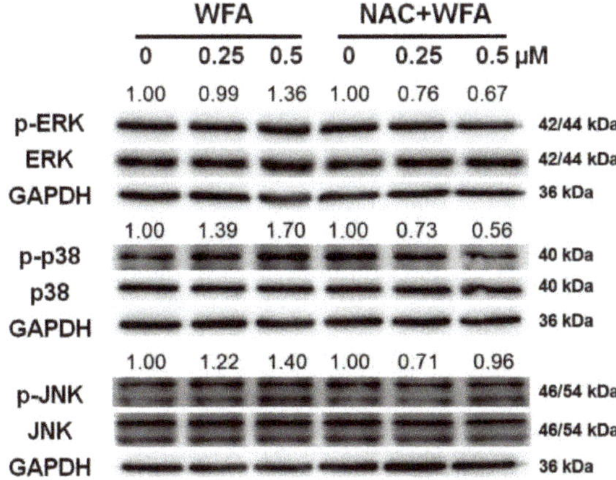

Figure 7. MAPK changes of low concentrations of WFA in oral cancer cells. Ca9-22 cells were pretreated with or without NAC (2 mM, 1 h) and post-treated with different concentrations of WFA for 24 h. ERK1/2, JNK1/2, p38, p-ERK1/2, p-JNK1/2, and p-p38 expressions were detected by Western blotting. The intensity ratio for each p-MAPK expression was adjusted to its matched MAPK and GAPDH intensities. Similar experiments were repeated 3 times.

4. Discussion

Previously, we discovered that high cytotoxic concentration of WFA, which was larger than IC_{50}, selectively killed oral cancer cells but rarely damaged normal oral cells [12], i.e., IC_{50} value of WFA in oral cancer Ca9-22 cells is 3 µM at 24 h MTS assay. In the current study, we focus on the evaluation of the migration regulating effects of low concentration (within 0.5 µM) of WFA in oral cancer Ca9-22 cells, which show 95% viability. This low concentration of WFA inhibits 2D/3D migration, 3D invasion, MMP-2 and MMP-9 activities, whereas it induces ROS generation, antioxidant related gene mRNA expressions and MAPK phosphorylation. The detailed mechanisms for low concentration of WFA inducing inhibition of migration and invasion are discussed below.

4.1. Low Cytotoxic Concentration of Drugs Is Suitable for Migration Study

The standard criteria for studying the migration effect of drugs is based on measurements using low cytotoxic concentrations [35–37]. With a high cytotoxic concentration (higher than IC_{50}), WFA had been reported to show migration inhibitory effects against gastric [14] and breast [15] cancer cells, though it may be attributed to apoptosis and cell death. Alternatively, low concentration of WFA with no cytotoxicity avoided side effect of cell death and provided a clear observation for migration response in the current study.

4.2. MMP-2 and MMP-9 Activity Changes in WFA-Treated Oral Cancer Cells

MMP-2 and MMP-9 are important mediators for cell migration, invasion, and metastasis in carcinogenesis [24]. A WFA-derived compound such as 3-azido WFA inhibits MMP-2 activity and migration of prostate PC-3 and cervical HeLa cancer cells [38]. Low concentration of WFA (>95% viability) inhibits MMP-9 activity of cervical Caski and liver SK-Hep-1 cancer cells by downregulating Akt phosphorylation [39].

In agreement with the inhibitory effect on MMP-9 activity [39], we further found that low concentration of WFA (>95% viability) exhibits inhibitory effects on MMP-2 activity in oral cancer Ca9-22 cells. Accordingly, WFA inhibits migration of oral cancer cells by inactivating MMP-2 and MMP-9. Moreover, MMP-2 and MMP-9 are overexpressed in the biopsy specimens of oral squamous cell carcinoma compared to the adjacent normal tissues [40,41]. Therefore, a low concentration of WFA has the potential to inhibit the MMP-2 and MMP-9 activities in order to inhibit migration or metastasis of oral cancer cells.

4.3. ROS Changes in WFA-Treated Oral Cancer Cells

As mentioned above, WFA exhibits a concentration-effect on apoptosis and migration, i.e., high concentration of WFA induces apoptosis while low concentration of WFA inhibits migration. Our previous study [12] demonstrated that the cytotoxic concentrations (>IC_{50}) of WFA induced 90% (+) ROS in oral cancer Ca9-22 cells. In the current study, the low concentration (>95% viability) of WFA induces 70% (+) ROS generation in Ca9-22 cells. It is possible that low concentration of WFA induces a ROS level lower than the redox threshold and leads to cell survival with inhibitory migration. In contrast, high cytotoxic concentration of WFA induces a ROS level higher than the redox threshold and leads to apoptosis and cell death. Accordingly, the differential ROS induction by WFA may lead to distinct fate of oral cancer cells, i.e., migration inhibition or inducible apoptosis.

4.4. Antioxidant Genes Changes in WFA-Treated Oral Cancer Cells

In cancer cells, ROS overproduction is counterbalanced by overexpression of antioxidant activity for redox homeostasis [42]. Moreover, antioxidant genes have the potential to regulate cellular migration. For example, knockdown of *HMOX1* and/or *NFE2L2* reversed the migration inhibitory effect of semaphorin 6A (SEMA6A) and the SEMA6A-driven downregulation of MMP-9 [43]. Knockdown of *NQO1* increases the invasion of human cutaneous squamous cancer SCC12 and SCC13 cells but it is reverted by *NQO1* overexpression [44]. Consistently, we found that low concentrations of WFA induced mRNA expressions of *NFE2L2*, *HMOX1*, and *NQO1* genes, which may lead to inhibitory migration of oral cancer cells.

4.5. MAPK Changes in WFA-Treated Oral Cancer Cells

As mentioned above, both high [12] and low (the current study) concentrations of WFA induced ROS. Moreover, ROS can regulate MAPK signaling [45], which is associated with tumor cell invasion [46]. Cytotoxic concentration of WFA induces apoptosis by phosphorylating p38 and ERK1/2 in leukemic [47] and glioblastomas cells [48], respectively. Similarly, we found that low concentration of WFA induces mild phosphorylation for ERK, JNK, and p38 MAPK.

4.6. The Role of ROS in Low Concentration of WFA Induced Migration Changes and Signaling in Oral Cancer Cells

Under low concentration of WFA, the changes of ROS generation, 2D migration, 3D migration/invasion, MMP-2/MMP-9 activities, antioxidant gene expression, and MAPK phosphorylation are reverted by NAC pretreatment. These results indicate that a low concentration of WFA inhibits migration and induces antioxidant signaling in a ROS-dependent manner in oral cancer cells.

5. Conclusions

Our study focuses on low concentrations of WFA to evaluate its inhibitory effects on migration and invasion in oral cancer Ca9-22 cells. Under low concentrations of WFA, Ca9-22 cells are grown with high viability and retained anti-migration and anti-invasion. Mechanically, this safe treatment of WFA inhibits MMP-2 and MMP-9 activities and induces antioxidant gene expression as well as MAPK activation in oral cancer cells. All these inhibitory migration changes and mechanisms after WFA treatment were suppressed by NAC pretreatment, suggesting that ROS plays an important role in WFA induced inhibitory migration in oral cancer cells. In conclusion, we provide here the first finding that supports low concentration of WFA could be a potent inhibitor for metastasis in oral cancer therapy.

Author Contributions: Conceptualization, L.-C.L. and H.-W.C.; Data curation, J.-Y.T., F.O.-Y., Y.-Y.W. and S.-S.F.Y.; Formal analysis, T.-J.Y.; Investigation, T.-J.Y.; Methodology, J.-Y.T., F.O.-Y., Y.-Y.W. and S.-S.F.Y.; Supervision, L.-C.L. and H.-W.C.; Writing—original draft, T.-J.Y. and H.-W.C.; Writing—review & editing, K.T., L.-C.L. and H.-W.C. All authors have read and agreed to the published version of the manuscript.

Funding: This work was partly supported by funds of the Ministry of Science and Technology (MOST 108-2320-B-037-015-MY3, MOST 108-2314-B-037-020, MOST 108-2314-B-037-080), the National Sun Yat-sen University-KMU Joint Research Project (#NSYSUKMU 109-I002), the Chimei-KMU jointed project (109CM-KMU-007), the Kaohsiung Medical University Research Center (KMU-TC108A04), and the Health and welfare surcharge of tobacco products, the Ministry of Health and Welfare, Taiwan, Republic of China (MOHW109-TDU-B-212-134016).

Conflicts of Interest: The authors declare no conflict of interest.

Abbreviations

WFA: withaferin A; ROS, reactive oxygen species; MMP, matrix metalloproteinase; *NFE2L2*, nuclear factor, erythroid 2-like 2; *HMOX1*, heme oxygenase 1; *GSR*, glutathione-disulfide reductase; *NQO1*, NAD(P)H quinone dehydrogenase 1; MAPK, mitogen-activated protein kinase; ERK1/2, extracellular signal-regulated kinases 1/2; JNK, c-Jun N-terminal kinase; NAC, *N*-acetylcysteine; DCFH-DA, 2′,7′-Dichlorodihydrofluorescein diacetate; qRT-PCR, quantitative RT-PCR; *NFE2L2*, nuclear factor erythroid 2-like 2; *GCLC*, glutamate-cysteine ligase catalytic subunit; *GPX1*, glutathione peroxidase 1; *TXN*, thioredoxin; *CAT*, catalase; *SOD1*, superoxide dismutase 1.

References

1. Bezivin, C.; Tomasi, S.; Lohezic-Le Devehat, F.; Boustie, J. Cytotoxic activity of some lichen extracts on murine and human cancer cell lines. *Phytomedicine* **2003**, *10*, 499–503. [CrossRef] [PubMed]
2. Jimenez, L.; Jayakar, S.K.; Ow, T.J.; Segall, J.E. Mechanisms of invasion in head and neck cancer. *Arch. Pathol. Lab. Med.* **2015**, *139*, 1334–1348. [CrossRef]

3.   Ghate, N.B.; Chaudhuri, D.; Sarkar, R.; Sajem, A.L.; Panja, S.; Rout, J.; Mandal, N. An antioxidant extract of tropical lichen, *Parmotrema reticulatum*, induces cell cycle arrest and apoptosis in breast carcinoma cell line MCF-7. *PLoS ONE* **2013**, *8*, e82293. [CrossRef] [PubMed]

4.   Noguti, J.; De Moura, C.F.; De Jesus, G.P.; Da Silva, V.H.; Hossaka, T.A.; Oshima, C.T.; Ribeiro, D.A. Metastasis from oral cancer: An overview. *Cancer Genom. Proteom.* **2012**, *9*, 329–335.

5.   Malik, F.; Kumar, A.; Bhushan, S.; Khan, S.; Bhatia, A.; Suri, K.A.; Qazi, G.N.; Singh, J. Reactive oxygen species generation and mitochondrial dysfunction in the apoptotic cell death of human myeloid leukemia HL-60 cells by a dietary compound withaferin A with concomitant protection by N-acetyl cysteine. *Apoptosis* **2007**, *12*, 2115–2133. [CrossRef] [PubMed]

6.   Munagala, R.; Kausar, H.; Munjal, C.; Gupta, R.C. Withaferin A induces p53-dependent apoptosis by repression of HPV oncogenes and upregulation of tumor suppressor proteins in human cervical cancer cells. *Carcinogenesis* **2011**, *32*, 1697–1705. [CrossRef]

7.   Li, X.; Zhu, F.; Jiang, J.; Sun, C.; Wang, X.; Shen, M.; Tian, R.; Shi, C.; Xu, M.; Peng, F.; et al. Synergistic antitumor activity of withaferin A combined with oxaliplatin triggers reactive oxygen species-mediated inactivation of the PI3K/AKT pathway in human pancreatic cancer cells. *Cancer Lett.* **2015**, *357*, 219–230. [CrossRef]

8.   Wang, H.C.; Hu, H.H.; Chang, F.R.; Tsai, J.Y.; Kuo, C.Y.; Wu, Y.C.; Wu, C.C. Different effects of 4beta-hydroxywithanolide E and withaferin A, two withanolides from Solanaceae plants, on the Akt signaling pathway in human breast cancer cells. *Phytomedicine* **2019**, *53*, 213–222. [CrossRef]

9.   Hsu, J.H.; Chang, P.M.; Cheng, T.S.; Kuo, Y.L.; Wu, A.T.; Tran, T.H.; Yang, Y.H.; Chen, J.M.; Tsai, Y.C.; Chu, Y.S.; et al. Identification of withaferin A as a potential candidate for anti-cancer therapy in non-small cell lung cancer. *Cancers* **2019**, *11*, 1003. [CrossRef]

10.   Xia, S.; Miao, Y.; Liu, S. Withaferin A induces apoptosis by ROS-dependent mitochondrial dysfunction in human colorectal cancer cells. *Biochem. Biophys. Res. Commun.* **2018**, *503*, 2363–2369. [CrossRef]

11.   Yang, I.-H.; Kim, L.-H.; Shin, J.-A.; Cho, S.-D. Chemotherapeutic effect of withaferin A in human oral cancer cells. *J. Cancer Ther.* **2015**, *6*, 735–742. [CrossRef]

12.   Chang, H.W.; Li, R.N.; Wang, H.R.; Liu, J.R.; Tang, J.Y.; Huang, H.W.; Chan, Y.H.; Yen, C.Y. Withaferin A induces oxidative stress-mediated apoptosis and DNA damage in oral cancer cells. *Front. Physiol.* **2017**, *8*, 634. [CrossRef] [PubMed]

13.   Ghosh, K.; De, S.; Mukherjee, S.; Das, S.; Ghosh, A.N.; Sengupta, S.B. Withaferin A induced impaired autophagy and unfolded protein response in human breast cancer cell-lines MCF-7 and MDA-MB-231. *Toxicol. In Vitro* **2017**, *44*, 330–338. [CrossRef] [PubMed]

14.   Kim, G.; Kim, T.H.; Hwang, E.H.; Chang, K.T.; Hong, J.J.; Park, J.H. Withaferin A inhibits the proliferation of gastric cancer cells by inducing G2/M cell cycle arrest and apoptosis. *Oncol. Lett.* **2017**, *14*, 416–422. [CrossRef]

15.   Liu, X.; Li, Y.; Ma, Q.; Wang, Y.; Song, A.L. Withaferin-A inhibits growth of drug-resistant breast carcinoma by inducing apoptosis and autophagy, endogenous reactive oxygen species (ROS) production, and inhibition of cell migration and nuclear factor kappa B (Nf-kappaB)/mammalian target of rapamycin (m-TOR) signalling pathway. *Med. Sci. Monit.* **2019**, *25*, 6855–6863.

16.   Hurd, T.R.; DeGennaro, M.; Lehmann, R. Redox regulation of cell migration and adhesion. *Trends Cell Biol.* **2012**, *22*, 107–115. [CrossRef]

17.   Chang, Y.T.; Huang, C.Y.; Li, K.T.; Li, R.N.; Liaw, C.C.; Wu, S.H.; Liu, J.R.; Sheu, J.H.; Chang, H.W. Sinuleptolide inhibits proliferation of oral cancer Ca9-22 cells involving apoptosis, oxidative stress, and DNA damage. *Arch. Oral Biol.* **2016**, *66*, 147–154. [CrossRef]

18.   Huang, C.H.; Yeh, J.M.; Chan, W.H. Hazardous impacts of silver nanoparticles on mouse oocyte maturation and fertilization and fetal development through induction of apoptotic processes. *Environ. Toxicol.* **2018**, *33*, 1039–1049. [CrossRef]

19.   Wang, T.S.; Lin, C.P.; Chen, Y.P.; Chao, M.R.; Li, C.C.; Liu, K.L. CYP450-mediated mitochondrial ROS production involved in arecoline N-oxide-induced oxidative damage in liver cell lines. *Environ. Toxicol.* **2018**, *33*, 1029–1038. [CrossRef]

20.   Yen, Y.H.; Farooqi, A.A.; Li, K.T.; Butt, G.; Tang, J.Y.; Wu, C.Y.; Cheng, Y.B.; Hou, M.F.; Chang, H.W. Methanolic extracts of *Solieria robusta* inhibits proliferation of oral cancer Ca9-22 cells via apoptosis and oxidative stress. *Molecules* **2014**, *19*, 18721–18732. [CrossRef]

21. Yen, C.Y.; Chiu, C.C.; Haung, R.W.; Yeh, C.C.; Huang, K.J.; Chang, K.F.; Hseu, Y.C.; Chang, F.R.; Chang, H.W.; Wu, Y.C. Antiproliferative effects of goniothalamin on Ca9-22 oral cancer cells through apoptosis, DNA damage and ROS induction. *Mutat. Res.* **2012**, *747*, 253–258. [CrossRef] [PubMed]

22. Chiu, C.C.; Liu, P.L.; Huang, K.J.; Wang, H.M.; Chang, K.F.; Chou, C.K.; Chang, F.R.; Chong, I.W.; Fang, K.; Chen, J.S.; et al. Goniothalamin inhibits growth of human lung cancer cells through DNA damage, apoptosis, and reduced migration ability. *J. Agric. Food Chem.* **2011**, *59*, 4288–4293. [CrossRef] [PubMed]

23. Peng, S.Y.; Hsiao, C.C.; Lan, T.H.; Yen, C.Y.; Farooqi, A.A.; Cheng, C.M.; Tang, J.Y.; Yu, T.J.; Yeh, Y.C.; Chuang, Y.T.; et al. Pomegranate extract inhibits migration and invasion of oral cancer cells by downregulating matrix metalloproteinase-2/9 and epithelial-mesenchymal transition. *Environ. Toxicol.* **2020**, *35*, 673–682. [CrossRef] [PubMed]

24. Wang, C.Y.; Lin, C.S.; Hua, C.H.; Jou, Y.J.; Liao, C.R.; Chang, Y.S.; Wan, L.; Huang, S.H.; Hour, M.J.; Lin, C.W. Cis-3-O-p-hydroxycinnamoyl ursolic acid induced ROS-dependent p53-mediated mitochondrial apoptosis in oral cancer cells. *Biomol. Ther. (Seoul)* **2019**, *27*, 54–62. [CrossRef]

25. Chang, H.W.; Yen, C.Y.; Chen, C.H.; Tsai, J.H.; Tang, J.Y.; Chang, Y.T.; Kao, Y.H.; Wang, Y.Y.; Yuan, S.F.; Lee, S.Y. Evaluation of the mRNA expression levels of integrins alpha3, alpha5, beta1 and beta6 as tumor biomarkers of oral squamous cell carcinoma. *Oncol. Lett.* **2018**, *16*, 4773–4781.

26. Yen, C.Y.; Huang, C.Y.; Hou, M.F.; Yang, Y.H.; Chang, C.H.; Huang, H.W.; Chen, C.H.; Chang, H.W. Evaluating the performance of fibronectin 1 (FN1), integrin alpha4beta1 (ITGA4), syndecan-2 (SDC2), and glycoprotein CD44 as the potential biomarkers of oral squamous cell carcinoma (OSCC). *Biomarkers* **2013**, *18*, 63–72. [CrossRef]

27. Stagos, D.; Balabanos, D.; Savva, S.; Skaperda, Z.; Priftis, A.; Kerasioti, E.; Mikropoulou, E.V.; Vougogiannopoulou, K.; Mitakou, S.; Halabalaki, M.; et al. Extracts from the mediterranean food plants *Carthamus lanatus*, *Cichorium intybus*, and *Cichorium spinosum* enhanced GSH levels and increased Nrf2 expression in human endothelial cells. *Oxid. Med. Cell Longev.* **2018**, *2018*, 6594101. [CrossRef]

28. Livak, K.J.; Schmittgen, T.D. Analysis of relative gene expression data using real-time quantitative PCR and the 2(-Delta Delta C(T)) Method. *Methods* **2001**, *25*, 402–408. [CrossRef]

29. Brunn, G.J.; Williams, J.; Sabers, C.; Wiederrecht, G.; Lawrence, J.C., Jr.; Abraham, R.T. Direct inhibition of the signaling functions of the mammalian target of rapamycin by the phosphoinositide 3-kinase inhibitors, wortmannin and LY294002. *EMBO J.* **1996**, *15*, 5256–5267. [CrossRef]

30. Fujii, Y.; Yoshihashi, K.; Suzuki, H.; Tsutsumi, S.; Mutoh, H.; Maeda, S.; Yamagata, Y.; Seto, Y.; Aburatani, H.; Hatakeyama, M. CDX1 confers intestinal phenotype on gastric epithelial cells via induction of stemness-associated reprogramming factors SALL4 and KLF5. *Proc. Natl. Acad. Sci. USA* **2012**, *109*, 20584–20589. [CrossRef]

31. Baribeau, S.; Chaudhry, P.; Parent, S.; Asselin, E. Resveratrol inhibits cisplatin-induced epithelial-to-mesenchymal transition in ovarian cancer cell lines. *PLoS ONE* **2014**, *9*, e86987. [CrossRef] [PubMed]

32. Chien, Y.C.; Liu, L.C.; Ye, H.Y.; Wu, J.Y.; Yu, Y.L. EZH2 promotes migration and invasion of triple-negative breast cancer cells via regulating TIMP2-MMP-2/-9 pathway. *Am. J. Cancer Res.* **2018**, *8*, 422–434. [PubMed]

33. Chen, J.; Zhang, Z.; Cai, L. Diabetic cardiomyopathy and its prevention by nrf2: Current status. *Diabetes Metab. J.* **2014**, *38*, 337–345. [CrossRef] [PubMed]

34. Hardingham, G.E.; Do, K.Q. Linking early-life NMDAR hypofunction and oxidative stress in schizophrenia pathogenesis. *Nat. Rev. Neurosci.* **2016**, *17*, 125–134. [CrossRef] [PubMed]

35. Wang, H.C.; Chu, Y.L.; Hsieh, S.C.; Sheen, L.Y. Diallyl trisulfide inhibits cell migration and invasion of human melanoma a375 cells via inhibiting integrin/facal adhesion kinase pathway. *Environ. Toxicol.* **2017**, *32*, 2352–2359. [CrossRef] [PubMed]

36. Shih, Y.L.; Au, M.K.; Liu, K.L.; Yeh, M.Y.; Lee, C.H.; Lee, M.H.; Lu, H.F.; Yang, J.L.; Wu, R.S.; Chung, J.G. Ouabain impairs cell migration, and invasion and alters gene expression of human osteosarcoma U-2 OS cells. *Environ. Toxicol.* **2017**, *32*, 2400–2413. [CrossRef]

37. Yeh, C.M.; Hsieh, M.J.; Yang, J.S.; Yang, S.F.; Chuang, Y.T.; Su, S.C.; Liang, M.Y.; Chen, M.K.; Lin, C.W. Geraniin inhibits oral cancer cell migration by suppressing matrix metalloproteinase-2 activation through the FAK/Src and ERK pathways. *Environ. Toxicol.* **2019**, *34*, 1085–1093. [CrossRef]

38. Rah, B.; Amin, H.; Yousuf, K.; Khan, S.; Jamwal, G.; Mukherjee, D.; Goswami, A. A novel MMP-2 inhibitor 3-azidowithaferin A (3-azidoWA) abrogates cancer cell invasion and angiogenesis by modulating extracellular Par-4. *PLoS ONE* **2012**, *7*, e44039. [CrossRef]

39. Lee, D.H.; Lim, I.H.; Sung, E.G.; Kim, J.Y.; Song, I.H.; Park, Y.K.; Lee, T.J. Withaferin A inhibits matrix metalloproteinase-9 activity by suppressing the Akt signaling pathway. *Oncol. Rep.* **2013**, *30*, 933–938. [CrossRef]

40. Tsai, C.H.; Hsieh, Y.S.; Yang, S.F.; Chou, M.Y.; Chang, Y.C. Matrix metalloproteinase 2 and matrix metalloproteinase 9 expression in human oral squamous cell carcinoma and the effect of protein kinase C inhibitors: Preliminary observations. *Oral Surg. Oral Med. Oral Pathol. Oral Radiol. Endod.* **2003**, *95*, 710–716. [CrossRef]

41. Patel, B.P.; Shah, P.M.; Rawal, U.M.; Desai, A.A.; Shah, S.V.; Rawal, R.M.; Patel, P.S. Activation of MMP-2 and MMP-9 in patients with oral squamous cell carcinoma. *J. Surg. Oncol.* **2005**, *90*, 81–88. [CrossRef] [PubMed]

42. Gorrini, C.; Harris, I.S.; Mak, T.W. Modulation of oxidative stress as an anticancer strategy. *Nat. Rev. Drug Discov.* **2013**, *12*, 931–947. [CrossRef] [PubMed]

43. Chen, L.H.; Liao, C.Y.; Lai, L.C.; Tsai, M.H.; Chuang, E.Y. Semaphorin 6A attenuates the migration capability of lung cancer cells via the NRF2/HMOX1 axis. *Sci. Rep.* **2019**, *9*, 13302. [CrossRef] [PubMed]

44. Zhang, Q.L.; Li, X.M.; Lian, D.D.; Zhu, M.J.; Yim, S.H.; Lee, J.H.; Jiang, R.H.; Kim, C.D. Tumor suppressive function of NQO1 in cutaneous squamous cell carcinoma (SCC) cells. *Biomed. Res. Int.* **2019**, *2019*, 2076579. [CrossRef] [PubMed]

45. Chen, Y.C.; Lu, M.C.; El-Shazly, M.; Lai, K.H.; Wu, T.Y.; Hsu, Y.M.; Lee, Y.L.; Liu, Y.C. Breaking down leukemia walls: Heteronemin, a sesterterpene derivative, induces apoptosis in leukemia Molt4 cells through oxidative stress, mitochondrial dysfunction and induction of talin Expression. *Mar. Drugs* **2018**, *16*, 212. [CrossRef]

46. Peng, Q.; Deng, Z.; Pan, H.; Gu, L.; Liu, O.; Tang, Z. Mitogen-activated protein kinase signaling pathway in oral cancer. *Oncol. Lett.* **2018**, *15*, 1379–1388. [CrossRef]

47. Mandal, C.; Dutta, A.; Mallick, A.; Chandra, S.; Misra, L.; Sangwan, R.S.; Mandal, C. Withaferin A induces apoptosis by activating p38 mitogen-activated protein kinase signaling cascade in leukemic cells of lymphoid and myeloid origin through mitochondrial death cascade. *Apoptosis* **2008**, *13*, 1450–1464. [CrossRef]

48. Grogan, P.T.; Sleder, K.D.; Samadi, A.K.; Zhang, H.; Timmermann, B.N.; Cohen, M.S. Cytotoxicity of withaferin A in glioblastomas involves induction of an oxidative stress-mediated heat shock response while altering Akt/mTOR and MAPK signaling pathways. *Investig. New Drugs* **2013**, *31*, 545–557. [CrossRef]

 *biomolecules*

Article

A Biochemometric Approach for the Identification of In Vitro Anti-Inflammatory Constituents in Masterwort

Julia Zwirchmayr [1], Ulrike Grienke [1], Scarlet Hummelbrunner [1], Jacqueline Seigner [2], Rainer de Martin [2], Verena M. Dirsch [1] and Judith M. Rollinger [1,*]

[1] Department of Pharmacognosy, Faculty of Life Sciences, University of Vienna, Althanstraße 14, 1090 Vienna, Austria; julia.zwirchmayr@univie.ac.at (J.Z.); ulrike.grienke@univie.ac.at (U.G.); scarlet.hummelbrunner@univie.ac.at (S.H.); verena.dirsch@univie.ac.at (V.M.D.)

[2] Department of Vascular Biology and Thrombosis Research, Medical University of Vienna, Schwarzspanierstaße. 17, 1090 Vienna, Austria; jacqueline.seigner@meduniwien.ac.at (J.S.); rainer.demartin@meduniwien.ac.at (R.d.M.)

* Correspondence: judith.rollinger@univie.ac.at; Tel.: +43-1-4277-55255; Fax: +43-1-4277-855255

Received: 26 March 2020; Accepted: 21 April 2020; Published: 28 April 2020

Abstract: *Peucedanum ostruthium* (L.) Koch, commonly known as masterwort, has a longstanding history as herbal remedy in the Alpine region of Austria, where the roots and rhizomes are traditionally used to treat disorders of the gastrointestinal and respiratory tract. Based on a significant NF-κB inhibitory activity of a *P. ostruthium* extract (PO-E), this study aimed to decipher those constituents contributing to the observed activity using a recently developed biochemometric approach named ELINA (Eliciting Nature's Activities). This -omics tool relies on a deconvolution of the multicomponent mixture, which was employed by generating microfractions with quantitative variances of constituents over several consecutive fractions. Using an optimized and single high-performance counter-current chromatographic (HPCCC) fractionation step 31 microfractions of PO-E were obtained. ^{1}H NMR data and bioactivity data from three in vitro cell-based assays, i.e., an NF-kB reporter-gene assay and two NF-κB target-gene assays (addressing the endothelial adhesion molecules E-selectin and VCAM-1) were collected for all microfractions. Applying heterocovariance analyses (HetCA) and statistical total correlation spectroscopy (STOCSY), quantitative variances of ^{1}H NMR signals of neighboring fractions and their bioactivities were correlated. This revealed distinct chemical features crucial for the observed activities. Complemented by LC-MS-CAD data this biochemometric approach differentiated between active and inactive constituents of the complex mixture, which was confirmed by NF-κB reporter-gene testing of the isolates. In this way, four furanocoumarins (imperatorin, ostruthol, saxalin, and 2′-O-acetyloxypeucedanin), one coumarin (ostruthin), and one chromone (peucenin) were identified as NF-κB inhibiting constituents of PO-E contributing to the observed NF-kB inhibitory activity. Additionally, this approach also enabled the disclose of synergistic effects of the PO-E metabolites imperatorin and peucenin. In sum, prior to any isolation an early identification of even minor active constituents, e.g. peucenin and saxalin, ELINA enables the targeted isolation of bioactive constituents and, thus, to effectively accelerate the NP-based drug discovery process.

Keywords: *Peucedanum ostruthium*; Apiaceae; ELINA; HetCA, STOCSY; coumarines; NF-kB; VCAM-1; E-selectin; inflammation

---

1. Introduction

Natural products (NPs) have been used for millennia as part of herbal remedies to alleviate and treat various types of diseases. Their unique chemical diversity provides a vast source of drug-like molecules endowed with various biological activities, which account for their significant contributions in drug discovery [1–4]. Until now, anti-inflammatory activities of NPs are one of the most frequently reported effects and have been described for many traditionally used herbal drugs in vitro and partly also in vivo, e.g., from *Zingiber officinale* Roscoe [5–7], *Symphytum officinale* L. [8,9], *Vaccinium myrtillus* L. [10–12], and *Calendula officinalis* L. [13,14], as well as for constituents of herbal remedies, such as curcumin [15–19] and resveratrol [20–22]. However, because of the intrinsic complexity of the inflammation process, the search for new anti-inflammatory compounds remains a challenging task [23]. Inflammation underlies a wide range of physiological and pathological changes and is triggered by microbial, chemical or physical stimuli, such as infection, tissue injury, and traumata. Activation of NF-kB leads to an enhanced expression of the cell adhesion molecules ICAM-1, VCAM-1, and E-selectin. Here, the activation of endothelial cells in blood vessels near the site of the injury plays a crucial role during the inflammation cascade as it promotes the chemoattraction, adhesion, and transmigration of leucocytes in the affected tissue. Hence, NF-κB plays a central role in the transcriptional regulation of inflammatory mediators and represents a rational target for intervention [8,24,25].

Nature is a prolific source of novel secondary metabolites, able to interfere with key players in inflammation [23] and their multi-target properties play an advantageous role when dealing with complex diseases. The structural diversity of NPs allows for identifying novel bioactives or the recognition of similar congeners with potential activity [26]. A previously published ethnopharmacological study on 71 Austrian herbal drugs, traditionally used to treat inflammation, revealed NF-κB inhibition for a detannified methanol extract and a dichloromethane extract generated from the roots of *Peucedanum ostruthium* (L.) W. D. J. Koch, also known as masterwort [27]. Masterwort, belonging to the Apiaceae family, has a longstanding history as a herbal remedy in the Alpine region of Austria. The rhizomes and roots (Radix Imperatoriae) are traditionally used to treat disorders of the respiratory tract, the cardiovascular system [28] as well as gastrointestinal diseases like stomach pain or ulcer [29]. One main compound class present in *P. ostruthium* are coumarins, such as osthole and ostruthin, as well as furanocoumarins such as oxypeucedanin, ostruthol, or imperatorin [28]. Although several studies on the isolation and identification of coumarins from *P. ostruthium* have been performed, the secondary metabolites responsible for the observed NF-kB inhibition of masterwort root extracts remain elusive.

One goal in NP-based drug discovery is to disclose constituents from complex mixtures responsible for a certain biological effect. Traditionally, bioassay-guided fractionation is applied as the gold standard to simplify these complex mixtures (i.e., extracts) and to isolate the comprised bioactive constituents. Although this approach has led to the discovery of drugs with great significance (e.g., taxol and artemisinin) [30], bioassay-guided fractionation has its restrictions: (i) bioactive compounds in low quantity are easily overlooked because of the presence of highly abundant compounds, (ii) bioactivity can get lost due to degradation or adhesion of compounds to chromatographic materials during repetitive fractionation steps [31], and (iii) synergistic or additive effects among several compounds are difficult to recognize [1,30,32–34]. Due to its intrinsic complexity, NP research is hampered by tedious fractionation and isolation steps that often results in the repeated isolation and identification of already known constituents. This can be avoided by the early identification of known NPs in complex mixtures (i.e., dereplication) [35]. In the recent past, there has been a great interest in the implementation of bioactivity data with chemical profiles to unveil the biologically active constituent(s) in multicomponent mixtures [33,36,37]. This so-called biochemometric approach is achieved by the correlation of biological and chemical datasets via multivariate statistics [33,36,37], as recently implemented by the NMR- and MS-based ELINA approach (Eliciting Nature's Activities) [32]. ELINA offers significant benefits over bioassay-guided fractionation, namely (i) the semi-quantitative estimation of secondary metabolite levels in mixtures by LC hyphenated to MS and a charged aerosol detector (CAD) [35]; (ii) the quantitative composition of secondary metabolites via ^{1}H NMR, and (iii) the in situ structural characterization of bioactives and inactive constituents prior to isolation by statistically correlating ^{1}H NMR profiles to bioactivity data [36,38]. Whereas in previous similar biochemometric approaches, structural data were correlated with bioactivity data derived from enzyme assays [36–38], we here probed the robustness and predictive power of ELINA when using the readout from a cell-based assay. This application study accordingly aimed to disclose those metabolites in the extract of masterwort, which contributes to its previously found NF-kB inhibitory activity. To achieve this aim, we applied the following workflow:

(i)   microfractionation of a bioactive extract prepared from the roots of *P. ostruthium* using high-performance counter-current chromatography (HPCCC);

(ii)   investigation of the masterwort extract and its generated microfractions for their ability to interfere with the NF-kB signaling pathway, and thus the expression of pro-inflammatory target genes (E-selectin, VCAM-1) in cell-based models;

(iii) recording of ^{1}H NMR and LC-MS-CAD data of all microfractions;

(iv)   correlation of structural data with bioactivity data to structurally identify and distinguish the bioactive/inactive metabolites from the extract, and

(v)   isolation and assaying of ELINA predicted active and inactive constituents from the masterwort root extract for validation of the used approach.

2. Materials and Methods

2.1. General Experimental Procedures

HPCCC fractionation was performed on a Spectrum HPCCC instrument (Dynamic Extractions Ltd) connected with an isocratic solvent pump (ecom Alpha 10+), a fraction collector and a chiller. UPLC analysis was performed on a Waters Acquity UPLC system (H-class) equipped with a binary solvent manager, a sample manager, a column manager, a PDA detector, an ELSD, and a fraction collector using a Waters Acquity UPLC BEH Phenyl column (1.7 µm, 2.1 × 100 mm). PO01_01-PO01_31 were chromatographed over a Dionex HPLC connected to a charged aerosol detector (CAD) and an MS Iontrap (LTQ XL™ Linear Ion Trap Mass Spectrometer, Thermo Fisher Scientific Inc. Bremen, Germany) equipped with ESI with a Waters Acquity UPLC BEH Phenyl column (1.7 µm, 2.1 × 100 mm). Flash chromatography was performed on an Interchim puriFlash 4250 system (Montluçon, France), equipped with an evaporative light scattering detector (ELSD), a photodiode array (PDA) detector, and a fraction collector, controlled by Interchim Software. Sephadex column chromatography (CC) was performed with Sephadex LH-20 in 100% MeOH. TLC analyses were performed with toluol:ether:10% aqueous acetic acid (1:1:1, upper layer) as mobile phase. Stationary phase: Merck silica gel 60 PF254, detection under both visible light and UV254 and UV366. NMR experiments were performed on a Bruker Avance 500 NMR spectrometer (UltraShield) (Bruker, Billerica, MA) with a 5 mm probe (TCI Prodigy CryoProbe, 5 mm, triple resonance inverse detection probe head) with z-axis gradients and automatic tuning and matching accessory (Bruker BioSpin). The resonance frequency for ^{1}H NMR was 500.13 MHz and for ^{13}C NMR 125.75 MHz. Standard 1D and gradient-enhanced (ge) 2D experiments, like double quantum filtered (DQF) COSY, HSQC, and HMBC, were used as supplied by the manufacturer. (Ultrahigh-)gradient grade solvents from Merck (Darmstadt, Germany) and deuterated solvents from Deutero GmbH (Kastellaun, Germany) were used.

2.2. Plant Material

Dried *Peucedanum ostruthium* roots and rhizomes were purchased from Kottas Pharma GmbH (Ch.Nr.: P17301770), Vienna. A voucher specimen (JR-20180119-A2) is deposited at the Department of Pharmacognosy, University of Vienna, Austria.

2.3. Extraction

For the extraction of *P. ostruthium* the protocol of [39,40] was used and modified as follows: 1 kg dried plant material was defatted with 2 L *n*-hexane (*n*-hex). The flasks were shaken for three days. The obtained *n*-hex extract was discarded. The remaining defatted plant material was extracted with

2.6 L CH_2Cl_2 and shaken for two days. The CH_2Cl_2 extract was transferred into a round bottom flask and the solvent was evaporated on a rotary evaporator. The remaining plant material was extracted with 2.6 L CH_2Cl_2 for a second time and shaken for one day. All CH_2Cl_2 extracts were combined and dried. The remaining plant material was extracted with 2.6 L MeOH for 7 days and filtrated. For an exhaustive extraction, this procedure was repeated twice. CH_2Cl_2/MeOH extracts were combined and concentrated to dryness on a rotary evaporator. The extraction yielded 348.96 g extract (PO-E; 35.81%).

2.4. UPLC Analysis of PO-E

PO-E was chromatographed over UPLC using a binary mobile phase system consisting of A) H_2O and B) CH_3CN. The gradient was from 13%–98% B in 12 min followed by 5 min re-equilibration. Method in detail: 13% B for 0.5 min, 13%–18% B in 0.5 min, 18%–45% B in 1 min, isocratic 45% B for 1.7 min, 45%–73% B in 2.8 min, 73%–98% B in 0.3 min, isocratic 98% B for 5 min, 98%–13% B in 0.1 min, isocratic 13% B for 0.1 min; Conditions: temperature, 40 °C; flow rate, 0.300 mL/min; injection volume, 1 µL. Detection of compounds using PDA and ELSD. PDA conditions: 210 nm and full range spectra 192–400 nm.

2.5. HPCCC Separation Procedure

2.5.1. Selection of Two-phase Solvent System for HPCCC

Mixtures of *n*-hex, ethyl acetate (EtOAc), methanol (MeOH), and H_2O (HEMWat) with various volume ratios were used for the two-phase solvent system selection. Briefly, a small amount of selected HEMWat solvent mixtures was prepared in small test-tubes with ground joint, whereas one aliquot of the MeOH ratio was replaced by an aliquot of a sample solution of PO-E (5 mg/mL in MeOH). The mixture was shaken vigorously followed by a 10 min equilibration time at room temperature. 1.00 mL of each phase was taken, dried and re-dissolved in 500 µL MeOH for UHPLC analysis. To find the ideal solvent system(s) for the micro-fractionation, the partition coefficient K_D was determined for each solvent system, as described by Garrard [41]: This was done for both, reversed-phase and normal-phase mode. Partition coefficients were expressed as the peak area of selected peaks in the stationary phase divided by the peak area of the corresponding peak in the mobile phase (data not shown). HEMWat systems resulting in K_D values for the main constituents in the range of 0.5 to 5.0 (i.e., systems 22, 21, 19, 17, 15, and 10) were selected for the semi-preparative HPCCC analysis.

2.5.2. Microfractionation with HPCCC

Microfractionation of PO-E was performed in a semi-preparative, normal-phase mode with gradient elution, starting with HEMWat system 22 and gradually increasing the polarity of the mobile phases by subsequently applying the mobile phases (i.e., upper layer) of HEMWat system 21, 20, 19, 17, 15, and 10 (Supporting Information, Table S1). The semi-preparative column was initially filled with the stationary phase (i.e., lower layer) of HEMWat system 22 at 200 rpm with a flow rate of 10 mL/min. After rotating up to 1600 rpm the mobile phase of HEMWat system 22 was pumped through the column with a flow rate of 6.0 mL/min. After the equilibrium was reached, the sample solution of PO-E (269.82 mg dissolved in 10 mL of HEMWat system 22) was injected and the fraction manager was set to 1 min time count (i.e., 6 mL/fraction). For each solvent system, 250 mL of the upper layer were used as the mobile phase for the fractionation, starting with 250 mL of the mobile phase of system 22. When the volume of system 22 was reduced to 60 mL, 50 mL of the upper layer of system 21 were added on top. As the solvent mixture of system 22 and 21 was reduced to 60 mL the remaining 200 mL of the upper layer of system 21 were added. This procedure was repeated for each selected solvent system, resulting in an even fractionation. Elution extrusion was performed with the lower, stationary phase of system 10 at 200 rpm and 10 mL/min. 285 fractions were collected. A second run with 329.40 mg was performed, resulting in 280 fractions. All HPCCC fractions were monitored by TLC and pooled to obtain 31 final microfractions, i.e., PO01_01–PO01_31. (Table S2, Figure S1).

2.6. NMR Measurements of PO01_01-PO01_31

The samples were measured at 298 K in fully deuterated methanol referenced to the residual non-deuterated solvent signal at 3.31 ppm (MeOH). Dry weighted samples (between 2.3 and 2.4 mg) of PO01_01-PO01_31 were dissolved in methanol-d4 to reach a concentration of 3.11 mg/mL. To avoid precipitation in the NMR tube, an aliquot of 750 μL of each fraction was put into an Eppendorf tube and centrifuged at 3000 rpm for 5 min. From the supernatants, 645 μL were transferred to NMR tubes.

Standard ^{1}H NMR spectra with 16 scans and relaxation delay of 6 s were recorded for all fractions using pulse sequences included in the standard pulse program library of Bruker. TopSpin 4.0, controlling a 60-position autosampler, was used for fully automated NMR operation, i.e., temperature control, sample loading, tuning and matching, shimming, lock phase optimization, 90° pulse calibration, and data recording. For the pure compounds, both standard 1D and 2D experiments were performed.

2.7. LC-MS-CAD Measurements of PO01_01–PO01_31

PO01_01-PO01_31 (c = 2 mg/mL in MeOH) were chromatographed using a binary mobile phase system consisting of A) H_2O: formic acid (100:0.01) and B) CH_3CN. The gradient was from 13%–98% B in 20 min. Isocratic 13% B for 0.5 min, 13%–18% B in 0.5 min, 18%–45% B in 1 min, isocratic 45% B for 1.7 min, 45%–73% B in 2.8 min, 73%–98% B in 0.3 min, isocratic 98% B for 5.2 min, 98%–13% B in 0.5 min, isocratic 13% B for 7.5 min; conditions: temperature, 40 °C; flow rate, 0.300 mL/min; injection volume, 10 μL. Detection of compounds using PDA and CAD. PDA conditions: 210 nm and full range spectra 192–400 nm. CAD nebulizer temperature, 35 °C. Mass conditions: source heater temperature, 300 °C; source voltage, 3.7 kV; sheath gas flow rate, 40; aux gas flow rate, 10.

2.8. ^{1}H NMR Spectra Processing and Statistical Correlation with Bioactivity Data

For spectral alignment, all ^{1}H NMR spectra of the 31 microfractions were subjected to chemical shift scale calibration by referencing to the MeOH resonance at 3.31 ppm. Then, a baseline correction factor was applied using a simple polynomial curve fitting of the mathematical equation A + Bx + Cx2 + Dx3 + Ex4. Baseline correction was carried out manually using the appropriate factors. To detect structural features of the active components prior to any purification, the previously described heterocovariance (HetCA) analysis was applied [36]. Briefly, ^{1}H NMR spectra of relevant fraction packages were bucketed (covered range: δ_H 0.5–10; bucket width: 0.0005 ppm). This means that spectra were reduced in their complexity by summation of all the data points per bucket. Since each bucket width was 0.0005 ppm, this procedure gave a total of 19,000 spectroscopic buckets. The intensities of ^{1}H NMR resonances of each bucket were calculated and served as variables for subsequent analyses. Covariance as a measure of the joint variability between the two variables (i) ^{1}H NMR resonance intensity and (ii) percentage of NF-kB inhibition at 10 μg/mL was calculated. Additionally, the normalized version of covariance, i.e., the correlation coefficient, was calculated for color coding. Thus, the resulting bucket-specific covariance values were plotted as ^{1}H NMR pseudo-spectrum and color coded according to the respective correlation coefficients. This procedure allowed for the straightforward identification of ^{1}H NMR resonances which are either positively (red) or negatively (blue) correlated with NF-κB inhibition. HetCA analysis was carried out with spectra of selected sets of fractions which showed a distinct variation in bioactivity and concentration of contained secondary metabolites. Additionally, the statistical total correlation spectroscopy (STOCSY) was applied [42], using the multicolinearity of the intensity variables over a set of spectra to give the correlation among the intensities of the various resonances across the whole set of spectra. STOCSY displays also covariance as a function of spectroscopic position and is color-coded according to the respective correlation coefficients (i.e., the intensities of the various resonances across the whole fraction package). STOCSY allows for the detection of multiple ^{1}H NMR signals from the same molecule based on the multi-collinearity of their intensities in the selected set of ^{1}H NMR spectra. The calculations for HetCA and STOCSY analyses were performed using the multi-paradigm numerical computing environment MATLAB.

2.9. Targeted Isolation of Bioactives

For the isolation of peucenin (**2**), PO-E was fractionated via isocratic HPCCC in semi-preparative normal-phase mode, using an optimized HEMWat system consisting of *n*-hex:EtOAc:MeOH:H_2O (volume ratios 1.8:1.2:2.1:0.9). The sample solution of PO-E (310 mg dissolved in 10 mL of optimized HEMWat system) was injected and fractionation was performed at 1600 rpm with a flow rate of 6.0 mL/min. The fraction manager was set to 1 min time count (i.e., 6 mL/fraction). In total, 79 tubes were collected, monitored with TLC and compared with the TLC pattern of the active microfraction PO01_16. Tubes 42–49 were pooled (5.38 mg) and purified via Sephadex CC collecting 60 sub-fractions. Sub-fractions 31–50 were combined yielding 3.12 mg of **2**. Fraction PO01_22 (3.33 mg) was fractionated via Sephadex CC yielding 79 tubes. Saxalin (**4**) (1.06 mg) was isolated by pooling fraction 46–57. Ostruthol (**5**) and oxypeucedanin methanolate (**6**) were isolated by combining PO01_24 and PO01_25 (~19 mg). The combined microfractions were separated via Sephadex CC, yielding five fractions PO02_01-PO02_05. PO02_02 (3.83 mg), PO02_03 (3.47 mg), and PO02_04 (5.31 mg) were further fractionated with flash chromatography via direct injection (PuriFlash C_{18} HQ column (6 g); flow rate: 2 mL/min) applying a gradient system of CH_3CN/water as mobile phase (0 min 40%/60%, 3 min 40%/60%, 13 min 60%/40%, 16 min 98%/2%, 20 min 98%/2%, 20 min 40%/60%, 23 min 40%/60%). 3 × 40 tubes were collected and chromatographed with TLC. Tubes were pooled according to their TLC pattern yielding five fractions PO03_01–PO03_05. Fractions PO03_01, PO03_03, and PO03_05 were combined, yielding 4.27 mg of **5**. Fractions PO03_02 and PO03_04 were combined, yielding 1.05 mg of **6**.

The purity of all isolated compounds was determined by UPLC-ELSD analysis to be >98%.

2.10. Cell-Lines, Chemicals and Biochemicals

2.10.1. Reporter-Gene Assay

HEK293 cells stably transfected with the NF-κB-driven luciferase reporter gene NF-κB-luc (293/NF-κB-luc cells, Panomics, RC0014) were stained with 2 μM cell tracker green (CTG, Thermo Scientific). After one hour, 4×10^4 cells per well were seeded in a 96 well plate in serum-free DMEM (4.5 g/L Glucose) obtained from Lonza and supplemented with 2 mM glutamine, 100 U/mL benzylpenicillin and 100 μg/mL streptomycin. After incubation at 37 °C, 5% CO_2 overnight, the cells were pre-treated on the next day with the samples for 1 h. Thereafter, cells were stimulated with 2 ng/mL human recombinant TNF-α (Sigma) for 3.5 h to activate the NF-κB signaling pathway. Then the medium was removed and cells were lysed with luciferase reporter lysis buffer (E3971, Promega, Madison, USA). PO-E and its microfractions were tested at a concentration of 10 μg/mL in at least three independent experiments, if not otherwise indicated. The sesquiterpene lactone parthenolide, an effective inhibitor of the NF-κB pathway [43], was used as a positive control at a concentration of 10 μM and 0.1% DMSO served as vehicle control. The luminescence of the firefly luciferase product and the CTG-derived fluorescence were quantified on a Tecan Spark plate reader (Tecan, Männedorf, Switzerland). The ratio of luminescence units to fluorescence units was calculated to account for differences in cell number. Results were expressed as fold changed relative to the vehicle control with TNFα, which was set to 1 [44]. CTG-fluorescence values used to estimate cell viability were also normalized to the vehicle control with TNFα. Compared to the vehicle control, treatments with fluorescence values below 0.75 were considered as toxic.

2.10.2. Target-Gene Assay

Primary human venous endothelial cells (HUVEC) were isolated from umbilical cords as described previously [8] and maintained in M199 medium (Lonza) supplemented with 20% FCS (Sigma), 2 mM L-glutamine (Sigma), penicillin (100 units/mL), streptomycin (100 mg/mL), 5 units/mL heparin, and 25 mg/mL ECGS (Promocell), and were used up to passage 5. HUVEC were grown to post-confluency in 12-well plates, pre-incubated with PO-E and its microfractions for 30 min, followed by stimulation with 5 ng/mL IL-1β for 90 min. The resorcylic acid lactone of fungal origin (5Z)-7-oxozeaenol [45], TAK1 inhibitor, was used as positive control at a concentration of 5 μM. Total RNA was isolated using the PeqGold Total RNA Isolation Kit (VWR International) according to the manufacturer's instructions. One μg RNA was reverse transcribed using random hexamers (Fermentas) and murine leukemia virus reverse transcriptase (Thermo Fisher). Real-time PCR was performed with the SsoAdvanced Universal SYBR Green Supermix (BioRad) using the StepOnePlus instrument (Applied Biosystems). The following primer pairs were used (forward/reverse, 5′-3′): E-selectin: CCTGTGAAGCTCCCACTGA/GGCTTTTGGTAGCTTCCATCT; VCAM-1: CCGG CTGGAGATATTAC/TGTATCTCTGGGGGCAA CAT; GAPDH: AGAAGGCTGGGGCTCATTT/CTAA GCAGTTGGTGGTGCAG. Relative mRNA levels were normalized to GAPDH and fold changes calculated according to the 2-ΔΔCT method. Results are shown as mean fold induction of averaged Ct values of triplicates. Cytotoxicity was judged by morphological examination.

2.11. Statistical Analysis

NF-κB inhibition data of pure compounds were expressed as the means ± SD of at least three independent biological experiments if not otherwise indicated. All statistical analyses were performed using GraphPad Prism 4.03 software. IC_{50} values were determined by non-linear regression with the sigmoidal dose-response settings (variable slope). Analysis of variance (ANOVA) with Dunnett's multiple comparison test was used to assess the significant differences between the control and treatment groups. A *p*-value < 0.05 was considered significant.

3. Results

As previously reported, a detannified MeOH extract and a CH_2Cl_2 extract prepared from the roots of *P. ostruthium* (both tested at 10 μg/mL) showed an NF-kB inhibition of >75% [27]. To unravel the compounds responsible for the pronounced inhibitory activity on this transcription factor, an optimized large-scale extract from the roots and rhizomes of *P. ostruhtium* was prepared. Briefly, the dried plant material was defatted with *n*-hex and the remaining material was subsequently extracted with CH_2Cl_2 and MeOH. Both extracts were combined to pool the arsenal of putatively bioactive molecules in one extract labeled as PO-E. The biochemometric approach ELINA [32] was applied to enable a straightforward identification and isolation of the active principles of PO-E. The first objective was to simplify and thus expand the structural complexity of the bioactive extract by the generation of microfractions with quantitative variances of constituents over several consecutive fractions. These were thereupon equally prepared for

(i) ^{1}H NMR analysis to obtain quantitative information on structural features of the constituents independent of their ability to ionize (in contrast to an MS-based approach),

(ii) LC-MS-CAD investigation for semi-quantitative information and dereplication of constituents present in each microfraction in both, positive and negative mode, and

(iii) bioactivity testing in three cell-based assays: an NF-kB reporter-gene assay and two functional assays quantifying mRNA expression of NF-kB target-genes (E-selectin and VCAM-1).

For microfractionation, an HPCCC with gradient elution in normal-phase mode was employed. An HPCCC method was developed allowing a high-resolution efficiency able to fractionate the crude extract in one single fractionation step. PO-E was fractionated by applying seven different two-phase solvent systems composed of *n*-hex/EtOAc/MeOH/H_2O with increasing polarity without stopping the apparatus (Table S1). By applying this optimized technique, an efficient fractionation of *P. ostruthium* constituents was guaranteed in a single operation by HPCCC. In total, more than 565 tubes were collected and pooled to 31 microfractions (PO01_01–PO01_31) according to their TLC fingerprint. Aliquots of PO01_01-PO01_31 were forwarded to ^{1}H NMR analysis, LC-MS-CAD measurements and bioactivity testing in an NF-kB reporter-gene assay using HEK293 cells, and two functional target-gene assays (E-selectin and VCAM-1) using endothelial cells. In parallel to a quantitative variance of ^{1}H NMR signals over consecutive microfractions (Figure 1), bioactivity patterns of three cell-based assays relating to this variation were obtained for PO01_01 to PO01_31 (Figure 2).

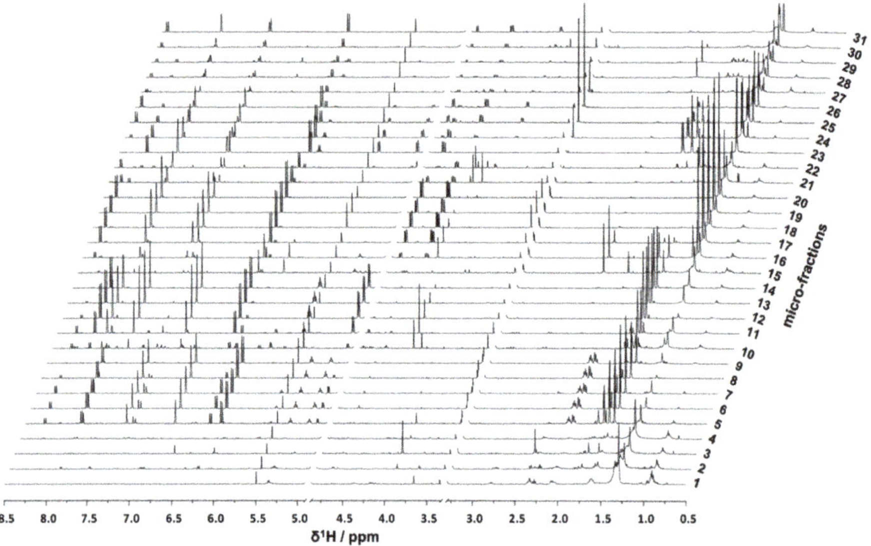

Figure 1. Stack Plot of PO01_01–PO01_31; ^{1}H NMR data (δ_H 0.5–8.5) of obtained *Peucedanum ostruthium* microfractions after fractionation with high-performance counter-current chromatographic (HPCCC). For clarity reasons, both, the water signal (at 4.9 ppm) and the signal of the solvent (at 3.31 ppm) are not shown.

By taking a closer look at the bioactivity results, it became obvious that the activity profile of the microfractions in the NF-kB reporter gene assay shows a very similar pattern as the activity profiles in the two NF-κB target-gene assays. For instance, PO01_06 to PO01_09 showed cytotoxicity in all assays. Further, an increase in activity was observed from PO01_11 to PO01_14, whereas a decrease was obvious from PO01_27 to PO01_29. Increasing activity was shown for PO01_01 to PO01_03 in the NF-kB reporter-gene assay and the NF-kB target-gene assay (E-selectin). Likewise, a similar bioactivity pattern was observed for PO01_15 to PO01_17 in the NF-kB reporter-gene assay and the NF-kB target-gene assay (VCAM-1). Here, a decreasing activity can be observed throughout these fractions. Interestingly, by perceiving the bioactivity results for PO01_22 to PO01_25, a decreasing activity could be shown for the reporter-gene assay and the target-gene assay (VCAM-1). In this package, the active principle seems to be comprised of PO01_22 rather than in PO01_24. On the contrary, in the target-gene assay quantifying E-selectin mRNA expression an explicit decrease in activity from PO01_24 to PO01_27 is shown, indicating that the active constituent(s) is/are accumulated in PO01_24 rather than in PO01_27. This decrease in activity for PO01_24 to PO01_26 was also shown in the target-gene assay when tested at 50 µg/mL (Figure S2). Because of the high correlation between the inhibition of NF-kB transactivation activity and the expression of both adhesion molecules on the mRNA level, the biochemometric correlations were predominantly elaborated based on the bioactivity data of the NF-kB reporter-gene assay. For ELINA, the concentration variances of compounds comprised in the microfractions of a chosen package (e.g., also reflected in their ^{1}H NMR data) were correlated with the bioactivity data by using the multivariate statistical tool HetCA as described before [32]: HetCA plots were generated to visualize the correlation between ^{1}H NMR spectra with their corresponding bioactivity data. Therefore, packages of three to four consecutive microfractions with a variance in activity were selected and depicted as HetCA plots. By applying this method, structural features of molecule(s) correlating to bioactivity could already be seen at the early stage of phytochemical workup, i.e., after the single fractionation step of PO-E. Hence, the HetCA plot of package I (i.e., PO01_11–PO01_14) (Figure 3A) displays structural features of molecule(s) correlating with NF-kB inhibition. Features belonging to ^{1}H NMR signals exhibiting a positive correlation with NF-kB inhibition (red) were assigned as "hot" features whereas, features belonging to ^{1}H NMR signals with a negative correlation with NF-kB inhibition (blue) were assigned as "cold" features. Thus, it became evident that aromatic compounds giving resonances in the downfield chemical shift area (such as coumarins or furanocoumarins) contribute to the inhibition of the transcription factor NF-kB. Further, statistical total correlation spectroscopy (STOCSY) analysis was implemented to deliver information in which molecule(s) share specific "hot" features [42]. For instance, the HetCA plot of package I (Figure 3A) resembles the STOCSY plot of package I (Figure 3B) indicating that there is only one molecule responsible for the observed anti-inflammatory activity. By taking a closer look on the generated STOCSY plot, it can be seen that the molecule with "hot" features gives five aromatic proton signals (between δ_H 6.00–9.00) with an aliphatic side chain: at δ_H 1.67 and 1.71 two singlets can be seen typical for methyl groups, a triplet at δ_H 5.55 typically given by a vinylic proton as well as a doublet at δ_H 4.97. Further, the signal of water and the solvent are present at 4.87 and 3.31 ppm. LC-MS was used to facilitate the identification of the bioactive molecule via dereplication. By the additional use of a charged aerosol detector (CAD), a semi-quantitative analysis could be performed and allowed for the visualization of increasing or decreasing peak areas under the curve (AUC) within a package. By implementing this information, further correlation with bioactivity could be achieved. For instance, an overlay of the four chromatograms of package I revealed a continuous increase of the AUC of the peak at the LC retention time (t_R) 7.6 min from the least active microfraction PO01_11 to the most active PO01_14 (Figure S3). A dereplication of the selected peak with a *m/z* value of 271.13 g/mol in the positive mode identified the furanocoumarin imperatorin (**1**) [28] as the bioactive constituent without any preceding isolation efforts. Figure 3C shows that the structural predictions delivered by the STOCSY plot match with the actual structure of the predicted molecule. Here, no targeted isolation was performed as **1** was already in-house available as a pure compound. LC-MS-CAD analysis was performed to compare the LC chromatogram and MS spectra of the microfraction PO01_14 and compound **1** (Figure S4A,B).

Package II was composed of the microfractions PO01_15 to PO01_17, since an explicit decline in activity was observed in both, the NF-kB reporter-gene assay and the target-gene assay on VCAM1. The HetCA pseudo spectrum and the STOCSY plot of package II revealed that at least two molecules contribute to the observed activity (compare Figure 4A,B): the STOCSY plot at δ_H 5.21 displays a molecule with two aromatic signals (singlets at δ_H 6.35 and 6.05), whereas the signal at δ_H 5.21 is given by a vinylic proton. Two methyl protons are further shown at δ_H 1.66 and 1.77 as well as a doublet at δ_H 3.29. The prominent signal at δ_H 2.35 indicates either the presence of a benzylic proton or a carbonyl methyl group. Apart from this, further signals in the downfield chemical shift area are present in the HetCA pseudo spectrum given by four doublets at δ_H 8.03, 7.89, 6.96, and 6.38 as well as a singlet at δ_H 7.57 (compare with Figure 3B). Besides the NMR-bioactivity correlation, a dereplication via LC-MS-CAD of package II was performed. In the most active fraction, PO01_15, two peaks at t_R 7.3 min and 7.6 min (red squares) are shown (Figure S5). The peak at t_R 7.6 min is the previously identified compound **1**, responsible for the activity of package I. The second peak in the most active microfraction PO01_15 present at t_R 7.30 min contains a *m/z* value of 261.28 in the positive mode. A dereplication via a literature search was performed under consideration of (i) the molecular weight of the compound at t_R 7.30 min and (ii) the structural information derived from the STOCSY plot. By this, the chromone peucenin (**2**) was identified as an active principle in package II. A targeted isolation and structure elucidation of **2** confirmed the ELINA prediction of package II (compare Figure 4B,C). Likewise, the furanocoumarin oxypeucedanin (**3**) [28] was identified as an inactive principle of package II (Figure S5; blue square).

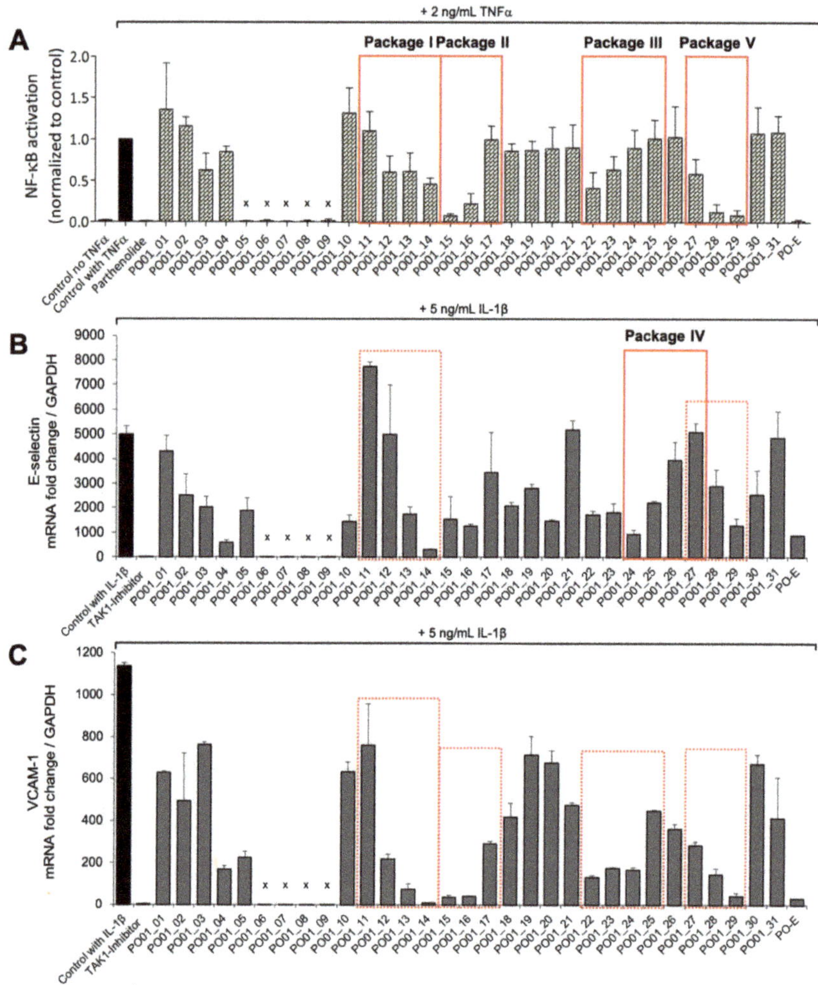

Figure 2. Bioactivity data of PO01_01–PO01_31 in (**A**) an NF-kB reporter-gene assay, a real-time PCR analysis of (**B**) E-selectin and (**C**) VCAM-1. Packages I–V are shown as red squares (solid lines, bioactivity data used for HetCA analysis; dotted lines, packages for comparison of bioactivity data); (**A**) PO01_01–PO01_31, the crude extract PO-E (all 10 µg/mL) and the positive control parthenolide (10 µM) were tested in a cell-based in vitro NF-κB-driven luciferase reporter assay in HEK293–NF-κB-luc cells stimulated with 2 ng/mL TNFα as indicated. The luciferase signal derived from the NF-κB reporter was normalized to the CTG-fluorescence and expressed as fold change normalized to the vehicle control (DMSO 0.1%) with TNFα. Bar charts represent (residual) NF-κB transactivation activity expressed as mean ± SD, n = 3. Microfractions marked with an "x" showed cytotoxicity; (**B**) PO01_01–PO01_31, the crude extract PO-E (all 50 µg/mL), the positive control TAK1-inhibitor (5 µM (5Z)-7-oxozeaenol) and vehicle control (DMSO 0.5%) were assayed for E-selectin expression in primary human venous endothelial cells (HUVEC). Real-time PCR of cDNA obtained from IL-1β (5 ng/mL, 90 min)-stimulated HUVEC which were either untreated (DMSO) or pre-treated (30 min) with 50 µg/mL PO01_01-PO01_31. Relative mRNA levels of E-selectin were normalized to GAPDH and expression levels are depicted as mean fold change ± SD compared to non-stimulated cells. n = 3. Microfractions marked with an "x"

showed cytotoxicity; (**C**) PO01_01-PO01_31, PO-E (all 50 µg/mL), the positive control TAK1-inhibitor (5 µM) and vehicle control (DMSO 0.5%) were assayed for VCAM-1 expression in primary HUVEC. Real-time PCR of cDNA obtained from IL-1β (5 ng/mL, 90 min)-stimulated HUVEC, which were either untreated (DMSO) or pre-treated (30 min) with 50 µg/mL PO01_01–PO01_31. Relative mRNA levels of VCAM-1 were normalized to GAPDH and expression levels are depicted as mean fold change ± SD compared to non-stimulated cells, n = 3. Microfractions marked with an "x" showed cytotoxicity.

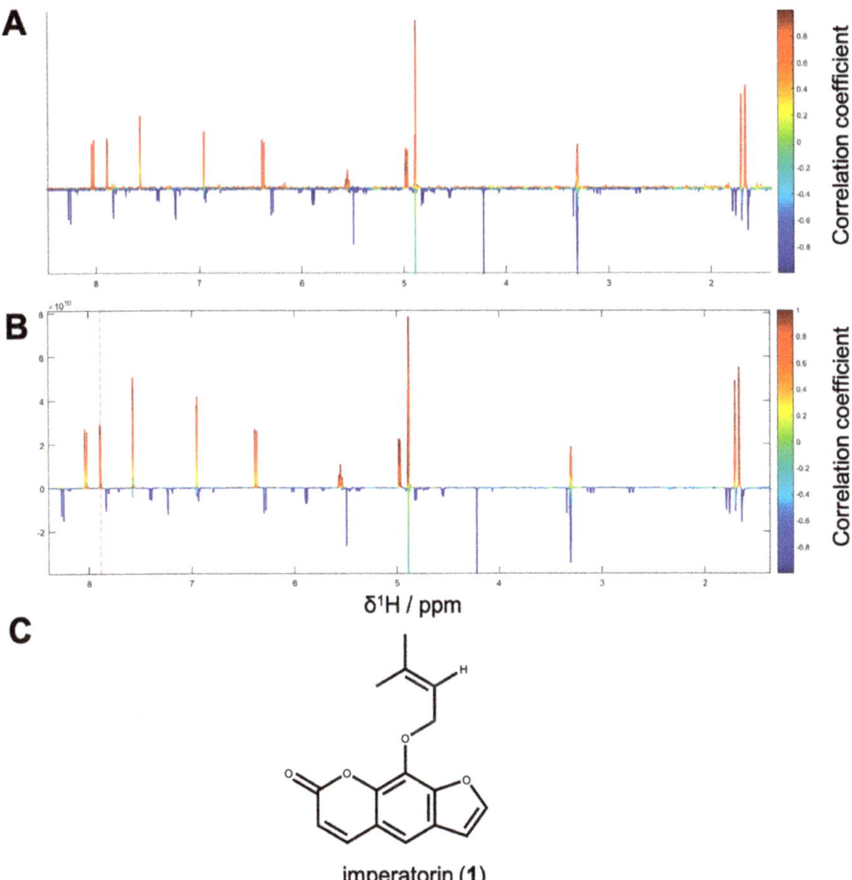

Figure 3. (**A**) HetCA plot of package I. The color code is based on the correlation coefficient: red signals ("hot features") are positive, blue signals ("cold features") are negatively correlated with bioactivity; (**B**) STOCSY plot of package I. The signal at δ_H 7.88 was chosen to obtain the information which molecule(s) share this hot feature. The plot is color coded based on the correlation coefficient: red = signals belonging to molecule(s) that have a signal at δ_H 7.88; blue = signals belong to molecule(s) that do not have the signal at δ_H 7.88; (**C**) structure of the positively correlated and identified compound **1**.

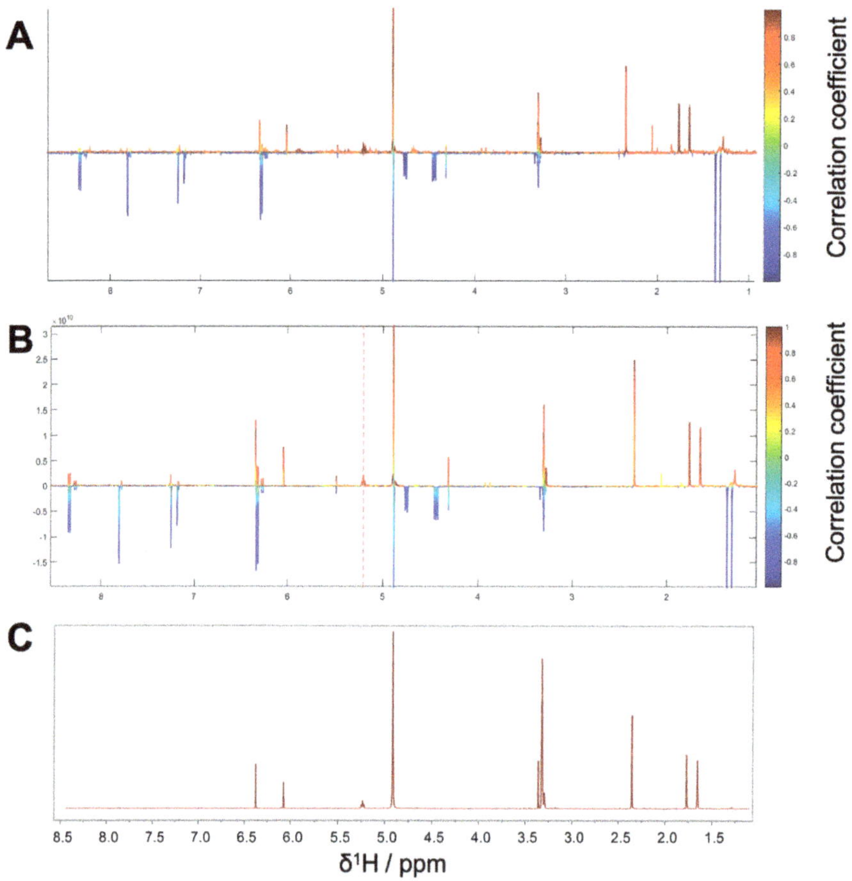

Figure 4. Package II (**A**) ^{1}H NMR pseudo-spectrum showing the HetCA of ^{1}H NMR spectra and NF-kB reporter-gene inhibition data of the microfractions PO01_15–PO01_17. (**B**) STOCSY plot of package II. The signal at δ_H 5.21 was chosen to obtain the information in which molecule(s) share this hot feature. (**C**) ^{1}H NMR spectra of the isolated active compound **2**.

The microfractions PO01_22 to PO01_25 were selected for the generation of package III. Here, the STOCSY plot (δ_H 4.0; Figure 5B) strongly resembles the HetCA plot of package III (Figure 5A), indicating that there is mainly one molecule responsible for the observed activity in the reporter-gene assay and target-gene assay (VCAM-1). The bioactive molecule shows resonances in the downfield chemical shift area, typical for the aromatic backbone of furanocoumarins (four doublets at δ_H 8.42, 7.81, 7.20, and 6.31 as well as a singlet at δ_H 7.24), two resonances in the upfield resonance area given by aliphatic protons such as methyl groups (δ_H 1.67 and 1.63). Further signals are present at δ_H 4.62 (singlet), 4.47 and 4.02 (doublet of doublet). A semi-quantitative LC-MS-CAD analysis allowed for the visualization of the concentration differences of the respective peaks within package III (Figure S6). The chromatogram revealed an increase of the AUC for the peak at t_R 7.33 min from the most active microfraction PO01_22 to PO01_24 (thereupon decreasing from PO01_24 to PO01_25). The peak at t_R 6.01 min is not present in PO01_22, whereas the peak at t_R 6.98 min is only present in the most active microfraction PO01_22. A dereplication for the respective peak revealed a *m/z* value of 323.12 in the positive mode. Further, an MS^2 fragmentation pattern of –35 g/mol was observed for this selected peak, typical for the halogene chlorine. Under consideration of the molecular weight, the fragmentation pattern and the hot features from the STOCSY plot, literature research was performed and unveiled the furanocoumarin saxalin, however with undefined stereochemistry (**4**) [46] as bioactive compound. Targeted isolation and structure elucidation with 1D and 2D NMR experiments (expect for NOESY) were performed and confirmed the presence of the chlorinated compound **4** within the most active microfraction PO01_22.

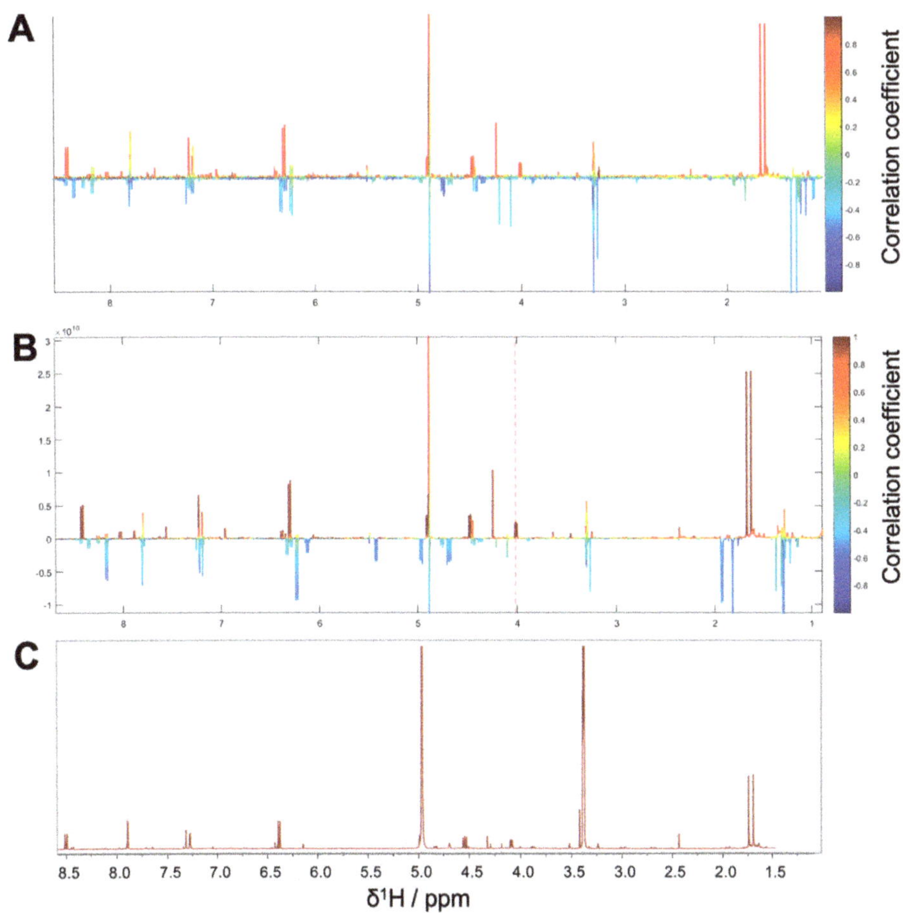

Figure 5. Package III (**A**) HetCA plot of PO01_22–PO01_25. (**B**) STOCSY plot; the signal at δ_H 4.03 was chosen to obtain the information which molecule(s) share this hot feature and (**C**) ^{1}H NMR spectra of the isolated active compound **4**.

As the microfractions PO01_24 to PO01_27 exhibited a decreasing activity (i) on the expression of E-selectin in the target-gene assay, (ii) in the reporter-gene assay at 50 µg/mL (Figure S2) and (iii) PO01_24 showed a similar activity than the preceding microfractions PO01_22 and PO01_23 on the expression of VCAM-1, these microfractions were used for the generation of package IV. Here, the ELINA approach unveiled the furanocoumarin ostruthol (**5**) [28] as the active principle (Figure 6A,B; red signals). Targeted isolation of **5** was performed; additionally, a second compound (**6**) was co-isolated showing cold features. Following the structure elucidation confirmed the isolated compound as **5** by using 1D and 2D NMR experiments. Compound **6** was identified as oxypeucedanin methanolate [47] which was negatively correlated with activity (Figure 6C,D).

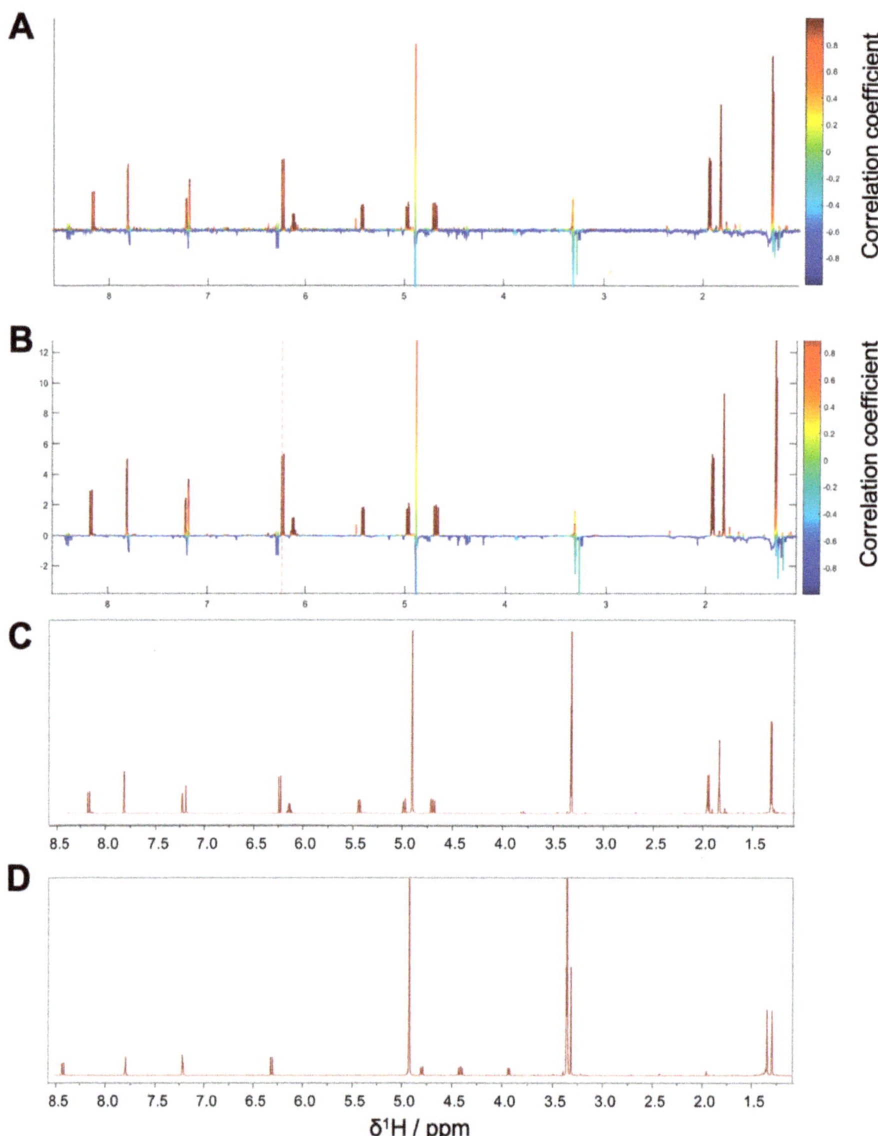

Figure 6. (**A**) ^{1}H NMR pseudo-spectrum showing the HetCA of ^{1}H NMR spectra and NF-kB target-gene assay on E-selectin of selected fractions PO01_24–PO01_27. (**B**) STOCSY plot of package IV. The signal at δ_H 6.24 was chosen to obtain the information in which molecule(s) share this hot feature. (**C**) ^{1}H NMR spectra of the isolated active compound **5** and (**D**) ^{1}H NMR spectra of the isolated compound **6** predicted as inactive.

The last package analyzed was generated from PO01_27 to PO01_29, i.e., package V. Here, ELINA unveiled the presence of two compounds responsible for the observed bioactivities in all three assays, i.e., the furanocoumarin 2′-O-acetyloxypeucedanin (**7**) [46] and the coumarin ostruthin (**8**). Further, the furanocoumarin oxypeucedanin hydrate (**9**) [28] was predicted as an inactive compound (Figure 7A,B; cold features). Because of the low quantity of the microfraction PO01_29 no isolation was performed of **7** and **8**, whereas compound **9** was available as an in-house pure compound (Figure 8).

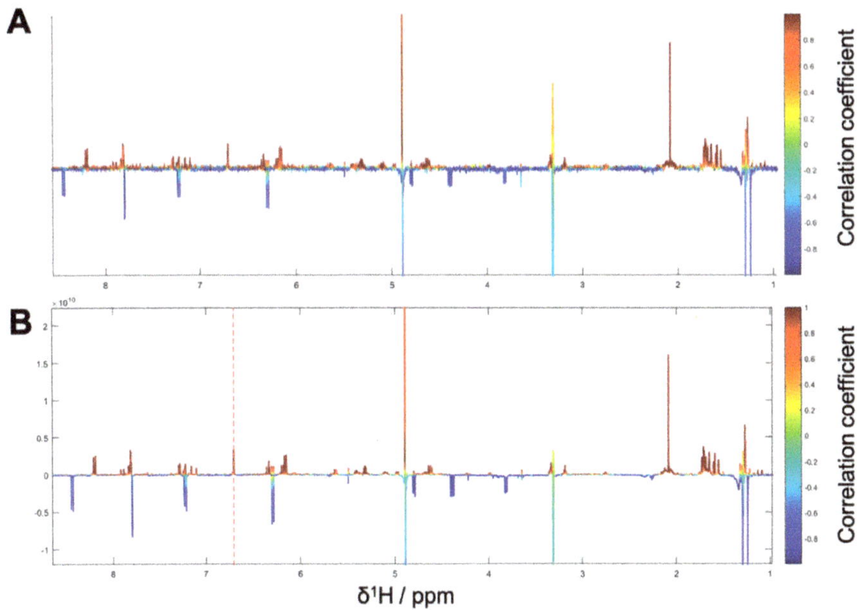

Figure 7. (**A**) ^{1}H NMR pseudo-spectrum showing the HetCA of ^{1}H NMR spectra and NF-kB reporter-gene assay of selected fractions PO01_27–PO01_29 (i.e., package V); (**B**) STOCSY plot of package V. The signal at δ_H 6.71 was chosen to obtain the information which molecule(s) share this hot feature.

As proof of concept, the isolated pure compounds **2**, **4**, **5**, and **6**, as well as the in-house available compounds **1**, **3**, and **9** were tested in the NF-κB reporter gene assay. Additionally, **1** and **2** were also tested as a mixture (1:1) as they were both predicted to contribute to the activity in the most active microfraction PO01_15 (Figure 9). Further, dose-response experiments were performed with positively correlated compounds and isolates **1**, **2**, **4**, and **5**, respectively (Figure S7). Compounds **1** and **2**, when tested separately, showed only weak NF-kB inhibitory activities with an IC_{50} value of 49.0 μM for **1**. For compound **2** dose-response experiments revealed hardly any concentration-dependent activity resulting in a rather flat curve. However, when tested as a mixture, a significant increase of the inhibitory effect was shown ($p < 0.01$) in the NF-kB reporter-gene assay (Figure 9). Whereas **5** exerted a moderate inhibitory activity (with an IC_{50} value of about 20 μM), **4** was identified as potent NF-κB inhibiting compound ($p < 0.01$) with an IC_{50} value of 8.08 μM. In line with the ELINA prediction, the negatively correlated compounds **3**, **6**, and **9**, showed no bioactivity, when tested in the NF-κB reporter-gene assay at 10 μg/mL, and thus confirmed the accuracy of the presented biochemometric approach.

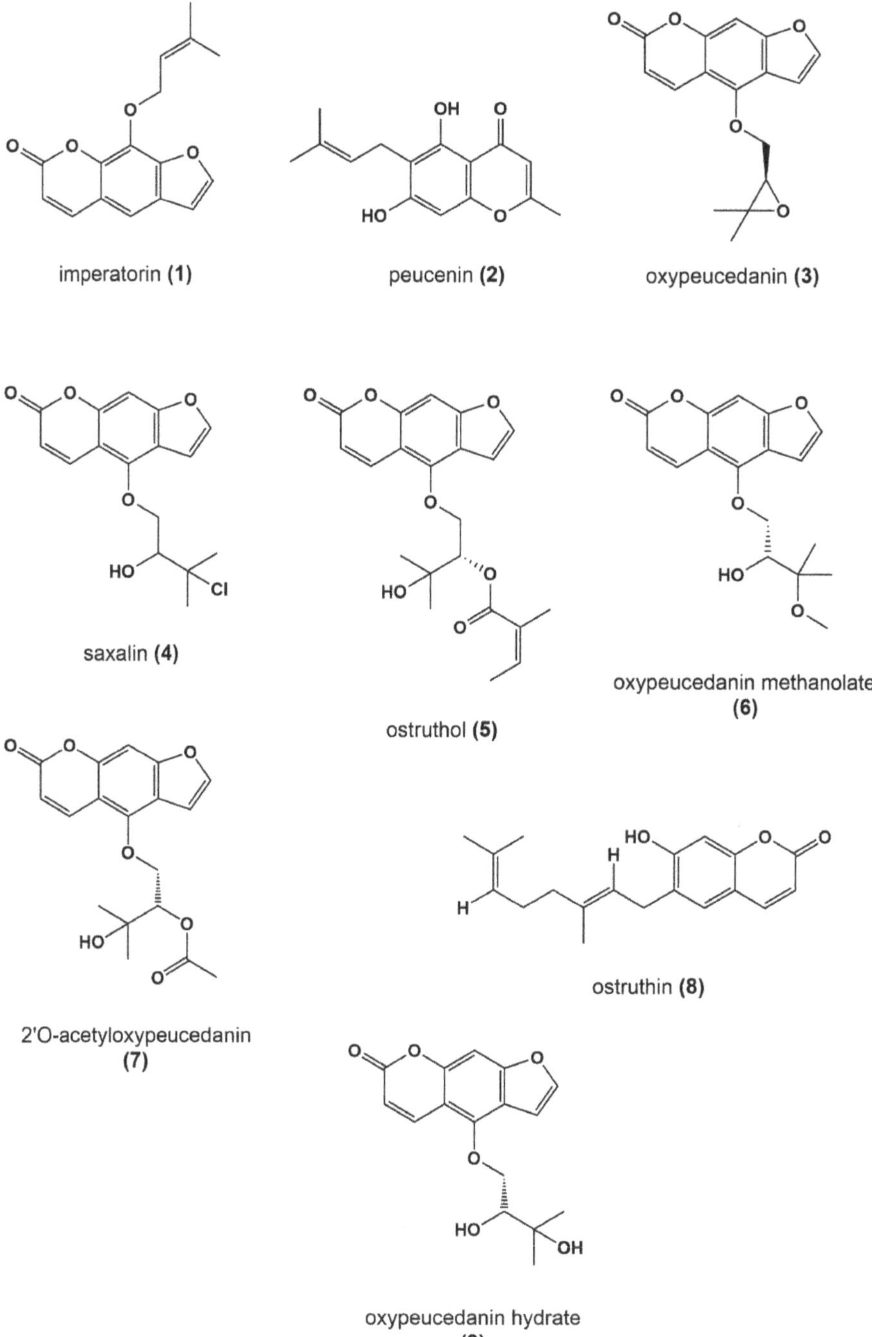

Figure 8. Chemical structures of identified *P. ostruthium* constituents.

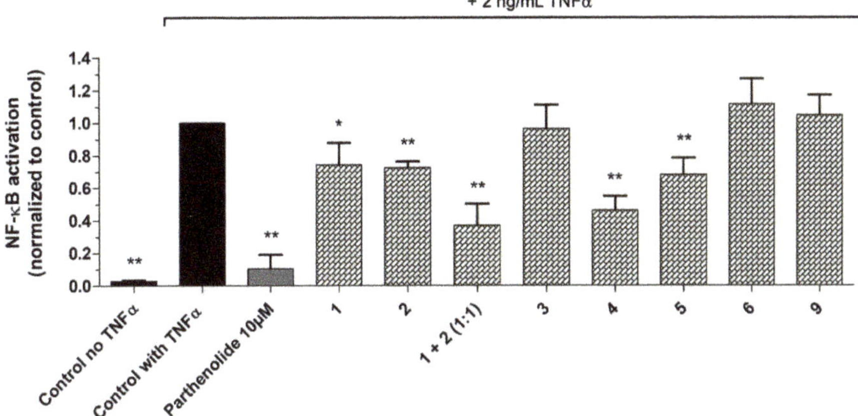

Figure 9. Residual NF-kB activity (normalized to the vehicle control with TNFα) in TNFα (2 ng/mL)-activated HEK293–NF-kB-luc cells after treatment with the compounds **1–6**, **9** and a 1:1 mixture of **1** and **2** (all tested at 10 µg/mL) and the positive control parthenolide (10 µM). The luciferase signal derived from the NF-kB reporter was normalized to CTG-fluorescence and expressed as fold change normalized to the vehicle control (0.1% DMSO) with TNFα. Bar charts represent (residual) NF-kB transactivation activity expressed as mean ± SD. $n = 4$; $* p < 0.05$ and $** p < 0.01$ as compared to the control with TNFα.

4. Discussion

In this study, we applied a biochemometric approach to unravel those constituents, which contribute to the NF-kB inhibitory activity in the masterwort extract PO-E using the recently established ELINA approach [37]. As a first and crucial step, a newly elaborated protocol for a comprehensive gradient-elution HPCCC was developed for the microfractionation of PO-E. This enabled an appropriate deconvolution of the crude extract without compound(s) adhesion to any chromatographic material, and accordingly without the risk of putatively losing bioactives. In this way, the complexity of PO-E was broken down to 31 microfractions with envisaged concentration variations of constituents.

Bioactivity data of three in vitro cell-based assays addressing the NF-kB inhibitory activity in masterwort were acquired for all 31 microfractions. Here, the implementation of in vitro cell-based assays in the biochemometric ELINA approach has substantial benefits. First, in contrast to cell-free assays, they offer the advantage to model the biology of intact cells. Thus, not only the activity data of a sample can be shown, but also biologically relevant information like the effect of the sample on cell viability and cytotoxicity can be elucidated. Second, molecular pathway interactions can be exposed [34,48,49]. Although some biochemometric studies have previously been performed with e.g. enzyme-based assays to decipher bioactives in complex mixtures [32,36–38], the present study aimed to evaluate the robustness of the ELINA approach applying a cell-based readout to correlate with big metabolite data derived from NMR and MS. Indeed, all the positively correlated constituents tested in the NF-kB reporter-gene assay (i.e., **1**, **2**, **4**, **5**) showed significant inhibitory activities (Figure 9), whereas the negatively correlated compounds (i.e., **3**, **6**, and **9**) revealed as inactive (tested at 10 µM). In addition, all the above tested compounds were well tolerated in the assay and showed no in vitro cytotoxicity (Figure S8). With the scope of this study, the application of a biochemometric approach in cell-based in vitro assays on the transcription factor NF-kB was successfully demonstrated.

Pharmacological effects given by a crude extract are often a result of the combination of constituents rather than individual chemical entities out of that mixture. Traditional medicinal systems like the European phytotherapy, traditional Chinese medicine, and Ayurveda, however, rely on multi-component mixtures instead of single compounds to treat pleiotropic diseases [50] such as inflammation. Identifying multiple compounds that contribute additively, synergistically or antagonistically to a biological effect remains a challenging task in NP drug discovery [34,51]. To unravel bioactive compound(s) from a complex mixture, metabolomics approaches have been introduced in the past few years. They offer a more holistic perspective on bioactives [36,52] by profiling multiple mixture components simultaneously [30]. Using the example of masterwort, additive effects of compounds **1** and **2** were observed within the most active microfraction PO01_15. This effect was imitated when **1** and **2** were tested as 1:1 mixture (Figure 9). On the contrary, when taking a closer look at the LC-MS chromatograms of package I (i.e., PO01_11-PO01_14; Figure S3), the detrimental effects of several compounds (present at t_R 7.8, 8.3, and 9.3 min) within PO01_11 are assumed. Several compounds seem to antagonize the activity of **1** (present at t_R 7.6 min in PO01_11-PO01_14). This conclusion is in line with the quantitative ^{1}H NMR data of package I (Figure 1; for more details see Figure S9): in all microfractions resonances given by **1** are present. These findings emphasize a unique strength of the ELINA approach: combinatorial effects of constituents that would probably have been missed using the classical bioactivity-guided isolation approach are disclosed from mixtures. Moreover, these positively correlated compounds can be identified prior to any isolation, thus avoiding unnecessary and tedious isolation procedures.

5. Conclusion

The biochemometric approach ELINA enabled a targeted identification of the NF-kB inhibiting metabolites of the traditionally used masterwort extract after one single fractionation step. Concentration variances of HPCCC microfractions unmasked the anti-inflammatory constituents probed in an NF-κB reporter-gene assay and two NF-κB target-gene assays quantifying VCAM-1 and E-selectin mRNA. In sum, three furanocoumarins (**1**, **4**, and **7**), one chromone (**2**) and one coumarin (**8**) were pinpointed as NF-kB inhibitory agents present in *P. ostruthium* even prior to any isolation. As proof of concept, the positively correlated compounds **1**, **2**, **4** and **5** as well as the negatively correlated compounds **3**, **6**, and **9** were tested in the cell-based NF-kB reporter-gene assay and confirmed the prediction. Intriguingly, ELINA enabled to unravel additive effects of compounds **1** and **2** that would have probably been missed in a classic bioactivity-guided isolation procedure. By applying this biochemometric approach, early identification and dereplication of even minor bioactives were achieved. Thus, this holistic –omics-based tool not only offers insight into the chemical features and those metabolites contributing to the bioactivity of multicomponent NPs but also has the potential to effectively accelerate the NP based drug discovery process.

Supplementary Materials: The following are available online at http://www.mdpi.com/2218-273X/10/5/679/s1. **Table S1:** HEMWat system table for HPCCC fractionation with gradient elution. **Table S2:** Pooling of microfractions. **Figure S1:** TLC of microfractions PO01_01-PO01_31; detection UV$_{254}$ (top) and UV$_{366}$ (bottom). **Figure S2** Bioactivity results of PO01_11-PO01_29, the crude extract PO-E (all 50 µg/mL) and the positive control parthenolide (10 µM) tested in a cell-based in vitro NF-kB-driven luciferase reporter assay in HEK293–NFKB-luc cells stimulated with 2 ng/mL TNFα as indicated. The luciferase signal derived from the NF-kB reporter was normalized to the vehicle control (0.1% DMSO) with TNFα. Bar charts represent (residual) NF-kB transactivation activity expressed as mean ± SD; n = 3. **Figure S3:** CAD chromatogram of package I (PO01_11–PO01_14); t_R 6.00–9.60 min. **Figure S4:** LC-MS-CAD analysis of the microfraction PO01_14 and the pure compound imperatorin; (**A**) LC chromatogram of PO01_14 and imperatorin at t_R 7.6 min; (**B**) Mass spectra of PO01_14 and imperatorin showing the m/z value of 271.1 g/mol. **Figure S5:** CAD chromatogram of package II (PO01_14-PO01_17); t_R 6.0–9.0 min. **Figure S6:** CAD chromatogram of package III (PO01_22-PO01_25); t_R 5.0–8.5 min. **Figure S7:** Concentration-response curves showing the positively correlated compounds 1, 2, 4 and 5 assayed in a cell-based NF-κB-reporter gene assay at different concentrations. Shown are the residual NF-kB activities in TNFα (2 ng/mL)-activated HEK293–NF-kB-luc cells after treatment with the respective compounds. The luciferase signal derived from the NF-kB reporter was normalized to the CTG-fluorescence and expressed as fold change normalized to the TNFα signal. IC$_{50}$ values were determined by non-linear regression with the sigmoidal dose response settings (variable slope) using GraphPad Prism 4.03 software. **Figure S8:** CTG fluorescence as a measure for cell viability after treatment with pure compounds 1–6, 9 and a 1:1 mixture of 1 and 2 (all tested at 10 µg/mL) and the positive control parthenolide (10 µM) in TNFα (2 ng/mL)-activated HEK293–NF-kB-luc cells; data are normalized to vehicle control (0.1% DMSO) with TNFα. Data represent mean ± SD. n = 4. **Figure S9:** ^{1}H NMR data of package I (PO01_11–PO01_14).

Author Contributions: Conceptualization, J.Z., U.G., and J.M.R.; investigation, J.Z., S.H., and J.S.; methodology, U.G. and J.M.R.; resources, R.d.M., V.M.D., and J.M.R.; software, J.Z. and U.G.; supervision, J.M.R.; validation, R.d.M., V.M.D., and J.M.R.; visualization, J.Z.; writing—original draft, J.Z.; Writing—review and editing, J.Z., U.G., R.d.M., V.M.D., and J.M.R. All authors have read and agreed to the published version of the manuscript.

Funding: This research received no external funding.

Acknowledgments: The authors thank B. Braunböck-Müller, C. Sykora (Department of Pharmacognosy, University of Vienna, Austria) and C. Lammel (Department of Vascular Biology and Thrombosis Research, Medical University of Vienna) for their technical support, E. Urban (Department of Pharmaceutical Chemistry, University of Vienna, Austria) for the recording of NMR spectra, and E. Mikros (Department of Pharmacy, Division of Pharmaceutical Chemistry, National and Kapodistrian University of Athens, Greece) for fruitful discussions regarding heterocovariance analyses. Open Access Funding by the University of Vienna.

Conflicts of Interest: The authors declare no conflict of interest.

References

1. Harvey, A.L.; Edrada-Ebel, R.; Quinn, R.J. The re-emergence of natural products for drug discovery in the genomics era. *Nat. Rev. Drug Discov.* **2015**, *14*, 111–129. [CrossRef] [PubMed]

2. Yuan, H.; Ma, Q.; Ye, L.; Piao, G. The traditional medicine and modern medicine from natural products. *Molecules* **2016**, *21*, 559. [CrossRef]

3. Atanasov, A.G.; Waltenberger, B.; Pferschy-Wenzig, E.M.; Linder, T.; Wawrosch, C.; Uhrin, P.; Temml, V.; Wang, L.; Schwaiger, S.; Heiss, E.H.; et al. Discovery and resupply of pharmacologically active plant-derived natural products: A review. *Biotechnol. Adv.* **2015**, *33*, 1582–1614. [CrossRef] [PubMed]

4. Boufridi, A.Q.R.J. Harnessing the properties of natural products. *Annu. Rev. Pharmacol. Toxicol.* **2018**, *58*, 451–470. [CrossRef] [PubMed]

5. Ezzat, S.M.; Ezzat, M.I.; Okba, M.M.; Menze, E.T.; Abdel-Naim, A.B. The hidden mechanism beyond ginger (*Zingiber officinale* Rosc.) potent In Vivo and In Vitro anti-inflammatory activity. *J. Ethnopharmacol.* **2018**, *214*, 113–123. [CrossRef] [PubMed]

6. Dongare, S.; Gupta, S.K.; Mathur, R.; Saxena, R.; Mathur, S.; Agarwal, R.; Nag, T.C.; Srivastava, S.; Kumar, P. *Zingiber officinale* attenuates retinal microvascular changes in diabetic rats via anti-inflammatory and antiangiogenic mechanisms. *Mol. Vision* **2016**, *22*, 599–609.

7. Dugasani, S.; Pichika, M.R.; Nadarajah, V.D.; Balijepalli, M.K.; Tandra, S.; Korlakunta, J.N. Comparative antioxidant and anti-inflammatory effects of [6]-gingerol, [8]-gingerol, [10]-gingerol and [6]-shogaol. *J. Ethnopharmacol.* **2010**, *127*, 515–520. [CrossRef]

8. Seigner, J.; Junker-Samek, M.; Plaza, A.; D'Urso, G.; Masullo, M.; Piacente, S.; Holper-Schichl, Y.M.; de Martin, R. A *Symphytum officinale* root extract exerts anti-inflammatory properties by affecting two distinct steps of NF-kappaB signaling. *Front. Pharmacol.* **2019**, *10*, 289. [CrossRef]

9. Sowa, I.; Paduch, R.; Strzemski, M.; Zielinska, S.; Rydzik-Strzemska, E.; Sawicki, J.; Kocjan, R.; Polkowski, J.; Matkowski, A.; Latalski, M.; et al. Proliferative and antioxidant activity of *Symphytum officinale* root extract. *Nat. Prod. Res.* **2018**, *32*, 605–609. [CrossRef]

10. Mykkanen, O.T.; Huotari, A.; Herzig, K.H.; Dunlop, T.W.; Mykkanen, H.; Kirjavainen, P.V. Wild blueberries (*Vaccinium myrtillus*) alleviate inflammation and hypertension associated with developing obesity in mice fed with a high-fat diet. *PLoS ONE* **2014**, *9*, e114790. [CrossRef]

11. Triebel, S.; Trieu, H.-L.; Richling, E. Modulation of inflammatory gene expression by a bilberry (*Vaccinium myrtillus* L.) extract and single anthocyanins considering their limited stability under cell culture conditions. *J. Agric. Food Chem.* **2012**, *60*, 8902–8910. [CrossRef] [PubMed]

12. Luo, H.; Lv, X.-D.; Wang, G.-E.; Li, Y.-F.; Kurihara, H.; He, R.-R. Anti-inflammatory effects of anthocyanins-rich extract from bilberry (*Vaccinium myrtillus* L.) on croton oil-induced ear edema and *Propionibacterium acnes* plus LPS-induced liver damage in mice. *Int. J. Food Sci. Nutr.* **2014**, *65*, 594–601. [CrossRef] [PubMed]

13. Ukiya, M.; Akihisa, T.; Yasukawa, K.; Tokuda, H.; Suzuki, T.; Kimura, Y. Anti-inflammatory, anti-tumor-promoting, and cytotoxic activities of constituents of marigold (*Calendula officinalis*) flowers. *J. Nat. Prod.* **2006**, *69*, 1692–1696. [CrossRef]

14. Colombo, E.; Sangiovanni, E.; D'Ambrosio, M.; Bosisio, E.; Ciocarlan, A.; Fumagalli, M.; Guerriero, A.; Harghel, P.; Dell'Agli, M. A bio-guided fractionation to assess the inhibitory activity of *Calendula officinalis* L. on the NF-kappaB driven transcription in human gastric epithelial cells. *J. Evid. Based Complementary Altern. Med.* **2015**, *2015*, 727342.

15. Kim, Y.S.; Ahn, Y.; Hong, M.H.; Joo, S.Y.; Kim, K.H.; Sohn, I.S.; Park, H.W.; Hong, Y.J.; Kim, J.H.; Kim, W.; et al. Curcumin attenuates inflammatory responses of TNF-alpha-stimulated human endothelial cells. *J. Cardiovasc. Pharmacol.* **2007**, *50*, 41–49. [CrossRef]

16. Kuhad, A.; Pilkhwal, S.; Sharma, S.; Tirkey, N.; Chopra, K. Effect of curcumin on inflammation and oxidative stress in cisplatin-induced experimental nephrotoxicity. *J. Agric. Food Chem.* **2007**, *55*, 10150–10155. [CrossRef] [PubMed]

17. Ghasemi, F.; Shafiee, M.; Banikazemi, Z.; Pourhanifeh, M.H.; Khanbabaei, H.; Shamshirian, A.; Amiri Moghadam, S.; ArefNezhad, R.; Sahebkar, A.; Avan, A.; et al. Curcumin inhibits NF-kB and Wnt/beta-catenin pathways in cervical cancer cells. *Pathol. Res. Pract.* **2019**, *215*, 152556. [CrossRef]

18. Jin, M.; Park, S.Y.; Shen, Q.; Lai, Y.; Ou, X.; Mao, Z.; Lin, D.; Yu, Y.; Zhang, W. Anti-neuroinflammatory effect of curcumin on pam3CSK4-stimulated microglial cells. *Int. J. Mol. Med.* **2018**, *41*, 521–530. [CrossRef]

19. Das, S.; Mandal, S.K. Current developments on anti-inflammatory natural medicines. *Asian J. Pharm. Clin. Res.* **2018**, *11*, 61. [CrossRef]

20. Jiang, T.; Gu, J.; Chen, W.; Chang, Q. Resveratrol inhibits high-glucose-induced inflammatory "metabolic memory" in human retinal vascular endothelial cells through SIRT1-dependent signaling. *Can. J. Physiol. Pharmacol.* **2019**, *97*, 1141–1151. [CrossRef]

21. Alharris, E.; Alghetaa, H.; Seth, R.; Chatterjee, S.; Singh, N.P.; Nagarkatti, M.; Nagarkatti, P. Resveratrol attenuates allergic asthma and associated inflammation in the lungs through regulation of miRNA-34a that targets FoxP3 in mice. *Front. Immunol.* **2018**, *9*, 2992. [CrossRef] [PubMed]

22. Chung, E.Y.; Kim, B.H.; Hong, J.T.; Lee, C.K.; Ahn, B.; Nam, S.Y.; Han, S.B.; Kim, Y. Resveratrol down regulates interferon-gamma-inducible inflammatory genes in macrophages: Molecular mechanism via decreased STAT-1 activation. *J. Nutr. Biochem.* **2011**, *22*, 902–909. [CrossRef] [PubMed]

23. Andrade, P.B.; Valentao, P. Insights into natural products in inflammation. *Int. J. Mol. Sci.* **2018**, *19*, 644. [CrossRef] [PubMed]

24. Medzhitov, R. Origin and physiological roles of inflammation. *Nature* **2008**, *454*, 428–435. [CrossRef]

25. Makarov, S.S.; Johnston, W.N.; Olsen, J.C.; Watson, J.M.; Mondal, K.; Rinehart, C.; Haskill, J.S. NF-κB as a target for anti-inflammatory gene therapy: Suppression of inflammatory responses in monocytic and stromal cells by stable gene transfer of IκBα cDNA. *Gene Ther.* **1997**, *4*, 846–852. [CrossRef] [PubMed]

26. Koeberle, A.; Werz, O. Multi-target approach for natural products in inflammation. *Drug Discov. Today* **2014**, *19*, 1871–1882. [CrossRef]

27. Vogl, S.; Picker, P.; Mihaly-Bison, J.; Fakhrudin, N.; Atanasov, A.G.; Heiss, E.H.; Wawrosch, C.; Reznicek, G.; Dirsch, V.M.; Saukel, J.; et al. Ethnopharmacological In Vitro studies on Austria's folk medicine—An unexplored lore In Vitro anti-inflammatory activities of 71 Austrian traditional herbal drugs. *J. Ethnopharmacol.* **2013**, *149*, 750–771. [CrossRef]

28. Vogl, S.; Zehl, M.; Picker, P.; Urban, E.; Wawrosch, C.; Reznicek, G.; Saukel, J.; Kopp, B. Identification and quantification of coumarins in *Peucedanum ostruthium* (L.) koch by HPLC-DAD and HPLC-DAD-MS. *J. Agric. Food Chem.* **2011**, *59*, 4371–4377. [CrossRef]

29. Schmiderer, C.; Ruzicka, J.; Novak, J. DNA-based identification of *Peucedanum ostruthium* specimens and detection of common adulterants by high-resolution melting curve analysis. *Mol. Cell. Probes* **2015**, *29*, 343–350. [CrossRef]

30. Kellogg, J.J.; Todd, D.A.; Egan, J.M.; Raja, H.A.; Oberlies, N.H.; Kvalheim, O.M.; Cech, N.B. Biochemometrics for natural products research: Comparison of data analysis approaches and application to identification of bioactive compounds. *J. Nat. Prod.* **2016**, *79*, 376–386. [CrossRef]

31. Nothias, L.F.; Nothias-Esposito, M.; da Silva, R.; Wang, M.; Protsyuk, I.; Zhang, Z.; Sarvepalli, A.; Leyssen, P.; Touboul, D.; Costa, J.; et al. Bioactivity-based molecular networking for the discovery of drug leads in natural product bioassay-guided fractionation. *J. Nat. Prod.* **2018**, *81*, 758–767. [CrossRef] [PubMed]

32. Grienke, U.; Foster, P.A.; Zwirchmayr, J.; Tahir, A.; Rollinger, J.M.; Mikros, E. ^{1}H NMR-MS-based heterocovariance as a drug discovery tool for fishing bioactive compounds out of a complex mixture of structural analogues. *Sci. Rep.* **2019**, *9*, 11113. [CrossRef] [PubMed]

33. Caesar, L.K.; Kellogg, J.J.; Kvalheim, O.M.; Cech, N.B. Opportunities and limitations for untargeted mass spectrometry metabolomics to identify biologically active constituents in complex natural product mixtures. *J. Nat. Prod.* **2019**, *82*, 469–484. [CrossRef]

34. Caesar, L.K.; Cech, N.B. Synergy and antagonism in natural product extracts: When 1 + 1 does not equal 2. *Nat. Prod. Rep.* **2019**, *36*, 869–888. [CrossRef] [PubMed]

35. Wolfender, J.L.; Litaudon, M.; Touboul, D.; Queiroz, E.F. Innovative omics-based approaches for prioritisation and targeted isolation of natural products—New strategies for drug discovery. *Nat. Prod. Rep.* **2019**, *36*, 855–868. [CrossRef] [PubMed]

36. Aligiannis, N.; Halabalaki, M.; Chaita, E.; Kouloura, E.; Argyropoulou, A.; Benaki, D.; Kalpoutzakis, E.; Angelis, A.; Stathopoulou, K.; Antoniou, S.; et al. Heterocovariance based metabolomics as a powerful tool accelerating bioactive natural product identification. *ChemistrySelect* **2016**, *1*, 2531–2535. [CrossRef]

37. Michalea, R.; Stathopoulou, K.; Polychronopoulos, P.; Benaki, D.; Mikros, E.; Aligiannis, N. Efficient identification of acetylcholinesterase and hyaluronidase inhibitors from *Paeonia parnassica* extracts through a heterocovariance approach. *J. Ethnopharmacol.* **2018**, 111547. [CrossRef]

38. Boka, V.-I.; Stathopoulou, K.; Benaki, D.; Gikas, E.; Aligiannis, N.; Mikros, E.; Skaltsounis, A.-L. Could multivariate statistics exploit HPTLC and NMR Data to reveal bioactive compounds? *Case Paeonia Mascula. Phytochem. Lett.* **2017**, *20*, 379–385. [CrossRef]

39. Camp, D.; Davis, R.A.; Campitelli, M.; Ebdon, J.; Quinn, R.J. Drug-like properties: Guiding principles for the design of natural product libraries. *J. Nat. Prod.* **2012**, *75*, 72–81. [CrossRef]

40. Kratz, J.M.; Mair, C.E.; Oettl, S.K.; Saxena, P.; Scheel, O.; Schuster, D.; Hering, S.; Rollinger, J.M. hERG channel blocking ipecac alkaloids identified by combined *in silico*—In Vitro screening. *Planta Med.* **2016**, *82*, 1009–1015. [CrossRef]

41. Garrard, I.J. Simple approach to the development of a CCC solvent selection protocol suitable for automation. *J. Liq. Chromatrogr. Relat. Technol.* **2007**, *28*, 1923–1935. [CrossRef]

42. Cloarec, O.; Dumas, M.E.; Craig, A.; Barton, R.H.; Trygg, J.; Hudson, J.; Blancher, C.; Gauguier, D.; Lindon, J.C.; Holmes, E.; et al. Statistical total correlation spectroscopy: An exploratory approach for latent biomarker identification from metabolic ^{1}H NMR data sets. *Anal. Chem.* **2005**, *77*, 1282–1289. [CrossRef] [PubMed]

43. Hehner, S.P.; Hofmann, T.G.; Droge, W.; Schmitz, M.L. The antiinflammatory sesquiterpene lactone parthenolide inhibits NF-kappa B by targeting the I kappa B kinase complex. *J Immunol.* **1999**, *163*, 5617–5623. [PubMed]

44. Fakhrudin, N.; Waltenberger, B.; Cabaravdic, M.; Atanasov, A.G.; Malainer, C.; Schachner, D.; Heiss, E.H.; Liu, R.; Noha, S.M.; Grzywacz, A.M.; et al. Identification of plumericin as a potent new inhibitor of the NF-kappaB pathway with anti-inflammatory activity In Vitro and In Vivo. *Br. J. Pharmacol.* **2014**, *171*, 1676–1686. [CrossRef]

45. Ninomiya-Tsuji, J.; Kajino, T.; Ono, K.; Ohtomo, T.; Matsumoto, M.; Shiina, M.; Mihara, M.; Tsuchiya, M.; Matsumoto, K. A resorcylic acid lactone, 5Z-7-oxozeaenol, prevents inflammation by inhibiting the catalytic activity of TAK1 MAPK kinase kinase. *J. Biol. Chem.* **2003**, *278*, 18485–18490. [CrossRef]

46. Harkar, S.; Razdan, T.K.; Waight, E.S. Steroids, chromone, and coumarins from Angelica officinalis. *Phytochemistry* **1984**, *23*, 419–426. [CrossRef]

47. Reisch, J.; Khaled, S.A.; Szendrei, K.; Novak, I. Natural product chemistry. 51. 5-Alkoxy-furanocoumarins from Peucedanum ostruthium. *Phytochemistry* **1975**, *14*, 1889–1890. [CrossRef]

48. Zimmermann, G.R.; Lehár, J.; Keith, C.T. Multi-target therapeutics: When the whole is greater than the sum of the parts. *Drug Discov. Today* **2007**, *12*, 34–42. [CrossRef]

49. Borisy, A.A.; Elliott, P.J.; Hurst, N.W.; Lee, M.S.; Lehar, J.; Price, E.R.; Serbedzija, G.; Zimmermann, G.R.; Foley, M.A.; Stockwell, B.R.; et al. Systematic discovery of multicomponent therapeutics. *Proc. Natl. Acad. Sci. USA* **2003**, *100*, 7977–7982. [CrossRef]

50. Thomas, E.; Egon, K. Complex interactions between phytochemicals. The multi-target therapeutic concept of phytotherapy. *Curr. Drug Targets* **2011**, *12*, 122–132.

51. Britton, E.R.; Kellogg, J.J.; Kvalheim, O.M.; Cech, N.B. Biochemometrics to identify synergists and additives from botanical medicines: A case study with *Hydrastis canadensis* (Goldenseal). *J. Nat. Prod.* **2018**, *81*, 484–493. [CrossRef] [PubMed]

52. Halabalaki, M.; Vougogiannopoulou, K.; Mikros, E.; Skaltsounis, A.L. Recent advances and new strategies in the NMR-based identification of natural products. *Curr. Opin. Biotechnol.* **2014**, *25*, 1–7. [CrossRef] [PubMed]

Review

Phenolic Compounds Exerting Lipid-Regulatory, Anti-Inflammatory and Epigenetic Effects as Complementary Treatments in Cardiovascular Diseases

Laura Toma [1], Gabriela Maria Sanda [1], Loredan Stefan Niculescu [1], Mariana Deleanu [1,2], Anca Volumnia Sima [1] and Camelia Sorina Stancu [1,*]

[1] Lipidomics Department, Institute of Cellular Biology and Pathology "Nicolae Simionescu" of the Romanian Academy, 8, B.P. Hasdeu Street, 050568 Bucharest, Romania; laura.toma@icbp.ro (L.T.); gabriela.sanda@icbp.ro (G.M.S.); loredan.niculescu@icbp.ro (L.S.N.); mariana.deleanu@icbp.ro (M.D.); anca.sima@icbp.ro (A.V.S.)

[2] Faculty of Biotechnology, University of Agronomical Sciences and Veterinary Medicine, 59, Marasti Blvd., 011464 Bucharest, Romania

* Correspondence: camelia.stancu@icbp.ro

Received: 30 March 2020; Accepted: 17 April 2020; Published: 21 April 2020

Abstract: Atherosclerosis is the main process behind cardiovascular diseases (CVD), maladies which continue to be responsible for up to 70% of death worldwide. Despite the ongoing development of new and potent drugs, their incomplete efficacy, partial intolerance and numerous side effects make the search for new alternatives worthwhile. The focus of the scientific world turned to the potential of natural active compounds to prevent and treat CVD. Essential for effective prevention or treatment based on phytochemicals is to know their mechanisms of action according to their bioavailability and dosage. The present review is focused on the latest data about phenolic compounds and aims to collect and correlate the reliable existing knowledge concerning their molecular mechanisms of action to counteract important risk factors that contribute to the initiation and development of atherosclerosis: dyslipidemia, and oxidative and inflammatory-stress. The selection of phenolic compounds was made to prove their multiple benefic effects and endorse them as CVD remedies, complementary to allopathic drugs. The review also highlights some aspects that still need clear scientific explanations and draws up some new molecular approaches to validate phenolic compounds for CVD complementary therapy in the near future.

Keywords: cardiovascular diseases; inflammation; lipid metabolism; non-coding RNA; oxidative stress; phenolic compounds

---

1. Introduction

Cardiovascular diseases (CVD) continue to be the leading cause of mortality and morbidity worldwide, despite the various therapies developed for their treatment, therapies that have an important global economic impact [1,2]. The main cause of CVD is atherosclerosis, a multifactorial disorder induced and augmented by risk factors such as dyslipidemia, oxidative and inflammatory stress, diabetes mellitus, hypertension, smoking, ageing and genetic mutations [3]. Many therapies designed to treat atherosclerosis either have failed completely or partially (cholesteryl ester transfer protein (CETP) inhibitors, antioxidants, vitamins), or were too expensive to be applied to the entire population at CVD risk (apoA-I Milano); others, though successful, induced considerable side-effects (statins) [4,5]. Thus, the quest for strategies to prevent or treat CVD is of high and continued interest.

In the last decade the scientific researchers turned their attention to phytochemicals, as effective, safe and low-cost natural bioactive compounds for CVD treatment.

Dyslipidemia consists of increased blood concentrations of total cholesterol (TC), low density lipoproteins–cholesterol (LDL-C) and/or triglycerides (TG), and decreased high density lipoproteins–cholesterol (HDL-C) [6]. The lipid metabolism is complex and the candidate mechanisms that could generate dyslipidemia include: (i) excessive dietary lipid absorption in the small intestine; (ii) packing of exogenous lipids with cholesterol and fatty acids produced de novo in the liver and their secretion as very low density lipoproteins (VLDL); (iii) hydrolysis of TG from VLDL by lipases and their conversion into LDL, which are taken up by the peripheral tissues through LDL receptor (LDL-R) and scavenger receptors; (iv) diminished production of HDL by the liver and small intestine, thereby decreasing reverse cholesterol transport (RCT) from the peripheral tissues to the liver; (v) lowered excess cholesterol excretion from the liver into gallbladder or to the intestinal lumen through the ATP-binding cassette G5 and G8 transporters (ABCG5/G8) that facilitate trans-intestinal cholesterol efflux (TICE). Dyslipidemia is associated with the accumulation of LDL in the sub-endothelium of the artery wall. At this site, LDL undergoes oxidative modifications (oxLDL) that trigger inflammatory responses, and is taken up by the monocyte-derived macrophages infiltrated in the sub-endothelium which thus become lipid-loaded foam cells, the hallmark of atheroma development [7]. Until now, the most effective lipid-lowering treatment for hyperlipidemic patients was the statin therapy. But recent recommendations have extended it to the asymptomatic adults at low cardiovascular risk (Systematic COronary Risk Evaluation <1%) and with lower LDL-C levels (<4.9 mmol/L), according to their levels of predicted CVD risk [8]. This extension will be accompanied by a higher number of subjects manifesting side effects and the need for replacement of the statin therapy with new effective and better tolerated medication.

Under physiological conditions, reactive oxygen species (ROS) are generated in low concentrations and serve as mediators that regulate vascular function [9,10]. Increased generation of ROS, due to the increased production of free radicals or paucity of antioxidants, generates oxidative stress and causes peroxidation of cellular lipids and lipoproteins that contribute to CVD progression [11]. Oxidative stress affects the bioavailability of nitric oxide (NO), an important vascular relaxing factor. When the endothelial nitric oxide synthase (eNOS) is uncoupled, NO and the superoxide radicals react and generate the oxidizing peroxynitrite anions, thereby inducing endothelial dysfunction [12]. Mitochondria control cell energy production and the respiratory cycle, processes closely related to oxidative stress [13]. To counteract or prevent the detrimental effect of ROS, numerous antioxidant systems have been developed, such as glutathione, flavonoids, superoxide dismutase (SOD), catalase, glutathione peroxidase (GSH-Px) and hemeoxygenase-1 (HO-1) [14]. The nuclear erythroid 2-related transcription factor (Nrf2) is redox-sensitive and plays an important role in regulating the production of the antioxidant enzymes [15]. Under physiological conditions, the enzyme paraoxonase 1 (PON1) accounts for HDL antioxidant properties, while the pro-oxidant enzyme myeloperoxidase (MPO) becomes attached to HDL under hyperlipidemic conditions [16]. In the last few decades, the quality and function of HDL became a consistent indicator of CVD risk [17,18]. Under augmented oxidative stress, HDL becomes dysfunctional, presenting an increased number of MPO molecules that replace the PON1 molecules [17,19]. Thus, dysfunctional HDL can no longer protect LDL from oxidation, these alterations being the key risk factors for initiation and progression of the atherosclerotic plaques.

Exposure of endothelial cells (EC) to risk factors induces their activation which consists in the production of pro-inflammatory molecules such as selectins (E-selectin, P-selectin, L-selectin), and of adhesion molecules, such as intracellular adhesion molecule 1 (ICAM-1) and vascular cell adhesion 1 (VCAM-1) promoting monocyte adhesion and their transmigration into the sub-endothelium [20,21]. The macrophages produce pro-inflammatory cytokines, such as interleukin-1β (IL-1β), IL-12 and tumor necrosis factor-α (TNF-α), or chemokines, such as monocyte chemoattractant protein-1 (MCP-1) to attract more monocytes, thereby stimulating the inflammatory process [22]. A central role in the stimulation of the inflammatory stress in atherosclerosis is played by the nuclear factor kappa B

(NF-κB) [23]. A strong inhibitor of NF-κB is sirtuin-1 (SIRT-1), and toll-like receptor 4 (TLR4) is an activator of NF-κB [21,24]. Another important pro-inflammatory intracellular protein complex is the nucleotide binding and domain-like receptor 3 (NLRP3) inflammasome, which stimulates the primary pro-inflammatory response determining the activation of the secondary inflammatory mediators, such as IL-6 or C reactive protein (CRP), thereby amplifying the inflammatory stress [25].

Since the completion of the Human Genome Project, it has become evident that several additional mechanisms related to epigenomic, transcriptomic, epitranscriptomic, proteomic and metabolomic regulations are crucial for determinations of the phenotypes of many human disorders, including CVD [26]. The emerging non-coding RNAs (ncRNAs) have major regulatory roles in gene expression, and thus could function as potential targets for personalized treatment of CVD patients [27–29]. They can be divided into small ncRNAs (<200 nt), such as microRNA (miRNAs), transfer RNAs and small nucleolar RNAs, and longer ncRNAs (lncRNA), that include ribosomal and natural antisense transcripts. Members of a class of small (~22 nt) ncRNA, single stranded in mature form miRNAs have been identified as potent post-transcriptional regulators of genes, including those involved in lipid metabolism [30]. MiRNAs control their target gene's expression by either imperfect base pairing to 3′ untranslated regions (3′UTR) of mRNA [31] or by binding to other regions, including 5′ UTRs or protein-coding exons [32], thereby inducing the repression of their target mRNAs. Circulating miRNAs are found in the blood closely associated with proteins, lipoproteins and extracellular vesicles (EVs) [33]. The levels of circulating miRNAs vary and have specific profiles in diverse pathophysiological states [34,35], leading to the possibility of using these molecules as promising markers for diagnosis and prognosis [36,37]. Published studies demonstrate that plasma miRNAs may be used as markers in the diagnosis of myocardial infarction [38], while EVs and HDL-associated miRNAs levels can discriminate between stable and vulnerable coronary artery disease patients [39–41]. It was reported that circulating and hepatic miRNAs expression could be modulated by hypolipidemic dietary interventions, such as probiotics administration [42]. For all these reasons, miRNAs could be very useful targets in nutritional science and could be used to test the pathways modulated by dietary treatments in healthy and/or diseased populations [43].

Phytochemicals such as phenolic compounds have been extensively studied, and important data exist on their use in CVD patients. In the present review we aim to focus on the molecular mechanisms of action that prove the multiple benefic effects of the phenolic compounds and endorse them as CVD remedies, complementary to allopathic drugs. We explore the active compounds that effectively target dyslipidemia, and oxidative and inflammatory stress, the main risk factors in atherosclerosis evolution. Thus, we discuss below the representative compounds for which a consistent number of mechanisms of action have been described. They belong to the following groups: hydroxycinnamic acids (curcumin, caffeic acid), stilbenes (resveratrol), flavonols (quercetin), flavones (apigenin, luteolin), flavanones (naringenin, hesperetin), flavanols (catechins, gallocatechins), isoflavones (genistein), anthocyanins/anthocyanidines and guaiacols (gingerols, shogaols).

2. Biologically Active Phenolic Compounds and Their Mechanisms of Action

Phytochemicals are biologically active compounds from plants that can regulate physiological and pathological processes with benefic consequences for human health. Phenolic compounds are one of the largest groups of phytochemicals. Their bioavailability is decisive to exerting beneficial effects in vivo and is influenced by the molecular size and complexity of their chemical structure, including conjugation with other phenols, polymerization, glycosylation, acylation or hydroxylation [44].

2.1. Hydroxycinnamic Acids Group

Curcumin (diferuloylmethane) (Figure 1) is the most important bioactive compound from *Curcuma longa*, belonging to the *Zingiberaceae* family. The curcuminoids comprise several compounds, such as curcumin, desmethoxycurcumin and bis-demethoxy-curcumin. The source of curcumin is turmeric, a yellow-colored spice [45]. Pharmacokinetic studies revealed that curcumin is poorly soluble in water; has low absorption in the gut, rapid metabolism and systemic elimination, and consequently, has low bioavailability after oral administration. The clinical efficacy of curcumin could be improved by formulations that enhance its solubility and stability and diminish the first-pass metabolism. To that end, certain strategies have been elaborated, such as the development of curcumin–piperine complexes, curcumin nanoparticles, cyclodextrin inclusions, curcumin liposomes and curcumin phospholipids' complexes, part of these systems exhibiting a 100-fold increase of bioavailability relative to unformulated curcumin [46].

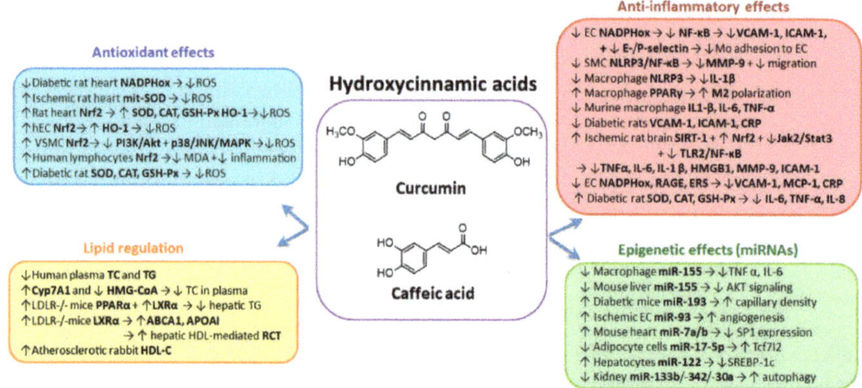

Figure 1. Chemical structure and protective effects exerted by curcumin and caffeic acid to improve cardiovascular diseases outcomes as demonstrated by experimental and clinical evidence.

In vitro and in vivo studies demonstrate curcumin's pleiotropic effects, due to its ability to interact with numerous molecular targets in different cell types. The anti-atherogenic potential of curcumin consists in its capacity to lower blood cholesterol and TG in healthy subjects (80 mg/day for 4 weeks) [47]. It is reported that turmeric inhibits LDL oxidation in atherosclerotic rabbits and increases serum HDL-C levels in diabetic rats [48]. The hypolipidemic effect of curcumin is based on the inhibition of the intestinal cholesterol absorption and increased activity of cholesterol-7alpha-hydroxylase (CYP7A1), the rate-limiting enzyme in the synthesis of bile acids [48]. In addition, curcumin inhibits 3-hydroxy-3-methyl-glutaryl-coenzyme A (HMG-CoA) reductase. Administration of 0.02% *w/w* curcumin for 18 weeks to ldlr$^{-/-}$ mice fed a high-cholesterol diet induces an inhibition of the hepatic TG accumulation by upregulation of peroxisome proliferator-activated receptors alpha (PPARα) and liver X receptor alpha (LXRα) expression [49]. PPARα is an important activator of fatty acid oxidation and inhibitor of hepatic fatty acid synthase (FAS) activity. LXRα regulates the gene expression of the key enzyme involved in cholesterol conversion to bile acid (CYP7A1), and increases expression of liver apolipoprotein A-I (apoA-I) and ATP-binding cassette A1 (ABCA1), which facilitates the HDL-mediated RCT [50]. It is known that an increase of 1 mg/dL in HDL-C level reduces coronary heart disease risk by 2–3%, and CVD mortality risk by 3.7–4.7% [51].

The antioxidant action of curcumin resides in the inhibition of ROS production by repressing the catalytic subunits p67phox, p47phox and p22phox of nicotinamide adenine dinucleotide phosphate (NADPH) oxidase [52,53]. In parallel, curcumin reduces ROS by upregulating the expression of endogenous antioxidant enzymes, such as SOD, catalase, GSH-Px and HO-1 [54,55]. Another interesting mechanism by which curcumin exerts antioxidant effects is the preservation of the mitochondrial redox potential [13] that has been evidenced in vivo in rat hearts subjected to ischemia-reperfusion. Curcumin pretreatment increases mitochondrial SOD activity and decreases mitochondrial H_2O_2 and malondialdehyde (MDA) levels [13]. HO-1 is an enzyme activated by oxidative stress which reduces inflammation by inhibiting the expression of endothelial adhesion molecules [56]. It was reported that curcumin can induce HO-1 in TNF-α-treated EA.hy926 cells in a dose-dependent manner and through activation of the transcription factor Nrf2. It also decreases ICAM-1 expression in a mouse model of lung injury [57,58]. As in EC, in vascular smooth muscle cells (SMC) curcumin activates Nrf2 which increases aldose reductase, an important enzyme that reduces oxidative stress in phosphatidylinositol 3-kinase (PI3K)/protein kinase B (Akt) and p38 mitogen-activated protein kinase (p38 MAPK)/c-Jun *N*-terminal kinase (JNK)-dependent manners [59]. These effects of curcumin on Nrf2 pathway have also been seen in human clinical trials involving patients with type 2 diabetes mellitus who each received an oral dose of curcumin of 500 mg/day for 15–30 days. This treatment induced an upregulation of Nrf2 in lymphocytes that controls other proteins, such as NAD(P)H quinone dehydrogenase 1 (NQO1), reduces inflammatory markers, and reduces plasma MDA levels [60]. In addition, curcumin supplementation of the diet increases NO bioavailability and inhibits the expression of pro-oxidative NADPH oxidase 2 (NOX-2) enzyme in rats [61]. Curcumin inhibits NF-κB and further decreases the expression of pro-inflammatory cytokines such as TNF-α, IL-1β and IL-6 both in vitro and in vivo [62].

There is growing interest in using curcumin to reduce vascular disease, in part due to its demonstrated anti-inflammatory effects. Using the in vitro model of TNF-α-activated EC, it was demonstrated that curcumin reduces the expression of VCAM-1, ICAM-1, E-selectin [63], fraktalkine and P-selectin [64], thereby inhibiting significantly monocyte adhesion through mechanisms involving the reduction of NADPH oxidase activation and consequently of the intracellular ROS production [64]. These results were confirmed in a recent study by Monfoulet et al., who demonstrated that curcumin pre-exposure reduces endothelial permeability and monocyte adhesion in both static and flow conditions [65]. In addition, most of the in vitro and in vivo studies confirm that curcumin administration determines the lowering of MCP-1 levels by downregulation of the MAPK and NF-κB signaling pathway [66–68]. The anti-inflammatory effects of curcumin on EC inflammatory markers were demonstrated also in vivo in various animal models. Tsai et al. evidenced the decrease of VCAM-1, ICAM-1 and CRP levels after curcumin supplementation to rats fed with a high-sucrose and high fat diet. The mechanisms that determine the improvement of the vascular function in this animal model involved an enhanced NO production and a reduction of the oxidative stress due to the increase of antioxidant enzyme activities [69]. SMC, an important component of the vascular wall, participates in the characteristic inflammatory process of atherosclerosis. In an in vitro study, Han et al. demonstrated that migration of human aortic SMC isolated from spontaneously hypertensive rats (SHR) or normal rats exposed to angiotensin II (AngII) was significantly inhibited by curcumin through the reduction of NLRP3/NF-κB signaling pathway [70]. The results were confirmed in vivo; it was demonstrated that administration of curcumin in SHR reduced intima-media thickness due to the inhibition of NF-κB and NLRP3 inflammasome and matrix metallopeptidase 9 (MMP-9) expression [70,71]. Yin et al. showed that curcumin inhibited caspase-1 activation and IL-1β secretion through suppressing lipopolysaccharide (LPS) priming and NLRP3 inflammasome in mouse bone marrow-derived macrophages, and confirmed these results in vivo in a model of high-fat diet-induced insulin resistance in wild-type C57BL/6 mice [72]. Interestingly, in an in vitro study, Chen et al. demonstrated that curcumin inhibits the M1 inflammatory phenotype of RAW264.7 macrophages as a result of the direct activation of the inhibitor protein κB-α (IkB-α) and polarizes the macrophages to become anti-inflammatory M2 phenotype through the activation of PPARγ [73]. Later on, the same group reported that curcumin dramatically reduced oxLDL-induced IL-1β, IL-6 and TNF-α cytokine production in M1 derived from M0 RAW264.7 macrophages [74].

In vivo studies demonstrated the benefic effects of curcumin to reduce the inflammatory burden present in the ischemia/reperfusion (I/R) in different organs by modulating the expression of different signaling pathways. In the brain tissue of rats after cerebral ischemia, curcumin decreased TNF-α and IL-6 levels via activation of SIRT-1 [75]; reduced TNF-α, IL-1β, IL-6, and high mobility group box 1 (HMGB1) by inhibition of the Janus kinase 2 (JAK2)/signal transducers and activator of transcription 3 (STAT3) signaling pathway [76]; and inhibited ICAM-1 and MMP-9 by downregulating NF-kB and elevating Nrf2 [77] or reduced TNF-α and IL-1β by inhibiting the TLR2/4/NF-kB signaling pathway [78].

It was reported that curcumin inhibits the activation of JNK and activator protein 1 (AP-1) transcription factor, and also the phosphorylation and degradation of IκB-α [79].

In a recent meta-analysis that proposed to evaluate the therapeutic effect of curcumin in mouse models of atherosclerosis, Lin et al. found that curcumin significantly decreases the aortic atherosclerotic lesion area, and the serum lipid levels (TC, TG and LDL-C) and inflammatory markers (TNF-α and IL-1β) [80]. In addition, Lin et al. highlighted the dose-response relation between curcumin and its protective effect on atherosclerosis, showing that the effect on decreasing the aortic lesion area is stronger in low and medium dosages (< 207 mg/kg BW/day) and weaker when the dose was more than 207 mg/kg BW/day, becoming pro-atherogenic when the dose reached 347 mg/kg BW/day [80]. Clinical trials focusing on curcumin effects in atherosclerosis progression gave dissimilar results, some of them evidencing that curcumin has no effect on risk factors of atherosclerosis [81,82], while others reporting an atheroprotective effect of curcumin by improvement of the lipidic profile in patients with metabolic syndrome, patients taking curcumin extract capsules (630 mg thrice daily) for 12 weeks [83]. In a pilot study, Panahi et al. demonstrated that administration of curcuminoids (500 mg/day, for 4 weeks) to subjects with pulmonary problems reduces the inflammatory mediators' expression: IL-6, IL-8, TNF-α, transforming growth factor β (TGF-β), high sensitivity C-reactive protein (hsCRP) and MCP-1 [84]. There are many studies supporting the anti-inflammatory properties of curcumin; however, of great importance is the establishment of a proper treatment with regard to the dose and time of administration.

To improve curcumin stability in vivo, different types of nanocarriers have been described for encapsulation. Thus, lipid nanoemulsions loaded with curcumin and functionalized with a cell-penetrating peptide were better taken up and internalized by human EC compared to the non-functionalized lipid nanoemulsions. This formulation of curcumin demonstrated anti-inflammatory effects by reducing the monocytes adhesion to TNF-α activated human EC [85].

In the last decade, published data evidenced that some phenolic compounds are able to perform a fine-tuning regulation of the mechanisms underlying oxidative and inflammatory stress by modulating the expression of epigenetic factors associated with RNA function, such as ncRNAs. Accordingly, recent in vitro and in vivo experiments showed that specific miRNAs mediate the molecular mechanisms affected by curcumin. Thus, treatment of murine Raw264.7 and human THP-1 macrophages with curcumin significantly reduced miR-155 expression through the modulation of PI3K/Akt pathway [86]. In the same report, septic mice obtained by LPS intraperitoneal injection were treated orally with curcumin, and a significant reduction of miR-155 expression and Akt phosphorylation in the liver and kidney was observed [86]. In a recent in vivo study of Zhang et al., mice with peripheral arterial disease (PAD) and a normal glycaemic profile were treated with curcumin [87]. They reported that curcumin treatment improved perfusion recovery, increased capillary density and increased miR-93 expression in ischemic muscle tissue. Moreover, in cultured EC under simulated ischemia, curcumin improved cell viability and enhanced tube formation. These data proved that curcumin may have beneficial effects in non-diabetic PAD by improving angiogenesis, which may have been achieved partially via the promotion of miR-93 expression [87]. In another in vivo experiment, after curcumin administration, C57BL/6 mice were subjected to left anterior descending coronary artery occlusion [88]. Geng et al. showed that curcumin administration significantly reduced the infarct size compared with control animals, increased miR-7a/b expression and downregulated the expression of transcription factor specific protein 1 (SP1). In hypoxia-induced mouse cardiac myocytes (MCM), curcumin led to the decrease of cell apoptosis. The authors suggested that curcumin pretreatment protected against hypoxia-induced MCM apoptosis through the upregulation of miR-7a/b and the downregulation of SP1 expression [88].

Curcumin was reported to modulate some miRNAs that are dysregulated in diabetes. In a study conducted by Tian et al., miR-17-5p was proven to stimulate adipogenic differentiation of mouse 3T3-L1 cells [89]. This gene encodes a key Wnt signaling pathway effector, and its human homologue transcription factor 7 like-2 (TCF7L2) is a highly diabetes risk gene. After treatment with curcumin, a decrease of miR-17-5p expression was observed, together with an increase of its target gene, Tcf7l2, in 3T3-L1 adipocyte cells. The authors also reported an elevation of miR-17-5p expression in mouse epididymal fat tissue in response to high fat diet. The authors suggested that miR-17-5p is among the central switches of adipogenic differentiation, activating adipogenesis via repressing the Wnt signaling pathway effector Tcf7l2, and its own expression is nutritionally regulated by curcumin [89].

Caffeic acid (3,4-dihydroxycinnamic acid) (Figure 1) is the major dietary hydroxycinnamic acid and is found in food, mainly as caffeic acid phenethyl ester (CAPE) or chlorogenic acid (5-O-caffeoylquinic acid), which results from its conjugation with quinic acid. The chlorogenic acid is one of the most widely consumed polyphenols, being present in many fruits (blueberries, apples, pears), vegetables (lettuce, potatoes, eggplants), and beverages, including coffee (caffeinated or decaffeinated), wine and tea. Regular consumption of coffee results in the ingestion of 0.5–1 g of chlorogenic acid and 250–500 mg of caffeic acid/day [90–92]. CAPE has poor bioavailability attributed to its low aqueous solubility, and in the plasma undergoes rapid hydrolysis to caffeic acid as the major metabolite. To overcome the poor bioavailability of CAPE, different formulations such as chemical modifications or microencapsulation in cyclodextrins were developed with success, the aqueous solubility of CAPE was notably increased [93]. The low bioavailability of caffeic acid (14.7%) is due to the low intestinal absorption and low permeability across the intestinal cells [94]. No formulations of caffeic acid to be used for CVD treatment have been developed at present.

CAPE has been reported to have antioxidant and anti-inflammatory properties [95]. It was demonstrated that CAPE induces the expression of redox-sensitive HO-1 [96] through activation of the Kelch-like ECH-associated protein 1 (Keap1)/Nrf2/antioxidant response element (ARE) pathway [97], and consequently generates the transcription and translation of detoxifying and antioxidant phase II cytoprotective enzymes. Nrf2 is bound to Keap1 in the cytoplasm before activation, and once inducers react with the sulfhydryl groups of Keap1, Nrf2 is released and eventually translocated into the nucleus, where it binds to and activates ARE, which acts as a promoter/enhancer regulating the genes of the mentioned antioxidant enzymes [98]. The Nrf2/ARE pathway can be also activated through Nrf2 phosphorylation by PI3K/Akt, extracellular signal-regulated kinases (ERK) or MAPK [97]. In addition, administration of CAPE to rats stimulates PON1 expression in the lung exposed to inflammatory stimuli [99].

The benefic effect of caffeic acid on inflammatory stress is not very well documented. It was reported that caffeic acid can reduce monocyte adhesion to TNF-α-activated human umbilical vein EC (HUVECs) by reducing the expression of VCAM-1, ICAM-1, E-selectin and MCP-1. These beneficial effects were attributed to the reduction of NF-κB p65 translocation from cytosol to nucleus, thereby decreasing the formation of NF-κB–DNA complex [100]. A relatively recent in vitro study confirmed these results and added new data, demonstrating that caffeic acid exerts anti-inflammatory effects on HUVEC exposed to glycated LDL (gLDL) by reducing the secretion of CRP, VCAM-1 and MCP-1. The downregulation of all these pro-inflammatory molecules is possible due to the inhibition of the receptor for advanced glycation end products (RAGE) expression and diminution of the oxidative stress (by reducing NOX4 and p22phox subunits of NADPH oxidase) and of the endoplasmic reticulum stress [101].

In vivo, it was shown that the caffeic acid reduces the plasma TNF-α, IL-6 and IL-8 in rats receiving a high-fructose diet by decreasing the oxidative stress due to restoring of the antioxidant enzymes concentration (SOD, catalase, GSH-Px, glutathione reductase and glucose 6-phosphate dehydrogenase) [102]. In a very interesting in silico study, the caffeic acid was identified as a potential therapeutic agent having anti-inflammatory potential due to its interactions with cyclooxygenase-1 (COX-1) and 2 (COX-2), coagulation factor Xa (FXa) and integrin αIIbβ3 proteins, which are directly or indirectly participants in the thrombosis pathways [103].

Another molecular mechanism associated with the positive effects of caffeic acid in ameliorating the lipid metabolism and oxidative and inflammatory stress is the regulation of epigenetic factors, in particular, miRNAs. Murase et al. showed that in vitro treatment of Hepa1-6 hepatocytes with coffee polyphenols significantly increased cellular miR-122 expression, and reduced sterol regulatory element-binding transcription factor 1c (SREBP-1c) expression [104]. Using high-fat diet/streptozotocin-induced diabetic rats, Matboli et al. showed that caffeic acid intake induced improvement in albumin excretion, blood glucose, reduced renal mesangial matrix extension with increased vacuolation and reappearance of autophagosomes [105]. Additionally, they demonstrated that caffeic acid treatment stimulates autophagy genes with simultaneous reduction in their epigenetic regulators: miR-133b, miR-342 and miR-30a. These data suggest that caffeic acid can modulate the autophagy pathway through inhibition of autophagy regulatory miRNAs that could explain its curative properties against diabetic kidney disease [105].

2.2. Stilbenes

Resveratrol (trans-3,5,4'-trihydroxystilbene) has a C6–C2–C6 structure containing three hydroxyl groups, which functions as a UV protectant of plants and defender against pathogenic infections (Figure 2). The food sources of resveratrol (RSV) are red wine, grapes, peanuts, passion fruit, white tea, plums and raspberries [106]. Oral ingestion of RSV is the most feasible route of administration, but the compound has low water solubility (~30 mg/L); it is rapidly metabolized, and consequently has a poor bioavailability. A slight increase in RSV solubility considerably enhances its bioavailability [107]. After ingestion, RSV is absorbed in the small intestine and then released into the bloodstream where it can bind to albumin and lipoproteins that further deliver RSV to the cells of the peripheral tissue. RSV is well tolerated and its plasma concentration depends on the dose consumed, but not in a linear relation. High oral doses (1 g/kg) may generate side effects (nausea, abdominal pain) [107]. Thus, improving RSV bioavailability will straighten its potential as a therapeutic agent. Recently the researchers have tried to increase RSV bioavailability by nanoencapsulation in lipid nanocarriers or liposomes, nanoemulsions, micelles, insertion into polymeric particles, solid dispersions and nanocrystals [108]. The results are promising, but further studies are needed to improve these methodological approaches and to compare effects of the most valuable strategies in the same trial.

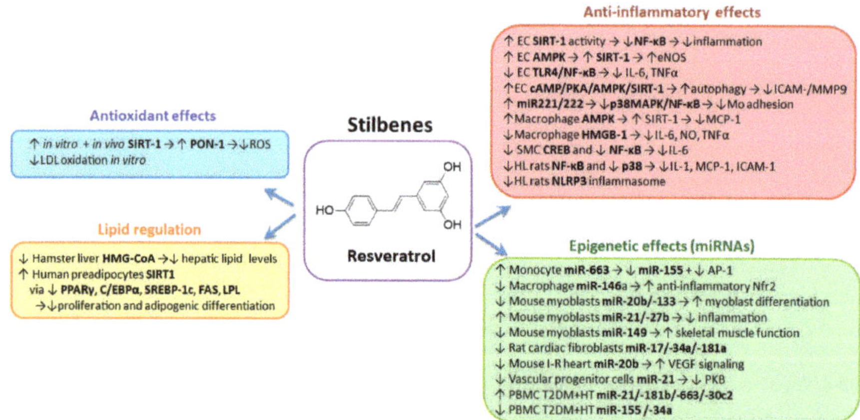

Figure 2. Resveratrol's chemical structure and its demonstrated effects of improving cardiovascular disease outcomes.

The lipid-lowering action of RSV is based on its capacity to reduce the level and activity of HMG-CoA reductase that has been demonstrated in the livers of hamsters with diet-induced dyslipidaemia [109]. Studies conducted in humans are controversial, some of them evidencing a reduction of LDL-C and an increase of HDL-C [110,111], and others showing no effect of RSV on plasma lipid profile [112]. RSV has important effects on adipose tissue due to its ability to inhibit differentiation of preadipocytes and stimulation of lipolysis. Thus, RSV inhibits proliferation and adipogenic differentiation of human preadipocytes by a SIRT1-dependent mechanism and in the 3T3 cells downregulate the expression of PPARγ, CCAAT-enhancer-binding proteins α (C/EBPα), SREBP-1c, FAS and lipoprotein lipase (LPL), which are important regulators of the lipolysis [113]. RSV has protective effects in terms of CVD risk due to its capacity to improve PON1 activity. Experiments performed in vitro and in vivo evidenced the positive effects of quercetin/RSV on PON1 activity [95,114–116]. The mechanisms responsible for this effect might be the activation of SIRT-1, which is known as activator of LXR, important regulator of PON1 gene [117]. The antioxidant potential of RSV was evidenced by inhibiting macrophages-induced in vitro oxidation of LDL [118]. The ability of RSV to exert antioxidant effects in humans is still under investigation.

Numerous studies reveal that the benefic effects of RSV are due to its potential to activate several important anti-inflammatory targets [119], although the exact mechanisms of action have not been clearly elucidated [120]. It was shown that the main target of RSV is SIRT-1, and its binding determines modifications of the SIRT-1 structure that enhances the binding of SIRT-1 to its substrates [121]. An important substrate of SIRT-1 is p65 of NF-κB (RelA) [122], the key transcription factor involved in regulation of inflammatory cytokines [123]. SIRT-1 activation by RSV determines the inhibition of RelA acetylation, which in turn decreases NF-κB expression [124]. In addition, RSV inhibits p300 expression and promotes the IκB-α degradation [125]. Another molecular target of RSV is AMP-activated protein kinase (AMPK), a protein that controls the activity of SIRT-1 by regulating the available cellular levels of NAD$^+$ [126]. Beside SIRT-1, AMPK is known to activate eNOS in EC. A recent clinical trial involving primary hypertensive patients evidenced that addition of a micronized formulation of RSV to standard antihypertensive therapy is sufficient to normalize the blood pressure, without additional antihypertensive drugs [127]. Other targets of RSV were identified by different other groups: TLR4 [128,129], miR-221/222 [130] and p38 MAPK [131,132].

RSV induces the decrease of endothelial activation and vascular inflammation, and improves the endothelial function. It was demonstrated that RSV determines the decrease of IL-6 and TNF-α via the TLR4/myeloid differentiation primary response gene 88 (MyD88)/NF-κB signal transduction pathway in HUVECs exposed to LPS [128]. In addition, it was reported that pre-incubation with RSV reduced the TNF-α-induced ICAM-1 secretion, as well as the intracellular expression of ICAM-1 and MMP-9 in EC by inducing autophagy, mediated in part through the activation of the cAMP/protein kinase A (PKA)/AMPK/SIRT-1 signaling pathway [133]. Liu et al. demonstrated that RSV decreases ICAM-1 expression and monocyte adhesion to TNF-α-exposed HUVECs by stimulating miR-221/-222 production, which determines p38 MAPK/NF-κB inhibition [130].

In THP-1 human macrophages stimulated with LPS, RSV pretreatment inhibited foam cells formation and reduced MCP-1 secretion, while increasing SIRT-1 and AMPK [134]. RSV significantly reduced the levels of secreted IL-6, NO and TNF-α in RAW264.7 cells exposed to LPS by attenuating HMGB-1 expression [135]. Other mechanisms of RSV action in LPS-exposed macrophages involve the attenuation of TLR4- TNF receptor-associated factor 6 (TRAF6), MAPK and Akt pathways [136].

The anti-inflammatory properties of RSV were demonstrated also in SMC. Inanaga et al. showed that RSV attenuates Ang II-induced IL-6 protein in the supernatant of vascular SMC in a dose-dependent manner. These effects were attributed to the ability of RSV to reduce the activity of cAMP-response element-binding protein (CREB) and NF-κB, two transcription factors which are critical for Ang II-induced IL-6 gene expression [137]. In addition, Zhang et al. demonstrated that RSV reduced the proliferation of vascular SMC exposed to Ang II by inhibiting ERK1/2 phosphorylation and NF-kB transcriptional activity [138].

The anti-inflammatory properties of RSV were demonstrated also in vivo. Using hyperlipidemic rats, Deng et al. demonstrated that RSV decreases the serum levels of IL-1β and reduces MCP-1, ICAM-1, p65 NF-κB and p38 MAPK mRNA and protein expression in the thoracic aortas samples [139]. In addition, NLRP3 inflammasome oligomerization was also decreased in the aortic tissue, in parallel with the upregulation of SIRT-1 expression [139]. Interestingly, Chang and colleagues previously demonstrated that RSV reduces inflammation, such as aortic macrophage infiltration and NF-κB expression in apoE-deficient mice fed with a high-cholesterol diet [140]. Using the model of an I/R-injured rat, Cong et al. demonstrated that RSV reduced the myocardial infarct area, in parallel with a reduction of serum and myocardial TNF-α levels through a mechanism dependent on NO production [141]. Li et al. confirmed the previous in vivo study, demonstrating that RSV significantly reduces myocardial infarct size and myocardial apoptosis, serum and myocardial TNF-α production by a mechanism dependent on TLR4/NF-κB attenuation and NO production [142].

The results concerning anti-inflammatory potential of RSV in humans were contradictory, some studies evidencing a positive effect in healthy people [143], others showing no effects in postmenopausal women [144]. The contrasting results suggest that in order to obtain beneficial effects, the dose and the way of administration have to be carefully analyzed. It was suggested that a moderate (>450 mg) continuous intake is better than a single, higher dose administration [145].

Recent in vitro and in vivo experiments proved that RSV positively regulates the mechanisms underlying oxidative and inflammatory stress by modulating the expression of a set of specific miRNAs. Tili et al. showed that RSV upregulates miR-663 in human THP-1 and circulating monocytes, this miRNA being proven as anti-inflammatory by inducing the decrease of AP-1 transcriptional activity [146]. Moreover, they showed that RSV impairs AP-1 upregulation induced by LPS at least in part by targeting JunB and JunD transcripts. In contrast, RSV impairs the LPS-induced upregulation of pro-inflammatory miR-155 in a manner dependent of increasing miR-663 levels [147]. These data suggest the potential modulation of miR-663 levels to stimulate the anti-inflammatory effects of RSV in metabolic disorders associated with elevated levels of miR-155. Since many in vitro experiments use high concentrations of phenolic compounds and do not reproduce their physiological in vivo plasma levels, Bigagli et al. incubated RAW264.7 macrophages with corresponding plasma physiological concentrations of RSV, hydroxytyrosol and oleuropein [148]. They showed that only RSV and hydroxytyrosol (at 10 μM) decreased miR-146a, which is known to target Nrf2 responsible for inhibiting pro-inflammatory mediators. In addition, the authors showed that Nrf2 was increased by RSV and hydroxytyrosol after in vitro stimulation of murine macrophages with LPS [148]. Lançon et al. reported that of 26 miRNAs were increased (miR-21 and miR-27b) in prevalence by RSV in mouse C2C12 skeletal myoblasts, while other 20 miRNAs (miR-20b and miR-133, a muscle-specific miRNA known to target genes involved in myoblast differentiation) were downregulated [149,150]. Additionally, miR-149 was downregulated by RSV, this miRNA having potential role in the regulation of skeletal muscle functionality. Recently, Zhang et al. demonstrated that RSV can inhibit in vitro the TGF-β1-induced proliferation of rat cardiac fibroblasts (CF) and collagen secretion [151]. RSV also decreased miR-17, miR-34a and miR-181a in TGF-β1-treated CF. The authors suggested that the inhibitory effect of RSV is mediated by the downregulation of miR-17 and the regulation of Smad7 [151].

In an in vivo study, Mukhopadhyay et al. reported the cardioprotective effect of RSV and proposed a RSV-induced miRNAs profile in a rat I/R model [152]. They reported that RSV significantly downregulated miR-20b, which might modulate vascular endothelial growth factor (VEGF) signaling. This downregulation of miR-20b was proposed to be linked with the potent anti-angiogenic action of RSV in the ischemic myocardium and with the synergic effects of RSV and γ-tocotrienol. An elegant and complex study performed by Campagnolo et al. showed that RSV can induce the expression of endothelial markers, such as CD31, VE-cadherin and eNOS in vascular resident progenitor cells and embryonic stem cells [153]. They also demonstrated that RSV significantly reduced miR-21 expression in these cells, which in turn diminished protein kinase B (PKB) phosphorylation. This signaling cascade reduced the nuclear β-catenin, inducing endothelial marker expression and increasing tube-like formation by progenitor cells. Additionally, the authors showed that vascular progenitor cells treated ex vivo with RSV produced better endothelialization of the decellularized vessels. Moreover, they demonstrated that RSV-enriched diet reduces lesion formation in a mouse model of vessel graft [153].

Tomé-Carneiro et al. performed a randomized placebo-controlled study with type-2 diabetic and hypertensive men, who received capsules containing either placebo (maltodextrin), grape extract (laking RSV) (GE) or grape extract with over 8 mg of RSV (GE-RES) during one year [154]. Their results show that supplementation with GE or GE-RES did not affect body weight, blood pressure, glucose, HbA1c or lipids, beyond the values regulated by gold standard medication in these patients. They also found molecular changes in peripheral blood mononuclear cells (PBMC), evidenced by the significantly reduced expression of the pro-inflammatory cytokines—macrophage inflammatory protein 1α (MIP1α), CCL3, IL-1β and TNF-α—and increased expression of transcriptional repressor leucine-rich repeat flightless-interacting protein 1 (LRRFIP-1) in PBMC from patients taking the GE-RES extract for 12 months [154]. Additionally, a GE-RES treatment-associated modulation of miRNAs involved in the inflammatory response was noticed, demonstrated by the increase of a set of miRNAs miR-21, miR-181b, miR-663 and miR-30c2, together with a decrease of miR-155 and miR-34a in PBMC after GE-RES treatment, as compared to the control group. These data provide evidence for the in vivo modulation of inflammatory miRNAs in PBMC by RSV in circulating immune cells of diabetic hypertensive medicated patients and support a beneficial immunomodulatory effect in these patients [154].

2.3. Flavonoids Group

Flavonoids (including flavonols, flavones, flavanones, flavanols, isoflavones and anthocyanidins) exert multiple beneficial effects. The food sources of flavonoids are berries, black tea, celery, citrus fruits, green tea, olives, onions, oregano, purple grapes, purple grape juice, soybean, soy products, vegetables, whole wheat and wine [155].

2.3.1. Flavonols

Quercetin (3,3′,4′,5,7-pentahydroxylflavone) (Figure 3) is administered as quercetin-3-glucoside (isoquercetin) which is hydrolyzed to quercetin in the small intestine, rapidly absorbed and then transferred into the blood. Vegetables that are important sources of quercetin are apples, grapes, berries, black tea, green tea, red onions, kale, leeks, broccoli, apricots, pepper, red wine and tomatoes [155]. A very recent study shows that quercetin alters the gut microbiota and reduces the atherogenic lipids, such as cholesterol and lysophosphatidic acids, all these effects being associated with the diminution of atherosclerotic lesions area [156]. Unlike most phenolic compounds, quercetin has a relatively high bioavailability. It is absorbed in the small intestine, then undergoes different transformations in the small intestine, colon, liver and kidney. Quercetin that is not intestinally absorbed is further subjected to colon microflora metabolization. Ingested quercetin is rapidly eliminated as metabolites through feces and urine. The bioavailability of quercetin orally administered to humans was estimated at ~45% with 3.3–5.7% of the dose found in the urine and 0.2–4.6% in the feces [113,157].

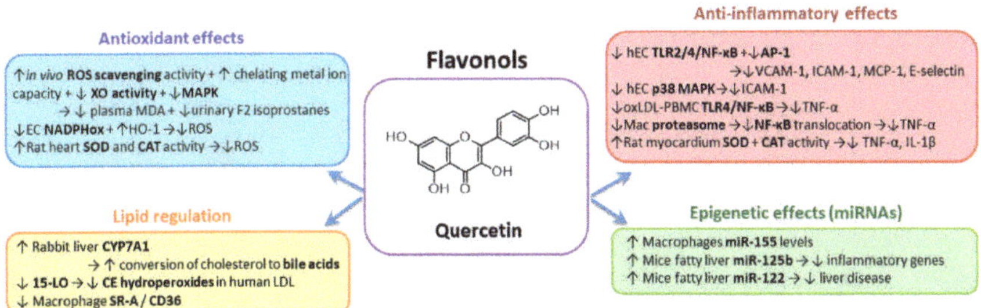

Figure 3. Quercetin chemical structure and beneficial effects in the context of cardiovascular diseases.

Quercetin is efficient at stimulating cytochrome P450 and CYP7A1 levels, and the conversion of cholesterol to bile acids in the liver of rabbits fed with a high fat diet [158]. By inhibition of 15-lipoxygenase, quercetin and its monoglucoside derivatives inhibits cholesteryl ester hydroperoxides formation in human LDL [158]. Antioxidant quercetin metabolites like quercetin-3-glucuronide (Q3GA) are taken-up by the human macrophages present in the intima and convert them to methylated derivatives, which suppress the gene expression of scavenger receptors SR-A and CD36 [159].

The data supporting the antioxidant potential of quercetin are contradictory. Thus, administration of quercetin at 10 mg/kg bodyweight for 13 weeks was shown to downregulate NADPH oxidase and increase eNOS activity, improving the endothelial function in hypertensive male rats [160], while a higher dose (1.5 g quercetin/kg diet for 5–11 weeks) was not associated with a reduced risk of developing CVD in hypertensive rats [161]. In such situation it is necessary to test the compound in the same experimental model, but at several concentrations to determine the optimum and identify the harmful one. In vivo, the antioxidant potential of quercetin was expressed as lower levels of urinary isoprostane F2 and plasma MDA, that could be due to the ability of quercetin to scavenge ROS, chelate metal ions, reduce xanthine oxidase activity and to inhibit the MAPK pathway [162,163]. The mechanism by which quercetin increases eNOS activity in a dose-dependent manner involves the phosphorylation on Ser1179 by cAMP/PKA pathway [164]. The antioxidant potential of quercetin is also reflected in the decrease of the oxidation levels of LDL [165]. In addition, quercetin reduces the activities of SIRT-1 and AMPK, upregulates HO-1 and decreases the expression of oxLDL-induced NOX2 and NOX4 in human EC [166]. Kaempferol (50 or 100 mg/kg for 4 weeks), another member of the flavonols group, reduces atherosclerotic lesions area in apoE-deficient mice through mechanisms involving reduction of the aortic ROS production and osteopontin downregulation [167]. Furthermore, kaempferol diminishes oxLDL-enhanced apoptosis of EC based on the upregulation of autophagy by inhibition of PI3K/Akt/mammalian target of rapamycin (mTOR) pathway [168].

Extensive studies using in vitro or in vivo models clearly indicate that quercetin manifests anti-atherosclerotic effects, in part due to its anti-inflammatory properties. In vitro studies demonstrate that quercetin reduces the expression of VCAM-1, ICAM-1, E-selectin or MCP-1 in cultured human EC exposed to different pro-inflammatory stimuli. The molecular mechanisms involve modulation of TLR2/4/NF-kB and AP-1 transcription factor. In addition, it was demonstrated that quercetin reduces ICAM-1 in EC exposed to uremic media by downregulating p38 MAPK [169–172]. Anti-inflammatory properties of quercetin were demonstrated also in monocytes/macrophages. It was demonstrated that TNF-α released by oxLDL-exposed human PBMC is reduced by quercetin through modulation of TLR/NF-κB signaling pathway [170]. In LPS-stimulated macrophages isolated from C57BL/6 and BALB/c mice, quercetin reduces the secretion of TNF-α and NO produced by the inducible NO synthase (iNOS) by a mechanism involving the inhibition of proteasome, which determines a diminished proteolytic degradation of phospho-IκB protein, resulting in the decreased translocation of activated NF-κB to the nucleus [173].

The decrease of the atherosclerotic lesions was not associated with the improvement of the lipid profile, but with the decreased inflammatory stress, measured as decreased IL-1 receptor, IKK and STAT3 [169,174]. In the rat model of acute myocardial infarction, quercetin administration determined the reduction of TNF-α, IL-1β expression in the myocardial tissue, in parallel with an increase of the antioxidant SOD and catalase activities [175]. Recently, a few clinical studies regarding the anti-inflammatory effects of quercetin in human subjects were published, but the results are controversial. A double-blind randomized clinical trial designed to measure the anti-inflammatory effects of quercetin (500 mg) administered for 10 weeks to women with type 2 diabetes reported no differences between groups [176]. Another study conducted on stable angina patients receiving 120 mg/day quercetin for two months showed a statistically significant reduction of IL-1β levels and attenuated TNF-α and IL-10 levels in treated patients' plasma, and decreased transcriptional activity of NF-kB in PBMC [177]. Several studies were done to evaluate the safety of quercetin administration to humans [178]. The conclusion of these studies was that up to 5000 mg/day quercetin supplementation for four weeks does not cause adverse effects.

In addition, quercetin can exert its positive effects on the regulation of oxidative and inflammatory stress by modulating the expression of specific miRNAs. Thus, Boesch-Saadatmandi et al. reported that quercetin and isorhamnetin upregulated miR-155 levels in LPS-activated macrophages, but quercetin metabolites, such as quercetin-3-glucoronide, did not modify miR-155 expression [179]. An in vivo study of the same group showed that addition of quercetin to high-fat diet fed C57BL/6J mice significantly increases hepatic expression of miR-125b, a negative regulator of inflammatory genes, and miR-122, known to be involved in lipid metabolism and pathogenesis of liver diseases [180]. These data suggest that miRNAs could represent potential targets of quercetin in the CVD prevention or treatment.

2.3.2. Flavones

Apigenin (4′,5,7-trihydroxy-flavone) is one of the major monomeric flavonoids existing in the diet and is found in a glycosylated form, with the tricyclic core structure linked to a sugar moiety through hydroxyl groups (O-glycosides) or directly to carbon (C-glycosides) (Figure 4). Apigenin is present in fresh parsley, vine spinach, celery seed, green celery heart, chinese celery, dried oregano, chamomile tea, red and white sorghum, rutabagas, oranges, kumquats, onions, wheat sprouts, tea and cilantro [181]. Apigenin-glycosides can be hydrolyzed in vivo into apigenin or chrysin. Oral bioavailability of apigenin is relatively low due to its poor solubility and because the main part of the ingested apigenin is either excreted unabsorbed or is rapidly metabolized after absorption. In vivo, after ingestion, apigenin is subjected to sulfation and glucuronidation, the absorbed apigenin being present in tissues (mainly hepatic and small intestin) as glucuronide, sulfate conjugates or luteolin [182].

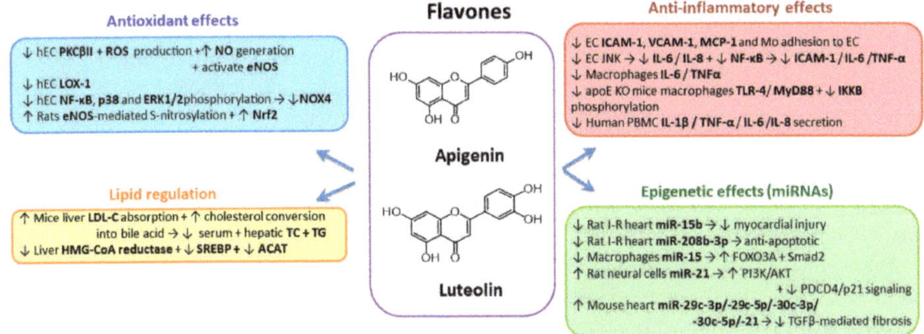

Figure 4. Antioxidant, anti-inflammatory, lipid-lowering and epigenetic mechanisms to improve cardiovascular diseases outcomes demonstrated by apigenin and luteolin.

Apigenin decreased the serum and hepatic levels of TC and TG in hyperlipidemic mice by promoting liver LDL-C absorption and increasing the conversion of hepatic cholesterol into bile acid [183]. In addition, apigenin markedly lowered the levels of hepatic enzymes involved in the synthesis of TG and cholesterol esters in HFD-induced obese mice, thereby ameliorating hepatic steatosis [184].

Data supporting the protective role of apigenin on the vascular wall in diabetes reveal the impeding of the endothelial dysfunction by inhibition of high-glucose-mediated protein kinase CβII (PKCβII) upregulation and ROS production, while stimulating NO generation [185]. The molecular mechanism of eNOS activation by apigenin involves continuous eNOS Ser1179 phosphorylation by the PI3K/Akt pathways [186]. In addition, apigenin inhibits oxLDL receptor 1 (LOX-1) expression after stimulation of EC with high glucose and TNF-α, LOX-1 being an important receptor involved in the uptake of modified lipoproteins and atherosclerotic plaque progression [187].

The anti-inflammatory properties of apigenin have been investigated in studies conducted in vitro and in vivo. Using cultured human EC activated with di-(2-ethylhexyl) phthalate, Wang et al. showed that apigenin suppressed the expression of ICAM-1 and inhibited THP-1 monocytic cells adhesion to HUVECs [188]. In addition, a dose-dependent inhibition of endothelial IL-6 and IL-8 expression was observed, and these inhibitory effects of apigenin are mediated by the JNK pathway, but not by IκBα/NF-κB or ERK pathways [188]. Apigenin inhibited NF-κB activation and ICAM-1 expression in EC exposed to palmitic acid [189]. In EC exposed to uremic plasma, p38 MAPK is another molecular target for apigenin [172]. Regarding the effect of apigenin on cytokine secretion, it has been shown that apigenin reduced IL-6 and TNF-α secretion in LPS-stimulated RAW 264.7 macrophages [190] and decreased TNF-α release in the media of LPS-activated macrophages [191]. Apigenin in LPS-exposed macrophages reduces TLR-4, MyD88 and phosphorylation of IκKB levels through nuclear NF-κB p65 signaling pathway [192]. Ren et al. confirmed these results in vivo, demonstrating that in LPS-challenged apoE deficient mice, treatment with apigenin determined the reduction of TLR-4 and NF-κB p65 levels and lessened the macrophages and SMC number in atherosclerotic regions [192]. Apigenin inhibited the expression of VCAM-1 and IκKB kinase and prevented the adhesion of U937 monocytes to EC exposed to high-glucose (30 mM) concentrations [193]. The beneficial effects of intra-gastric administration of apigenin to type 2 diabetic (T2D) rats were expressed as decreases of the blood glucose concentration and ICAM-1 levels and improved impaired glucose tolerance [189].

Luteolin (3′,4′,5,7-tetrahydroxyflavone) (Figure 4) is a member of the flavones family and is found in carrots, cabbage, artichokes, tea, celery and apples [194]. Luteolin and its glucosides are absorbed quickly in the intestine. The time of maximum blood concentration is under one hour, and the maximum plasma concentration is of 1–100 μmol/L, depending on the dose ingested and the type of food consumed; the purer the luteolin used, the faster the absorption [195].

Luteolin exerts lipid-lowering effects due to the interaction with HMG-CoA reductase, the SREBPs and acyl-CoA cholesterol acyltransferase (ACAT) in the liver [3]. In all studies available, no major side effects have been detected, confirming the good tolerability and safety of artichoke extract, but long-term safety studies are needed [196,197]. There are no data about luteolin effects on the EC function or atherosclerotic plaque. Luteolin was shown to impede TNF-α–induced NOX4, which generates a decrease of ROS production in human EC. The mechanism involves the inhibition of the TNF-α–induced transcriptional activity of NF-κB, p38 and ERK1/2phosphorylation [198]. Recently, it was shown that luteolin (100 mg/kg/d) reduces cardiac I/R injury by enhancing eNOS-mediated S-nitrosylation and Nrf2 redox function in diabetic rats [199].

The effects of luteolin on monocyte adhesion to EC, a key event in triggering vascular inflammation, were evaluated in a few studies. Jia et al. demonstrated that physiological concentrations (0.5 μM) of luteolin suppress TNF-α-induced expression of MCP-1, VCAM-1 and ICAM-1 [200]. In addition, luteolin inhibited endothelial NF-κB transcriptional activity, IκB-α degradation and IκB-β expression, and thus reduced NF-κB p65 nuclear translocation. The in vitro results were confirmed in vivo using C57BL/6 mice fed with luteolin diet supplementation. The authors show that luteolin suppresses TNF-α-induced MCP-1 and soluble ICAM-1 in the plasma, as well as VCAM-1 in the aorta of mice [200]. In a recent study, Zhang et al. demonstrated that a combination of luteolin (0.5 μM) and curcumin (1 μM) inhibits, synergistically, VCAM-1 and MCP-1, and the subsequent monocyte adhesion to EC exposed to TNF-α by suppressing NF-kB translocation [201]. These results were confirmed in vivo in C57BL/6 mice that received luteolin and curcumin [201]. In human monocytes exposed to high glucose concentrations, luteolin significantly reduced IL-6 and TNF-α by inhibiting NF-κB activity [202]. Recently, it was demonstrated that luteolin effectively suppresses IL-1β, TNF-α, IL-6 and IL-8 secretion from PBMC from healthy donors incubated with LPS [203], confirming reports showing that luteolin reduces IL-6 and TNF-α secretion in LPS-stimulated RAW 264.7 macrophages [190]. Hong et al. evaluated the anti-inflammatory effects of luteolin in renal I/R injury in Sprague–Dawley rats and demonstrated that this natural compound attenuated serum and renal TNF-α, IL-1β and IL-6, and the renal HMGB1 and NF-κB expression levels in I/R rats [204]. In addition, luteolin significantly reduced the endoplasmic reticulum stress and renal cell apoptosis caused by renal I/R injury [204]. Recently, Ding et al. demonstrated that luteolin attenuates atherosclerosis in high-fat fed apoE-/- mice by alleviating inflammation through inhibition of signal transducer and STAT3 [205].

Recent studies evidenced that apigenin and luteolin modulate the expression of certain epigenetic factors, in particular miRNAs, that constitute fine regulators of the oxidative and inflammatory stress. Thus, an in vivo study by Wang et al. shows that the increased cardiac miR-15b expression observed during myocardial I/R injury in rats correlates with the decreased expression of JAK2 and activity of JAK2/STAT3 pathway, with augmented myocardial apoptosis and ROS production, and aggravated heart injury [206]. Apigenin treatment of I/R rats downregulates miR-15b expression in the heart, improves the altered mechanisms and alleviates myocardial I/R injury [206]. In an in vivo report, Bian et al. showed that luteolin pretreatment induces anti-apoptotic effects by decreasing miR-208b-3p expression in myocardial tissue of I/R rats [207]. Arango et al. performed a high-throughput PCR screening of 312 miRNAs in RAW 264.7 murine macrophages and evidenced that apigenin reduces LPS-induced miR-155 expression by transcriptional regulation [208]. They further demonstrated that apigenin-reduced expression of miR-155 led to the increase of the anti-inflammatory mediators forkhead box O3a (FOXO3A) and α smooth-muscle-actin (α-SMA) and MAD-related protein 2 (Smad2) in LPS-treated murine macrophages. Arango et al. also demonstrated that in vivo apigenin or a celery-based apigenin-rich diet reduced LPS-induced expression of miR-155 and decreased TNF-α in lungs from LPS-treated mice, thereby diminishing the inflammatory process [208].

Zhang et al. reported recently that low concentrations of luteolin presented protective effects on H_2O_2-induced ischemic cerebrovascular disease cell viability loss, proliferation inhibition, ROS generation, oxidative stress increase and apoptosis, together with an increase of miR-21 expression level [209]. Furthermore, they showed that luteolin alleviated H_2O_2-induced inactivation of PI3K/Akt pathway and activated programmed cell death protein 4 (PDCD4)/p21 pathway in PC-12 cells by upregulating miR-21 [209]. Ning et al. showed that pretreatment and post-treatment with luteolin-7-diglucuronide (L7DG), a naturally occurring flavonoid glycoside found in leaves of basil or *Verbena officinalis*, significantly attenuated isoproterenol-induced myocardial injury and fibrosis in mice [210]. Furthermore, L7DG pretreatment blocked isoproterenol-stimulated expression of genes encoding the enzymatic subunits of NADPH oxidase (Cyba, Cybb, Ncf1, Ncf4 and Rac2). In addition, the authors showed that L7DG pretreatment almost reversed isoproterenol-altered expression of miRNAs which were cross-talking with TGF-β-mediated fibrosis, including miR-29c-3p, miR-29c-5p, miR-30c-3p, miR-30c-5p and miR-21 [210].

2.3.3. Flavanones

Naringenin (4,5,7-trihydroxy-flavanone) and *hesperetin* (3,5,7-trihydroxy-4'-methoxyflavanone) (Figure 5) are the representative flavanones that are found in glycoside form in nature; that form favors their intestinal absorption. They are present in citrus fruits, tomatoes and cherries [211], but unfortunately have limited solubility in water. Kanaze et al. reports that oral administration of naringenin results in a low bioavailability (5.81%) in human subjects [212]. To overcome the poor solubility and to increase naringenin use in clinical applications, several drug delivery systems have been developed. As a result, naringenin was formulated into liposomes, nanoparticles, self-nanoemulsifying drug delivery systems or nano-suspensions to assure a friendly and efficient delivery system to be used in the future [213].

It was reported that the alcoholic extracts of bergamot enriched in flavanones reduce the intestinal absorption of cholesterol and increase cholesterol excretion based on the bile acids secretion pathway [214]. Supplementation with naringenin (3%) of the high-fat diet in LDL-R-/- mice prevented hepatic steatosis by diminishing the SREBP1c expression and fatty acid synthesis, stimulated hepatic fatty acid oxidation and progression of atherosclerosis in the aortic sinus [215]. In addition, naringenin inhibited microsomal triglyceride transfer protein (MTTP) activity in HepG2 cells, reduced TG accumulation and decreased apoB100 secretion by 50–70% [216]. Naringenin (0.05%) and naringin (0.1%) reduced aortic fatty streaks in rabbits fed high-cholesterol diets by a mechanism involving reduction of hepatic ACAT activity [217]. In a clinical trial involving hypercholesterolemic patients, administration of naringin (400 mg/day, for 8 weeks) reduced plasma LDL-C and serum apoB by more than 14%, without altering plasma TG or HDL-C levels [218]. Naringenin found in bergamot inhibits LDL oxidation, initiates AMPK, modulates the activation of redox-sensitive transcription factors NF-κB induced by TNF-α and acts as ROS scavenger [219]. There are reports showing that hesperetin-induced eNOS expression increased NO production via phosphorylation of Src, Akt and AMPK in cultured EC, and consequently prevented hypertension by improving endothelial dysfunction in the hypertensive rat [220,221]. Naringenin increased the NO production diminished by the high glucose concentrations and reduced ROS production via decreased PKCβII expression in EC [187]. Certain epidemiologic studies evidence that the intake of drinks made from flavanone-rich citrus fruits improve endothelial function in humans [222].

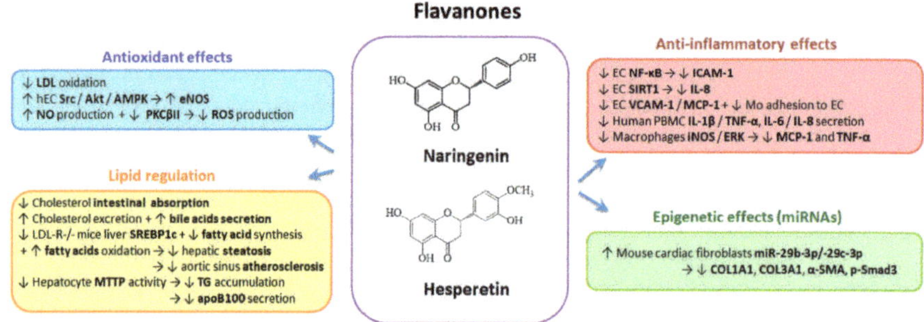

Figure 5. Chemical structure and cardioprotective mechanisms of action demonstrated by naringenin and hesperetin in experimental and clinical studies.

An in vitro study demonstrated that naringenin reduces ICAM-1 in palmitic acid exposed EC by alleviating NF-κB signaling [189]. The anti-inflammatory effects of naringenin were confirmed in vivo by the same group in a model of T2D rats [189]. It was suggested that the anti-inflammatory action of flavanones from bergamot juice is due to the activation of SIRT-1 that further inhibits the transcription of pro-inflammatory cytokine IL-8 induced by LPS in EC [223,224]. Interestingly, Testai et al. observed recently that naringenin presents structural similarity with RSV, the natural SIRT-1 activator [225]. Thus, they performed an in silico study in which they detailed the binding mode of naringenin to SIRT-1. In addition, the same group demonstrated in cultured H9c2 cardiomyocytes and in vivo in 6-month-old mice that naringenin activates SIRT-1 [225].

Hesperetin attenuated VCAM-1 upregulation and adhesion of monocytes to cultured EC induced by TNF-α stimulation [220]. Administration of glucosyl hesperidin (500 mg/day) to hypertriglyceridemic patients for 24 weeks significantly reduced plasma TG and apoB levels [226]. Other human studies using capsules of hesperidin (800 mg) or naringin (500 mg) administered for 4 weeks to moderately hypercholesterolemic individuals evidenced no alterations in plasma TC, LDL-C or TG concentrations [227]. These studies reveal that the effects of flavonoids in clinical studies depend on the metabolite used, the dose, the patient population and the length of study [228].

Using LPS-stimulated PBMC from healthy subjects, Zaragoza et al. demonstrated that naringenin decreases IL-1β, TNF-α, IL-6 and IL-8 production [203]. Hsu et al. showed that naringenin extracted from *Nymphaea mexicana Zucc.* has an important inhibitory effect on MCP-1 and TNF-α production in LPS-activated RAW264.7 macrophages by decreasing the expression of iNOS and ERK phosphorylation [229]. The anti-inflammatory effects of naringenin in vivo were evaluated by Raza et al., demonstrating that naringenin treatment downregulates the NF-κB expression levels in the brains of Wistar rats after the cerebral I/R injury [230].

Although demonstrating anti-inflammatory effects, the poor water solubility and the reduced bioavailability of naringenin restricted its therapeutic use. To overcome this limitation, naringenin formulations have been developed. In a recent study, Fuior et al. encapsulated naringenin into lipid nanoemulsions (LNs), targeted to VCAM-1 exposed on the surface of activated EC [231]. They found that encapsulated naringenin decreased THP-1 monocytes adhesion and transmigration to/through activated ECs by mechanisms involving the reduction of MCP-1 and diminished nuclear translocation of NF-κB [231].

A very recent study showed that naringenin modulates the expression of miRNAs involved in fine regulation of certain oxidative and inflammatory processes. It is known that the migration inhibitory factor (MIF) has antioxidant properties and is markedly increased in Ang-II-infused mouse myocardium. Starting from these data, Liang et al. showed that miR-29b-3p and miR-29c-3p are decreased in the myocardium of Ang-II-infused MIF-KO mice, but upregulated in mouse CF with MIF overexpression or by treatment with MIF protein [232]. Interestingly, miR-29b-3p and miR-29c-3p could suppress the expression of collagen type I α1 (COL1A1), collagen type III α1 (COL3A1) and α-SMA in mouse CF by a mechanism involving the repression of the pro-fibrosis genes TGFβ2 and MMP2. Further, the authors show that naringenin could markedly reverse Ang-II-induced downregulation of miR-29b-3p and miR-29c-3p expression. Moreover, COL1A1, COL3A1, α-SMA and p-Smad3 expression were significantly decreased in Ang-II-treated CF by pretreatment with naringenin [232].

2.3.4. Flavanols

Catechins are a family of flavonoids, subgroup flavan-3-ols (flavanol) (Figure 6). The most abundant component of catechins is epigallocatechin gallate [155]. Epigallocatechin-3-gallate (EGCG) is a catechin conjugated with gallic acid. The two or more aromatic rings of these polyphenols have at least one hydroxyl group linked by a carbon bridge which is the main source for electron donor and efficiently scavenging reactive species (singlet oxygen). Catechins are present in fresh tea leaves, red wine, broad beans, apples, pears, black grapes, apricots, strawberries, blackberries, cherries, raspberries and chocolate. Cocoa is the richest source of EGCG [155]. The poor bioavailability of catechins is a consequence of their rapid degradation under physiological conditions and their low absorption in the intestinal tract by passive diffusion. To mitigate the reduced bioavailability of EGCG and to increase its effectiveness, the consumption of 8–16 cups of green tea daily is needed, but the excessive consumption of green tea has been demonstrated to be toxic [233]. Thus, other methods to increase catechins bioavailability were developed, including encapsulation into nanostructure-based drug delivery systems, molecular modification and co-administration with some other bioactive ingredients to produce a synergistic effect [234].

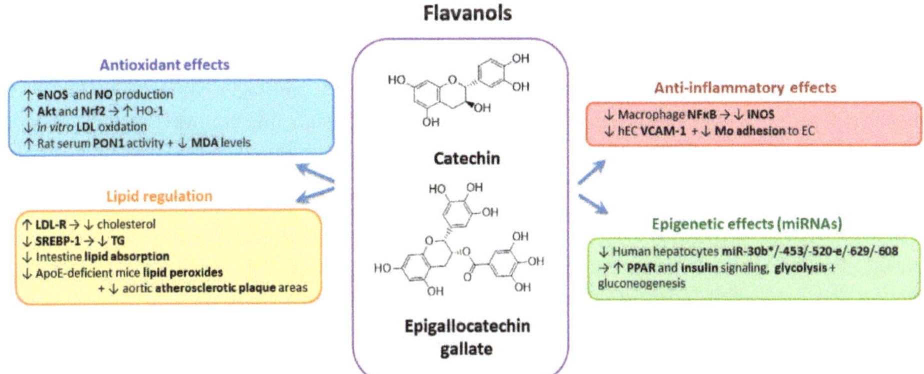

Figure 6. Chemical structures of catechin and epigallocatechin gallate and their molecular mechanisms of action to combat cardiovascular diseases.

Catechins present in the green tea are responsible for its cholesterol-lowering properties. It was shown that consumption of green tea in supplement formulation has cholesterol-lowering effects due to upregulation of liver LDL-R, thereby modulating the intracellular processing of lipids [235–237].

Their mechanisms of action involve inhibition of lipid absorption in the intestine by interfering with the micelle formation, emulsification, hydrolysis, solubilization and inhibition of squalene oxidase, a key enzyme in the hepatic cholesterol biosynthesis [238,239]. EGCG was described as the most potent inhibitor of lipid absorption in the intestine [48]. The EGCG ability to lower plasma lipids was associated with the consequential reduction of lipid peroxides levels and aortic atherosclerotic plaque areas in apoE-deficient mice [240]. Cocoa is a polyphenol-rich fruit that has been investigated for its potential to regulate the lipid metabolism. A special attention was given to the raise of HDL levels by cocoa ingestion. The results are controversial; the beneficial effects observed in vitro for cocoa epicatechins are not all confirmed by those in humans [241]. Besides the cholesterol lowering potential, catechins were described as having a TG lowering effect [242]. The mechanism responsible for decreasing TG is inhibition of hepatic lipogenesis, and more specifically, the inhibition of SREBP-1 [243].

A meta-analysis study concerning green tea consumption has indicated that it decreases the risk of CVD [244]. Many cardiovascular health benefits of flavanols have been reported, and one possible mechanism could be the modulation of homocysteine, its elevated concentrations being associated with increased CVD risk [245]. It was reported that in vitro EGCG reduces homocysteine-enhanced apoptosis by modulating mitochondrial-dependent signaling and PI3K/Akt/eNOS signaling in human EC [246]. In addition, epicatechin activates eNOS and increases NO production by inducing Ser633 and Ser1177 phosphorylation and Thr495 dephosphorylation. EGCG induces HO-1 expression in EC exposed to H_2O_2 through activation of Akt and Nrf2, thereby diminishing the effects of oxidative stress [247]. Many other catechins exert antioxidant properties expressed as prevention of in vitro LDL oxidation and in humans [248]. Studies on green tea found that its administration (2% in water) to diabetic rats increased serum PON1 activity [249]. These antioxidant mechanisms driven by catechins are responsible for the significant reduction of MDA levels measured in vivo and the protective action exerted on LDL and HDL. The molecular mechanisms by which EGCG or other catechins exert antioxidant effects need to be further explored in vivo.

Catechins from black tea are important scavengers of peroxyl, hydroxyl and superoxide radicals, singlet oxygen and lipid peroxides, NO and peroxynitrite radicals [250]. EGCG exerts also anti-inflammatory effects on macrophages pre-exposed to pro-inflammatory stimuli, such as LPS. EGCG blocks the disappearance of IκB from the cytosolic fraction, thereby obstructing NF-κB activation, which in turn decreases the transcription of iNOS [240]. Another in vitro study evidenced that EGCG inhibits VCAM-1 expression induced by IL-1 or TNF-α, thereby diminishing the monocytes' adhesion to cultured human EC [250].

Catechins and their derivatives were proven to contribute to beneficial health effects by the modulation of miRNAs. In human HepG2 hepatocytes, EGCG isolated from green tea was shown to differentially inhibit the expression of a set of five miRNAs (miR-30b*, miR-453, miR-520-e, miR-629 and miR-608) that are involved in inflammatory pathways, the PPAR signaling pathway, insulin signaling, glycolysis and gluconeogenesis, oxidative phosphorylation and glutathione metabolism [251]. It was demonstrated by using 1H NMR spectroscopy, that there was direct binding of EGCG and RSV to miR-33a and miR-122 [252]. While RSV binds miR-33a and miR-122 through an A ring interaction and increases their expression levels, EGCG decreases miR-33a and miR-122 expression by direct binding through an interaction with all rings in the molecule. Wang et al. demonstrated that EGCG binds hypoxia-inducible factor 1α (HIF-1α) protein, a known transcriptional activator of miR-210, and interferes with Proline residues hydroxylation in the oxygen-dependent degradation domain [253]. While the hydroxylation of Proline residues is essential for the proteasome-mediated degradation of HIF-1α [254], EGCG binding increases HIF-1α expression and enhances miR-210 levels.

2.3.5. Isoflavones

Genistein (4′,5,7-trihydroxyisoflavone) (Figure 7) is an isoflavone found in high quantities in soybeans and in many products based on soy. Genistein is also present in alfalfa and clover sprouts, barley meal, broccoli, cauliflower and sunflowers, caraway and clover seeds [255]. In humans, the plasma concentrations of genistein are dependent on the food type consumed and are highest between 2 and 12 h after ingestion of isoflavone-rich foods. Genistein has an absorption rate around 30% of the ingested dose. A small part of the intake of isoflavone aglycones (10%) is absorbed from the small intestine and metabolized in the liver. Most of the ingested isoflavone (90%) undergoes different transformations under gut microbiota action in the colon [113].

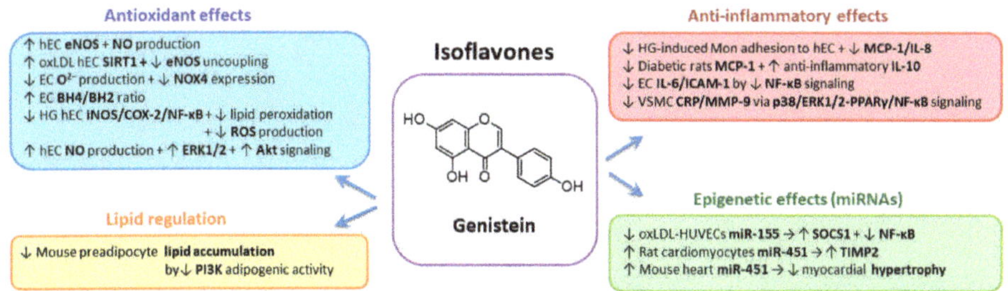

Figure 7. Chemical structure and protective effects of genistein in the context of cardiovascular diseases.

Genistein (80 mM) induced a 67% reduction of lipid accumulation in 3T3-L1 mouse preadipocytes, by inhibiting the adipogenic activity of PI3K, thereby exerting an anti-adipogenic action [113]. Opinions on the potential of isoflavones to reduce CVD risk are divided. Some epidemiologic studies highlight their protective effects on vascular EC, while other reports found no correlation between isoflavones consumption and reduction of CVD risk [256,257]. Exposure of human EC to 1–10 μM of genistein upregulated eNOS expression and enhanced NO production [258]. In addition, genistein prevented eNOS uncoupling by stimulating SIRT-1 pathway in human EC incubated with oxLDL. Furthermore, genistein diminished superoxide anion production and NOX4 expression, and improved the tetrahydrobiopterin (BH4)/dihydrobiopterin (BH2) ratio [259]. An intake of isoflavone of 50–99 mg/day was found to increase endothelial function measured as brachial flow-mediated dilation [260]. In another in vivo study, an intake of 80 mg of soy isoflavone decreased the aortic pulse-wave velocity, another marker of CVD risk [261]. Daidzein (40 μM), another member of the isoflavones group, inhibited high-glucose–induced iNOS, COX-2 and NF-κB expression in human EC, in parallel with the reduction of lipid peroxidation and ROS production [262]. Equol is a metabolite of daidzein produced by the intestinal microflora in the gut [263] and was described as activator of eNOS in EC by modulating the epidermal growth factor receptor (EGFR), the G-protein–coupled receptor GPR30 and mitochondrial ROS production [264]. Equol induced the relaxation of the rat aortic rings and stimulated endothelial NO production by activation of ERK1/2 and Akt signaling in human fetal EC [265]. In addition, soy isoflavones consumption by peri-menopausal women induced the increase of serum PON1 activity [95].

Studies from literature show that genistein exerts anti-inflammatory effects in vitro and in vivo by regulating different pro-inflammatory signaling pathways. Babu et al. showed that physiological concentrations of genistein significantly inhibit high glucose-induced adhesion of monocytes to human aortic EC and suppress MCP-1 and IL-8 endothelial production. These effects were due to genistein promoting PKA activity and were confirmed in diabetic db/db mice [266]. Genistein supplementation in diabetic rats determined a statistically significant reduction of MCP-1, while increasing the anti-inflammatory IL-10 [266]. In EC exposed to homocysteine, genistein diminished the expression of IL-6 and ICAM-1 via NF-kB inhibition [267]. In a recent study, Xu et al. evaluated the effect of genistein on inflammation induced by Ang II in vascular SMC [268]. This study demonstrates that genistein decreases CRP and MMP-9 levels in SMC by regulating p38/ERK1/2-PPARγ/NF-kB signaling pathway [268]. Besides, it was shown that genistein inhibits TNF-α secretion in LPS-activated macrophages [191].

Recent studies show that genistein modulates the expression of miRNAs involved in inflammatory processes. It was reported that genistein pretreatment of HUVECs reduces in a dose-dependent manner the oxLDL-induced expressions of E-selectin, P-selectin, MCP-1, IL-8, VCAM-1 and ICAM-1. Further analyses established that the mechanism of action consists of genistein inducing reduction of miR-155 levels and elevation of the suppressor of cytokine signaling 1 (SOCS1) expression that further induces the inhibition of NF-kB signaling pathway in HUVECs [269]. A recent study of Gan et al. showed that genistein can inhibit isoproterenol-induced cardiac hypertrophy by increasing miR-451 and the tissue inhibitor of metalloproteinases 2 (TIMP2) expression (a miR-451 target gene), both in vitro in H9C2 embryonic rat cardiomyocytes and in vivo in isoproterenol-induced myocardial hypertrophy in mice [270].

2.3.6. Anthocyanidins

Anthocyanins and *anthocyanidins* are glycosylated, poly-hydroxy or poly-methoxy derivatives of flavylium cations (2-phenylchromenylium) (Figure 8). In nature, about 702 different anthocyanins and 27 anthocyanidins are present. They are water-soluble plant pigments that give red, purple or blue coloration to many fruits, flowers and leaves. The widely distributed anthocyanidins in human foods are: cyanidin, delphinidin, pelargonidin, peonidin, malvidin and petunidin. They are found in many berry fruits, eggplantss, red onion, purple cabbage and black rice [271]. Anthocyanins' bioavailability has quite large inter-individual variability, and is influenced by the food processing, availability of the enzymes involved in anthocyanins metabolism and the composition of the gut microbiota that metabolize anthocyanins [272]. After ingestion, anthocyanins appear rapidly in the circulation, reach the maximal concentration of ~100 nM within 1.5h and disappear from the bloodstream by 6h post consumption. The bioavailability of anthocyanins is very low (1%), but it is increased by the absorption of their active metabolites (12.4%) that were recently identified [271].

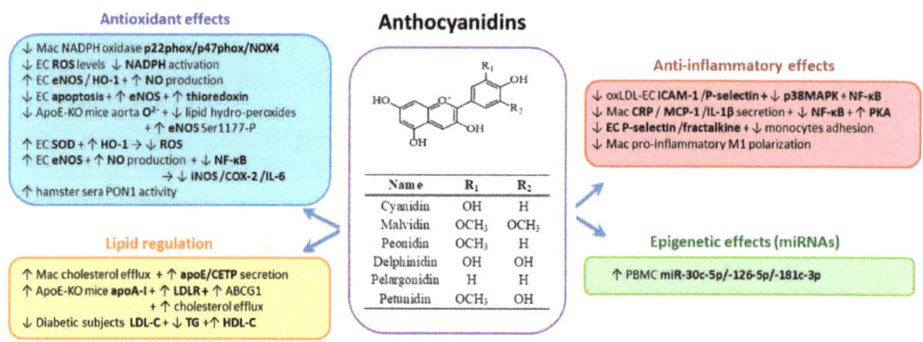

Figure 8. Anthocyanidins' classifications, chemical structures and beneficial effects for improving cardiovascular disease outcomes.

Anthocyanines/anthocyanidines and their metabolites ameliorate endothelial dysfunction and diminish the CVD risk [273]. Anthocyanins and flavonoids from bilberry (*Vaccinium myrtillus*) alcoholic extract can reduce the lipid deposits from the artery wall by inducing the cholesterol efflux from lipid-loaded macrophages through a mechanism involving increased secretion of apoE and CETP [274]. Anthocyanins from black elderberry (*Sambucus nigra*) extract induce expression of apoA-I and LDL-R in apoE-deficient mice [275]. Administration of 320 mg/day anthocyanins to diabetic patients for 24 weeks improved their lipid profiles by significantly decreasing LDL-C and TG, and increasing HDL-C [276].

In parallel with the lipid-lowering effects, anthocyanins show antioxidant properties. Thus, the alcoholic extract of bilberries reduces the expression of NADPH oxidase subunits (p22phox, p47phox and NOX4) in lipid-loaded macrophages derived from THP-1 monocytes [274]. Mulberry

(*Morus alba*) leaves have considerable amounts of flavonoids and anthocyanins. Treatment of human EC, pre-exposed to pro-inflammatory stimuli, with *Morus Alba* extract, inhibited the intracellular ROS levels due to reduction of NADPH oxidase activation [64]. Cyanidin-3-O-b-glucoside (C3G), a metabolite of cyanidin, has been shown to upregulate eNOS and HO-1 expression in a dose-dependent manner in EC, in parallel with an increase of NO production. The mechanism involves eNOS phosphorylation at Ser1179 and dephosphorylation of Ser116 [277,278]. Thioredoxin is one of the key regulators of intracellular redox status and it protects EC against oxidative stress. It was reported that nonaglycone cyanidin reduces TNFα–induced apoptosis and upregulates eNOS and thioredoxin in EC [278]. Other pathways involved in EC protection by cyanidins are Akt, ERK1/2 and Src kinase positive regulation. In vivo studies show that C3G intake (2 g/kg diet for 8 weeks) diminished the area of atherosclerotic plaques and alleviated the endothelium-dependent relaxation in fat-fed apoE-deficient mice. The underlying mechanisms of these benefic effects are: the decrease of superoxide and lipid hydroperoxide generation; the increase of Ser1177 phosphorylation in eNOS protein from the aorta; and the increase of ABCG1 expression that facilitates cholesterol efflux [279,280]. It was demonstrated that malvidin and its metabolites reduce ROS levels by upregulating SOD and HO-1 in EC [281]. In parallel, they induce eNOS expression, increase NO production and reduce peroxynitrite-induced NF-kB activation that further decreases the levels of pro-inflammatory mediators such as iNOS, COX-2 and IL-6 [282]. Anthocyanins combined with gallocatechins from *Hibiscus sabdariffa* extract, increase PON1 activity in hamster sera in a dose-response manner [283]. Alcoholic extract of maqui berry, another black fruit, decreases the levels of lipid peroxides and F2-isoprostanes in humans [284]. A study conducted in women shows that the dietary intake of anthocyanins is accompanied by a lower carotid-femoral pulse wave velocity and carotid intima-media thickness [285].

Chen et al. demonstrated that pretreatment with delphidin determines the reduction of ICAM-1 and P-selectin expression in oxLDL-activated EC, resulting in an inhibition of monocytes adhesion and transmigration [286]. These anti-inflammatory effects were determined by the inhibition of oxidative stress, mitigation of p38 MAPK expression and inhibition of NF-κB [286]. The bilberry extract diminishes the secretion of CRP, MCP-1 and IL-1β in lipid–loaded macrophages. These effects are driven by the inhibition of NF-κB and activation of PKA signaling pathways [274]. The *Morus alba* extract downregulates the pro-inflammatory molecules P-selectin and fractalkine, decreasing monocytes adhesion to EC [64]. In a recent study, Lee et al. demonstrated that anthocyanin-rich blackcurrant extract exerts anti-inflammatory action by repressing the pro-inflammatory M1 polarization of mouse bone marrow-derived macrophages and human THP-1 cells [287].

The effects of anthocyanins and their metabolites on miRNAs expression are still largely unknown, only a few studies having been published. Rodriguez-Mateos et al. performed, recently, a nutrigenomic study to explore the mechanism of action of anthocyanins in vivo [288]. They analyzed mRNAs and miRNAs in PBMC isolated from healthy volunteers at the beginning and the end of a 28-day period of anthocyanins-enriched vitamin mix or blueberry consumption. The results of the microarray analysis in PBMC showed that a daily blueberry consumption led to differential expression (>1.2-fold) of 608 genes and three miRNAs (miR-30c-5p, miR-126-5p and miR-181c-3p). The most striking finding was a 13-fold increase of miR-181c expression evidenced in PMBC. Specific patterns of 13 metabolites were proven as independent predictors of mRNA expression alteration, and pathway enrichment analysis revealed significantly modulated biological processes involved in cell adhesion, migration, immune response and cell differentiation [288].

Anthocyanins are well tolerated and have no side effects, but their bioavailability is rather low, and they are rapidly transformed into phenolic acid derivates [289]. For this reason, additional experiments should be performed to develop new formulation of these natural active compounds to increase their protection, bioavailability and efficiency in vivo. Further studies are also desirable to assess the clinical efficiency of anthocyanins in different populations and to evaluate their benefic effects exerted on arteries affected by atherosclerosis.

2.4. Guaiacols Group

Gingerols and *shogaols* are the most abundant active compounds of ginger (*Zingiber officinale*) rhizomes that are used since old times in the treatment of various symptoms and as dietary supplement in drinks and food products [290,291]. Gingerols differ in the length of their unbranched alkyl side chains, [6]-gingerol being the most abundant type in fresh ginger root, followed by [10]-gingerol and [8]-gingerol (Figure 9). Dehydration of these major gingerols generates the corresponding shogaols (Shao et al., 2010). The low solubility of the orally ingested gingerols generates their low bioavailability. These compounds are not completely free of side effects due to their interactions with other pharmaceutical active compounds, the gingerols acting as bioenhancers of certain drugs. Possible technological solutions for enhancing gingerols solubility and bioavailability, and preventing their harmful interactions comprise the microemulsions and nanocarriers/nanoparticles formulations of gingerols. Some liposomal ginger products were developed to increase their bioavailability. These structures are not degraded in the stomach, can enter liver cells, but their benefic effects remain to be demonstrated by in vivo studies [292].

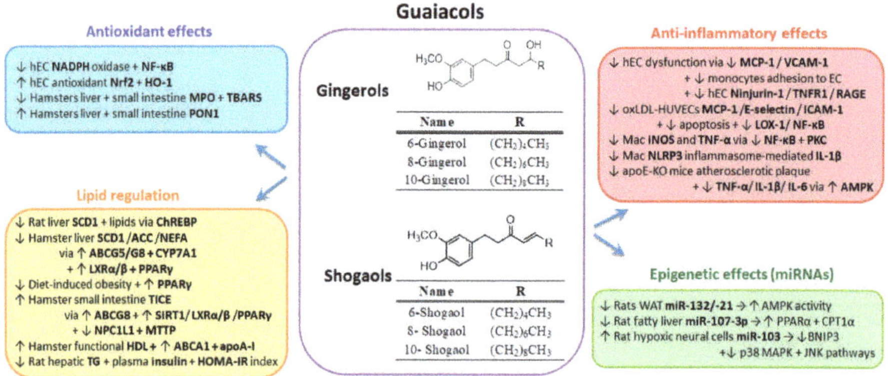

Figure 9. Guaiacols' classifications, chemical structures and molecular mechanisms of action to reduce cardiovascular diseases.

Ginger extract (GEx) exerts lipid–lowering effects, reduces stearoyl CoA desaturase 1 (SCD1) gene expression and accumulation of lipids in the liver of rats fed a fructose diet through a pathway mediated by hepatic carbohydrate response element-binding protein [293]. Recently, it was shown that a GEx, with very well characterized composition in gingerols and shogaols, diminishes the fatty acid production in hyperlipidemic conditions by reducing SCD1, ACC and non-esterified fatty acids levels in the liver and plasma of hyperlipidemic hamster [294]. The lipid-lowering properties of gingerols and shogaols can be also explained by the enhancement of hepatic cholesterol excretion into bile fluids, based on the induction of ABCG5/G8 and CYP7A1 expression in the liver of hyperlipidemic hamster. Lei et al. reported that the treatment with gingerol and shogaol-enriched GEx upregulates hepatic CYP7A1 [295]. The increase of ABCG5/G8 and CYP7A1 levels can be due to the induction of their transcription regulators LXRα/β and PPARγ which are stimulated by the decrease of endoplasmic reticulum stress [294]. The upregulation of PPARγ by GEx was also detected by Misawa et al., who showed that GEx attenuates diet-induced obesity and improves exercise endurance capacity by activation of the PPARγ pathway [296]. Gingerols and shogaols stimulate TICE through stimulation of ABCG8 gene and protein expression in the small intestine, due to the upregulation of SIRT1-LXRα/β-PPARγ pathway [117,295]. In addition, the treatment with gingerol and shogaol-enriched GEx induces the downregulation of Niemann-Pick C1-Like 1 (NPC1L1) and MTTP in the small intestine of hamster [295]. In parallel, GEx stimulates the small intestine to produce functional HDL by restoring ABCA1 levels and apoA-I quality and quantity through inhibition of the oxidative stress. It was reported that these processes happening in the small intestine and liver are associated with the reduction of the aortic valves lipid-deposits [117].

GEx active constituents have proven to be effective in exerting antioxidant effects on dysfunctional EC in culture by decreasing the expression of NADPH oxidase subunits, inhibiting NF-κB, activating the antioxidant Nrf2 and HO-1. GEx acts in vivo as antioxidant under dyslipidemic conditions by reducing MPO and thiobarbituric acid reactive substances (TBARS) levels and increasing PON1 levels in the liver, small intestine and plasma of hyperlipidemic hamsters [117,294].

Ginger is known since ancient times for its anti-inflammatory effects that ameliorate various diseases [297]. Recent data evidence the GEx potential and its major components, 6-gingerol and shogaol, to exert anti-inflammatory effects in EC by reversing TNFα-induced EC dysfunction. Their mechanisms of action involve reduction of MCP-1, VCAM-1 and monocytes adhesion to EC, due to the decrease of Ninjurin-1, TNFα receptor 1 and RAGE expression [298]. Wang et al. demonstrated that 6-shogaol decreases MCP-1, E-selectin and ICAM-1 and the apoptosis of HUVECs exposed to oxLDL by inhibiting LOX-1, oxidative stress and NF-κB signaling [299]. Beneficial effects of 6-gingerol and 6-shogaol were demonstrated also in macrophages. In LPS-stimulated macrophages, 6-gingerol decreased iNOS and TNF-α expression by inhibiting NF-κB and PKC signaling [300]. In another study, 6-shogaol significantly inhibited the canonical NLRP3 inflammasome-mediated IL-1β secretion in THP-1 macrophages stimulated with LPS [301]. In a recent in vivo study, Wang et al. demonstrated that administration of 6-gingerol to fat-fed apoE-deficient mice determines the reduction of atherosclerosis, expressed as decreased plaque formation and reduced levels of pro-inflammatory cytokines (TNF-α, IL-1β, and IL-6) by a mechanism mediated in part by AMPK activation [302].

Knowing the GEx pleiotropic effects, several groups investigated its potential to modulate epigenetic factors associated with ncRNA. As the main active ingredient of ginger, 6-gingerol, was shown to significantly improve lipid metabolism abnormalities in adult rodents. A few studies have reported its molecular effects on age-related non-alcoholic fatty liver disease (NAFLD), as well as on epigenetic factors. In an in vivo recent study in aged rats, Li et al. demonstrated that 6-gingerol brought to normal the hepatic TG content, plasma insulin and HOMA-IR index of ageing rats [303]. Mechanistically, they showed that 6-gingerol modulates lipid metabolism by increasing β-oxidation and decreasing lipogenesis, through activation of liver PPARα and carnitine palmitoyl-transferase 1α (CPT1α), and inhibition of diacylglycerol O-acyltransferase 2 (DGAT-2) expression at translational level, but not at transcriptional level, to ameliorate ageing-associated hepatosteatosis. The authors further analyzed miRNAs targeting PPARα and CPT1α genes and reported that ageing significantly increased hepatic miR-34a expression in rats. Interestingly, 6-gingerol showed minimal effect on hepatic miR-34a. Additionally, 6-gingerol significantly decreased hepatic miR-107-3p level, thereby increasing PPARα and CPT1α genes [303]. Kim et al. investigated molecular factors involved in lipid metabolism and inflammation of the white adipose tissue (WAT), including miRNAs, modulated by GEx in Sprague–Dawley rats fed a high-fat diet [304]. They reported that GEx reduced body weight and WAT mass, mRNA levels of adipogenic genes, PPARγ, adipocyte protein 2 (aP2), as well as pro-inflammatory cytokines (TNFα, IL-6 and MCP-1). Obese rats expressed increased expression of miR-21 and miR-132 in WAT. Additionally, this report showed downregulated expression of miR-132 and miR-21 in WAT of rats after GEx administration, in parallel with a greater AMPK activity. The authors assumed that the reduced miR-21 and miR-132 levels could be associated with the post-translational regulation of genes involved in adipogenesis and inflammation in high-fat diet fed rats [304].

It has been reported that some phenolic compounds such as curcumin, RSV, quercetin, EGCG and genistein can interfere with assays via a number of different mechanisms (are potential pan-assay interference compounds-PAINS) and/or are considered invalid metabolic panaceas (IMPs) [305,306]. We recommend that this "warning bell" should be very seriously taken in consideration by the researchers that intend to extend the research in the field of phytochemicals therapeutic potential. The present review has taken in consideration only the phenolic compounds with known molecular mechanisms in vitro and in vivo, precisely to support their specific actions, and does not propose them as panacea.

3. Conclusions

This review amassed consistent knowledge concerning the mechanisms of action of the phenolic compounds, which ascertain their therapeutic potential to prevent or treat CVD. Phenolic compounds can regulate lipid metabolism and balance the oxidative and inflammatory stress through various epigenetic, transcriptional and translational mechanisms that have been demonstrated in vitro and in animal models; some of them have been confirmed in humans.

Phenolic compounds have various pharmacological properties and the present review highlighted those having robust potential to amend dyslipidemia or to diminish the oxidative and inflammatory stress, important risk factors for CVD. Thus, remarkable lipid-regulatory properties have the phenols from hydroxycinnamic acid, flavanols, anthocyanidines and guaiacols groups. They inhibit the lipid absorption in the small intestine; stimulate the cholesterol efflux from atheroma and cholesterol excretion through gallbladder or small intestine; and imped de novo lipid synthesis in the liver. The molecular mechanisms responsible for these effects involve activation of transcription regulators (SIRT-1, LXRs, PPARs). Important antioxidant and anti-inflammatory effects are described for almost all presented phenolic compounds, the main mechanisms of action being inhibition of NLRP3 inflammasome and NF-κB; activation of Nrf2 and Akt, and the consequent stimulation of eNOS and antioxidant enzymes, and inhibition of NADPH oxidase. All these processes are susceptible to be regulated at the epigenetic level, an increasing number of miRNAs modulating them being depicted. Interestingly, certain phenolic compounds, such as resveratrol and naringenin, bind directly to the genes to regulate various proteins synthesis.

Most of these findings come from experiments performed either in vitro or in vivo (experimental animals). Many of these outcomes have been consistently achieved from clinical studies, but some of them still need to be proved in humans. More studies are needed to widen the knowledge concerning the molecular actions of the phenolic compounds; some of them are not completely understood or are still controversial. Since the results obtained until now on improving the human health are promising, more studies in humans are needed to validate the beneficial effects of phenolic compounds in CVD. Additional clinical trials are warranted, but with improvement and standardization of the study design, formulations and doses to be assessed. Further investigations are necessary using larger and longer studies to better define the therapeutic role of phenolic compounds.

Herbal medication has lower adverse effects compared to synthetic drugs and thus presents the advantage that they can be used for a longer period of time. Many of the plant phenolic compounds exert their effects through different mechanisms, and we might expect additive effects when taken in combination, but this has to be scientifically proven. In many cases, it was demonstrated that the biologically active compounds are more efficient in plant extracts compared to the purified molecular components, the data supporting the synergistic action of plants' active compounds. Analyses comparing intake of phenolic compounds in different formulation (powder capsules, alcoholic extracts or encapsulation in nanocarriers) have produced conflicting results. Thus, greater attention should be given to combinations of phenolic compounds, their doses and formulations for administration. Further studies are needed to develop new forms of delivery for such natural bioactive compounds to improve compliance of patients at risk for CVD, and ensure their increased bioavailability, thereby creating more efficient products than the commercially available ones at present.

In the present review we have focused on the molecular mechanisms of action of the phenolic compounds that target dyslipidemia, oxidative and inflammatory stress, the main risk factors in the atherosclerotic process. We hope that we convinced the reader that their multiple benefic effects warrant their use as CVD remedies, complementarily to allopathic drugs. Consumption of foods containing natural phenolic compounds should be encouraged for people with low to moderate CVD

Biomolecules **2020**, *10*, 641

risks; those who are preventive, and those who are at the same time on the prescription drugs to lower the incidence of fatal cardiovascular events.

Author Contributions: Conceptualization, L.T., A.V.S. and C.S.S.; writing—original draft preparation, L.T., G.M.S., L.S.N., M.D. and C.S.S.; writing—review and editing, L.T., C.S.S. and A.V.S. All authors have read and agreed to the published version of the manuscript.

Funding: This work was supported by the Romanian Academy.

Conflicts of Interest: The authors declare no conflict of interest.

References

1. European Society of Cardiology. 2019 ESC/EAS guidelines for the management of dyslipidaemias: Lipid modification to reduce cardiovascular risk. *Atherosclerosis* **2019**, *290*, 140–205. [CrossRef]

2. Bloom, D.E.; Cafiero, E.T.; Jané-Llopis, E.; Abrahams-Gessel, S.; Bloom, L.R.; Fathima, S.; Feigl, A.B.; Gaziano, T.; Mowafi, M.; Pandya, A.; et al. *The Global Economic Burden of Non-Communicable Diseases*; World Economic Forum: Geneva, Switzerland, 2011.

3. Cicero, A.F.G.; Colletti, A.; Bajraktari, G.; Descamps, O.; Djuric, D.M.; Ezhov, M.; Fras, Z.; Katsiki, N.; Langlois, M.; Latkovskis, G.; et al. Lipid-lowering nutraceuticals in clinical practice: Position paper from an International Lipid Expert Panel. *Nutr. Rev.* **2017**, *75*, 731–767. [CrossRef] [PubMed]

4. Sando, K.R.; Knight, M. Nonstatin therapies for management of dyslipidemia: A review. *Clin. Ther.* **2015**, *37*, 2153–2179. [CrossRef] [PubMed]

5. Uehara, Y.; Chiesa, G.; Saku, K. High-Density Lipoprotein-Targeted Therapy and Apolipoprotein A-I Mimetic Peptides. *Circ. J.* **2015**, *79*, 2523–2528. [CrossRef] [PubMed]

6. Musunuru, K. Atherogenic dyslipidemia: Cardiovascular risk and dietary intervention. *Lipids* **2010**, *45*, 907–914. [CrossRef]

7. Sima, A.V.; Stancu, C.S.; Simionescu, M. Vascular endothelium in atherosclerosis. *Cell Tissue Res.* **2009**, *335*, 191–203. [CrossRef]

8. Pavlovic, J.; Kavousi, M.; Ikram, M.A.; Leening, M.J.G. Updated treatment thresholds in the 2019 ESC/EAS dyslipidaemia guidelines substantially expand indications for statin use for primary prevention at population level: Results from the Rotterdam Study. *Atherosclerosis* **2020**. [CrossRef]

9. Wolin, M.S. Interactions of oxidants with vascular signaling systems. *Arterioscler. Thromb. Vasc. Biol.* **2000**, *20*, 1430–1442. [CrossRef]

10. Puddu, P.; Puddu, G.M.; Cravero, E.; De Pascalis, S.; Muscari, A. The emerging role of cardiovascular risk factor-induced mitochondrial dysfunction in atherogenesis. *J. Biomed. Sci.* **2009**, *16*, 112. [CrossRef]

11. Wattanapitayakul, S.K.; Bauer, J.A. Oxidative pathways in cardiovascular disease: Roles, mechanisms, and therapeutic implications. *Pharmacol. Ther.* **2001**, *89*, 187–206. [CrossRef]

12. Valko, M.; Leibfritz, D.; Moncol, J.; Cronin, M.T.; Mazur, M.; Telser, J. Free radicals and antioxidants in normal physiological functions and human disease. *Int. J. Biochem. Cell Biol.* **2007**, *39*, 44–84. [CrossRef] [PubMed]

13. Yang, Y.; Duan, W.; Lin, Y.; Yi, W.; Liang, Z.; Yan, J.; Wang, N.; Deng, C.; Zhang, S.; Li, Y.; et al. SIRT1 activation by curcumin pretreatment attenuates mitochondrial oxidative damage induced by myocardial ischemia reperfusion injury. *Free Radic. Biol. Med.* **2013**, *65*, 667–679. [CrossRef] [PubMed]

14. Martindale, J.L.; Holbrook, N.J. Cellular response to oxidative stress: Signaling for suicide and survival. *J. Cell. Physiol.* **2002**, *192*, 1–15. [CrossRef] [PubMed]

15. Lee, J.H.; Khor, T.O.; Shu, L.; Su, Z.Y.; Fuentes, F.; Kong, A.N. Dietary phytochemicals and cancer prevention: Nrf2 signaling, epigenetics, and cell death mechanisms in blocking cancer initiation and progression. *Pharmacol. Ther.* **2013**, *137*, 153–171. [CrossRef]

16. Ortiz-Munoz, G.; Couret, D.; Lapergue, B.; Bruckert, E.; Meseguer, E.; Amarenco, P.; Meilhac, O. Dysfunctional HDL in acute stroke. *Atherosclerosis* **2016**, *253*, 75–80. [CrossRef]

17. Carnuta, M.G.; Stancu, C.S.; Toma, L.; Sanda, G.M.; Niculescu, L.S.; Deleanu, M.; Popescu, A.C.; Popescu, M.R.; Vlad, A.; Dimulescu, D.R.; et al. Dysfunctional high-density lipoproteins have distinct composition, diminished anti-inflammatory potential and discriminate acute coronary syndrome from stable coronary artery disease patients. *Sci. Rep.* **2017**, *7*, 7295. [CrossRef]

18. Parhofer, K.G. Increasing HDL-cholesterol and prevention of atherosclerosis: A critical perspective. *Atheroscler. Suppl.* **2015**, *18*, 109–111. [CrossRef]

19. Variji, A.; Shokri, Y.; Fallahpour, S.; Zargari, M.; Bagheri, B.; Abediankenari, S.; Alizadeh, A.; Mahrooz, A. The combined utility of myeloperoxidase (MPO) and paraoxonase 1 (PON1) as two important HDL-associated enzymes in coronary artery disease: Which has a stronger predictive role? *Atherosclerosis* **2019**, *280*, 7–13. [CrossRef] [PubMed]

20. Nguyen, M.T.; Fernando, S.; Schwarz, N.; Tan, J.T.; Bursill, C.A.; Psaltis, P.J. Inflammation as a Therapeutic Target in Atherosclerosis. *J. Clin. Med.* **2019**, *8*, 1109. [CrossRef] [PubMed]

21. Zhu, Y.; Xian, X.; Wang, Z.; Bi, Y.; Chen, Q.; Han, X.; Tang, D.; Chen, R. Research Progress on the Relationship between Atherosclerosis and Inflammation. *Biomolecules* **2018**, *8*, 80. [CrossRef]

22. Bi, Y.; Chen, J.; Hu, F.; Liu, J.; Li, M.; Zhao, L. M2 Macrophages as a Potential Target for Antiatherosclerosis Treatment. *Neural Plast.* **2019**, *2019*, 6724903. [CrossRef] [PubMed]

23. Pamukcu, B.; Lip, G.Y.; Shantsila, E. The nuclear factor—kappa B pathway in atherosclerosis: A potential therapeutic target for atherothrombotic vascular disease. *Thromb. Res.* **2011**, *128*, 117–123. [CrossRef] [PubMed]

24. Sosnowska, B.; Mazidi, M.; Penson, P.; Gluba-Brzozka, A.; Rysz, J.; Banach, M. The sirtuin family members SIRT1, SIRT3 and SIRT6: Their role in vascular biology and atherogenesis. *Atherosclerosis* **2017**, *265*, 275–282. [CrossRef] [PubMed]

25. Grebe, A.; Hoss, F.; Latz, E. NLRP3 Inflammasome and the IL-1 Pathway in Atherosclerosis. *Circ. Res.* **2018**, *122*, 1722–1740. [CrossRef]

26. Botker, H.E.; Hausenloy, D.; Andreadou, I.; Antonucci, S.; Boengler, K.; Davidson, S.M.; Deshwal, S.; Devaux, Y.; Di Lisa, F.; Di Sante, M.; et al. Practical guidelines for rigor and reproducibility in preclinical and clinical studies on cardioprotection. *Basic Res. Cardiol.* **2018**, *113*, 39. [CrossRef]

27. Jacquier, A. The complex eukaryotic transcriptome: Unexpected pervasive transcription and novel small RNAs. *Nat. Rev. Genet.* **2009**, *10*, 833–844. [CrossRef]

28. Papageorgiou, N.; Tslamandris, S.; Giolis, A.; Tousoulis, D. MicroRNAs in Cardiovascular Disease: Perspectives and Reality. *Cardiol. Rev.* **2016**, *24*, 110–118. [CrossRef]

29. Gomes, C.P.C.; Schroen, B.; Kuster, G.M.; Robinson, E.L.; Ford, K.; Squire, I.B.; Heymans, S.; Martelli, F.; Emanueli, C.; Devaux, Y.; et al. Regulatory RNAs in Heart Failure. *Circulation* **2020**, *141*, 313–328. [CrossRef]

30. Moore, K.J.; Rayner, K.J.; Suarez, Y.; Fernandez-Hernando, C. microRNAs and cholesterol metabolism. *Trends Endocrinol. Metab. TEM* **2010**, *21*, 699–706. [CrossRef]

31. Bartel, D.P. MicroRNAs: Target recognition and regulatory functions. *Cell* **2009**, *136*, 215–233. [CrossRef]

32. Forman, J.J.; Coller, H.A. The code within the code: MicroRNAs target coding regions. *Cell Cycle* **2010**, *9*, 1533–1541. [CrossRef] [PubMed]

33. Vickers, K.C.; Palmisano, B.T.; Shoucri, B.M.; Shamburek, R.D.; Remaley, A.T. MicroRNAs are transported in plasma and delivered to recipient cells by high-density lipoproteins. *Nat. Cell Biol.* **2011**, *13*, 423–433. [CrossRef] [PubMed]

34. Arroyo, J.D.; Chevillet, J.R.; Kroh, E.M.; Ruf, I.K.; Pritchard, C.C.; Gibson, D.F.; Mitchell, P.S.; Bennett, C.F.; Pogosova-Agadjanyan, E.L.; Stirewalt, D.L.; et al. Argonaute2 complexes carry a population of circulating microRNAs independent of vesicles in human plasma. *Proc. Natl. Acad. Sci. USA* **2011**, *108*, 5003–5008. [CrossRef]

35. Weber, J.A.; Baxter, D.H.; Zhang, S.; Huang, D.Y.; Huang, K.H.; Lee, M.J.; Galas, D.J.; Wang, K. The microRNA spectrum in 12 body fluids. *Clin. Chem.* **2010**, *56*, 1733–1741. [CrossRef] [PubMed]

36. Devaux, Y. Transcriptome of blood cells as a reservoir of cardiovascular biomarkers. *Biochim. Biophys. Acta. Mol. Cell Res.* **2017**, *1864*, 209–216. [CrossRef] [PubMed]

37. Cortez, M.A.; Bueso-Ramos, C.; Ferdin, J.; Lopez-Berestein, G.; Sood, A.K.; Calin, G.A. MicroRNAs in body fluids–the mix of hormones and biomarkers. *Nat. Rev. Clin. Oncol.* **2011**, *8*, 467–477. [CrossRef] [PubMed]

38. Devaux, Y.; Vausort, M.; Goretti, E.; Nazarov, P.V.; Azuaje, F.; Gilson, G.; Corsten, M.F.; Schroen, B.; Lair, M.L.; Heymans, S.; et al. Use of circulating microRNAs to diagnose acute myocardial infarction. *Clin. Chem.* **2012**, *58*, 559–567. [CrossRef]

39. Jansen, F.; Yang, X.; Proebsting, S.; Hoelscher, M.; Przybilla, D.; Baumann, K.; Schmitz, T.; Dolf, A.; Endl, E.; Franklin, B.S.; et al. MicroRNA expression in circulating microvesicles predicts cardiovascular events in patients with coronary artery disease. *J. Am. Heart Assoc.* **2014**, *3*, e001249. [CrossRef]

40. Niculescu, L.S.; Simionescu, N.; Sanda, G.M.; Carnuta, M.G.; Stancu, C.S.; Popescu, A.C.; Popescu, M.R.; Vlad, A.; Dimulescu, D.R.; Simionescu, M.; et al. MiR-486 and miR-92a Identified in Circulating HDL Discriminate between Stable and Vulnerable Coronary Artery Disease Patients. *PLoS ONE* **2015**, *10*, e0140958. [CrossRef]

41. Simionescu, N.; Niculescu, L.S.; Carnuta, M.G.; Sanda, G.M.; Stancu, C.S.; Popescu, A.C.; Popescu, M.R.; Vlad, A.; Dimulescu, D.R.; Simionescu, M.; et al. Hyperglycemia Determines Increased Specific MicroRNAs Levels in Sera and HDL of Acute Coronary Syndrome Patients and Stimulates MicroRNAs Production in Human Macrophages. *PLoS ONE* **2016**, *11*, e0161201. [CrossRef]

42. Niculescu, L.S.; Dulceanu, M.D.; Stancu, C.S.; Carnuta, M.G.; Barbalata, T.; Sima, A.V. Probiotics administration or the high-fat diet arrest modulates microRNAs levels in hyperlipidemic hamsters. *J. Funct. Foods* **2019**, *56*, 295–302. [CrossRef]

43. Afman, L.; Milenkovic, D.; Roche, H.M. Nutritional aspects of metabolic inflammation in relation to health–insights from transcriptomic biomarkers in PBMC of fatty acids and polyphenols. *Mol. Nutr. Food Res.* **2014**, *58*, 1708–1720. [CrossRef] [PubMed]

44. Sobhani, M.; Farzaei, M.H.; Kiani, S.; Khodarahmi, R. Immunomodulatory; Anti-inflammatory/antioxidant Effects of Polyphenols: A Comparative Review on the Parental Compounds and Their Metabolites. *Food Rev. Int.* **2020**, 1–53. [CrossRef]

45. Subramaniam, S.; Selvaduray, K.R.; Radhakrishnan, A.K. Bioactive Compounds: Natural Defense Against Cancer? *Biomolecules* **2019**, *9*, 758. [CrossRef]

46. Jamwal, R. Bioavailable curcumin formulations: A review of pharmacokinetic studies in healthy volunteers. *J. Integr. Med.* **2018**, *16*, 367–374. [CrossRef]

47. DiSilvestro, R.A.; Joseph, E.; Zhao, S.; Bomser, J. Diverse effects of a low dose supplement of lipidated curcumin in healthy middle aged people. *Nutr. J.* **2012**, *11*, 79. [CrossRef]

48. Mollazadeh, H.; Mahdian, D.; Hosseinzadeh, H. Medicinal plants in treatment of hypertriglyceridemia: A review based on their mechanisms and effectiveness. *Phytomed. Int. J. Phytother. Phytopharmacol.* **2019**, *53*, 43–52. [CrossRef]

49. Asai, A.; Nakagawa, K.; Miyazawa, T. Antioxidative effects of turmeric, rosemary and capsicum extracts on membrane phospholipid peroxidation and liver lipid metabolism in mice. *Biosci. Biotechnol. Biochem.* **1999**, *63*, 2118–2122. [CrossRef]

50. Shin, S.K.; Ha, T.Y.; McGregor, R.A.; Choi, M.S. Long-term curcumin administration protects against atherosclerosis via hepatic regulation of lipoprotein cholesterol metabolism. *Mol. Nutr. Food Res.* **2011**, *55*, 1829–1840. [CrossRef]

51. Hamer, M.; O'Donovan, G.; Stamatakis, E. High-Density Lipoprotein Cholesterol and Mortality: Too Much of a Good Thing? *Arterioscler. Thromb. Vasc. Biol.* **2018**, *38*, 669–672. [CrossRef]

52. Soetikno, V.; Sari, F.R.; Sukumaran, V.; Lakshmanan, A.P.; Mito, S.; Harima, M.; Thandavarayan, R.A.; Suzuki, K.; Nagata, M.; Takagi, R.; et al. Curcumin prevents diabetic cardiomyopathy in streptozotocin-induced diabetic rats: Possible involvement of PKC-MAPK signaling pathway. *Eur. J. Pharm. Sci.* **2012**, *47*, 604–614. [CrossRef] [PubMed]

53. Yu, W.; Wu, J.; Cai, F.; Xiang, J.; Zha, W.; Fan, D.; Guo, S.; Ming, Z.; Liu, C. Curcumin alleviates diabetic cardiomyopathy in experimental diabetic rats. *PLoS ONE* **2012**, *7*, e52013. [CrossRef]

54. Tanwar, V.; Sachdeva, J.; Kishore, K.; Mittal, R.; Nag, T.C.; Ray, R.; Kumari, S.; Arya, D.S. Dose-dependent actions of curcumin in experimentally induced myocardial necrosis: A biochemical, histopathological, and electron microscopic evidence. *Cell Biochem. Funct.* **2010**, *28*, 74–82. [CrossRef] [PubMed]

55. Nazam Ansari, M.; Bhandari, U.; Pillai, K.K. Protective role of curcumin in myocardial oxidative damage induced by isoproterenol in rats. *Hum. Exp. Toxicol.* **2007**, *26*, 933–938. [CrossRef] [PubMed]

56. Soares, M.P.; Seldon, M.P.; Gregoire, I.P.; Vassilevskaia, T.; Berberat, P.O.; Yu, J.; Tsui, T.Y.; Bach, F.H. Heme oxygenase-1 modulates the expression of adhesion molecules associated with endothelial cell activation. *J. Immunol.* **2004**, *172*, 3553–3563. [CrossRef] [PubMed]

57. Olszanecki, R.; Gebska, A.; Korbut, R. The role of haem oxygenase-1 in the decrease of endothelial intercellular adhesion molecule-1 expression by curcumin. *Basic Clin. Pharmacol. Toxicol.* **2007**, *101*, 411–415. [CrossRef]

58. Nabavi, S.F.; Barber, A.J.; Spagnuolo, C.; Russo, G.L.; Daglia, M.; Nabavi, S.M.; Sobarzo-Sanchez, E. Nrf2 as molecular target for polyphenols: A novel therapeutic strategy in diabetic retinopathy. *Crit. Rev. Clin. Lab. Sci.* **2016**, *53*, 293–312. [CrossRef]

59. Kang, E.S.; Woo, I.S.; Kim, H.J.; Eun, S.Y.; Paek, K.S.; Kim, H.J.; Chang, K.C.; Lee, J.H.; Lee, H.T.; Kim, J.H.; et al. Up-regulation of aldose reductase expression mediated by phosphatidylinositol 3-kinase/Akt and Nrf2 is involved in the protective effect of curcumin against oxidative damage. *Free Radic. Biol. Med.* **2007**, *43*, 535–545. [CrossRef]

60. Yang, H.; Xu, W.; Zhou, Z.; Liu, J.; Li, X.; Chen, L.; Weng, J.; Yu, Z. Curcumin attenuates urinary excretion of albumin in type II diabetic patients with enhancing nuclear factor erythroid-derived 2-like 2 (Nrf2) system and repressing inflammatory signaling efficacies. *Exp. Clin. Endocrinol. Diabetes* **2015**, *123*, 360–367. [CrossRef]

61. Wongeakin, N.; Bhattarakosol, P.; Patumraj, S. Molecular mechanisms of curcumin on diabetes-induced endothelial dysfunctions: Txnip, ICAM-1, and NOX2 expressions. *BioMed Res. Int.* **2014**, *2014*, 161346. [CrossRef]

62. Zeng, C.; Zhong, P.; Zhao, Y.; Kanchana, K.; Zhang, Y.; Khan, Z.A.; Chakrabarti, S.; Wu, L.; Wang, J.; Liang, G. Curcumin protects hearts from FFA-induced injury by activating Nrf2 and inactivating NF-kappaB both in vitro and in vivo. *J. Mol. Cell. Cardiol.* **2015**, *79*, 1–12. [CrossRef] [PubMed]

63. Coban, D.; Milenkovic, D.; Chanet, A.; Khallou-Laschet, J.; Sabbe, L.; Palagani, A.; Vanden Berghe, W.; Mazur, A.; Morand, C. Dietary curcumin inhibits atherosclerosis by affecting the expression of genes involved in leukocyte adhesion and transendothelial migration. *Mol. Nutr. Food Res.* **2012**, *56*, 1270–1281. [CrossRef] [PubMed]

64. Pirvulescu, M.M.; Gan, A.M.; Stan, D.; Simion, V.; Calin, M.; Butoi, E.; Tirgoviste, C.I.; Manduteanu, I. Curcumin and a Morus alba extract reduce pro-inflammatory effects of resistin in human endothelial cells. *Phytother. Res. PTR* **2011**, *25*, 1737–1742. [CrossRef] [PubMed]

65. Monfoulet, L.E.; Mercier, S.; Bayle, D.; Tamaian, R.; Barber-Chamoux, N.; Morand, C.; Milenkovic, D. Curcumin modulates endothelial permeability and monocyte transendothelial migration by affecting endothelial cell dynamics. *Free Radic. Biol. Med.* **2017**, *112*, 109–120. [CrossRef] [PubMed]

66. Karimian, M.S.; Pirro, M.; Majeed, M.; Sahebkar, A. Curcumin as a natural regulator of monocyte chemoattractant protein-1. *Cytokine Growth Factor Rev.* **2017**, *33*, 55–63. [CrossRef]

67. Hao, Q.; Chen, X.; Wang, X.; Dong, B.; Yang, C. Curcumin Attenuates Angiotensin II-Induced Abdominal Aortic Aneurysm by Inhibition of Inflammatory Response and ERK Signaling Pathways. *Evid.-Based Complement. Altern. Med. eCAM* **2014**, *2014*, 270930. [CrossRef]

68. Wang, J.; Dong, S. ICAM-1 and IL-8 are expressed by DEHP and suppressed by curcumin through ERK and p38 MAPK in human umbilical vein endothelial cells. *Inflammation* **2012**, *35*, 859–870. [CrossRef]

69. Tsai, I.J.; Chen, C.W.; Tsai, S.Y.; Wang, P.Y.; Owaga, E.; Hsieh, R.H. Curcumin supplementation ameliorated vascular dysfunction and improved antioxidant status in rats fed a high-sucrose, high-fat diet. *Appl. Physiol. Nutr. Metab.* **2018**, *43*, 669–676. [CrossRef]

70. Han, Y.; Sun, H.J.; Tong, Y.; Chen, Y.Z.; Ye, C.; Qiu, Y.; Zhang, F.; Chen, A.D.; Qi, X.H.; Chen, Q.; et al. Curcumin attenuates migration of vascular smooth muscle cells via inhibiting NFkappaB-mediated NLRP3 expression in spontaneously hypertensive rats. *J. Nutr. Biochem.* **2019**, *72*, 108212. [CrossRef]

71. Sun, H.J.; Ren, X.S.; Xiong, X.Q.; Chen, Y.Z.; Zhao, M.X.; Wang, J.J.; Zhou, Y.B.; Han, Y.; Chen, Q.; Li, Y.H.; et al. NLRP3 inflammasome activation contributes to VSMC phenotypic transformation and proliferation in hypertension. *Cell Death Dis.* **2017**, *8*, e3074. [CrossRef]

72. Yin, H.; Guo, Q.; Li, X.; Tang, T.; Li, C.; Wang, H.; Sun, Y.; Feng, Q.; Ma, C.; Gao, C.; et al. Curcumin Suppresses IL-1beta Secretion and Prevents Inflammation through Inhibition of the NLRP3 Inflammasome. *J. Immunol.* **2018**, *200*, 2835–2846. [CrossRef] [PubMed]

73. Chen, F.; Guo, N.; Cao, G.; Zhou, J.; Yuan, Z. Molecular analysis of curcumin-induced polarization of murine RAW264.7 macrophages. *J. Cardiovasc. Pharmacol.* **2014**, *63*, 544–552. [CrossRef] [PubMed]

74. Chen, F.Y.; Zhou, J.; Guo, N.; Ma, W.G.; Huang, X.; Wang, H.; Yuan, Z.Y. Curcumin retunes cholesterol transport homeostasis and inflammation response in M1 macrophage to prevent atherosclerosis. *Biochem. Biophys. Res. Commun.* **2015**, *467*, 872–878. [CrossRef] [PubMed]

75. Miao, Y.; Zhao, S.; Gao, Y.; Wang, R.; Wu, Q.; Wu, H.; Luo, T. Curcumin pretreatment attenuates inflammation and mitochondrial dysfunction in experimental stroke: The possible role of Sirt1 signaling. *Brain Res. Bull.* **2016**, *121*, 9–15. [CrossRef] [PubMed]

76. Wu, Y.; Xu, J.; Xu, J.; Zheng, W.; Chen, Q.; Jiao, D. Study on the mechanism of JAK2/STAT3 signaling pathway-mediated inflammatory reaction after cerebral ischemia. *Mol. Med. Rep.* **2018**, *17*, 5007–5012. [CrossRef] [PubMed]

77. Li, W.; Suwanwela, N.C.; Patumraj, S. Curcumin by down-regulating NF-kB and elevating Nrf2, reduces brain edema and neurological dysfunction after cerebral I/R. *Microvasc. Res.* **2016**, *106*, 117–127. [CrossRef]

78. Tu, X.K.; Yang, W.Z.; Chen, J.P.; Chen, Y.; Ouyang, L.Q.; Xu, Y.C.; Shi, S.S. Curcumin inhibits TLR2/4-NF-kappaB signaling pathway and attenuates brain damage in permanent focal cerebral ischemia in rats. *Inflammation* **2014**, *37*, 1544–1551. [CrossRef] [PubMed]

79. Hatcher, H.; Planalp, R.; Cho, J.; Torti, F.M.; Torti, S.V. Curcumin: From ancient medicine to current clinical trials. *Cell. Mol. Life Sci. CMLS* **2008**, *65*, 1631–1652. [CrossRef]

80. Lin, K.; Chen, H.; Chen, X.; Qian, J.; Huang, S.; Huang, W. Efficacy of Curcumin on Aortic Atherosclerosis: A Systematic Review and Meta-Analysis in Mouse Studies and Insights into Possible Mechanisms. *Oxid. Med. Cell. Longev.* **2020**, *2020*, 1520747. [CrossRef]

81. Mirzabeigi, P.; Mohammadpour, A.H.; Salarifar, M.; Gholami, K.; Mojtahedzadeh, M.; Javadi, M.R. The Effect of Curcumin on some of Traditional and Non-traditional Cardiovascular Risk Factors: A Pilot Randomized, Double-blind, Placebo-controlled Trial. *Iran. J. Pharm. Res. IJPR* **2015**, *14*, 479–486.

82. Baum, L.; Cheung, S.K.; Mok, V.C.; Lam, L.C.; Leung, V.P.; Hui, E.; Ng, C.C.; Chow, M.; Ho, P.C.; Lam, S.; et al. Curcumin effects on blood lipid profile in a 6-month human study. *Pharmacol. Res.* **2007**, *56*, 509–514. [CrossRef] [PubMed]

83. Yang, Y.S.; Su, Y.F.; Yang, H.W.; Lee, Y.H.; Chou, J.I.; Ueng, K.C. Lipid-lowering effects of curcumin in patients with metabolic syndrome: A randomized, double-blind, placebo-controlled trial. *Phytother. Res. PTR* **2014**, *28*, 1770–1777. [CrossRef]

84. Panahi, Y.; Ghanei, M.; Bashiri, S.; Hajihashemi, A.; Sahebkar, A. Short-term Curcuminoid Supplementation for Chronic Pulmonary Complications due to Sulfur Mustard Intoxication: Positive Results of a Randomized Double-blind Placebo-controlled Trial. *Drug Res.* **2015**, *65*, 567–573. [CrossRef]

85. Simion, V.; Stan, D.; Constantinescu, C.A.; Deleanu, M.; Dragan, E.; Tucureanu, M.M.; Gan, A.M.; Butoi, E.; Constantin, A.; Manduteanu, I.; et al. Conjugation of curcumin-loaded lipid nanoemulsions with cell-penetrating peptides increases their cellular uptake and enhances the anti-inflammatory effects in endothelial cells. *J. Pharm. Pharmacol.* **2016**, *68*, 195–207. [CrossRef]

86. Ma, F.; Liu, F.; Ding, L.; You, M.; Yue, H.; Zhou, Y.; Hou, Y. Anti-inflammatory effects of curcumin are associated with down regulating microRNA-155 in LPS-treated macrophages and mice. *Pharm. Biol.* **2017**, *55*, 1263–1273. [CrossRef] [PubMed]

87. Zhang, J.; Wang, Q.; Rao, G.; Qiu, J.; He, R. Curcumin improves perfusion recovery in experimental peripheral arterial disease by upregulating microRNA-93 expression. *Exp. Ther. Med.* **2019**, *17*, 798–802. [CrossRef] [PubMed]

88. Geng, H.H.; Li, R.; Su, Y.M.; Xiao, J.; Pan, M.; Cai, X.X.; Ji, X.P. Curcumin protects cardiac myocyte against hypoxia-induced apoptosis through upregulating miR-7a/b expression. *Biomed. Pharmacother.* **2016**, *81*, 258–264. [CrossRef]

89. Tian, L.; Song, Z.; Shao, W.; Du, W.W.; Zhao, L.R.; Zeng, K.; Yang, B.B.; Jin, T. Curcumin represses mouse 3T3-L1 cell adipogenic differentiation via inhibiting miR-17-5p and stimulating the Wnt signalling pathway effector Tcf7l2. *Cell Death Dis.* **2017**, *8*, e2559. [CrossRef]

90. El-Seedi, H.R.; El-Said, A.M.; Khalifa, S.A.; Goransson, U.; Bohlin, L.; Borg-Karlson, A.K.; Verpoorte, R. Biosynthesis, natural sources, dietary intake, pharmacokinetic properties, and biological activities of hydroxycinnamic acids. *J. Agric. Food Chem.* **2012**, *60*, 10877–10895. [CrossRef]

91. David, I.G.; Bizgan, A.M.; Popa, D.E.; Buleandra, M.; Moldovan, Z.; Badea, I.A.; Tekiner, T.A.; Basaga, H.; Ciucu, A.A. Rapid determination of total polyphenolic content in tea samples based on caffeic acid voltammetric behaviour on a disposable graphite electrode. *Food Chem.* **2015**, *173*, 1059–1065. [CrossRef]

92. Rebelo, M.J.; Rego, R.; Ferreira, M.; Oliveira, M.C. Comparative study of the antioxidant capacity and polyphenol content of Douro wines by chemical and electrochemical methods. *Food Chem.* **2013**, *141*, 566–573. [CrossRef] [PubMed]

93. Garrido, E.; Cerqueira, A.S.; Chavarria, D.; Silva, T.; Borges, F.; Garrido, J. Microencapsulation of caffeic acid phenethyl ester and caffeic acid phenethyl amide by inclusion in hydroxypropyl-beta-cyclodextrin. *Food Chem.* **2018**, *254*, 260–265. [CrossRef] [PubMed]

94. Wang, S.J.; Zeng, J.; Yang, B.K.; Zhong, Y.M. Bioavailability of caffeic acid in rats and its absorption properties in the Caco-2 cell model. *Pharm. Biol.* **2014**, *52*, 1150–1157. [CrossRef] [PubMed]

95. Moya, C.; Manez, S. Paraoxonases: Metabolic role and pharmacological projection. *Naunyn-Schmiedeberg's Arch. Pharmacol.* **2018**, *391*, 349–359. [CrossRef]

96. Wang, X.; Stavchansky, S.; Zhao, B.; Bynum, J.A.; Kerwin, S.M.; Bowman, P.D. Cytoprotection of human endothelial cells from menadione cytotoxicity by caffeic acid phenethyl ester: The role of heme oxygenase-1. *Eur. J. Pharmacol.* **2008**, *591*, 28–35. [CrossRef]

97. Kim, J.K.; Jang, H.D. Nrf2-mediated HO-1 induction coupled with the ERK signaling pathway contributes to indirect antioxidant capacity of caffeic acid phenethyl ester in HepG2 cells. *Int. J. Mol. Sci.* **2014**, *15*, 12149–12165. [CrossRef]

98. Dinkova-Kostova, A.T.; Holtzclaw, W.D.; Cole, R.N.; Itoh, K.; Wakabayashi, N.; Katoh, Y.; Yamamoto, M.; Talalay, P. Direct evidence that sulfhydryl groups of Keap1 are the sensors regulating induction of phase 2 enzymes that protect against carcinogens and oxidants. *Proc. Natl. Acad. Sci. USA* **2002**, *99*, 11908–11913. [CrossRef]

99. Taylan, M.; Kaya, H.; Demir, M.; Evliyaoglu, O.; Sen, H.S.; Firat, U.; Keles, A.; Yilmaz, S.; Sezgi, C. The Protective Effects of Caffeic Acid Phenethyl Ester on Acetylsalicylic Acid-induced Lung Injury in Rats. *J. Investig. Surg.* **2016**, *29*, 328–334. [CrossRef]

100. Moon, M.K.; Lee, Y.J.; Kim, J.S.; Kang, D.G.; Lee, H.S. Effect of caffeic acid on tumor necrosis factor-alpha-induced vascular inflammation in human umbilical vein endothelial cells. *Biol. Pharm. Bull.* **2009**, *32*, 1371–1377. [CrossRef]

101. Toma, L.; Sanda, G.M.; Niculescu, L.S.; Deleanu, M.; Stancu, C.S.; Sima, A.V. Caffeic acid attenuates the inflammatory stress induced by glycated LDL in human endothelial cells by mechanisms involving inhibition of AGE-receptor, oxidative, and endoplasmic reticulum stress. *BioFactors* **2017**, *43*, 685–697. [CrossRef]

102. Ibitoye, O.B.; Ajiboye, T.O. Dietary phenolic acids reverse insulin resistance, hyperglycaemia, dyslipidaemia, inflammation and oxidative stress in high-fructose diet-induced metabolic syndrome rats. *Arch. Physiol. Biochem.* **2018**, *124*, 410–417. [CrossRef] [PubMed]

103. Shahbazi, S.; Sahrawat, T.R.; Ray, M.; Dash, S.; Kar, D.; Singh, S. Drug Targets for Cardiovascular-Safe Anti-Inflammatory: In Silico Rational Drug Studies. *PLoS ONE* **2016**, *11*, e0156156. [CrossRef] [PubMed]

104. Murase, T.; Misawa, K.; Minegishi, Y.; Aoki, M.; Ominami, H.; Suzuki, Y.; Shibuya, Y.; Hase, T. Coffee polyphenols suppress diet-induced body fat accumulation by downregulating SREBP-1c and related molecules in C57BL/6J mice. *Am. J. Physiol. Endocrinol. Metab.* **2011**, *300*, E122–E133. [CrossRef]

105. Matboli, M.; Eissa, S.; Ibrahim, D.; Hegazy, M.G.A.; Imam, S.S.; Habib, E.K. Caffeic Acid Attenuates Diabetic Kidney Disease via Modulation of Autophagy in a High-Fat Diet/Streptozotocin- Induced Diabetic Rat. *Sci. Rep.* **2017**, *7*, 2263. [CrossRef]

106. Hou, C.Y.; Tain, Y.L.; Yu, H.R.; Huang, L.T. The Effects of Resveratrol in the Treatment of Metabolic Syndrome. *Int. J. Mol. Sci.* **2019**, *20*, 535. [CrossRef]

107. Amri, A.; Chaumeil, J.C.; Sfar, S.; Charrueau, C. Administration of resveratrol: What formulation solutions to bioavailability limitations? *J. Controll. Release* **2012**, *158*, 182–193. [CrossRef]

108. Chimento, A.; De Amicis, F.; Sirianni, R.; Sinicropi, M.S.; Puoci, F.; Casaburi, I.; Saturnino, C.; Pezzi, V. Progress to Improve Oral Bioavailability and Beneficial Effects of Resveratrol. *Int. J. Mol. Sci.* **2019**, *20*, 1381. [CrossRef]

109. Cho, I.J.; Ahn, J.Y.; Kim, S.; Choi, M.S.; Ha, T.Y. Resveratrol attenuates the expression of HMG-CoA reductase mRNA in hamsters. *Biochem. Biophys. Res. Commun.* **2008**, *367*, 190–194. [CrossRef]

110. Kumar, B.J.; Joghee, N. Resveratrol supplementation in patients with type 2 diabetes mellitus: A prospective, open label, randomized controlled trial. *Int. Res. J. Pharm.* **2013**, *4*, 245–249. [CrossRef]

111. Movahed, A.; Nabipour, I.; Lieben Louis, X.; Thandapilly, S.J.; Yu, L.; Kalantarhormozi, M.; Rekabpour, S.J.; Netticadan, T. Antihyperglycemic effects of short term resveratrol supplementation in type 2 diabetic patients. *Evid.-Based Complement. Altern. Med. eCAM* **2013**, *2013*, 851267. [CrossRef]

112. Sahebkar, A. Effects of resveratrol supplementation on plasma lipids: A systematic review and meta-analysis of randomized controlled trials. *Nutr. Rev.* **2013**, *71*, 822–835. [CrossRef] [PubMed]

113. Luca, S.V.; Macovei, I.; Bujor, A.; Miron, A.; Skalicka-Wozniak, K.; Aprotosoaie, A.C.; Trifan, A. Bioactivity of dietary polyphenols: The role of metabolites. *Crit. Rev. Food Sci. Nutr.* **2020**, *60*, 626–659. [CrossRef] [PubMed]

114. Gong, M.; Garige, M.; Varatharajalu, R.; Marmillot, P.; Gottipatti, C.; Leckey, L.C.; Lakshman, R.M. Quercetin up-regulates paraoxonase 1 gene expression with concomitant protection against LDL oxidation. *Biochem. Biophys. Res. Commun.* **2009**, *379*, 1001–1004. [CrossRef] [PubMed]

115. Noll, C.; Hamelet, J.; Matulewicz, E.; Paul, J.L.; Delabar, J.M.; Janel, N. Effects of red wine polyphenolic compounds on paraoxonase-1 and lectin-like oxidized low-density lipoprotein receptor-1 in hyperhomocysteinemic mice. *J. Nutr. Biochem.* **2009**, *20*, 586–596. [CrossRef] [PubMed]

116. Garige, M.; Gong, M.; Varatharajalu, R.; Lakshman, M.R. Quercetin up-regulates paraoxonase 1 gene expression via sterol regulatory element binding protein 2 that translocates from the endoplasmic reticulum to the nucleus where it specifically interacts with sterol responsive element-like sequence in paraoxonase 1 promoter in HuH7 liver cells. *Metab. Clin. Exp.* **2010**, *59*, 1372–1378. [CrossRef]

117. Barbalata, T.; Deleanu, M.; Carnuta, M.G.; Niculescu, L.S.; Raileanu, M.; Sima, A.V.; Stancu, C.S. Hyperlipidemia Determines Dysfunctional HDL Production and Impedes Cholesterol Efflux in the Small Intestine: Alleviation by Ginger Extract. *Mol. Nutr. Food Res.* **2019**, *63*, e1900029. [CrossRef]

118. Guo, R.; Su, Y.; Liu, B.; Li, S.; Zhou, S.; Xu, Y. Resveratrol suppresses oxidised low-density lipoprotein-induced macrophage apoptosis through inhibition of intracellular reactive oxygen species generation, LOX-1, and the p38 MAPK pathway. *Cell. Physiol. Biochem. Int. J. Exp. Cell. Physiol. Biochem. Pharmacol.* **2014**, *34*, 603–616. [CrossRef]

119. Bonnefont-Rousselot, D. Resveratrol and Cardiovascular Diseases. *Nutrients* **2016**, *8*, 250. [CrossRef]

120. Wicinski, M.; Socha, M.; Walczak, M.; Wodkiewicz, E.; Malinowski, B.; Rewerski, S.; Gorski, K.; Pawlak-Osinska, K. Beneficial Effects of Resveratrol Administration-Focus on Potential Biochemical Mechanisms in Cardiovascular Conditions. *Nutrients* **2018**, *10*, 1813. [CrossRef]

121. Borra, M.T.; Smith, B.C.; Denu, J.M. Mechanism of human SIRT1 activation by resveratrol. *J. Biol. Chem.* **2005**, *280*, 17187–17195. [CrossRef]

122. Yeung, F.; Hoberg, J.E.; Ramsey, C.S.; Keller, M.D.; Jones, D.R.; Frye, R.A.; Mayo, M.W. Modulation of NF-kappaB-dependent transcription and cell survival by the SIRT1 deacetylase. *EMBO J.* **2004**, *23*, 2369–2380. [CrossRef]

123. Bonizzi, G.; Karin, M. The two NF-kappaB activation pathways and their role in innate and adaptive immunity. *Trends Immunol.* **2004**, *25*, 280–288. [CrossRef]

124. Malaguarnera, L. Influence of Resveratrol on the Immune Response. *Nutrients* **2019**, *11*, 946. [CrossRef]

125. Shakibaei, M.; Buhrmann, C.; Mobasheri, A. Resveratrol-mediated SIRT-1 interactions with p300 modulate receptor activator of NF-kappaB ligand (RANKL) activation of NF-kappaB signaling and inhibit osteoclastogenesis in bone-derived cells. *J. Biol. Chem.* **2011**, *286*, 11492–11505. [CrossRef]

126. Price, N.L.; Gomes, A.P.; Ling, A.J.; Duarte, F.V.; Martin-Montalvo, A.; North, B.J.; Agarwal, B.; Ye, L.; Ramadori, G.; Teodoro, J.S.; et al. SIRT1 is required for AMPK activation and the beneficial effects of resveratrol on mitochondrial function. *Cell Metab.* **2012**, *15*, 675–690. [CrossRef]

127. Theodotou, M.; Fokianos, K.; Mouzouridou, A.; Konstantinou, C.; Aristotelous, A.; Prodromou, D.; Chrysikou, A. The effect of resveratrol on hypertension: A clinical trial. *Exp. Ther. Med.* **2017**, *13*, 295–301. [CrossRef]

128. Chen, J.; Cao, X.; Cui, Y.; Zeng, G.; Chen, J.; Zhang, G. Resveratrol alleviates lysophosphatidylcholine-induced damage and inflammation in vascular endothelial cells. *Mol. Med. Rep.* **2018**, *17*, 4011–4018. [CrossRef]

129. Zhang, Z.; Chen, N.; Liu, J.B.; Wu, J.B.; Zhang, J.; Zhang, Y.; Jiang, X. Protective effect of resveratrol against acute lung injury induced by lipopolysaccharide via inhibiting the myd88-dependent Toll-like receptor 4 signaling pathway. *Mol. Med. Rep.* **2014**, *10*, 101–106. [CrossRef]

130. Liu, C.W.; Sung, H.C.; Lin, S.R.; Wu, C.W.; Lee, C.W.; Lee, I.T.; Yang, Y.F.; Yu, I.S.; Lin, S.W.; Chiang, M.H.; et al. Resveratrol attenuates ICAM-1 expression and monocyte adhesiveness to TNF-alpha-treated endothelial cells: Evidence for an anti-inflammatory cascade mediated by the miR-221/222/AMPK/p38/NF-kappaB pathway. *Sci. Rep.* **2017**, *7*, 44689. [CrossRef]

131. Lin, J.W.; Yang, L.H.; Ren, Z.C.; Mu, D.G.; Li, Y.Q.; Yan, J.P.; Wang, L.X.; Chen, C. Resveratrol downregulates TNF-alpha-induced monocyte chemoattractant protein-1 in primary rat pulmonary artery endothelial cells by P38 mitogen-activated protein kinase signaling. *Drug Des. Dev. Ther.* **2019**, *13*, 1843–1853. [CrossRef]

132. Pan, W.; Yu, H.; Huang, S.; Zhu, P. Resveratrol Protects against TNF-alpha-Induced Injury in Human Umbilical Endothelial Cells through Promoting Sirtuin-1-Induced Repression of NF-KB and p38 MAPK. *PLoS ONE* **2016**, *11*, e0147034. [CrossRef]

133. Chen, M.L.; Yi, L.; Jin, X.; Liang, X.Y.; Zhou, Y.; Zhang, T.; Xie, Q.; Zhou, X.; Chang, H.; Fu, Y.J.; et al. Resveratrol attenuates vascular endothelial inflammation by inducing autophagy through the cAMP signaling pathway. *Autophagy* **2013**, *9*, 2033–2045. [CrossRef]

134. Park, D.W.; Baek, K.; Kim, J.R.; Lee, J.J.; Ryu, S.H.; Chin, B.R.; Baek, S.H. Resveratrol inhibits foam cell formation via NADPH oxidase 1- mediated reactive oxygen species and monocyte chemotactic protein-1. *Exp. Mol. Med.* **2009**, *41*, 171–179. [CrossRef]

135. Yang, Y.; Li, S.; Yang, Q.; Shi, Y.; Zheng, M.; Liu, Y.; Chen, F.; Song, G.; Xu, H.; Wan, T.; et al. Resveratrol reduces the proinflammatory effects and lipopolysaccharide- induced expression of HMGB1 and TLR4 in RAW264.7 cells. *Cell. Phys. Biochem. Int. J. Exp. Cell. Physiol. Biochem. Pharmacol.* **2014**, *33*, 1283–1292. [CrossRef]

136. Jakus, P.B.; Kalman, N.; Antus, C.; Radnai, B.; Tucsek, Z.; Gallyas, F., Jr.; Sumegi, B.; Veres, B. TRAF6 is functional in inhibition of TLR4-mediated NF-kappaB activation by resveratrol. *J. Nutr. Biochem.* **2013**, *24*, 819–823. [CrossRef]

137. Inanaga, K.; Ichiki, T.; Matsuura, H.; Miyazaki, R.; Hashimoto, T.; Takeda, K.; Sunagawa, K. Resveratrol attenuates angiotensin II-induced interleukin-6 expression and perivascular fibrosis. *Hypertens. Res.* **2009**, *32*, 466–471. [CrossRef]

138. Zhang, J.; Chen, J.; Yang, J.; Xu, C.W.; Pu, P.; Ding, J.W.; Jiang, H. Resveratrol attenuates oxidative stress induced by balloon injury in the rat carotid artery through actions on the ERK1/2 and NF-kappa B pathway. *Cell. Phys. Biochem. Int. J. Exp. Cell. Physiol. Biochem. Pharmacol.* **2013**, *31*, 230–241. [CrossRef]

139. Deng, Z.Y.; Hu, M.M.; Xin, Y.F.; Gang, C. Resveratrol alleviates vascular inflammatory injury by inhibiting inflammasome activation in rats with hypercholesterolemia and vitamin D2 treatment. *Inflamm. Res.* **2015**, *64*, 321–332. [CrossRef]

140. Chang, G.R.; Chen, P.L.; Hou, P.H.; Mao, F.C. Resveratrol protects against diet-induced atherosclerosis by reducing low-density lipoprotein cholesterol and inhibiting inflammation in apolipoprotein E-deficient mice. *Iran. J. Basic Med. Sci.* **2015**, *18*, 1063–1071.

141. Cong, X.; Li, Y.; Lu, N.; Dai, Y.; Zhang, H.; Zhao, X.; Liu, Y. Resveratrol attenuates the inflammatory reaction induced by ischemia/reperfusion in the rat heart. *Mol. Med. Rep.* **2014**, *9*, 2528–2532. [CrossRef]

142. Li, J.; Xie, C.; Zhuang, J.; Li, H.; Yao, Y.; Shao, C.; Wang, H. Resveratrol attenuates inflammation in the rat heart subjected to ischemia-reperfusion: Role of the TLR4/NF-kappaB signaling pathway. *Mol. Med. Rep.* **2015**, *11*, 1120–1126. [CrossRef] [PubMed]

143. Espinoza, J.L.; Trung, L.Q.; Inaoka, P.T.; Yamada, K.; An, D.T.; Mizuno, S.; Nakao, S.; Takami, A. The Repeated Administration of Resveratrol Has Measurable Effects on Circulating T-Cell Subsets in Humans. *Oxid. Med. Cell. Longev.* **2017**, *2017*, 6781872. [CrossRef] [PubMed]

144. Yoshino, J.; Conte, C.; Fontana, L.; Mittendorfer, B.; Imai, S.; Schechtman, K.B.; Gu, C.; Kunz, I.; Rossi Fanelli, F.; Patterson, B.W.; et al. Resveratrol supplementation does not improve metabolic function in nonobese women with normal glucose tolerance. *Cell Metab.* **2012**, *16*, 658–664. [CrossRef] [PubMed]

145. Ramirez-Garza, S.L.; Laveriano-Santos, E.P.; Marhuenda-Munoz, M.; Storniolo, C.E.; Tresserra-Rimbau, A.; Vallverdu-Queralt, A.; Lamuela-Raventos, R.M. Health Effects of Resveratrol: Results from Human Intervention Trials. *Nutrients* **2018**, *10*, 1892. [CrossRef] [PubMed]

146. Tili, E.; Michaille, J.J.; Adair, B.; Alder, H.; Limagne, E.; Taccioli, C.; Ferracin, M.; Delmas, D.; Latruffe, N.; Croce, C.M. Resveratrol decreases the levels of miR-155 by upregulating miR-663, a microRNA targeting JunB and JunD. *Carcinogenesis* **2010**, *31*, 1561–1566. [CrossRef] [PubMed]

147. Michaille, J.J.; Piurowski, V.; Rigot, B.; Kelani, H.; Fortman, E.C.; Tili, E. MiR-663, a MicroRNA Linked with Inflammation and Cancer That Is under the Influence of Resveratrol. *Medicines* **2018**, *5*, 74. [CrossRef]

148. Bigagli, E.; Cinci, L.; Paccosi, S.; Parenti, A.; D'Ambrosio, M.; Luceri, C. Nutritionally relevant concentrations of resveratrol and hydroxytyrosol mitigate oxidative burst of human granulocytes and monocytes and the production of pro-inflammatory mediators in LPS-stimulated RAW 264.7 macrophages. *Int. Immunopharmacol.* **2017**, *43*, 147–155. [CrossRef]

149. Lancon, A.; Kaminski, J.; Tili, E.; Michaille, J.J.; Latruffe, N. Control of MicroRNA expression as a new way for resveratrol to deliver its beneficial effects. *J. Agric. Food Chem.* **2012**, *60*, 8783–8789. [CrossRef]

150. Chen, J.F.; Mandel, E.M.; Thomson, J.M.; Wu, Q.; Callis, T.E.; Hammond, S.M.; Conlon, F.L.; Wang, D.Z. The role of microRNA-1 and microRNA-133 in skeletal muscle proliferation and differentiation. *Nat. Genet.* **2006**, *38*, 228–233. [CrossRef]

151. Zhang, Y.; Lu, Y.; Ong'achwa, M.J.; Ge, L.; Qian, Y.; Chen, L.; Hu, X.; Li, F.; Wei, H.; Zhang, C.; et al. Resveratrol Inhibits the TGF-beta1-Induced Proliferation of Cardiac Fibroblasts and Collagen Secretion by Downregulating miR-17 in Rat. *BioMed Res. Int.* **2018**, *2018*, 8730593. [CrossRef]

152. Mukhopadhyay, P.; Mukherjee, S.; Ahsan, K.; Bagchi, A.; Pacher, P.; Das, D.K. Restoration of altered microRNA expression in the ischemic heart with resveratrol. *PLoS ONE* **2010**, *5*, e15705. [CrossRef]

153. Campagnolo, P.; Hong, X.; di Bernardini, E.; Smyrnias, I.; Hu, Y.; Xu, Q. Resveratrol-Induced Vascular Progenitor Differentiation towards Endothelial Lineage via MiR-21/Akt/beta-Catenin Is Protective in Vessel Graft Models. *PLoS ONE* **2015**, *10*, e0125122. [CrossRef]

154. Tome-Carneiro, J.; Larrosa, M.; Yanez-Gascon, M.J.; Davalos, A.; Gil-Zamorano, J.; Gonzalvez, M.; Garcia-Almagro, F.J.; Ruiz Ros, J.A.; Tomas-Barberan, F.A.; Espin, J.C.; et al. One-year supplementation with a grape extract containing resveratrol modulates inflammatory-related microRNAs and cytokines expression in peripheral blood mononuclear cells of type 2 diabetes and hypertensive patients with coronary artery disease. *Pharmacol. Res.* **2013**, *72*, 69–82. [CrossRef]

155. Cione, E.; La Torre, C.; Cannataro, R.; Caroleo, M.C.; Plastina, P.; Gallelli, L. Quercetin, Epigallocatechin Gallate, Curcumin, and Resveratrol: From Dietary Sources to Human MicroRNA Modulation. *Molecules* **2019**, *25*, 63. [CrossRef]

156. Nie, J.; Zhang, L.; Zhao, G.; Du, X. Quercetin reduces atherosclerotic lesions by altering the gut microbiota and reducing atherogenic lipid metabolites. *J. Appl. Microbiol.* **2019**, *127*, 1824–1834. [CrossRef]

157. Rasouli, H.; Farzaei, M.H.; Khodarahmi, R. Polyphenols and their benefits: A review. *Int. J. Food Prop.* **2017**, *20*, 1700–1741. [CrossRef]

158. Basu, A.; Das, A.S.; Majumder, M.; Mukhopadhyay, R. Antiatherogenic Roles of Dietary Flavonoids Chrysin, Quercetin, and Luteolin. *J. Cardiovasc. Pharmacol.* **2016**, *68*, 89–96. [CrossRef]

159. Kawai, Y.; Nishikawa, T.; Shiba, Y.; Saito, S.; Murota, K.; Shibata, N.; Kobayashi, M.; Kanayama, M.; Uchida, K.; Terao, J. Macrophage as a target of quercetin glucuronides in human atherosclerotic arteries: Implication in the anti-atherosclerotic mechanism of dietary flavonoids. *J. Biol. Chem.* **2008**, *283*, 9424–9434. [CrossRef]

160. Sanchez, M.; Galisteo, M.; Vera, R.; Villar, I.C.; Zarzuelo, A.; Tamargo, J.; Perez-Vizcaino, F.; Duarte, J. Quercetin downregulates NADPH oxidase, increases eNOS activity and prevents endothelial dysfunction in spontaneously hypertensive rats. *J. Hypertens.* **2006**, *24*, 75–84. [CrossRef]

161. Carlstrom, J.; Symons, J.D.; Wu, T.C.; Bruno, R.S.; Litwin, S.E.; Jalili, T. A quercetin supplemented diet does not prevent cardiovascular complications in spontaneously hypertensive rats. *J. Nutr.* **2007**, *137*, 628–633. [CrossRef] [PubMed]

162. Duarte, J.; Galisteo, M.; Ocete, M.A.; Perez-Vizcaino, F.; Zarzuelo, A.; Tamargo, J. Effects of chronic quercetin treatment on hepatic oxidative status of spontaneously hypertensive rats. *Mol. Cell. Biochem.* **2001**, *221*, 155–160. [CrossRef]

163. Min, Z.; Yangchun, L.; Yuquan, W.; Changying, Z. Quercetin inhibition of myocardial fibrosis through regulating MAPK signaling pathway via ROS. *Pak. J. Pharm. Sci.* **2019**, *32*, 1355–1359.

164. Li, P.G.; Sun, L.; Han, X.; Ling, S.; Gan, W.T.; Xu, J.W. Quercetin induces rapid eNOS phosphorylation and vasodilation by an Akt-independent and PKA-dependent mechanism. *Pharmacology* **2012**, *89*, 220–228. [CrossRef] [PubMed]

165. Bhaskar, S.; Sudhakaran, P.R.; Helen, A. Quercetin attenuates atherosclerotic inflammation and adhesion molecule expression by modulating TLR-NF-kappaB signaling pathway. *Cell. Immunol.* **2016**, *310*, 131–140. [CrossRef] [PubMed]

166. Hung, C.H.; Chan, S.H.; Chu, P.M.; Tsai, K.L. Quercetin is a potent anti-atherosclerotic compound by activation of SIRT1 signaling under oxLDL stimulation. *Mol. Nutr. Food Res.* **2015**, *59*, 1905–1917. [CrossRef] [PubMed]

167. Xiao, H.B.; Lu, X.Y.; Sun, Z.L.; Zhang, H.B. Kaempferol regulates OPN-CD44 pathway to inhibit the atherogenesis of apolipoprotein E deficient mice. *Toxicol. Appl. Pharmacol.* **2011**, *257*, 405–411. [CrossRef]

168. Che, J.; Liang, B.; Zhang, Y.; Wang, Y.; Tang, J.; Shi, G. Kaempferol alleviates ox-LDL-induced apoptosis by up-regulation of autophagy via inhibiting PI3K/Akt/mTOR pathway in human endothelial cells. *Cardiovasc. Pathol.* **2017**, *31*, 57–62. [CrossRef]

169. Kleemann, R.; Verschuren, L.; Morrison, M.; Zadelaar, S.; van Erk, M.J.; Wielinga, P.Y.; Kooistra, T. Anti-inflammatory, anti-proliferative and anti-atherosclerotic effects of quercetin in human in vitro and in vivo models. *Atherosclerosis* **2011**, *218*, 44–52. [CrossRef]

170. Bhaskar, S.; Helen, A. Quercetin modulates toll-like receptor-mediated protein kinase signaling pathways in oxLDL-challenged human PBMCs and regulates TLR-activated atherosclerotic inflammation in hypercholesterolemic rats. *Mol. Cell. Biochem.* **2016**, *423*, 53–65. [CrossRef]
171. Calabriso, N.; Scoditti, E.; Massaro, M.; Pellegrino, M.; Storelli, C.; Ingrosso, I.; Giovinazzo, G.; Carluccio, M.A. Multiple anti-inflammatory and anti-atherosclerotic properties of red wine polyphenolic extracts: Differential role of hydroxycinnamic acids, flavonols and stilbenes on endothelial inflammatory gene expression. *Eur. J. Nutr.* **2016**, *55*, 477–489. [CrossRef]
172. Vera, M.; Torramade-Moix, S.; Martin-Rodriguez, S.; Cases, A.; Cruzado, J.M.; Rivera, J.; Escolar, G.; Palomo, M.; Diaz-Ricart, M. Antioxidant and Anti-Inflammatory Strategies Based on the Potentiation of Glutathione Peroxidase Activity Prevent Endothelial Dysfunction in Chronic Kidney Disease. *Cell. Phys. Biochem. Int. J. Exp. Cell. Physiol. Biochem. Pharmacol.* **2018**, *51*, 1287–1300. [CrossRef]
173. Qureshi, A.A.; Tan, X.; Reis, J.C.; Badr, M.Z.; Papasian, C.J.; Morrison, D.C.; Qureshi, N. Suppression of nitric oxide induction and pro-inflammatory cytokines by novel proteasome inhibitors in various experimental models. *Lipids Health Dis.* **2011**, *10*, 177. [CrossRef] [PubMed]
174. Lu, X.L.; Zhao, C.H.; Yao, X.L.; Zhang, H. Quercetin attenuates high fructose feeding-induced atherosclerosis by suppressing inflammation and apoptosis via ROS-regulated PI3K/AKT signaling pathway. *Biomed. Pharmacother.* **2017**, *85*, 658–671. [CrossRef] [PubMed]
175. Li, B.; Yang, M.; Liu, J.W.; Yin, G.T. Protective mechanism of quercetin on acute myocardial infarction in rats. *Genet. Mol. Res. GMR* **2016**, *15*, 15017117. [CrossRef]
176. Zahedi, M.; Ghiasvand, R.; Feizi, A.; Asgari, G.; Darvish, L. Does Quercetin Improve Cardiovascular Risk factors and Inflammatory Biomarkers in Women with Type 2 Diabetes: A Double-blind Randomized Controlled Clinical Trial. *Int. J. Prev. Med.* **2013**, *4*, 777–785. [PubMed]
177. Chekalina, N.; Burmak, Y.; Petrov, Y.; Borisova, Z.; Manusha, Y.; Kazakov, Y.; Kaidashev, I. Quercetin reduces the transcriptional activity of NF-kB in stable coronary artery disease. *Indian Heart J.* **2018**, *70*, 593–597. [CrossRef] [PubMed]
178. Dabeek, W.M.; Marra, M.V. Dietary Quercetin and Kaempferol: Bioavailability and Potential Cardiovascular-Related Bioactivity in Humans. *Nutrients* **2019**, *11*, 2288. [CrossRef] [PubMed]
179. Boesch-Saadatmandi, C.; Loboda, A.; Wagner, A.E.; Stachurska, A.; Jozkowicz, A.; Dulak, J.; Doring, F.; Wolffram, S.; Rimbach, G. Effect of quercetin and its metabolites isorhamnetin and quercetin-3-glucuronide on inflammatory gene expression: Role of miR-155. *J. Nutr. Biochem.* **2011**, *22*, 293–299. [CrossRef] [PubMed]
180. Boesch-Saadatmandi, C.; Wagner, A.E.; Wolffram, S.; Rimbach, G. Effect of quercetin on inflammatory gene expression in mice liver in vivo—Role of redox factor 1, miRNA-122 and miRNA-125b. *Pharmacol. Res.* **2012**, *65*, 523–530. [CrossRef]
181. Wang, M.; Firrman, J.; Liu, L.; Yam, K. A Review on Flavonoid Apigenin: Dietary Intake, ADME, Antimicrobial Effects, and Interactions with Human Gut Microbiota. *BioMed Res. Int.* **2019**, *2019*, 7010467. [CrossRef]
182. Tang, D.; Chen, K.; Huang, L.; Li, J. Pharmacokinetic properties and drug interactions of apigenin, a natural flavone. *Expert Opin. Drug Metab. Toxicol.* **2017**, *13*, 323–330. [CrossRef] [PubMed]
183. Zhang, K.; Song, W.; Li, D.; Jin, X. Apigenin in the regulation of cholesterol metabolism and protection of blood vessels. *Exp. Ther. Med.* **2017**, *13*, 1719–1724. [CrossRef] [PubMed]
184. Jung, U.J.; Cho, Y.Y.; Choi, M.S. Apigenin Ameliorates Dyslipidemia, Hepatic Steatosis and Insulin Resistance by Modulating Metabolic and Transcriptional Profiles in the Liver of High-Fat Diet-Induced Obese Mice. *Nutrients* **2016**, *8*, 305. [CrossRef] [PubMed]
185. Qin, W.; Ren, B.; Wang, S.; Liang, S.; He, B.; Shi, X.; Wang, L.; Liang, J.; Wu, F. Apigenin and naringenin ameliorate PKCbetaII-associated endothelial dysfunction via regulating ROS/caspase-3 and NO pathway in endothelial cells exposed to high glucose. *Vasc. Pharmacol.* **2016**, *85*, 39–49. [CrossRef] [PubMed]
186. Chen, C.C.; Ke, W.H.; Ceng, L.H.; Hsieh, C.W.; Wung, B.S. Calcium- and phosphatidylinositol 3-kinase/Akt-dependent activation of endothelial nitric oxide synthase by apigenin. *Life Scie.* **2010**, *87*, 743–749. [CrossRef]
187. Pothineni, N.V.K.; Karathanasis, S.K.; Ding, Z.; Arulandu, A.; Varughese, K.I.; Mehta, J.L. LOX-1 in Atherosclerosis and Myocardial Ischemia: Biology, Genetics, and Modulation. *J. Am. Coll. Cardiol.* **2017**, *69*, 2759–2768. [CrossRef]

188. Wang, J.; Liao, Y.; Fan, J.; Ye, T.; Sun, X.; Dong, S. Apigenin inhibits the expression of IL-6, IL-8, and ICAM-1 in DEHP-stimulated human umbilical vein endothelial cells and in vivo. *Inflammation* **2012**, *35*, 1466–1476. [CrossRef]

189. Ren, B.; Qin, W.; Wu, F.; Wang, S.; Pan, C.; Wang, L.; Zeng, B.; Ma, S.; Liang, J. Apigenin and naringenin regulate glucose and lipid metabolism, and ameliorate vascular dysfunction in type 2 diabetic rats. *Eur. J. Pharmacol.* **2016**, *773*, 13–23. [CrossRef]

190. Mueller, M.; Hobiger, S.; Jungbauer, A. Anti-inflammatory activity of extracts from fruits, herbs and spices. *Food Chem.* **2010**, *122*, 987–996. [CrossRef]

191. Comalada, M.; Ballester, I.; Bailon, E.; Sierra, S.; Xaus, J.; Galvez, J.; de Medina, F.S.; Zarzuelo, A. Inhibition of pro-inflammatory markers in primary bone marrow-derived mouse macrophages by naturally occurring flavonoids: Analysis of the structure-activity relationship. *Biochem. Pharmacol.* **2006**, *72*, 1010–1021. [CrossRef]

192. Ren, K.; Jiang, T.; Zhou, H.F.; Liang, Y.; Zhao, G.J. Apigenin Retards Atherogenesis by Promoting ABCA1-Mediated Cholesterol Efflux and Suppressing Inflammation. *Cell. Phys. Biochem. Int. J. Exp. Cell. Physiol. Biochem. Pharmacol.* **2018**, *47*, 2170–2184. [CrossRef] [PubMed]

193. Yamagata, K.; Miyashita, A.; Matsufuji, H.; Chino, M. Dietary flavonoid apigenin inhibits high glucose and tumor necrosis factor alpha-induced adhesion molecule expression in human endothelial cells. *J. Nutr. Biochem.* **2010**, *21*, 116–124. [CrossRef] [PubMed]

194. Luo, Y.; Shang, P.; Li, D. Luteolin: A Flavonoid that Has Multiple Cardio-Protective Effects and Its Molecular Mechanisms. *Front. Pharmacol.* **2017**, *8*, 692. [CrossRef]

195. Hostetler, G.L.; Ralston, R.A.; Schwartz, S.J. Flavones: Food Sources, Bioavailability, Metabolism, and Bioactivity. *Adv. Nutr.* **2017**, *8*, 423–435. [CrossRef]

196. Rangboo, V.; Noroozi, M.; Zavoshy, R.; Rezadoost, S.A.; Mohammadpoorasl, A. The Effect of Artichoke Leaf Extract on Alanine Aminotransferase and Aspartate Aminotransferase in the Patients with Nonalcoholic Steatohepatitis. *Int. J. Hepatol.* **2016**, *2016*, 4030476. [CrossRef] [PubMed]

197. Sahebkar, A.; Pirro, M.; Banach, M.; Mikhailidis, D.P.; Atkin, S.L.; Cicero, A.F.G. Lipid-lowering activity of artichoke extracts: A systematic review and meta-analysis. *Crit. Rev. Food Sci. Nutr.* **2018**, *58*, 2549–2556. [CrossRef]

198. Xia, F.; Wang, C.; Jin, Y.; Liu, Q.; Meng, Q.; Liu, K.; Sun, H. Luteolin protects HUVECs from TNF-alpha-induced oxidative stress and inflammation via its effects on the Nox4/ROS-NF-kappaB and MAPK pathways. *J. Atheroscler. Thromb.* **2014**, *21*, 768–783. [CrossRef]

199. Xiao, C.; Xia, M.L.; Wang, J.; Zhou, X.R.; Lou, Y.Y.; Tang, L.H.; Zhang, F.J.; Yang, J.T.; Qian, L.B. Luteolin Attenuates Cardiac Ischemia/Reperfusion Injury in Diabetic Rats by Modulating Nrf2 Antioxidative Function. *Oxid. Med. Cell. Longev.* **2019**, *2019*, 2719252. [CrossRef]

200. Jia, Z.; Nallasamy, P.; Liu, D.; Shah, H.; Li, J.Z.; Chitrakar, R.; Si, H.; McCormick, J.; Zhu, H.; Zhen, W.; et al. Luteolin protects against vascular inflammation in mice and TNF-alpha-induced monocyte adhesion to endothelial cells via suppressing IKappaBalpha/NF-kappaB signaling pathway. *J. Nutr. Biochem.* **2015**, *26*, 293–302. [CrossRef]

201. Zhang, L.; Wang, X.; Zhang, L.; Virgous, C.; Si, H. Combination of curcumin and luteolin synergistically inhibits TNF-alpha-induced vascular inflammation in human vascular cells and mice. *J. Nutr. Biochem.* **2019**, *73*, 108222. [CrossRef]

202. Kim, H.J.; Lee, W.; Yun, J.M. Luteolin inhibits hyperglycemia-induced proinflammatory cytokine production and its epigenetic mechanism in human monocytes. *Phytother. Res. PTR* **2014**, *28*, 1383–1391. [CrossRef] [PubMed]

203. Zaragoza, C.; Villaescusa, L.; Monserrat, J.; Zaragoza, F.; Alvarez-Mon, M. Potential Therapeutic Anti-Inflammatory and Immunomodulatory Effects of Dihydroflavones, Flavones, and Flavonols. *Molecules* **2020**, *25*, 1017. [CrossRef] [PubMed]

204. Hong, X.; Zhao, X.; Wang, G.; Zhang, Z.; Pei, H.; Liu, Z. Luteolin Treatment Protects against Renal Ischemia-Reperfusion Injury in Rats. *Mediat. Inflamm.* **2017**, *2017*, 9783893. [CrossRef] [PubMed]

205. Ding, X.; Zheng, L.; Yang, B.; Wang, X.; Ying, Y. Luteolin Attenuates Atherosclerosis Via Modulating Signal Transducer And Activator Of Transcription 3-Mediated Inflammatory Response. *Drug Des. Dev. Ther.* **2019**, *13*, 3899–3911. [CrossRef] [PubMed]

206. Wang, P.; Sun, J.; Lv, S.; Xie, T.; Wang, X. Apigenin Alleviates Myocardial Reperfusion Injury in Rats by Downregulating miR-15b. *Med. Sci. Monit. Int. Med. J. Exp. Clin. Res.* **2019**, *25*, 2764–2776. [CrossRef] [PubMed]

207. Bian, C.; Xu, T.; Zhu, H.; Pan, D.; Liu, Y.; Luo, Y.; Wu, P.; Li, D. Luteolin Inhibits Ischemia/Reperfusion-Induced Myocardial Injury in Rats via Downregulation of microRNA-208b-3p. *PLoS ONE* **2015**, *10*, e0144877. [CrossRef]

208. Arango, D.; Diosa-Toro, M.; Rojas-Hernandez, L.S.; Cooperstone, J.L.; Schwartz, S.J.; Mo, X.; Jiang, J.; Schmittgen, T.D.; Doseff, A.I. Dietary apigenin reduces LPS-induced expression of miR-155 restoring immune balance during inflammation. *Mol. Nutr. Food Res.* **2015**, *59*, 763–772. [CrossRef]

209. Zhang, Z.; Xu, P.; Yu, H.; Shi, L. Luteolin protects PC-12 cells from H2O2-induced injury by up-regulation of microRNA-21. *Biomed. Pharmacother.* **2019**, *112*, 108698. [CrossRef]

210. Ning, B.B.; Zhang, Y.; Wu, D.D.; Cui, J.G.; Liu, L.; Wang, P.W.; Wang, W.J.; Zhu, W.L.; Chen, Y.; Zhang, T. Luteolin-7-diglucuronide attenuates isoproterenol-induced myocardial injury and fibrosis in mice. *Acta Pharmacol. Sin.* **2017**, *38*, 331–341. [CrossRef]

211. Kara, S.; Gencer, B.; Karaca, T.; Tufan, H.A.; Arikan, S.; Ersan, I.; Karaboga, I.; Hanci, V. Protective effect of hesperetin and naringenin against apoptosis in ischemia/reperfusion-induced retinal injury in rats. *Sci. World J.* **2014**, *2014*, 797824. [CrossRef]

212. Kanaze, F.I.; Bounartzi, M.I.; Georgarakis, M.; Niopas, I. Pharmacokinetics of the citrus flavanone aglycones hesperetin and naringenin after single oral administration in human subjects. *Eur. J. Clin. Nutr.* **2007**, *61*, 472–477. [CrossRef]

213. Joshi, R.; Kulkarni, Y.A.; Wairkar, S. Pharmacokinetic, pharmacodynamic and formulations aspects of Naringenin: An update. *Life Sci.* **2018**, *215*, 43–56. [CrossRef] [PubMed]

214. Miceli, N.; Mondello, M.R.; Monforte, M.T.; Sdrafkakis, V.; Dugo, P.; Crupi, M.L.; Taviano, M.F.; De Pasquale, R.; Trovato, A. Hypolipidemic effects of Citrus bergamia Risso et Poiteau juice in rats fed a hypercholesterolemic diet. *J. Agric. Food Chem.* **2007**, *55*, 10671–10677. [CrossRef] [PubMed]

215. Mulvihill, E.E.; Assini, J.M.; Sutherland, B.G.; DiMattia, A.S.; Khami, M.; Koppes, J.B.; Sawyez, C.G.; Whitman, S.C.; Huff, M.W. Naringenin decreases progression of atherosclerosis by improving dyslipidemia in high-fat-fed low-density lipoprotein receptor-null mice. *Arterioscler. Thromb. Vasc. Biol.* **2010**, *30*, 742–748. [CrossRef] [PubMed]

216. Borradaile, N.M.; de Dreu, L.E.; Barrett, P.H.; Behrsin, C.D.; Huff, M.W. Hepatocyte apoB-containing lipoprotein secretion is decreased by the grapefruit flavonoid, naringenin, via inhibition of MTP-mediated microsomal triglyceride accumulation. *Biochemistry* **2003**, *42*, 1283–1291. [CrossRef] [PubMed]

217. Lee, C.H.; Jeong, T.S.; Choi, Y.K.; Hyun, B.H.; Oh, G.T.; Kim, E.H.; Kim, J.R.; Han, J.I.; Bok, S.H. Anti-atherogenic effect of citrus flavonoids, naringin and naringenin, associated with hepatic ACAT and aortic VCAM-1 and MCP-1 in high cholesterol-fed rabbits. *Biochem. Biophys. Res. Commun.* **2001**, *284*, 681–688. [CrossRef]

218. Jung, U.J.; Kim, H.J.; Lee, J.S.; Lee, J.F.; Lee, M.K.; Lee, M.F.; Kim, H.O.; Kim, H.F.; Park, E.J.; Park, E.F.; et al. Naringin supplementation lowers plasma lipids and enhances erythrocyte antioxidant enzyme activities in hypercholesterolemic subjects. *Clin. Nutr.* **2003**, *22*, 561–568. [CrossRef]

219. Perna, S.; Spadaccini, D.; Botteri, L.; Girometta, C.; Riva, A.; Allegrini, P.; Petrangolini, G.; Infantino, V.; Rondanelli, M. Efficacy of bergamot: From anti-inflammatory and anti-oxidative mechanisms to clinical applications as preventive agent for cardiovascular morbidity, skin diseases, and mood alterations. *Food Sci. Nutr.* **2019**, *7*, 369–384. [CrossRef]

220. Rizza, S.; Muniyappa, R.; Iantorno, M.; Kim, J.A.; Chen, H.; Pullikotil, P.; Senese, N.; Tesauro, M.; Lauro, D.; Cardillo, C.; et al. Citrus polyphenol hesperidin stimulates production of nitric oxide in endothelial cells while improving endothelial function and reducing inflammatory markers in patients with metabolic syndrome. *J. Clin. Endocrinol. Metab.* **2011**, *96*, E782–E792. [CrossRef]

221. Yamamoto, M.; Jokura, H.; Hashizume, K.; Ominami, H.; Shibuya, Y.; Suzuki, A.; Hase, T.; Shimotoyodome, A. Hesperidin metabolite hesperetin-7-O-glucuronide, but not hesperetin-3'-O-glucuronide, exerts hypotensive, vasodilatory, and anti-inflammatory activities. *Food Funct.* **2013**, *4*, 1346–1351. [CrossRef]

222. Rendeiro, C.; Dong, H.; Saunders, C.; Harkness, L.; Blaze, M.; Hou, Y.; Belanger, R.L.; Altieri, V.; Nunez, M.A.; Jackson, K.G.; et al. Flavanone-rich citrus beverages counteract the transient decline in postprandial endothelial function in humans: A randomised, controlled, double-masked, cross-over intervention study. *Br. J. Nutr.* **2016**, *116*, 1999–2010. [CrossRef] [PubMed]

223. Borgatti, M.; Mancini, I.; Bianchi, N.; Guerrini, A.; Lampronti, I.; Rossi, D.; Sacchetti, G.; Gambari, R. Bergamot (Citrus bergamia Risso) fruit extracts and identified components alter expression of interleukin 8 gene in cystic fibrosis bronchial epithelial cell lines. *BMC Biochem.* **2011**, *12*, 15. [CrossRef] [PubMed]

224. Xie, J.; Zhang, X.; Zhang, L. Negative regulation of inflammation by SIRT1. *Pharmacol. Res.* **2013**, *67*, 60–67. [CrossRef]

225. Testai, L.; Piragine, E.; Piano, I.; Flori, L.; Da Pozzo, E.; Miragliotta, V.; Pirone, A.; Citi, V.; Di Cesare Mannelli, L.; Brogi, S.; et al. The Citrus Flavonoid Naringenin Protects the Myocardium from Ageing-Dependent Dysfunction: Potential Role of SIRT1. *Oxid. Med. Cell. Longev.* **2020**, *2020*, 4650207. [CrossRef]

226. Miwa, Y.; Mitsuzumi, H.; Sunayama, T.; Yamada, M.; Okada, K.; Kubota, M.; Chaen, H.; Mishima, Y.; Kibata, M. Glucosyl hesperidin lowers serum triglyceride level in hypertriglyceridemic subjects through the improvement of very low-density lipoprotein metabolic abnormality. *J. Nutr. Sci. Vitaminol.* **2005**, *51*, 460–470. [CrossRef]

227. Demonty, I.; Lin, Y.; Zebregs, Y.E.; Vermeer, M.A.; van der Knaap, H.C.; Jakel, M.; Trautwein, E.A. The citrus flavonoids hesperidin and naringin do not affect serum cholesterol in moderately hypercholesterolemic men and women. *J. Nutr.* **2010**, *140*, 1615–1620. [CrossRef] [PubMed]

228. Assini, J.M.; Mulvihill, E.E.; Huff, M.W. Citrus flavonoids and lipid metabolism. *Curr. Opin. Lipidol.* **2013**, *24*, 34–40. [CrossRef] [PubMed]

229. Hsu, C.L.; Fang, S.C.; Yen, G.C. Anti-inflammatory effects of phenolic compounds isolated from the flowers of Nymphaea mexicana Zucc. *Food Funct.* **2013**, *4*, 1216–1222. [CrossRef]

230. Raza, S.S.; Khan, M.M.; Ahmad, A.; Ashafaq, M.; Islam, F.; Wagner, A.P.; Safhi, M.M.; Islam, F. Neuroprotective effect of naringenin is mediated through suppression of NF-kappaB signaling pathway in experimental stroke. *Neuroscience* **2013**, *230*, 157–171. [CrossRef]

231. Fuior, E.V.; Deleanu, M.; Constantinescu, C.A.; Rebleanu, D.; Voicu, G.; Simionescu, M.; Calin, M. Functional Role of VCAM-1 Targeted Flavonoid-Loaded Lipid Nanoemulsions in Reducing Endothelium Inflammation. *Pharmaceutics* **2019**, *11*, 391. [CrossRef]

232. Liang, J.N.; Zou, X.; Fang, X.H.; Xu, J.D.; Xiao, Z.; Zhu, J.N.; Li, H.; Yang, J.; Zeng, N.; Yuan, S.J.; et al. The Smad3-miR-29b/miR-29c axis mediates the protective effect of macrophage migration inhibitory factor against cardiac fibrosis. *(BBA)-Mol. Basis Dis.* **2019**, *1865*, 2441–2450. [CrossRef] [PubMed]

233. Murakami, A. Dose-dependent functionality and toxicity of green tea polyphenols in experimental rodents. *Arch. Biochem. Biophys.* **2014**, *557*, 3–10. [CrossRef] [PubMed]

234. Cai, Z.Y.; Li, X.M.; Liang, J.P.; Xiang, L.P.; Wang, K.R.; Shi, Y.L.; Yang, R.; Shi, M.; Ye, J.H.; Lu, J.L.; et al. Bioavailability of Tea Catechins and Its Improvement. *Molecules* **2018**, *23*, 2346. [CrossRef] [PubMed]

235. Onakpoya, I.; Spencer, E.; Heneghan, C.; Thompson, M. The effect of green tea on blood pressure and lipid profile: A systematic review and meta-analysis of randomized clinical trials. *Nutr. Metab. Cardiovasc. Dis. NMCD* **2014**, *24*, 823–836. [CrossRef] [PubMed]

236. Kim, A.; Chiu, A.; Barone, M.K.; Avino, D.; Wang, F.; Coleman, C.I.; Phung, O.J. Green tea catechins decrease total and low-density lipoprotein cholesterol: A systematic review and meta-analysis. *J. Am. Diet. Assoc.* **2011**, *111*, 1720–1729. [CrossRef] [PubMed]

237. Zheng, X.X.; Xu, Y.L.; Li, S.H.; Liu, X.X.; Hui, R.; Huang, X.H. Green tea intake lowers fasting serum total and LDL cholesterol in adults: A meta-analysis of 14 randomized controlled trials. *Am. J. Clin. Nutr.* **2011**, *94*, 601–610. [CrossRef]

238. Koo, S.I.; Noh, S.K. Green tea as inhibitor of the intestinal absorption of lipids: Potential mechanism for its lipid-lowering effect. *J. Nutr. Biochem.* **2007**, *18*, 179–183. [CrossRef]

239. Abe, I.; Seki, T.; Umehara, K.; Miyase, T.; Noguchi, H.; Sakakibara, J.; Ono, T. Green tea polyphenols: Novel and potent inhibitors of squalene epoxidase. *Biochem. Biophys. Res. Commun.* **2000**, *268*, 767–771. [CrossRef]

240. Naveed, M.; BiBi, J.; Kamboh, A.A.; Suheryani, I.; Kakar, I.; Fazlani, S.A.; FangFang, X.; Kalhoro, S.A.; Yunjuan, L.; Kakar, M.U.; et al. Pharmacological values and therapeutic properties of black tea (Camellia sinensis): A comprehensive overview. *Biomed. Pharmacother.* **2018**, *100*, 521–531. [CrossRef]

241. Santos, H.O.; Macedo, R.C.O. Cocoa-induced (Theobroma cacao) effects on cardiovascular system: HDL modulation pathways. *Clin. Nutr. ESPEN* **2018**, *27*, 10–15. [CrossRef]

242. Snoussi, C.; Ducroc, R.; Hamdaoui, M.H.; Dhaouadi, K.; Abaidi, H.; Cluzeaud, F.; Nazaret, C.; Le Gall, M.; Bado, A. Green tea decoction improves glucose tolerance and reduces weight gain of rats fed normal and high-fat diet. *J. Nutr. Biochem.* **2014**, *25*, 557–564. [CrossRef] [PubMed]

243. Shrestha, S.; Ehlers, S.J.; Lee, J.Y.; Fernandez, M.L.; Koo, S.I. Dietary green tea extract lowers plasma and hepatic triglycerides and decreases the expression of sterol regulatory element-binding protein-1c mRNA and its responsive genes in fructose-fed, ovariectomized rats. *J. Nutr.* **2009**, *139*, 640–645. [CrossRef] [PubMed]

244. Pang, J.; Zhang, Z.; Zheng, T.Z.; Bassig, B.A.; Mao, C.; Liu, X.; Zhu, Y.; Shi, K.; Ge, J.; Yang, Y.J.; et al. Green tea consumption and risk of cardiovascular and ischemic related diseases: A meta-analysis. *Int. J. Cardiol.* **2016**, *202*, 967–974. [CrossRef] [PubMed]

245. Mangels, D.R.; Mohler, E.R., 3rd. Catechins as Potential Mediators of Cardiovascular Health. *Arterioscler. Thromb. Vasc. Biol.* **2017**, *37*, 757–763. [CrossRef] [PubMed]

246. Liu, S.; Sun, Z.; Chu, P.; Li, H.; Ahsan, A.; Zhou, Z.; Zhang, Z.; Sun, B.; Wu, J.; Xi, Y.; et al. EGCG protects against homocysteine-induced human umbilical vein endothelial cells apoptosis by modulating mitochondrial-dependent apoptotic signaling and PI3K/Akt/eNOS signaling pathways. *Apoptosis* **2017**, *22*, 672–680. [CrossRef]

247. Wu, C.C.; Hsu, M.C.; Hsieh, C.W.; Lin, J.B.; Lai, P.H.; Wung, B.S. Upregulation of heme oxygenase-1 by Epigallocatechin-3-gallate via the phosphatidylinositol 3-kinase/Akt and ERK pathways. *Life Sci.* **2006**, *78*, 2889–2897. [CrossRef]

248. Suzuki-Sugihara, N.; Kishimoto, Y.; Saita, E.; Taguchi, C.; Kobayashi, M.; Ichitani, M.; Ukawa, Y.; Sagesaka, Y.M.; Suzuki, E.; Kondo, K. Green tea catechins prevent low-density lipoprotein oxidation via their accumulation in low-density lipoprotein particles in humans. *Nutr. Res.* **2016**, *36*, 16–23. [CrossRef]

249. Tas, S.; Sarandol, E.; Ziyanok, S.; Aslan, K.; Dirican, M. Effects of green tea on serum paraoxonase/arylesterase activities in streptozotocin-induced diabetic rats. *Nutr. Res.* **2005**, *25*, 1061–1074. [CrossRef]

250. Ludwig, A.; Lorenz, M.; Grimbo, N.; Steinle, F.; Meiners, S.; Bartsch, C.; Stangl, K.; Baumann, G.; Stangl, V. The tea flavonoid epigallocatechin-3-gallate reduces cytokine-induced VCAM-1 expression and monocyte adhesion to endothelial cells. *Biochem. Biophys. Res. Commun.* **2004**, *316*, 659–665. [CrossRef]

251. Arola-Arnal, A.; Blade, C. Proanthocyanidins modulate microRNA expression in human HepG2 cells. *PLoS ONE* **2011**, *6*, e25982. [CrossRef]

252. Baselga-Escudero, L.; Blade, C.; Ribas-Latre, A.; Casanova, E.; Suarez, M.; Torres, J.L.; Salvado, M.J.; Arola, L.; Arola-Arnal, A. Resveratrol and EGCG bind directly and distinctively to miR-33a and miR-122 and modulate divergently their levels in hepatic cells. *Nucleic Acids Res.* **2014**, *42*, 882–892. [CrossRef] [PubMed]

253. Wang, H.; Bian, S.; Yang, C.S. Green tea polyphenol EGCG suppresses lung cancer cell growth through upregulating miR-210 expression caused by stabilizing HIF-1alpha. *Carcinogenesis* **2011**, *32*, 1881–1889. [CrossRef]

254. Kaelin, W.G., Jr.; Ratcliffe, P.J. Oxygen sensing by metazoans: The central role of the HIF hydroxylase pathway. *Mol. Cell* **2008**, *30*, 393–402. [CrossRef] [PubMed]

255. Jaiswal, N.; Akhtar, J.; Singh, S.P.; Badruddeen; Ahsan, F. An Overview on Genistein and its Various Formulations. *Drug Res.* **2019**, *69*, 305–313. [CrossRef] [PubMed]

256. Beavers, D.P.; Beavers, K.M.; Miller, M.; Stamey, J.; Messina, M.J. Exposure to isoflavone-containing soy products and endothelial function: A Bayesian meta-analysis of randomized controlled trials. *Nutr. Metab. Cardiovasc. Dis. NMCD* **2012**, *22*, 182–191. [CrossRef]

257. Yu, D.; Shu, X.O.; Li, H.; Yang, G.; Cai, Q.; Xiang, Y.B.; Ji, B.T.; Franke, A.A.; Gao, Y.T.; Zheng, W.; et al. Dietary isoflavones, urinary isoflavonoids, and risk of ischemic stroke in women. *Am. J. Clin. Nutr.* **2015**, *102*, 680–686. [CrossRef]

258. Si, H.; Liu, D. Genistein, a soy phytoestrogen, upregulates the expression of human endothelial nitric oxide synthase and lowers blood pressure in spontaneously hypertensive rats. *J. Nutr.* **2008**, *138*, 297–304. [CrossRef]

259. Zhang, H.P.; Zhao, J.H.; Yu, H.X.; Guo, D.X. Genistein ameliorated endothelial nitric oxidase synthase uncoupling by stimulating sirtuin-1 pathway in ox-LDL-injured HUVECs. *Environ. Toxicol. Pharmacol.* **2016**, *42*, 118–124. [CrossRef]

260. Ras, R.T.; Streppel, M.T.; Draijer, R.; Zock, P.L. Flow-mediated dilation and cardiovascular risk prediction: A systematic review with meta-analysis. *Int. J. Cardiol.* **2013**, *168*, 344–351. [CrossRef]

261. Hazim, S.; Curtis, P.J.; Schar, M.Y.; Ostertag, L.M.; Kay, C.D.; Minihane, A.M.; Cassidy, A. Acute benefits of the microbial-derived isoflavone metabolite equol on arterial stiffness in men prospectively recruited according to equol producer phenotype: A double-blind randomized controlled trial. *Am. J. Clin. Nutr.* **2016**, *103*, 694–702. [CrossRef]

262. Park, M.H.; Ju, J.W.; Kim, M.; Han, J.S. The protective effect of daidzein on high glucose-induced oxidative stress in human umbilical vein endothelial cells. *Z. für Naturforschung C J. Biosci.* **2016**, *71*, 21–28. [CrossRef] [PubMed]

263. Setchell, K.D.; Clerici, C.; Lephart, E.D.; Cole, S.J.; Heenan, C.; Castellani, D.; Wolfe, B.E.; Nechemias-Zimmer, L.; Brown, N.M.; Lund, T.D.; et al. S-equol, a potent ligand for estrogen receptor beta, is the exclusive enantiomeric form of the soy isoflavone metabolite produced by human intestinal bacterial flora. *Am. J. Clin. Nutr.* **2005**, *81*, 1072–1079. [CrossRef] [PubMed]

264. Rowlands, D.J.; Chapple, S.; Siow, R.C.; Mann, G.E. Equol-stimulated mitochondrial reactive oxygen species activate endothelial nitric oxide synthase and redox signaling in endothelial cells: Roles for F-actin and GPR30. *Hypertension* **2011**, *57*, 833–840. [CrossRef] [PubMed]

265. Joy, S.; Siow, R.C.; Rowlands, D.J.; Becker, M.; Wyatt, A.W.; Aaronson, P.I.; Coen, C.W.; Kallo, I.; Jacob, R.; Mann, G.E. The isoflavone Equol mediates rapid vascular relaxation: Ca2+-independent activation of endothelial nitric-oxide synthase/Hsp90 involving ERK1/2 and Akt phosphorylation in human endothelial cells. *J. Biol. Chem.* **2006**, *281*, 27335–27345. [CrossRef]

266. Babu, P.V.; Si, H.; Fu, Z.; Zhen, W.; Liu, D. Genistein prevents hyperglycemia-induced monocyte adhesion to human aortic endothelial cells through preservation of the cAMP signaling pathway and ameliorates vascular inflammation in obese diabetic mice. *J. Nutr.* **2012**, *142*, 724–730. [CrossRef]

267. Han, S.; Wu, H.; Li, W.; Gao, P. Protective effects of genistein in homocysteine-induced endothelial cell inflammatory injury. *Mol. Cell. Biochem.* **2015**, *403*, 43–49. [CrossRef]

268. Xu, L.; Liu, J.T.; Li, K.; Wang, S.Y.; Xu, S. Genistein inhibits Ang II-induced CRP and MMP-9 generations via the ER-p38/ERK1/2-PPARgamma-NF-kappaB signaling pathway in rat vascular smooth muscle cells. *Life Sci.* **2019**, *216*, 140–146. [CrossRef]

269. Zhang, H.; Zhao, Z.; Pang, X.; Yang, J.; Yu, H.; Zhang, Y.; Zhou, H.; Zhao, J. Genistein Protects Against Ox-LDL-Induced Inflammation Through MicroRNA-155/SOCS1-Mediated Repression of NF-kB Signaling Pathway in HUVECs. *Inflammation* **2017**, *40*, 1450–1459. [CrossRef]

270. Gan, M.; Zheng, T.; Shen, L.; Tan, Y.; Fan, Y.; Shuai, S.; Bai, L.; Li, X.; Wang, J.; Zhang, S.; et al. Genistein reverses isoproterenol-induced cardiac hypertrophy by regulating miR-451/TIMP2. *Biomed. Pharmacother.* **2019**, *112*, 108618. [CrossRef]

271. Krga, I.; Milenkovic, D. Anthocyanins: From Sources and Bioavailability to Cardiovascular-Health Benefits and Molecular Mechanisms of Action. *J. Agric. Food Chem.* **2019**, *67*, 1771–1783. [CrossRef]

272. Eker, M.E.; Aaby, K.; Budic-Leto, I.; Brncic, S.R.; El, S.N.; Karakaya, S.; Simsek, S.; Manach, C.; Wiczkowski, W.; Pascual-Teresa, S. A Review of Factors Affecting Anthocyanin Bioavailability: Possible Implications for the Inter-Individual Variability. *Foods* **2019**, *9*, 2. [CrossRef] [PubMed]

273. Hollands, W.J.; Armah, C.N.; Doleman, J.F.; Perez-Moral, N.; Winterbone, M.S.; Kroon, P.A. 4-Week consumption of anthocyanin-rich blood orange juice does not affect LDL-cholesterol or other biomarkers of CVD risk and glycaemia compared with standard orange juice: A randomised controlled trial. *Br. J. Nutr.* **2018**, *119*, 415–421. [CrossRef] [PubMed]

274. Niculescu, L.S.; Sanda, G.M.; Simionescu, N.; Sima, A.V. Bilberries exert an anti-atherosclerotic effect in lipid-loaded macrophages. *Cent. Eur. J. Biol.* **2014**, *9*, 268–276. [CrossRef]

275. Farrell, N.; Norris, G.; Lee, S.G.; Chun, O.K.; Blesso, C.N. Anthocyanin-rich black elderberry extract improves markers of HDL function and reduces aortic cholesterol in hyperlipidemic mice. *Food Funct.* **2015**, *6*, 1278–1287. [CrossRef] [PubMed]

276. Wallace, T.C.; Slavin, M.; Frankenfeld, C.L. Systematic Review of Anthocyanins and Markers of Cardiovascular Disease. *Nutrients* **2016**, *8*, 32. [CrossRef] [PubMed]

277. Sorrenti, V.; Mazza, F.; Campisi, A.; Di Giacomo, C.; Acquaviva, R.; Vanella, L.; Galvano, F. Heme oxygenase induction by cyanidin-3-O-beta-glucoside in cultured human endothelial cells. *Mol. Nutr. Food Res.* **2007**, *51*, 580–586. [CrossRef]

278. Xu, J.W.; Ikeda, K.; Yamori, Y. Upregulation of endothelial nitric oxide synthase by cyanidin-3-glucoside, a typical anthocyanin pigment. *Hypertension* **2004**, *44*, 217–222. [CrossRef]

279. Kennedy, M.A.; Barrera, G.C.; Nakamura, K.; Baldan, A.; Tarr, P.; Fishbein, M.C.; Frank, J.; Francone, O.L.; Edwards, P.A. ABCG1 has a critical role in mediating cholesterol efflux to HDL and preventing cellular lipid accumulation. *Cell Metab.* **2005**, *1*, 121–131. [CrossRef]

280. Terasaka, N.; Yu, S.; Yvan-Charvet, L.; Wang, N.; Mzhavia, N.; Langlois, R.; Pagler, T.; Li, R.; Welch, C.L.; Goldberg, I.J.; et al. ABCG1 and HDL protect against endothelial dysfunction in mice fed a high-cholesterol diet. *J. Clin. Investig.* **2008**, *118*, 3701–3713. [CrossRef]

281. Huang, W.; Zhu, Y.; Li, C.; Sui, Z.; Min, W. Effect of Blueberry Anthocyanins Malvidin and Glycosides on the Antioxidant Properties in Endothelial Cells. *Oxid. Med. Cell. Longev.* **2016**, *2016*, 1591803. [CrossRef]

282. Paixao, J.; Dinis, T.C.; Almeida, L.M. Malvidin-3-glucoside protects endothelial cells up-regulating endothelial NO synthase and inhibiting peroxynitrite-induced NF-kB activation. *Chem.-Biol. Interact.* **2012**, *199*, 192–200. [CrossRef] [PubMed]

283. Huang, T.W.; Chang, C.L.; Kao, E.S.; Lin, J.H. Effect of Hibiscus sabdariffa extract on high fat diet-induced obesity and liver damage in hamsters. *Food Nutr. Res.* **2015**, *59*, 29018. [CrossRef] [PubMed]

284. Davinelli, S.; Bertoglio, J.C.; Zarrelli, A.; Pina, R.; Scapagnini, G. A Randomized Clinical Trial Evaluating the Efficacy of an Anthocyanin-Maqui Berry Extract (Delphinol(R)) on Oxidative Stress Biomarkers. *J. Am. Coll. Nutr.* **2015**, *34* (Suppl. S1), 28–33. [CrossRef] [PubMed]

285. Jennings, A.; Welch, A.A.; Fairweather-Tait, S.J.; Kay, C.; Minihane, A.M.; Chowienczyk, P.; Jiang, B.; Cecelja, M.; Spector, T.; Macgregor, A.; et al. Higher anthocyanin intake is associated with lower arterial stiffness and central blood pressure in women. *Am. J. Clin. Nutr.* **2012**, *96*, 781–788. [CrossRef]

286. Chen, C.Y.; Yi, L.; Jin, X.; Zhang, T.; Fu, Y.J.; Zhu, J.D.; Mi, M.T.; Zhang, Q.Y.; Ling, W.H.; Yu, B. Inhibitory effect of delphinidin on monocyte-endothelial cell adhesion induced by oxidized low-density lipoprotein via ROS/p38MAPK/NF-kappaB pathway. *Cell Biochem. Biophys.* **2011**, *61*, 337–348. [CrossRef]

287. Lee, Y.; Lee, J.Y. Blackcurrant (Ribes nigrum) Extract Exerts an Anti-Inflammatory Action by Modulating Macrophage Phenotypes. *Nutrients* **2019**, *11*, 975. [CrossRef] [PubMed]

288. Rodriguez-Mateos, A.; Istas, G.; Boschek, L.; Feliciano, R.P.; Mills, C.E.; Boby, C.; Gomez-Alonso, S.; Milenkovic, D.; Heiss, C. Circulating Anthocyanin Metabolites Mediate Vascular Benefits of Blueberries: Insights From Randomized Controlled Trials, Metabolomics, and Nutrigenomics. *J. Gerontol. Ser. A* **2019**, *74*, 967–976. [CrossRef] [PubMed]

289. Marques, C.; Fernandes, I.; Norberto, S.; Sa, C.; Teixeira, D.; de Freitas, V.; Mateus, N.; Calhau, C.; Faria, A. Pharmacokinetics of blackberry anthocyanins consumed with or without ethanol: A randomized and crossover trial. *Mol. Nutr. Food Res.* **2016**, *60*, 2319–2330. [CrossRef] [PubMed]

290. Singletary, K. Ginger: An Overview of Health Benefits. *Nutr. Today* **2010**, *45*, 171–183. [CrossRef]

291. Arablou, T.; Aryaeian, N.; Valizadeh, M.; Sharifi, F.; Hosseini, A.; Djalali, M. The effect of ginger consumption on glycemic status, lipid profile and some inflammatory markers in patients with type 2 diabetes mellitus. *Int. J. Food Sci. Nutr.* **2014**, *65*, 515–520. [CrossRef]

292. Santos Braga, S. Ginger: Panacea or Consumer's Hype? *Appl. Sci.* **2019**, *9*, 1570. [CrossRef]

293. Gao, H.; Guan, T.; Li, C.; Zuo, G.; Yamahara, J.; Wang, J.; Li, Y. Treatment with ginger ameliorates fructose-induced Fatty liver and hypertriglyceridemia in rats: Modulation of the hepatic carbohydrate response element-binding protein-mediated pathway. *Evid.-Based Complement. Altern. Med. eCAM* **2012**, *2012*, 570948. [CrossRef] [PubMed]

294. Carnuta, M.G.; Deleanu, M.; Barbalata, T.; Toma, L.; Raileanu, M.; Sima, A.V.; Stancu, C.S. Zingiber officinale extract administration diminishes steroyl-CoA desaturase gene expression and activity in hyperlipidemic hamster liver by reducing the oxidative and endoplasmic reticulum stress. *Phytomedicine* **2018**, *48*, 62–69. [CrossRef] [PubMed]

295. Lei, L.; Liu, Y.; Wang, X.; Jiao, R.; Ma, K.Y.; Li, Y.M.; Wang, L.; Man, S.W.; Sang, S.; Huang, Y.; et al. Plasma cholesterol-lowering activity of gingerol- and shogaol-enriched extract is mediated by increasing sterol excretion. *J. Agric. Food Chem.* **2014**, *62*, 10515–10521. [CrossRef] [PubMed]

296. Misawa, K.; Hashizume, K.; Yamamoto, M.; Minegishi, Y.; Hase, T.; Shimotoyodome, A. Ginger extract prevents high-fat diet-induced obesity in mice via activation of the peroxisome proliferator-activated receptor delta pathway. *J. Nutr. Biochem.* **2015**, *26*, 1058–1067. [CrossRef]

297. Semwal, R.B.; Semwal, D.K.; Combrinck, S.; Viljoen, A.M. Gingerols and shogaols: Important nutraceutical principles from ginger. *Phytochemistry* **2015**, *117*, 554–568. [CrossRef]

298. Toma, L.; Raileanu, M.; Deleanu, M.; Stancu, C.S.; Sima, A.V. Novel molecular mechanisms by which ginger extract reduces the inflammatory stress in TNFα—Activated human endothelial cells; decrease of Ninjurin-1, TNFR1 and NADPH oxidase subunits expression. *J. Funct. Foods* **2018**, *48*, 654–664. [CrossRef]

299. Wang, Y.K.; Hong, Y.J.; Yao, Y.H.; Huang, X.M.; Liu, X.B.; Zhang, C.Y.; Zhang, L.; Xu, X.L. 6-Shogaol Protects against Oxidized LDL-Induced Endothelial Injruries by Inhibiting Oxidized LDL-Evoked LOX-1 Signaling. *Evid.-Based Complement. Altern. Med. eCAM* **2013**, *2013*, 503521. [CrossRef]

300. Lee, T.Y.; Lee, K.C.; Chen, S.Y.; Chang, H.H. 6-Gingerol inhibits ROS and iNOS through the suppression of PKC-alpha and NF-kappaB pathways in lipopolysaccharide-stimulated mouse macrophages. *Biochem. Biophys. Res. Commun.* **2009**, *382*, 134–139. [CrossRef]

301. Ho, S.C.; Chang, Y.H. Comparison of Inhibitory Capacities of 6-, 8- and 10-Gingerols/Shogaols on the Canonical NLRP3 Inflammasome-Mediated IL-1beta Secretion. *Molecules* **2018**, *23*, 466. [CrossRef]

302. Wang, S.; Tian, M.; Yang, R.; Jing, Y.; Chen, W.; Wang, J.; Zheng, X.; Wang, F. 6-Gingerol Ameliorates Behavioral Changes and Atherosclerotic Lesions in ApoE(-/-) Mice Exposed to Chronic Mild Stress. *Cardiovasc. Toxicol.* **2018**, *18*, 420–430. [CrossRef]

303. Li, J.; Wang, S.; Yao, L.; Ma, P.; Chen, Z.; Han, T.L.; Yuan, C.; Zhang, J.; Jiang, L.; Liu, L.; et al. 6-gingerol ameliorates age-related hepatic steatosis: Association with regulating lipogenesis, fatty acid oxidation, oxidative stress and mitochondrial dysfunction. *Toxicol. Appl. Pharmacol.* **2019**, *362*, 125–135. [CrossRef] [PubMed]

304. Kim, S.; Lee, M.S.; Jung, S.; Son, H.Y.; Park, S.; Kang, B.; Kim, S.Y.; Kim, I.H.; Kim, C.T.; Kim, Y. Ginger Extract Ameliorates Obesity and Inflammation via Regulating MicroRNA-21/132 Expression and AMPK Activation in White Adipose Tissue. *Nutrients* **2018**, *10*, 1567. [CrossRef]

305. Nelson, K.M.; Dahlin, J.L.; Bisson, J.; Graham, J.; Pauli, G.F.; Walters, M.A. The Essential Medicinal Chemistry of Curcumin. *J. Med. Chem.* **2017**, *60*, 1620–1637. [CrossRef] [PubMed]

306. Baell, J.; Walters, M.A. Chemistry: Chemical con artists foil drug discovery. *Nature* **2014**, *513*, 481–483. [CrossRef] [PubMed]

Article

Green Tea Catechin (-)-Epigallocatechin-3-Gallate (EGCG) Facilitates Fracture Healing

Sung-Yen Lin [1,2,3,4,5,6], Jung Yu Kan [6,7], Cheng-Chang Lu [1,2,3,5,8], Han Hsiang Huang [9], Tsung-Lin Cheng [1,5,10], Hsuan-Ti Huang [1,2,3,4], Cheng-Jung Ho [1,2,3], Tien-Ching Lee [1,2,3,4,5], Shu-Chun Chuang [1,5], Yi-Shan Lin [1,5], Lin Kang [11,*] and Chung-Hwan Chen [1,2,3,4,5,12,*]

[1]   Orthopaedic Research Center, Kaohsiung Medical University, Kaohsiung 80708, Taiwan; tony8501031@gmail.com (S.-Y.L.); cclu0880330@gmail.com (C.-C.L.); junglecc@gmail.com (T.-L.C.); hthuang@kmu.edu.tw (H.-T.H.); rick_free@mail2000.com.tw (C.-J.H.); tn916943@gmail.com (T.-C.L.); hawayana@gmail.com (S.-C.C.); 327lin@gmail.com (Y.-S.L.)

[2]   Department of Orthopedics, Kaohsiung Medical University Hospital, Kaohsiung Medical University, Kaohsiung 80708, Taiwan

[3]   Departments of Orthopedics, College of Medicine, Kaohsiung Medical University, Kaohsiung 80708, Taiwan

[4]   Department of Orthopedics, Kaohsiung Municipal Ta-Tung Hospital, Kaohsiung City 80145, Taiwan

[5]   Regeneration Medicine and Cell Therapy Research Center, Kaohsiung Medical University, Kaohsiung 80708, Taiwan

[6]   Graduate Institute of Medicine, College of Medicine, Kaohsiung Medical University, Kaohsiung 80708, Taiwan; kan890043@gmail.com

[7]   Division of Breast Surgery, Department of Surgery, Kaohsiung Medical University Hospital, Kaohsiung 80708, Taiwan

[8]   Department of Orthopedics, Kaohsiung Municipal Siaogang Hospital, Kaohsiung Medical University, Kaohsiung 80708, Taiwan

[9]   Department of Veterinary Medicine, National Chiayi University, Chiayi City 60054, Taiwan; hhuang@mail.ncyu.edu.tw

[10] Department of Physiology, College of Medicine, Kaohsiung Medical University, Kaohsiung 80708, Taiwan

[11] Department of Obstetrics and Gynecology, National Cheng Kung University Hospital, College of Medicine, National Cheng Kung University, Tainan 70101, Taiwan

[12] Institute of Medical Science and Technology, National Sun Yat-Sen University, Kaohsiung 80424, Taiwan

*   Correspondence: kanglin@mail.ncku.edu.tw (L.K.); hwan@kmu.edu.tw (C.-H.C.); Tel.: +886-7-3209209 (C.-H.C.)

Received: 16 March 2020; Accepted: 14 April 2020; Published: 16 April 2020

Abstract: Green tea drinking can ameliorate postmenopausal osteoporosis by increasing the bone mineral density. (-)-Epigallocatechin-3-gallate (EGCG), the abundant and active compound of tea catechin, was proven to be able to reduce bone loss and ameliorate microarchitecture in female ovariectomized rats. EGCG can also enhance the osteogenic differentiation of murine bone marrow mesenchymal stem cells and inhibit the osteoclastogenesis in RAW264.7 cells by modulation of the receptor activator of Nuclear factor-kB (RANK)/RANK ligand (RANKL)/osteoprotegrin (OPG) (RANK/RANKL/OPG) pathway. Our previous study also found that EGCG can promote bone defect healing in the distal femur partially via bone morphogenetic protein-2 (BMP-2). Considering the osteoinduction property of BMP-2, we hypothesized that EGCG could accelerate the bone healing process with an increased expression of BMP-2. In this manuscript, we studied whether the local use of EGCG can facilitate tibial fracture healing. Fifty-six 4-month-old rats were randomly assigned to two groups after being weight-matched: a control group with vehicle treatment (Ctrl) and a study group with 10 µmol/L, 40 µL, EGCG treatment (EGCG). Two days after the operation, the rats were treated daily with EGCG or vehicle by percutaneous local injection for 2 weeks. The application of EGCG enhanced callus formation by increasing the bone volume and subsequently improved the mechanical properties of the tibial bone, including the maximal load, break load, stiffness, and Young's modulus. The results of the histology and BMP-2 immunohistochemistry staining showed that EGCG treatment accelerated the bone matrix formation and produced a stronger expression of BMP-2. Taken together, this study for the first time demonstrated that local treatment of EGCG can accelerate the fracture healing process at least partly via BMP-2.

Keywords: (-)-epigallocatechin-3-gallate (EGCG); bone morphogenetic protein-2 (BMP-2); catethin; fracture healing; local use

1. Introduction

Long-bone fracture is a common musculoskeletal trauma representing a considerable burden of disease. Bone fracture healing is a long process usually requiring several months. The delayed union or Nonunion rate is approximately 5%–10% [1]. Acceleration the fracture healing process and shortening the recovery time is important. As a result, different types of Nourishment and medications have been intensively investigated in order to identify the optimal strategies to promote fracture healing.

The beneficial effects of green tea on bone health have been widely reported. Numerous clinical studies indicated a positive effect between tea consumption and bone health [2]. Those who drink tea habitually have better bone mineral density (BMD) and a lower hip fracture rate [3,4]. Hegarty's study reported approximately 5% higher lumbar spine BMD in frequent tea drinkers than in those who did Not drink tea [5]. A cross-sectional analysis from a 5-year prospective trial related to oral calcium supplements and osteoporotic fractures also demonstrated that tea drinkers had a lower rate of BMD loss compared to Non-tea drinkers (1.6% vs. 4% in 4 years) [6]. (-)-Epigallocatechin-3-gallate (EGCG) is widely studied because of its potent antioxidant effects [7]. Previous animal studies found that green tea polyphenol extracts improved several bone loss models related to aging, estrogen-deficiency, and chronic inflammation [8–14]. Although Numerous clinical and animal studies indicated that tea can increase bone volume and diminish osteoporotic fractures, the relationship between tea consumption and fracture healing remained unclear.

The fracture healing process is a specialized healing process that depends on the coordination of different growth factors to stimulate bone formation. Bone morphogenetic protein-2 (BMP-2) is a powerful osteogenic factor that induces osteoblast differentiation and promotes bone formation [15]. Enhanced BMP-2 expression is known to play an essential role in the initiation of fracture healing [16]. We found that EGCG enhanced BMP-2 mRNA expression in human bone marrow derived mesenchymal stem cells (BMSCs) [17]. EGCG can facilitate osteogenic differentiation of both murine and human BMSCs and eventually increase mineralization in vitro [17,18]. We previously reported that intraperitoneal injection of EGCG for 3 months in ovariectomized rats could increase bone volume and microarchitecture at a dosage of 3.4 mg/kg/day, which achieved 10 μmol/L in serum. The effect of EGCG may rely on BMP-2 [19]. We also found that local treatment of EGCG could improve the healing of femoral bone defects and that this effect might be mediated at least in part by BMP-2 [20]. In this study, we hypothesized that local percutaneous injection of EGCG could promote fracture healing by enhancing BMP-2 expression.

2. Materials and Methods

2.1. Chemicals

High purity (>99%) grade EGCG (No. E4143) was purchased from Sigma-Aldrich (St. Louis, MO, USA) and was dissolved in dimethylsulfoxide (DMSO).

2.2. Experiment Animals

This animal study was approved by the Institutional Animal Care and Use Committee. Male Sprague–Dawley (SD) rats at 12 weeks of age were obtained from the National Laboratory Center (Taipei, Taiwan) with a mean body weight of 350 ± 25 g, and were provided with free food and water in a temperature-controlled room (25 ± 1 °C) and kept on a 12:12 light–dark cycle during the experiment. Fifty-six 4-month-old rats were randomly divided into two groups after being weight-matched: fracture with vehicle treatment as control group (Ctrl) (n = 28) and fracture with treatment of EGCG (EGCG) (n = 28). We used EGCG, 40 µL at 10 µmol/L, with a total dose of 0.52 µg/kg/time [20]. Isolated right tibia fractures were made with a saw. Percutaneous local injection of either vehicle or EGCG were given locally and daily 2 days after the fracture creation for 2 weeks (n = 14 at each group). Micro-computed tomography (µ-CT), biomechanical analysis (n = 7 at each group), and histological study (n = 7 at each group) were done 2 and 4 weeks after treatment [19–21].

2.3. Tibial Fracture Model

An isolated right tibia fracture with intramedullary Needle fixation was selected as the bone fracture healing model. Each animal was anesthetized with ketamine (60 mg/kg) administered via intraperitoneal injection. The right hind limb was prepped and draped in a sterile manner. A 1 cm longitudinal skin incision was made over the antero-medial aspect of the lower limb. The tibia was exposed without elevating the periosteum, and osteotomized using an oscillating saw under continuous irrigation. After open osteotomy, a 23-gauge syringe Needle was inserted into the bone marrow cavity of the tibia to stabilize the fracture. The surgical wound was sutured and covered with sterile dressings for two days. The proper alignment after fixation was radiographically confirmed.

2.4. Radiographic and µCT Analyses

During radiologic examination and micro-CT image scanning, the animals were sedated. For investigation of the fracture repair process in living animals, small animal micro-CT (Skyscan 1076, Bruker, Belgium) and image software (CTAn) were used and calculated at the indicated time point (n = 14 in each time point). For scanning, the scan settings were an aluminum filter of 0.5 mm, 9 µm scanning resolution, X-ray voltage of 50 kV, X-ray current of 200 mA, and exposure time of 600 ms. The analysis began from the proximal tibia and through the whole fracture site until the distal tibia [19,21–26].

2.5. Histological Study

Rats were sacrificed at 2 weeks and 4 weeks for histological analysis after surgery (n = 7 in each time point). The bone samples were harvested after sacrifice and fixed with 10% Neutral buffered formaldehyde for 2 days, decalcified in 14% ethylenediaminetetraacetic acid (EDTA)/phosphate buffered saline (PBS) for 14 days and then embedded in paraffin. The 5-µm bone sections were hematoxylin and eosin stained for histomorphometry of the bone volume after decalcification. At a magnification of 40×, we defined the counted callus area as the 1-mm regions proximal and distal to the bone graft ends. The area of callus formation and the intact tibia bone was measured using Image-Pro Plus 5.0 software (Media Cybernetics, Inc., Rockville, MD, USA). We calculated the percentage of bone matrix of the callus and the intact tibia and compared with the results from the control group [27–30].

2.6. Immunohistochemistry (IHC) of BMP-2

The samples were prepared for indirect immune detection using a rabbit polyclonal anti-BMP-2 (Abcam, Cambridge, MA, USA) and mouse and rabbit specific horseradish peroxidase/ diaminobenzidine detection IHC kit (Abcam, Cambridge, MA, USA) by protocols provided by manufacture. The sections were then counterstained with hematoxylin to visualize cell Nuclei. BMP-2 were stained brown [20]. Under high power magnification, the BMP-2 stain in the callus area was measured and quantified [20,29].

2.7. Three Point Bending for Biomechanical Testing

Instron 4466 (model 4465; Instron, Canton, Massachusetts) was used for the mechanical property of the repaired fracture in the tibia bone samples after the soft tissues were removed ($n = 7$ in each time point). For three-point bending, the tibia bone was placed between two metal supports, and a single-pronged loading device was applied to the opposite surface at a point precisely in the middle between the two supports. The distance between the two supports was 4 cm. The loading force was 1 N at a rate of 1 mm/min and directed to the mid-diaphyseal region [21]. We measured the deflection of the bone at the point of load application and the simultaneous measurement of the load, yielding a force–deflection graph. The biomechanical parameters, including yield point, maximal load, fracture load, and whole-bone stiffness (defined as the slope of the early, linear portion of the load–deflection curve) were recorded. The Young's modulus of the material was also measured by the geometry of the loading device and the stiffness of the bone.

2.8. Statistical Analysis

All data were expressed as mean ± standard error by at least three independent experiments. Comparisons of the data were analyzed by one-way ANOVA, and the Scheffe post hoc test using SPSS (version 22 for Windows, SPSS Inc, Chicago, IL, USA). Statistical significance was defined as $p < 0.05$.

3. Results

3.1. X-ray and Microarchitecture Assessment by μ-CT

The images of x-ray and μ-CT are revealed in Figure 1. The results of the radiographic analysis showed that the administration of EGCG enhanced bone callus formation. With EGCG treatment, the fracture gap decreased at the end of weeks 2 and 4 after treatment both in the X-ray images (Figure 1A) and in the μ-CT images (Figure 1B). Compared to the control group, the fracture gap decreased gradually in the EGCG treated group at weeks 2 and 4. The quantification results of the callus in μ-CT are shown in Table 1. Furthermore, there was No fracture gap visible at week 4 in the EGCG treated group, confirming the beneficial role of EGCG on the fracture healing process.

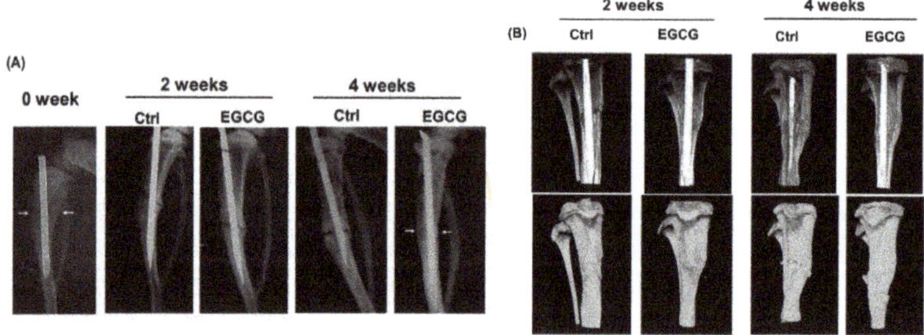

Figure 1. (-)-Epigallocatechin-3-gallate (EGCG) promotion of bone fracture healing in rats. The images of X-ray and micro-computed tomography (μ-CT). (**A**) The X-ray radiographic analysis showed a clear fracture gap at weeks 0, 2, and 4 in the control group, and this fracture was gradually blurred at weeks 2 and 4 in the EGCG treated group. Arrows indicate the fracture site. (**B**) The results of μ-CT analysis also showed a gradual union of fracture gap at the end of weeks 2 and 4 after treatment. Compared to the control group, the fracture was united at the end of week 4 after the EGCG treatment, suggesting that fracture healing process was accelerated by EGCG.

Table 1. The quantification results of callus in μ-CT.

	Week 2		Week 4	
	Control	**EGCG**	**Control**	**EGCG**
Tissue volume (mm^3)	440.87 ± 24.82	640.32 ± 58.44 *	447.03 ± 38.63	622.64 ± 40.95 **
Bone volume (mm^3)	58.30 ± 11.65	70.28 ± 4.81	57.29 ± 12.13	66.35 ± 11.20

* $p < 0.05$ versus control group after treatment. ** $p < 0.01$ versus control after treatment.

3.2. Three-Point Bending Test for the Mechanical Properties of the Bone

After harvesting the tibia, the three-point bending test was used for the mechanical strength of fracture healing. The results of the biomechanical testing showed that the treatment with EGCG enhanced the bone mechanical strength (Figure 2A–D). The maximal load increased from 42.3 ± 5.5 N in the control group to 75.2 ± 9.9 N in the EGCG group at the end of week 2 ($p < 0.05$) and from 68.1 ± 12.0 N in the control group to 99.1 ± 0.7 N in the EGCG group at the end of week 4 ($p < 0.05$) (Figure 2A). The break point improved from 31.2 ± 4.0 N in the control group to 72.2 ± 11.4 N in the EGCG group at the end of week 2 ($p < 0.05$) and from 62.1 ± 11.2 N in the control group to 91.5 ± 1.1 N in the EGCG group at the end of week 4 ($p < 0.05$) (Figure 2B). The stiffness also improved from 149.1 ± 23.9 N/mm^2 in the control group to 213.5 ± 8.8 N/mm^2 in the EGCG group at the end of week 2 ($p < 0.05$) and from 171.4 ± 18.4 N/mm^2 in the control group to 226.6 ± 12.9 N/mm^2 in the EGCG group at the end of week 4 ($p < 0.05$) (Figure 2C). Young's modulus improved from 1.4 ± 0.06 GPa in the control group to 2.5 ± 0.25 GPa in the EGCG group at the end of week 2 ($p < 0.001$) and 2.2 ± 0.5 GPa in the control group to 3.4 ± 0.18 GPa in the EGCG group at the end of week 4 ($p < 0.05$) (Figure 2D).

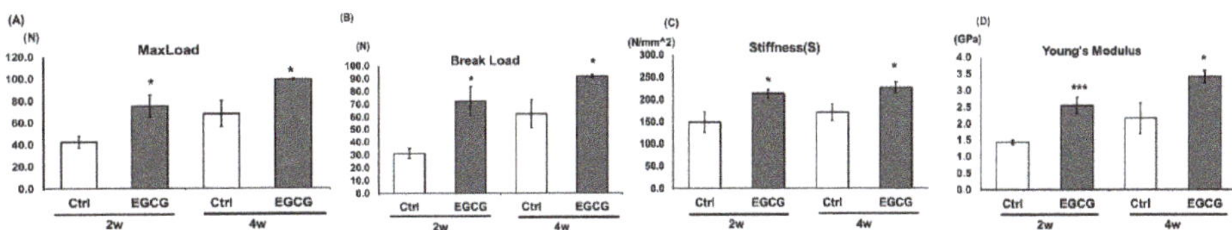

Figure 2. EGCG increased the bone mechanical properties of the fractured tibia. The results of mechanical testing showed that the maximal load (**A**), break load (**B**), stiffness (**C**), and Young's modulus (**D**) (at weeks 2 and 4) were significantly increased after the treatment of EGCG. All data are expressed as mean ± standard error. * $p < 0.05$ versus control group after treatment. *** $p < 0.001$ versus control after treatment.

3.3. Histological Study

Histological analysis showed that fracture healing was accelerated by EGCG treatment. More New matrix formation in EGCG groups was Noted than that in the control group at the end of both weeks 2 and 4 (Figure 3A). The bone matrix ratio increased from 44.0% ± 0.9% to 70.6% ± 4.7% at the end of week 2 ($p < 0.01$) and 58.6% ± 3.4% to 82.6% ± 7.1% at the end of week 4 ($p < 0.01$) (Figure 3B). The histological study further approved the X-ray and μ-CT results.

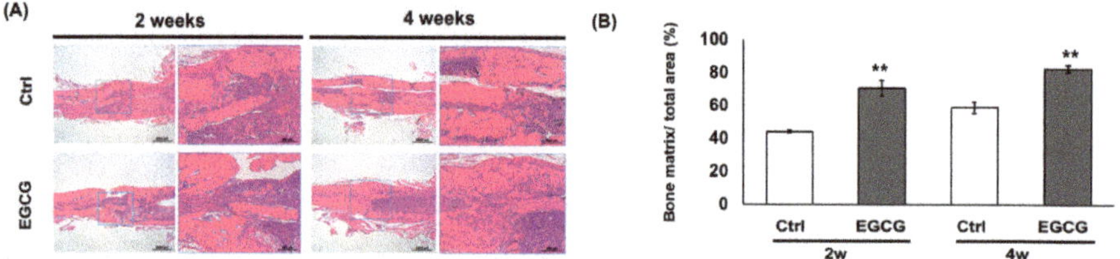

Figure 3. (**A**) Histological analysis. There was significantly more New matrix formation in the EGCG groups than in the control groups. (**B**) The New bone matrix formation in callus tissues was significantly enhanced by EGCG treatment at weeks 2 and 4. ** $p < 0.01$ versus control after treatment.

3.4. IHC Analysis

In the IHC study, immunolocalized BMP-2 in New form matrices were quantified (Figure 4A). The brown stained ratio in the callus tissue increased from 0.20 ± 0.04 to 0.36 ± 0.02 and from 0.24 ± 0.03 to 0.51 ± 0.02 at the end of weeks 2 and 4, respectively, with the treatment of EGCG ($p < 0.01$) (Figure 4B). Our findings indicated that EGCG can facilitate bone fracture healing by increasing the BMP-2 expression.

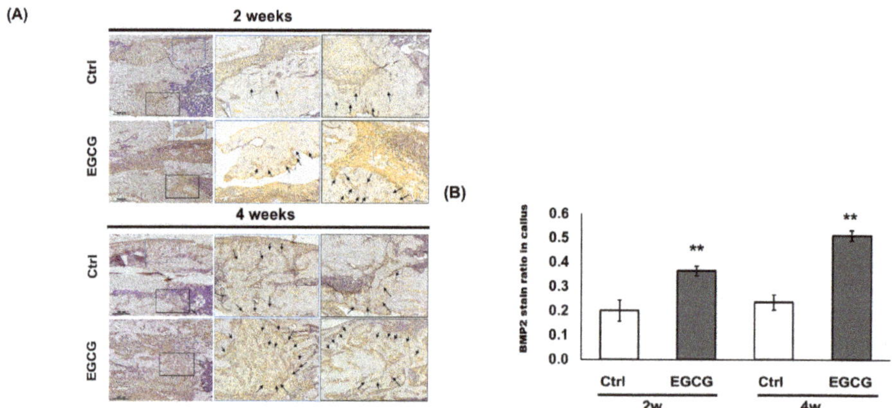

Figure 4. Immunohistochemistry study. (**A**) Immunolocalized bone morphogenetic protein-2 (BMP-2) in New form matrices were stained brown. The arrows indicate positive BMP-2 stains. (**B**) Quantification of immunolocalized BMP-2. EGCG facilitated bone fracture healing by increasing the BMP-2 expression in callus tissue. ** $p < 0.01$ versus control after treatment.

4. Discussion

EGCG, one of the most bioactive catechins derived from green tea, can stimulate bone formation both in vitro and in vivo [17,19,20,31]. The effects of EGCG on the promotion of the fracture healing process, however, remain unclear. This is the first study to indicate the beneficial effects of local percutaneous EGCG treatment in facilitating the fracture healing of tibia established by X-ray, μCT images, and eventually mechanical property tests. The study results provided supportive evidence that the local EGCG improved callus formation and increased the bone mechanical strength in the fractured tibia. Significantly, the current study results also indicated that the effect on accelerating fracture healing may be partially related to EGCG's role in the upregulation of BMP-2.

The healing process of fractured bone is regulated by the differentiation of osteoblasts and BMSCs. In our previous study, we indicated that EGCG can facilitate the osteogenic differentiation of murine and human BMSCs especially at 10 μmol/L [17]. At the same concentration, EGCG can also decrease osteoclastogenesis via the regulation of osteoprotegrin (OPG) and the receptor activator of Nuclear factor-kB (RANK) ligand (RANKL) in pre-osteoclast feeder cells, ST2 [32]. Qiu's study also showed that 10–40 µmol/L of EGCG significantly inhibited hypoxia-induced apoptosis in BMSCs and increased the level of runt-related transcription factor 2 (Runx2), BMP-2, type I procollagen, and alkaline phosphatase activity [33]. We found No published data regarding the effect dose of EGCG on fracture healing. As a result, we chose 10 µmol/L of EGCG in this study. We inferred that EGCG may increase the osteogenic differentiation of BMSCs around the fracture site at this concentration. Moreover, the callus size may be further increased due to decreased osteoclastogenesis.

BMPs are crucial regulators of both chondrocyte and osteoblast differentiation and thus can promote bone formation by improving both intramembranous and endochondral ossification [34]. BMP-2 can strongly induce New bone formation and is one of the most powerful osteoinductive biofactors. BMP-2 can be applied for delayed union, Non-union, and bone defects [35]. The role of BMP-2 in fracture healing has been investigated using BMP2 limb bud mesenchyme conditional knockout mice. Researchers demonstrated that the fracture healing process was slower in the heterozygotes compared to the wild-type mice and No fracture healing was Noted in the knockout mice [34].

The direct application of BMP-2 was also proven to accelerate the healing process in small animal, large animal, and even human clinical trials [36–38]. Our previously reported intraperitoneal EGCG use mitigated the BMD decline and ameliorated the bone architecture. Moreover, EGCG treatment also increased the immunolocalized BMP-2 in the bone matrix [19]. In another in vivo study, we locally applied EGCG on femoral bone defects and found that the local use of EGCG at femoral defects could facilitate New bone formation by enhancing the bone volume and subsequently recovered the mechanical strength of the bone. The expression of the immunolocalized BMP-2 was shown to increase up to 3-fold in EGCG treated rats [20]. With this concept, we used EGCG to facilitate tibia fracture healing. Our results revealed that treatment with EGCG, at 10 µmol/L, significantly accelerated the fracture healing and thus improved the mechanical strength of the fractured tibia.

The modulation of osteoclast activity may be another function of EGCG to increase the callus size. Chen et al. reported that EGCG can inhibit osteoclastogenesis in RAW264.7 cell and ST2 cell co-cultures by modulation of the RANK/RANKL/OPG pathway [32]. Similar results were also reported in Xu's study, where EGCG and its oxidation product were both found to effectively inhibit RANKL-induced osteoclastogenesis in RAW 264.7 cells [39]. A study by Song et al. also found that EGCG supplementation at a dose of 10 mg/kg/day for 12 weeks markedly increased both BMD and total bone volume and also improved the microarchitecture of the trabecular bone in the femur.

In an IHC study, researchers reported that EGCG diminished semaphorin 4D expression [40]. Semaphorin 4D is highly expressed in osteoclasts and may potently inhibit bone formation by binding to Plexin-B1 receptors, which are highly expressed by osteoblasts [41]. Semaphorin 4D specific antibody increased osteoblastic bone formation but did Not disturb osteoclastic bone resorption and eventually diminished bone loss after ovariectomies [41]. As a result, EGCG could regulate the RANKL-induced osteoclastogenesis and possibly reduce the resorption of woven bones and increase the callus size, which may be another possible mechanism for increased tibia fracture healing.

Previous studies indicated that taking one cup of green tea can achieve a serum level of 1 µmol/L EGCG [42,43]. An oral dose of 1600 mg EGCG can achieved a serum level of 7.6 µmol/L [44]. The effective concentration of EGCG to enhance fracture healing was 10 µmol/L in this study, which can be reached by daily tea consumption. Adequate oral EGCG was used in the study for fracture healing. The molecular mechanisms of EGCG on fracture healing are likely due, at least in part, to the increased BMP-2 expression. Further studies are required to find more molecular mechanisms.

5. Conclusions

In conclusion, we found that locally administered EGCG, to a level of 10 μmol/L with 40 μL at a dose of 0.52 μg/kg/time, at a tibia fracture could enhance the fracture healing by improving the callus size and then aid in recovering the mechanical strength of the bone, including the max load, modulus, stiffness, and break load. The promotion effect of EGCG on fracture healing might be in part due to the effects on BMP-2.

Author Contributions: Conceptualization, S.-Y.L., J.Y.K., C.-C.L., L.K. and C.-H.C.; methodology, S.-Y.L., J.Y.K., H.H.H., T.-L.C., S.-C.C. and C.-H.C.; software, H.-T.H., C.-J.H., T.-C.L. and J.Y.K.; validation, S.-Y.L., C.-C.L., L.K. and C.-H.C.; formal analysis, S.-Y.L., J.Y.K., H.H.H., T.-L.C., S.-C.C., Y.-S.L., J.Y.K. and C.-H.C.; investigation, S.-Y.L., J.Y.K., H.H.H., T.-L.C., S.-C.C., Y.-S.L. and C.-H.C.; data curation, S.-Y.L., J.Y.K., H.H.H., T.-L.C., H.-T.H., C.-J.H., T.-C.L., S.-C.C., Y.-S.L. and C.-H.C.; writing—original draft preparation, S.-Y.L., C.-C.L. and L.K.; writing—review and editing, S.-Y.L., C.-C.L., L.K. and C.-H.C.; project administration, L.K. and C.-H.C.; funding acquisition, C.-H.C. All authors have read and agreed to the published version of the manuscript.

Funding: This study was supported in part by the National Health Research Institute (NHRI-EX101-9935EI) of Taiwan, Kaohsiung Medical University Hospital (KMUH-107-7R54), Kaohsiung Medical University (KMU-TC108A02-1, NCTUKMU108-BIO-04 and KMU-DK105009), Kaohsiung Municipal Ta-Tung Hospital (KMTTH-100-004) and the Minister of Science and Technology of Taiwan (MOST 106-2314-B-037 -050 -MY3 and 108-2314-B-037 -059 -MY3). The funders had No role in the study design, data collection, and analysis, decision to publish, or preparation of the manuscript.

Acknowledgments: We appreciate the support from members of our orthopedic research center and Department of Physiology, Kaohsiung Medical University, Kaohsiung City, Taiwan.

Conflicts of Interest: The authors declare No conflict of interest. The funders had No role in the design of the study; in the collection, analyses, or interpretation of data; in the writing of the manuscript, or in the decision to publish the results.

References

1. Gómez-Barrena, E.; Rosset, P.; Lozano, D.; Stanovici, J.; Ermthaller, C.; Gerbhard, F. Bone fracture healing: Cell therapy in delayed unions and Nonunions. *Bone* **2015**, *70*, 93–101. [CrossRef] [PubMed]
2. Muraki, S.; Yamamoto, S.; Ishibashi, H.; Oka, H.; Yoshimura, N.; Kawaguchi, H.; Nakamura, K. Diet and lifestyle associated with increased bone mineral density: Cross-sectional study of Japanese elderly women at an osteoporosis outpatient clinic. *J. Orthop. Sci.* **2007**, *12*, 317–320. [CrossRef]
3. Kanis, J.; Johnell, O.; Gullberg, B.; Allander, E.; Elffors, L.; Ranstam, J.; Dequeker, J.; Dilsen, G.; Gennari, C.; Vaz, A.L.; et al. Risk factors for hip fracture in men from southern Europe: The MEDOS study. Mediterranean Osteoporosis Study. *Osteoporos. Int.* **1999**, *9*, 45–54. [CrossRef]
4. Johnell, O.; Gullberg, B.; Kanis, J.A.; Allander, E.; Elffors, L.; Dequeker, J.; Dilsen, G.; Gennari, C.; Lopes Vaz, A.; Lyritis, G.; et al. Risk factors for hip fracture in European women: The MEDOS Study. Mediterranean Osteoporosis Study. *J. Bone Miner. Res.* **1995**, *10*, 1802–1815. [CrossRef] [PubMed]
5. Hegarty, V.M.; May, H.M.; Khaw, K.T. Tea drinking and bone mineral density in older women. *Am. J. Clin. Nutr.* **2000**, *71*, 1003–1007. [CrossRef] [PubMed]
6. Devine, A.; Hodgson, J.M.; Dick, I.M.; Prince, R.L. Tea drinking is associated with benefits on bone density in older women. *Am. J. Clin. Nutr.* **2007**, *86*, 1243–1247. [CrossRef] [PubMed]
7. Fujiki, H. Green tea: Health benefits as cancer preventive for humans. *Chem. Rec.* **2005**, *5*, 119–132. [CrossRef]
8. Shen, C.L.; Yeh, J.K.; Stoecker, B.J.; Chyu, M.C.; Wang, J.S. Green tea polyphenols mitigate deterioration of bone microarchitecture in middle-aged female rats. *Bone* **2009**, *44*, 684–690. [CrossRef]
9. Shen, C.L.; Yeh, J.K.; Cao, J.J.; Tatum, O.L.; Dagda, R.Y.; Wang, J.S. Synergistic effects of green tea polyphenols and alphacalcidol on chronic inflammation-induced bone loss in female rats. *Osteoporos. Int.* **2010**, *21*, 1841–1852. [CrossRef]
10. Shen, C.L.; Yeh, J.K.; Cao, J.J.; Tatum, O.L.; Dagda, R.Y.; Wang, J.S. Green tea polyphenols mitigate bone loss of female rats in a chronic inflammation-induced bone loss model. *J. Nutr. Biochem.* **2010**, *21*, 968–974. [CrossRef]
11. Shao, C.; Chen, L.; Lu, C.; Shen, C.L.; Gao, W. A gel-based proteomic analysis of the effects of green tea polyphenols on ovariectomized rats. *Nutrition* **2011**, *27*, 681–686. [CrossRef] [PubMed]

12. Shen, C.L.; Yeh, J.K.; Cao, J.J.; Wang, J.S. Green tea and bone metabolism. *Nutr. Res.* **2009**, *29*, 437–456. [CrossRef] [PubMed]

13. Shen, C.L.; von Bergen, V.; Chyu, M.C.; Jenkins, M.R.; Mo, H.; Chen, C.H.; Kwun, I.S. Fruits and dietary phytochemicals in bone protection. *Nutr. Res.* **2012**, *32*, 897–910. [CrossRef] [PubMed]

14. Shen, C.L.; Kwun, I.S.; Wang, S.; Mo, H.; Chen, L.; Jenkins, M.; Brackee, G.; Chen, C.H.; Chyu, M.C. Functions and mechanisms of green tea catechins in regulating bone remodeling. *Curr. Drug Targets* **2013**, *14*, 1619–1630. [CrossRef]

15. Salazar, V.S.; Gamer, L.W.; Rosen, V. BMP signalling in skeletal development, disease and repair. *Nat. Rev. Endocrinol.* **2016**, *12*, 203–221. [CrossRef]

16. Yu, Y.H.; Wilk, K.; Waldon, P.L.; Intini, G. In vivo identification of Bmp2-correlation Networks during fracture healing by means of a limb-specific conditional inactivation of Bmp2. *Bone* **2018**, *116*, 103–110. [CrossRef]

17. Lin, S.Y.; Kang, L.; Wang, C.Z.; Huang, H.H.; Cheng, T.L.; Huang, H.T.; Lee, M.J.; Lin, Y.S.; Ho, M.L.; Wang, G.J.; et al. (-)-Epigallocatechin-3-Gallate (EGCG) Enhances Osteogenic Differentiation of Human Bone Marrow Mesenchymal Stem Cells. *Molecules* **2018**, *23*, 3221. [CrossRef]

18. Chen, C.H.; Ho, M.L.; Chang, J.K.; Hung, S.H.; Wang, G.J. Green tea catechin enhances osteogenesis in a bone marrow mesenchymal stem cell line. *Osteoporos. Int.* **2005**, *16*, 2039–2045. [CrossRef]

19. Chen, C.H.; Kang, L.; Lin, R.W.; Fu, Y.C.; Lin, Y.S.; Chang, J.K.; Chen, H.T.; Chen, C.H.; Lin, S.Y.; Wang, G.J.; et al. (-)-Epigallocatechin-3-gallate improves bone microarchitecture in ovariectomized rats. *Menopause* **2013**, *20*, 687–694. [CrossRef]

20. Lin, S.Y.; Kang, L.; Chen, J.C.; Wang, C.Z.; Huang, H.H.; Lee, M.J.; Cheng, T.L.; Chang, C.F.; Lin, Y.S.; Chen, C.H. (-)-Epigallocatechin-3-gallate (EGCG) enhances healing of femoral bone defect. *Phytomedicine* **2019**, *55*, 165–171. [CrossRef]

21. Wang, Y.H.; Rajalakshmanan, E.; Wang, C.K.; Chen, C.H.; Fu, Y.C.; Tsai, T.L.; Chang, J.K.; Ho, M.L. PLGA-linked alendronate enhances bone repair in diaphysis defect model. *J. Tissue Eng. Regen. Med.* **2017**, *11*, 2603–2612. [CrossRef] [PubMed]

22. Chen, C.H.; Kang, L.; Lo, H.C.; Hsu, T.H.; Lin, F.Y.; Lin, Y.S.; Wang, Z.J.; Chen, S.T.; Shen, C.L. Polysaccharides of Trametes versicolor Improve Bone Properties in Diabetic Rats. *J. Agric. Food Chem.* **2015**, *63*, 9232–9238. [CrossRef] [PubMed]

23. Teong, B.; Kuo, S.M.; Tsai, W.H.; Ho, M.L.; Chen, C.H.; Huang, H.H. Liposomal Encapsulation for Systemic Delivery of Propranolol via Transdermal Iontophoresis Improves Bone Microarchitecture in Ovariectomized Rats. *Int. J. Mol. Sci.* **2017**, *18*, 822. [CrossRef] [PubMed]

24. Fu, Y.C.; Wang, Y.H.; Chen, C.H.; Wang, C.K.; Wang, G.J.; Ho, M.L. Combination of calcium sulfate and simvastatin-controlled release microspheres enhances bone repair in critical-sized rat calvarial bone defects. *Int. J. Nanomedicine* **2015**, *10*, 7231–7240. [PubMed]

25. Bouxsein, M.L.; Boyd, S.K.; Christiansen, B.A.; Guldberg, R.E.; Jepsen, K.J.; Muller, R. Guidelines for assessment of bone microstructure in rodents using micro-computed tomography. *J. Bone Miner. Res.* **2010**, *25*, 1468–1486. [CrossRef] [PubMed]

26. Chou, L.Y.; Chen, C.H.; Lin, Y.H.; Chuang, S.C.; Chou, H.C.; Lin, S.Y.; Fu, Y.C.; Chang, J.K.; Ho, M.L.; Wang, C.Z. Discoidin domain receptor 1 regulates endochondral ossification through terminal differentiation of chondrocytes. *FASEB J.* **2020**, *34*, 5767–5781. [CrossRef] [PubMed]

27. Lee, T.C.; Wang, Y.H.; Hung, S.H.; Chen, C.H.; Ho, M.L.; Fu, Y.C. Evaluations of Clinical-Grade Bone Substitute-Combined Simvastatin Carriers to Enhance Bone Growth: In vitro and in vivo analyses. *J. Bioact. Compat. Polym.* **2018**, *33*, 160–177. [CrossRef]

28. Wang, C.Z.; Wang, Y.H.; Lin, C.W.; Lee, T.C.; Fu, Y.C.; Ho, M.L.; Wang, C.K. Combination of a Bioceramic Scaffold and Simvastatin Nanoparticles as a Synthetic Alternative to Autologous Bone Grafting. *Int. J. Mol. Sci.* **2018**, *19*, 99. [CrossRef]

29. Fu, Y.C.; Lin, C.C.; Chang, J.K.; Chen, C.H.; Tai, I.C.; Wang, G.J.; Ho, M.L. A Novel single pulsed electromagnetic field stimulates osteogenesis of bone marrow mesenchymal stem cells and bone repair. *PLoS ONE* **2014**, *9*, e91581. [CrossRef]

30. Wang, C.; Ho, M.; Wang, G.; Chang, J.; Chen, C.; Fu, Y.; Fu, H. Controlled-release of rhBMP-2 carriers in the regeneration of osteonecrotic bone. *Biomaterials* **2009**, *30*, 4178–4186. [CrossRef]

31. Huang, A.; Honda, Y.; Li, P.; Tanaka, T.; Baba, S. Integration of Epigallocatechin Gallate in Gelatin Sponges Attenuates Matrix Metalloproteinase-Dependent Degradation and Increases Bone Formation. *Int. J. Mol. Sci.* **2019**, *20*, 42. [CrossRef] [PubMed]

32. Chen, S.T.; Kang, L.; Wang, C.Z.; Huang, P.J.; Huang, H.T.; Lin, S.Y.; Chou, S.H.; Lu, C.C.; Shen, P.C.; Lin, Y.S.; et al. (-)-Epigallocatechin-3-Gallate Decreases Osteoclastogenesis via Modulation of RANKL and Osteoprotegrin. *Molecules* **2019**, *24*, 156. [CrossRef] [PubMed]

33. Qiu, Y.; Chen, Y.; Zeng, T.; Guo, W.; Zhou, W.; Yang, X. EGCG ameliorates the hypoxia-induced apoptosis and osteogenic differentiation reduction of mesenchymal stem cells via upregulating miR-210. *Mol. Biol. Rep.* **2016**, *43*, 183–193. [CrossRef]

34. Hankenson, K.D.; Gagne, K.; Shaughnessy, M. Extracellular signaling molecules to promote fracture healing and bone regeneration. *Adv. Drug Deliv. Rev.* **2015**, *94*, 3–12. [CrossRef]

35. Geiger, M.; Li, R.H.; Friess, W. Collagen sponges for bone regeneration with rhBMP-2. *Adv. Drug Deliv. Rev.* **2003**, *55*, 1613–1629. [CrossRef]

36. Starr, A.J. Recombinant human bone morphogenetic protein-2 for treatment of open tibial fractures. *J. Bone Joint Surg. Am.* **2003**, *85*, 2049–2050. [CrossRef] [PubMed]

37. Tölli, H.; Kujala, S.; Jämsä, T.; Jalovaara, P. Reindeer bone extract can heal the critical-size rat femur defect. *Int. Orthop.* **2011**, *35*, 615–622. [CrossRef]

38. Morishita, Y.; Naito, M.; Miyazaki, M.; He, W.; Wu, G.; Wei, F.; Sintuu, C.; Hymanson, H.; Brochmann, E.J.; Murray, S.S.; et al. Enhanced effects of BMP-binding peptide combined with recombinant human BMP-2 on the healing of a rodent segmental femoral defect. *J. Orthop. Res.* **2010**, *28*, 258–264. [CrossRef] [PubMed]

39. Xu, H.; Liu, T.; Li, J.; Xu, J.; Chen, F.; Hu, L.; Zhang, B.; Zi, C.; Wang, X.; Sheng, J. Oxidation derivative of (-)-epigallocatechin-3-gallate (EGCG) inhibits RANKL-induced osteoclastogenesis by suppressing RANK signaling pathways in RAW 264.7 cells. *Biomed. Pharmacother.* **2019**, *118*, 109237. [CrossRef]

40. Song, D.; Gan, M.; Zou, J.; Zhu, X.; Shi, Q.; Zhao, H.; Luo, Z.; Zhang, W.; Li, S.; Niu, J.; et al. Effect of (-)-epigallocatechin-3-gallate in preventing bone loss in ovariectomized rats and possible mechanisms. *Int J. Clin. Exp. Med.* **2014**, *7*, 4183–4190.

41. Negishi-Koga, T.; Shinohara, M.; Komatsu, N.; Bito, H.; Kodama, T.; Friedel, R.H.; Takayanagi, H. Suppression of bone formation by osteoclastic expression of semaphorin 4D. *Nat. Med.* **2011**, *17*, 1473–1480. [CrossRef]

42. Liao, S.; Kao, Y.H.; Hiipakka, R.A. Green tea: Biochemical and biological basis for health benefits. *Vitam. Horm.* **2001**, *62*, 1–94. [PubMed]

43. Van het Hof, K.H.; Wiseman, S.A.; Yang, C.S.; Tijburg, L.B. Plasma and lipoprotein levels of tea catechins following repeated tea consumption. *Proc. Soc. Exp. Biol Med.* **1999**, *220*, 203–209. [PubMed]

44. Ullmann, U.; Haller, J.; Decourt, J.P.; Girault, N.; Girault, J.; Richard-Caudron, A.S.; Pineau, B.; Weber, P. A single ascending dose study of epigallocatechin gallate in healthy volunteers. *J. Int. Med. Res.* **2003**, *31*, 88–101. [CrossRef] [PubMed]

 *biomolecules*

Article

Bacopa monnieri and Their Bioactive Compounds Inferred Multi-Target Treatment Strategy for Neurological Diseases: A Cheminformatics and System Pharmacology Approach

Rajendran Jeyasri [1,†], **Pandiyan Muthuramalingam** [1,†] , **Vellaichami Suba** [1],
Manikandan Ramesh [1,*] and **Jen-Tsung Chen** [2]

[1] Department of Biotechnology, Science Campus, Alagappa University, Karaikudi 630 003, India;
    jeyasri8220@gmail.com (R.J.); pandianmuthuramalingam@gmail.com (P.M.); subahrj@gmail.com (V.S.)
[2] Department of Life Sciences, National University of Kaohsiung, Kaohsiung 811, Taiwan;
    jentsung@nuk.edu.tw
* Correspondence: mrbiotech.alu@gmail.com; Tel.: +91-04565-225215
† Author's contributed equally to this article.

Received: 14 February 2020; Accepted: 31 March 2020; Published: 2 April 2020

Abstract: Neurological diseases (NDs), especially Alzheimer's and Spinocerebellar ataxia (SCA), can severely cause biochemical abnormalities in the brain, spinal cord and other nerves of human beings. Their ever-increasing prevalence has led to a demand for new drug development. Indian traditional and Ayurvedic medicine used to combat the complex diseases from a holistic and integrative point of view has shown efficiency and effectiveness in the treatment of NDs. *Bacopa monnieri* is a potent Indian medicinal herb used for multiple ailments, but is significantly known as a nootropic or brain tonic and memory enhancer. This annual herb has various active compounds and acts as an alternative and complementary medicine in various countries. However, system-level insights of the molecular mechanism of a multiscale treatment strategy for NDs is still a bottleneck. Considering its prominence, we used cheminformatics and system pharmacological approaches, with the aim to unravel the various molecular mechanisms represented by Bacopa-derived compounds in identifying the active human targets when treating NDs. First, using cheminformatics analysis combined with the drug target mining process, 52 active compounds and their corresponding 780 direct receptors were retrieved and computationally validated. Based on the molecular properties, bioactive scores and comparative analysis with commercially available drugs, novel and active compounds such as asiatic acid (ASTA) and loliolide (LLD) to treat the Alzheimer's and SCA were identified. According to the interactions among the active compounds, the targets and diseases were further analyzed to decipher the deeper pharmacological actions of the drug. NDs consist of complex regulatory modules that are integrated to dissect the therapeutic effects of compounds derived from Bacopa in various pathological features and their encoding biological processes. All these revealed that Bacopa compounds have several curative activities in regulating the various biological processes of NDs and also pave the way for the treatment of various diseases in modern medicine.

Keywords: Alzheimer's disease; *Bacopa monnieri*; bioactive compounds; cheminformatics; neurological diseases; spinocerebellar ataxia; system pharmacology

---

1. Introduction

Medicinal plants are a pivotal reservoir of plenty of pharmacologically active compounds, and have been used as a therapeutic medicine for several diseases since the ancient period. Medicinal plants are the essential backbone of traditional medicine; around 3.3 billion people in the least-developed countries are still utilizing them on a regular basis [1]. Medicinal plants offer rich resources of pharmacological ingredients that can be used in drug development. Above and beyond that, these plants play a significant role in the development of human cultures around the globe. According to the International Union for Conservation of Nature (IUCN) report, worldwide, almost 80,000 flowering plant species are used for pharmaceutical purposes.

Bacopa monnieri (L.) is an important medicinal plant in Indian traditional Ayurvedic medicines. It is a small perennial herbaceous plant commonly known as 'Brahmi', belonging to the family Scrophulariaceae. It is a renowned Indian medicinal plant that has been used as a memory booster in the Ayurvedic medicinal system for more than 3000 years [2,3]. It is used in traditional medicine to treat various nervous disorders, digestive aid, improve learning, memory, and concentration and to provide relief to patients with anxiety, and skin disorders; specific uses include the treatment of asthma, insanity and epilepsy [4–6]. The Bacopa herb, also called nootropic herb, helps in the repair of damaged neurons, neuronal synthesis, and the restoration of synaptic activity, and improves brain function. *B. monnieri* contains alkaloid brahmine, nicotinine, herpestine, bacosides A and B, saponins A, B and C, triterpenoid saponins, stigmastanol, β-sitosterol, betulinic acid, D-mannitol, stigmasterol, α-alanine, aspartic acid, glutamic acid, and serine and pseudojujubogenin glycoside [7]. The plant possesses a wide variety of pharmacologically active principles including memory enhancing, tranquillizing, sedative, antidepressant, antioxidant, cognitive, anticancer, antianxiety, adaptogenic, antiepileptic, gastrointestinal effects, endocrine, gastrointestinal, smooth muscle relaxant effects, cardiovascular, analgesic, antipyretic, antidiabetic, antiarthritic, anticancer, antihypertensive, antimicrobial, antilipidemia, anti-inflammatory, neuroprotective, and hepatoprotective activities [8–10].

It has a very important role in Ayurvedic therapies for the treatment of cognitive disorders of aging [8,11]. Diverse mechanisms of action for its cognitive effects have been proposed, including acetylcholinesterase (AChE) inhibition, antioxidant neuroprotection, β-amyloid reduction, neurotransmitter modulation (acetylcholine [ACh], 5-hydroxytryptamine [5-HT], dopamine [DA]), choline acetyltransferase activation, and increased cerebral blood flow [12]. Theoretically, paring *B. monnieri* with a stimulant would ward off malaise, but this combination has not been tested [10]. Metabolites or active compounds from *B. monnieri* interact with the dopamine and serotonergic systems, but its main molecular mechanism concerns promoting neuron communication. It does this by increasing the growth of nerve endings, also called dendrites. Characteristics of saponins called "bacosides", particularly bacoside A, have been considered to be the major bioactive constituents responsible for the cognitive effects of *B. monnieri* [13–15]. In addition, herbal medicine contains a plethora of active chemical compounds/molecules evolved through a long-term evolutionary process for safeguarding *B. monnieri* from various environmental stresses such as physio-chemical changes, the attack of pathogens, climatic changes, etc. The drug activity of herbal medicine may be due to the individual and synergistic action of phytomolecules. To the best of our knowledge, the molecular mechanism of such actions and the identification of the novel lead molecules against neurological diseases (NDs) are more difficult to study.

In light of this, and despite the significance of traditional Indian medicine, how the metabolites/compounds work and what their active human targets are are still vague. The present study aimed to explore the holistic molecular mechanism and the biological properties/pharmacological activity of phytomolecules derived from *B. monnieri* against NDs. The following issues need to be solved immediately: (i) Which active phytomolecules are involved in the regulatory processes for the treatment NDs? (ii) Which human targets are linked and modulated by the phytochemicals to achieve the biological activity and the purpose of curing NDs? With the emerging prosperity of pharmacological systems and powerful analytical tools such as cheminformatics, network pharmacology investigation allows us to understand the holistic mechanisms of Indian traditional medicine in treating complex NDs. The current study reveals deeper insights into the molecular mechanisms of active phytomolecules in treating NDs. The cheminformatics analysis was used to filter out the active and novel phytomolecules with potent pharmacological activity; also, the reliability of compound/drug–human target interactions was evaluated. The obtained potential human targets were then inputted into specific repositories to figure out their encoding NDs, pathways and holistic mechanisms of active phytomolecules. We hope that with the help of systems pharmacology and the exploration of molecular cross-talks of Indian traditional medicine, we will positively promote the development of new therapies for a mixture of neurological and other diseases in the near future.

2. Materials and Methods

An integrated cheminformatics and system pharmacology approach has been applied for the first time, to unveil the curative effects of Bacopa-derived, pharmacologically active compounds—consisting of: (1) target fishing and functional analysis to identify the compounds—direct target network; (2) network construction and analysis to illustrate the molecular mechanisms of Bacopa-derived compounds in treating NDs, particularly Alzheimer's and SCA; (3) gene ontology enrichment for targets will pave the way for pathway integration analysis to reveal the regulatory mode of target players in multiple functional nodes from a signaling pathway level.

2.1. Collection of Phytomolecules

Detailed information on 52 phytochemicals from *B. monnieri* was collected from the literature and other web sources [16]. A list of biologically phytomolecules is depicted in Table 1.

Table 1. List of Phytomolecuels.

S. No	Compounds	Abbreviations/Acronyms
1	Nicotine	Nt
2	D-Mannitol	D-Mn
3	Bacoside A	BCS A
4	Bacopasaponin A	BCSN A
5	Bacopasaponin B	BCSN B
6	Bacopasaponin C	BCSN C
7	Bacopasaponin D	BCSN D
8	Bacopasaponin E	BCSN E
9	Bacopasaponin F	BCSN F
10	Bacopasaponin G	BCSN G
11	Bacopaside I	BPS I
12	Bacopaside II	BPS II
13	Bacopaside III	BPS III
14	Bacopaside IV	BPS IV
15	Bacopaside V	BPS V
16	Bacopaside VIII	BPS VIII
17	Bacopaside XII	BPS XII
18	Plantainoside B	PTS B
19	Betulinic acid	BTA
20	Cucurbitacin A	CCB A
21	Cucurbitacin B	CCB B

Table 1. *Cont.*

S. No	Compounds	Abbreviations/Acronyms
22	Cucurbitacin C	CCB C
23	Cucurbitacin D	CCB D
24	Cucurbitacin E	CCB E
25	Stearic acid	STRA
26	Rosavin	RSV
27	3,4Dimethoxycinnamic acid	3,4DMCA
28	Ascorbic acid	ASBA
29	Asiatic acid	ASTA
30	Brahmic acid	BMA
31	Wogonin	WG
32	Oroxindin	OX
33	Loliolide	LLD
34	Stigmasterol	SGMS
35	β-sitosterol	β-SS
36	Ebelin lactone	EBL
37	Stigmastanol	SGMSL
38	Bacosterol	BCSt
39	Bacosine	BCSn
40	Heptacosane	HCS
41	Octacosane	OCS
42	Nonacosane	NCS
43	Triacontane	TC
44	Hentriacontane	HTA
45	Dotriacontane	DOT
46	Apigenin	AG
47	Quercetin	QR
48	Ursolic acid	USA
49	Luteolin	LT
50	Asiaticoside	ASTS
51	Bacopaside VI	BPS VI
52	Bacopaside VII	BPS VII

2.2. Phytochemical Information Retrieval

In total, 52 biologically active plant-derived molecules from *B. monnieri* were procured from the PubChem database [17]. The 3D structures and Canonical SMILIES were collected from Molinspiration tool [18] and the PubChem database [17].

2.3. Human Target Imputations

The identified compounds with their canonical SMILIES were searched against human in Swiss TargetPrediction tool to retrieve the compounds in combating human targets (www.swisstargetprediction.ch/).

2.4. Mining of Human Targets and Its Features

Identified potential human targets were subjected onto the Expression Atlas database for retrieving human targets and their features, such as corresponding genomic transcripts, chromosome numbers, the start and end positions of the targets, and UniProt IDs [19].

2.5. Gene Ontology (GO) Analysis

Identified targets and their corresponding official gene symbols were subjected to the GOnet database (https://tools.dice-database.org/GOnet/) [20], to obtain GO against humans with a significant q-value threshold level of <0.05. Targets were also categorized as per GO molecular function, cellular component, and biological process according to the GOnet functional enrichment classification.

2.6. Network Construction

2.6.1. Compound–Target-Network (C-T-N)

A C-T-N was constructed to combat NDs by expounding the multi-target therapeutic feature of the pharmacologically active compounds. In this interaction, a possible target protein and a candidate compound were linked if the protein was targeted by the phyto-compound.

2.6.2. Target-Disease-Network (T-D-N)

The specific targets correlated with NDs were sorted out to explore a detailed interrelationship between potential targets and diseases, and those corresponding diseases with their potential targets were obtained from Expression Atlas database. Finally, T-D-N were built by linking all the target proteins with their relevant diseases using the previously obtained target associated disease information.

Visualization of all networks was accomplished by Cytoscape 3.7.2 [21]. In the obtained result network, nodes represent compounds, diseases and targets, whereas edges indicate the interactions between them.

2.6.3. Identification of Properties of Active Compounds

Phytochemicals with their corresponding canonical SMILIES were uploaded on to the Molinspiration tool to extract the significant calculation on phytomolecules with their molecular properties, and also to predict a bioactive score for the vital targets such as GPCR ligand activity (GPCR), Kinase inhibitor activity (Ki), protease inhibitor activity (Pi), enzymes and nuclear receptors (Ncr), and the number of violations (nvio) (http://www.molinspiration.com) [18].

2.6.4. Compound Comparison

Identified potential compounds with their significant molecular and bioactive properties were compared with commercially available drugs which are responsible for two different neurological diseases. The comparison revealed the pharmacologically active compounds with the help of GPCR, nvio, Ki, Pi, Ei, and Ncr properties.

3. Results

3.1. Compound Information Retrieval

Fifty-two numbers of phytomolecules were used as a query in the PubChem database and the Molinspiration tool to retrieve the Canonical SMILIES (Table S1) and 3D structure of the compounds (Figure 1). This collected information was used for further analyses.

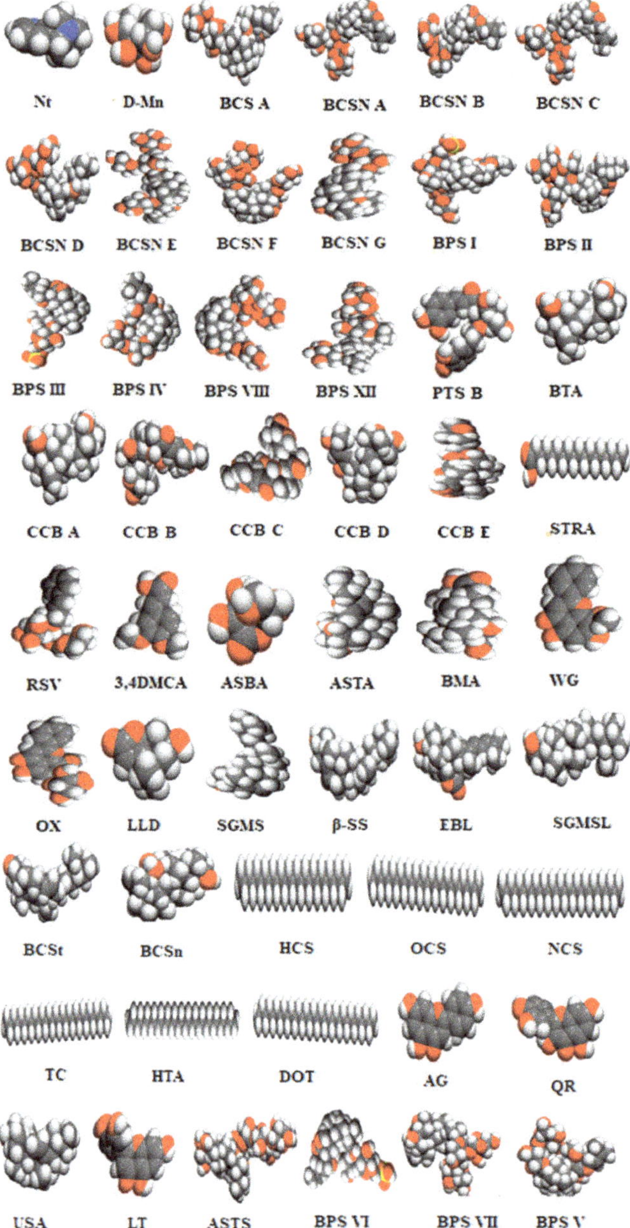

Figure 1. Phytomolecules and their 3D structures.

3.2. Identification Compounds Combating Human Targets

Phytomolecules targeting human receptors were identified using the Swiss TargetPrediction tool. The study identified 52 phytomolecules targeting 780 human direct receptors. A list of compound and human target information was given in Table S2. These human receptors/targets involved in various functions, especially on the signal transducer and activator, neuronal acetylcholine receptor, apoptosis regulator Bcl–2, transcription factor activities, controling the microtubule associated protein functions, and so forth. In addition, a functional description of the target proteins is given in Table S3.

3.3. Properties of Human Targets

Fifty-two numbers of potential compounds and 52 active human targets with their corresponding information of UniProt ID, chromosome number, start and end position and ortholog information were retrieved, and are given in Table 2. These attributes will pave the way for deciphering their detailed molecular function.

Table 2. Features of human active targets.

Compound Name	Target	Uniprot ID	Chr. No	Start	End	Orthologs
Nt	CHRNB1	P11230	17	7445061	7457707	Chrnb1 (Mus musculus)
D-Mn	TDP1	Q9NUW8	14	89954939	90044768	Tdp1 (Mus musculus)
BCS A	STAT3	P40763	17	42313324	42388568	Stat3 (Mus musculus)
BCSN A	MBNL1	Q9NR56	3	152243828	152465780	Mbnl1 (Mus musculus)
BCSN B	STAT3	P40763	17	42313324	42388568	Stat3 (Mus musculus)
BCSN C	PTPN2	P17706	18	12785478	12929643	Ptpn2 (Mus musculus)
BCSN D	PTPN1	P18031	20	50510321	50585241	Ptpn1 (Mus musculus)
BCSN E	MBNL1	Q9NR56	3	152243828	152465780	Mbnl1 (Mus musculus)
BCSN F	MBNL1	Q9NR56	3	152243828	152465780	Mbnl1 (Mus musculus)
BCSN G	MAPT	P10636	17	45894382	46028334	Mapt (Mus musculus)
BPS I	FGF1	P05230	5	142592178	142698070	Fgf1 (Mus musculus)
BPS II	PTPN1	P18031	20	50510321	50585241	Ptpn1 (Mus musculus)
BPS III	FGF1	P05230	5	142592178	142698070	Fgf1 (Mus musculus)
BPS IV	MBNL1	Q9NR56	3	152243828	152465780	Mbnl1 (Mus musculus)
BPS V	STAT3	P40763	17	42313324	42388568	Stat3 (Mus musculus)
BPS VIII	MBNL1	Q9NR56	3	152243828	152465780	Mbnl1 (Mus musculus)
BPS XII	STAT3	P40763	17	42313324	42388568	Stat3 (Mus musculus)

Table 2. *Cont.*

Compound Name	Target	Uniprot ID	Chr. No	Start	End	Orthologs
PTS B	PRKCG	P05129	19	53879190	53907652	Prkcg (Mus musculus),
BTA	AKR1B10	O60218	7	134527592	134541408	Akr1b10 (Mus musculus)
CCB A	ITGAL	P20701	16	30472658	30523185	Itgal (Mus musculus)
CCB B	ITGAL	P20701	16	30472658	30523185	Itgal (Mus musculus)
CCB C	ITGAL	P20701	16	30472658	30523185	Itgal (Mus musculus)
CCB D	ITGAL	P20701	16	30472658	30523185	Itgal (Mus musculus)
CCB E	ITGAL	P20701	16	30472658	30523185	Itgal (Mus musculus)
STRA	FABP3	P05413	1	31365625	31376850	Fabp3 (Mus musculus)
RSV	MMP1	P03956	11	102789920	102798160	Mmp1a (Mus musculus)
3,4DMCA	CA1	P00915	8	85327608	85379014	Car1 (Mus musculus)
ASBA	TDP1	Q9NUW8	14	89954939	90044768	Tdp1 (Mus musculus)
ASTA	AKR1B10	AKR1B10	7	134527592	134541408	Akr1b10 (Mus musculus)
BMA	AKR1B10	O60218	7	134527592	134541408	Akr1b10 (Mus musculus)
WG	PTGS1	P23219	9	122370530	122395703	Ptgs1 (Mus musculus)
OX	ADORA1	P30542	1	203090654	203167405	Adora1 (Mus musculus)
LLD	TDP1	Q9NUW8	14	89954939	90044768	Tdp1 (Mus musculus)
SGMS	AR	P10275	10	67544032	67730619	Ar (Mus musculus)
β-SS	TDP1	Q9NUW8	14	89954939	90044768	Tdp1 (Mus musculus)
EBL	DRD2	DRD2	11	113409615	113475691	Drd2 (Mus musculus)
SGMSL	AR	P10275	10	67544032	67730619	Ar (Mus musculus)
BCSt	TDP1	Q9NUW8	14	89954939	90044768	Tdp1 (Mus musculus)
BCSn	POLB	P06746	8	42338454	42371808	Polb (Mus musculus)
HCS	CES2	O00748	16	66934444	66945096	Ces2c (Mus musculus)
OCS	CES2	O00748	16	66934444	66945096	Ces2c (Mus musculus)
NCS	CDC25A	P30304	3	48157146	48188402	Cdc25a (Mus musculus)
TC	CES2	O00748	16	66934444	66945096	Ces2c (Mus musculus)
HTA	CES2	O00748	16	66934444	66945096	Ces2c (Mus musculus)

Table 2. *Cont.*

Compound Name	Target	Uniprot ID	Chr. No	Start	End	Orthologs
DOT	CES2	O00748	16	66934444	66945096	*Ces2c (Mus musculus)*
AG	AKR1B10	O60218	7	134527592	134541408	*Akr1b10 (Mus musculus)*
QR	CA12	O43570	15	63321378	63382161	*Car12 (Mus musculus)*
USA	POLB	P06746	8	42338454	42371808	*Polb (Mus musculus)*
LT	MMP1	P03956	11	102789920	102798160	*Mmp1a (Mus musculus)*
ASTS	STAT3	P40763	17	42313324	42388568	*Stat3 (Mus musculus)*
BPS VI	FGF1	P05230	5	142592178	142698070	*Fgf1 (Mus musculus)*
BPS VII	MAPT	P10636	17	45894382	46028334	*Mapt (Mus musculus)*

3.4. GO Annotation

Human targets with their characteristic features were analyzed by official gene symbols using the GOnet database, and showed a significant involvement of these proteins in diverse molecular functions, cellular components and biological processes. The target gene-encoding proteins were predicted to be involved in biological regulation, including in synaptic regulation, immune system processes, cell–cell signaling, responses to stimuli, and developmental growth (Figure 2). In cellular components, targets were present in the synapse, organelle, membrane, and protein-containing complex (Figure 3). The inherent molecular functions of these proteins corresponded to different types of catalytic activity, binding activity and transcription regulator activity (Figure 4). The activation effect of these compounds on their encoding human targets would definitely reduce the risk of NDs and their associated diseases. For example, these active compounds serve as an inhibitor of Aβ – peptide accumulation in the brain in the development of Alzheimer's disease.

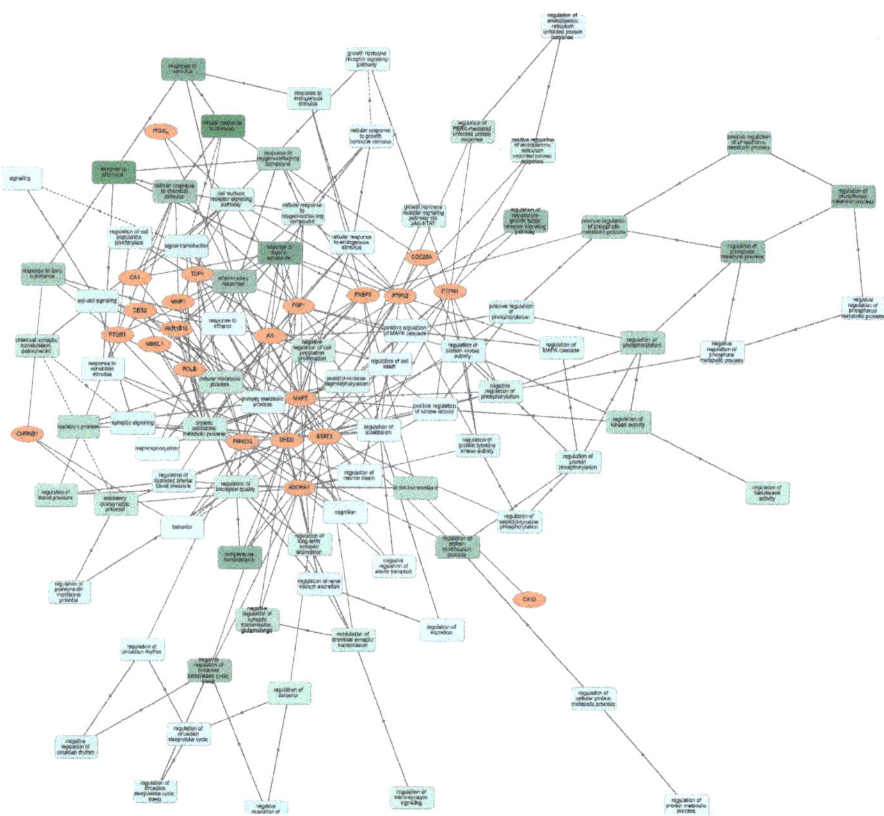

Figure 2. Classification human targets with their biological processes. Orange color encodes human targets; green color represents biological processes.

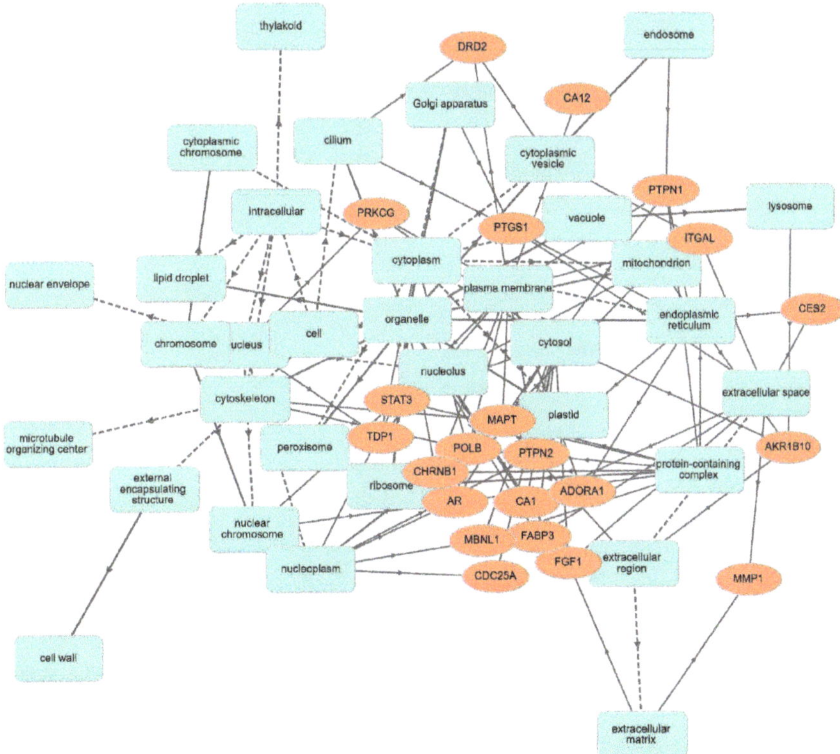

Figure 3. Classification human targets with corresponding cellular component. Orange color encodes human targets; green color represents cellular components.

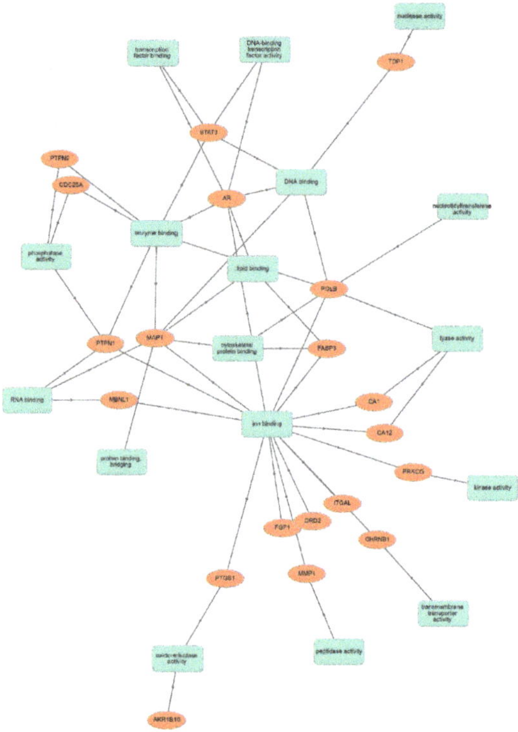

Figure 4. Classification human targets with encoding molecular functions. Orange color encodes human targets; green color represents various molecular functions.

3.5. C-T-N Analysis

The C-T-N result was displayed in Figure 5, consisting of 52 compounds and 780 human targets. Network interaction between the compounds and targets revealed the multi-target properties of elements, which is the essence of the plausible action mode of herbal drugs. As for the human targets, 52 compounds (Table 2) had a high probability, which showed the potential therapeutic effect of each compound present in Bacopa for combatting NDs, particularly Alzheimer's and SCAby—modulating, inhibiting and or transducing the signals of these possible proteins.

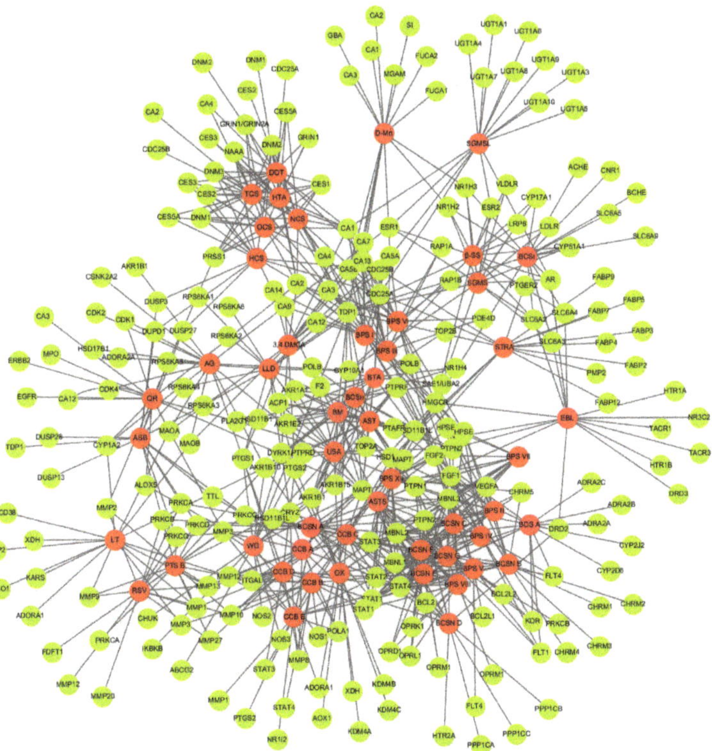

Figure 5. Compound target network (C-T-N). The red color represents compounds and the green color indicates targets.

3.6. T-D-N Analysis

T-D-N analysis illustrated the particular mechanisms of the potential drug case by case, which were most relevant to NDs (especially Alzheimer's and SCA) and their associated targets, and were constructed into T–D interaction (Figure 6). The results showed that the treatment of NDs has the multi-target therapeutic efficacy present in Bacopa-derived compounds. These compounds may alter the targets (yet to be characterized), and their disease-associated pathways deserve more attention in continuous therapy.

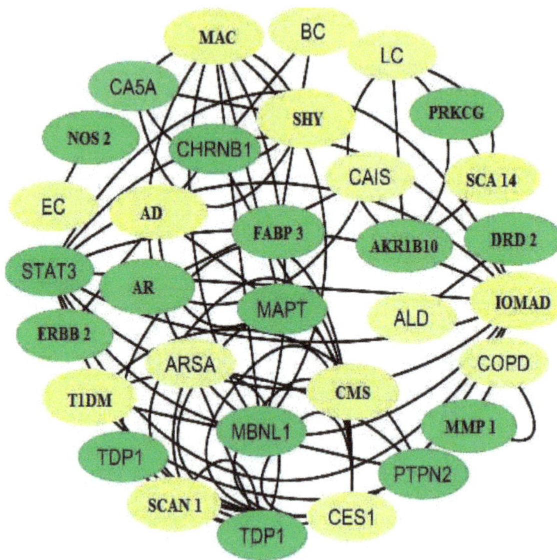

Figure 6. Target disease network (T-D-N). The dark color represents targets and the light color indicates diseases.

3.7. Features of Active Molecules and Novel Compounds

The significant calculated properties of GPCR, Ki, Pi, Ei, Ncr, and nVio were retrieved and are given in Table 3. Based on the number of violations, enzyme inhibitor activity feature scores above 0.5 were considered to be of a significant level. According to this, commercially available drugs (responsible for Alzheimer's and SCA) and compounds were compared, and novel compounds were identified and are listed in Table 4.

Table 3. Active Compounds and its Features.

Compound	GPCR lg	Ki	Ncr	Pi	Ei	Nvio
Nt	−0.32	−0.79	−1.57	−0.8	−0.31	0
D-Mn	−0.64	−0.88	−0.88	−0.72	−0.01	1
BCS A	−1.05	−2	−1.64	−0.74	−1.14	3
BCSN A	−0.77	−1.61	−1.13	−0.48	−0.68	3
BCSN B	−0.57	−1.51	−1.21	−0.36	−0.58	3
BCSN C	−2.69	−3.53	−3.34	−2.21	−2.67	3
BCSN D	−0.89	−1.87	−1.54	−0.59	−0.97	3
BCSN E	−3.6	−3.79	−3.73	−3.53	−3.51	3
BCSN F	−3.65	−3.82	−3.77	−3.6	−3.57	3
BCSN G	−0.62	−1.61	−1.24	−0.46	−0.52	3
BPS I	−3.13	−3.68	−3.61	−2.75	−2.9	3
BPS II	−2.99	−3.6	−3.48	−2.6	−2.95	3
BPS III	−1.65	−2.85	−2.47	−1.05	−1.7	3

Table 3. *Cont.*

Compound	GPCR lg	Ki	Ncr	Pi	Ei	Nvio
BPS IV	−1.04	−1.99	−1.52	−0.76	−1.04	3
BPS V	−1.04	−1.99	−1.52	−0.76	−1.04	3
BPS VIII	−3.65	−3.82	−3.77	−3.6	−3.57	3
BPS XII	−3.65	−3.8	−3.77	−3.58	−3.59	3
PTS B	0.21	−0.04	0.1	0.13	0.36	2
BTA	0.31	−0.5	0.93	0.14	0.55	1
CCB A	0.45	−0.42	0.79	0.1	0.52	1
CCB B	0.52	−0.4	0.88	0.1	0.56	1
CCB C	0.42	−0.36	0.8	0.12	0.61	1
CCB D	0.54	−0.33	0.91	0.06	0.65	1
CCB E	0.47	−0.38	0.69	0.04	0.47	1
STRA	0.11	−0.2	0.17	0.06	0.2	1
RSV	0.14	−0.08	−0.08	0.02	0.34	1
3,4DMCA	−0.42	−0.66	−0.15	−0.68	−0.13	0
ASBA	−0.53	−1.09	−0.01	−0.81	0.2	0
ASTA	0.2	−0.46	0.91	0.28	0.66	0
BMA	0.25	−0.45	0.93	0.29	0.75	1
WG	−0.14	0.12	0.13	−0.31	0.23	0
OX	0.02	−0.06	0.24	−0.06	0.39	1
LLD	−0.45	−0.91	−0.04	−0.33	0.56	0
SGMS	0.12	−0.48	0.74	−0.02	0.53	1
β-SS	0.14	−0.51	0.73	0.07	0.51	1
EBL	0.33	−0.12	0.92	0.13	0.68	1
SGMSL	0.21	−0.35	0.65	0.24	0.48	1
BCSt	0.21	−0.37	0.56	0.15	0.47	1
BCSn	0.29	−0.49	0.91	0.19	0.56	1
HCS	0.04	−0.04	0.04	0.04	0.03	1
OCS	0.04	−0.04	0.04	0.04	0.03	1
NCS	0.04	−0.04	0.04	0.03	0.02	1
TC	0.04	−0.04	0.04	0.03	0.02	1
HTA	0.03	−0.03	0.04	0.03	0.02	1
DOT	0.03	−0.03	0.03	0.03	0.02	1
AG	−0.07	0.18	0.34	−0.25	0.26	0
QR	−0.06	0.28	0.36	−0.25	0.28	0
USA	0.28	−0.5	0.89	0.23	0.69	1
LT	−0.02	0.26	0.39	−0.22	0.28	0
ASTS	−3.38	−3.7	−3.55	−2.96	−3.26	3
BPS VI	−1.65	−2.85	−2.47	−1.05	−1.7	3
BPS VII	−2.73	−3.56	−3.37	−2.28	−2.62	3

Table 4. Comparison of Drug and Novel Compounds.

Drug	GPCR lg	Ki	Ncr	Pi	Ei	Nvio
Alzheimer's Disease						
Tacrine	−0.11	−0.37	−0.93	−0.59	0.43	0
Edrophonium	−0.64	−1.59	−1.42	−0.87	0.69	0
Neostigmine	−0.22	−0.82	−0.35	−0.07	−0.66	0
Donepezil	0.22	−0.16	0.03	0.03	0.25	0
Pyriostigmine	−0.19	−0.86	−2.19	−0.46	0.59	0
Compound	GPCR lg	Ki	Ncr	Pi	Ei	nvio
WG	−0.14	0.12	0.13	−0.31	0.23	0
ASTA	0.2	−0.46	0.91	0.28	0.66	0
AG	−0.07	0.18	0.34	−0.25	0.26	0
QR	−0.06	0.28	0.36	−0.25	0.28	0
LLD	−0.45	−0.91	-0.04	−0.33	0.56	0
CCB A	0.45	−0.42	0.79	0.1	0.52	1
CCB B	0.52	−0.4	0.88	0.1	0.56	1
CCB C	0.42	−0.36	0.8	0.12	0.61	1
BMA	0.25	−0.45	0.93	0.29	0.75	1
OX	0.02	−0.06	0.24	−0.06	0.39	1
USA	0.28	−0.5	0.89	0.23	0.69	1
BCSn	0.29	−0.49	0.91	0.19	0.56	1
BTA	0.31	−0.5	0.93	0.14	0.55	1
Spinocerebellar Ataxia						
Drug	GPCR lg	Ki	Ncr	Pi	Ei	nvio
Phenytoin	0.07	−0.47	−0.32	0.01	−0.02	0
Primidone	−0.06	−0.58	−0.64	−0.38	−0.06	0
Valporic Acid	−0.83	−1.55	−0.78	−0.74	−0.39	0
Trimethadione	−0.59	−1.44	−1.53	−0.54	−0.6	0
Mephenytoin	−0.65	−1.25	−1.68	−0.76	−0.51	0
Lamotrigine	−0.16	0.36	−1.12	−0.84	0.08	0
Ethosuximide	−0.76	−2.1	−1.7	−1.1	−0.38	0
Ethotoin	−0.22	−0.98	−1.32	−0.43	−0.16	0
Oxcarbazepine	−0.02	0.1	−0.35	−0.26	−0.2	0
Compound	GPCR lg	Ki	Ncr	Pi	Ei	nvio
ASTA	0.2	−0.46	0.91	0.28	0.66	0
WG	−0.14	0.12	0.13	−0.31	0.23	0
LLD	−0.45	−0.91	−0.04	−0.33	0.56	0
OX	0.02	−0.06	0.24	−0.06	0.39	1
BCSn	0.29	−0.49	0.91	0.19	0.56	1
BTA	0.31	−0.5	0.93	0.14	0.55	1
D-Mn	−0.64	−0.88	−0.88	−0.72	−0.01	1

Table 4. *Cont.*

Drug	GPCR lg	Ki	Ncr	Pi	Ei	Nvio
SGMS	0.12	−0.48	0.74	−0.02	0.53	1
β-SS	0.14	−0.51	0.73	0.07	0.51	1
BCSt	0.21	−0.37	0.56	0.15	0.47	1
HCS	0.04	−0.04	0.04	0.04	0.03	1
OCS	0.04	−0.04	0.04	0.04	0.03	1
NCS	0.04	−0.04	0.04	0.03	0.02	1
TC	0.04	−0.04	0.04	0.03	0.02	1
HTA	0.03	−0.03	0.04	0.03	0.02	1
DOT	0.03	−0.03	0.03	0.03	0.02	1

4. Discussion

The prevalence of NDs and the ineffectiveness of allopathic medicine in dealing with multiple trait disorders mean it is crucial that we research and develop successful curative systems. The significant efficacy of Indian traditional medicine has been well established over 1000 years of practice. Though the pharmaceutical components of drugs have been extracted and purified for new drug development, this method always ends in failure due to the violation of functional drug regulation. Traditional Indian medicine treating multiple complex diseases can be also seen as a complexity fronting another complexity, which primarily speculates through the equilibrium of the entire human body system by controlling the molecular interactions between all the elements within the species. Nevertheless, the exact mode of action on the target protein and pathway stage of traditional or herbal drug mechanisms remains a roadblock to us. Therefore, in the present study, we applied a cheminformatics and integrated pharmacology approach to human systems to unravel the pharmacological role of Bacopa-derived phytomolecules in the treatment of NDs at a molecular system level. This finding also described the novel compounds for treating Alzheimer's and SCA by inhibiting AChE, β-amyloid accumulation in the brain and degenerative changes in the cerebellum and spinal cord, respectively.

In this study, with the help of cheminformatics and the PubChem database, 52 active compounds [16] were identified. All 52 compounds strongly interact with 780 direct human targets by drug targeting. Interestingly, it was predicted to be involved in diverse biological activities against NDs (Table S3) that have not been previously reported, demonstrating the reliability of the Swiss TargetPrediction, UniProtKB and GOnet evaluation methods. A Cytoscape v3.7.2 was then executed to get the molecular interaction straight to C-T-N. The analytical results revealed the various biological processes and modes of action used by drug compounds to achieve their curative effects. Finally, in further analyses to decipher their therapeutic potential of Bacopa, a T-D-N and an NDs-associated pathway (especially for Alzheimer's and SCA/long-term depression) were constructed (Figures S1 and S2).

Previous reports have detailed that even though the "unique/single target" compounds utilize their utmost suppressive effects on their direct human targets, they may not always produce appropriate results [22]. To the best of our knowledge, various compounds acting on multiple or the same targets hit by the same compounds gain more efficacy on binding opportunities with each other and gain more chances of affecting the entire interaction equilibrium, making the Indian traditional medicine therapy more efficacious and fruitful to society. The C-T-N analytical results showed that there are multi-target functional modules in a single active compound. Of the 52 compounds with corresponding human targets, all of them were able to act on more than 15 active targets. For example, CHRNB1 (acetylcholine ion-channel activation receptor), indicated for the activation of the cholinergic receptor nicotinic β1 subunit, has shown that myasthenic syndrome is usually the result of the prolonged activation of CHRNB1, which is mechanistically correlated with the development of NDs. Accordingly, active compounds can exert their various biological effects in this complex human system with numerous compound–target–disease interactions by controlling the associated goals and pathways and achieving curative outcomes for NDs.

NDs are not caused by a single factor, affecting different types of NDs and aggravating one another. Some stressors involved in the development of NDs are complex, but at the same time drugs have not yet emerged specifically for those conditions. Therefore, the multi-level curative effects of Bacopa phytomolecules include controlling various physiological functional regulation, transcriptional reprogramming—particularly on cerebral blood flow, and their important effectiveness in the battle against NDs from a panoramic perspective. Since modern medicine is incapable of making a breakthrough so far in the treatment of NDs, why should we not transfer our focus to Indian traditional medicine, which is well documented as confronting complex NDs?

5. Conclusions

B. monnieri showed various potential actions in the amelioration of cognitive disorders and cognitive enhancement in healthy people. Biomedical research on *B. monnieri* is still at a roadblock. Notably, our results on novel compounds—with their encoding properties, active targets, biological processes, and interactions—have opened the research floodgates with the integration of Ayurveda to the modern medicine era. This study also hypothesizes that Bacopa compounds and their combination with other substances—as is recommended by the Ayurvedic and modern medicine system—may result in synergistic effects and need to be studied further. The ethical implications of drugs which enhance cognition are vital, but should be appropriately mitigated with social and ethical considerations as field researchers enter the brave, advancing world of neural enhancement.

Supplementary Materials: The following are available online at http://www.mdpi.com/2218-273X/10/4/536/s1, Figure S1: title, Table S1: title, Video S1: title.

Author Contributions: M.R. and P.M. conceived and designed the experiments. R.J., P.M. and V.S. performed the experiments. R.J., P.M., V.S. and J.-T.C. analyzed the results. R.J., P.M. wrote the manuscript. M.R. and. J.-T.C. revised the manuscript. M.R. approved the final version of the manuscript. All authors have read and approved the final manuscript.

Funding: This research received no extrenal funding.

Acknowledgments: The authors thank RUSA 2.0 [F. 24-51/2014-U, Policy (TN Multi-Gen), Dept of Edn, GoI]. RJ and PM sincerely acknowledge the computational and bioinformatics facility provided by the Alagappa University Bioinformatics Infrastructure Facility (funded by DBT, GOI; File No. BT/BI/25/012/2012,BIF). The authors also thankfully acknowledge DST-FIST (Grant No. SR/FST/LSI-639/2015(C)), UGC-SAP (Grant No.F.5-1/2018/DRS-II (SAP-II)) and DST-PURSE (Grant No. SR/PURSE Phase 2/38 (G)) for providing instrumentation facilities.

Conflicts of Interest: The authors declare no conflict of interest.

References

1. Davidson-Hunt, I. Ecological ethnobotany: Stumbling toward new practices and paradigms. *MASA J.* **2000**, *16*, 1–13.

2. Anonymous. *The Ayurvedic Pharmacopoeia of India*; Government of India: New Delhi, India; Ministry of Health and Family Welfare: New Delhi, India, 2001; Part-I; Volume III.

3. Gohil, K.J.; Patel, J.A. A review on Bacopa monniera: Current research and future prospects. *Int. J. Green Pharm.* **2010**, *4*, 1–9. [CrossRef]

4. Jyoti, A.; Sharma, D. Neuroprotective role of Bacopa monniera extract against aluminium-induced oxidative stress in the hippocampus of rat brain. *Neuro Toxicol.* **2006**, *27*, 457. [CrossRef] [PubMed]

5. Calabrese, C.; Gregory, W.L.; Leo, M.; Kraemer, D.; Bone, K.; Oken, B. Effects of a standardized Bacopa monnieri extract on cognitive performance, anxiety, and depression in the elderly: A randomized, double-blind, placebo-controlled trial. *J. Altern. Complement. Med.* **2008**, *14*, 707–713. [CrossRef] [PubMed]

6. Kamkaew, N.; Scholfield, C.N.; Ingkaninan, K.; Taepavarapruk, N.; Chootip, K. Bacopa monnieri increases cerebral blood flow in rat independent of blood pressure. *Phytother. Res.* **2013**, *27*, 135–138. [CrossRef] [PubMed]

7. Devishree, R.A.; Kumar, S.; Jain, A.R. Short term effect of Bacopa monnieri on memory- A brief review. *J. Pharm. Res.* **2017**, *11*, 1447–1450.

8. Russo, A.; Borrelli, F. Bacopa monniera, a reputed nootropic plant: An overview. *Phytomedicine* **2005**, *12*, 305–317. [CrossRef]

9. Sinha, S.; Saxena, R. Effect of iron on lipid peroxidation, and enzymatic and non-enzymatic antioxidants and bacoside-A content in medicinal plant Bacopa monnieri L. *Chemosphere* **2006**, *62*, 1340–1350. [CrossRef]

10. Ramasamy, S.; Chin, S.P.; Sukumaran, S.D.; Buckle, M.J.C.; Kiew, L.V.; Chung, L.Y. In silico and in vitro analysis of bacoside a aglycones and its derivatives as the constituents responsible for the cognitive effects of Bacopa monnieri. *PLoS ONE* **2015**, *10*, e0126565. [CrossRef]

11. Ernst, E. Herbal remedies for anxiety–a systematic review of controlled clinical trials. *Phytomedicine* **2006**, *13*, 205–208. [CrossRef]

12. Aguiar, S.; Borowski, T. Neuropharmacological review of the nootropic herb Bacopa monnieri. *Rejuvenation Res.* **2013**, *16*, 313–326. [CrossRef] [PubMed]

13. Singh, H.K.; Rastogi, R.P.; Srimal, R.C.; Dhawan, B.N. Effect of bacosides A and B on avoidance responses in rats. *Phytother. Res.* **1988**, *2*, 70–75. [CrossRef]

14. Dhawan, B.N.; Singh, H.K. Pharmacological studies on Bacopa monnieri, an Ayurvedic nootropic agent. *Eur. Neuropsychopharmacol.* **1996**, *6*, 144. [CrossRef]

15. Singh, H.K.; Dhawan, B.N. Neuropsychopharmacological effects of the Ayurvedic nootropic Bacopa monniera Linn. (Brahmi). *Indian J. Pharm.* **1997**, *29*, 359.

16. Shefin, B.; Sreekumar, S.; Biju, C.K. Identification of lead molecules with anti-hepatitis B activity in Bacopa monnieri (L.) Wettst. and Cassia fistula L. through in silico method. *J. Pharm. Biol. Sci.* **2016**, *11*, 16–21.

17. Bolton, E.E.; Wang, Y.; Thiessen, P.A.; Bryant, S.H. PubChem: Integrated platform of small molecules and biological activities. In Annual reports in computational chemistry. *Annu. Rep. Comput. Chem.* **2008**, *4*, 217–241.

18. Sharma, C.S.; Mishra, S.S.; Singh, H.P.; Kumar, N. In silico ADME and toxicity study of some selected antineoplastic drugs. *Int. J. Pharm. Sci. Drug Res.* **2016**, *8*, 65–67. [CrossRef]

19. Petryszak, R.; Keays, M.; Tang, Y.A.; Fonseca, N.A.; Barrera, E.; Burdett, T.; Füllgrabe, A.; Fuentes, A.M.; Jupp, S.; Koskinen, S.; et al. Expression Atlas update—An integrated database of gene and protein expression in humans, animals and plants. *Nucleic Acids Res.* **2016**, *44*, D746–D752. [CrossRef]

20. Pomaznoy, M.; Ha, B.; Peters, B. GOnet: A tool for interactive Gene Ontology analysis. *BMC Bioinform.* **2018**, *19*, 470. [CrossRef]

21. Smoot, M.E.; Ono, K.; Ruscheinski, J.; Wang, P.L.; Ideker, T. Cytoscape 2.8: New features for data integration and network visualization. *Bioinformatics* **2011**, *27*, 431–432. [CrossRef]

22. Zhang, B.; Wang, X.; Li, S. An integrative platform of TCM network pharmacology and its application on a herbal formula, Qing-Luo-Yin. *Evid. Based Complementary Altern. Med.* **2013**, *2013*, 456747. [CrossRef] [PubMed]

Article

Luteolin-7-O-Glucoside Inhibits Oral Cancer Cell Migration and Invasion by Regulating Matrix Metalloproteinase-2 Expression and Extracellular Signal-Regulated Kinase Pathway

Bharath Kumar Velmurugan [1], Jen-Tsun Lin [2,3], B. Mahalakshmi [4], Yi-Ching Chuang [5], Chia-Chieh Lin [5], Yu-Sheng Lo [5], Ming-Ju Hsieh [5,6,7,8,*] and Mu-Kuan Chen [9,*]

[1] Toxicology and Biomedicine Research Group, Faculty of Applied Sciences, Ton Duc Thang University, Ho Chi Minh City 700000, Vietnam; bharath.kumar.velmurugan@tdtu.edu.vn
[2] Division of Hematology and Oncology, Department of Medicine, Changhua Christian Hospital, Changhua 500, Taiwan; 111227@cch.org.tw
[3] School of Medicine, Chung Shan Medical University, Taichung 402, Taiwan
[4] Institute of Research and Development, Duy Tan University, Da Nang 550000, Vietnam; mahalakshmibharath05@gmail.com
[5] Oral Cancer Research Center, Changhua Christian Hospital, Changhua 500, Taiwan; 177267@cch.org.tw (Y.-C.C.); 181327@cch.org.tw (C.-C.L.); 165304@cch.org.tw (Y.-S.L.)
[6] Institute of Medicine, Chung Shan Medical University, Taichung 402, Taiwan
[7] Department of Holistic Wellness, MingDao University, Changhua 52345, Taiwan
[8] Graduate Institute of Biomedical Sciences, China Medical University, Taichung 404, Taiwan
[9] Department of Otorhinolaryngology, Head and Neck Surgery, Changhua Christian Hospital, Changhua 500, Taiwan
* Correspondence: 170780@cch.org.tw (M.-J.H.); 53780@cch.org.tw (M.-K.C.); Tel.: +886-4-7238595 (ext. 4966) (M.-K.C.); Fax: +886-4-7232942 (M.-K.C.)

Received: 28 February 2020; Accepted: 23 March 2020; Published: 26 March 2020

Abstract: Oral squamous cell carcinoma is the sixth most common type of cancer globally, which is associated with high rates of cancer-related deaths. Metastasis to distant organs is the main reason behind worst prognostic outcome of oral cancer. In the present study, we aimed at evaluating the effects of a natural plant flavonoid, luteolin-7-O-glucoside, on oral cancer cell migration and invasion. The study findings showed that in addition to preventing cell proliferation, luteolin-7-O-glucoside caused a significant reduction in oral cancer cell migration and invasion. Mechanistically, luteolin-7-O-glucoside caused a reduction in cancer metastasis by reducing p38 phosphorylation and downregulating matrix metalloproteinase (MMP)-2 expression. Using a p38 inhibitor, SB203580, we proved that luteolin-7-O-glucoside exerts anti-migratory effects by suppressing p38-mediated increased expression of MMP-2. This is the first study to demonstrate the luteolin-7-O-glucoside inhibits cell migration and invasion by regulating MMP-2 expression and extracellular signal-regulated kinase pathway in human oral cancer cell. The study identifies luteolin-7-O-glucoside as a potential anti-cancer candidate that can be utilized clinically for improving oral cancer prognosis.

Keywords: luteolin-7-O-glucoside; oral cancer; migration; invasion; MMP-2

---

1. Introduction

Oral squamous cell carcinoma that primarily affects the lips, oral cavity, and tongue represents about 90% of all head and neck cancers. Together with pharyngeal cancer, oral carcinoma has become the sixth most common type of cancer globally [1]. Despite advancement in diagnostic processes and therapeutic interventions, oral cancer is associated with a high rate of morbidity and mortality [2]. As a result of the initial asymptomatic nature, oral cancer is often diagnosed in the later stage, resulting in distant metastasis and poor prognosis [3–6].

Metastasis that occurs in the advanced stage of cancer is the leading cause of cancer-related deaths [5]. The extra cellular matrix degradation by proteases, including matrix metalloproteinase (MMPs), is a vital phenomenon associated with cancer cell migration [7]. An overexpression or increased activity of MMPs is associated with higher cancer aggressiveness and poor survival rate [8]. MMPs are a target for developing treatment strategies against cancer. In this context, several studies have shown that selective inhibition of MMP can be associated with better cancer management [9–11].

Mitogen-activated protein kinases (MAPKs), including p38 MAP kinase, Jun N-terminal kinase 1/2/3 (JNK1/2/3), and extracellular signal-regulated kinase 1/2 (ERK1/2), are the primary regulators of proliferation, differentiation, apoptosis, migration, and invasion of cancer cells [12]. Introducing a gap between MAPK signaling pathway and MMP activity, it is known that the stimulation of p38 causes increased metastasis of neoplastic squamous epithelial cells by regulating the expression of MMPs [13]. In prostate cancer, p38 MAPK is known to trigger invasion by increasing MMP-2 expression and activity [14]. Moreover, breast cancer cells lacking p38 MAPK expression have been shown to have reduced MMP-9 activity and significantly lower rate of bone metastasis [15]. Interestingly, ERK1/2 and p38 MAPK maintains an inverse relationship wherein a high ERK activity and a low p38 activity result in increased cell growth and survival [16]. Regarding cancer metastasis, it has been found that stimulation of ERK phosphorylation leads to increased cancer metastasis [17–20]. Moreover, previous studies have shown that human lung cancer cell migration and invasion can be suppressed by pharmacologically downregulated ERK/p38 signaling pathway and inhibited MMP-2 and MMP-9 activities [21]. Similarly, in ovarian cancer, increased cell growth and migration by Rap1A, a Ras-associated protein, has been shown to be associated with elevated ERK/p38 and notch signaling [22]. Taken together, it is well-evidenced that the complex crosstalk between MAPK signaling pathway components and MMPs plays an immensely important role in regulating cancer metastasis and progression. Thus, selective targeting of any of these components through pharmacological interventions can be an effective strategy to ameliorate cancer-related burdens, especially metastasis.

Luteolin (3′,4′,5,7-hydroxyl-flavone) is a naturally occurring plant flavonoid that has been used extensively in Chinese traditional medicine because of several health benefits, including anti-inflammatory, antioxidative, and anti-cancer effects [23,24]. Mostly, luteolin is present in plants as a glycosylated component (glucoside), which is hydrolyzed in the gut to produce free luteolin during absorption [25]. Several studies evaluating the therapeutic properties of luteolin have potentiated its anti-cancer effects, which are primarily associated with increased cancer cell death, reduced proliferation, and angiogenesis, and increased cancer cell sensitization to chemotherapies [23,24,26]. Moreover, luteolin has been shown to have preventive effects against cytotoxicity produced by chemotherapeutic agents, such as cisplatin [27].

Despite having a large number of evidences on chemopreventive and anti-proliferative properties of luteolin, its effect on cancer metastasis has not been studied extensively. In the present study, we aimed at evaluating the effects of luteolin-7-O-glucoside on oral cancer cell migration and invasion. A detailed mechanism of action of luteolin-7-O-glucoside was also studied.

2. Materials and Methods

2.1. Cell Culture

Human oral squamous carcinoma cell lines (FaDu and HSC-3) were received from ATCC (Manas, VA, USA). In addition, Ca9-22 cell line was obtained from Japanese Collection of Research Bioresources Cell Bank (JCRB, Shinjuku, Japan) and cultured in the in Dulbecco's Modified Eagle Medium (DMEM; Life Technologies, Grand Island, NY, USA): Ham's F12 Nutrient Mixture (Life Technologies, Grand Island, NY, USA) supplemented with 10% FBS (Invitrogen, Waltham, MA, USA) [28]. The cells were cultured in 5% CO_2 at 37 °C.

2.2. Luteolin-7-O-Glucoside Treatments

Luteolin-7-O-glucoside (≥97% purity) was obtained from Sigma-Aldrich (St. Louis, MO, USA). Luteolin-7-O-glucoside stock solution (100 mM) was prepared using dimethyl sulfoxide (DMSO) and stored at −20 °C. The DMSO concentration was less than 0.2% for each experiment. For luteolin-7-O-glucoside treatments, appropriate amounts of stock solution were administered to the medium to get the final experimental doses.

2.3. MTT Assay

To study the effects of luteolin-7-O-glucoside on cell viability, MTT (3-(4,5-dimethylthiazol-2-yl)-2,5-diphenyltetrazolium bromide) assay was conducted using HSC-3, FaDu, and CA9-22 cells. The cells were seeded onto 24-well plates and treated with 0 to 40 µM of luteolin-7-O-glucoside solutions at 37 °C for 24 h. Cell viability was determined following methods described previously [29].

2.4. Wound Healing Assay

Oral cancer cells FaDu, CA9-22, and HSC-3 were seeded into 12-well plates and cultured to 90% confluence. A cell monolayer was scratched with a 1-mL micropipette tip in each well, and treated luteolin-7-O-glucoside (0, 10, 20, and 40 µM) at 37 °C for 0, 2, 4, 6, 24 h. The cells were photographed under a microscope for its migration ability and its mean crawling distance of cells was measured.

2.5. Cell Migration and Invasion Assay

Migration and invasion assays were performed as described previously [28,30]. Oral cancer cells were mixed in serum free medium and seeded into the upper chambers of the inserts (Greiner Bio-One, Monroe, NC, USA). The inserts were placed in 24-well plates containing complete medium with various concentration of luteolin-7-O-glucoside in lower wells. Migration and invasion ability was observed and captured under light microscope.

2.6. Western Blot Assay

Western blot analysis was performed as previously described [31]. The membranes were blocked with 5% non-fat milk in TBST (Tris-Buffered Saline, 0.1% Tween 20 Detergent) for 1 h and incubated with indicated primary antibodies for 24 h at 4 °C. After washing, the blots were incubated with peroxidase-conjugated secondary antibodies for 1 h. Finally, bands were monitored using were visualized through an ECL detection system.

2.7. Statistical Analysis

All experiments were repeated at least 3 times. All statistical analysis was performed by Student's *t*-test (Sigma Plot, version 10.0). $p < 0.05$ is considered as statistically significant.

3. Results

3.1. Luteolin-7-O-Glucoside Exerts Cytotoxic Effects on Human Oral Cancer Cells

To investigate whether luteolin-7-O-glucoside can alter cell proliferation, three oral cancer cell lines, FaDu, HSC-3, and CA9-22, were treated with various concentrations (0, 10, 20, and 40 μM) of luteolin-7-O-glucoside for 24 h, and the cell viability was determined by MTT assay. As observed in Figure 1a–c, the treatment with 20 and 40 μM of luteolin-7-O-glucoside significantly reduced the viability of oral cancer cells as compared to untreated controls, indicating the growth inhibition potential of the compound.

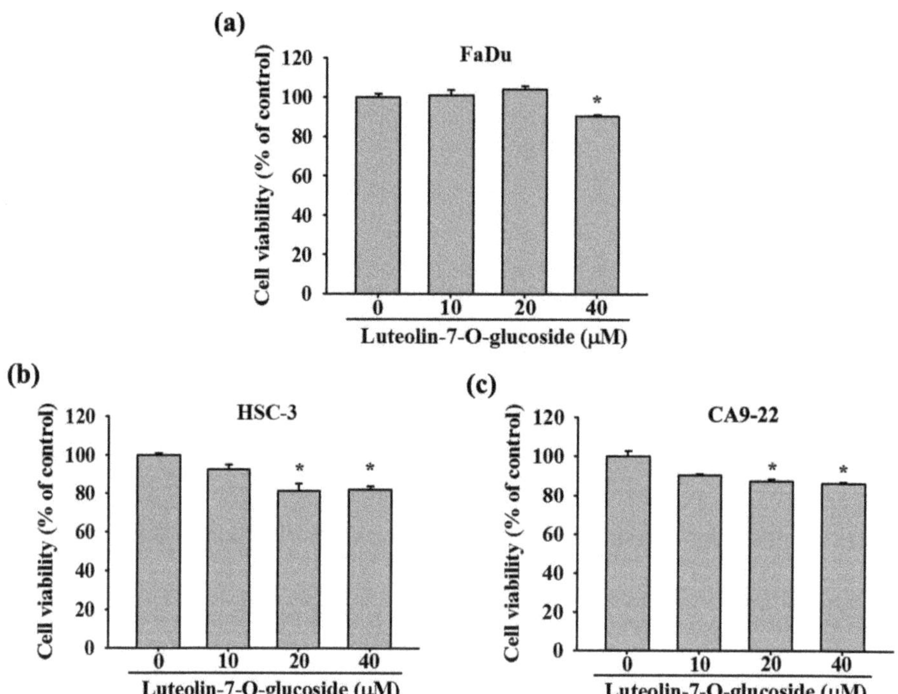

Figure 1. Cytotoxicity of Luteolin-7-O-glucosidein human oral cancer cells. (**a**) FaDu, (**b**) HSC-3, and (**c**) CA9-22 cell lines were treated with various concentrations (0, 10, 20, and 40 μM) of luteolin-7-O-glucoside for 24 h, and the cell viability was determined by MTT assay. The values are represented as mean ± SD of at least three independent experiments. *$p < 0.05$, compared with the control group.

3.2. Luteolin-7-O-Glucoside Inhibits Motility of Human Oral Cancer Cells

In addition to cell proliferation, we investigated the effects of luteolin-7-O-glucoside on cell motility. As mentioned previously, FaDu, HSC-3, and CA9-22 cells were treated with various concentrations (0, 10, 20, and 40 μM) of luteolin-7-O-glucoside for 3, 6, and 24 h. The analysis of cell motility using wound closer assay revealed that luteolin-7-O-glucoside significantly reduced the cancer cell migration in a dose-dependent manner (Figure 2a–f). These findings indicate that luteolin-7-O-glucoside possesses anti-migratory activity.

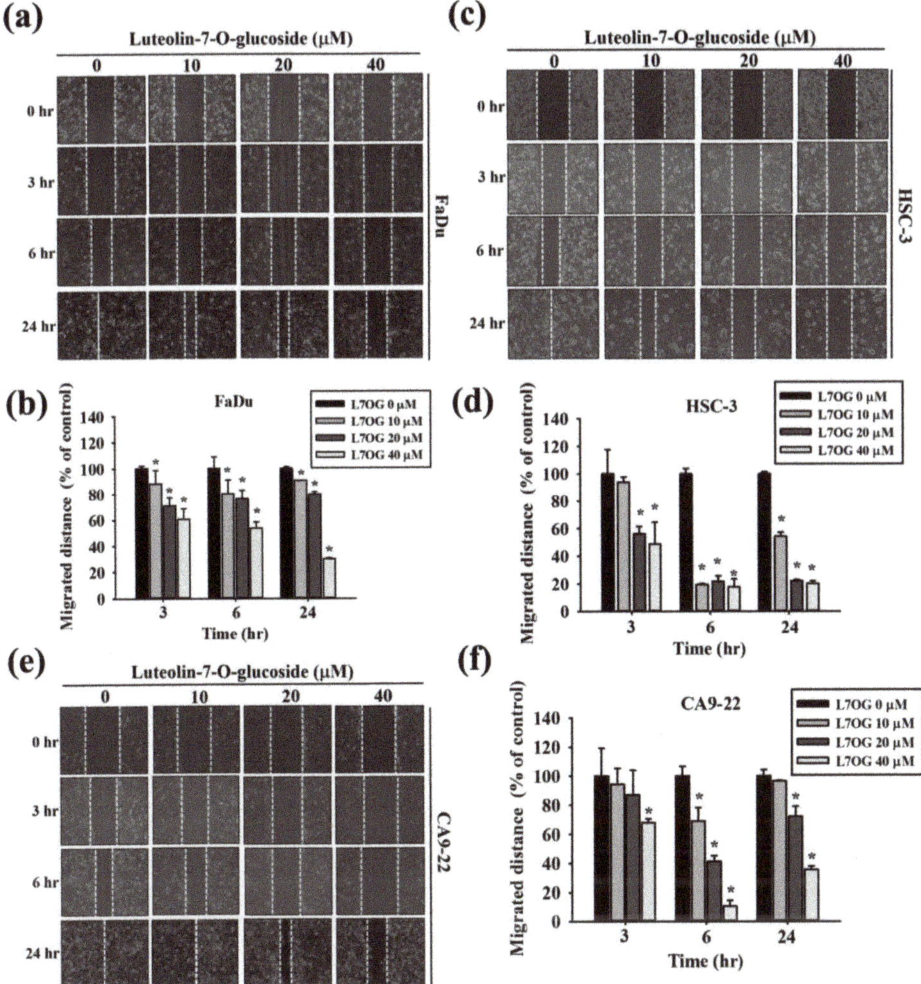

Figure 2. Luteolin-7-O-glucoside inhibits cell motility inhuman oral cancer cells. The effect of luteolin-7-O-glucoside treatment on cell motility was analyzed in (**a,b**) FaDu, (**c,d**) HSC-3, and (**e,f**) CA9-22 cells. The values are represented as mean ± SD of at least three independent experiments. *$p < 0.05$, compared with the control group.

3.3. Luteolin-7-O-Glucoside Inhibits Migration and Invasion of Human Oral Cancer Cells

To investigate whether luteolin-7-O-glucoside can inhibit both migration and invasion, FaDu, HSC-3, and CA9-22 cells were treated similarly as the previous experiments. As observed in Figure 3a,b, luteolin-7-O-glucoside significantly reduced the migration of all three oral cancer cells in a dose-dependent manner. Similarly, a significant reduction in invasion was observed after 24 h treatment with luteolin-7-O-glucoside (Figure 3c,d). Taken together, all these findings clearly indicate that luteolin-7-O-glucoside is capable of significantly ameliorating the metastatic profile of oral cancer cells.

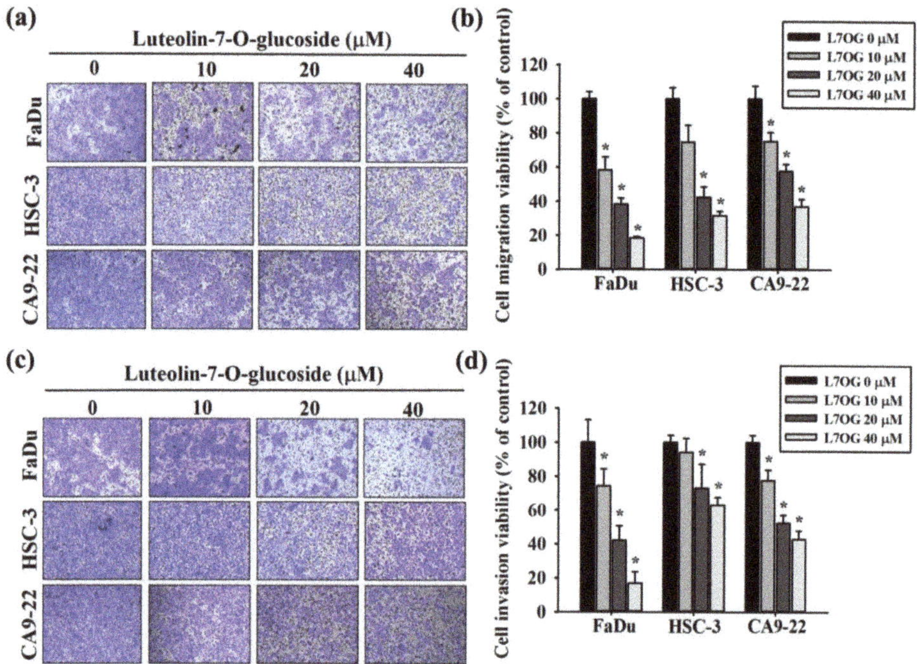

Figure 3. Luteolin-7-O-glucoside inhibits cell migration and invasion inhuman oral cancer cells. The effect of luteolin-7-O-glucoside treatment on cell migration (**a**) and invasion (**c**) was measured using trans-well assay in FaDu, HSC-3, and CA9-22 cells. The percentages of cells in migration and invasion assays are shown in (**b**) and (**d**), respectively. The values are represented as mean ± SD of at least three independent experiments. *$p < 0.05$, compared with the control group.

3.4. Luteolin-7-O-Glucoside Reduces the Protein Expression of MMP-2 in Human Oral Cancer Cells

Next, we investigated the molecular mechanism responsible for the luteolin-7-O-glucoside activity. Given the significant involvement of MMPs in cancer cell migration and invasion, we checked the protein expression of MMP-2 in FaDu and HSC-3 cells after the treatment with various concentrations (0, 10, 20, and 40 µM) of luteolin-7-O-glucoside for 24 h. As observed in Western blot analysis, all the doses of luteolin-7-O-glucoside significantly reduced the expression of MMP-2 in both the cell lines (Figure 4a–d). These findings indicate that luteolin-7-O-glucoside reduces oral cancer cell migration and invasion by decreasing the cellular level of MMP-2.

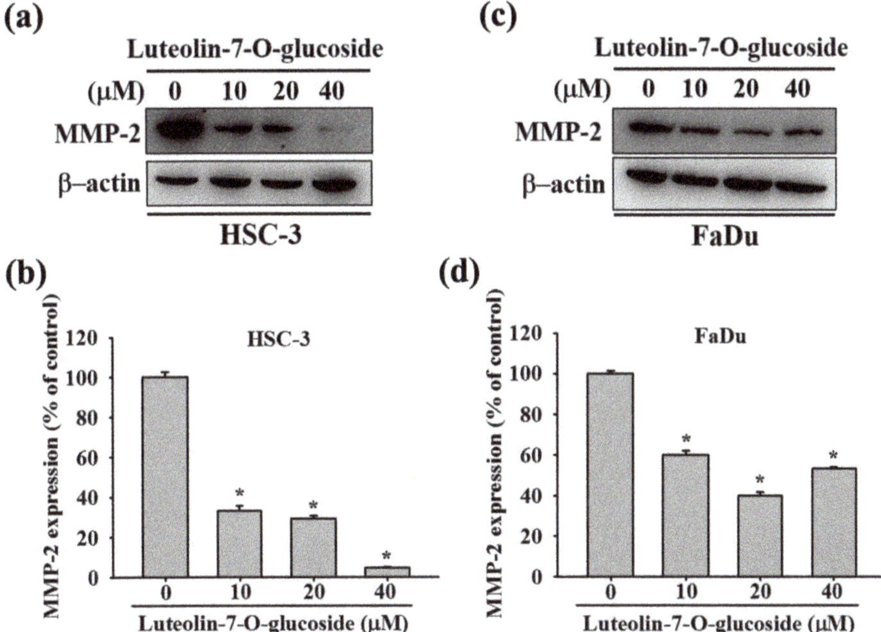

Figure 4. Luteolin-7-O-glucoside reduces the protein expression of matrix metalloproteinase (MMP)-2 in human oral cancer cells. The protein expression of MMP-2 was determined using Western blot in (**a**) HSC-3 and (**c**) FaDu cells. The quantitative results are shown in (**b**) and (**d**). The values are represented as mean± SD of at least three independent experiments. *$p < 0.05$, compared with the control group.

3.5. Luteolin-7-O-Glucoside Inhibits the p38 Pathway in Human Oral Cancer Cells

Since the expressions and activities of MMPs are tightly regulated by MAPK pathway components, we next investigated the protein phosphorylation status of ERK, p38, and JNK in FaDu and HSC-3 cells after employing similar treatment as before. As observed in Figure 5a,b, luteolin-7-O-glucoside treatment caused a significantly reduction in p38 and JNK phosphorylation in HSC-3 cells. In contrast, the phosphorylation of ERK increased significantly after the treatment. In case of FaDu cells, the treatment caused similar reduction in p38 phosphorylation as observed in HSC-3 cells (Figure 5c,d). However, the phosphorylation of JNK increased significantly after the treatment. Moreover, the phosphorylation of ERK reduced significantly only after the treatment with 40 μM of luteolin-7-O-glucoside (Figure 5c,d).

3.6. Effect of SB203580 and Luteolin-7-O-Glucoside Co-Treatment on MMP-2 Protein Expression in Human Oral Cancer Cells

Since we observed a consistent change in p38 phosphorylation status in all the oral cancer cells, we next thought of using a p38 inhibitor, SB203580, to further evaluate the mechanistic details of luteolin-7-O-glucoside action. Both FaDu and HSC-3 cells were pretreated with SB203580 for 1 h, followed by treatment with 40 μM of luteolin-7-O-glucoside for 24 h. As observed in Figure 6a–d, the collective effects of SB203580 and luteolin-7-O-glucoside caused a further reduction in MMP-2 expression, indicating that luteolin-7-O-glucoside mediates anti-migratory effects by altering p38-induced activation of MMP-2.

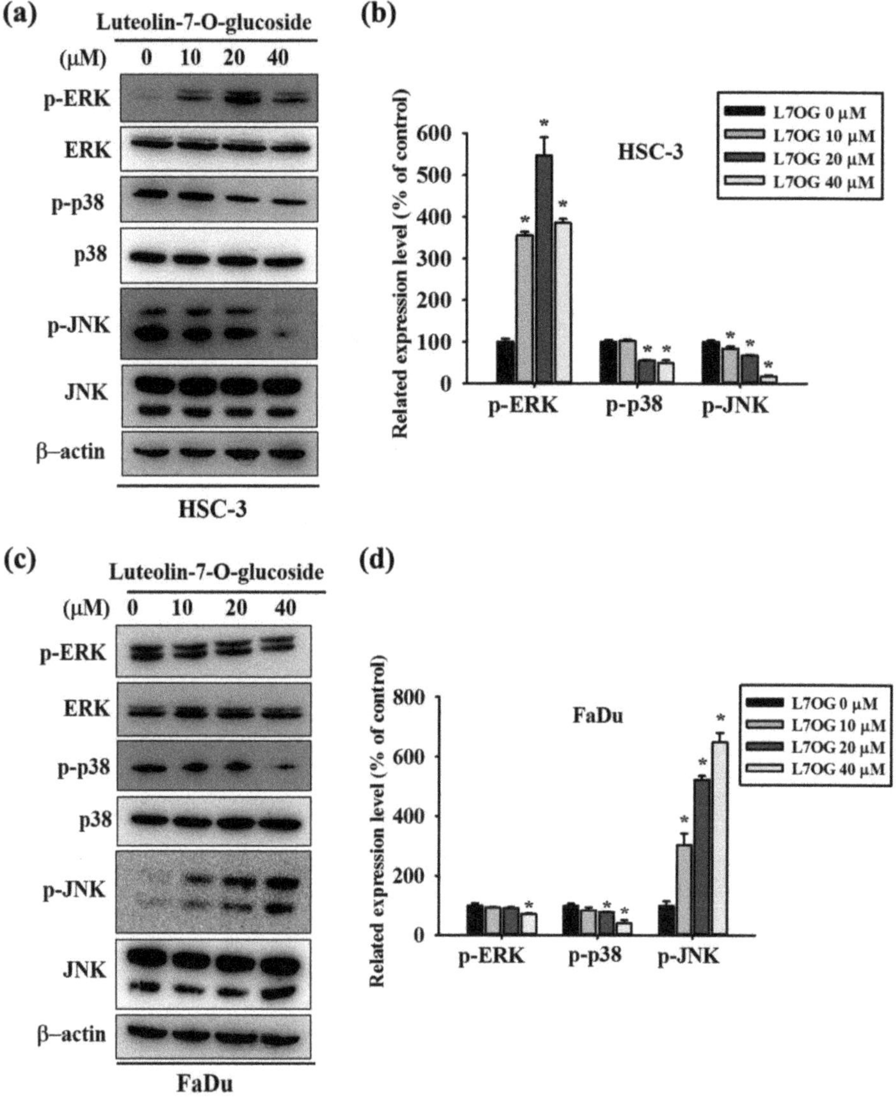

Figure 5. Luteolin-7-O-glucoside inhibits p38 pathway in human oral cancer cells. The phosphorylation as well as the total protein expressions of extracellular signal-regulated kinase 1/2 (ERK1/2), Jun N-terminal kinase 1/2 (JNK1/2), and p38 were measured after luteolin-7-O-glucoside treatment for 24 h in (**a,b**) HSC-3 and (**c,d**) FaDu cell lines. The values are represented as mean ± SD of at least three independent experiments. *$p < 0.05$, compared with the control group.

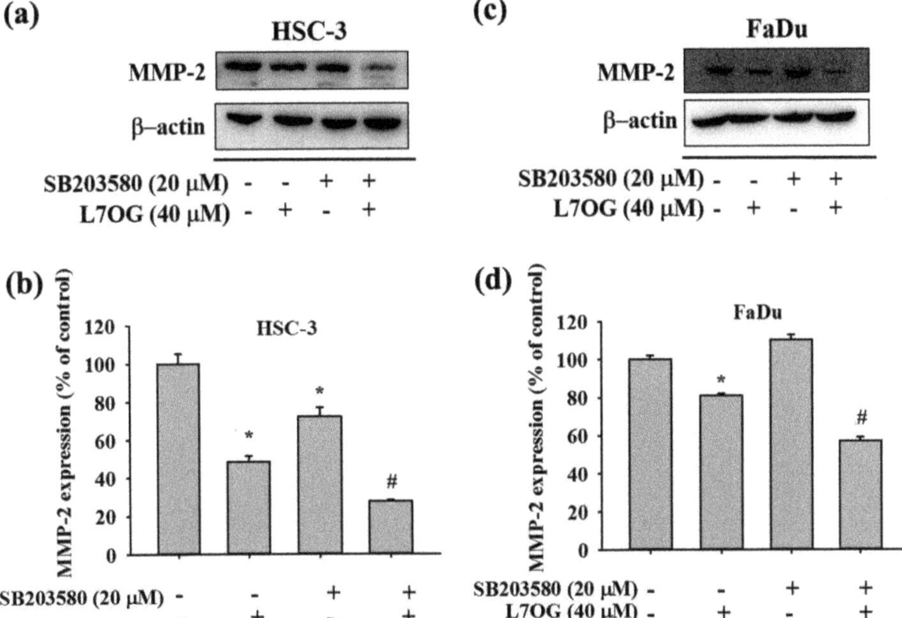

Figure 6. Effect of SB203580 and luteolin-7-O-glucoside co-treatment on MMP-2 protein expression in HSC-3 and FaDu cell lines. (**a,b**) HSC-3 and (**c,d**) FaDu cell lines were pre-treated with SB203580 for 1h, followed by treatment with Luteolin-7-O-glucoside for 24 h. Next, the culture medium was subjected to western blot assay to determine the MMP-2 expression. The values are represented as mean ± SD of at least three independent experiments. *$p < 0.05$, compared to the control group; #$p < 0.05$, compared to the luteolin-7-O-glucoside treated group.

3.7. Effect of SB203580 and Luteolin-7-O-Glucoside Co-Treatment on Cell Motility in Human Oral Cancer Cells

To check the effects of SB203580 and luteolin-7-O-glucoside co-treatment on cell motility, we performed wound closer assay in FaDu and HSC-3 cells using the similar treatment as mentioned before. As observed in Figure 7a–d, the co-treatment caused further reduction in cell motility as compared to the luteolin-7-O-glucoside treatment alone. These findings further confirm the involvement of p38 pathway in mediating the effects of luteolin-7-O-glucoside in oral cancer cells.

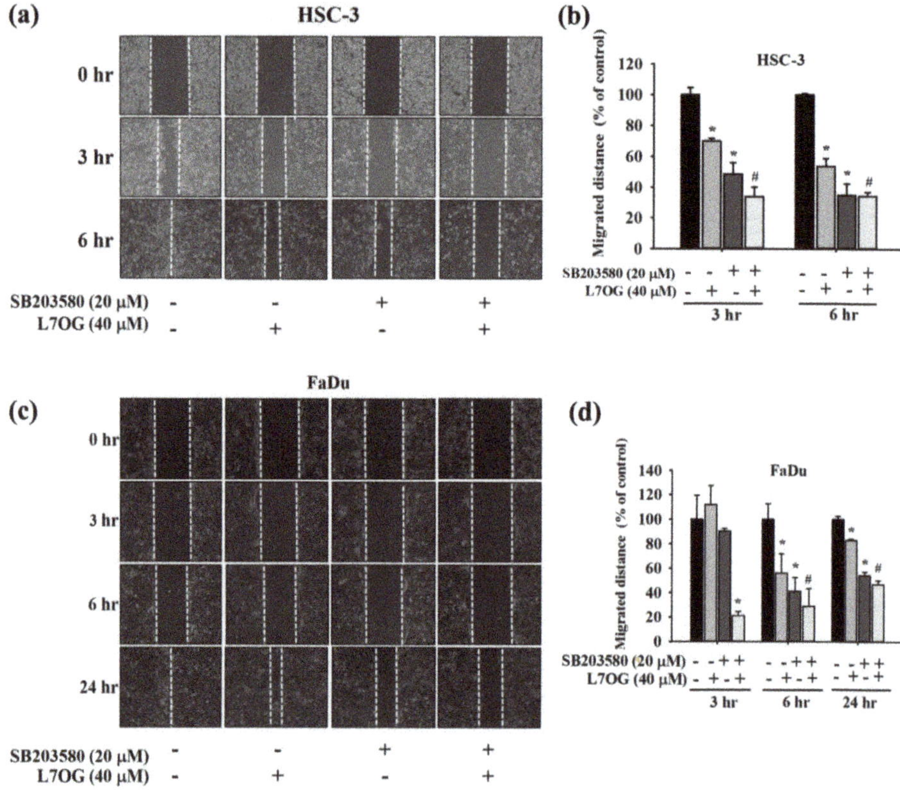

Figure 7. Effect of SB203580 and luteolin-7-O-glucoside co-treatment on cell motility in HSC-3 and FaDu cell lines. The cell motility was measured using wound healing assay after the co-treatment with SB203580 and luteolin-7-O-glucosidein (**a,b**) HSC-3 and (**c,d**) FaDu cell lines. The values are represented as mean ± SD of at least three independent experiments. *$p < 0.05$, compared to the control group; #$p < 0.05$, compared to the luteolin-7-O-glucoside treated group.

3.8. Effect of SB203580 and Luteolin-7-O-Glucoside Co-Treatment on Cell Migration in Human Oral Cancer Cells

Given the significant effect of SB203580 and luteolin-7-O-glucoside co-treatment on cell motility, we performed trans-well migration assay to finally prove the involvement of p38 pathway and MMP-2 in mediating luteolin-7-O-glucoside-induced reduction in oral cancer cell migration. As clearly observed in Figure 8a–c, the co-treatment resulted in a significantly higher reduction in cell migration as compared to the luteolin-7-O-glucoside treatment alone. Taken together, all these findings clearly indicate that luteolin-7-O-glucoside-induced inhibition of oral cancer cell metastasis is mediated by the combined action of p38 and MMP-2.

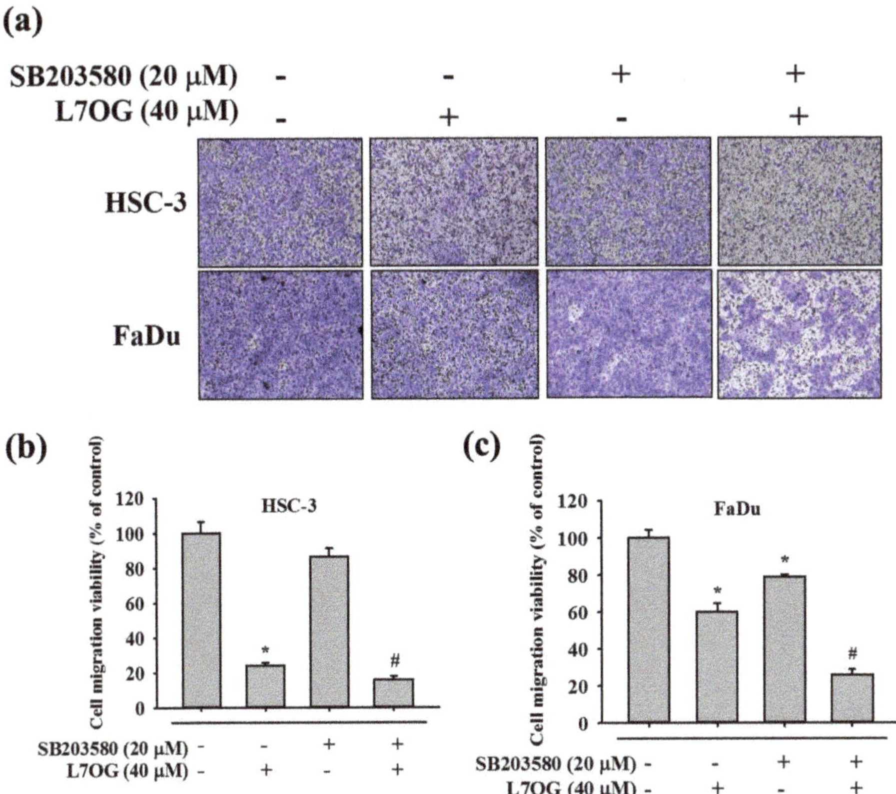

Figure 8. Effect of SB203580 and luteolin-7-O-glucoside co-treatment on cell migration in HSC-3 and FaDu cell lines. The cell migration (**a**) was measured using trans-well assay after the co-treatment with SB203580 and luteolin-7-O-glucosidein HSC-3 and FaDu cell lines. The quantitative results are shown for (**b**) HSC-3 and (**c**) FaDu cells. The values are represented as mean ± SD of at least three independent experiments. *$p < 0.05$, compared to the control group; #$p < 0.05$, compared to the luteolin-7-O-glucoside treated group.

4. Discussion

In this study, we evaluated the anti-migratory and anti-invasive effects of luteolin-7-O-glucoside, a natural flavonoid, on oral cancer cells. To the best of our knowledge, this is the first study describing the potential role of luteolin-7-O-glucoside as an anti-metastatic agent. The findings of the study can open up a new path toward developing novel therapeutic candidates for preventing distant metastasis of oral cancer, which is known to be responsible for higher rates of cancer-related deaths worldwide.

The study was designed to evaluate the effects of different doses of luteolin-7-O-glucoside (10, 20, and 40 µM) on oral cancer cells, as well as to compare the findings with untreated controls. Of these doses, 20 and 40 µM have shown that luteolin-7-O-glucoside significant anti-proliferative or cytotoxic effects (Figure 1). These findings are in line with previous studies showing the pro-apoptotic effects of luteolin-7-O-glucoside on cancer cells [27,32,33]. A study using human liver cancer cells has shown thatluteolin-7-O-glucoside causes cancer cell death by inducing cell cycle arrest at G2/M phase, and the effect is mediated by increased production of free radical and phosphorylation of JNK [34]. Moreover, a recent study has shown that luteolin-7-O-glucoside extracted from *Cuminum cyminum* has cytotoxic effects against breast cancer cells, which makes luteolin-7-O-glucoside a potential chemotherapeutic agent [35]. In contrast, it has been found that luteolin-7-O-glucoside exerts anti-apoptotic effects on cardiomyocytes that are treated with angiotensin II to develop cardiac hypertrophy [36]. The cardio-protective effects of luteolin-7-O-glucoside has been found to be associated with reduced free radical level, improved antioxidative capacity, and decreased hypoxia/reperfusion-induced cell death. Luteolin-7-O-glucoside has been found to mediate all these effects by modulating the MAPK signaling pathway. These findings clearly suggest that luteolin-7-O-glucosideexerts cytoprotective effects on normal, non-cancerous cells, such as cardiomyocytes.

Regarding the impact of luteolin-7-O-glucoside on cancer cell motility, we found that the compound is capable of significantly reducing the migration and invasion of oral cancer cells in a dose-dependent manner (Figures 2 and 3). Although there is no direct evidence on the anti-migratory activity of luteolin-7-O-glucoside, luteolin has been shown to reduce migration by inhibiting the production and secretion of pro-inflammatory cytokines, such as TNFα and IL-6 [37], which are known to induce cell migration by increasing the expression of migration-promoting proteins, including MMPs [38]. Moreover, luteolin has been found to inhibit the migration and invasion of pancreatic cancer cells by trans-inactivating EGFR activity and suppressing the phosphorylation of focal adhesion kinase (FAK) and MMP secretion [39]. In case of liver cancer cells, luteolin has been found to inhibit cell migration and invasion by reducing the phosphorylation of c-Met (hepatocyte growth factor receptor), as well as modulating the MAPK and PI3K-AKT pathways [40]. A recent study using human squamous carcinoma cell lines has shown that luteolin ameliorates cancer metastasis by reducing the expression and activity of ribosomal protein S19 and inhibiting the AKT/mTOR/c-Myc signaling pathway [41]. Luteolin suppress vascular endothelial growth factor receptor (VEGF) 2 and mediated human prostate tumor growth and angiogenesis [42]. In addition, the migration of squamous carcinoma cells has been found to be modulated by luteolin-induced reduction in expressions and activities of S100A7, Src, and STAT3 [43].

In addition, we also investigated the mode of action of luteolin-7-O-glucoside. According to the results in Figure 5, luteolin-7-O-glucoside treatment caused a significantly reduction in p38 and JNK phosphorylation in HSC-3 cells. In case of FaDu cells, the treatment caused similar reduction in p38 phosphorylation as observed in HSC-3 cells (Figure 5c,d). The study findings demonstrated that luteolin-7-O-glucoside mediates anti-metastatic effects by reducing MMP-2 expression and p38 phosphorylation (Figures 4 and 5). Using a p38 inhibitor, SB203580, we confirmed that luteolin-7-O-glucoside reduces oral cancer cell migration by altering p38-induced activation of MMP-2 (Figures 6–8). The anti-migratory effects of natural flavonoids such as luteolin has been well-documented in the literature [44–47]. Moreover, previous studies have shown that anti-metastatic effects of natural bioactive compounds are primarily mediated by modulation of MAPK pathways and MMP expression [48–50].

Several MMPs such as MT1-MMP, MMP-2, and MMP-9 were found correlated to the stages of cancer cell invasion progression. Specifically, our study findings are in line with a recent study showing that two bioactive compounds of pomegranate namely ellagic acid and luteolin prevent proliferation and migration of ovarian cancer cells by reducing the expression of MMP-2 and MMP-9 [51]. Similarly, luteolin has been found to reduce the proliferation, migration, and invasion of human melanoma cells by suppressing the expression of MMP-2 and MMP-9 and altering the PI3K/AKT signaling pathway [52]. Regarding the effect of luteolin on the p38 signaling pathway, a recent study using gastric cancer cells has shown that luteolin exerts anti-proliferative and anti-migratory effects by inhibiting the phosphorylation of PI3K, AKT, mTOR, ERK, and p38, as well as reducing the expression of MMP-2 and MMP-9 [53].

5. Conclusions

In conclusion, the present study clearly demonstrates that luteolin-7-O-glucoside can significantly reduce the oral cancer metastasis by mitigating p38-induced increased expression of MMP-2. The study identifies luteolin-7-O-glucoside as a potential anti-cancer candidate that can be utilized clinically for improving oral cancer prognosis.

Author Contributions: Conceptualization, M.-J.H. and J.-T.L.; methodology, Y.-S.L., C.-C.L., Y.-C.C.; software, Y.-S.L., C.-C.L., Y.-C.C.; writing—original draft preparation, B.K.V. and B.M.; writing—review and editing, M.-J.H. and M.-K.C. All authors have read and agreed to the published version of the manuscript.

Acknowledgments: This research was funded by National Science Council, Taiwan (MOST 106-2314-B-371-006-MY3; MOST 106-2314-B-371-005-MY3; MOST 108-2314-B-371-010) and Changhua Christian Hospital (108-CCH-ICO-145). The authors of the manuscript do not have a direct financial relation with the commercial identity mentioned in this paper.

Conflicts of Interest: The authors declare no conflict of interest.

References

1. Smriti, K.; Ray, M.; Chatterjee, T.; Shenoy, R.-P.; Gadicherla, S.; Pentapati, K.-C.; Rustaqi, N. Salivary MMP-9 as a Biomarker for the Diagnosis of Oral Potentially Malignant Disorders and Oral Squamous Cell Carcinoma. *Asian Pac. J. Cancer Prev.* **2020**, *21*, 233–238. [CrossRef] [PubMed]

2. Su, S.-C.; Lin, C.-W.; Liu, Y.-F.; Fan, W.-L.; Chen, M.-K.; Yu, C.-P.; Yang, W.-E.; Su, C.-W.; Chuang, C.-Y.; Li, W.-H.; et al. Exome Sequencing of Oral Squamous Cell Carcinoma Reveals Molecular Subgroups and Novel Therapeutic Opportunities. *Theranostics* **2017**, *7*, 1088–1099. [CrossRef] [PubMed]

3. Hsu, I.; Yen, C.-W.; Huang, K.-F.; Lin, Y.-S. Distant Intramuscular Metastases of Head and Neck Squamous Cell Carcinoma. *Ann. Plast. Surg.* **2020**, *84* (Suppl. S1), S11–S16. [CrossRef]

4. Haigentz, M., Jr.; Hartl, D.M.; Silver, C.E.; Langendijk, J.A.; Strojan, P.; Paleri, V.; De Bree, R.; Machiels, J.P.; Hamoir, M.; Rinaldo, A.; et al. Distant metastases from head and neck squamous cell carcinoma. Part III. Treatment. *Oral Oncol.* **2012**, *48*, 787–793. [CrossRef] [PubMed]

5. Takes, R.P.; Rinaldo, A.; Silver, C.E.; Haigentz, M., Jr.; Woolgar, J.A.; Triantafyllou, A.; Mondin, V.; Paccagnella, D.; De Bree, R.; Shaha, A.R.; et al. Distant metastases from head and neck squamous cell carcinoma. Part I. Basic aspects. *Oral Oncol.* **2012**, *48*, 775–779. [CrossRef] [PubMed]

6. De Bree, R.; Haigentz, M., Jr.; Silver, C.E.; Paccagnella, D.; Hamoir, M.; Hartl, D.M.; Machiels, J.P.; Paleri, V.; Rinaldo, A.; Shaha, A.R.; et al. Distant metastases from head and neck squamous cell carcinoma. Part II. Diagnosis. *Oral Oncol.* **2012**, *48*, 780–786. [CrossRef] [PubMed]

7. Tahtamouni, L.H.; Ahram, M.; Koblinski, J.; Rolfo, C.C. Molecular Regulation of Cancer Cell Migration, Invasion, and Metastasis. *Anal. Cell. Pathol. (Amst)* **2019**, *2019*, 1356508. [CrossRef] [PubMed]

8. Warnakulasuriya, S. Global epidemiology of oral and oropharyngeal cancer. *Oral Oncol.* **2009**, *45*, 309–316. [CrossRef]

9. Gonzalez-Avila, G.; Sommer, B.; Mendoza-Posada, D.A.; Ramos, C.; García-Hernández, A.A.; Falfán-Valencia, R.; Falfán-Valencia, R. Matrix metalloproteinases participation in the metastatic process and their diagnostic and therapeutic applications in cancer. *Crit. Rev. Oncol. Hematol.* **2019**, *137*, 57–83. [CrossRef]

10. Winer, A.; Adams, S.; Mignatti, P. Matrix Metalloproteinase Inhibitors in Cancer Therapy: Turning Past Failures Into Future Successes. *Mol. Cancer Ther.* **2018**, *17*, 1147–1155. [CrossRef]

11. Radisky, E.S.; Raeeszadeh-Sarmazdeh, M.; Radisky, D.C. Therapeutic Potential of Matrix Metalloproteinase Inhibition in Breast Cancer. *J. Cell. Biochem.* **2017**, *118*, 3531–3548. [CrossRef] [PubMed]

12. Dhillon, A.S.; Hagan, S.; Rath, O.; Kolch, W. MAP kinase signalling pathways in cancer. *Oncogene* **2007**, *26*, 3279–3290. [CrossRef] [PubMed]

13. Koul, H.K.; Pal, M.; Koul, S. Role of p38 MAP Kinase Signal Transduction in Solid Tumors. *Genes Cancer* **2013**, *4*, 342–359. [CrossRef] [PubMed]

14. Xu, L.; Chen, S.; Bergan, R.C. MAPKAPK2 and HSP27 are downstream effectors of p38 MAP kinase-mediated matrix metalloproteinase type 2 activation and cell invasion in human prostate cancer. *Oncogene* **2006**, *25*, 2987–2998. [CrossRef] [PubMed]

15. Buck, M.B.; Pfizenmaier, K.; Knabbe, C. Antiestrogens induce growth inhibition by sequential activation of p38 mitogen-activated protein kinase and transforming growth factor-beta pathways in human breast cancer cells. *Mol. Endocrinol.* **2004**, *18*, 1643–1657. [CrossRef] [PubMed]

16. Chelsky, Z.L.; Yue, P.; Kondratyuk, T.P.; Paladino, D.; Pezzuto, J.M.; Cushman, M.; Turkson, J. A Resveratrol Analogue Promotes ERKMAPK-Dependent Stat3 Serine and Tyrosine Phosphorylation Alterations and Antitumor Effects In Vitro against Human Tumor Cells. *Mol. Pharmacol.* **2015**, *88*, 524–533. [CrossRef]

17. Tang, B.; Liang, W.; Liao, Y.; Li, Z.; Wang, Y.; Yan, C. PEA15 promotes liver metastasis of colorectal cancer by upregulating the ERK/MAPK signaling pathway. *Oncol. Rep.* **2018**, *41*, 43–56. [CrossRef]

18. Chowdhury, P.; Dey, P.; Ghosh, S.; Sarma, A.; Ghosh, U. Reduction of metastatic potential by inhibiting EGFR/Akt/p38/ERK signaling pathway and epithelial-mesenchymal transition after carbon ion exposure is potentiated by PARP-1 inhibition in non-small-cell lung cancer. *BMC Cancer* **2019**, *19*, 829. [CrossRef]

19. Shen, T.; Cheng, X.; Xia, C.; Li, Q.; Gao, Y.; Pan, D.; Zhang, X.; Zhang, C.; Li, Y. Erlotinib inhibits colon cancer metastasis through inactivation of TrkB-dependent ERK signaling pathway. *J. Cell. Biochem.* **2019**, *120*, 11248–11255. [CrossRef]

20. Nokin, M.-J.; Bellier, J.; Durieux, F.; Peulen, O.; Rademaker, G.; Gabriel, M.; Monseur, C.; Charloteaux, B.; Verbeke, L.; Van Laere, S.; et al. Methylglyoxal, a glycolysis metabolite, triggers metastasis through MEK/ERK/SMAD1 pathway activation in breast cancer. *Breast Cancer Res.* **2019**, *21*, 11. [CrossRef]

21. Yue, S.-J.; Zhang, P.-X.; Zhu, Y.; Li, N.-G.; Chen, Y.-Y.; Li, J.; Zhang, S.; Jin, R.-Y.; Yan, H.; Shi, X.-Q.; et al. A Ferulic Acid Derivative FXS-3 Inhibits Proliferation and Metastasis of Human Lung Cancer A549 Cells via Positive JNK Signaling Pathway and Negative ERK/p38, AKT/mTOR and MEK/ERK Signaling Pathways. *Molecules* **2019**, *24*, 2165. [CrossRef] [PubMed]

22. Lu, L.; Wang, J.; Wu, Y.; Wan, P.; Yang, G. Rap1A promotes ovarian cancer metastasis via activation of ERK/p38 and notch signaling. *Cancer Med.* **2016**, *5*, 3544–3554. [CrossRef]

23. Lin, Y.; Shi, R.; Wang, X.; Shen, H.-M. Luteolin, a flavonoid with potential for cancer prevention and therapy. *Curr. Cancer Drug Targets* **2008**, *8*, 634–646. [CrossRef]

24. Park, C.M.; Song, Y.-S. Luteolin and luteolin-7-O-glucoside protect against acute liver injury through regulation of inflammatory mediators and antioxidative enzymes in GalN/LPS-induced hepatitic ICR mice. *Nutr. Res. Pr.* **2019**, *13*, 473–479. [CrossRef] [PubMed]

25. Richelle, M.; Pridmore-Merten, S.; Bodenstab, S.; Enslen, M.; A Offord, E. Hydrolysis of isoflavone glycosides to aglycones by beta-glycosidase does not alter plasma and urine isoflavone pharmacokinetics in postmenopausal women. *J. Nutr.* **2002**, *132*, 2587–2592. [CrossRef] [PubMed]

26. Seelinger, G.; Merfort, I.; Wölfle, U.; Schempp, C.M. Anti-carcinogenic effects of the flavonoid luteolin. *Molecules* **2008**, *13*, 2628–2651. [CrossRef] [PubMed]

27. Maatouk, M.; Abed, B.; Bouhlel, I.; Krifa, M.; Khlifi, R.; Ioannou, I.; Ghedira, K.; Ghedira, L.C. Heat treatment and protective potentials of luteolin-7-O-glucoside against cisplatin genotoxic and cytotoxic effects. *Environ. Sci. Pollut. Res.* **2020**. [CrossRef]

28. Hsieh, M.-J.; Chin, M.-C.; Lin, C.-C.; His, Y.-T.; Lo, Y.-S.; Chuang, Y.-C.; Chen, M.-K. Pinostilbene Hydrate Suppresses Human Oral Cancer Cell Metastasis by Downregulation of Matrix Metalloproteinase-2 Through the Mitogen-Activated Protein Kinase Signaling Pathway. *Cell. Physiol. Biochem.* **2018**, *50*, 911–923. [CrossRef]

29. Wei, C.-W.; Lin, C.-C.; Yu, Y.-L.; Lin, C.-Y.; Lin, P.-C.; Wu, M.-T.; Chen, C.-J.; Chang, W.; Lin, S.-Z.; Chen, Y.-L.S.; et al. n-Butylidenephthalide induced apoptosis in the A549 human lung adenocarcinoma cell line by coupled down-regulation of AP-2alpha and telomerase activity. *Acta Pharmacol. Sin.* **2009**, *30*, 1297–1306. [CrossRef]

30. Ho, H.-Y.; Ho, Y.-C.; Hsieh, M.-J.; Yang, S.-F.; Chuang, C.-Y.; Lin, C.-W.; Hsin, C.-H. Hispolon suppresses migration and invasion of human nasopharyngeal carcinoma cells by inhibiting the urokinase-plasminogen activator through modulation of the Akt signaling pathway. *Environ. Toxicol.* **2016**, *32*, 645–655. [CrossRef]

31. Chen, A.; Tseng, Y.-S.; Lin, C.-C.; Hsi, Y.-T.; Lo, Y.-S.; Chuang, Y.-C.; Lin, S.-H.; Yu, C.-Y.; Hsieh, M.-J.; Chen, M.-K. Norcantharidin induce apoptosis in human nasopharyngeal carcinoma through caspase and mitochondrial pathway. *Environ. Toxicol.* **2017**, *33*, 343–350. [CrossRef] [PubMed]

32. Rub, R.A.; Pati, M.J.; Siddiqui, A.A.; Moghe, A.S.; Shaikh, N.N. Characterization of Anticancer Principles of Celosia argentea (Amaranthaceae). *Pharmacogn. Res.* **2016**, *8*, 97–104. [CrossRef] [PubMed]

33. Liu, Q.-M.; Zhao, H.-Y.; Zhong, X.-K.; Jiang, J.-G. Eclipta prostrata L. phytochemicals: Isolation, structure elucidation, and their antitumor activity. *Food Chem. Toxicol.* **2012**, *50*, 4016–4022. [CrossRef] [PubMed]

34. Hwang, Y.J.; Lee, E.J.; Kim, H.R.; Hwang, K.A. Molecular mechanisms of luteolin-7-O-glucoside-induced growth inhibition on human liver cancer cells: G2/M cell cycle arrest and caspase-independent apoptotic signaling pathways. *BMB Rep.* **2013**, *46*, 611–616. [CrossRef]

35. Goodarzi, S.; Tabatabaei, M.J.; Jafari, R.M.; Shemirani, F.; Tavakoli, S.; Mofasseri, M.; Tofighi, Z. Cuminum cyminum fruits as source of luteolin-7-O-glucoside, potent cytotoxic flavonoid against breast cancer cell lines. *Nat. Prod. Res.* **2018**, 1–5. [CrossRef]

36. Chen, S.; Yang, B.; Xu, Y.; Rong, Y.; Qiu, Y. Protection of Luteolin-7-O-glucoside against apoptosis induced by hypoxia/reoxygenation through the MAPK pathways in H9c2 cells. *Mol. Med. Rep.* **2018**, *17*, 7156–7162. [CrossRef]

37. Chen, C.-Y.; Peng, W.-H.; Tsai, K.-D.; Hsu, S.-L. Luteolin suppresses inflammation-associated gene expression by blocking NF-kappaB and AP-1 activation pathway in mouse alveolar macrophages. *Life Sci.* **2007**, *81*, 1602–1614. [CrossRef]

38. Chen, C.C.; Chow, M.P.; Huang, W.C.; Lin, Y.C.; Chang, Y.J. Flavonoids inhibit tumor necrosis factor-alpha-induced up-regulation of intercellular adhesion molecule-1 (ICAM-1) in respiratory epithelial cells through activator protein-1 and nuclear factor-kappaB: Structure-activity relationships. *Mol. Pharmacol.* **2004**, *66*, 683–693.

39. Lee, L.-T.; Huang, Y.-T.; Hwang, J.-J.; Lee, A.Y.-L.; Ke, F.-C.; Huang, C.-J.; Kandaswami, C.; Lee, P.-P.H.; Lee, M.-T. Transinactivation of the epidermal growth factor receptor tyrosine kinase and focal adhesion kinase phosphorylation by dietary flavonoids: Effect on invasive potential of human carcinoma cells. *Biochem. Pharmacol.* **2004**, *67*, 2103–2114. [CrossRef]

40. Lee, W.-J.; Wu, L.-F.; Chen, W.-K.; Wang, C.-J.; Tseng, T.-H. Inhibitory effect of luteolin on hepatocyte growth factor/scatter factor-induced HepG2 cell invasion involving both MAPK/ERKs and PI3K-Akt pathways. *Chem. Biol. Interact.* **2006**, *160*, 123–133. [CrossRef]

41. Chen, K.-C.; Hsu, W.-H.; Ho, J.-Y.; Lin, C.; Chu, C.-Y.; Kandaswami, C.C.; Lee, M.-T.; Cheng, C.-H. Flavonoids Luteolin and Quercetin Inhibit RPS19 and contributes to metastasis of cancer cells through c-Myc reduction. *J. Food Drug Anal.* **2018**, *26*, 1180–1191. [CrossRef] [PubMed]

42. Pratheeshkumar, P.; Son, Y.; Budhraja, A.; Wang, X.; Ding, S.; Wang, L.; Hitron, A.; Lee, J.-C.; Kim, D.; Divya, S.P.; et al. Luteolin inhibits human prostate tumor growth by suppressing vascular endothelial growth factor receptor 2-mediated angiogenesis. *PLoS ONE* **2012**, *7*, e52279. [CrossRef] [PubMed]

43. Fan, J.J.; Hsu, W.H.; Lee, K.H.; Chen, K.C.; Lin, C.W.; Lee, Y.L.A.; Ko, T.P.; Lee, L.T.; Lee, M.T.; Chang, M.S.; et al. Dietary Flavonoids Luteolin and Quercetin Inhibit Migration and Invasion of Squamous Carcinoma through Reduction of Src/Stat3/S100A7 Signaling. *Antioxidants* **2019**, *8*, 557. [CrossRef] [PubMed]

44. Wang, Q.; Wang, H.; Jia, Y.; Ding, H.; Zhang, L.; Pan, H. Luteolin reduces migration of human glioblastoma cell lines via inhibition of the p-IGF-1R/PI3K/AKT/mTOR signaling pathway. *Oncol. Lett.* **2017**, *14*, 3545–3551. [CrossRef]

45. Lin, Y.-C.; Tsai, P.-H.; Lin, C.-Y.; Cheng, C.-H.; Lin, T.-H.; Lee, K.P.H.; Huang, K.-Y.; Chen, S.-H.; Hwang, J.-J.; Kandaswami, C.C.; et al. Impact of flavonoids on matrix metalloproteinase secretion and invadopodia formation in highly invasive A431-III cancer cells. *PLoS ONE* **2013**, *8*, e71903. [CrossRef]

46. Wang, L.; Li, W.; Lin, M.; Garcia, M.; Mulholland, D.; Lilly, M.; Martins-Green, M. Luteolin, ellagic acid and punicic acid are natural products that inhibit prostate cancer metastasis. *Carcinogenesis* **2014**, *35*, 2321–2330. [CrossRef]

47. Naso, L.G.; Badiola, I.; Clavijo, J.M.; Valcarcel, M.; Salado, C.; Ferrer, E.G.; Williams, P.A. Inhibition of the metastatic progression of breast and colorectal cancer in vitro and in vivo in murine model by the oxidovanadium(IV) complex with luteolin. *Bioorganic Med. Chem.* **2016**, *24*, 6004–6011. [CrossRef]

48. Chen, M.-K.; Liu, Y.-T.; Lin, J.-T.; Lin, C.-C.; Chuang, Y.-C.; Lo, Y.-S.; Hsi, Y.-T.; Hsieh, M.-J. Pinosylvin reduced migration and invasion of oral cancer carcinoma by regulating matrix metalloproteinase-2 expression and extracellular signal-regulated kinase pathway. *Biomed. Pharmacother.* **2019**, *117*, 109160. [CrossRef]

49. Tseng, P.-Y.; Liu, Y.-T.; Lin, C.-C.; Chuang, Y.-C.; Lo, Y.-S.; Hsi, Y.-T.; Hsieh, M.-J.; Chen, M.-K. Pinostilbene Hydrate Inhibits the Migration and Invasion of Human Nasopharyngeal Carcinoma Cells by Downregulating MMP-2 Expression and Suppressing Epithelial-Mesenchymal Transition Through the Mitogen-Activated Protein Kinase Signaling Pathways. *Front. Oncol.* **2019**, *9*, 1364. [CrossRef]

50. Yeh, C.; Hsieh, M.-J.; Yang, J.; Yang, S.; Chuang, Y.; Su, S.; Liang, M.; Chen, M.; Lin, C.-W. Geraniin inhibits oral cancer cell migration by suppressing matrix metalloproteinase-2 activation through the FAK/Src and ERK pathways. *Environ. Toxicol.* **2019**, *34*, 1085–1093. [CrossRef]

51. Liu, H.; Zeng, Z.; Wang, S.; Li, T.; Mastriani, E.; Li, Q.-H.; Bao, H.-X.; Zhou, Y.-J.; Wang, X.; Liu, Y.; et al. Main components of pomegranate, ellagic acid and luteolin, inhibit metastasis of ovarian cancer by down-regulating MMP2 and MMP9. *Cancer Biol. Ther.* **2017**, *18*, 990–999. [CrossRef] [PubMed]

52. Yao, X.; Jiang, W.; Yu, D.; Yan, Z. Luteolin inhibits proliferation and induces apoptosis of human melanoma cells in vivo and in vitro by suppressing MMP-2 and MMP-9 through the PI3K/AKT pathway. *Food Funct.* **2019**, *10*, 703–712. [CrossRef] [PubMed]

53. Pu, Y.; Zhang, T.; Wang, J.; Mao, Z.; Duan, B.; Long, Y.; Xue, F.; Liu, N.; Liu, S.; Gao, Z. Luteolin exerts an anticancer effect on gastric cancer cells through multiple signaling pathways and regulating miRNAs. *J. Cancer* **2018**, *9*, 3669–3675. [CrossRef] [PubMed]

 *biomolecules*

Article

UV-B Radiation Largely Promoted the Transformation of Primary Metabolites to Phenols in *Astragalus mongholicus* Seedlings

Yang Liu [1,†], Jia Liu [2,†], Ann Abozeid [3,4] , Ke-Xin Wu [3], Xiao-Rui Guo [3,*], Li-Qiang Mu [1,*] and Zhong-Hua Tang [3]

1 School of Forestry, Northeast Forestry University, Harbin 150040, China; liuyang00612@163.com
2 Material Science and Engineering College, Northeast Forestry University, Harbin 150040, China; liujia19880906@163.com
3 Key Laboratory of Plant Ecology, Northeast Forestry University, Harbin 150040, China; annabozeid@yahoo.com (A.A.); wukexin94@126.com (K.-X.W.); tangzh@nefu.edu.cn (Z.-H.T.)
4 Botany Department, Faculty of Science, Menoufia University, Shebin El-koom 32511, Egypt
* Correspondence: xruiguo@nefu.edu.cn (X.-R.G.); mlq041@163.com (L.-Q.M.); Tel.: +86-451-8219-2098 (X.-R.G.); +86-451-8219-1317 (L.-Q.M.)
† Yang Liu and Jia Liu contributed equally to this work.

Received: 28 November 2019; Accepted: 20 March 2020; Published: 26 March 2020

Abstract: Ultraviolet-B (UV-B) radiation (280–320 nm) may induce photobiological stress in plants, activate the plant defense system, and induce changes of metabolites. In our previous work, we found that between the two *Astragalus* varieties prescribed by the Chinese Pharmacopoeia, *Astragalus mongholicus* has better tolerance to UV-B. Thus, it is necessary to study the metabolic strategy of *Astragalus* under UV-B radiation further. In the present study, we used untargeted gas chromatography-mass spectrometry (GC-MS) and targeted liquid chromatography-mass spectrometry (LC-MS techniques) to investigate the profiles of primary and secondary metabolic. The profiles revealed the metabolic response of *Astragalus* to UV-B radiation. We then used real-time polymerase chain reaction (RT-PCR) to obtain the transcription level of relevant genes under UV-B radiation (UV-B supplemented in the field, λ_{max} = 313 nm, 30 W, lamp-leaf distance = 60 cm, 40 min·day^{-1}), which annotated the responsive mechanism of phenolic metabolism in roots. Our results indicated that supplemental UV-B radiation induced a stronger shift from carbon assimilation to carbon accumulation. The flux through the phenylpropanoids pathway increased due to the mobilization of carbon reserves. The response of metabolism was observed to be significantly tissue-specific upon the UV-B radiation treatment. Among phenolic compounds, C6C1 carbon compounds (phenolic acids in leaves) and C6C3C6 carbon compounds (flavones in leaves and isoflavones in roots) increased at the expense of C6C3 carbon compounds. Verification experiments show that the response of phenolics in roots to UV-B is activated by upregulation of relevant genes rather than phenylalanine. Overall, this study reveals the tissues-specific alteration and mechanism of primary and secondary metabolic strategy in response to UV-B radiation.

Keywords: *Astragalus mongholicus*; ultraviolet-B radiation; phenolics; untargeted gas chromatography-mass spectrometry; targeted liquid chromatography-mass spectrometry

1. Introduction

Astragalus mongholicus (*A. mongholicus*) is an important perennial herb of the Legumes family [1,2]. Its dried roots (Radix astragali) are one of the most popular Chinese herbal medicines in East Asia and are considered as healthy food in Western countries [3,4]. Radix astragali is often used as an antiperspirant, a vimmuno-stimulant, and a supplementary medicine during cancer therapy [5–7]. Calycosin-7-*O*-β-D-glucoside (CAG) can potentially be used as a "marker compound" for the chemical evaluation or product standardization of Radix astragali [8–10].

The need for *A. mongholicus* to capture sunlight for photosynthesis inevitably exposes the plant to ultraviolet (UV) radiation. Ultraviolet radiation is the general term for three radiation wavelengths: UV-C (200–280 nm), UV-B (280–320 nm), and UV-A (320–400 nm). Ultraviolet-C is completely absorbed by atmospheric gases. Ultraviolet-Ais barely absorbed by atmospheric ozone, but UV-A impact on plants is small. Ultraviolet-B is potentially harmful to plants but is partially absorbed by ozone [11–13]. It was reported that the influx of UV-B radiation will likely increase as a result of the depletion of stratospheric ozone [14]. Increased amounts of UV-B radiation affect plant development, morphology, and physiology [15]. The sessile lifestyle of plants forces them to adapt to dynamic environmental conditions. To counteract these problems, plants use a range of strategies, including increases in leaf thickness, UV-B reflective properties, and the accumulation of UV-B-absorbing secondary metabolites. The most common protective mechanism against potentially damaging irradiation is the biosynthesis of UV-absorbing compounds [16]. These secondary metabolites mostly consist of phenolic compounds, flavonoids, and hydroxycinnamate esters, which accumulate in the vacuoles of epidermal cells in response to UV-B radiation and attenuate the penetration of the UV-B portion of the solar spectrum into deeper cell layers with little effect on the visible region. These responses to UV-B radiation may result in the reprogramming of metabolites in *A. mongholicus* and even altered accumulation of bioactive compounds. Given this fact, it is essential to better understand the adaptive responses of metabolites in *A. mongholicus* to increased UV-B radiation.

Phenolic compounds play diverse roles in plants. These compounds provide structural support of the cell wall or protect plants against pathogens, herbivores, and UV radiation [17]. Of all classes of secondary metabolites, phenolics, specifically flavonoids, are the most relevant for UV protection. Plant phenolics are compounds having at least one aromatic ring substituted with at least one hydroxyl group. The hydroxyl group can be free or engaged in another function as an ether, ester, or glycoside [18–20]. Phenolics exhibit a large variety of structures in nature and can be divided into three groups according to their chemical structure: (1) compounds having a C6C1 carbon skeleton, such as 4-hydroxybenzoic acid, vanillic acid, and salicylic acid; (2) compounds having a C6C3 carbon skeleton, such as caffeic acid, *p*-coumaric acid, and ferulic acid; and (3) compounds having a C6C3C6 carbon skeleton (flavonoids are typical C6C3C6 phenolic compounds) [21,22]. Phenolic compounds are generally synthesized via the shikimate pathway. The shikimate pathway is a major biosynthetic route for both primary and secondary metabolism. This pathway begins with phosphoenolpyruvate and erythrose-4-phosphate and ends with chorismate [23,24]. Phenylalanine, a key metabolite, is synthesized by chorismate [25]. Phenylalanine is considered the general precursor of C6C1-, C6C3-, and C6C3C6 compounds and their polymers in plant metabolism [26]. Many investigations have closely focused on the accumulation of phenolic compounds, which is regulated by biotic and abiotic stress [27]. *Astragalus mongholicus* is considered a rich source of natural phenolic compounds [28,29].

Regarding the complex primary and secondary metabolites fluctuations of higher plants, metabolomics plays a vital role as a suitable tool [15,30]. Metabolomics approaches are becoming more widely used in modern biology [28,31,32]. In plant metabolomics studies, common analytical technologies include liquid chromatography-mass spectrometry (LC-MS), gas chromatography-mass spectrometry (GC-MS), and nuclear magnetic resonance spectroscopy (NMR), among others [33–35]. Gas chromatography-mass spectrometry is widely used in metabolomic studies due to its high quality and reproducibility, wide dynamic range, universal mass spectral library and ability to detect hydrophilic metabolites after derivatization [36]. A reference pool of GC-MS for many primary metabolites, such as sugars, organic acids, fatty acids, and amino acids, has been established [37–39]. The short run time and relatively low running cost are strong advantages of GC-MS [40]. In contrast, LC-MS can potentially analyze a wide variety of large hydrophobic metabolites predominant in secondary compounds in plants; these compounds include phenolic compounds, terpenoids and alkaloids [41–43]. Moreover, the development of the ultra-performance liquid chromatography-mass spectrometry (UPLC-MS) methods leads to more powerful analyses of metabolites because of its higher throughput and shorter run time than those of conventional HPLC-MS [44,45]. Accordingly, the combination of GC-MS and LC-MS could clearly enhance the coverage of metabolites, allowing both a comprehensive overview and detailed analysis of metabolic changes in plants [46]. Recently, these two methods have been jointly adopted for many botanical studies [30,47–49].

Many of the biosynthetic enzymes of phenolics, such as phenylalanine ammonia lyase (*PAL*, E.C. 4.3.1.25) and chalcone synthase (*CHS*, E.C. 2.3.1.74), are activated by UV-B [1,50]. Flavonoid biosynthesis and their regulation have been thoroughly investigated. As the major bioactive compound, the calycosin 7-*O*-glucoside (CAG) biosynthesis pathway has been completely elucidated based on its presence in other legumes [10,51–53]. Calycosin 7-*O*-glucoside is synthesized from L-phenylalanine via the isoflavonoid branch of phenylpropanoid metabolism. *PAL*, cinnamate 4-hydroxylase (*C4H*, E.C. 1.14.13.11) and 4-coumarate-coenzyme A ligase (*4CL*, E.C. 6.2.1.12) are enzymes involved in the upstream general phenylalanine pathway. Isoliquiritigenin, an isoflavonoid skeleton, is synthesized via *CHS* and chalcone reductase (*CHR*), co-catalyzed by the condensation of 4-coumaroyl-CoA and three molecules of malonyl-CoA. Afterward, a series of chemical reactions is performed under the catalysis of chalcone isomerase (*CHI*, E.C. 5.5.1.6), isoflavone synthase (*IFS*, E.C. 1.14.13.86), isoflavone *O*-methyltransferase (*IOMT*) and isoflavone 3′-hydroxylase (*I3′H*). At the last step, the formation of CAG from calycosin (CA) is catalyzed by uridine diphosphate glucose (UDP)-glucose: calycosin 7-*O*-glucosyltransferase (*UCGT*). For the last step, the formation of CAG from CAis catalyzed by uridine diphosphate glucose (UDP)-glucose: calycosin 7-*O*-glucosyltransferase (*UCGT*) [54,55].

Due to the release of anthropogenic pollutants such as chlorofluorocarbons, a larger proportion of the UV-B spectrum reaches the surface of the earth and affects all living organisms [56,57]. In general, the most common protective mechanism against potentially damaging irradiation in plants is the biosynthesis of UV-absorbing compounds, such as phenolic compounds, flavonoids, and hydroxycinnamate esters [16]. The use of UV-B radiation is expected to induce the accumulation of CAG in *A. mongholicus* [55,58]. This could satisfy consumer demand for these naturally derived health-promoting products. Our goal is to understand the response of the metabolites in *A. mongholicus* to UV-B radiation. To the best of our knowledge, no information on the metabolites response mechanism of *A. mongholicus* to UV-B radiation is currently available. To deeply explore the response and mechanism explaining how metabolic reprogramming can achieve a new steady state under increased UV-B radiation, we investigated the influence on the specific tissue accumulation of primary and phenolic metabolites using untargeted (GC-MS) and targeted (LC-MS) metabolomics in *A. mongholicus* under increased UV-B radiation.

2. Materials and Methods

2.1. Plant Materials, Growth Conditions and Treatments

Seeds of *A. mongholicus* were sown in Botanical Garden of the Key Laboratory of Plant Ecology, Northeast Forestry University, Harbin, northeastern of China (natural environment at east longitude 126°38′, north latitude 45°43′). All plants grew under standard field conditions, with an average temperature of 16 °C during the day (14 h) and 4 °C at night (10 h), and water was applied every two days. After three months, plants of uniform height and stem thickness were transferred to containers (37 cm × 28 cm × 10 cm) and used for this study. All containers ($n = 6$) contained equal bulk density of potting mix and 12 plants, half of the containers for UV-B treated plants and the other for controlled plants. Plants were treated in the field with UV irradiation (λ_{max} = 313 nm, 30 W, lamp-leaf distance = 60 cm, 40 min·day^{-1}). The UV-B irradiation intensity (33.5 µW/cm^2) was monitored by a UV light meter (UV340B, Xin Bao Technologies, ShenZhen, China). That is, the irradiation doses of UV-B treated group was 804 J m^{-2}, the dose chosen was based on our previous study [57]. Control plants were maintained in the previous environment. Sampling was performed for both plants treated with UV and controlled plants at 10 days after treatment. Morphological indicators (plant height, fresh weight of root and whole plant, leaf area) were recorded for 20 biologic repetitions, firstly. Then, these plants were sampled. Collected plants were rinsed briefly in deionized water and separated into roots, stems, leaves and petioles. The samples were then immediately frozen in liquid nitrogen and stored at −80 °C for biochemical analysis. Six biological replicates were prepared for each sample. The analysis of H_2O_2 and antioxidant enzymes was performed based on the methods of Liu et al. [59–61]. Chlorophyll analysis was performed in accordance with the methods of Arnon [62,63].

2.2. Primary Metabolite Extraction and Gas Chromatography-Mass Spectrometry Analysis

Plant samples were pulverized to 5 μm using a grinding instrument (MM 400, Retsch GmbH, Haan, Germany), and 90 mg aliquots of roots, stems, leaves, and petioles were extracted. A mixed solution of 40 μL 2-chloro-L-phenylalanine (0.3 mg/mL, internal standard) in 360 μL methanol (pre-cooled at −20 °C) was prepared for metabolite extraction. Metabolites were extracted with the mixed solution by sonication for 30 min, after which they were sonicated again with a chloroform and water solution. Subsequently, the solution was centrifuged at 14,000 rpm (revolutions per minute). The supernatant was collected in a derivatized glass bottle, evaporated to dryness. A derivatization method oximation reaction was then performed to increase the volatility of the metabolites [30]. The dried residue was redissolved in 80 μL methoxyamine pyridine solution (15 mg/mL) and incubated at 37 °C (90 min). Subsequently, an 80 μL aliquot of *N,O*-bis(trimethylsilyl)trifluoroacetamide (BSTFA) (including 1% trimethyl-cholorosilane), and 20 μL were added to the mixture, vortexing for 2 min, which were then derivatized for 60 min in 70 °C. After pretreatment, the solution was centrifuged (12,000 rpm at 4 °C for 10 min), after which the supernatant was transferred to a glass vial for GC-MS analysis.

The GC-MS analysis was performed using a 7890A-5975C system (Agilent, Santa Clara, CA, USA). An Agilent HP-5 (5%-phenyl) -methyl polysiloxane non-polar column was used for GC-MS. The GC process was as follows: an initial temperature of 60 °C; oven temperature increments of 8 °C·min^{-1} to 125 °C, 4 °C·min^{-1} increments to 210 °C, 5 °C·min^{-1} to 270 °C, and 10 °C·min^{-1} to 305 °C; and a final holding temperature of 305 °C for 3 min. Mass spectra acquisition parameters were as follows: ion source temperature, 260 °C; filament bias, −70 V; mass range, *m/z* of 50–600; and acquisition rate, 20 spectra·sec^{-1} [30].

2.3. Phenolic Metabolite Extraction and Liquid Chromatography-Mass Spectrometry Targeted Analysis

Plant samples were pulverized to 5 μm using a grinding instrument (MM 400, Retsch GmbH), and 500 mg aliquots of roots, stems, leaves, and petioles were extracted. Samples were extracted with 80% ethanol in water (10 mL) containing 0.1 mg/L lidocaine (internal standard) by sonication twice, each for 45 min. The sample extractions were filtered, and the filtrates were merged. The filtrates were then dried under low pressure using a vacuum cavitation instrument. The resultant extracted material was dissolved in the mobile phase (1 mL) and filtered through 0.22 μm-diameter micropores. The purified solution was analyzed by ultra-performance liquid chromatography quadrupole time-of-flight mass spectrometry (UPLC/Q-TOF-MS).

Separation was performed on an Acquity UPLC BEH C18 column (1.7 μm, 2.1 mm × 50 mm) with a VanGuard precolumn (BEH C18, 1.7 μm, 2.1 × 5 mm; Waters, Shang-Hai, China) and maintained at 30 °C. The volume injected was 2 μL. Gradient elution was performed at a flow rate of 0.25 mL·min^{-1} using the following solvent system: 0.05% acetic acid-water (A), 0.05% acetic acid-acetonitrile (B); 5% B-95% B from 0–23 min; 95% B-5% B from 23–25 min; and 5% B from 25–31 min. Analyses were performed using a UPLC/Q-TOF-MS system (Waters). The MS conditions were set as follows: positive ion mode, capillary voltage of 3.0 kV, cone voltage of 45 V, source temperature of 400 °C, desolvation temperature of 500 °C, cone gas flow of 50 L/h, and desolvation gas flow of 800 L/h. Detection was performed in positive ion mode in the *m/z* range of 50–1000.

2.4. Multivariate Analysis

The GC-MS raw data were transformed into NetCDF format by data analysis software (Agilent GC-MS 5975) and were later processed using the software R. Each compound was displayed as the peak area normalized to that of the internal standard. For further analysis, the treated R output data were exported to Microsoft Excel. National Institute of Standards and Technology (NIST) 14 library was searched to compare the structures of the compounds with that of the NIST database. Compounds were then identified based on the retention index and mass spectra with already known compounds in the NIST library. Peak detection, retention time alignment, and library matching were performed using the TargetSearch package from Bioconductor (Solvusoft Corporation, Los Angeles, CA, USA) [64], after which the normalized data were imported into SIMCA-P version 11.0 software (Umetrics, Umea, Sweden) for multivariate statistical analysis. The supervised method of partial least-squares discriminant analysis (PLS-DA) was used to compare tissue-specific differences between control and UV-B treatment regarding the identification of significant metabolites, and *t*-test combinatory approaches were used to screen for important metabolites ($p < 0.05$). The LC-MS data were analyzed using the software MassLynx version 4.1. This software detected peaks and listed the detected and matched peaks with the retention time and m/z pair and their corresponding intensities. The relative signal intensities of compounds were standardized by dividing them by the intensities of internal standard. The relationships between 15 primary metabolites and phenolic compounds were used for hierarchical clustering analysis (HCA) by R (www.r-project.org/) for both species. Pearson's correlation coefficients were calculated for these metabolites, and the Tukey test was performed using Statistical Product and Service Solutions (SPSS) version 17.0. Metabolic pathways were analyzed using the Metaboanalyst web portal (www.metaboanalyst.ca) and MBRole (http://csbg.cnb.csic.es/mbrole). The pathways of metabolites were analyzed using database sources including the Kyoto Encyclopedia of Genomes and Genomes (KEGG) (http://www.genome.jp/kegg/) to identify the most affected metabolic pathways and facilitate further metabolite interpretation. The metabolites and corresponding pathways were imported into Cytoscape software version 3.1.0 to visualize the network models. A metabolic correlation distribution network was created from the 144 primary metabolite data using the WGCNA package (Solvusoft Corporation, Los Angeles, CA, USA) [65,66].

2.5. RNA Extraction and Real-Time Polymerase Chain Reaction Analysis

The extraction and derivatization of RNA from roots, leaves, stems and petioles were performed as described previously by Liu et al. [67]. Amplification, detection, and data analysis were performed using a Rotor-Gene 6000 real-time rotary analyzer (Corbett Life Science, Sydney, Australia). Primer sequences for PCR were as follows (Table 1):

Table 1. Primer sequences for polymerase chain reaction (PCR).

Gene Name		Primer Sequence (5′ to 3′)
CHS	Forward	CCTTCTTTGGATGCTAGACAAGACA
	Reverse	CGAAGACCCAAGAGTTTGGTTAGTT
PAL	Forward	CATCAAATCTCTCTGGCAGTAGGAA
	Reverse	AGTTCACATCTTGGTTATGCTGCTC
C4H	Forward	AACAAAGTGAGGGATGAAATTGACA
	Reverse	GGATTGCCATTCTTAGCCTTAGTGT
4CL	Forward	TGTCCCTCCTATTGTTTTGGCTATT
	Reverse	CTTTGGGGAATTTAGCTCTGACAGT
CHR	Forward	AAACAAGGTTACAGGCATTTTGACA
	Reverse	GGAAGAACGAGATGAGGATGATTTT

Table 1. *Cont.*

Gene Name		Primer Sequence (5′ to 3′)
CHI	Forward	ATCGAGTTTTTCCACCAGGATCTAC
	Reverse	ATCATAGTCTCCAACACAGCCTCAG
IFS	Forward	CCTTCACCTATTGGACAAACCTCTT
	Reverse	CCTGGTATTAAAGGAAGAAGCCTCA
IOMT	Forward	GCACAAAACACAAGATCAAACTTC
	Reverse	GCATTACGGCCATTGATTG
I3′H	Forward	GGATGTTAAAGAAGCGAAGCAATTT
	Reverse	ATCAAACAATCTCAACAAAGGCAAA
UCGT	Forward	AGGTTTTGAAGATTTATGCACCA
	Reverse	TCCTTTCTGAGTTCCAGGACA
18S	Forward	TGCAGAATCCCGTGAACCATC
	Reverse	AGGCATCGGGCAACGATATG

To determine relative fold differences in template abundance for each sample, the cycle threshold (Ct) values for each of the gene-specific primers were normalized to the Ct value for *18S* RNA.

3. Results

3.1. Morphological and Physiological Changes Induced by Ultraviolet-B Radiation

The effect of UV-B radiation on morphology can be seen in Figure 1 and Table 2. Compared with the control group, stem growth was relatively short and the leaf more compact under UV-B radiation. Both plant height and leaf area decreased under UV irradiation. The whole-plant proportion of roots (n = 10) was 19.07% for the control group and 20.70% for the UV-B treatment group. In contrast to the control group, the chlorophyll content decreased in the UV-B treated group. The activity of antioxidant enzymes and content of H_2O_2 increased under UV-B radiation.

Table 2. Morphological and physiological indices of control and Ultraviolet treatments (n = 20).

	Height (cm)	Root/Whole Fresh Weight (FW) (%)	Leaf Area (cm^2)	Chlorophyll ($mg\cdot g^{-1}$)
Control	24.2 ± 0.6	19.07 ± 0.31	0.292 ± 0.004	1.242 ± 0.008
UV-treated	23.4 ± 0.6	20.7 ± 0.67	0.230 ± 0.006	0.799 ± 0.011
	CAT ($U\cdot g^{-1}\cdot min^{-1}$, FW)	**POD** ($U\cdot g^{-1}\cdot min^{-1}$, FW)	**APX** ($U\cdot g^{-1}\cdot min^{-1}$, FW)	H_2O_2 ($\mu mol\cdot g^{-1}$)
Control	26.25 ± 0.01	1800 ± 1	0.06 ± 0.001	3.59 ± 0.01
UV-treated	36.75 ± 0.01	7050 ± 1	0.15 ± 0.003	3.79 ± 0.01

Abbreviations: CAT: catalase; POD: peroxidase; APX: ascorbate peroxidase; H_2O_2: hydrogen peroxide.

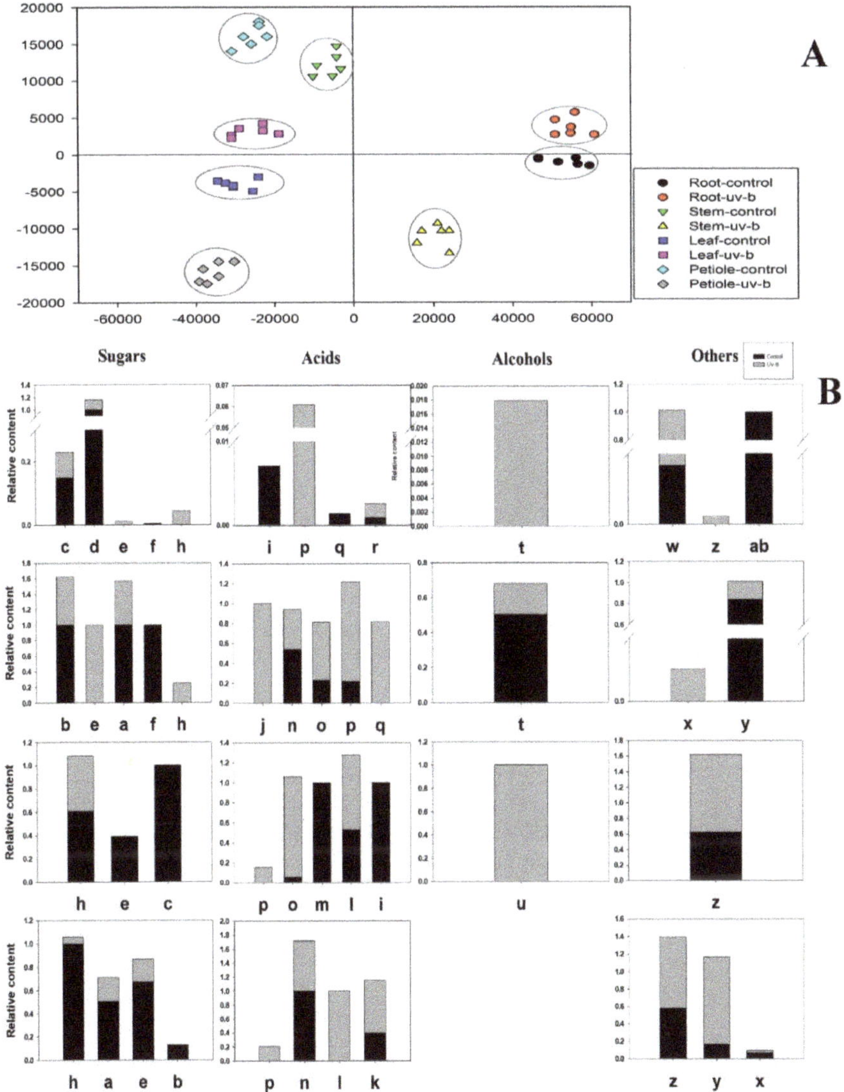

Figure 1. Metabolic analysis of specific tissues of *A. mongholicus* under control and UV-B treatments analyzed by GC-MS. (**A**) Partial least-squares discriminant analysis (PLS-DA); (**B**) The expression pattern of potential differences. **a**, mannose; **b**, 3-mannobiose; **c**, fructose; **d**, talose; **e**, glucose; **f**, psicose; **g**, sorbofuranose; **h**, myo-inositol; **i**, 2,3,4-trihydroxybutyric acid; **j**, 4-aminobutyric acid; **k**, citric acid; **l**, hexonic acid; **m**, isothiocyanic acid; **n**, malic acid; **o**, oxalic acid; **p**, phosphonic acid; **q**, palmitic acid; **r**, propanedioic acid; **s**, stearic acid; **t**, pinitol; **u**, erythritol; **w**, D-glucopyranoside; **x**, 2-O-glycerol-D-galactopyranoside; **y**, methyl galactoside; **z**, glycerol; **ab**, dotriacontane; **ac**, 1-monopalmitin.

3.2. Primary Metabolism Reprogramming between the below- and Aboveground Organs in Response to Ultraviolet-B Radiation

In the present study, approximately 184 metabolites comprising sugars, acids, alcohols, and other compounds were identified using a GC-MS platform in four tissues of *A. mongholicus* under control and UV-B treatments (Supplementary Table S1).

To identify the metabolites that contributed to these variances, 27 significantly differential compounds were identified by their variables of importance in the projection (VIP) values >1, and their *p*-values < 0.05 (Figure 1B). There were eight significant differences regarding sugars; most of the sugars were monosaccharides. The 3-α-mannobiose was the only disaccharide that significantly changed under UV-B, and its monosaccharide (mannose) constituents have a common tendency to decrease in aboveground tissue (stems and petioles). In contrast to the control, the levels of D-glucose and myo-inositol increased in roots and stems but decreased in leaves and petioles under UV-B treatment. The D-talose was significantly reduced by UV-B radiation in belowground tissues (roots) but did not significantly change in aboveground tissues. Regarding acids, including eight organic acids, one inorganic acid and two fatty acids, diverse trends were observed. Phosphonic acid level decreased under UV-B radiation in all organs. Propanedioic acid increased only in roots, and there was no significant difference in aboveground tissue under control or UV-B treatments. Palmitic acid decreased in belowground tissue (roots) but increased in aboveground tissues (stems) under UV-B radiation. Oxalic acid increased in leaves and stems while hexonic acid increased in leaves and petioles under UV-B radiation. Significant increases in alcoholic compounds in belowground tissues (roots) were observed for D-pinitol, whereas significant increases in aboveground tissues (leaves) were observed for erythritol. In addition, the levels of six other compounds significantly differed from those of the control group under UV-B radiation. Among these compounds, α-D-glucopyranoside levels significantly differed only in belowground tissues (roots).

The PLS-DA was used to profile these metabolites. In the PLS-DA score plot, separation of the two treatments and their different tissues was obtained. The samples were clustered into separate groups (Figure 1A). A PLS-DA model was created with two principal components: PLS1 (47.9%) and PLS2 (64.7%). A clear classification trend was observed among roots, stems, leaves, and petioles for all samples in the score plot. The control and UV-B treatment groups of the aboveground organs (stems, leaves and petioles) were better separated than those of the belowground organs (roots).

3.3. Phenolic Compounds Were Concentrated in Leaves Compared to Roots under Ultraviolet-B Radiation

To understand the effects of increased UV-B radiation on the level of phenolic metabolites, the targeted phenolic compounds content of the collected samples was analyzed via LC-MS. A total of 29 standard reference compounds were used, all provided by Shanghai yuanye Bio-Technology Co. Ltd (ShangHai, China), with a purity of ≥98%. The targeted compounds in the samples were identified in accordance with the mass spectral information and retention time of the corresponding standard reference compounds. The mass spectral information has been added as Supplementary Table S2; the retention time and corresponding standard reference compounds were shown in the chromatogram of Figure 2. In total, 29 phenolic metabolites were analyzed.

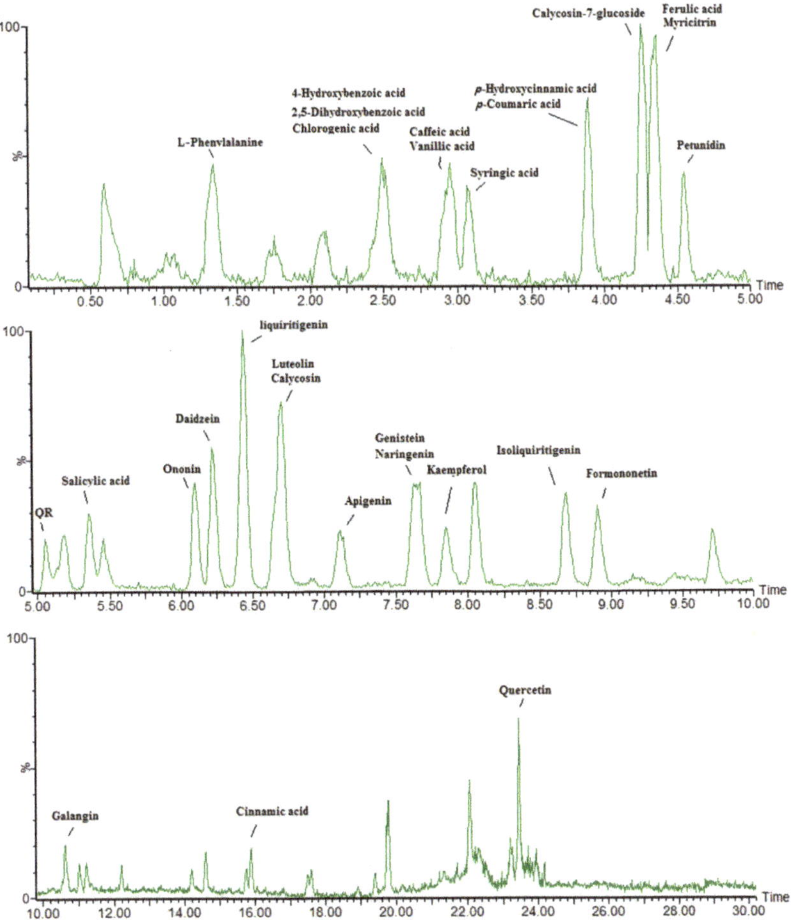

Figure 2. Chromatogram of the target compounds from ultra-performance liquid chromatography quadrupole time-of-flight mass spectrometry (UPLC-TOF-MS). Targeted compounds in samples were determined in accordance with the mass spectral information and retention time of the corresponding standard reference compounds. QR: Quercetin-3-*O*-rhamnoside.

To further clarify phenolic accumulation patterns in different tissues under the control and UV-B treatments, the roots, stems, leaves, and petioles of *A. mongholicus* were determined separately. Visualization of the phenolic profile was performed using HCA (Figure 3). The accumulation of phenolics displayed a clear phenotypic variation in terms of phenolic abundance in the different tissues and treatments. Roots contained the highest levels of the majority of tested phenolics, followed by the leaves, petioles and stems. Based on their tissue- and UV-B specific accumulation patterns, phenolics clearly grouped into two clusters.

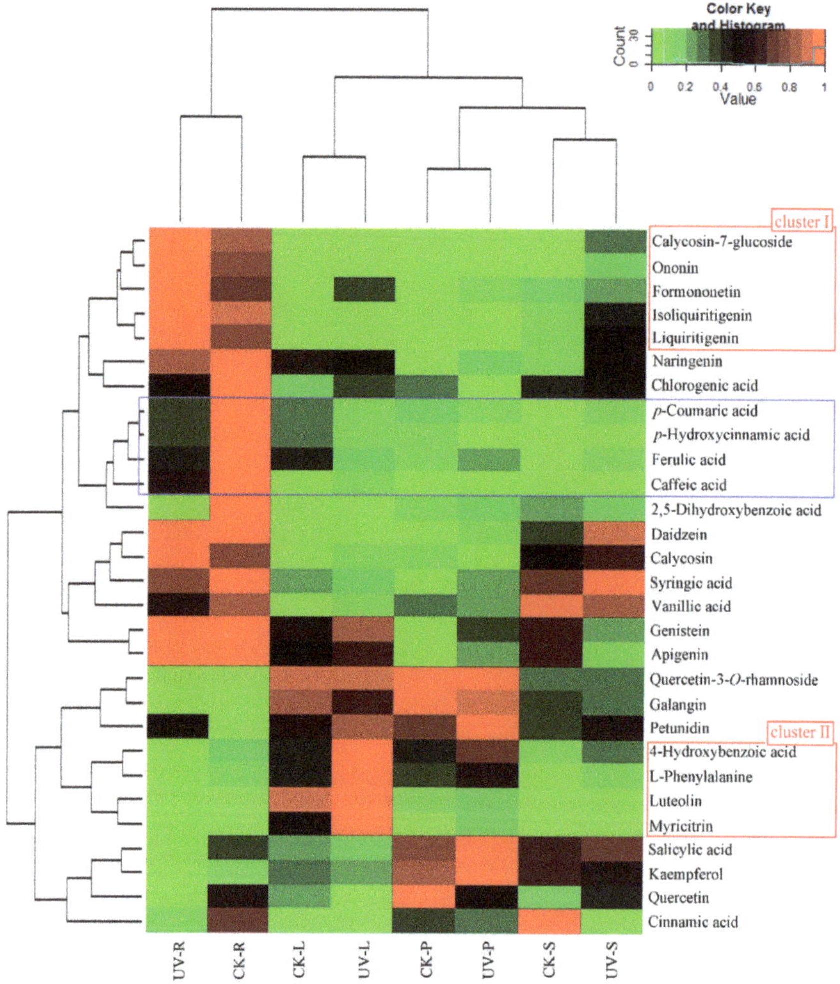

Figure 3. Distribution of potential biomarkers in specific tissues from the control and UV-B treatment groups. Heat map visualization of the relative differences of potential biomarkers in the different samples. The data of the content value of each compound were normalized to complete the linkage hierarchical clustering. Each tissue type is visualized in a single column, and each metabolite is represented by a single row: CK-R, CK-S, CK-L, and CK-P correspond to roots, stems, leaves, and petioles of the control group; UV-R, UV-S, UV-L, and UV-P correspond to roots, stems, leaves, and petioles from the UV-B-treated group. Red indicates high abundance of compounds, whereas green indicates low abundance (color key scale above heat map).

Phenolics in cluster I showed higher levels in the roots of the UV-B treatment group than in those of the control group and mostly consisted of calycosin-7-glucoside, ononin, formononetin, isoliquiritigenin, and liquiritigenin. Phenolics in cluster II consisted of 4-hydroxybenzoic acid, L-phenylalanine, luteolin and myricitrin; higher levels of these compounds were detected in the leaves of the UV-B radiation group than in those of the control group. In addition, UV-B radiation altered the distribution of phenolic compounds between above- and belowground tissues (Figure 4). The ordinate of Figure 4 is set to the ratio of compound contents in leaves and roots.

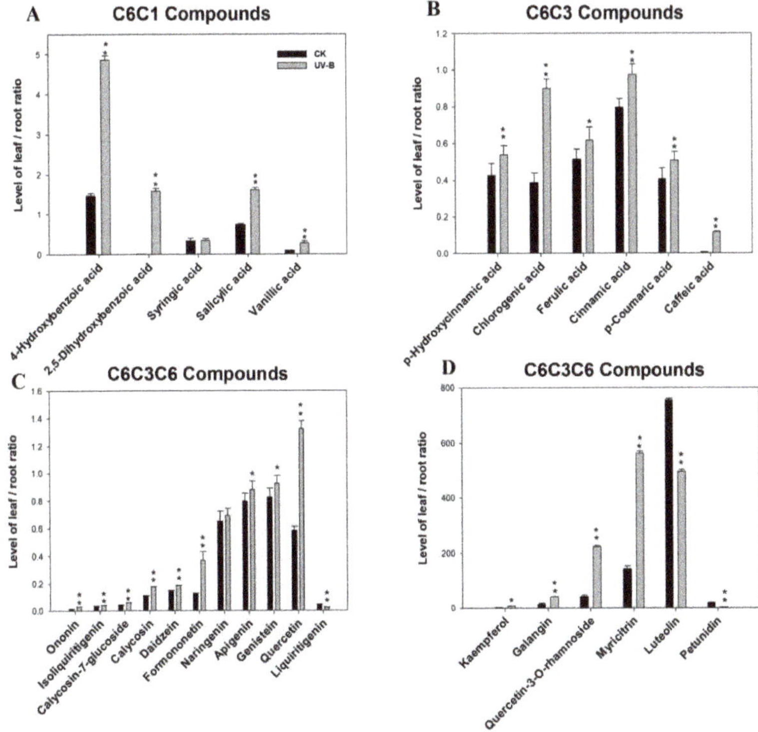

Figure 4. Distribution of phenolics that have various carbon skeletons in above- and belowground tissues under UV-B radiation. The ordinate is set to the ratio of compound contents in leaves and roots. (**A**) Phenolics with C6C1 carbon skeletons; (**B**) phenolics with C6C3 carbon skeletons; (**C**) and (**D**) phenolics with C6C3C6 carbon skeletons.

According to Figure 4, the majority of the ratios of phenolic levels in leaves/roots under UV-B treatment were higher than those of the control. In other words, compared with those of the control, phenolic compounds were more concentrated in leaves under UV-B radiation. Moreover, all the C6C3 carbon skeleton phenolics were translocated from the roots to the leaves.

3.4. Construction of an Integrative View of the Primary and Phenolic Metabolite Network for Specific Tissues and Growth Conditions

To provide better overview of these data, a simplified primary and phenolic metabolites pathway was used to show the metabolic responses to UV-B radiation among different tissues. As shown in Figure 5A,B, for primary metabolites, the accumulation of soluble sugars, such as glucose, fructose, and mannose, decreased in response to UV-B radiation in both roots and leaves. In difference with leaves the level of erythrytiol and sorbitol increased in the roots. Some amino acids including valine and aspartate increased in response to UV-B radiation in both leaves and roots. The levels of acids involved in the tricarboxylic acid (TCA) cycle, including fumarate in roots, malate, and succinate in leaves, increased in the UV-B treatment group. For phenolic metabolites, most of the isoflavones in roots were upregulated by UV-B induction. Unlike roots, the upregulated phenolic metabolites in leaves mainly belong to phenolic acids and flavones. We only observed that the C6C3 phenolics decreased both in roots and leaves. In addition, phenylalanine, which is the key node in the synthesis of phenolic compounds from the shikimate pathway, may be responsible for the increase in the majority of phenolic compounds under UV-B radiation.

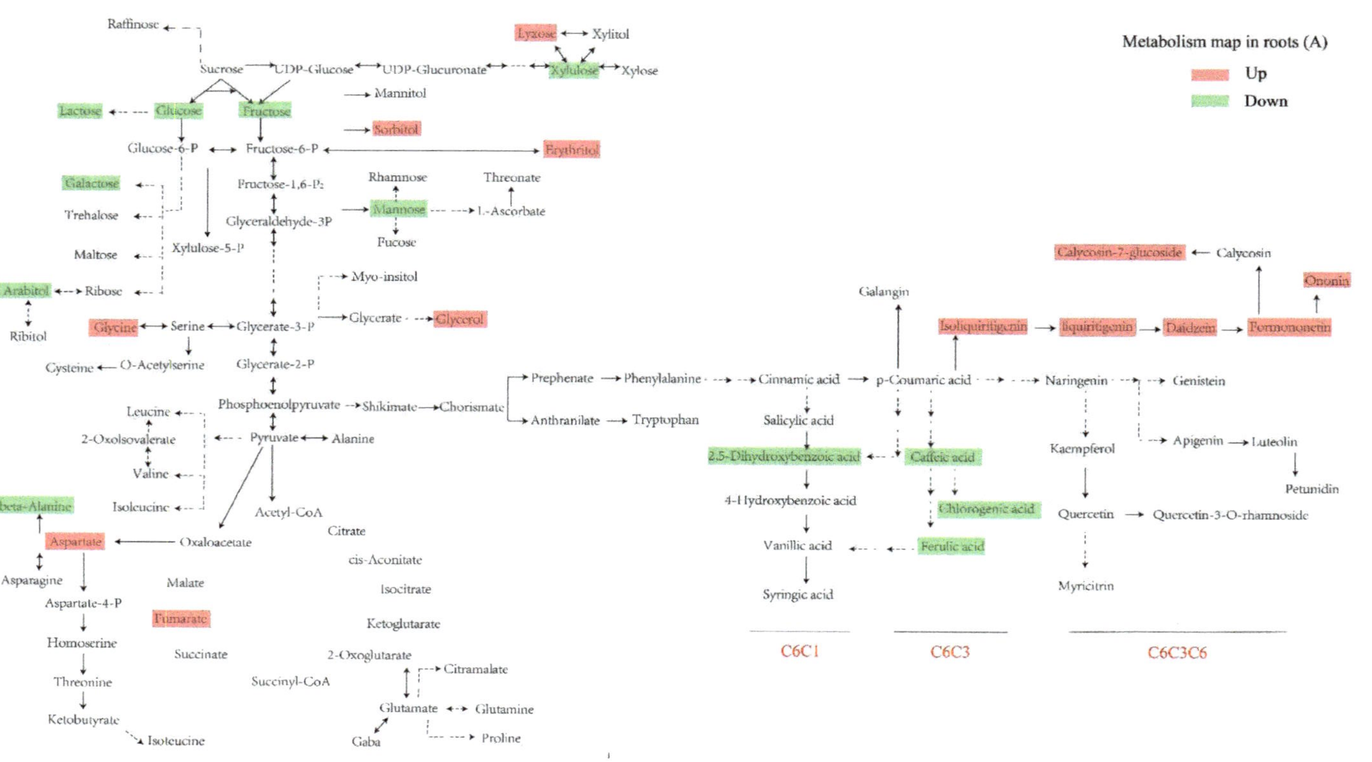

Figure 5. *Cont.*

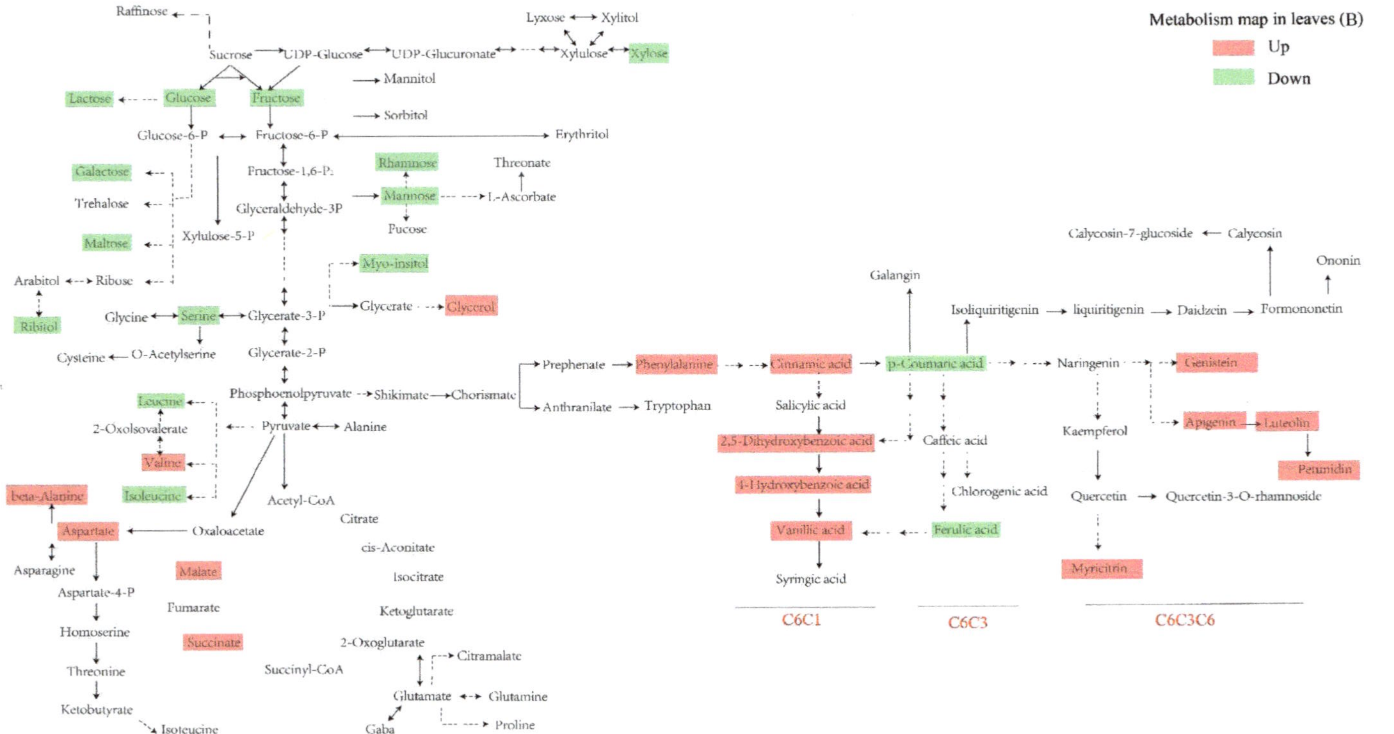

Figure 5. Visualization of primary and phenolic metabolites dynamics on a biochemical pathway map. (**A**) Metabolism map in roots. (**B**) Metabolism map in leaves. Compounds in the red box are upregulated by UV-B radiation, and compounds in the green box are downregulated.

3.5. Expression of Genes Involved in Isoflavonoids Pathway in Leaves and Roots

To further explore the metabolic mechanisms of CAG in the enhanced UV-B environment, the CAG biosynthesis pathway and the expression levels of relevant genes were visualized, as shown in Figure 6. The expression of the genes encoding enzymes that directly participate in the formation of CAG in samples collected from different treatments and different tissues was analyzed using RT-PCR. Gene expression levels were normalized using the *18S* RNA reference gene as an internal standard. The transcription level of the synthetase genes involved in the CAG pathway, including *CHI*, *IFS*, *IOMT*, *I3'H*, and *UCGT*, were upregulated in roots but downregulated in aboveground tissues in response to UV-B induction. This variation is strikingly similar to the accumulation pattern of their corresponding compounds. The *CHS* and *CHR* co-catalyze the condensation of *p*-coumaryl-CoA with three malonyl-CoA molecules toward the formation of the isoliquiritigenin, an isoflavonoid skeleton. The increased expression of *CHS* was induced by UV-B in the roots. Compared with *CHR*, *CHS* is clearly more responsive to UV-B.

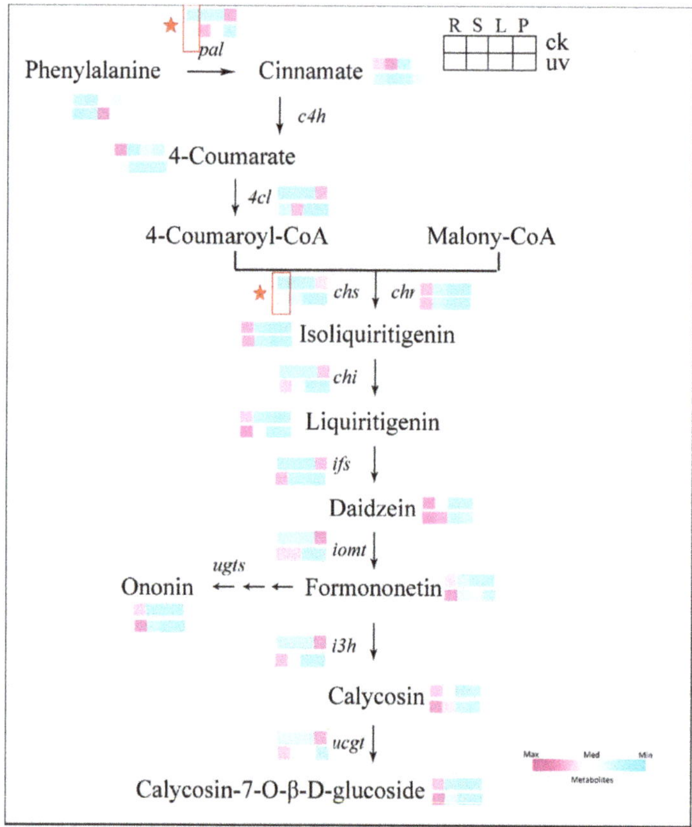

Figure 6. Visualization of isoflavonoid metabolites and relative gene dynamics on a biochemical pathway map. The levels of isoflavonoid metabolites and relative genes were both averaged over three biological replicates after normalization. Within each box, each column is a specific tissue (from left to right: root, R; stem, S; leaf, L; petiole, P), as shown in the upper-right corner. The intensity of the ratio of the UV-B group to the control (CK) group is indicated by the color scale key in the upper-right corner.

4. Discussion

Increased UV-B radiation affects plant development, morphology, physiology and metabolism. At the same time, environmental vicissitude will greatly impact *A. mongholicus* in the field. Therefore, the metabolic response of *A. mongholicus* to UV-B strongly influences the quality of Radix astragali that is provided to the market. In this study, we used untargeted GC-MS and targeted LC-MS techniques at primary and secondary compound levels to investigate the abundance and identity of compounds, revealing different metabolic profiles for specific tissues under different conditions. To explore the reprogramming of primary and secondary metabolic responses to UV-B and the formation of adaptive stable metabolism in above- and under- ground tissues. We also evaluated the level of isoflavones and their related genes to explore the mechanism of CAG accumulation under UV-B.

Based on the observation of morphological indicators and the detection of physiological indicators (Figure 1 and Table 1), leaves were more compact and plant height, leaf area, and chlorophyll content were reduced. These results indicate that growth was inhibited and that energy might be transferred to the accumulation of more functional metabolites to adapt to the changing environment [68]. The relative root biomass increased slightly, suggesting that energy was transferred to the roots in response to the UV-B radiation. Moreover, an increase antioxidant enzymes and H_2O_2 (Table 1) indicated that the stress response of plants to UV-B was actively operating [69,70].

We based our study on GC-MS-metabolomics to analyze tissue-specific changes under UV-B radiation and control conditions of *A. mongholicus*. Sugars, which are energy sources for growth, decreased in aboveground tissues but increased in belowground tissues. This explains the decrease in plant height and leaf area and the increase in root specific gravity. Additionally, the reduction in glucose in the leaves could indicate that a stronger shift from carbon assimilation to carbon accumulation occurred after UV-B radiation. Moreover, increased levels of sugar alcohols (e.g., erythritol) derived from the pentose phosphate pathway (PPP) were also observed in the UV-B treatment group, likely reflecting the higher respiratory rate of the PPP pathway in these plants. In addition, PLS-DA plots of the primary metabolites for four tissues under different conditions were generated (Figure 1A). The analysis of the score plots revealed a clear clustering of biological replicates for each sample.

The dynamic patterns of targeted phenolic compounds in four tissues of control and UV-B treatment plants was detected by UPLC-q/TOF-MS. Interestingly, the content of flavanones and flavonols increased in the leaves in response to UV-B radiation, whereas the isoflavonoids increased in the roots. Several early physiological experiments provided circumstantial evidence that phenolics are involved in UV-B protection [71]. Among these phenolics, flavones and flavonols protect cells because these compounds accumulate in the epidermal layers of leaves and stems, acting as filters and absorbing radiation in the UV-B portion of the spectrum [72,73]. Ryan et al. reported that UV radiation induces the synthesis of flavonols that have higher hydroxylation levels in *Petunia* and *Arabidopsis*. Those authors suggested that flavonols may play as yet uncharacterized roles in the UV stress response because flavonols have UV-absorbing properties, and the hydroxylation of flavonols might positively affect antioxidant capacity [74,75]. Regarding compounds involved in UV-B protection, the structures of flavones and flavonols are the most advantageous [15]. Thus, the most significant changes occurred regarding the levels of flavone and flavonol compounds (such as luteolin and myricitrin) under UV-B radiation were in leaves. At the same time, the accumulation of isoflavonoids (such as calycosin-7-glucoside, ononin, and formononetin) in roots constitutes a reserve of active phenolics that can easily be mobilized at any given time, especially under UV-B stress conditions. Furthermore, a common feature of the changes in phenolic compounds under UV-B exposure in roots and leaves was observed. The C6C3C6 carbon compounds increasingly accumulated at the expense of C6C3 carbon compounds such as chlorogenic acids, ferulic acid, and cinnamic acid. In view of this phenomenon, the increase in C6C3C6 carbon compounds (flavone and flavonol in the leaves and isoflavonoid in the roots) at the cost of C6C3 carbon compounds may be due to the stronger UV-B absorptive and antioxidant capacity of C6C3C6 carbon compounds than that of C6C3 ones in *A. mongholicus*. Under UV-B radiation, the phenolic compounds of various carbon skeletons are concentrated in the leaves (Figure 4), probably because the leaves are most directly affected by UV-B radiation. The decreasing of sugars and enrichment of phenolics in leaves could also result in reduced energy for growth and reduced leaf area.

To obtain a more detailed overview regarding the tissue-specific differences of the identified metabolites between the UV-B and control treatments, we built a primary and phenolic metabolites network (Figure 5A,B). The TCA cycle intermediates such as succinate and fumarate, which are major regulators of carbon and nitrogen interactions, increased under UV-B radiation. At the same time, this increase also occurred for shikimate and L-phenylalanine. It seems highly likely that shikimate would be elevated following the mobilization of carbon reserves stored in plants to increase the flux through the phenylpropanoid pathway under UV-B radiation [76]. In plants, phenylalanine is thought to be the general precursor of C6C1-, C6C3-, and C6C3C6 compounds and their polymers such as tannins and lignins [77]. As such, when considered together, these data suggest that in response to UV-B the plant cell is "primed" at the level of primary metabolism by a mechanism that involves the reprogramming of metabolism to efficiently divert carbon toward the aromatic amino acid precursors of the phenylpropanoid pathway [78]. Tryptophan and phenylalanine compete for chorismate to synthesize alkaloids and phenols, respectively. The results showed more accumulation of phenylalanine under UV-B radiation. This suggests that the increase in phenolic compounds under UV-B radiation is caused by the transfer of carbon from primary metabolism and involves metabolic reprogramming. For phenylpropanoid pathway under UV-B radiation, a significant difference appears between the above- and the underground tissues. Flavones were significantly increased in leaves, probably because of the roles of flavones accumulation in the epidermal layers of leaves acting as filters and absorbing radiation in the UV-B portion of the spectrum [79]. The phenolic acids enhanced accumulation in leaves might be acting as antioxidant supplements. Different from the metabolic response in leaves, enhanced accumulation of isoflavones were observed in roots. These isoflavones might constitute a reserve of active phenolics that can easily be mobilized at any given time, especially under UV-B stress conditions. Phenylalanine, as a key node in phenolic metabolism, showed a significant increase only in leaves under UV-B radiation. A bold and reasonable assumption is that the upregulation of the relevant genes rather than the synthetic precursor (phenylalanine) might be the most important contributor for the activation of phenolic metabolism in the roots.

In order to verify the possibility of this assumption for further, and due to the therapeutic potential of these isoflavones (especially CAG), the metabolic mechanisms of these compounds in the enhanced UV-B environment need to be further explored. The CAG biosynthesis pathway and relevant genes are visualized in Figure 6. The promoters of a series of genes involved in the flavonoid biosynthetic pathway contain a specific recognized domain that can interact with the MYB family of transcription factors via light-responsive elements [80]. In the present study, targeted isoflavonoid biosynthetic genes have transcriptional activation; as suggested, these transcription factors could be activated by UV-B radiation and could affect the investigated genes (Figure 6). In general, *PAL* genes have been proposed as a dominant control point of phenylpropanoids, flavonoids, and isoflavonoids biosynthesis in response to various biotic and abiotic stresses inclusive of pathogen attack, UV radiation, and mechanical wounding [81,82]. The no-significant of phenylalanine and increased transcription levels of *PAL* indicates that the response of phenolics in roots to UV-B is activated by relevant genes rather than phenylalanine. The increased transcription levels of *PAL* in roots, stems, and leaves indicated that the response of phenolics to UV-B is stimulated in different tissues. *CHS* and *CHR* co-catalyze the condensation of *p*-coumaryl-CoA with three malonyl-CoA molecules toward the formation of the Isoliquiritigenin, an isoflavonoid skeleton. The increased expression of *CHS* was induced by UV-B in the roots. Compared with *CHR*, *CHS* is clearly more responsive to UV-B, which suggests that the increased accumulation of isoflavonoids might be due to the elevated levels of *CHS* in the UV-B environment. The transcription level of the synthetase genes involved the CAG pathway, including *CHI*, *IFS*, *IOMT*, *I3'H*, and *UCGT*, were upregulated in the roots in response to UV-B induction but downregulated in aboveground tissues. This variation is strikingly similar to the accumulation pattern of the corresponding compounds of these enzymes. These results suggest that in the enhanced UV-B environment, tissue-specific increases in the levels of active isoflavones such as CAG are due to the regulation of the elevated levels of synthesis genes in the roots.

5. Conclusions

A metabolic profile was revealed using untargeted GC-MS and targeted LC-MS combined techniques to investigate the tissue-specific metabolic mechanism of *A. mongholicus* in an enhanced UV-B environment. We found that in response to UV-B the plant cell is "primed" at the level of primary metabolism by a mechanism that involves the reprogramming of metabolism to efficiently divert carbon toward the aromatic amino acid precursors of the phenylpropanoid pathway. A stronger shift from carbon assimilation to carbon accumulation has occurred. Among the accumulation of phenolics, C6C1 carbon compounds (phenolic acids in leaves) and C6C3C6 carbon compounds (flavones and flavonols in leaves and isoflavones in roots) increased at the expense of C6C3 carbon compounds in order to obtain the stronger UV-B absorptive and antioxidant capacity. Compared with the control treatment, the response of phenolics has a significant tissue-specific in the UV-B radiation treatment. Notably, the response of phenolics in roots to UV-B is activated by upregulation of relevant genes rather than phenylalanine.

Supplementary Materials: The following are available online at http://www.mdpi.com/2218-273X/10/4/504/s1, Table S1: Compounds identified in *Astragalus mongholicus* measured by Gas Chromatography-Mass Spectrometry; Table S2: Metabolite reporting checklist and recommendations for Liquid Chromatography-Mass Spectrometry.

Author Contributions: Conceptualization, Y.L., J.L., L.-Q.M.; methodology, Y.L., K.-X.W.; software, J.L.; validation, K.-X.W., Y.L.; formal analysis, A.A.; investigation, K.-X.W., X.-R.G.; resources, L.-Q.M.; data curation, Y.L.; writing—original draft preparation, Y.L.; writing—review and editing, Y.L., A.A.; visualization, J.L.; supervision, X.-R.G., Z.-H.T.; project administration, X.-R.G., Z.-H.T.; funding acquisition, Z.-H.T., X.-R.G., Y.L. All authors have read and agreed to the published version of the manuscript.

Funding: This research was funded by the Heilongjiang Province Postdoctoral Fund (415556), the Fundamental Research Funds for the Central Universities (2572018BU01), the China Postdoctoral Science Foundation (2019M651239) and Northeast Forestry University Postdoctoral Fund (60201105).

Conflicts of Interest: The authors declare no conflict of interest.

References

1. Pan, H.; Fang, C.; Zhou, T.; Wang, Q.; Chen, J. Accumulation of calycosin and its 7-O-beta-D-glucoside and related gene expression in seedlings of Astragalus membranaceus Bge. var. mongholicus (Bge.) Hsiao induced by low temperature stress. *Plant Cell Rep.* **2007**, *26*, 1111–1120. [CrossRef] [PubMed]

2. Napolitano, A.; Akay, S.; Mari, A.; Bedir, E.; Pizza, C.; Piacente, S. An analytical approach based on ESI-MS, LC–MS and PCA for the quali–quantitative analysis of cycloartane derivatives in Astragalus spp. *J. Pharm. Biomed. Anal.* **2013**, *85*, 46–54. [CrossRef] [PubMed]

3. Kuo, Y.-H.; Tsai, W.-J.; Loke, S.-H.; Wu, T.-S.; Chiou, W.-F. Astragalus membranaceus flavonoids (AMF) ameliorate chronic fatigue syndrome induced by food intake restriction plus forced swimming. *J. Ethnopharmacol.* **2009**, *122*, 28–34. [CrossRef] [PubMed]

4. Wu, J.-H.; Li, Q.; Wu, M.-Y.; Guo, D.-J.; Chen, H.-L.; Chen, S.-L.; Seto, S.W.; Au, A.L.; Poon, C.C.; Leung, G.P.-H. Formononetin, an isoflavone, relaxes rat isolated aorta through endothelium-dependent and endothelium-independent pathways. *J. Nutr. Biochem.* **2010**, *21*, 613–620. [CrossRef]

5. Fan, Y.; Wu, D.-Z.; Gong, Y.-Q.; Zhou, J.-Y.; Hu, Z.-B. Effects of calycosin on the impairment of barrier function induced by hypoxia in human umbilical vein endothelial cells. *Eur. J. Pharmacol.* **2003**, *481*, 33–40. [CrossRef]

6. Cho, W.C.S.; Leung, K.N. In vitro and in vivo immunomodulating and immunorestorative effects of Astragalus membranaceus. *J. Ethnopharmacol.* **2007**, *113*, 132–141. [CrossRef]

7. Cho, W.C.; Leung, K.N. In vitro and in vivo anti-tumor effects of Astragalus membranaceus. *Cancer Lett.* **2007**, *252*, 43–54. [CrossRef]

8. Ma, X.; Tu, P.; Chen, Y.; Zhang, T.; Wei, Y.; Ito, Y. Preparative isolation and purification of two isoflavones from Astragalus membranaceus Bge. var. mongholicus (Bge.) Hsiao by high-speed counter-current chromatography. *J. Chromatogr. A* **2003**, *992*, 193–197. [CrossRef]

9. Nakamura, T.; Hashimoto, A.; Nishi, H.; Kokusenya, Y. Investigation on the Marker Substances of Crude Drugs in Formulations. I.: Marker Substances for the Identification of Astragali Radix in Kampo and Drinkable Preparations. *Yakugaku Zasshi* **1999**, *119*, 391–400. [CrossRef]

10. Yu, O.; Shi, J.; Hession, A.O.; Maxwell, C.A.; Mcgonigle, B.; Odell, J.T. Metabolic engineering to increase isoflavone biosynthesis in soybean seed. *Phytochem.* **2003**, *63*, 753–763. [CrossRef]

11. Frohnmeyer, H.; Staiger, D. Ultraviolet-B Radiation-Mediated Responses in Plants. Balancing Damage and Protection. *Plant Physiol.* **2003**, *133*, 1420–1428. [CrossRef] [PubMed]

12. Jansen, M.A.K.; Gaba, V.; Greenberg, B.M. Higher plants and UV-B radiation: Balancing damage, repair and acclimation. *Trends Plant Sci.* **1998**, *3*, 131–135. [CrossRef]

13. Rozema, J.; Van De Staaij, J.; Björn, L.O.; Caldwell, M. UV-B as an environmental factor in plant life: Stress and regulation. *Trends Ecol. Evol.* **1997**, *12*, 22–28. [CrossRef]

14. Kusano, M.; Tohge, T.; Fukushima, A.; Kobayashi, M.; Hayashi, N.; Otsuki, H.; Kondou, Y.; Goto, H.; Kawashima, M.; Matsuda, F.; et al. Metabolomics reveals comprehensive reprogramming involving two independent metabolic responses of Arabidopsis to UV-B light. *Plant J.* **2011**, *67*, 354–369. [CrossRef]

15. Matsuura, H.N.; De Costa, F.; Yendo, A.C.A.; Fett-Neto, A.G. Photoelicitation of Bioactive Secondary Metabolites by Ultraviolet Radiation: Mechanisms, Strategies, and Applications. *Biotechnol. Med. Plants* **2012**, 171–190.

16. Hahlbrock, K.; Scheel, D. Physiology and Molecular Biology of Phenylpropanoid Metabolism. *Annu. Rev. Plant Biol.* **1989**, *40*, 347–369. [CrossRef]

17. Frolov, A.; Henning, A.; Böttcher, C.; Tissier, A.; Strack, D. An UPLC-MS/MS Method for the Simultaneous Identification and Quantitation of Cell Wall Phenolics in Brassica napus Seeds. *J. Agric. Food Chem.* **2013**, *61*, 1219–1227. [CrossRef]

18. Parr, A.J.; Bolwell, G.P. Phenols in the plant and in man. The potential for possible nutritional enhancement of the diet by modifying the phenols content or profile. *J. Sci. Food Agric.* **2000**, *80*, 985–1012. [CrossRef]

19. Beckman, C.H. Phenolic-storing cells: Keys to programmed cell death and periderm formation in wilt disease resistance and in general defence responses in plants? *Physiol. Mol. Plant Pathol.* **2000**, *57*, 101–110. [CrossRef]

20. Valcarcel, J.; Reilly, K.; Gaffney, M.; O'Brien, N.M. Antioxidant Activity, Total Phenolic and Total Flavonoid Content in Sixty Varieties of Potato (Solanum tuberosum L.) Grown in Ireland. *Potato Res.* **2015**, *58*, 221–244. [CrossRef]

21. García, B.A.; Berrueta, L.A.; Garmón-Lobato, S.; Gallo, B.; Vicente, F. A general analytical strategy for the characterization of phenolic compounds in fruit juices by high-performance liquid chromatography with diode array detection coupled to electrospray ionization and triple quadrupole mass spectrometry. *J. Chromatogr. A* **2009**, *1216*, 5398–5415. [CrossRef] [PubMed]

22. Caravaca, G.; López-Cobo, A.; Verardo, V.; Segura-Carretero, A.; Gutierrez, A.F. HPLC-DAD-Q-TOF-MS as a powerful platform for the determination of phenolic and other polar compounds in the edible part of mango and its by-products (peel, seed and seed husk). *Electrophoresis* **2016**, *37*, 1072–1084. [CrossRef] [PubMed]

23. Herrmann, K.M.; Weaver, L.M. The shikimate pathway. *Annu. Rev. Plant Boil.* **1999**, *50*, 473–503. [CrossRef] [PubMed]

24. Maeda, H.; Dudareva, N. The Shikimate Pathway and Aromatic Amino Acid Biosynthesis in Plants. *Annu. Rev. Plant Boil.* **2012**, *63*, 73–105. [CrossRef] [PubMed]

25. Babst, B.A.; Harding, S.A.; Tsai, C.-J. Biosynthesis of Phenolic Glycosides from Phenylpropanoid and Benzenoid Precursors in Populus. *J. Chem. Ecol.* **2010**, *36*, 286–297. [CrossRef]

26. Wink, M. *Biochemistry of Plant Secondary Metabolism*, 2nd ed.; Wiley Blackwell: Hoboken, NJ, USA, 2010; Volume 40, pp. 484–485.

27. Marković, S.; Tošović, J. Application of Time-Dependent Density Functional and Natural Bond Orbital Theories to the UV–vis Absorption Spectra of Some Phenolic Compounds. *J. Phys. Chem. A* **2015**, *119*, 9352–9362. [CrossRef]

28. Song, J.-Z.; Mo, S.-F.; Yip, Y.-K.; Qiao, C.-F.; Han, Q.-B.; Xu, H. Development of microwave assisted extraction for the simultaneous determination of isoflavonoids and saponins in radix astragali by high performance liquid chromatography. *J. Sep. Sci.* **2007**, *30*, 819–824. [CrossRef]

29. Wu, T.; Bligh, S.A.; Gu, L.-H.; Wang, Z.-T.; Liu, H.-P.; Cheng, X.-M.; Branford-White, C.J.; Hu, Z.-B. Simultaneous determination of six isoflavonoids in commercial Radix Astragali by HPLC-UV. *Fitoterapia* **2005**, *76*, 157–165. [CrossRef]

30. Liu, J.; Liu, Y.; Wang, Y.; Abozeid, A.; Yuan-Gang, Z.; Tang, Z.-H. The integration of GC–MS and LC–MS to assay the metabolomics profiling in Panax ginseng and Panax quinquefolius reveals a tissue- and species-specific connectivity of primary metabolites and ginsenosides accumulation. *J. Pharm. Biomed. Anal.* **2017**, *135*, 176–185. [CrossRef]

31. Park, H.-W.; In, G.; Kim, J.-H.; Cho, B.-G.; Han, G.-H.; Chang, I.-M. Metabolomic approach for discrimination of processed ginseng genus (Panax ginseng and Panax quinquefolius) using UPLC-QTOF MS. *J. Ginseng Res.* **2013**, *38*, 59–65. [CrossRef]

32. Li, L.; Luo, G.-A.; Liang, Q.; Hu, P.; Wang, Y.-M. Rapid qualitative and quantitative analyses of Asian ginseng in adulterated American ginseng preparations by UPLC/Q-TOF-MS. *J. Pharm. Biomed. Anal.* **2010**, *52*, 66–72. [CrossRef]

33. Fernie, A.R.; Trethewey, R.N.; Krotzky, A.J.; Willmitzer, L. Metabolite profiling: From diagnostics to systems biology. *Nat. Rev. Mol. Cell Boil.* **2004**, *5*, 763–769. [CrossRef] [PubMed]

34. Fiehn, O. Combining Genomics, Metabolome Analysis, and Biochemical Modelling to Understand Metabolic Networks. *Comp. Funct. Genom.* **2001**, *2*, 155–168. [CrossRef]

35. Obata, T.; Fernie, A.R. The use of metabolomics to dissect plant responses to abiotic stresses. *Cell. Mol. Life Sci.* **2012**, *69*, 3225–3243. [CrossRef] [PubMed]

36. Ye, G.; Zhu, B.; Yao, Z.; Yin, P.; Lu, X.; Kong, H.; Fan, F.; Jiao, B.; Xu, G. Analysis of Urinary Metabolic Signatures of Early Hepatocellular Carcinoma Recurrence after Surgical Removal Using Gas Chromatography–Mass Spectrometry. *J. Proteome Res.* **2012**, *11*, 4361–4372. [CrossRef]

37. Li, Y.; Ruan, Q.; Li, Y.; Ye, G.; Lü, X.; Lin, X.; Xu, G. A novel approach to transforming a non-targeted metabolic profiling method to a pseudo-targeted method using the retention time locking gas chromatography/mass spectrometry-selected ions monitoring. *J. Chromatogr. A* **2012**, *1255*, 228–236. [CrossRef] [PubMed]

38. Lisec, J.; Schauer, N.; Kopka, J.; Willmitzer, L.; Fernie, A.R. Gas chromatography mass spectrometry–based metabolite profiling in plants. *Nat. Protoc.* **2006**, *1*, 387–396. [CrossRef]

39. Lisec, J.; Schauer, N.; Kopka, J.; Willmitzer, L.; Fernie, A.R. Corrigendum: Gas chromatography mass spectrometry-based metabolite profiling in plants. *Nat Protoc.* **2015**, *10*, 1457. [CrossRef]

40. Schauer, N.; Steinhauser, D.; Strelkov, S.; Schomburg, D.; Allison, G.; Moritz, T.; Lundgren, K.; Roessner, U.; Forbes, M.G.; Willmitzer, L.; et al. GC-MS libraries for the rapid identification of metabolites in complex biological samples. *FEBS Lett.* **2005**, *579*, 1332–1337. [CrossRef]

41. Ye, G.; Liu, Y.; Yin, P.; Zeng, Z.; Huang, Q.; Kong, H.; Lu, X.; Zhong, L.; Zhang, Z.; Xu, G. Study of Induction Chemotherapy Efficacy in Oral Squamous Cell Carcinoma Using Pseudotargeted Metabolomics. *J. Proteome Res.* **2014**, *13*, 1994–2004. [CrossRef] [PubMed]

42. Imperlini, E.; Santorelli, L.; Orrù, S.; Scolamiero, E.; Ruoppolo, M.; Caterino, M. Mass Spectrometry-Based Metabolomic and Proteomic Strategies in Organic Acidemias. *BioMed Res. Int.* **2016**, *2016*, 1–13. [CrossRef] [PubMed]

43. Allwood, J.W.; Goodacre, R. An introduction to liquid chromatography–mass spectrometry instrumentation applied in plant metabolomic analyses. *Phytochem. Anal.* **2010**, *21*, 33–47. [CrossRef]

44. Farag, M.A.; Huhman, D.; Lei, Z.; Sumner, L. Metabolic profiling and systematic identification of flavonoids and isoflavonoids in roots and cell suspension cultures of Medicago truncatula using HPLC–UV–ESI–MS and GC–MS. *Phytochemistry* **2007**, *68*, 342–354. [CrossRef] [PubMed]

45. Rogachev, I.; Aharoni, A. UPLC-MS-Based Metabolite Analysis in Tomato. *Adv. Struct. Saf. Stud.* **2011**, *860*, 129–144.

46. Doerfler, H.; Lyon, D.; Nägele, T.; Sun, X.; Fragner, L.; Hadaček, F.; Egelhofer, V.; Weckwerth, W. Granger causality in integrated GC–MS and LC–MS metabolomics data reveals the interface of primary and secondary metabolism. *Metabolomics* **2012**, *9*, 564–574. [CrossRef] [PubMed]

47. Han, J.S.; Lee, S.; Kim, H.Y.; Lee, C.H. MS-Based Metabolite Profiling of Aboveground and Root Components of Zingiber mioga and Officinale. *Molecules* **2015**, *20*, 16170–16185. [CrossRef]

48. Cuadros-Inostroza, A.; Ruiz-Lara, S.; González, E.; Eckardt, A.; Willmitzer, L.; Peña-Cortés, H. GC–MS metabolic profiling of Cabernet Sauvignon and Merlot cultivars during grapevine berry development and network analysis reveals a stage- and cultivar-dependent connectivity of primary metabolites. *Metabolomics* **2016**, *12*, 39. [CrossRef]

49. Moing, A.; Aharoni, A.; Biais, B.; Rogachev, I.; Meir, S.; Brodsky, L.; Allwood, J.W.; Erban, A.; Dunn, W.; Kay, L.; et al. Extensive metabolic cross-talk in melon fruit revealed by spatial and developmental combinatorial metabolomics. *New Phytol.* **2011**, *190*, 683–696. [CrossRef]

50. Logemann, E.; Tavernaro, A.; Schulz, W.; Somssich, I.; Hahlbrock, K. UV light selectively coinduces supply pathways from primary metabolism and flavonoid secondary product formation in parsley. *Proc. Natl. Acad. Sci. USA* **2000**, *97*, 1903–1907. [CrossRef]

51. Liu, C.-J.; Deavours, B.E.; Richard, S.B.; Ferrer, J.-L.; Blount, J.W.; Huhman, D.; Dixon, R.A.; Noel, J.P. Structural Basis for Dual Functionality of Isoflavonoid O-Methyltransferases in the Evolution of Plant Defense Responses[OA]. *Plant Cell* **2006**, *18*, 3656–3669. [CrossRef]

52. Liu, C.-J.; Huhman, D.; Sumner, L.; Dixon, R.A. Regiospecific hydroxylation of isoflavones by cytochrome p450 81E enzymes from Medicago truncatula. *Plant J.* **2003**, *36*, 471–484. [CrossRef] [PubMed]

53. Dhaubhadel, S.; McGarvey, B.D.; Williams, R.; Gijzen, M. Isoflavonoid biosynthesis and accumulation in developing soybean seeds. *Plant Mol. Boil.* **2003**, *53*, 733–743. [CrossRef] [PubMed]

54. Graham, M.Y. The Diphenylether Herbicide Lactofen Induces Cell Death and Expression of Defense-Related Genes in Soybean1. *Plant Physiol.* **2005**, *139*, 1784–1794. [CrossRef] [PubMed]

55. Jiao, J.; Gai, Q.-Y.; Wang, W.; Luo, M.; Gu, C.-B.; Fu, Y.-J.; Ma, W. Ultraviolet Radiation-Elicited Enhancement of Isoflavonoid Accumulation, Biosynthetic Gene Expression, and Antioxidant Activity in Astragalus membranaceus Hairy Root Cultures. *J. Agric. Food Chem.* **2015**, *63*, 8216–8224. [CrossRef]

56. Caldwell, M.M.; Ballare, C.L.; Bornman, J.F.; Flint, S.D.; Bjorn, L.O.; Teramura, A.H.; Kulandaivelu, G.; Tevini, M. Terrestrial ecosystems, increased solar ultraviolet radiation and interactions with other climatic change factors. *Photochem. Photobiol. Sci.* **2003**, *2*, 29–38.

57. Xiong, F.S.; Day, T.A. Effect of solar ultraviolet-B radiation during springtime ozone depletion on photosynthesis and biomass production of Antarctic vascular plants. *Plant Physiol.* **2001**, *125*, 738–751.

58. Xu, R.Y.; Nan, P.; Yang, Y.; Pan, H.; Zhou, T.; Chen, J. Ultraviolet irradiation induces accumulation of isoflavonoids and transcription of genes of enzymes involved in the calycosin-7-O-beta-d-glucoside pathway in Astragalus membranaceus Bge. var. mongholicus (Bge.) Hsiao. *Physiol. Plant* **2011**, *142*, 265–273. [CrossRef]

59. Liu, Y.; Liu, J.; Wang, Y.; Abozeid, A.; Tian, D.; Zhang, X.-N.; Tang, Z. The Different Resistance of Two Astragalus Plants to UV-B Stress is Tightly Associated with the Organ-specific Isoflavone Metabolism. *Photochem. Photobiol.* **2017**, *94*, 115–125. [CrossRef]

60. Li, J.; Ou-Lee, T.M.; Raba, R.; Last, A.R.L. Arabidopsis flavonoid mutants are hypersensitive to UB-B irradiation. *Plant Cell.* **1993**, *5*, 171–179. [CrossRef] [PubMed]

61. Liu, L.; McClure, J.W. Effects of UV-B on activities of enzymes of secondary phenolic metabolism in barley primary leaves. *Physiol. Plant.* **1995**, *93*, 734–739. [CrossRef]

62. Arnon, D.I. Copper Enzymes in Isolated Chloroplasts. Polyphenoloxidase in Beta Vulgaris. *Plant Physiol.* **1949**, *24*, 1–15. [CrossRef]

63. Aiamla-Or, S.; Shigyo, M.; Ito, S.-I.; Yamauchi, N. Involvement of chloroplast peroxidase on chlorophyll degradation in postharvest broccoli florets and its control by UV-B treatment. *Food Chem.* **2014**, *165*, 224–231. [CrossRef]

64. Cuadros-Inostroza, Á.; Caldana, C.; Redestig, H.; Kusano, M.; Lisec, J.; Peña-Cortés, H.; Willmitzer, L.; Hannah, M. TargetSearch-a Bioconductor package for the efficient preprocessing of GC-MS metabolite profiling data. *BMC Bioinform.* **2009**, *10*, 428.

65. Langfelder, P.; Horvath, S. WGCNA: An R package for weighted correlation network analysis. *BMC Bioinform.* **2008**, *9*, 559. [CrossRef] [PubMed]

66. Dileo, M.V.; Strahan, G.D.; Bakker, M.D.; Hoekenga, O. Weighted Correlation Network Analysis (WGCNA) Applied to the Tomato Fruit Metabolome. *PLoS ONE* **2011**, *6*, e26683. [CrossRef] [PubMed]

67. Liu, J.; Liu, Y.; Wang, Y.; Zhang, Z.-H.; Zu, Y.-G.; Efferth, T.; Tang, Z.-H. The Combined Effects of Ethylene and MeJA on Metabolic Profiling of Phenolic Compounds in Catharanthus roseus Revealed by Metabolomics Analysis. *Front. Physiol.* **2016**, *7*, 2555. [CrossRef] [PubMed]

68. Zu, Y.-G.; Pang, H.-H.; Yu, J.-H.; Li, D.-W.; Wei, X.-X.; Gao, Y.-X.; Tong, L. Responses in the morphology, physiology and biochemistry of Taxus chinensis var. mairei grown under supplementary UV-B radiation. *J. Photochem. Photobiol. B Boil.* **2010**, *98*, 152–158. [CrossRef] [PubMed]

69. Soriano-Melgar, L.D.A.A.; Alcaraz-Meléndez, L.; Méndez-Rodríguez, L.C.; Puente, M.E.; Rivera-Cabrera, F.; Zenteno-Savin, T. Antioxidant responses of damiana (Turnera diffusa Willd) to exposure to artificial ultraviolet (UV) radiation in an in vitro model; part ii; UV-B radiation. *Nutr. Hosp.* **2014**, *29*, 1116–1122.

70. Dias, D.A.; Hill, C.B.; Jayasinghe, N.S.; Atieno, J.; Sutton, T.; Roessner, U. Quantitative profiling of polar primary metabolites of two chickpea cultivars with contrasting responses to salinity. *J. Chromatogr. B* **2015**, *1000*, 1–13. [CrossRef]

71. Harborne, J.B.; A Williams, C. Advances in flavonoid research since 1992. *Phytochemistry* **2000**, *55*, 481–504. [CrossRef]

72. Peer, W.; Murphy, A.S. Flavonoids and auxin transport: Modulators or regulators? *Trends Plant Sci.* **2007**, *12*, 556–563. [CrossRef]

73. Taylor, L.P.; Grotewold, E. Flavonoids as developmental regulators. *Curr. Opin. Plant Boil.* **2005**, *8*, 317–323. [CrossRef]

74. Ryan, K.G.; Swinny, E.E.; Markham, K.R.; Winefield, C. Flavonoid gene expression and UV photoprotection in transgenic and mutant Petunia leaves. *Phytochemistry* **2002**, *59*, 23–32. [CrossRef]

75. Ryan, K.G.; Swinny, E.E.; Winefield, C.; Markham, K.R. Flavonoids and UV photoprotection in Arabidopsis mutants. *Z. Nat. C* **2001**, *56*, 745–754. [CrossRef]

76. Gibon, Y.; Pyl, E.-T.; Sulpice, R.; Lunn, J.; Höhne, M.; Günther, M.; Stitt, M. Adjustment of growth, starch turnover, protein content and central metabolism to a decrease of the carbon supply when Arabidopsisis grown in very short photoperiods. *Plant Cell Environ.* **2009**, *32*, 859–874. [CrossRef] [PubMed]

77. Robbins, M.P. Biochemistry of plant secondary metabolism. Annual Plant Reviews, Volume 2. Edited by Michael Wink. *Eur. J. Plant Pathol.* **2000**, *106*, 487. [CrossRef]

78. Kusano, M.; Fukushima, A.; Redestig, H.; Saito, K. Metabolomic approaches toward understanding nitrogen metabolism in plants. *J. Exp. Bot.* **2011**, *62*, 1439–1453. [CrossRef]

79. Agati, G.; Cerovic, Z.G.; Pinelli, P.; Tattini, M. Light-induced accumulation of ortho-dihydroxylated flavonoids as non-destructively monitored by chlorophyll fluorescence excitation techniques. *Environ. Exp. Bot.* **2011**, *73*, 3–9. [CrossRef]

80. Zhang, Z.-Z.; Che, X.-N.; Pan, Q.-H.; Li, X.-X.; Duan, C.-Q. Transcriptional activation of flavan-3-ols biosynthesis in grape berries by UV irradiation depending on developmental stage. *Plant Sci.* **2013**, *208*, 64–74. [CrossRef]

81. Kanazawa, K.; Hashimoto, T.; Yoshida, S.; Sungwon, P.; Fukuda, S. Short Photoirradiation Induces Flavonoid Synthesis and Increases Its Production in Postharvest Vegetables. *J. Agric. Food Chem.* **2012**, *60*, 4359–4368. [CrossRef]

82. Singh, K.; Kumar, S.; Rani, A.; Gulati, A.; Ahuja, P.S. Phenylalanine ammonia-lyase (*PAL*) and cinnamate 4-hydroxylase (*C4H*) and catechins (flavan-3-ols) accumulation in tea. *Funct. Integr. Genom.* **2008**, *9*, 125–134. [CrossRef]

 *biomolecules*

Article

Network Pharmacology-Based Approaches of *Rheum undulatum* Linne and *Glycyrriza uralensis* Fischer Imply Their Regulation of Liver Failure with Hepatic Encephalopathy in Mice

Su Youn Baek [1], Eun Hye Lee [2], Tae Woo Oh [3], Hyun Ju Do [3], Kwang-Youn Kim [3], Kwang-Il Park [4] and Young Woo Kim [5,*]

[1] Institute for Phylogenomics and Evolution, Kyungpook National University, Daegu 41566, Korea; rhodeus@nate.com
[2] School of Medical Science, Kyungpook National University, Daegu 41566, Korea; eun90hye@nate.com
[3] Korea Institute of Oriental Medicine, Daegu 41062, Korea; taewoo2080@kiom.re.kr (T.W.O.); dododo@kiom.re.kr (H.J.D.); lokyve@kiom.re.kr (K.-Y.K.)
[4] College of Veterinary Medicine, Gyeongsang National University, Jinju 52828, Korea; kipark@gnu.ac.kr
[5] School of Korean Medicine, Dongguk University, Gyeongju 38066, Korea
*   Correspondence: ywk@dongguk.ac.kr; Fax: +82-31-961-5835

Received: 30 January 2020; Accepted: 6 March 2020; Published: 12 March 2020

Abstract: *Rheum undulatum* and *Glycyrrhiza uralensis* have been used as supplementary ingredients in various herbal medicines. They have been reported to have anti-inflammatory and antioxidant effects and, therefore, have potential in the treatment and prevention of various liver diseases. Considering that hepatic encephalopathy (HE) is often associated with chronic liver failure, we investigated whether an *R. undulatum* and *G. uralensis* extract mixture (RG) could reduce HE. We applied systems-based pharmacological tools to identify the active ingredients in RG and the pharmacological targets of RG by examining mechanism-of-action profiles. A CCl_4-induced HE mouse model was used to investigate the therapeutic mechanisms of RG on HE. We successfully identified seven bioactive ingredients in RG with 40 potential targets. Based on an integrated target–disease network, RG was predicted to be effective in treating neurological diseases. In animal models, RG consistently relieved HE symptoms by protecting blood–brain barrier permeability via downregulation of matrix metalloproteinase-9 (MMP-9) and upregulation of claudin-5. In addition, RG inhibited mRNA expression levels of both interleukin (IL)-1β and transforming growth factor (TGF)-β1. Based on our results, RG is expected to function various biochemical processes involving neuroinflammation, suggesting that RG may be considered a therapeutic agent for treating not only chronic liver disease but also HE.

Keywords: *Rheum undulatum*; *Glycyrriza uralensis*; hepatic encephalopathy; MMP-9; neuroinflammation

---

1. Introduction

Hepatic encephalopathy (HE) is a decline in neuropsychiatric function observed in patients with acute or chronic liver diseases [1,2]. The clinical symptoms are so diverse that it may appear as a subtle impairment in mental state, but it can lead to coma [1]. There are numerous explanations of why liver dysfunction can lead to encephalopathy; however, the most common clinical mechanism through which HE develops is an increase in the level of blood ammonia, which could be removed by sodium benzoate [1–4]. In a healthy body, the nitrogen-containing compounds that remain following the food digestion process are transported through the portal vein to the liver, where most such compounds are metabolized. If the liver cells or the metabolism processes are impaired, nitrogen wastes containing a large amount of ammonia can accumulate in systemic circulation. Small nitrogen-containing molecules, such as ammonia, can pass through the blood–brain barrier (BBB), and astrocytes in the cerebral cortex can absorb ammonia to convert glutamate into glutamine [3,5]. An excessively high glutamine level in astrocytes may increase osmotic pressure and the activity of the inhibitory GABA system, leading to a shortage in the energy supply to the brain [3,5].

Commonly utilized pharmacologic HE treatments tend to focus on lowering the blood ammonia level [3,4]. Clinically, two classes of medications—non-absorbed disaccharides, including lactulose and lactitol; and nonabsorbable antibiotics, including neomycin and rifaximin—are normally used [2–4]. However, both these two classes are known to show negative side-effects, such as diarrhea, electrolyte disturbances and resistances, which can limit long-term effectiveness and patient compliance [4,6]. Therefore, it is necessary to be simultaneously considered for effective treatment, including the presence of oxidative stress or neurotoxins, as well as changes in neurotransmission, GABA-ergic pathways and energy metabolism. During such consideration, it is wise to consider the potential of herbal medicine treatments, which have been widely applied in a variety of medical fields due to their high efficiency and stability [6,7].

Rhubarb (*Rheum undulatum*) is a perennial plant belonging to the Polygonaceae family, and the components of dried roots of *R. undulatum* are reported to have therapeutic efficacy in oriental traditional medicine against various diseases [8,9]. Herbal preparations of *R. undulatum* are used as a stimulant laxative with excellent efficacy in reducing inflammation in the liver, large intestine, and kidney. Previous studies have reported that its components have potent antitumor, antithrombotic, and antioxidant properties [10–12]. A review argued that rhubarb, alone or combined with other herbs, can be used to treat HE and to improve treatment efficacy, which was promising but not based on solid empirical evidence [8]. Ning Z. et al. (2017) reported that the rhubarb-based Chinese herbal formulae (RCHF) might be effective at treating hepatic encephalopathy using meta-analysis in China. Their data showed that the rhubarb-based Chinese herbal formulae (RCHF) patients had significant improvements in their ammonia and alanine aminotransferase levels in their blood when compared to the other groups of patients that received other treatments [13]. Moreover, a few studies demonstrated that rhubarb contained multiple biological activities probably associated with BBB function, including anti-oxidation, anti-inflammation, and inhibition of aquaporin expression [14,15]. However, no research has yet been reported on which molecular mechanisms rhubarb treats HE.

A previous our study showed a significant protective effect of an *R. undulatum* and *Glycyrrhiza uralensis* extract mixture (hereafter RG) against hepatic oxidative injury in two model systems: hepatocytes treated with arachidonic acid (AA) plus iron and mice with CCl_4-induced liver injury [16]. We found the best ratio of RG (1:10) in vitro and applied at mice model of acute hepatitis [16]. *Glycyrrhiza glabra* Fisch have been used to treat a wide array of illnesses in herbal and traditional medicines for thousands of years [16,17]. Triterpene saponins, such as glycyrrhizin, uralsaponin, liquorice saponins, and glycyrrhetinic acid; as well as flavonoids, such as liquiritin, liquiritigenin, Isoliquiritin, isoliquiritigenin, and isolicoflavonol, are reported to be responsible for the pharmacological properties of licorice [17]. In the present study, we analyzed RG to elucidate the mechanism-of-action of each component, to identify the components that are thought to have a key role and, ultimately, to predict the therapeutic effects of these components by constructing a drug-target-disease network model. We performed in vivo and in vitro experiments to determine whether RG has an effect on HE, and if so, by what molecular mechanisms. Our results are expected to provide theoretical and in vivo empirical evidence that RG may be used as an effective drug in the treatment of HE.

2. Materials and Methods

2.1. Reagents

Anti-claudin 5 was obtained from Santa Cruz Biotechnology (Santa Cruz, CA, USA). Horseradish peroxidase conjugated goat anti-rabbit and goat anti-mouse IgGs were purchased from Novus Biological (Centennial, CO, USA). Trizol was obtained from Invitrogen (Invitrogen, Carlsbad, CA, USA) and reverse transcriptional polymerase chain reaction (RT-PCR) kit was obtained from Promega (Promega, Madison WI, USA). Sennoside A, emodin, chrysophanol, aloe-emodin, rhein, glycyrrhizin acid, liquiritigenin, isoliquiritigenin, and other reagents were purchased from Sigma-Aldrich (St. Louis, MO, USA). The reagents for ultra-performance liquid chromatography (UPLC) analysis were methanol (Junsei for the high-performance liquid chromatography (HPLC), acetonitrile (JT BAKER for the HPLC), and then Water (Tertiary distilled water).

2.2. Preparation of RG Extracts

Rheum undulatum (200 g) and *Glycyrriza uralensis* (200 g), which are warranted as a standard herb for medicine by Korea Food and Drug Administration, were purchased from Sejong pharmacy (Daegu, South Korea) and were boiled in 1500 mL of distilled water for 3 h. This aqueous mixture was passed through a 0.2 mm filter (Nalgene, New York, NY, USA). The filtrate was concentrated under reduced pressure at 50 °C and lyophilized into powder. The extraction yields of *R. undulatum* and *Glycyrriza uralensis* were 11.33% and 15.61%, respectively.

2.3. Profiling the Chemical Contents of RUE and GUE by Ultra-Performance Liquid Chromatography

Ultra-performance liquid chromatography) analysis was performed using an ACQUITY ultraperformance LC system equipped with ACQUITY photodiode array (PDA) detector (Waters Corporation, Milford, CT, USA) based on the methods described in a previous study [18]. ACQUITY BEH C18 column (Waters Corporation; 1.7 μm, 2.1 × 100 mm) was used to elute the compounds off and Empower software was applied to tract and compare the peak data. The standard compounds were melted by methanol and DMSO. Next, they were prepared based on a standard undiluted solution containing 1 mg/mL. The PDA analysis wavelengths were sennoside A (340 nm), emodin (254 nm), chrysophanol (254 nm), aloe-emodin (254 nm), rhein (254 nm), glycyrrhizin acid (254 nm), liquiritigenin (280 nm), and isoliquiritigenin (280 nm), respectively. Samples were 2 mL, and the flow rate was 0.4 mL/min. Five and three potential ingredients were collected from *R. undulatum* and *G. uralensis*, respectively (Figure 1 and Table 1).

Table 1. Chemical formula and mass accuracy of potential ingredients.

Sample	Identity	Chemical Formula	Mass Accuracy (ppm)
RUE (*Rheum undulatum* Linne)	Sennoside A	$C_{42}H_{38}O_{20}$	145.85 ± 7.080
	Emodin	$C_{15}H_{10}O_5$	1.20 ± 0.012
	Chrysophanol	$C_{15}H_{10}O_4$	0.064 ± 0.001
	Aloe-emodin	$C_{15}H_{10}O_5$	2.11 ± 0.616
	Rhein	$C_{15}H_8O_6$	29.47 ± 0.447
GUE (*Glycyrriza uralensis*)	Glycyrrhizin acid	$C_{42}H_{62}O_{16}$	325.91 ± 6.8
	Liquiritigenin	$C_{15}H_{12}O_4$	124.25 ± 3.7
	Isoliquiritigenin	$C_{15}H_{12}O_4$	6.08 ± 0.7

A) RUE (*Rheum undulatum* Linne)

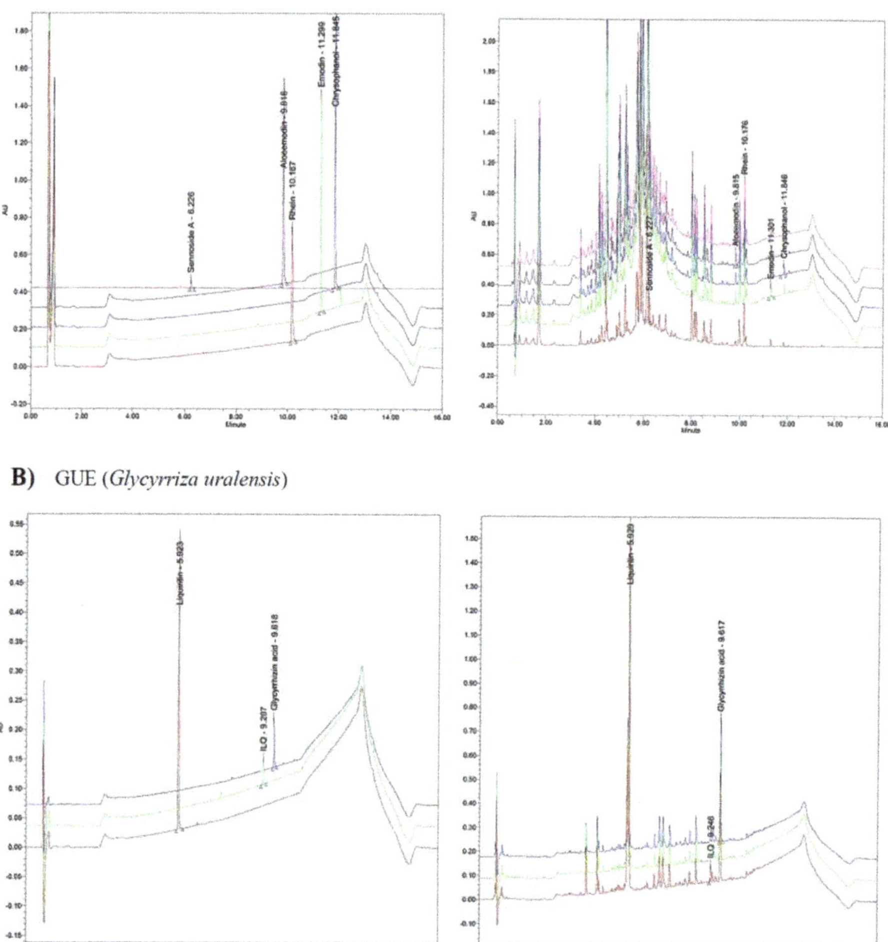

B) GUE (*Glycyrriza uralensis*)

Figure 1. Characterization of RUE (*Rheum undulatum* Linne) (**A**) and GUE (*Glycyrriza uralensis*) (**B**) by UPLC chromatogram. (**A**) UPLC chromatogram of five commercial standards (left) and marker compounds in GUE (right). Each peak represents sennoside A (340 nm), emodin (254 nm), chrysophanol (254 nm), aloe-emodin (254 nm), and rhein (254 nm), respectively. (**B**) UPLC chromatogram of theww commercial standards (left) and marker compounds in RUE (right). Each peak represents glycyrrhizin acid (254 nm), liquiritigenin (280 nm), and isoliquiritigenin (280 nm), respectively.

2.4. Systems Pharmacology-Based Analysis

Molecules with oral bioavailability ($\geq$30%) and drug-likeness indices ($\geq$0.18) were considered to select the target compounds that could be used for our study [19,20]. Oral bioavailability (OB), the capable of being delivered to systemic circulation after oral administration, is one of the most important pharmacokinetic parameters in drug screening [20]. The drug-likeness index was used to examine the herbal ingredients under consideration whether they are suitable as medicaments by quantifying the structural similarities to all drugs registered in the DrugBank database (http://www.drugbank.ca/) [19]. The potential targets and diseases of the candidate compounds were predicted by Traditional Chinese

Medicine Systems Pharmacology Database and Analysis Platform (TCMSP) (http://lsp.nwsuaf.edu.cn/) [21]. The protein targets predicted were examined for their functional annotation based on the sequence similarities to the proteins registered in UniProt (http://www.uniprot.org/). Finally, the potential targets were imported to the DAVID (http://david.abcc.ncifcrf.gov) for gene ontology (GO) enrichment analysis. The GO terms with p-values of less than 0.05 were filtered out, and functional annotation clustering analysis was implemented for the remaining terms to identify the pharmacological and biological processes. The likelihood networks connecting candidate compounds, potential targets and their related diseases were constructed by Cytoscape 3.2.1 (Bethesda, MD, USA). The topological properties of the networks generated were analyzed using the Plugins (Network Analysis, Analyze Network) of this software.

2.5. Animals

Balb/c male mice (6 weeks old) purchased from Charles River Orient Bio (Seongnam, Korea) were maintained under standard condition (23 ± 1 °C; 50 ± 5% humidity; 12-h light/dark cycle). All animal experimental procedures were approved by the Institutional Animal Care and Use Committee of Daegu Haany University and were conducted in accordance with the guidelines of the National Institutes of Health (Protocol # DHU2016-047). The mice were allocated randomly into three treatment groups: 'vehicle-treated control' ($n = 8$), 'CCl$_4$' ($n = 8$), and 'CCl$_4$ + RG' ($n = 8$). Mice were orally administered by either RG (10 mg/kg of RUE plus 100 mg/kg of GUE, dissolved in water; for CCl$_4$ + RG) or water (for vehicle-treated control and CCl$_4$) three times per week for 4 weeks. To induce liver damage in groups CCl$_4$ and CCl$_4$ + RG, CCl$_4$ (0.5 mL/kg of body weight and 1:9 diluted in corn oil) was intraperitoneally injected into mice twice a week for 4 weeks. All the mice were sacrificed on day 28 using carbon dioxide chamber.

2.6. Behavioral Tests

An open-field test was designed to evaluate our animals' basal activity and its change across time in response to our pharmacological treatments in accordance with the procedure described in a previous study [22]. Video tracking with recording was conducted under dim lighting condition to examine the movement pattern within the peripheral and central zones in an open field for 30 min. Movement patterns were quantified based on the distance traveled and time spent in each zone using a video tracking system (SMART V3.0; Panlab, Barcelona, Spain).

2.7. Brain Immunohistochemistry

The brain sections from three groups of mice (vehicle-treated control, CCl$_4$ and CCl$_4$ + RG) were fixed in 10% neutral buffered formalin, then embedded in paraffin by tissue processor and embedding center. Tissue blocks were sectioned in 4 μm thick ribbons and applied on slide glass. Two slides were randomly selected for each group, and hematoxylin and eosin (H&E) staining was performed on selected slides according to the standard protocol. After deparaffinization in xylene, tissue sections were hydrated in EtOH and stained in hematoxylin for 5 min. After bluing, tissue sections were stained with eosin for 10 s. Followed by eosin, tissue sections were dehydrated and cleared before mounting.

These slides were used for immunohistochemical detection of glial fibrillary acidic protein (GFAP), which was carried out according to the protocol described below. A BenchMark XT autostainer (Ventana Medical Systems, Tucson, AZ, USA) was used in immunohistochemistry (IHC). Coated glass slides were loaded with 4-μm-thick liver tissue sections and heated at 60 °C for 2 h. EZ prep. solution (Ventana Medical Systems) was used to deparaffinize followed by cell conditioning 1 solution. Cell conditioning 1 solution (Ventana Medical Systems) was used as heat-induced antigen retrieval agent. Anti-GFAP antibody (Abcam, Cambridge, UK) was diluted to 1:100 and applied on slides for 60 min. Universal DAB detection kit (Ventana Medical Systems) was used to detect specific reaction and nucleus was stained with hematoxylin followed by bluing reagent.

2.8. Blood Analysis

Collected whole blood was incubated at room temperature for 2 h and centrifuged for 5 min at 5000 rpm to isolate plasma. Plasma ALT was analyzed by using an automated blood analyzer (Fuji Dri-Chem NX500i, Fuji Medical System Co., Ltd., Tokyo, Japan).

2.9. Liver Histopathology

The largest lobe of each liver removed was fixed in 10% neutral buffered formalin, then embedded in paraffin by tissue processor and embedding center. Tissue block was sectioned in 4-μm-thick ribbon and applied on slide glass. After deparaffinization in xylene, tissue sections were hydrated in EtOH and stained with either H&E or with Masson's trichrome for collagen fibers. Following staining, tissue sections were dehydrated and cleared before mounting. The histological changes were observed under a light microscope (Nikon, Tokyo, Japan).

2.10. Western Blot Analysis

Western blot analysis was performed as previously described [18]. Briefly, the brain tissues were lysed with radioimmunoprecipitation assay (RIPA) buffer (Thermo Scientific, Rockford, IL, USA) at 4 °C and each protein sample (30–50 μg) was separated by 10% sodium dodecyl sulfate–polyacrylamide gel electrophoresis (SDS–PAGE) and transferred onto polyvinylidene fluoride (PVDF) membranes (Millipore, Bedford, MA, USA). The protein bands were detected with WesternBright ECL (Amersham Biosciences, Piscataway, NJ, USA) and a gel-doc image analyzer (Vilber Lourmat, France). The band intensity was quantified using Image J 1.42 software (NIH; Bethesda, MD, USA).

2.11. Real-Time RT-qPCR Analysis

Total RNA was extracted from the brain tissues using Trizol (Invitrogen, Carlsbad, CA, USA) following the manufacturer's instructions [18], and primer sets were shown in Table 2. The mRNA levels of matrix metalloproteinase-9 (MMP-9), transforming growth factor beta 1 (TGF-β1), and interleukin 1β (IL-1β) were compared by calculating the crossing point (C_p) value and were normalized by glyceraldehyde-3-phosphate dehydrogenase (GAPDH) using LightCycler 96 relative quantification software (Roche, München, Germany). A melting curve analysis was performed following the amplification to verify the accuracy of the amplicons.

Table 2. Primers used in real-time PCR analysis in this study.

Genes	Sense	Antisense
MMP-9	5′-TCCCTCTGAATAAAGTCGACA-3′	5′-AGGTGACAAGGTGGACCATG-3′
IL-1β	5′-CAGGATGAGGACATGAGC-3′	5′-CTCTGCAGACTCAAACTCCA-3′
TGF-β1	5′-GAGGTTTGCTGGGGTGAG-3′	5′-CAGCACGAGGAGGAGCAG-3′
GAPDH	5′-AACGACCCCTTCATTGAC-3′	5′-TCCACGACATACTCAGCAC-3′

2.12. Statistics

Mean ± standard deviation (SD) was provided for all quantitative data. All comparisons were examined based on either one-way ANOVA followed by Bonferroni's post hoc test or a two-tailed Student's *t*-test. The criterion for statistical significance was set at $p < 0.05$, but an indication was added when p was less than 0.05, 0.01, or 0.001.

3. Results

3.1. Systems Pharmacology-Based Approach

Based on our ultra-performance liquid chromatography (UPLC) results, five (sennoside A, emodin, aloe-emodin, chrysophanol, and rhein) and three (glycyrrhizin acid, liquiritigenin, and isoliquiritigenin) representative components were identified from the *R. undulatum* and *G. uralensis* extracts, respectively (Figure 1 and Table 1). We compared between two peaks of the commercial standards and of corresponding herbal component. All but sennoside A were considered target compounds to be tested in this study because they exceeded our oral bioavailability (OB) and drug-likeness evaluation criteria. Sennoside A was excluded as it had a very low OB value (less than 0.1). In addition, given that the absorption rate is very low in the small intestine when a component enters via the oral cavity and can be easily converted to rhein anthrone by various enzymes associated with intestinal microorganisms [23,24], it was not necessary to consider sennoside A as a target component.

To understand the mechanism associated with the RG components' HE-related medicinal action at a systems level, a compound–target network was constructed to uncover the underlying interactions (Figure 2). The compound–target network with colour-coded nodes to indicate components (green) and candidate targets (yellow) comprised 141 interactions connecting the 7 selected RG components to 40 targets (Figure 2A). Two components, emodin (degree = 34) and isoliquiritigenin (degree = 30), appeared to exhibit the most versatile functions, with many of their functions associated with the pharmacological regulation of HE (Figure 2A). As the network indicates, the RG components and their targets are intertwined; thus, it is better to interpret a component as having a complex function rather than having a specific function. In an effort to illuminate the therapeutic mechanism of the 40 selected targets, we applied ClueGO, a Cytoscape (https://cytoscape.org/) plugin, to perform a biologically relevant interpretation of the potential targets. The majority of targets were associated with oxidoreductase, nucleocytoplasmic transport and inflammatory response (Figure 2B).

All 40 of the targets were entered into pharmacology database platforms including DrugBank, Therapeutic Target, and PharmGkb to obtain a list of potentially related diseases. A total of 135 diseases, belonging to 18 groups suggested by the MeSH Browser (2017 MeSH), were identified. The target–disease interactions were visualized by using a network format (Figure 2C). Of the 135 diseases identified, 38 were found to be related to neoplasms, 17 to nervous system diseases, and 18 to cardiovascular diseases. These three disease groups are all reported to have the potential to be improved by the administration of RG.

3.2. RG Improves CCl$_4$-Induced Behavioral Damage

Because it had been reported that RG protects against liver damage, we focused on nervous system related diseases among the three key disease groups. To determine the effects of hepatic damage on behavior, we performed open-field tests on mice. Based on the methods described in a previous study [25], locomotion activities were quantified by measuring distance travelled and time spent in the centre or periphery of the open field for 30 min. The results showed that exposure to CCl$_4$ was associated with depressive-like symptoms (Figure 3A). Compared with the level of resting activities of the vehicle group, the group with CCl$_4$-induced hepatic damage (CCl$_4$ group) showed a significant increase in the amount of time spent resting (Figure 3B). In addition, the level of slow activities was decreased in the CCl$_4$ group compared to the level in the vehicle group (Figure 3A,B). Furthermore, the CCl$_4$ group travelled a shorter distance than the vehicle group (Figure 3A,C). In contrast, depression did not appear in the RG-treated group, and behavioral symptoms of that group were similar to those of the vehicle group that did nothing more than CCl$_4$ group (Figure 3). Furthermore, the group treated with CCl$_4$ plus RG (CCl$_4$ + RG group) exhibited significantly more activity and greater distance travelled than those in the CCl$_4$ group (Figure 3). Furthermore, in the RG treatment group, there was a greater distance travelled than that in the control group (Figure 3A).

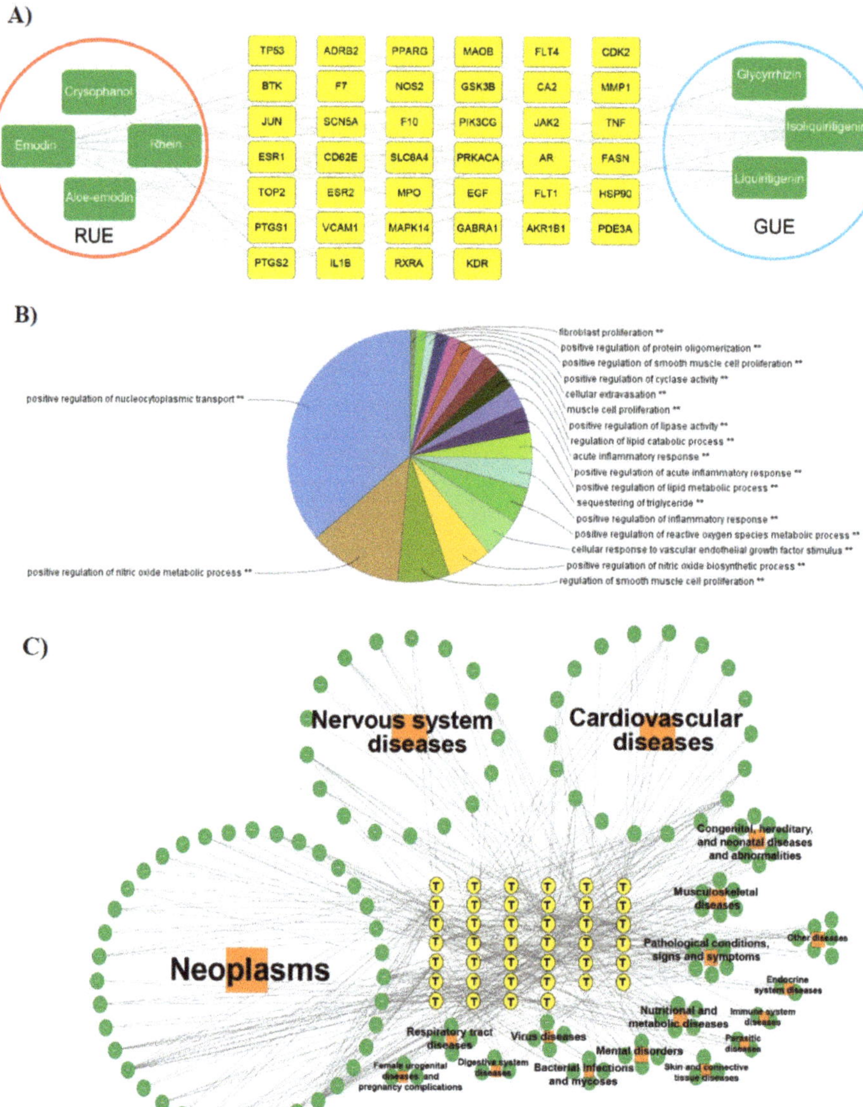

Figure 2. Systems pharmacology-based approach. (**A**) Compound–target network (C–T network). The C–T network was constructed by connecting the candidate compounds (green rectangles) of *R. undulatum* (red circle) and *G. uralensis* (blue circle) with their potential targets (yellow rectangles). The C–T network was composed of 141 compound–target links generated from the connection of the 7 candidate compounds to 40 targets. (**B**) ClueGO analysis of the predicted targets. The pie chart represents the molecular function, immune system processes and reactome pathways of the targets identified in the network analysis. (**C**) Target–disease network (T–D network). In the T–D network, candidate targets were connected to the related diseases. Target proteins (40, yellow ellipses) were connected to 135 diseases (green circles), which could be assigned into 18 separated groups (orange squares).

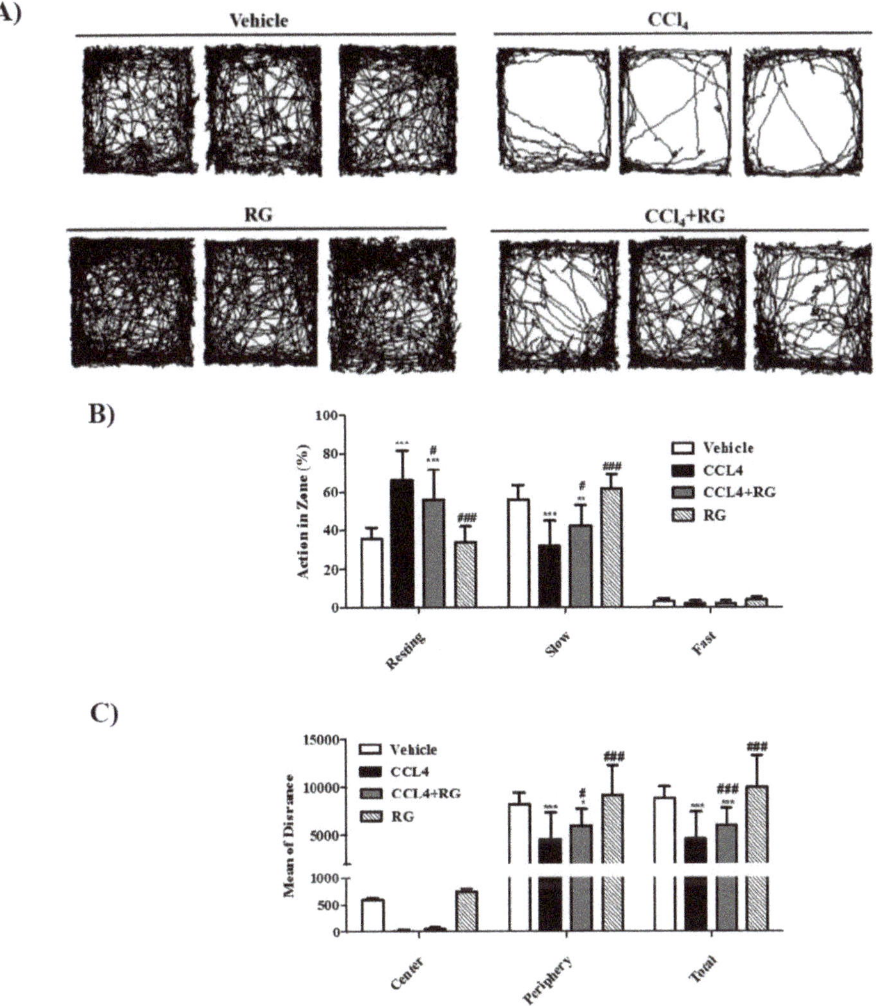

Figure 3. RG ameliorates neurobehavioral changes in CCl4-induced hepatic damage model. Neurobehavioral changes observed in the open field tests (OFTs) in CCl4-induced hepatic damage model. (**A**) Representative traces of mouse movement in the experimental field. (**B**) % action categories (resting, slow, and fast) in both center and periphery zones by mice. (**C**) Distance travelled in both center and periphery zones by mice. Data were analyzed for statistical significance using ANOVA with Tukey's test for ad hoc multiple comparison implemented in GraphPad Prism. Data were presented as mean ± SD for three independent experiments. $*p < 0.05$, $**p < 0.01$, $***p < 0.001$ when compared with vehicle group; $^{\#}p < 0.05$, $^{\#\#}p < 0.01$, $^{\#\#\#}p < 0.001$ when compared with group CCl4.

3.3. RG Ameliorates CCl4-Induced Histopathological Changes of Brain

We investigated histopathological changes in the cerebral cortex by examining H&E-stained samples (Figure 4). The CCl4 treatment produced lesions in the cerebral cortex that exhibited swollen astrocytes with large vesicular nuclei and prominent nucleoli (Figure 4). However, the RG combination treatment (10 mg/kg of R plus 100 mg/kg of G) protected astrocytes in the cerebral cortex from exhibiting the lesional damages that are observed following CCl4 treatment (Figure 4A).

A) H & E　　　　　　　　**B) GFAP-IHC**

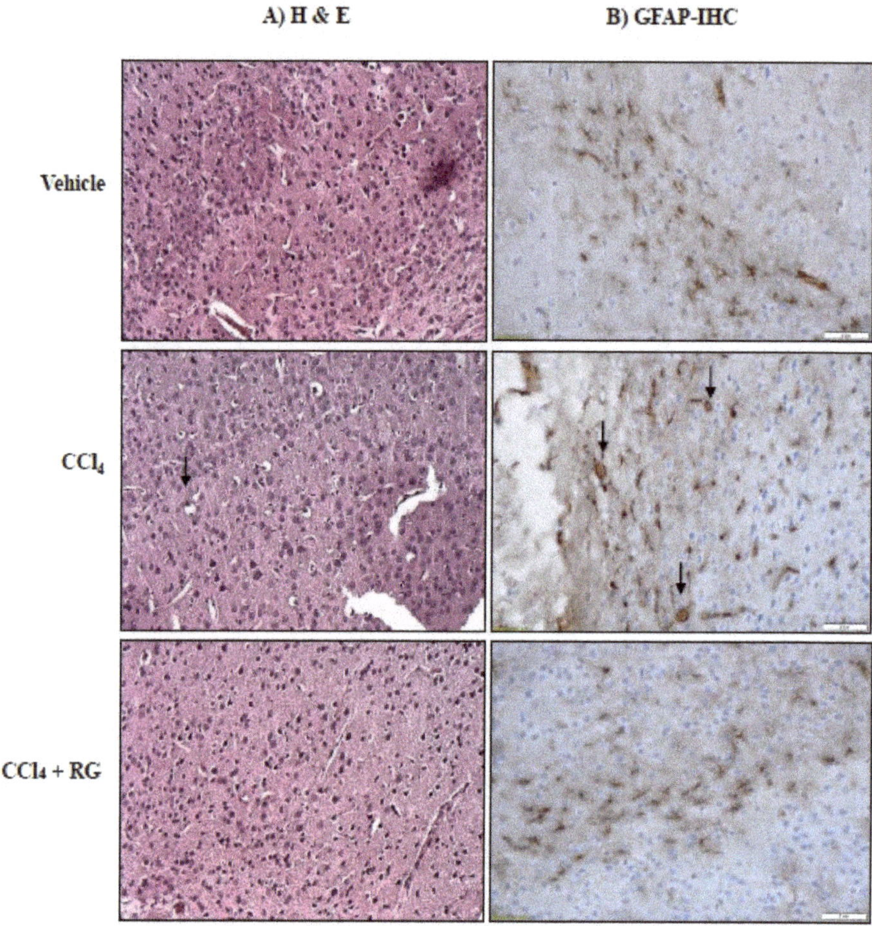

Figure 4. RG ameliorates CCl$_4$-induced histological changes in cerebral cortex. Mice were given with repeated injections of CCl$_4$ (5 mg/kg, 2 times/week, 4 weeks), whereas vehicle mice instead received saline (n = 4). The CCl$_4$ + RG group was treated with RG (10 mg/kg of R plus 100 mg/kg of G) with CCl$_4$. After four weeks of CCl$_4$ injection, mice were sacrificed. (**A**) Histopathological change (when stained with hematoxylin and eosin; H&E) and (**B**) astrogliosis (glial fibrillary acidic protein, GFAP) in cerebral cortex. Magnification: 400×.

Astrogliosis was examined by quantifying GFAP-immunoreactive cells in the area of the cerebral cortex (Figure 4B). CCl$_4$ seemed to induce intensive GFAP staining and to increase process complexity. However, the number of GFAP-positive cells in the CCl$_4$ + RG group was reduced when compared to that in the CCl$_4$ group (Figure 4B). Overall, it appears that CCl$_4$ administration increases the number of activated astrocytes and RG treatment can significantly reduce that effect.

3.4. RG Inhibits the BBB Disruption and Neuroinflammation

The effect of RG on the CCl$_4$-induced increase in brain endothelial cell permeability was determined by analysing MMP9 mRNA in mouse brain tissue samples. The mRNA expression of MMP9 was significantly upregulated in brain tissues of the CCl$_4$ group, but that upregulation could be blocked by pre-treatment with RG (Figure 5A). To verify the effectiveness of RG in preventing BBB disruption, the expression of claudin-5, a tight junction protein, was assessed by undertaking an immunoblotting analysis. The level of claudin-5 was significantly decreased in the CCl_4 group, compared to that in the control group (Figure 5B). However, the presence of RG led to a greater increase in the level of claudin-5 protein in the CCl_4 + RG group than in the CCl_4 group (Figure 5B), indicating that RG has a protective function in preventing BBB disruption.

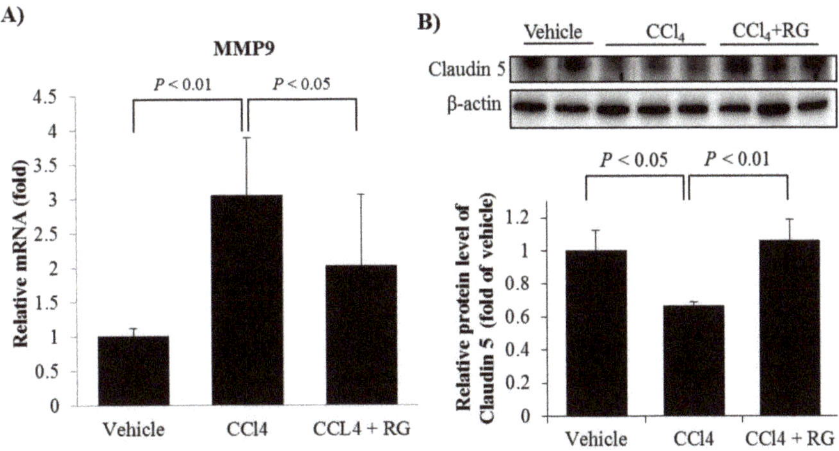

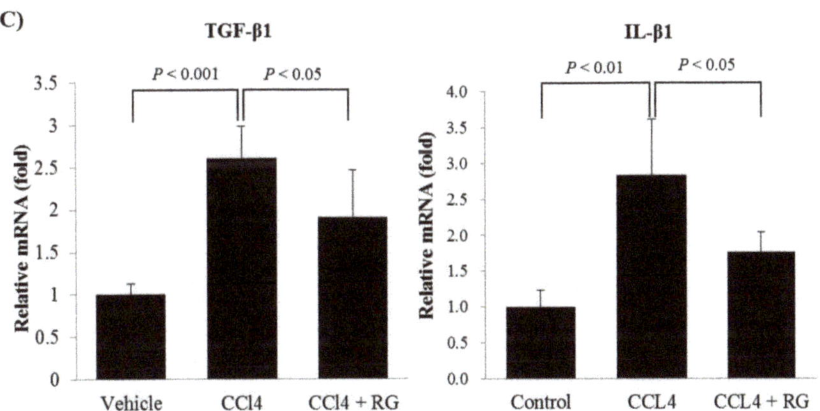

Figure 5. RG suppresses BBB disruption and inflammation in mice with CCl4-induced HE. Mice were treated with CCl_4, and RG as described in Figure 3. RT-qPCR analysis of the mRNA expression levels of (**A**) MMP-9, (**C**) TGF-β1, and (**D**) IL-1β. (**B**) The protein levels of claudin-5 were assessed by western blot, in which β-actin served as a loading control. Protein levels were presented as relative band intensities to control (vehicle treated) group. Data were represented by mean ± SD.

Neuroinflammation is a fundamental immune response in the central nervous system (CNS) through which the brain reacts to diverse pathogens passing through the damaged BBB. Therefore, we examined the mRNA expression levels of the representative pro-inflammatory cytokines IL-1β and TGF-β1 in the cerebral cortex. Real-time RT-qPCR analysis showed that the presence of RG effectively inhibited the CCl_4-induced expressions of IL-1β and TGF-β1 (Figure 5C).

3.5. RG Ameliorates CCl₄-Induced Liver Injury

To examine the hepatic protective effects of RG, we established a chronic hepatic cirrhosis mouse model by using injections of CCl_4. Repeated injections of CCl_4 (5 mg/kg twice per week for 4 weeks) significantly elevated the serum ALT level, indicating that liver damage had occurred (Figure 6A). However, RG pre-treatment markedly inhibited the elevation of the serum ALT level caused by the CCl_4 injections (Figure 6A). It is known that the increase of ammonia in serum is a representative indicator of hepatic encephalopathy as well as liver disease. Our results of reducing ammonia in serum increased by CCl_4 with RG treatment provide further evidence that RG can mitigate live injury caused by CCl_4 (Figure 6B).

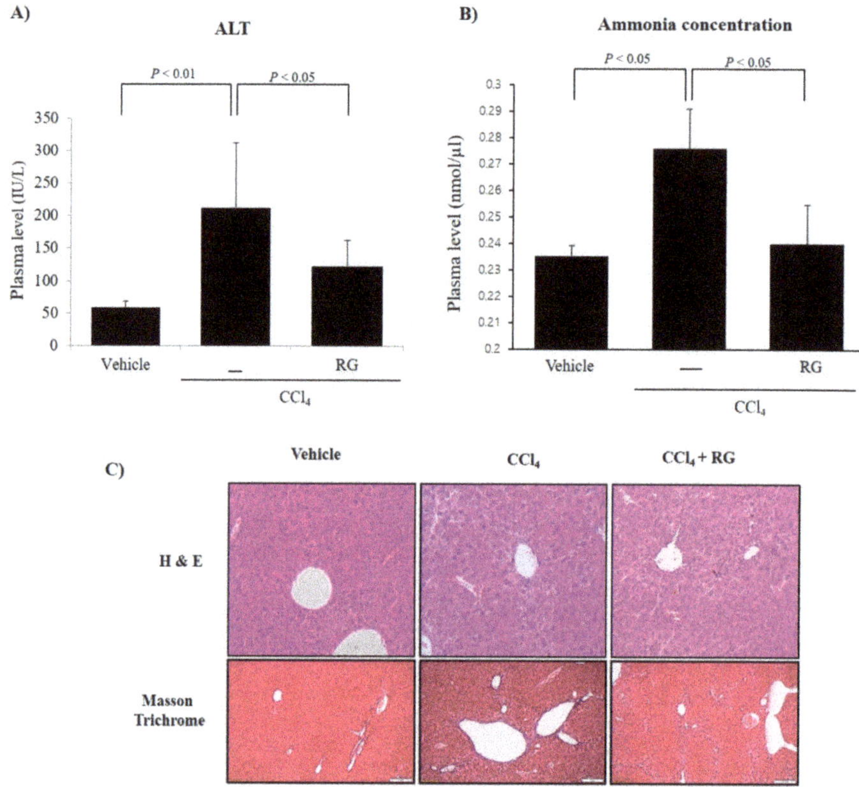

Figure 6. RG protects CCl_4-induced liver toxicity in mice. Mice were given with repeated injections of CCl_4 (5 mg/kg, 2 times/week, 4 weeks), whereas vehicle mice instead received saline ($n = 4$). The CCl_4 + RG group was treated with RG (10 mg/kg of R plus 100 mg/kg of G) with CCl_4. After four weeks of CCl_4 injection, mice were sacrificed. (**A**) The activities ALT and (**B**) ammonia were assayed by using semi-automated blood chemistry analyzer after 4 weeks of treatment. Data were represented by the mean ± SD. (**C**) Representative photomicrographs of liver sections processed for H&E (upper row) and Masson's trichrome (lower row) staining with vehicle, CCl_4, and CCl_4 + RG. Scale bars = 80 µm.

In addition, we observed histopathological differences between groups in livers that had been stained with H&E and Masson's trichrome (Figure 6B). CCl_4 treatment can produce degenerative damages to liver tissue, typically, centrilobular necroses such as ballooning and vacuolation (deposition of lipid droplets) of hepatocytes and the infiltration of inflammatory cells (Figure 6C). However, pre-treatment with RG markedly minimized the CCl_4-induced liver damage (Figure 6C).

4. Discussion

There are many reports indicating that a mixture of various medicines can be used to obtain a better effect than that when the medicines are used separately. Based on a previous report, extracts from *R. undulatum* and *G. uralensis* were able to provide synergistic antioxidant and hepatic protective effects against AA and iron-induced oxidative stress as well as against CCl_4-induced acute liver injury [16]. Considering that HE is a frequent disease associated with chronic liver failure [2], in this study, we investigated whether RG could directly reduce HE or its symptoms. The study consisted of several experimental and analytic steps.

First, we constructed a systems-based pharmacological model that integrated OB predictions, target predictions and interaction networks in order to provide insights into the effects and mechanisms of RG on HE. After identifying the components of RG and choosing among those components those that were deemed worth prescribing, we examined how these candidate ingredients interact with certain targets in human physiology. Based on our target–disease network results, it can be assumed that some of the components identified in RG could be used for the treatment of several diseases in addition to HE. This multi-target potential is a benefit of many herbal medicines, and it appears that RG is likely to treat HE in a manner that can interact with other important targets in the overall human physiology rather than concentrating its effects only on one disease. In addition, the discovery of a close relationship in the target–disease network with neoplasms and cardiovascular disease indicates another asset of RG as a therapeutic agent.

Second, we examined whether RG had sufficient efficacy to reduce symptoms in HE induced by CCl_4 in mice. Administration of CCl_4 to mice induces chronic liver failure, which can lead to CNS changes similar to those seen in HE. CCl_4 is one of the most widely used chemicals that cause liver injury. It is activated in liver microsomes by several functional enzymes dependent on the cytochrome P450 family, and that activation generates an abundance of free radicals, which leads to lipid peroxidation within the cells. In addition, CCl_4 can covalently bind to cellular lipids and proteins, which may lead to dissociation of cell membrane structure and increase membrane permeability [26]. CCl_4 is often used to create rat HE models [27]. In a previous study, it was reported that rats treated with CCl_4 or alcohol had very similar symptoms to those of HE [28]. It was also reported that BALB/c nude mice showed rapid and obvious liver injury and brain dysfunction following intraperitoneal injection of CCl_4 [29]. As predicted, we observed that CCl_4 administration resulted in a severely impaired neurological score, decreased activity, and diminished cognitive function. However, RG treatment recovered those effects to a considerable level, though not to the level in the control subjects. Likewise, astrocyte swelling in the brains of CCl_4-treated mice was remarkably moderated by RG treatment. Such results show that RG has a protective effect on the brain against the development of HE.

The major ingredients of RG and their potential targets shown in our results indicate that RG could ameliorate disruption of the BBB by regulating the molecules involved in forming tight junctions, such as MMP and vascular cell adhesion molecule 1 (VCAM1). MMP-9 can digest major proteins in the capillary endothelial BBB and in tight junctions. MMP-9 is known to degrade occludin and claudin-5 in focal cerebral ischemia [30], and it has been reported that it affects the expression of occludin, claudin-5, and ZO-1 and ZO-2 levels in early diabetic retinopathy [27,31,32]. MMP-9 upregulation was previously reported to open the BBB during the later stages of HE [31]. During ammonia-induced HE, for example, the upregulation of MMP-9 disrupted the expression of major tight junction proteins, and when MMP-9 was pharmacologically inhibited, their expressions were restored to near control level [33]. In our results, an increased level of MMP-9 mRNA expression in the cerebral cortex was observed following CCl_4 administration. However, RG treatment resulted in inhibition of MMP-9 gene expression. In addition, the inhibition of claudin-5 expression observed in the brains of CCl_4-treated mice was completely moderated by RG treatment. These results affirm the results of previous studies that showed that rhubarb could maintain BBB integrity in traumatic brain injury treatment [14] or intracerebral haemorrhage [15]. Our data suggest that RG can effectively attenuate CCl_4-induced BBB damage in mice, raising the possibility that RG or its active components may be considered as potential neuroprotective drugs useful in relieving the severe symptoms of HE.

If the BBB is breached and circulating neurotoxic agents and inflammatory mediators penetrate into the brain, astrocytes absorb them and, simultaneously, secrete a variety of pro-inflammatory cytokines including IL-1β, IL-6, and TNF-α [34]. Cytokines may directly induce astrocytes to diffuse into brain parenchyma and activate transcription factors in astrocytes in order to increase the production of cytokines. In a recent study, increased brain efflux of IL-1β, TNFα, and IL-6 was identified in patients with uncontrolled intracranial hypertension due to acute liver failure [35,36]. Intracerebral synthesis of cytokines may be a primary cause for the characteristic astrocyte swelling often seen in this kind of condition. In this study, we show that that CCl₄ markedly increased the levels of IL-1β and TGF-β1 in mouse cerebral cortex, and that increase was effectively inhibited by the presence of RG.

Sometimes, the general consumers expect that the some hepatoprotective agents could reduce the severity of extra-hepatic manifestation of liver failure (e.g., hepatic encephalopathy). Like this, it is inferred that reduced encehphalopathy after RG could be off-label uses, not a direct effect of RG on brain mechanism. It might be partly explained by a reduction in severity of liver failure.

5. Conclusions

Our results demonstrate the therapeutic effects on chronic HE models of a mixture of extracts from *R. undulatum* and *G. uralensis*. We applied theoretical and computational approaches to determine the active ingredients of those extracts and to identify their pharmacological targets by exploring their mechanism-of-action profiles. In the animal models, RG consistently relieved the symptoms of HE by protecting BBB permeability via downregulation of MMP-9 and upregulation of claudin-5. The effects of RG are mediated by a variety of biochemical processes involving neuroinflammation. In this regard, the use of RG appears to be advantageous as it can provide more effective and rapid treatment effect than other medicines that are involved in only one or two processes. Thus, RG may be considered as a therapeutic agent for treating not only chronic liver disease but also HE.

Author Contributions: Conceptualization, S.Y.B. and Y.W.K.; investigation, S.Y.B., E.H.L., T.W.O., H.J.D., K.-Y.K., and K.-I.P.; formal analysis, S.Y.B. and E.H.L.; writing—original draft preparation, Y.W.K. and S.Y.B.; writing—review and editing, Y.W.K. and S.Y.B.; supervision, Y.W.K.; funding acquisition, Y.W.K. and S.Y.B. All authors have read and agreed to the published version of the manuscript.

Funding: This work was supported by the National Research Foundation of Korea (NRF) Grant the Korea Government (MSIP) (No. 2019R1A2C1003200) and also partly by (No. 2017R1D1A3B03027847).

Conflicts of Interest: The authors declare no conflict of interest.

Abbreviations

ALT	alanine aminotransferase
BBB	blood–brain barrier
CCl4	carbon tetrachloride
C-T network	compound–target network
H&E	harri's hematoxylin and eosin
HPLC	High-performance liquid chromatography
GABA	γ-Aminobutyric acid
IHC	immunohistochemistry
IL-1β	Interleukin 1 beta
MMP-9	matrix metallopeptidase 9
OB	Oral bioavailability
TGF-β1	Transforming growth factor beta 1
UPLC	ultra-performance liquid chromatography

References

1. Ferenci, P. Hepatic encephalopathy. *Gastroenterol. Rep. (Oxf)*. **2017**, *5*, 138–147. [CrossRef]

2. Vilstrup, H.; Amodio, P.; Bajaj, J.; Cordoba, J.; Ferenci, P.; Mullen, K.D.; Weissenborn, K.; Wong, P. Hepatic encephalopathy in chronic liver disease: 2014 Practice Guideline by the American Association for the Study of Liver Diseases and the European Association for the Study of the Liver. *Hepatology* **2014**, *60*, 715–735. [CrossRef]

3. Butterworth, R.F. Pathophysiology of hepatic encephalopathy: A new look at ammonia. *Metab. Brain Dis.* **2002**, *17*, 221–227. [CrossRef]

4. Hadjihambi, A.; Arias, N.; Sheikh, M.; Jalan, R. Hepatic encephalopathy: A critical current review. *Hepatol. Int.* **2018**, *12*, 135–147. [CrossRef]

5. Dadsetan, S.; Waagepetersen, H.S.; Schousboe, A.; Bak, L.K. Glutamine and ammonia in hepatic encephalopathy. In *Glutamine in Clinical Nutrition*; Rajendram, R., Preedy, V.R., Patel, V.B., Eds.; Springer: Berlin/Heidelberg, Germany, 2015.

6. Stickel, F.; Schuppan, D. Herbal medicine in the treatment of liver diseases. *Dig. Liver Dis.* **2007**, *39*, 293–304. [CrossRef]

7. Ferrucci, L.M.; Bell, B.P.; Dhotre, K.B.; Manos, M.M.; Terrault, N.A.; Zaman, A.; Murphy, R.C.; Vanness, G.R.; Thomas, A.R.; Bialek, S.R.; et al. Complementary and alternative medicine use in chronic liver disease patients. *J. Clin. Gastroenterol.* **2010**, *44*, e40–e45. [CrossRef]

8. Yao, C.; Tang, N.; Xie, G.; Zheng, X.; Liu, P.; Fu, L.; Xie, W.; Yao, F.; Li, H.; Jia, W. Management of hepatic encephalopathy by traditional chinese medicine. *Evid. Based Complement. Alternat. Med.* **2012**, *2012*, 835686. [CrossRef]

9. Zhao, Y.L.; Wang, J.B.; Zhou, G.D.; Shan, L.M.; Xiao, X.H. Investigations of free anthraquinones from rhubarb against alpha-naphthylisothiocyanate-induced cholestatic liver injury in rats. *Basic Clin. Pharmacol. Toxicol.* **2009**, *104*, 463–469. [CrossRef]

10. Huang, Q.; Lu, G.; Shen, H.M.; Chung, M.C.; Ong, C.N. Anti-cancer properties of anthraquinones from rhubarb. *Med. Res. Rev.* **2007**, *27*, 609–630. [CrossRef]

11. Seo, E.J.; Ngoc, T.M.; Lee, S.M.; Kim, Y.S.; Jung, Y.S. Chrysophanol-8-O-glucoside, an anthraquinone derivative in rhubarb, has antiplatelet and anticoagulant activities. *J. Pharmacol. Sci.* **2012**, *118*, 245–254. [CrossRef]

12. Zhang, R.; Kang, K.A.; Piao, M.J.; Lee, K.H.; Jang, H.S.; Park, M.J.; Kim, B.J.; Kim, J.S.; Kim, Y.S.; Ryu, S.Y.; et al. Rhapontigenin from *Rheum undulatum* protects against oxidative-stress-induced cell damage through antioxidant activity. *J. Toxicol. Environ. Health A* **2007**, *70*, 1155–1166. [CrossRef]

13. Ning, Z.; Shuangnan, Z.; Xiaohe, X.; Zhen, W.; Yunfeng, B.; Tingting, H.; Chao, Z.; Yao, W.; Zhou, K.; Zhongxia, W.; et al. Rhubarb-based Chinese herbal formulae for hepatic encephalopathy: A systematic review and Meta-analysis. *J. Tradit. Chin. Med.* **2017**, *37*, 721–734. [CrossRef]

14. Wang, Y.; Fan, X.; Tang, T.; Fan, R.; Zhang, C.; Huang, Z.; Peng, W.; Gan, P.; Xiong, X.; Huang, W.; et al. Rhein and rhubarb similarly protect the blood–brain barrier after experimental traumatic brain injury via gp91^{phox} subunit of NADPH oxidase/ROS/ERK/MMP-9 signaling pathway. *Sci. Rep.* **2016**, *6*, 37098. [CrossRef]

15. Wang, Y.; Peng, F.; Xie, G.; Chen, Z.Q.; Li, H.G.; Tang, T.; Luo, J.K. Rhubarb attenuates blood–brain barrier disruption via increased zonula occludens-1 expression in a rat model of intracerebral hemorrhage. *Exp. Ther. Med.* **2016**, *12*, 250–256. [CrossRef]

16. Lee, E.H.; Baek, S.Y.; Kim, K.Y.; Lee, S.G.; Kim, S.C.; Lee, H.S.; Kim, Y.W. Effect of *Rheum undulatum* Linne extract and *Glycyrriza uralensis* Fischer extract against arachidonic acid and iron-induced oxidative stress in HepG2 cell and CCl$_4$-induced liver injury in mice. *Herb. Formula Sci.* **2016**, *24*, 163–174. [CrossRef]

17. Wang, Y.C.; Yang, Y.S. Simultaneous quantification of flavonoids and triterpenoids in licorice using HPLC. *J. Chromatogr. B Analyt. Technol. Biomed. Life Sci.* **2007**, *850*, 392–399. [CrossRef]

18. Sohn, K.H.; Jo, M.J.; Cho, W.J.; Lee, J.R.; Cho, I.J.; Kim, S.C.; Kim, Y.W.; Jee, S.Y. Bojesodok-eum, a Herbal Prescription, Ameliorates Acute Inflammation in Association with the Inhibition of NF-κB-Mediated Nitric Oxide and ProInflammatory Cytokine Production. *Evid. Based Complement. Alternat. Med.* **2012**, *2012*, 457370. [CrossRef]

19. Ma, C.; Wang, L.; Xie, X.Q. GPU accelerated chemical similarity calculation for compound library comparison. *J. Chem. Inf. Model.* **2011**, *51*, 1521–1527. [CrossRef]

20. Xu, X.; Zhang, W.; Huang, C.; Li, Y.; Yu, H.; Wang, Y.; Duan, J.; Ling, Y. A novel chemometric method for the prediction of human oral bioavailability. *Int. J. Mol. Sci.* **2012**, *13*, 6964–6982. [CrossRef]

21. Ru, J.; Li, P.; Wang, J.; Zhou, W.; Li, B.; Huang, C.; Li, P.; Guo, Z.; Tao, W.; Yang, Y.; et al. TCMSP: A database of systems pharmacology for drug discovery from herbal medicines. *J. Cheminform.* **2014**, *6*, 13. [CrossRef]

22. Monteggia, L.M.; Luikart, B.; Barrot, M.; Theobold, D.; Malkovska, I.; Nef, S.; Parada, L.F.; Nestler, E.J. Brain-derived neurotrophic factor conditional knockouts show gender differences in depression-related behaviors. *Biol. Psychiatry* **2007**, *61*, 187–197. [CrossRef] [PubMed]

23. Lee, J.H.; Kim, J.M.; Kim, C. Pharmacokinetic analysis of rhein in *Rheum undulatum* L. *J. Ethnopharmacol.* **2003**, *84*, 5–9. [CrossRef]

24. Yagi, T.; Miyawaki, Y.; Nishikawa, A.; Horiyama, S.; Yamauchi, K.; Kuwano, S. Prostaglandin E2-mediated stimulation of mucus synthesis and secretion by rhein anthrone, the active metabolite of sennosides A and B, in the mouse colon. *J. Pharm. Pharmacol.* **1990**, *42*, 542–545. [CrossRef]

25. Ströhle, A.; Höfler, M.; Pfister, H.; Müller, A.G.; Hoyer, J.; Wittchen, H.U.; Lieb, R. Physical activity and prevalence and incidence of mental disorders in adolescents and young adults. *Psychol. Med.* **2007**, *37*, 1657–1666. [CrossRef] [PubMed]

26. Ingawale, D.K.; Mandlik, S.K.; Naik, S.R. Models of hepatotoxicity and the underlying cellular, biochemical and immunological mechanism(s): A critical discussion. *Environ. Toxicol. Pharmacol.* **2014**, *37*, 118–133. [CrossRef] [PubMed]

27. Giebel, S.J.; Menicucci, G.; McGuire, P.G.; Das, A. Matrix metalloproteinases in early diabetic retinopathy and their role in alteration of the blood-retinal barrier. *Lab. Investig.* **2005**, *85*, 597–607. [CrossRef]

28. Wang, W.W.; Zhang, Y.; Huang, X.B.; You, N.; Zheng, L.; Li, J. Fecal microbiota transplantation prevents hepatic encephalopathy in rats with carbon tetrachloride-induced acute hepatic dysfunction. *World J. Gastroenterol.* **2017**, *23*, 6983–6994. [CrossRef]

29. Chang, H.M. Improvement of carbon tetrachloride-induced acute hepatic failure by transplantation of induced pluripotent stem cells without reprogramming factor c-Myc. *Int. J. Mol. Sci.* **2012**, *13*, 3598–3617. [CrossRef]

30. Yang, Y.; Rosenberg, G.A. MMP-mediated disruption of claudin-5 in the blood–brain barrier of rat brain after cerebral ischemia. *Methods Mol. Biol.* **2011**, *762*, 333–345. [CrossRef]

31. Dhanda, S.; Sandhir, R. Blood–brain Barrier Permeability Is Exacerbated in Experimental Model of Hepatic Encephalopathy via MMP-9 Activation and Downregulation of Tight Junction Proteins. *Mol. Neurobiol.* **2018**, *55*, 3642–3659. [CrossRef]

32. Spindler, K.R.; Hsu, T.H. Viral disruption of the blood–brain barrier. *Trends Microbiol.* **2012**, *20*, 282–290. [CrossRef] [PubMed]

33. Chen, F.; Ohashi, N.; Li, W.; Eckman, C.; Nguyen, J.H. Disruptions of occludin and claudin-5 in brain endothelial cells in vitro and in brains of mice with acute liver failure. *Hepatology* **2009**, *50*, 1914–1923. [CrossRef]

34. Bhalala, U.S.; Koehler, R.C.; Kannan, S. Neuroinflammation and neuroimmune dysregulation after acute hypoxic-ischemic injury of developing brain. *Front. Pediatr.* **2015**, *2*, 144. [CrossRef] [PubMed]

35. Bémeur, C.; Butterworth, R.F. Liver-brain proinflammatory signalling in acute liver failure: Role in the pathogenesis of hepatic encephalopathy and brain edema. *Metab. Brain Dis.* **2013**, *28*, 145–150. [CrossRef] [PubMed]

36. Butterworth, R.F. The liver-brain axis in liver failure: Neuroinflammation and encephalopathy. *Nat. Rev. Gastroenterol. Hepatol.* **2013**, *10*, 522–528. [CrossRef]

 *biomolecules*

Article

Targeted Isolation of Lignans from *Trachelospermum asiaticum* Using Molecular Networking and Hierarchical Clustering Analysis

Jiho Lee [1], Hong Seok Yang [1], Hyogeun Jeong [1], Jung-Hwan Kim [2] and Heejung Yang [1,*]

[1]   Laboratory of Natural Products Chemistry, College of Pharmacy, Kangwon National University, Chuncheon 24341, Korea; jiho3232@kangwon.ac.kr (J.L.); comboy10@naver.com (H.S.Y.); jung99gs@naver.com (H.J.)

[2]   Department of Pharmacology, Gyeongsang National University, Jinju 52727, Korea; junghwan.kim@gnu.ac.kr

*   Correspondence: heejyang@kangwon.ac.kr; Tel.: +82-33-250-6919

Received: 29 January 2020; Accepted: 24 February 2020; Published: 1 March 2020

Abstract: High-resolution-mass-spectrometry (HR-MS) methods rapidly provide extensive structural information for the isolation of metabolites in natural products. However, they may occasionally provide more information than required and interfere with the targeted analysis of natural products. In this study, we aimed to selectively isolate lignans from *Trachelospermum asiaticum* by applying the Global Natural Product Social Molecular Networking (GNPS) platform and hierarchical clustering analysis (HCA). *T. asiaticum*, which contains lignans, triterpenoids and flavonoids that possess various biological activities, was analyzed in a data-dependent acquisition (DDA) analysis mode using HR-MS. The preprocessed MS spectra were applied not only to GNPS for molecular networking but also to HCA based on similarity patterns between two nodes. The combination of these two methods reliably helped in the targeted isolation of lignan-type metabolites, which are expected to possess potent anti-cancer or anti-inflammatory activities.

Keywords: lignans; *Trachelospermum asiaticum*; GNPS; targeted isolation

---

1. Introduction

Trachelospermum asiaticum (Korean name: "Nagseogdeung", Apocynaceae), which is regionally distributed in East Asian countries such as Korea, China, and Japan, has been reported to contain lignans [1], triterpenoids [2] and flavonoids [3]. As such, it has been used in traditional medicine for treating hypertension and neuralgia. *T. asiaticum* has been reported to exert tuberculosis and bronchitis effects, and has also been used for the treatment of rheumatism [4].

High-resolution-mass-spectrometry (HR-MS), which employs various instruments such as a quadrupole time-of-flight (qTOF) mass spectrometer and Orbitrap, has become one of the most powerful techniques to obtain information on metabolites in natural products. HR-MS is most helpful for the identification, over many decades, of compounds from natural products [5–7]. The data-dependent analysis (DDA) mode of the tandem MS technique using HR-MS detects ions in two stages. It detects two or three ions of the most intense ions in the first stage (MS) and their fragmented ions (MS/MS) in the second stage. The two layers of MS data consisting of the parent ions and their fragmented ions are very useful for annotating unidentified peaks. Recently, the Global Natural Product Social (GNPS) platform has received increasing attention among natural product chemists [8]. GNPS helps to process numerous *m/z* and intensity values from raw MS spectral data acquired in the DDA mode. The GNPS also helps generate molecular networks (MNs) based on the similarity between processed MS spectra of single compounds. Nevertheless, MN only focuses on the structural similarity between two nodes. In the present study, we applied hierarchical clustering analysis (HCA) to obtain insights into the targeted isolation of lignans from nodes that are not directly connected by an MN but have the same chemical scaffolds. The combination of molecular networking and HCA was more reliable than the application of each method alone for the targeted isolation of five lignans (**1–5**) from *T. asiaticum* roots (Figure 1). We could isolate compounds **1–5** which have cytotoxic activities against four cell lines, namely, human lung cancer (A549), ovarian cancer (SKOV3), prostate cancer (PC3) and laryngeal carcinoma (Hep2) cells.

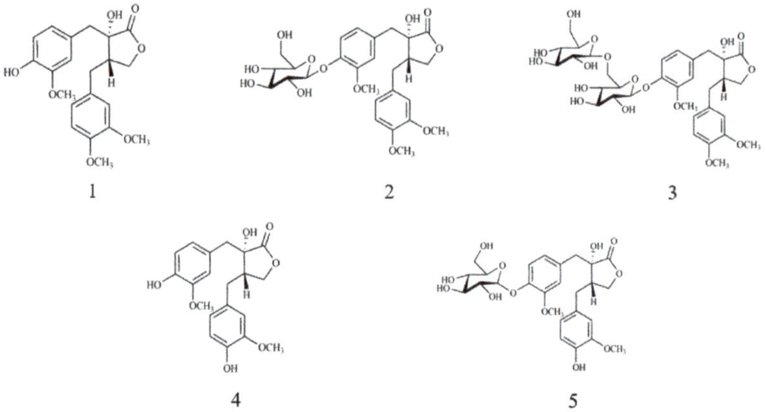

Figure 1. Chemical structures of compounds **1–5**.

2. Results and Discussion

2.1. Molecular Networking and Hierarchical Clustering Analysis of Mass Spectral Data from T. asiaticum

In the present study, we compared the HCA results between the structural similarities of the processed MS spectral data derived from single compounds with the molecular networking results obtained through GNPS (Figure 2). The molecular networking results showed connections between two nodes with similar spectral patterns (cosine similarity threshold >0.7). The molecular type of the cluster was identified by comparison with MS databases [8]. We attempted to selectively isolate specific types of compounds using the information on the nodes annotated by the MN. Although these nodes were derived from the same backbone, the different strong fragmented ions lowered the similarity score and hindered the connection between them. We applied HCA to improve the results of molecular networking. The matrix profile generated using similarity scores provided clear evidence on the nodes that were absent in the MN and the results were presented as a dendrogram [9]. The nodes with similar score profiles were more closely located in the smaller clusters.

Trachelospermum species contains dibenzylbutylrolactone-type lignans as a bioactive component. These lignans possess potent anti-estrogenic [10], antitumor [11] and anti-cancer activities [12]. In the present study, we focused on the isolation of dibenzylbutylrolactone-type lignans from the methanolic extract of *T. asiaticum* using two approaches, GNPS and HCA (Figure 3). The MN from the total methanolic extract and the four sub-fractions of *T. asiaticucm*, *n*-hexane, EtOAc, *n*-BuOH and H₂O, comprised 489 nodes, with 144 paired nodes and 345 non-cluster nodes. Based on the node information provided by the network annotation propagation (NAP) tool, an in silico node annotation tool, we could annotate the nodes from the total extract and four sub-fractions for the discovery of dibenzylbutylrolactone-type lignans using the MN (Figure 3a) (See Figure S6 in Supplementary Materials). In the MN, the nodes annotated as dibenzylbutylrolactone-type lignans mainly existed in the E6 sub-fraction from the EtOAc and the B3 sub-fraction from the *n*-BuOH sub-fractions, respectively (See Figures S7 and S8 in Supplementary Materials). Next, we found that the HCA results showed that the nodes for lignans and triterpenoids, which were not grouped together in the sub-cluster of the MN, were closely gathered together by the same backbones and distinctly separated by different ones (Figure 3b). Based on these two in silico results, we aimed for targeted isolation using the information on the nodes annotated as dibenzylbutylrolactone-type lignans. Consequently, we successfully isolated five dibenzylbutylrolactone-type lignans from *T. asiaticum* roots.

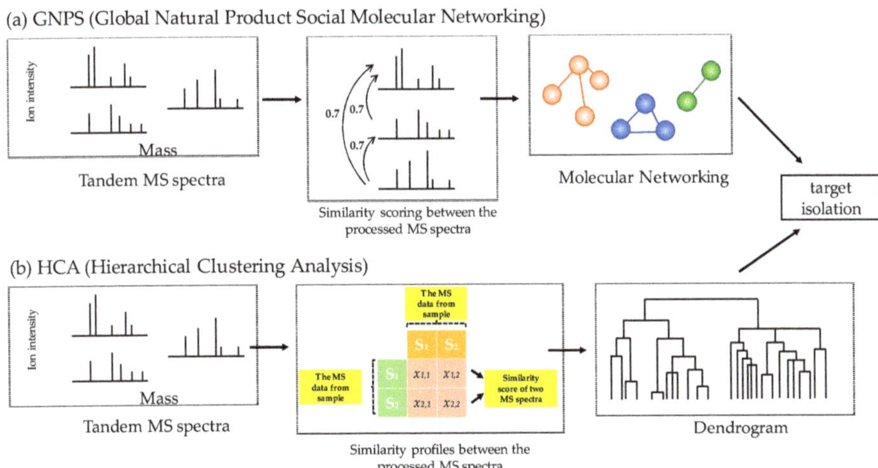

Figure 2. Workflow of molecular networking and hierarchical clustering analysis (HCA). MS spectral data from *Trachelospermum asiaticum* were acquired by high-resolution mass spectrometry (See Materials and Methods section) and preprocessed by the MZmine software (Version. 2.34). The molecular network was constructed using the Global Natural Product Social Molecular Networking (GNPS) platform. The processed MS spectra with similar patterns (cosine similarity >0.7) were grouped into clusters (**a**). The matrix profile was generated using similarity scores between the processed MS spectra which were calculated by the Pearson correlation coefficient and visualized in a dendrogram (**b**).

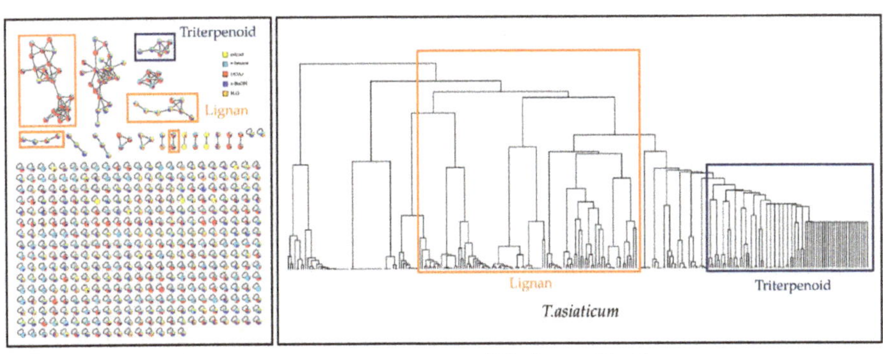

Figure 3. Molecular networking of *Trachelospermum asiaticum* using the GNPS platform (**a**) and Dendrogram of HCA results (**b**).

2.2. Targeted Isolation of Compounds **1–5** Using Molecular Networking and HCA Results

The five lignans were identified as trachelogenin (**1**) [13], tracheloside (**2**) [14], trachelogenin β-gentionbioside (**3**) [15], nortrachelogenin (**4**) [16] and nortracheloside (**5**) [17] by comparing them with previously reported spectroscopic data. Nodes from compounds **1–5** were easily displayed in the molecular networking and HCA results (Figure 4). Notably, in the MN, nodes 2 (*m/z* 387.14 [M–H]⁻),

1 (*m/z* 595.20 [M+Glc+2Na]⁻) and 28 (*m/z* 711.24 [M+2Glc]⁻) for compounds **1–3**, which have the same backbone except for the sequential attachment of glucose moieties, were not present in one cluster but they were closely located in the HCA results (Figure 4).

(a) Dendrogram of *T. asiaticum*

(b) Compounds 1-5 from *T. asiaticum*

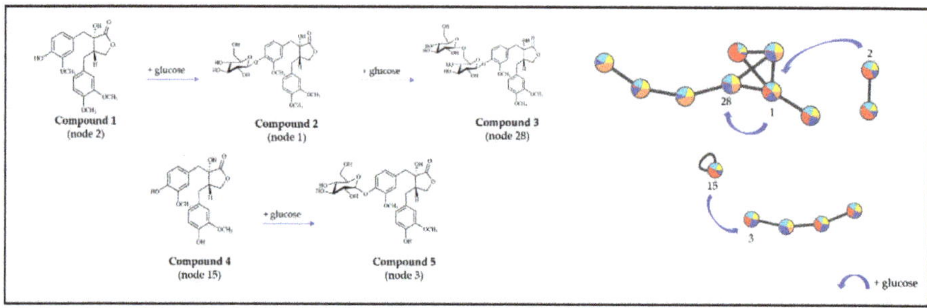

Figure 4. Compounds **1–5** identified based on molecular networking and HCA results.

Compound **5** had an additional glucose unit compared to compound **4**; however, compounds **4** (node 15) and **5** (node 3) were not clustered in the same sub-group in the MN. Contrastingly, nodes 3 and 15 in the HCA results were closely located together in the low-level cluster. It was confirmed that node 114 possesses an additional glucose compared to compound **5** and was identified as nortrachelogenin 4,4′-di-*O*-β-D-glucopyranoside [18,19]. Node 114, which was not clustered in GNPS, was found near compounds **4** and **5** in the HCA (See Figure S8 in Supplementary Materials).

In the present study, we found that some nodes (e.g., 15, 114, and 141) that were not connected in GNPS could be located in close clusters in the dendrogram. This approach helped to predict the structures corresponding to unknown nodes. Nodes 21 and 69, which were included in the same cluster in the MN and closely located in the dendrogram, were predicted as dibenzylbutylrolactone-type lignans. Node 21 with an *m/z* value of 519.19 [M+Glc]⁻ was proposed as the matairescinoside [20]. The MS spectral pattern corresponding to node 69 (*m/z* 682.24 [M+2Glc]⁻) showed one more glucose moiety attached to the compound at node 21, and the compound was suggested as matairesinol 4,40-di-*O*-β-D-glucopyranoside [2] (See Figure S9 in Supplementary Materials).

Recently, the GNPS platform has become one of the best choices for natural product chemists to discover novel chemicals from natural products and to determine their metabolic changes. Molecular networking using a GNPS platform can help to visualize the connectivity between similar structures. Because the connectivity of nodes in the MN can only be decided by similarity scoring between every two MS spectra, the MN occasionally failed to identify the clustering of the same types of backbone. In the present study, we found that HCA, which compares a single node against all the nodes, provided clearer results for the annotation of backbones than molecular networking. By combining the results of molecular networking and HCA, we could easily identify and selectively isolate dibenzylbutylrolactone-type lignans from the extract of *T. asiaticum* roots.

2.3. Cytotoxic Activity of Isolated Compounds 1–5

The cytotoxic activities of compounds **1–5** were tested against four cell lines (A549, SKOV3, PC3 and hep2) using a 3-(4,5-Dimethylthiazol-2-yl)-2,5-diphenyltetrazolium bromide (MTT) assay (Table 1). Compounds **1–5** exhibited different cytotoxic activities against different cell lines. Compounds **1** and **4** showed cytotoxic activities against the A549 cancer cell line with IC_{50} values of 19.5 and 20.6 μM, respectively. Compounds **2, 4,** and **5** exhibited cytotoxicity against the SKOV3 cell line with IC_{50} values of 23.8, 24.7 and 23.2 μM, respectively. Compound **1** exhibited cytotoxicity against HEP2 cells with an IC_{50} value of 18.3 μM, while compound **2** showed cytotoxicity against the PC3 cell line with an IC_{50} value of 19.3 μM.

Table 1. Cytotoxicity data of compounds **1–5** from *Trachelospermum asiaticum*.

Compound	IC_{50} (μM) [1]			
	A549	SKOV3	PC3	HEP2
1	19.5	63.3	27.8	18.3
2	67.9	23.8	19.3	>100
3	33.7	72.7	48.7	46.7
4	20.6	24.7	55.6	43.1
5	46.9	23.2	40.8	47.2
Etoposide [2]	6.56	6.57	7.78	4.14

[1] The results are IC_{50} values of compounds against each cancer cell line; [2] Etoposide was used as the positive control.

3. Materials and Methods

3.1. Plant Material

The roots of *T. asiaticum* were collected from Jinju, Korea in May 2018 and deposited in the Herbarium at the College of Pharmacy, Kangwon National University (KNUTA-01).

3.2. Apparatus and Reagents

^{1}H NMR data were recorded at 600 MHz on a Bruker Avance Neo 600 (Brucker, Billerica, MA, USA) spectrometer in the Central Laboratory of Kangwon National University (Chuncheon, Korea). The Mass experiments were performed on a Waters Xevo G2 qTOF mass spectrometer (Waters MS Technologies, Manchester, UK) with the UPLC system through an electrospray ionization (ESI) interface for the MN study, as well as the exact mass analysis. Silica gel Kieselgel 60 (40–60 μm, 230–400 mesh, Art. 9385, Merck, Darmstadt, Land Hessen, Germany) and Diaion HP-20 (Mitsubishi Chemical Industries Ltd., Chiyoda-ku, Tokyo, Japan) were used for column chromatography. Thin layer chromatography (TLC) was performed to monitor the different fractions of *T. asiaticum* using a Kieselgel 60 F_{254} (Art. 5715, Merck, Darmstadt, land Hessen, Germany) and an RP-C_{18} F_{254} (Art. 15389, Merck, Darmstadt, Land Hessen, Germany). Semipreparative HPLC was performed on an Agilent 1260 Infinity Quaternary LC (Agilent, Santa, CA, USA) with an Inspire^{TM} 5 μm C_{18} (250 × 21.2 mm, Dikima, Foothill Ranch, CA, USA). Methanol (MeOH), *n*-hexane, ethyl acetate (EtOAc) and *n*-butanol (*n*-BuOH) were purchased from Daejung (Si-heung, Korea). The other reagents were purchased from Sigma-Aldrich (St. Louis, MO, USA).

3.3. Extraction and Isolation

The dried roots of *T. asiaticum* (696.8 g) were extracted with 100% MeOH (each for 3 days) at room temperature. The methanol extract (71.7 g) was fractionated with *n*-hexane (3.0 g), EtOAc (6.5 g), *n*-BuOH (24.8 g) and H_2O (33.4 g), successively. The *n*-BuOH (24.8 g) fraction of *T. asiaticum* was applied to a HP-20 resin column, eluted with MeOH-H_2O (0:1, 2:3, 1:1, 3:2 and 1:0) to yield five fractions (B1: 2.9 g, B2: 1.6 g, B3: 5.7 g, B4: 2.6 g, B5: 3.5 g). The fraction B3 (5.7 g) was re-chromatographed by reversed-phase medium-pressure liquid chromatography (MPLC) silica gel with MeOH-H_2O (1:4, 1:3,

1:2, 1:1, 2:1, and 1:0) to yield seven fractions (B3-1 to 7), which were collected and monitored by TLC analysis. Fraction B3-5 was applied to reversed-phase MPLC silica gel eluted with MeOH-H_2O (1:4, 1:3, 1:2, 1:1, 2:1, and 1:0) to afford six fractions. Fraction B3-5-4 (374.2 mg) was applied to a preparative HPLC [mobile phase MeCN-H_2O with 0.1% formic acid (60:40)] to produce compounds **3** (18 mg) and **5** (128.2 mg).

The EtOAc (6.5 g) fraction of *T. asiaticum* was separated by MPLC silica gel with n-hexane and ethyl acetate (HE) (5:1, 1:1, 1:5), chloroform and methanol (CM) (5:1, 3:2, 2:3, 1:5), and MeOH 100% to yield compound **2** (1.8 g) and 12 fractions (E1 ~ 12). These fractions were monitored by TLC analysis. E6 (357.1 mg) was subjected to preparative HPLC [mobile phase MeCN-H_2O with 0.1% formic acid (50:50)] to produce compounds **1** (94.6 mg) and **4** (25 mg).

3.3.1. Trachelogenin (1)

Yellow gum, $[\alpha]$D − 36° (c = 0.43, CHCl$_3$); HR-MS m/z: 387.1480 [M−H]$^-$ (calcd for $C_{21}H_{23}O_7$, 387.1444); ^{1}H NMR (600 MHz, CDCl$_3$) δ_H: 2.51 (m, 1H, H-8), 2.54 (m, 1H, H-7), 2.93 (d, J = 5.0 Hz, 1H, H-7), 2.96 (m, 1H, H-7′), 3.11 (d, J = 13.7 Hz, 1H, H-7′), 3.84, 3.85, 3.86 (s, 3H each, -OMe), 4.00 (m, 1H, H-9), 4.04 (m, 1H, H-9), 6.62 (d, 1H, J = 2.0 Hz, H-2), 6.63 (d, 1H, J = 1.1 Hz, H-6′), 6.67 (dd, J = 8.1, 1.9 Hz, 1H, H-6), 6.71 (d, J = 1.9 Hz, 1H, H-2′), 6.79 (d, 1H, J = 8.1 Hz, H-5), 6.84 (d, 1H, J = 8.0 Hz, H-5′).

3.3.2. Tracheloside (2)

Amorphous powder, $[\alpha]_D^{27}$ − 12.6° (c = 1.6, MeOH); HR-MS m/z: 549.1970 [M−H]$^-$ (calcd for $C_{27}H_{33}O_{12}$, 549.1972); ^{1}H NMR (600 MHz, pyridine-d_5) δ_H: 2.77 (ddd, J = 17.3, 9.5, 5.0 Hz, 1H, H-8), 2.96 (dd, J = 13.8, 10.0 Hz, 1H, H-7), 3.26 (dd, J = 13.8, 4.8 Hz, 1H, H-7), 3.34 (d, J = 13.6 Hz, 1H, H-7′), 3.66 (d, J = 13.6 Hz, 1H, H-7′), 3.76, 3.77, 3.82 (s, 3H each, -OMe), 4.19 (t, J = 8.0 Hz, 1H, H-9), 4.36 (m, 1H, H-9), 5.67 (d, J = 6.2 Hz, 1H, H$_{glc}$-1), 6.89 (dd, J = 8.1, 1.6 Hz, 1H, H-6), 6.93 (s, 1H, H-2), 6.94 (d, J = 7.6 Hz, 1H, H-5), 7.06 (dd, J = 8.3, 1.7 Hz, 1H, H-6′), 7.17 (d, J = 1.7 Hz, 1H, H-2′), 7.60 (d, J = 8.3 Hz, 1H, H-5′).

3.3.3. Trachelogenin β-Gentionbioside (3)

Amorphous powder, $[\alpha]_D^{28}$ − 70.0° (c = 2.0, MeOH); HR-MS m/z: 711.2489 [M−H]$^-$ (calcd for $C_{33}H_{43}O_{17}$, 711.2500); ^{1}H NMR (600 MHz, pyridine-d_5) δ_H: 2.76 (ddd, J = 17.3, 9.4, 5.1 Hz, 1H, H-8), 2.93 (dd, J = 13.8, 10.1 Hz, 1H, H-7), 3.22 (dd, J = 13.9, 7.0 Hz, 1H, H-7), 3.30 (d, J = 13.6 Hz, 1H, H-7′), 3.61 (d, J = 13.9 Hz, 1H, H-7′), 3.75, 3.76, 3.81 (s, 3H each, -OMe), 3.85 (m, 1H, H-8), 4.04 (t, J = 8.0 Hz, 1H, H-9), 4.49 (dd, J = 11.8, 2.2 Hz, 1H, H-9), 5.06 (d, J = 7.8 Hz, 1H, H$_{glc}$-1′), 5.58 (d, J = 7.3 Hz, 1H, H$_{glc}$-1), 6.88 (dd, J = 8.1, 1.6 Hz, 1H, H-6), 6.92 (d, J = 1.6 Hz, 1H, H-2), 6.94 (d, J = 8.1 Hz, 1H, H-5), 7.14 (d, J = 1.7 Hz, 1H, H-2′), 7.19 (d, J = 1.6 Hz, 1H, H-6′), 7.70 (d, J = 8.3 Hz, 1H, H-5′).

3.3.4. Nortrachelogenin (4)

Yellow resin, $[\alpha]$D + 15.4° (c = 0.52, CHCl$_3$); HR-MS m/z: 373.1307 [M−H]$^-$ (calcd for $C_{20}H_{21}O_7$, 373.1287); ^{1}H NMR (600 MHz, CDCl$_3$) δ_H: 2.49 (m, 1H, H-8), 2.53 (m, 1H, H-7), 2.92 (d, J = 4.1 Hz, 1H, H-7), 2.93 (d, J = 7.1 Hz, 1H, H-7′), 3.12 (d, J = 13.7 Hz, 1H, H-7′), 3.84, 3.82 (s, 3H each, -OMe), 3.98 (dd, J = 15.6, 6.8 Hz, 1H, H-9), 4.03 (m, 1H, H-9), 6.60 (s, 1H, H-5′), 6.61 (s, 1H, H-2′), 6.63 (m, 1H, H-5), 6.70 (d, J = 1.3 Hz, 1H, H-2), 6.82 (s, 1H, H-6′), 6.83 (d, J = 3.7 Hz, 1H, H-6).

3.3.5. Nortracheloside (5)

White powder, $[\alpha]_D^{19}$ − 47.9° (c = 1.02, EtOH); HR-MS m/z: 534.1850 [M−H]$^-$ (calcd for $C_{26}H_{31}O_{12}$, 535.1816); ^{1}H NMR (600 MHz, DMSO-d_6) δ_H: 2.38 (m, 1H, H-8), 2.45 (dd, J = 13.6, 9.9 Hz, 1H, H-7), 2.69 (dd, J = 14.7, 4.4 Hz, 1H, H-7), 2.91 (d, J = 13.5 Hz, 1H, H-7′), 3.09 (d, J = 13.5 Hz, 1H, H-7′), 3.71, 3.73 (s, 3H each, -OMe), 3.97 (d, J = 7.7 Hz, 2H, H-9), 4.93 (d, J = 7.3 Hz, 1H, H$_{glc}$-1′), 6.56 (dd, J = 8.0,

1.6 Hz, 1H, H-5′), 6.69 (d, *J* = 1.7 Hz, 1H, H-6′), 6.72 (d, *J* = 8.0 Hz, 1H, H-2′), 6.75 (dd, *J* = 8.4, 1.5 Hz, 1H, H-5), 6.81 (d, *J* = 1.6 Hz, 1H, H-6), 7.05 (d, *J* = 8.4 Hz, 1H, H-2).

3.4. GNPS Analysis

The raw data of total methanolic extract and four sub-fractions from *T. asiaticum*, *n*-hexane, EtOAc, *n*-BuOH and H_2O, were acquired by performing UPLC-MS/MS. The mobile phases involved a mixture with H_2O (A) buffered with 0.1% formic acid and acetonitrile (B). 5%–95% B (0–13 min), 95% B (13–14.5 min), 95%–5% B (14.5–14.7 min) and 5% B (14.7–15 min). The flow rate was set at 300 µL/min. The temperatures in the autosampler and in the column oven were set at 10 and 45 °C, respectively. The ESI conditions for MS analyses were set as follows: negative ion mode, capillary voltage of 2.5 kV, cone voltage of 20 V, source temperature of 120 °C, desolvation temperature of 350 °C and a desolvation gas flow of 800 L/h. The ion acquisition rate was 0.3 s with a resolution in excess of 20,000 FWHM. The energy for the collision-induced dissociation (CID) was set to 4 V for the precursor ions in the MS1 scan, and MS/MS scans were acquired in negative ion automated data-dependent acquisition (DDA) mode, in which MS/MS scans for the three most intense ion were produced (scan time 100 ms). The MS/MS acquisition was set to be activated when the Total Ion Current (TIC) of the MS1 survey scan rose and switched back to survey scanning after two scans of MS/MS. The MS1 and MS/MS data were converted to XML format by MZmine software (Ver. 2.3.4). The MN was constructed using the website GNPS (https://gnps.ucsd.edu/ProteoSAFe/static/gnps-splash.jsp). The parent mass tolerance was 0.05 Da and the MS/MS fragment ion tolerance was set to 0.05 Da. Subsequently, consensus spectra containing less than three spectra were eliminated. The MN was created using a cosine score above 0.7 and more than three matched peaks. The mass spectral data in the MN were explored by comparing with the GNPS spectral libraries. The MN was visualized with Cytoscape 3.7.0 (http://www.cytoscape.org/) and the information on the nodes and edges in the MN were found at a GNPS repository (GNPS project ID: a81b6ca055a64428b42ca7971bac677b).

3.5. Dendrogram Analysis

HCA was created with R program version (3.6.1) using "dendextend" and "qplots" packages, which indicate the hierarchical similarity between mass data.

3.6. Cytotoxicity Assay

Compounds **1–5** were tested for their cytotoxic activity against A549, SKOV3, PC3 and Hep2 cells using an MTT assay. Etoposide was used as a positive control. A549, SKOV3, PC3, and Hep2 cells were seeded at a density of 5×10^3 cells/well in a 96-well plate. After overnight incubation, compounds **1–5** were dissolved in dimethyl sulfoxide and treated with different concentrations (10–100 µM). After 48 h of incubation in a 37 °C incubator, cell viability was evaluated at 490 nm. IC_{50} values were calculated as the mean of three-times treatment tests.

4. Conclusions

We constructed an MN of total methanolic extract fraction and sub-fractions (*n*-hexane, EtOAc, *n*-BuOH and H_2O) from *T. asiaticum* using the GNPS platform. In our study, some clusters in the MN included nodes annotated as lignans and triterpenoids, but many other nodes that were not connected in the cluster were not annotated. To overcome the general limitation of GNPS, we used the similarity score profiles of the MS spectral data of every single node against the similarity score profiles corresponding to all of the nodes. The nodes that had more similar backbones were closely located together in the lower branches. Using this proof of concept, we could identify the nodes for lignans and triterpenoids in the dendrogram. Some nodes that were not grouped in the MN were located in the lower branches of the dendrogram and were successfully utilized for the targeted isolation of five dibenzylbutylrolactone-type lignans (**1–5**). Thus, the combination of molecular networking and HCA improved the experimental efficiency of the targeted isolation of compounds possessing cytotoxic activities against four cancer cell lines. In the next study, we will try to automatically compare the MN and HCA results for the simple and rapid annotation of nodes.

Supplementary Materials: The following are available online at http://www.mdpi.com/2218-273X/10/3/378/s1: Figures S1–S5: are NMR data of compounds (**1–5**); Figure S6: presents the NAP data of *T. asiaticum*; Figure S7: presents the GNPS of B1-B5 fractions; Figure S8: presents the GNPS of E1-E12 fractions; Figures S9 and S10: are GNPS and HCA data of prediction nodes.

Author Contributions: Conceptualization, H.Y. and J.L.; methodology, H.Y. and J.L.; writing—original draft preparation, J.L.; performed the experiments, J.L., H.S.Y. and H.J.; writing—review and editing, H.Y.; visualization, H.Y. and J.L.; funding acquisition, J.-H.K., J.L. and H.Y. All authors have read and agreed to the published version of the manuscript.

Funding: This work was supported by the Basic Science Research Program through the National Research Foundation of Korea (NRF) grant funded by the Korean government (MEST) (NRF-2015R1A5A2008833, NRF-2018R1C1B6002574, NRF-2019R1A6A3A13096888).

Conflicts of Interest: The authors declare no conflict of interest.

References

1. Abe, F.; Yamauchi, T. Glycosides of 19α-hydroxyoleanane-type triterpenoids from *Trachelospermum asiaticum* (Trachelospermum. IV). *Chem. Pharm. Bull.* **1987**, *35*, 1833–1838. [CrossRef]

2. Nishibe, S.; Hisada, S.; Inagaki, I. The ether-soluble lignans of *Trachelospermum asiaticum* var. *intermedium*. *Phytochemistry* **1971**, *10*, 2231–2232. [CrossRef]

3. Zhang, J.; Yin, Z.; Liang, J. A new isoflavonoid glycoside from the aerial parts of *Trachelospermum jasminoides*. *Chin. J. Nat. Med.* **2013**, *11*, 274–276. [CrossRef] [PubMed]

4. Devi, N. Indian tribe's and villager's health and habits: Popularity of apocynaceae plants as medicine. *Int. J. Green Pharm.* **2017**, *11*, 256–259.

5. Bouslimani, A.; Sanchez, L.M.; Garg, N.; Dorrestein, P.C. Mass spectrometry of natural products: Current, emerging and future technologies. *Nat. Prod. Rep.* **2014**, *31*, 718–729. [CrossRef] [PubMed]

6. Zhang, H.; Li, S.; Zhang, H.; Wang, Y.; Zhao, Z.; Chen, S.; Xu, H. Holistic quality evaluation of commercial white and red ginseng using a UPLC-QTOF-MS/MS-based metabolomics approach. *J. Pharm. Biomed. Anal.* **2012**, *62*, 258–273. [CrossRef] [PubMed]

7. Zhang, J.; Chu, C.; Li, X.; Yao, S.; Yan, B.; Ren, H.; Xu, N.; Liang, Z.; Zhao, Z. Isolation and identification of antioxidant compounds in *Vaccinium Bracteatum* thunb. by UHPLC-Q-TOF LC/MS and their kidney damage protection. *J. Funct. Foods* **2014**, *11*, 62–70. [CrossRef]

8. Wang, M.; Carver, J.J.; Phelan, V.V.; Sanchez, L.M.; Garg, N.; Peng, Y.; Nguyen, D.D.; Watrous, J.; Kapono, C.A.; Luzzatto-Knaan, T. Sharing and community curation of mass spectrometry data with global natural products social molecular networking. *Nat. Biotechnol.* **2016**, *34*, 828–837. [CrossRef] [PubMed]

9. Galili, T. Dendextend: An R package for visualizing, adjusting and comparing trees of hierarchical clustering. *Bioinformatics* **2015**, *31*, 3718–3720. [CrossRef] [PubMed]

10. Yoo, H.H.; Park, J.H.; Kwon, S.W. An anti-estrogenic lignan glycoside, tracheloside, from seeds of *Carthamus tinctorius*. *Biosci. Biotechnol. Biochem.* **2006**, *70*, 2783–2785. [CrossRef] [PubMed]

11. Kitamura, Y.; Yamagishi, M.; Okazaki, K.; Son, H.; Imazawa, T.; Nishikawa, A.; Iwata, T.; Yamauchi, Y.; Kasai, M.; Tsutsumi, K.; et al. Lack of significant inhibitory effects of a plant lignan tracheloside on 2-amino-1-methyl-6-phenylimidazo[4,5-*b*]pyridine (PhIP)-induced mammary carcinogenesis in female sprague-dawley rats. *Cancer Lett.* **2003**, *200*, 133–139. [CrossRef]

12. Rajalekshmi, D.S.; Kabeer, F.A.; Madhusoodhanan, A.R.; Bahulayan, A.K.; Prathapan, R.; Prakasan, N.; Varughese, S.; Nair, M.S. Anticancer activity studies of cubebin isolated from *Piper Cubeba* and its synthetic derivatives. bioorg. *Med. Chem. Lett.* **2016**, *26*, 1767–1771. [CrossRef] [PubMed]

13. John, L.M.; Tinto, W.F. Revised 13C-NMR assignments for the biologically active butyrolactone (-)-trachelogenin. *J. Nat. Prod.* **1992**, *55*, 1313–1314. [CrossRef]

14. Liu, X.; Wang, Z.; Yang, Y.; Wang, L.; Sun, R.; Zhao, Y.; Yu, N. Active components with inhibitory activities on IFN-Γ/STAT1 and IL-6/STAT3 signaling pathways from caulis trachelospermi. *Molecules* **2014**, *19*, 11560–11571. [CrossRef] [PubMed]

15. Abe, F.; Yamauchi, T. Lignans from *Trachelospermum Asiaticum* (Tracheolospermum. II). *Chem. Pharm. Bull.* **1986**, *34*, 4340–4345. [CrossRef]

16. Kato, A.; Hashimoto, Y.; Kidokor, M. (+)-Nortrachelogenin, a new pharmacologically active lignan from *Wikstroemia indica*. *J. Nat. Prod.* **1979**, *42*, 159–162. [CrossRef] [PubMed]

17. Nishibe, S.; Hisada, S.; Inagaki, I. Structures of tracheloside and nortracheloside from *Trachelospermum asiaticum* NAKAI var. *intermedium* NAKAI. *Chem. Pharm. Bull.* **1971**, *19*, 866–868. [CrossRef]

18. Nishibe, S.; Hisada, S.; Inagaki, I. Lignan diglucosides from *Trachelospermum asiticum*. *Phytochemistry* **1972**, *11*, 3084–3085.

19. Nishibe, S.; Hisada, S.; Inagaki, I. Lignans of *Trachelospermum asiaticum*. var. *intermedium*. V. Isolation of nortrachelogenin-4, 4'-Di-O-β-D-glucopyranoside. *Chem. Pharm. Bull.* **1973**, *21*, 1114–1117. [CrossRef]

20. Inagaki, I.; Hisada, S.; Nishibe, S. Triterpenoids of Trachelospermum asiaticum var. *intermedium*. *Phytochemistry* **1970**, *9*, 2241. [CrossRef]

 *biomolecules*

Article

Acorus gramineusand and *Euodia ruticarpa* Steam Distilled Essential Oils Exert Anti-Inflammatory Effects Through Decreasing Th1/Th2 and Pro-/Anti-Inflammatory Cytokine Secretion Ratios In Vitro

Tzu-He Yeh and Jin-Yuarn Lin *

Department of Food Science and Biotechnology, National Chung Hsing University, 250 Kuokuang Road, Taichung 40227, Taiwan; http00000@yahoo.com.tw
* Correspondence: jinlin@dragon.nchu.edu.tw; Tel./Fax: + 886-4-22-851-857

Received: 31 December 2019; Accepted: 17 February 2020; Published: 19 February 2020

Abstract: To clarify the effects of steam distilled essential oils (SDEO) from herbs used in traditional Chinese medicine on immune functions, two potential herbs, *Acorus gramineusand* (AG) and *Euodia ruticarpa* (ER) cultivated in Taiwan, were selected to assess their immunomodulatory effects using mouse primary splenocytes and peritoneal macrophages. T helper type 1 lymphocytes (Th1) (IL-2), Th2 (IL-5), pro-inflammatory (TNF-α) and anti-inflammatory (IL-10) cytokines secreted by correspondent immune cells treated with SDEO samples were determined using enzyme-linked immunosorbent assay. The total amounts of potential phytochemicals, including total flavonoids, polyphenols and saponins, in these two selected SDEOs were measured and correlated with cytokine levels secreted by immune cells. Our results evidenced that ER SDEO is rich in total flavonoids, polyphenols and saponins. Treatments with AG and ER SDEO significantly ($p < 0.05$) increased IL-5/IL-2 (Th2/Th1) cytokine secretion ratios by splenocytes, suggesting that both AG and ER SDEO have the Th2-polarization property and anti-inflammatory potential. In addition, AG and ER SDEO, particularly ER SDEO, markedly decreased TNF-α/IL-10 secretion ratios by macrophages in the absence or presence of lipopolysaccharide (LPS), exhibiting substantial effects on spontaneous and LPS-induced inflammation. Significant correlations were found between the total polyphenols, flavonoids or saponins content in the two selected SDEOs and Th1/Th2 immune balance or anti-inflammatory ability in linear, non-linear or biphasic manners, respectively. In conclusion, our results suggest that AG and ER, particularly ER, SDEO have immunomodulatory potential in shifting the Th1/Th2 balance toward Th2 polarization in splenocytes and inhibiting inflammation in macrophages in the absence or presence of LPS.

Keywords: *Acorus gramineusand*; *Euodia ruticarpa*; pro-/anti-inflammatory cytokines; steam distillation essential oil; Th1/Th2 cytokines

---

1. Introduction

When a host is stimulated by microorganisms or other harmful substances, inflammation occurs immediately. Typically, acute inflammation will result in redness, swelling, warmth and pain reactions [1]. This occurs because leukocytes are recruited and penetrate blood vessels to the infection site to clear pathogens. Inflammatory cells, mainly macrophages and neutrophils, produce a large number of soluble inflammatory mediators, including pro-inflammatory cytokines (tumor necrosis factor (TNF)-α and interleukin (IL)-1, etc.), inflammatory mediators (prostaglandin E2 (PGE$_2$) and nitric oxide (NO), etc.), contributing to the inflammatory reactions [2]. Cytokines are small proteins secreted by particular immune cells like macrophages, dendritic cells, B lymphocytes, T lymphocytes and mast cells, as well as other non-immune cells like endothelial cells, fibroblasts and various stromal cells. Cytokines are similar to hormones that act on cells having the corresponding cytokine receptors on the cell membrane in autocrine, paracrine and endocrine manners. Among the cytokines, IL-10 is a cytokine synthesis inhibitor that inhibits the synthesis of pro-inflammatory cytokines secreted in the late stage of inflammation to effectively control the inflammation process [3]. Inflammation is the fixed mode of the body's defense against foreign harmful substances. However, chronic inflammation is recognized as a potential risk factor [4], resulting in continuous and repeated inflammatory reactions that can cause other diseases such as cancer, cardiovascular disease, obesity and rheumatoid arthritis [5,6]. One strategy includes natural or synthetic products that are non-toxic to normal cells that can suppress or prevent initial carcinogenesis or the progression of premalignant cells in invasive diseases [7–9].

Homeostasis between T helper type 1 (Th1) lymphocytes and T helper type 2 (Th2) lymphocytes is very important to human health and immunity [10,11]. Th1 cells that dominantly express IL-2 and IFN-γ are designated to fight viruses, intracellular pathogens, and cancerous cells, as well as induce delayed-type hypersensitivity skin reactions [10,12]. In contrast, Th2 cells uniquely secrete IL-4, IL-5 and IL-13 that fight extracellular organisms [10]. Overactivation of either Th1 or Th2 immune balance may cause diseases. Moreover, Th1/Th2 balance is also involved in many immune deficient diseases. Th1 polarization may induce autoimmune diseases including multiple sclerosis, inflammatory bowel disease, rheumatoid arthritis, etc., but Th2-inclination may result in allergic diseases and susceptibility to infection [10,13]. T lymphocyte subsets can be detected by changes in cytokine expression, unique surface markers and nuclear transcription factors [13,14]. The Th1/Th2 balance was found influenced by nutrients, hormones, omega-3 fatty acids, plant sterols/sterolins, particular minerals, probiotics, progesterone and melatonin, suggesting that Th1/Th2-based immunotherapies are promising to date [10].

Traditional Chinese medicine (TCM) is widely used for food supplements and plant-based medicine to treat chronic diseases. Some TCMs are rich in essential oils (EOs) that are complex mixtures of volatile compounds extracted from plants by steam distillation or various solvents [15]. Most EOs exude a special aroma that may be useful for aromatherapy. Increasingly more in vitro and in vivo studies have demonstrated EO bioactivities, including antioxidant, antimicrobial, anti-inflammatory and anti-cancer effects [15–17]. EOs plays multiple roles in immune-regulation, including immune activation, inhibition or regulation [18]. In general, EO components are plant secondary metabolites, consisting mostly of a mixture of terpenoids and phenylpropanoids, few aromatics, flavonoids as well as phenolic constituents [15]. The components in EOs are suggested to have antitumor, cytotoxic and chemopreventive properties; therefore, attention has been given to EOs over the last decade.

Among TCMs, *Acorus gramineus* (AG), which belongs to the Acoraceae family, has been reported for its chemical composition and bioactivity [18]. The major active components of *Acorus calamus* steam distillation essential oil (SDEO) have been found to be α-asarone and β-asarone, monoterpene hydrocarbons, sesquiterpenes, sesquiterpenoids, monoterpene alcohols, sesquiterpene alcohols and monoterpenes (including α- and β-pinenes) [19]. It is observed that AG EOs increase superoxide dismutase (SOD) and glutathione peroxidase (GSH-Px) activities, and block the peroxidatic injury induced by free radicals [20]. Both α- and β-asarone in different EOs have been reported to have numerous pharmacological activities such as acting as sedative, anti-Alzheimer's, anticonvulsant, antispasmodic, immunosuppressive, anti-inflammatory and anticancer agents [21–23].

To date, more than 100 kinds of active ingredients have been isolated and identified from *Tetradium ruticarpum* (synonym: *Euodia ruticarpa* (ER)), including alkaloids, terpenoids and phenols [24]. The main essential oil ingredients of *E. rutaecarpa* (Juss.) Benth are β-pinene, α-pinene and β-myrcene [24]. The study of Evodia extracts suggests that volatile compounds in these extracts possess anti-inflammatory properties [24,25]. Saponins, such as limonin from *Tetradium ruticarpum*, exhibit potency in antitumor and anti-inflammatory activities [26]. Herbal Zuojin Pill (ZJP), a traditional Chinese medicine formula, is composed of *Coptis chinensis* French. and *Evodia rutaecarpa* (Juss.) Benth. at a ratio of 6:1 (*w/w*), was found to attenuate the release of inflammatory factors including IL-6, IL-1β and TNF-α by regulating the NF- B signaling pathway in a gastric ulceration ICR mouse model [27].

Undoubtedly, particular EOs possess some bioactivities, however their regulatory functions in Th1/Th2 balance and inflammation are not fully understood, yet. To unravel this puzzle, two steam distillation essential oils (SDEO) from AG and ER herbs widely used in TCM in Taiwan were selected for this study. We hypothesized that potent SDEO may be rich in active phytochemicals such as polyphenols, flavonoids and saponins that are potentially valuable to regulate Th1/Th2 balance and decrease spontaneous inflammation status in the body through long-term daily low-dose supplementation. In the present study, we investigated the possible regulatory functions of these two selected SDEOs on Th1/Th2 balance and inflammation using murine splenocytes and peritoneal macrophages. Cytokine secretions, including Th1/Th2 and pro-/anti-inflammatory cytokines, were measured using the enzyme-linked immunosorbent assay (ELISA). Total amounts of potential phytochemicals, including total flavonoids, polyphenols and saponins, in these two selected SDEOs were measured and correlated with cytokine levels secreted by immune cells.

2. Materials and Methods

2.1. Isolation of Steam Distillation Essential Oils (SDEO) from Two Selected Herbs

Two herbs, *Acorus gramineusand* (AG) and *Euodia ruticarpa* (ER) cultivated in Taiwan and widely used in TCM, were purchased from a Chinese herbal medicine shop in Taichung, Taiwan. The dried herb, which has moisture content lower than 13%, was ground into a powder and then passed through a 40-mesh sieve for use to extract steam distillation essential oil. Briefly, an aliquot of 100 g sample powder was extracted with 10 volume deionized water, performed using a rotary evaporator at 90 °C for 8 h. The steam mixture was condensed and collected with a cooler. The collected steam mixture was further extracted with 400 mL ethyl acetate three times. The solvent in the steam mixture was removed by evaporation using a rotary evaporator at reduced pressure. Finally, AG and ER steam distillation essential oil (SDEO) was obtained. The extract experiment was performed in triplicate. The extract yield was expressed as the mean ± standard deviation (SD). AG and ER SDEO extract yields were 1.40 ± 0.10 and 0.05 ± 0.00 (%, *w/w*), respectively. These 2 selected SDEOs were stored at −80 °C until use. Before use, SDEO was dissolved in dimethyl sulfoxide (DMSO) to prepare a 50 mM stock solution and sterilized using a 0.22 μm pore size filter (Millipore).

2.2. Potential Phytochemicals Determination including Total Phenolic, Flavonoid and Saponin Contents in AG and ER SDEO

2.2.1. Total Phenolic Content

The total SDEO phenolic content was determined using the Folin–Ciocalteu reagent method with a slight modification [28]. Briefly, an aliquot of 0.1 mL sample solution was pipetted into a test tube. An aliquot of 2 mL of 2% Na_2CO_3 solution was added, mixed and allowed to stand for 2 min. An aliquot of 0.1 mL 50% Folin–Ciocalteu reagent solution was added to the reaction mixture. The resultant solution was mixed and allowed to stand at room temperature for 30 min. The resultant solution absorbance at 750 nm wavelength was measured using a UV–visible spectrophotometer (Hitachi-U2900 UV–vis spectrophotometer, Tokyo, Japan). Gallic acid with a serial dilution was chosen as a standard for phenolics. Total phenolic content in the sample was calculated using a standard gallic acid curve.

2.2.2. Total Flavonoid Content

Total flavonoid content in SDEO was determined as described by Shen et al. [28] with a slight modification. Briefly, an aliquot of 0.5 mL of the sample solution was pipetted into a test tube; 1.5 mL of

95% ethanol was added and mixed thoroughly. To the mixture was sequentially added 0.1 mL of 10% aluminum chloride solution, 0.1 mL of 1 M potassium acetate solution and 2.8 mL of deionized water. The resultant mixture was mixed thoroughly and allowed to stand at room temperature for 30 min. Finally, the resultant mixture absorbance at 415 nm wavelength was measured using a UV–visible spectrophotometer (Hitachi-U2900 UV–vis spectrophotometer, Tokyo, Japan). Quercetin with a serial dilution was chosen as a standard for flavonoids. Total flavonoid content in the sample was calculated using a quercetin standard curve.

2.2.3. Total Saponin Content

The total SDEO saponin content was measured using the method described by Yu et al. [29] with a slight modification. Briefly, an aliquot of 1 mL of the sample solution was pipetted into a test tube, with 0.2 mL of 5% vanillin–glacial acetic acid added, and 0.8 mL of perchloric acid. After mixing thoroughly, the reaction was carried out at 60 °C for 20 min in a water bath. After cooled on ice, an aliquot of 4 mL of glacial acetic acid was added to the reaction mixture. The resultant mixture absorbance at 550 nm wavelength was measured using a UV–visible spectrophotometer (Hitachi-U2900 UV–vis spectrophotometer, Tokyo, Japan). Oleanolic acid with a serial dilution was chosen as a standard for saponins. Total saponin content in the sample was calculated using an oleanolic acid standard curve.

2.3. Determination of Chemical Components of ER SDEO using Gas Chromatography-Mass Spectrometry (GC-MS)

Since ER SDEO had the higher amount of potential phytochemicals, it was subjected to GC-MS analysis. Chemical components of ER SDEO were analyzed using GC-MS (Agilent GC/Mass Selective Detector (MSD) 5973Network, Agilent Technologies, Inc., Santa Clara, CA, USA) connected a series with a DB-1 column (60 m × 250 mm; film thickness 0.25 μm); carrier gas (helium at a flow rate of 1 mL/min); injector temperature, 250 °C; sample injection, split; the operational temperature was kept at 50 °C for 1 min initially. The detector and injector temperature was set at 250 and 300 °C respectively, and the column temperature was raised from 50 to 270 °C at a rate of 4 °C/min; ionization voltage, 70 eV. An aliquot of 1.0 μL of steam essential oil sample (diluted in methane dichloride) was injected into the GC-MSD instrument for analysis. The chemical components of essential oil were identified by comparing their retention indices (RI) and mass fragmentation patterns with those stored on the Wiley and National Institute of Standards and Technology (NIST) Library.

2.4. Experimental Animals

Female BALB/c ByJNarl mice at 6–8 weeks of age were purchased from the National Laboratory Animal Center, National Applied Research Laboratory, Ministry of Science and Technology in Taipei, Taiwan. Groups of five mice were housed in a standard cage (25 ± 2 °C, 50%–75% ambient humidity) and maintained on a chow diet (laboratory standard pellet diet, Diet MF 18, Oriental Yeast Co., Ltd., Osaka, Japan) with free access to food and water under a 12-h light/dark cycle. At 11–12 weeks of age, the mice were weighed and anesthetized with 2% isoflurane (cat. no., 4900-1605, Panion & BF Biotech Inc., Taipei, Taiwan) using a vaporizer (CAS-01, Northern Vaporiser Limited, Cheshire, England, UK). Blood was taken from the experimental mice through retro-orbital venous plexus puncture. Animals were sacrificed with CO_2 inhalation immediately after the blood collection. The primary mice peritoneal macrophages and splenocytes were isolated aseptically. The experimental animal use protocol was examined and verified by the Institutional Animal Care and Use Committee (IACUC No: 103-119), National Chung Hsing University, Taiwan.

2.5. Isolation of Mouse Primary Peritoneal Macrophages and Splenocytes

2.5.1. Isolation of Primary Peritoneal Macrophages

Mouse primary peritoneal macrophages were isolated using the method as described [30]. Briefly, peritoneal macrophages were isolated by lavaging the peritoneal cavity of experimental mice with 2 aliquots of 5 mL sterile Hank's balanced salts solution (HBSS) (50 mL of 10× HBSS (Hyclone, SH30015.02, South Logan, UT, USA), 2.5 mL of penicillin–streptomycin–amphotericin solution (PSA, 100×, Biological Industries, 03-033-1B, Kibbutz Beit Haemek, Israel), 20 mL of 3% bovine serum albumin (BSA, Sigma-Aldrich Co., A9418, St. Louis, MO, USA) in phosphate-buffered saline (PBS, 137 mM NaCl, 2.7 mM KCl, 8.1 mM Na_2HPO_4, 1.5 mM KH_2PO_4, pH 7.4, 0.22 μm filtered), 2.0 mL of 7.5% $NaHCO_3$ (Wako, 191-01305, Osaka, Japan) and 425.5 mL sterile water) for a total of 10 mL through peritoneum. The peritoneal lavage fluid was collected and centrifuged at 400× g for 10 min. The cell pellet was isolated and resuspended in tissue culture medium (TCM, a serum substitute, Celox Laboratories, Lake Zurich, IL, USA). TCM medium consisted of 10 mL TCM, 500 mL Roswell Park Memorial Institute (RPMI) 1640 medium (Atlanta Biologicals Inc., Norcross, GA, USA) and 2.5 mL of antibiotic–antimycotic solution (100× PSA). Isolated peritoneal cells are macrophages that can serve as a cell culture model for assessing inflammation status in vitro. The viable cell number was counted under a microscope with a hemocytometer using the trypan blue exclusion method. The macrophages cell density was adjusted to 2×10^6 cells/mL TCM medium for use.

2.5.2. Primary Splenocytes Isolation

After peritoneal macrophages were collected, the mouse spleen was cut aseptically, immersed in TCM medium and ground to isolate splenocytes [31]. Splenocytes were then collected and centrifuged at 400× g for 7 min. Next, the splenocytes were resuspended in an aliquot of 10 mL of red blood cell (RBC) lysis buffer (pH 7.4, 0.22 μm filtered) consisting of 0.017 M Trizma Base (Sigma-Aldrich Co., St. Louis, MO, USA) and 0.144 M NH_4Cl (Sigma-Aldrich Co., St. Louis, MO, USA) in sterile water. After standing for 3 min, the cell solution was centrifuged at 400× g for 7 min. The splenocyte pellet was carefully washed with HBSS three times. Isolated splenocytes were resuspended in a 3 mL TCM medium. The splenocytes were computed using the trypan blue dye exclusion method using a hemocytometer. Finally, the splenocytes cell density was adjusted to 1×10^7 cells/mL TCM medium for use. Isolated splenocytes are approximately composed of 41.54% of B lymphocytes and 47.11% of T lymphocytes, as well as trace antigen-presenting cells that are suitable cell cultures for evaluating Th1/Th2 immune responses in vitro [32].

2.6. Determination of Optimal AG and ER SDEO Concentrations

To obtain non-cytotoxic optimal concentrations for treating immune cells, AG and ER SDEO at different concentrations were used to treat splenocytes, respectively. The cell viabilities were evaluated using the 3-(4,5-dimethylthiazol-2-diphenyl)-2,5-tetrazolium bromide (MTT) assay. Briefly, each individual SDEO stock solution was aseptically diluted into working solutions using TCM medium before use. Aliquots of 50 μL/well splenocytes (1×10^7 cells/mL) were pipetted into a 96-well plate. Aliquots of 50 μL/well SDEO at different concentrations or lipopolysaccharide (LPS, as a positive control, final concentration in the medium was 2.5 μg/mL) were added to the well and mixed thoroughly. The plate was incubated in an incubator with 5% CO_2 and 95% air at 37 °C for 72 h. After incubation, aliquots of 10 μL of MTT (Sigma M5655, St. Louis, MO), 5 mg/mL in PBS were added to each well in the 96-well plate and incubated in an incubator with 5% CO_2 and 95% air at 37 °C for another 4 h. The plate was centrifuged at 400× g for 10 min. The supernatant was decanted to remove excess MTT. Aliquots of 100 μL/well PBS buffer were added to each well to rinse the cells three times. Aliquots of 100 μL/well DMSO were added to each well. The plate was gently oscillated for 30 min to lyse the cells. The absorbance (A) at 550 nm was measured using an ELISA reader. The cell viability was expressed as the survival rate (%) compared to the control mean absorbance. The following equation was used to calculate the cell viability: cell viability (% of control) = [(A_{sample} - A_{blank})/($A_{control}$ - A_{blank})] × 100. A_{sample}: cells added with SDEO samples; A_{blank}: TCM medium alone; $A_{control}$: cells alone. Based on changes in cell viabilities, optimal non-cytotoxic concentrations of these 2 selected SDEOs were achieved and adopted for the following immune cell cultures.

2.7. Mouse Splenocytes Cultures with AR and ER SDEO at Different Optimal Concentrations

To assess the effects of the selected SDEOs on Th1/Th2 immune balance, isolated splenocytes (1×10^7 cells/mL, 500 µL/well) were cultured with SDEO (500 µL/well) samples at different optimal concentrations in a 24-well plate. The plate was incubated at 37 °C in a humidified incubator with 5% CO_2 and 95% air for 48 h. Lipopolysaccharide (LPS, Sigma-Aldrich Co., L-2654, St. Louis, MO, USA) at a final concentration of 2.5 µg/mL was selected as a positive control in each experiment. After incubation, the plate was centrifuged at 400× *g* for 10 min. The supernatant in the cell culture was collected and stored at −80 °C for Th1/Th2 cytokine assays. Based on changes in Th1/Th2 cytokine secretions, AG and ER SDEO exhibited the potential to regulate Th1/Th2 balance [31]. Thus, AG and ER SDEO were further selected to evaluate their anti-inflammatory potential.

2.8. Mouse Peritoneal Macrophages Cultures with AG and ER SDEO at Different Optimal Concentrations in the Absence or Presence of LPS

To assess the anti-inflammatory potential of AG and ER SDEO, peritoneal macrophages (2×10^6 cells/mL, 500 µL/well) were cultured with SDEO samples (500 µL/well) at different optimal concentrations in a 24-well plate in the absence or presence of LPS. The plate was incubated at 37 °C in a humidified incubator with 5% CO_2 and 95% air for 48 h. In each experiment, endotoxin LPS with a final concentration of 2.5 µg/mL in the culture was selected as a positive control or control, respectively. After incubation, the supernatant in the cell culture was collected and stored at −80 °C for pro-/anti-inflammatory cytokine assays. Based on changes in pro-/anti-inflammatory cytokine secretions, the anti-inflammatory potential of AG and ER SDEO were evaluated [30].

2.9. Th1/Th2 and pro-/anti-Inflammatory Cytokine Levels Secreted by Immune Cells Measured Using an Enzyme-Linked Immunosorbent Assay (ELISA)

Th1 (IL-2)/Th2 (IL-5) cytokines secreted by splenocytes and pro- (TNF-α)/anti-inflammatory (IL-10) cytokines secreted by macrophages were respectively measured using sandwich ELISA kits, and assayed according to the cytokine ELISA protocol from the manufacturer's instructions (mouse DuoSet ELISA Development system, R&D Systems, Minneapolis, MN, USA). Briefly, aliquots of 100 µL of anti-mouse captured antibodies (1:180 diluted with PBS) were added to 96-well plate and incubated overnight at 4 °C. After incubation, the plate was washed with ELISA wash buffer (0.05% Tween 20 in PBS, pH 7.4) three times. Aliquots of 200 µL of block buffer (1% bovine serum albumin (BSA, Sigma-Aldrich Corp., S-2002, St. Louis, MO, USA) and 0.05% NaN_3 in PBS) were added to each well to block non-specific binding. The plate was incubated at room temperature for 1 h. After incubation, the plate was washed with ELISA wash buffer three times. Aliquots of 100 µL of the cytokine test sample or standard in reagent diluent (0.1% BSA in Tris-buffered saline (TBS, 20 mM Trizma base, 150 mM NaCl, pH 7.4, 0.22 µm filtered)) were added to the wells and the plate was incubated for 2 h at room temperature. A seven-point (in duplicate) standard curve (1000–15.6 pg/mL) using 2-fold serial dilutions in reagent diluent was performed. After incubation, the plate was washed with ELISA wash buffer three times. Aliquots of 100 µL of the detection antibody (biotinylated goat anti-mouse monoclonal antibody at 1:180 dilution in reagent diluent) were added to each well. The plate was incubated at room temperature for 2 h. After incubation, the plate was washed with ELISA wash buffer three times. Aliquots of 100 µL of working streptavidin–horseradish peroxidase (HRP) dilution were added to each well. Then, the plate was incubated at room temperature for 20 min. After incubation, the plate was washed with ELISA wash buffer three times. Aliquots of 100 µL of substrate solution (tetramethylbenzidine, TMB, Clinical Science Products Inc., 01016-1-500, Mansfield, MA,

USA) were pipetted into each well. To develop color, the plate was incubated at room temperature for 20 min. Finally, aliquots of 50 µL of stop solution (2N H_2SO_4) were added to each well to cease the reaction. The plate was measured for absorbance at 450 nm using an ELISA reader (Microplate Reader FLUOstar-Omega, 415-1103, Ortenberg, Germany). The cytokine levels were calculated using the seven-point standard curves. The inter/intra coefficient of variability (CV, %) of the DuoSet ELISA kits used in this study were calculated based on the average absorbance from duplicate standards and/or plates. With the inter CV (%) assays, CV (%) of IL-2, IL-5, TNF-α, and IL-10 were 1.94 (1.14–2.81), 4.33 (2.07–8.67), 1.75 (0.90–3.26), and 2.25 (0.95–3.67), respectively. With the intra CV (%) assays, CV (%) of IL-2, IL-5, TNF-α, and IL-10 were 2.11 (0.00–3.08), 3.10 (1.64–4.00), 1.87 (0.95–4.53), and 2.30 (0.21–4.63), respectively. The detection sensitivity of the ELISA kits used in this study was <15.6 pg/mL.

2.10. Statistical Analysis

Values are expressed as mean ± standard deviation (SD) and statistically analyzed using repeated ANOVA measurements, if analyzed by statistical probability ($p < 0.05$), followed by Duncan's new multi-range test. Statistical analyses were performed using IBM SPSS statistics 20. The relationship between phytochemicals (total phenolic, total flavonoid and total saponin) contents in 2 selected SDEOs and cytokine secretion profiles by correspondent immune cells was described as the Spearman correlation coefficient (rho). The difference was considered statistically significant if $p < 0.05$.

3. Results and Discussion

3.1. Total Flavonoid, Phenolic and Saponin Contents in AG and ER Selected SDEO

As shown in Table 1, ER SDEO contained much higher total flavonoid, polyphenol and saponin contents than those of AG SDEO. In comparison with these three phytochemicals, total saponins had the greatest amounts in either AG or ER SDEO. We hypothesized that all three detected phytochemicals might contribute to anti-inflammatory effects. The potential phytochemical contents in different SDEOs varied dramatically, possibly reflecting their immunomodulatory functions. If merely depending on the quantity, total saponins in ER SDEO seemed to have the most contribution to modulate Th1/Th2 and pro-/anti-inflammatory cytokines secretions. However, the composition of ER SDEO is so complicated that other constituents could not be excluded the role to partake in the immunomodulatory effects. Chemical components of ER SDEO assayed with GC-MS were found to contain at least 51 compounds, including 2 alkenes, 22 alcohols, 2 ketones, 1 acid, 2 esters as well as other compounds (Table 2). Other compounds in ER SDEO such as phenolic compounds and alkaloids may also contribute their effects. Individual active compounds in ER SDEO should be further assayed and clarified in the future.

Table 1. Total flavonoid, polyphenol and saponin contents in AG and ER steam distilled essential oils (SDEO).

Samples	Total Flavonoids (mg Quercetin Equivalent/g Sample)	Total Polyphenols (mg Gallic Acid Equivalent/g Sample)	Total Saponins (mg Oleanolic Acid Equivalent/g Sample)
AG	0.1 ± 0.0 [C,b]	17.3 ± 2.3 [B,b]	161 ± 32 [A,b]
ER	20.3 ± 1.6 [B,a]	33.3 ± 0.7 [B,a]	601 ± 13 [A,a]

Value are means ± SD (n = 3 replications). AG, *Acorus gramineus* SDEO; ER, *Euodia ruticarpa* SDEO. Values within same row not sharing a common superscript capital letter are significantly different ($p < 0.05$) from each other. Values within same column not sharing a common superscript small letter are significantly different ($p < 0.05$) from one another analyzed using one-way ANOVA, followed by Duncan's multiple range test.

Table 2. Chemical components of ER SDEO assayed with Gas Chromatography-Mass Spectrometry (GC-MS).

NO.	RT	RI	Compounds	M.W.	Chemical Formula	CAS NO.
1	18.619	1092.33	Linalool	154.14	$C_{10}H_{18}O$	000078-70-6
2	19.54	1118	4-Isopropyl-1-methyl-2-cyclohexen-1-ol	154.14	$C_{10}H_{18}O$	029803-81-4
3	20.136	1135.1	Terpenene-1-ol	154.14	$C_{10}H_{18}O$	000586-82-3
4	21.321	1167.66	4-Isopropyl-1-methyl-2-cyclohexen-1-ol	138.10	$C_9H_{14}O$	000500-02-7
5	21.539	1173.45	p-Cymen-8-ol	150.10	$C_{10}H_{14}O$	001197-01-9
6	21.975	1184.86	β-Fenchyl alcohol	154.14	$C_{10}H_{18}O$	000470-08-6
7	22.344	1194.35	4-Methyl-1,4-heptadiene	110.11	C_8H_{14}	013857-55-1
8	22.967	1212.25	Trans-(+)-carveol	152.12	$C_{10}H_{16}O$	001197-07-5
9	23.328	1223.12	Carveol	152.12	$C_{10}H_{16}O$	000000-00-0
10	23.471	1227.38	Cyclobutanol	150.1	$C_{10}H_{14}O$	091531-61-2
11	24.201	1248.73	3-Ethyl-2-pentanone	114.10	$C_7H_{14}O$	006137-03-7
12	24.349	1252.98	2,5-Dimethyl-1,5-hexadiene-3,4-diol	142.10	$C_8H_{14}O_2$	004723-10-8
13	24.648	1261.49	3,3,5-Trimethyl-heptane	142.17	$C_{10}H_{22}$	007154-80-5
14	25.22	1277.48	3,3,6-Trimethyl-4,5-heptadien-2-one	152.12	$C_{10}H_{16}O$	081250-41-1
15	25.325	1280.38	Methyl (2E)-2,5-dimethylhexa-2,4-dienoate	154.10	$C_9H_{14}O_2$	000000-00-0
16	25.421	1283.01	Cuminic alcohol	150.10	$C_{10}H_{14}O$	000536-60-7
17	25.664	1289.65	Dill ether	152.12	$C_{10}H_{16}O$	000000-00-0
18	25.846	1294.57	Perilla alcohol	152.12	$C_{10}H_{16}O$	000536-59-4
19	25.986	1298.34	4-Methyl-2-(3-methyl-2-butenyl)-furan	150.10	$C_{10}H_{14}O$	000000-00-0
20	26.787	1323.45	Methyl anthranilate	151.06	$C_8H_9NO_2$	000134-20-3
21	27.423	1343.13	Eugenol	164.08	$C_{10}H_{12}O_2$	000097-53-0
22	27.825	1355.33	(±)-Eldanolide	168.12	$C_{10}H_{16}O_2$	092843-42-0
23	29.064	1391.85	(±)-Eldanolide	168.12	$C_{10}H_{16}O_2$	092843-42-0
24	29.863	1417.23	(E)-1-Cyclohexyl-3,3-dimethyl-1-butene	166.17	$C_{12}H_{22}$	109660-16-4
25	29.945	1419.95	1-(2-Hydroxy-4-methoxyphenyl)-ethanone	166.06	$C_9H_{10}O_3$	000552-41-0
26	32.435	1499.08	(E)-3,4-Epoxy-1-(1′,2′-epoxy-3′,3′-epoxymethano-2′,6′,6′-trimethyl-1′-cyclohexyl)-3-methyl-1-butene	250.16	$C_{15}H_{22}O_3$	091186-32-2
27	32.997	1518.69	β-myrcene	136.13	$C_{10}H_{16}$	000127-91-3
28	33.962	1551.85	cis-5-Dodecenoic acid	198.16	$C_{12}H_{22}O_2$	002430-94-6
29	34.239	1561.19	3,7,11-Trimethyl-, (Z,E)-1,3,6,10-Dodecatetraene	204.19	$C_{15}H_{24}$	026560-14-5
30	35.051	1588.15	(+)-Spathulenol	220.18	$C_{15}H_{24}O$	077171-55-2
31	35.153	1591.49	4-Hydroxy-β-ionone	208.15	$C_{13}H_{20}O_2$	015401-34-0
32	35.474	1602.23	Methyl 4-(4-methyl-3-pentenyl)-3-cyclohexen-1-yl ketone	206.17	$C_{14}H_{22}O$	038758-04-2
33	35.864	1616.64	6-Isopropenyl-4,8a-dimethyl-1,2,3,5,6,7,8,8a-octahydro-2-naphthalenol	220.18	$C_{15}H_{24}O$	000000-00-0
34	36.293	1632.31	(-)-Spathulenol	220.18	$C_{15}H_{24}O$	077171-55-2
35	36.636	1644.7	Caryophylla-4(12),8(13)-dien-5β-ol	220.18	$C_{15}H_{24}O$	000000-00-0
36	37.131	1662.39	(+)-β-Costol	220.18	$C_{15}H_{24}O$	065018-15-7
37	37.604	1679.07	Patchulane	206.20	$C_{15}H_{26}$	019078-35-4
38	37.943	1690.9	2-Pentadecanone	226.23	$C_{15}H_{30}O$	002345-28-0
39	38.502	1711.51	Cedr-8-en-13-ol	220.18	$C_{15}H_{24}O$	018319-35-2
40	38.819	1723.74	(E)-2-Methyl-4-(2′,6′,6′-trimethyl-3′-methyliden-1′,2′-epoxy-1′-cyclohexyl)-1,3-butadiene	218.17	$C_{15}H_{22}O$	077822-46-9
41	38.965	1729.33	Valerenol	220.18	$C_{15}H_{24}O$	084249-42-3
42	39.203	1738.41	(+)-Valencene	220.18	$C_{15}H_{24}O$	004630-07-3
43	39.462	1748.23	7-Isopropenyl-1,4a-dimethyl-4,4a,5,6,7,8-hexahydro-3H-naphthalen-2-one	218.17	$C_{15}H_{22}O$	000473-08-5
44	40.416	1783.85	6-Phenyl(deuterate)-2,3,4,5-tetrahydro-3-pyridazinone	179.11	$C_{10}H_5D_5N_2O$	055999-93-4
45	41.356	1820.32	cis-Z-α-Bisabolene epoxide	220.18	$C_{15}H_{24}O$	000000-00-0
46	42.828	1878.73	1-Isopropyl-4,8,12-trimethylcyclotetrtadeca-2,4,7,11-tetraene	272.25	$C_{20}H_{32}$	000000-00-0
47	42.917	1882.19	4,4,8-Trimethyltricyclo[6.3.1.0(1,5)]dodecane-2,9-diol	238.19	$C_{15}H_{26}O_2$	000000-00-0
48	43.43	1902.27	7,11-Dimethyl-3-methylene-(Z)-1,6,10-dodecatriene	204.19	$C_{15}H_{24}$	028973-97-9
49	44.732	1957.07	2,6,11,15-Tetramethyl-hexadeca-2,6,8,10,14-pentaene	272.25	$C_{20}H_{32}$	038259-79-9
50	45.437	1986.09	2,4a,5,6,7,8,9,9a-Octahydro-3,5,5-trimethyl-9-methylene-1H-Benzocycloheptene	204.19	$C_{15}H_{24}$	080923-88-2
51	45.544	1990.45	2,6,11,15-Tetramethyl-hexadeca-2,6,8,10,14-pentaene	272.25	$C_{20}H_{32}$	038259-79-9

RT: Retention time (min); RI: Retention indices.

3.2. Optimal AG and ER SDEO Concentrations using Murine Splenocytes

To achieve optimal concentrations for use, the possible cytotoxicity of AG or ER SDEO to murine splenocytes was assessed, respectively. The results showed that there are significant ($p < 0.05$)

cytotoxic effects on splenocytes at the higher concentrations of these two selected SDEO (Figure 1). Based on changes in cell viabilities, the optimal non-cytotoxic concentrations of AG and ER SDEO were 0.125–5 µg/mL and 0.25–25 µg/mL, respectively (Figure 1). In comparison, ER SDEO had less cytotoxicity than that of AG SDEO. The stronger cytotoxic effects of AG SDEO result possibly from its sesquiterpene composition such as the most notable component β-asarone [33]. To avoid cytotoxic effects on immune cells, non-cytotoxic optimal concentrations of these two selected SDEOs were adopted for the following immunomodulatory evaluation.

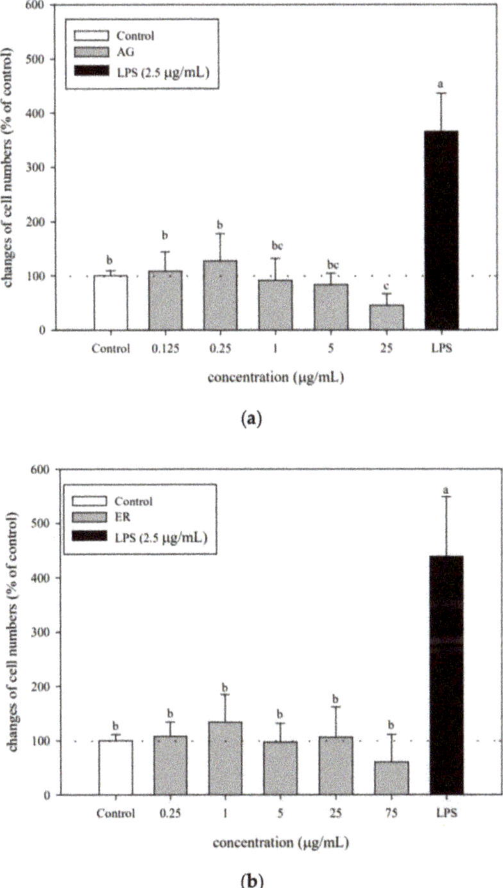

(a)

(b)

Figure 1. *Acorus gramineus* (AG) (**a**) and *Euodia ruticarpa* (ER) (**b**) SDEO treatment effects on splenocyte cell growth from female BALB/c mice. Values are means ± SD (n = 6 biological determination). Data are assayed using one-way ANOVA, followed by Duncan's multiple range test. Bars in the same plot not sharing a common small letter are significantly different ($p < 0.05$) from each other. The original cell density was 5×10^6 cells/mL. Lipopolysaccharides (LPS) at 2.5 µg/mL in each experiment was selected as a positive control.

3.3. AG and ER SDEO Effects on Th1/Th2 Cytokine Secretions by Mouse Primary Splenocytes

To evaluate the effects of different SDEOs on Th1/Th2 cytokine secretions, AG and ER SDEO at the indicated non-cytotoxic concentrations were administered to splenocyte cultures for 48 h, respectively. The results showed that AG SDEO significantly ($p < 0.05$) decreased IL-2 (Th1) secretions, but increased

IL-5 (Th2) secretions (Table 3). Importantly, IL-5/IL-2 (Th2/Th1) cytokine secretion ratios by splenocytes were significantly ($p < 0.05$) increased by AG SDEO in a dose-dependent manner, suggesting that AG SDEO has a Th2-inclination property even though both Th1 and Th2 secretion amounts were still low (Table 3).

In addition, ER SDEO significantly ($p < 0.05$) and dose-dependently decreased IL-2 (Th1) secretions, but markedly increased IL-5 secretions (Table 3). Importantly, IL-5/IL-2 (Th2/Th1) cytokine secretion ratios by splenocytes were significantly ($p < 0.05$) increased by ER SDEO (Table 3). Our results evidenced that ER SDEO had a Th2- inclination property (Table 3).

Taken together, these two selected SDEOs exhibited an obvious Th2-inclination property (Table 3), suggesting that AG and ER SDEO, particularly ER SDEO, have anti-inflammatory potential via their Th2-polarization property that inhibits pro-inflammatory Th1 immune balance. However, we caution that excess Th2-inclination in the adaptive immune system may cause adverse effects such as allergic diseases.

Table 3. AG and ER SDEO treatment effects on Th1/Th2 cytokine secretions using primary splenocytes from female BALB/c mice.

| | | Th1 | Th2 | Th2/Th1 Cytokines Ratio (pg/pg) |
| | | Cytokines (pg/mL) | | |
Samples	Treatments (µg/mL)	IL-2	IL-5	IL-5/IL-2
AG	control	23.9 ± 4.5 [B]	0.0 ± 0.0 [B]	0.00 ± 0.00 [E]
	0.125	21.3 ± 3.3 [B]	15.3 ± 4.9 [A]	0.73 ± 0.26 [C,D]
	0.25	16.8 ± 2.4 [C]	13.4 ± 3.4 [A]	0.82 ± 0.26 [B,C]
	1	9.7 ± 2.5 [D]	12.0 ± 4.0 [A]	1.28 ± 0.42 [A]
	5	11.1 ± 2.6 [D]	11.8 ± 4.6 [A]	1.13 ± 0.59 [A,B]
	LPS	35.3 ± 5.1 [A]	13.4 ± 3.1 [A]	0.39 ± 0.12 [D]
ER	control	23.9 ± 4.5 [b]	0.0 ± 0.0 [d]	0.00 ± 0.00 [c]
	0.25	9.8 ± 1.6 [c]	50.9 ± 6.1 [a]	5.50 ± 1.13 [b]
	1	10.3 ± 2.3 [c]	49.0 ± 2.5 [a,b]	4.90 ± 0.97 [b]
	5	9.4 ± 2.1 [c]	50.8 ± 8.3 [a]	5.43 ± 1.10 [b]
	25	3.0 ± 1.1 [d]	45.3 ± 2.5 [b]	16.8 ± 6.72 [a]
	LPS	35.3 ± 5.1 [a]	13.4 ± 3.1 [c]	0.39 ± 0.12 [c]

Values are means ± SD ($n = 6$ biological determinations). Values within same column in the same item (AG or ER) not sharing a common superscript letter are significantly different ($p < 0.05$) from each other assayed using one-way ANOVA, followed by Duncan's multiple range test. The detection sensitivity of cytokine ELISA kits used in this study was <15.6 pg/mL.

3.4. AG and ER SDEO Effects on pro-/anti-Inflammatory Cytokine Secretions by Mouse Peritoneal Macrophages in the Absence or Presence of LPS

To examine anti-inflammatory potential, AG and ER SDEO at the indicated non-cytotoxic optimal concentrations were administered to peritoneal macrophages in the absence or presence of LPS (2.5 µg/mL) for 48 h. The results showed that TNF-α secretions by macrophages in the absence of LPS were slightly but not significantly ($p > 0.05$) inhibited by AG SDEO (0.125–5 µg/mL), while IL-10 levels slightly but not significantly ($p > 0.05$) increased (Table 4). Importantly, TNF-α/IL-10 cytokine secretion ratios by macrophages in the absence of LPS were significantly ($p < 0.05$) decreased by AG SDEO, suggesting that AG SDEO has anti-inflammatory potential. In addition, ER SDEO markedly inhibited TNF-α secretions, but increased IL-10 secretions (Table 4). Importantly, TNF-α/IL-10 cytokine secretion ratios by macrophages in the absence of LPS were significantly ($p < 0.05$) and dose-dependently decreased by ER SDEO, suggesting that ER SDEO has strong anti-inflammatory potential. Our results suggest that these two selected SDEOs, particularly ER SDEO, have the potential to inhibit spontaneous inflammation in macrophages.

AG SDEO administration more or less increased pro-inflammatory cytokines TNF-α, and anti-inflammatory cytokine IL-10 by peritoneal macrophages in the presence of LPS (Table 5). However, AG SDEO administration dose-dependently, but not significantly ($p > 0.05$), decreased

TNF-α/IL-10 secretion ratios by LPS-stimulated macrophages. Our results suggest that AG SDEO enhances cytokine secretions but has mild anti-inflammatory potential. As to ER SDEO, ER SDEO administration at appropriate concentrations significantly ($p < 0.05$) decreased pro-inflammatory cytokine TNF-α, as well as anti-inflammatory cytokine IL-10 by LPS-stimulated macrophages (Table 5). However, ER SDEO administrations significantly ($p < 0.05$) decreased TNF-α/IL-10 secretion ratios by LPS-stimulated macrophages, suggesting that ER SDEO has anti-inflammatory potential via inhibiting both pro- and anti-inflammatory cytokine secretions and decreasing TNF-α/IL-10 secretion ratios by LPS-stimulated macrophages.

In comparison with AG and ER SDEO, we concluded that both AG and ER SDEO have the potential to inhibit spontaneous (Table 4) and LPS-stimulated inflammation (Table 5) in macrophages. However, the anti-inflammatory properties of these two selected SDEO in LPS-induced inflammation were quite different from each other (Table 5). ER SDEO seems to have a mild inhibitory property to IL-10 secretions by LPS-stimulated macrophages that are similar to glucocorticoid functions but it can regulate pro-/anti-inflammatory cytokine secretion ratio. In contrast, AG SDEO has the potential to inhibit spontaneous and LPS-induced inflammation through its potent immunomodulatory but not inhibitory property to immune cells.

Table 4. AG and ER SDEO treatment effects on pro- and anti-inflammatory cytokine secretions using mouse peritoneal macrophages in the absence of LPS.

Treatments (μg/mL)	Pro-Inflammatory Cytokines (pg/mL)	Anti-Inflammatory Cytokines (pg/mL)	Pro-/Anti-Inflammatory Cytokines Secretion Ratios (pg/pg)
	TNF-α	IL-10	TNF-α/IL-10
AG control	45 ± 20 [B]	78 ± 19 [B]	0.63 ± 0.34 [A]
0.125	26 ± 5 [B]	91 ± 31 [B]	0.31 ± 0.07 [B]
0.25	28 ± 7 [B]	84 ± 27 [B]	0.35 ± 0.16 [A,B]
1	33 ± 5 [B]	88 ± 27 [B]	0.40 ± 0.04 [A,B]
5	26 ± 4 [B]	102 ± 24 [B]	0.26 ± 0.04 [B]
LPS	248 ± 152 [A]	735 ± 148 [A]	0.33 ± 0.19 [A,B]
ER control	45 ± 20 [b]	78 ± 19 [c]	0.63 ± 0.34 [a]
0.25	26 ± 18 [b]	108 ± 14 [c]	0.25 ± 0.20 [b]
1	25 ± 14 [b]	104 ± 9 [c]	0.24 ± 0.13 [b]
5	23 ± 11 [b]	120 ± 7 [c]	0.19 ± 0.08 [b]
25	28 ± 14 [b]	339 ± 34 [b]	0.08 ± 0.04 [b]
LPS	248 ± 152 [a]	735 ± 148 [a]	0.33 ± 0.19 [b]

Values are means ± SD ($n = 4$ biological determinations). Values within same column in the same item (AG or ER) not sharing a common superscript letter are significantly different ($p < 0.05$) from each other assayed using one-way ANOVA, followed by Duncan's multiple range test. The detection sensitivity of cytokine ELISA kits used in this study was <15.6 pg/mL.

Table 5. AG and ER SDEO treatment effects on pro- and anti-inflammatory cytokine secretions using LPS-stimulated peritoneal macrophages from female BALB/c mice.

Treatments (μg/mL)	Pro-Inflammatory Cytokines (pg/mL)	Anti-Inflammatory Cytokines (pg/mL)	Pro-/Anti-Inflammatory Cytokines Secretion Ratios (pg/pg)
	TNF-α	IL-10	TNF-α/IL-10
VC	144 ± 30 [C]	190 ± 63 [B]	0.81 ± 0.26 [A]
control (LPS alone)	499 ± 81 [B]	1005 ± 282 [A]	0.52 ± 0.14 [B]
AG 0.125	592 ± 53 [A]	1292 ± 405 [A]	0.50 ± 0.16 [B]
0.25	580 ± 74 [A,B]	1274 ± 371 [A]	0.48 ± 0.10 [B]
1	546 ± 68 [A,B]	1234 ± 389 [A]	0.47 ± 0.14 [B]
5	557 ± 69 [A,B]	1308 ± 405 [A]	0.46 ± 0.15 [B]

Table 5. *Cont.*

Treatments (μg/mL)	Pro-Inflammatory Anti-Inflammatory Cytokines (pg/mL)		Pro-/Anti-Inflammatory Cytokines Secretion Ratios (pg/pg)
	TNF-α	IL-10	TNF-α/IL-10
VC	144 ± 30 [d]	190 ± 63 [c]	0.81 ± 0.26 [a]
control (LPS alone)	499 ± 81 [a]	1005 ± 282 [a]	0.52 ± 0.14 [b]
ER 0.25	239 ± 39 [c]	726 ± 193 [b]	0.34 ± 0.05 [c]
1	222 ± 39 [c]	712 ± 120 [b]	0.31 ± 0.02 [c]
5	235 ± 49 [c]	817 ± 149 [a,b]	0.29 ± 0.02 [c]
25	396 ± 47 [b]	922 ± 146 [a,b]	0.43 ± 0.03 [b,c]

Values are means ± SD (*n* = 6 biological determinations). Values within same column in the same item (AG or ER) not sharing a common superscript letter are significantly different (*p* < 0.05) from each other assayed using one-way ANOVA, followed by Duncan's multiple range test. The sensitivity of cytokine ELISA kits used in this study was <15.6 pg/mL.

3.5. The Correlation between Th1/Th2 Cytokine Secretion Levels in Mouse Primary Splenocyte Cultures and Total Flavonoid, Phenol or Saponin Contents in AG and ER SDEO

Correlations between Th1/Th2 cytokine secretion levels and potent phytochemicals, including total flavonoid, total phenol and total saponin contents in AG and ER SDEO, were determined using the Spearman correlation coefficient. The results showed that there are significant (*p* < 0.05) positive correlations between IL-5/IL-2 (Th2/Th1) cytokine secretion ratios by mouse splenocytes and total polyphenol, flavonoid or saponin contents in a linear manner (Figure 2). Our results further suggest that potent anti-inflammatory phytochemicals including polyphenols (Figure 2A), flavonoids (Figure 2B) and saponins (Figure 2C) may exert their anti-inflammatory ability via decreasing Th2/Th1 cytokine secretions in a linear manner.

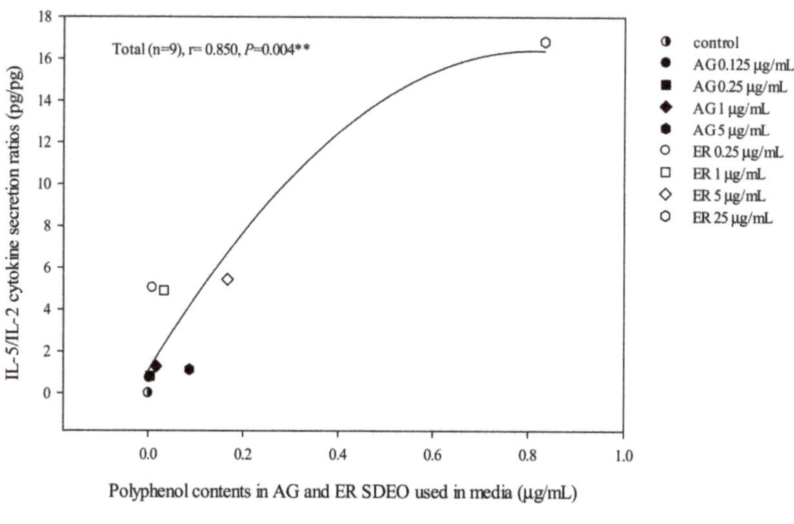

(**a**)

Figure 2. *Cont.*

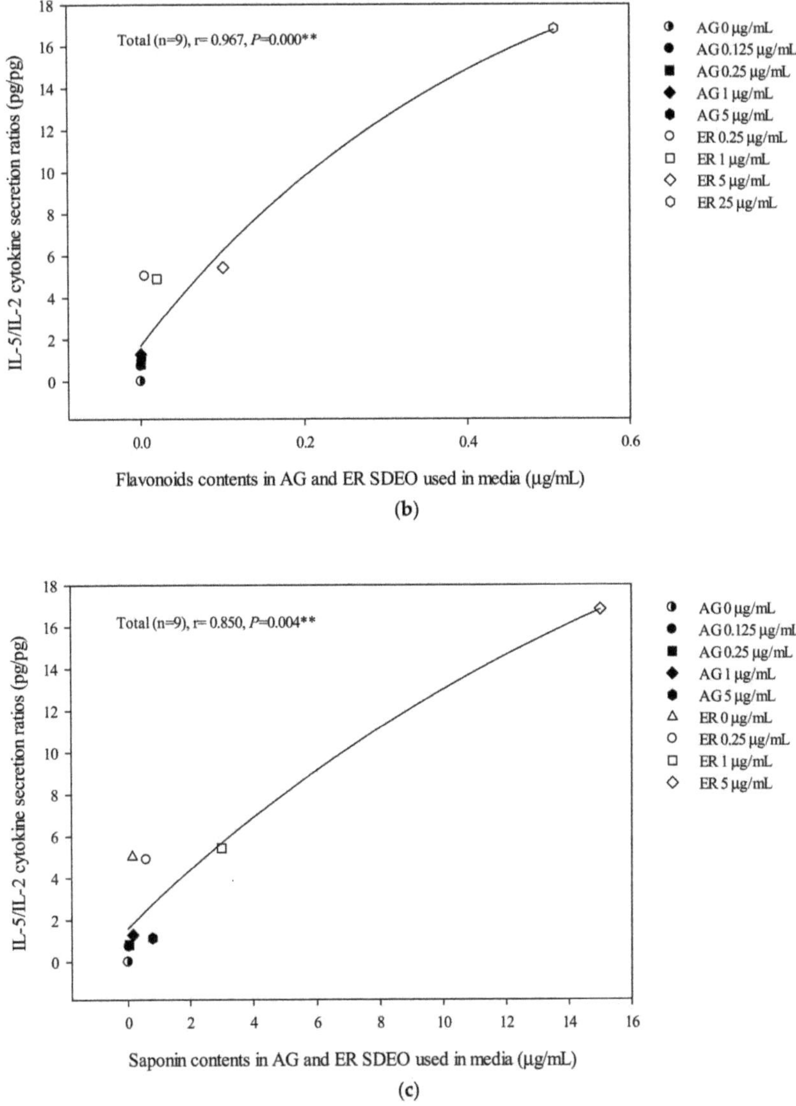

(b)

(c)

Figure 2. The correlation between total polyphenol (**a**), flavonoid (**b**) and saponin (**c**) contents in AG and ER SDEO used in the media and IL-5/IL-2 cytokine secretion ratios using mouse splenocytes. The correlation was expressed using the Spearman correlation coefficient. The correlation is considered statistically different if $p < 0.05$. **, $p < 0.01$.

3.6. The Correlation between pro-/anti-Inflammatory Cytokine Secretion Levels in Mouse Primary Peritoneal Macrophage Cultures in the Absence or Presence of LPS and Total Flavonoid, Phenol or Saponin Contents in AG and ER SDEO

The correlations between cytokine secretion levels and potent phytochemicals, including total flavonoid, phenol or saponin contents in AG and ER SDEO, were determined using the Spearman correlation coefficient. We found that there are significant ($p < 0.05$) positive correlations between anti-inflammatory IL-10 cytokine secretions by mouse peritoneal macrophages in the absence of

LPS and total polyphenol, flavonoid or saponin contents in a linear manner (Figure 3). Our results further suggest that potent anti-inflammatory phytochemicals including polyphenols (Figure 3A), flavonoids (Figure 3B) and saponins (Figure 3C) may exert their anti-inflammatory ability via increasing spontaneously anti-inflammatory cytokine secretions in a linear manner. In addition, there are significantly ($p < 0.05$) but non-linearly negative correlations between TNF-α/IL-10 secretions by peritoneal macrophages in the absence of LPS and total polyphenol, flavonoid or saponin contents (Figure 4). Our results suggest that potent anti-inflammatory phytochemicals including polyphenols (Figure 4A), flavonoids (Figure 4B) and saponins (Figure 4C) may exert their anti-inflammatory ability via decreasing spontaneous pro-/anti-inflammatory cytokine secretion ratios by macrophages in a non-linearly pharmacological manner. In the presence of LPS, there was a significant negative correlation between TNF-α/IL-10 cytokine secretion ratios in the LPS-stimulated peritoneal macrophage cultures and total flavonoid, polyphenol and saponin contents in the two selected SDEOs in a biphasic manner (Figure 5). Interestingly, higher doses of active anti-inflammatory ingredients might enhance LPS-induced inflammation status. Our results suggest that lower-dose daily supplements of active anti-inflammatory ingredients such as flavonoids, polyphenols and saponins, etc. may provide anti-inflammatory protections but avoid their adverse side effects.

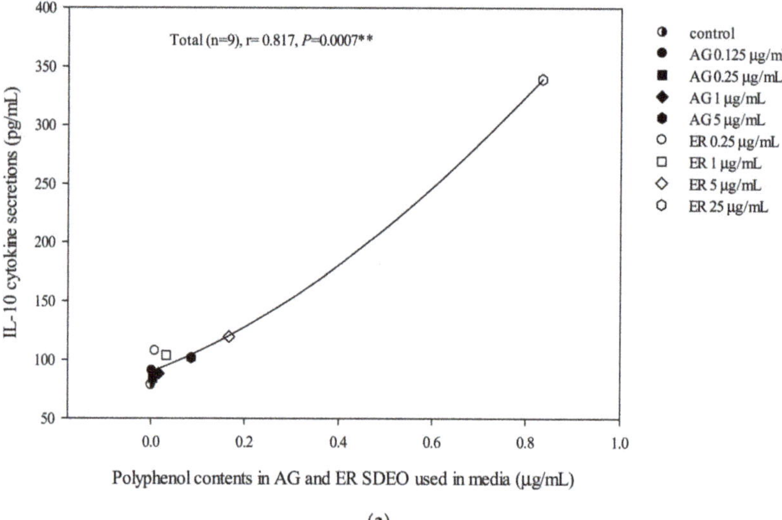

(a)

Figure 3. *Cont.*

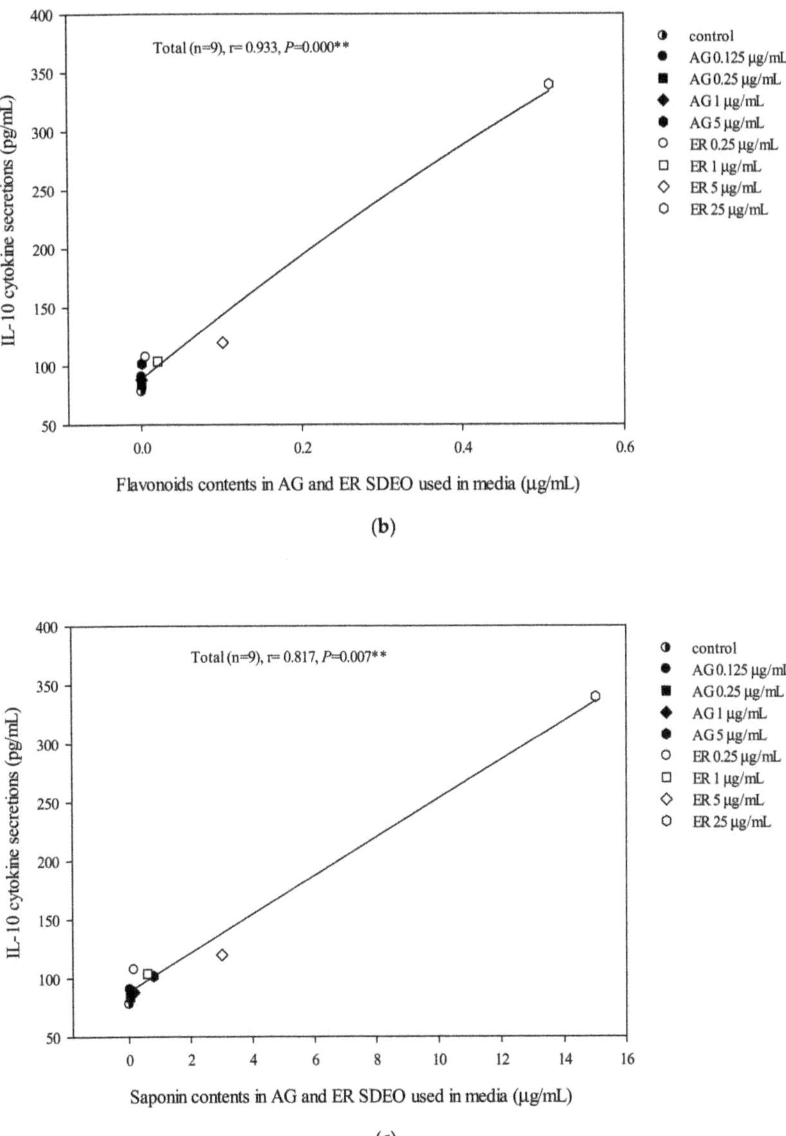

(b)

(c)

Figure 3. The correlation between total polyphenol (**a**), flavonoid (**b**) and saponin (**c**) contents in AG and ER SDEO used in the media and IL-10 cytokine secretion using mouse macrophages in the absence of LPS. The correlation was expressed using the Spearman correlation coefficient. The correlation is considered statistically different if $p < 0.05$. **, $p < 0.01$.

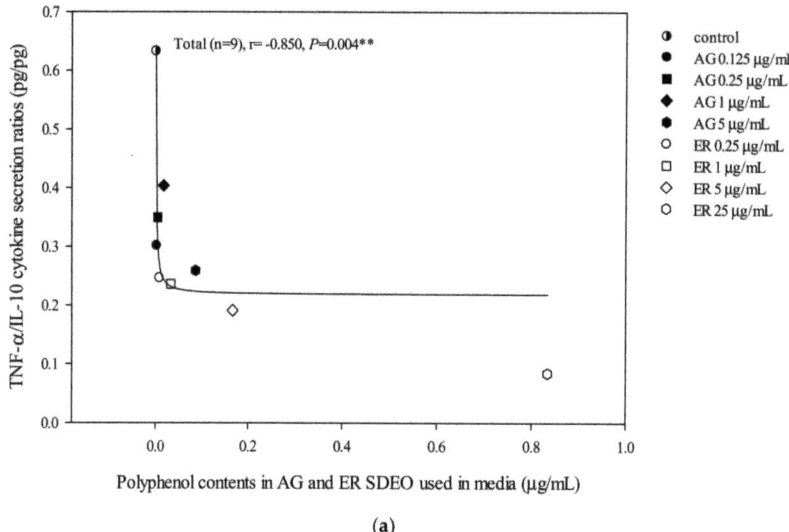

(**a**)

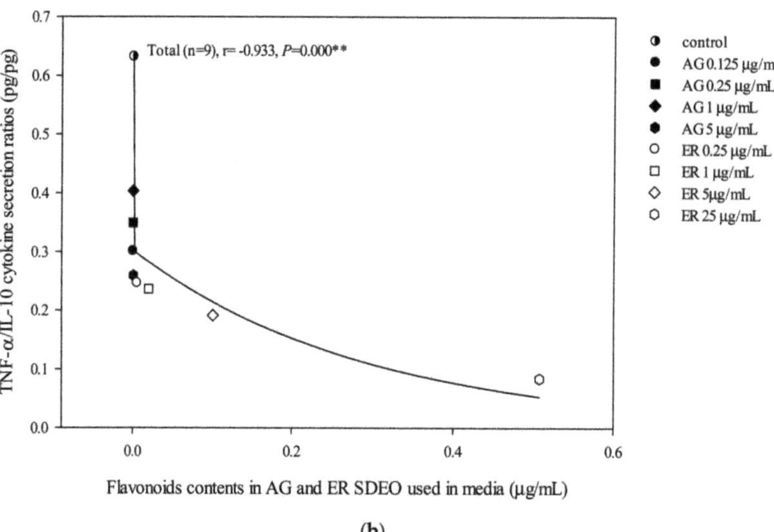

(**b**)

Figure 4. *Cont.*

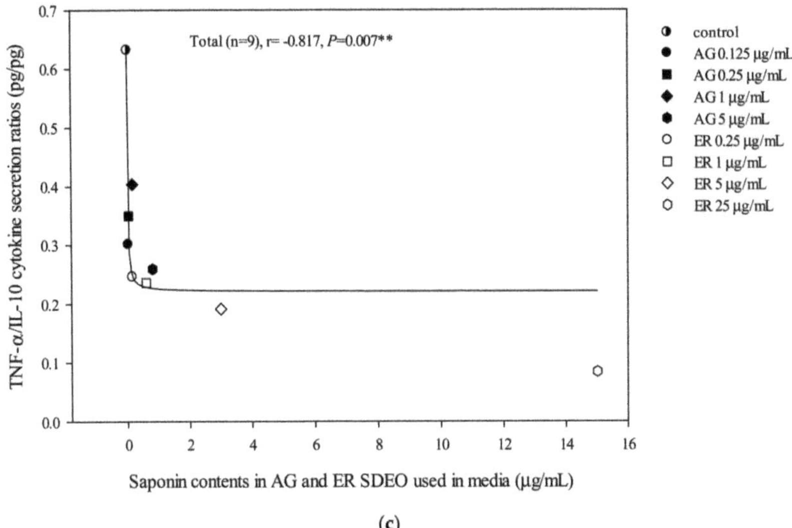

(**c**)

Figure 4. The correlation between total polyphenol (**a**), flavonoid (**b**) and saponin (**c**) contents in AG and ER SDEO used in the media and TNF-α/IL-10 cytokine secretion ratios using mouse macrophages in the absence of LPS. The correlation was expressed using the Spearman correlation coefficient. The correlation is considered statistically different if $p < 0.05$. **, $p < 0.01$.

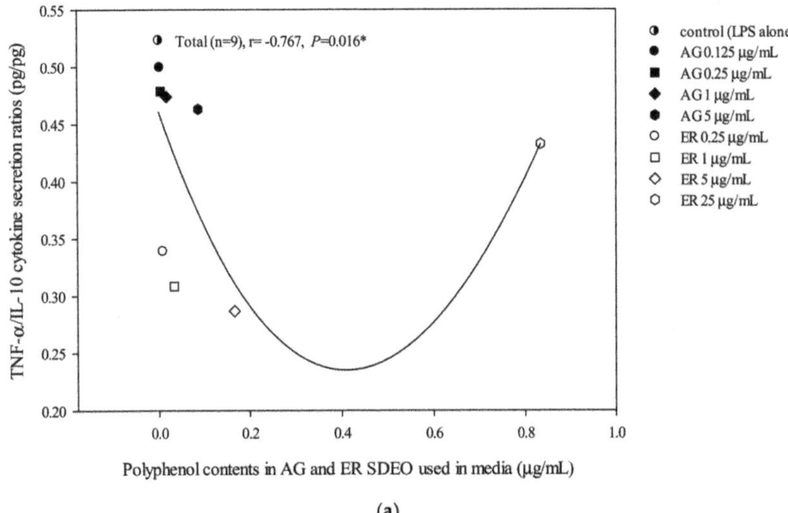

(**a**)

Figure 5. *Cont.*

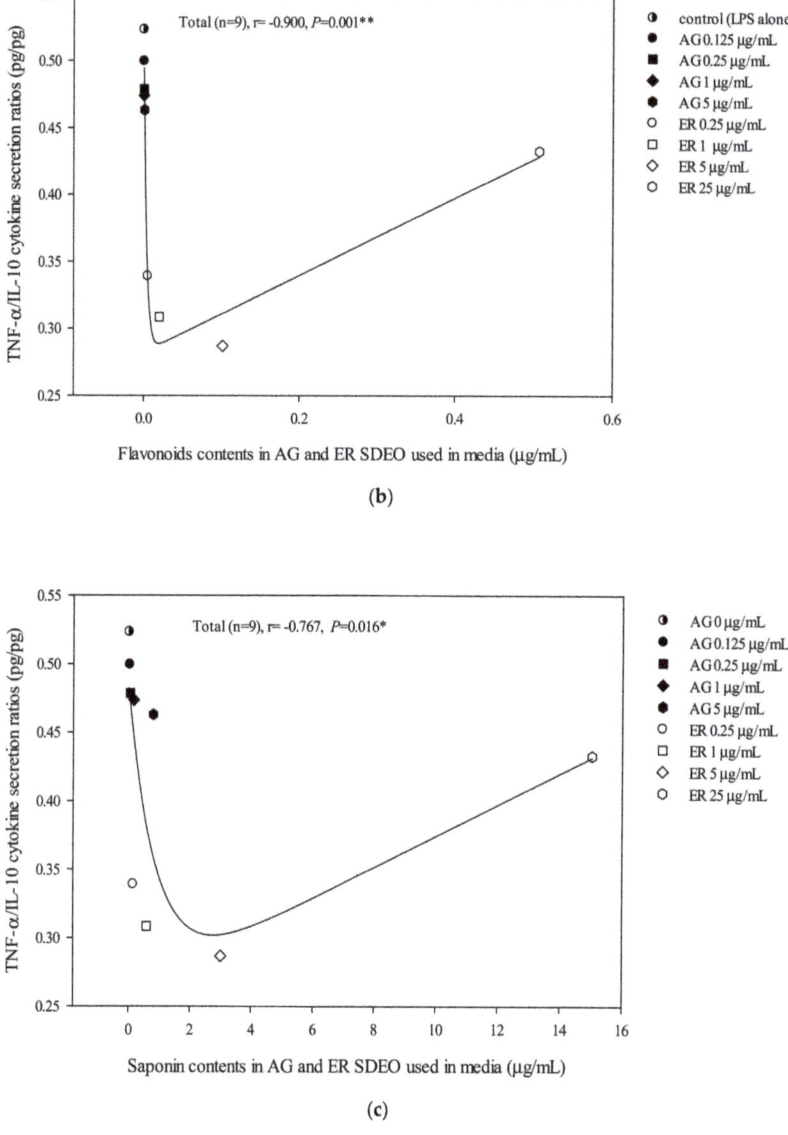

(b)

(c)

Figure 5. The correlation between total polyphenol (**a**), flavonoid (**b**) and saponin (**c**) contents in AG and ER SDEO used in the media and TNF-α/IL-10 cytokine secretion ratios using mouse macrophages in the presence of LPS. The correlation was expressed using the Spearman correlation coefficient. The correlation is considered statistically different if $p < 0.05$. **, $p < 0.01$.

In the present study, we reported on the immunomodulatory effects of AG and ER SDEO on Th1/Th2 and pro-/anti-inflammatory immune responses using mouse splenocytes and peritoneal macrophages, respectively. We evidenced that these two selected SDEOs demonstrate potent anti-inflammatory potential through their Th2-inclination property to splenocytes (Table 3) and decreasing pro-/anti-inflammatory cytokine secretion ratios by macrophages either in the absence (Table 4) or presence (Table 5) of LPS. Undoubtedly, essential oils have the potential to regulate cytokine secretions by immune cells. We hypothesized that SDEO is safer than traditional essential oils because of water steam distillation in place of traditional toxic solvents for essential oil extraction. Therefore, functional SDEOs including AG and ER SDEO may be used for the development of functional foods and pharmaceuticals including aromatherapy. The highest dose of ER SDEO used in this study was 25 μg/mL (versus 5×10^6 splenocytes/mL or 1×10^6 macrophages/mL) (Tables 3 and 5). It is still a low concentration, nevertheless, the dose is an effective concentration with little cytotoxicity. We supposed that there are 20 folds of the same tested cells in a 20 g mouse. Therefore, an aliquot of 0.5 mg (25 μg × 20 = 500 μg) is suggested to furnish a 20 g mouse daily. Moreover, dose conversion between animals and humans is based on body surface area. There is a conversion factor of 387.9 between a 70 kg person and a 20 g mouse based on body surface area ratio for pharmacological application. Thus, daily supplementation with 193.95 mg ER SDEO (0.5 mg × 387.9 = 193.95 mg) or 38.79 mg AG SDEO for a 70 kg person may achieve a therapeutic effect. The recommendatory doses may be useful for practical applicability. However, our suggestion still can't entirely and appropriately calculate the effective dose necessary for mouse/human body. Our assumption is just based on a simple multiplication of the cell numbers used in our experiment in proportion to the expected mass of mouse/human body. This assumption could not respect any continuous metabolic elimination by organs such as the liver or kidneys, as well as distribution volume for these active chemicals in the living body. Particularly, these chemicals could affect not only immune cells but also any other cell/tissue of the host organism. It is so complicated that it is difficult to propose a simple and perfect model to predict the effective dose in vivo. Nevertheless, we postulate a hypothesis which can be easily proved/disproved with the known chemical composition of these distilled essential oils in the future.

AG and ER SDEO regulated the secretion of cytokines by immune cells, but it was still unclear which active components in SDEO are responsible. To clarify this puzzle, potential phytochemicals, including total polyphenols, flavonoids and saponins, in the two selected SDEOs were measured and correlated with cytokine secretions by correspondent immune cells. Total polyphenols, flavonoids or saponins contents in the two selected SDEOs are significantly ($p < 0.05$) correlated with Th1/Th2 immune balance and anti-inflammatory ability in linear, non-linear or biphasic manners, respectively (Figures 2–5). We presumed that other phytochemicals such as plant alkaloids in addition to polyphenols, flavonoids or saponins may exist in these two selected SDEOs. Individual active phytochemicals in selected SDEOs may mutually interact with each other with synergism or antagonism effects, resulting in the linear, non-linear or biphasic manners. It has been reported that plant essential oils with biologically active functions either extracted with water distillation or organic solvents are usually attributed to the ability of low molecular weight molecules penetrating into cellular membranes (including mitochondrial membranes), to scavenge free radicals and inhibit the expression of pro-inflammatory *IL-1β* and *TNF-α* cytokine genes [34]. Essential oils (EOs) can be complex mixtures of low molecular weight molecules (less than 500 Da) with very variable concentrations extracted from plants by steam distillation, hydro-distillation or other various solvents [15,34]. These low molecular weight compounds are not practically soluble in water [35]. Monoterpenes, sesquiterpenes, oxygenated monoterpenes, oxygenated sesquiterpenes, phenolics and others [17,36] are the major constituents that provide the characteristic aroma and biological properties to EOs. To avoid the toxic effects of residual organic solvents, relatively less toxic SDEOs such as AG and ER SDEO (Tables 3–5) are a new and potential therapy to regulate Th1/Th2 immune balance and inhibit an inflammatory response without compromising an immune defense. A study on active components in AG and ER SDEOs is being performed to clarify their immunomodulatory mechanisms.

There is growing evidence that chronic inflammatory responses are important for cancer development involved in the cell differentiation of monocytes/macrophages [37]. The pro-inflammatory cytokine TNF-α stimulates macrophages to release nitric oxide (NO) when they make pinocytosis [38]. However, excess TNF-α can induce macrophages to secrete amounts of NO, causing damage to tissue cells and eventually leading to apoptosis [38,39]. In the present study, ER SDEO demonstrated potent anti-inflammatory capacity through decreasing TNF-α/IL10 cytokine secretion ratios by macrophages in the absence or presence of LPS (Tables 4 and 5). ER SDEO may be used for cancer immunotherapy via its potent anti-inflammatory potential. In comparison with phytochemical composition, Th1/Th2 immune balance property and anti-inflammatory potential, ER SDEO has better potential than AR SDEO in developing future functional foods and pharmaceuticals.

4. Conclusions

In the present study, our results evidenced ER SDEO was rich in phytochemicals including total flavonoids, polyphenols and saponins. AG and ER, particularly ER, SDEO have immunomodulatory potential in shifting the Th1/Th2 balance toward Th2 polarization in splenocytes and inhibiting inflammation in macrophages in the absence or presence of LPS. There are significant correlations between the total polyphenols, flavonoids or saponins content in AG and ER SDEO and Th1/Th2 immune balance or anti-inflammatory ability in linear, non-linear or biphasic manners.

Author Contributions: J.-Y.L.: Conceptualization, Methodology, Validation, Writing-Reviewing and Editing, Funding acquisition. T.-H.Y.: Data curation, Formal analysis, Investigation, Writing- Original draft preparation. All authors have read and agreed to the published version of the manuscript.

Funding: This study was kindly supported by research grants MOST 104-2320-B-005-006-MY3 and MOST 107-2320-B-005-010-MY3 from the Ministry of Science and Technology, Taipei, Taiwan.

Acknowledgments: We thank Li-Yun Lin, Department of Food Science and Technology, HungKuang University, Taichung, Taiwan, for kindly providing GC/MSD instrument for essential oils analysis.

Conflicts of Interest: The authors declare no conflict of interest.

References

1. Vasto, S.; Candore, G.; Balistreri, C.R.; Caruso, M.; Colonna-Romano, G.; Grimaldi, M.P.; Listi, F.; Nuzzo, D.; Lio, D.; Caruso, C. Inflammatory networks in ageing, age-related diseases and longevity. *Mech. Ageing Dev.* **2007**, *128*, 83–91. [CrossRef] [PubMed]

2. Adeyemi, O.O.; Yemitan, O.K.; Afolabi, L. Inhibition of chemically induced inflammation and pain by orally and topically administered leaf extract of *Manihot esculenta* Crantz in rodents. *J. Ethnopharmacol.* **2008**, *119*, 6–11. [CrossRef] [PubMed]

3. Liao, Y.R.; Lin, J.Y. Quercetin intraperitoneal administration ameliorates lipopolysaccharide-induced systemic inflammation in mice. *Life Sci.* **2015**, *137*, 89–97. [CrossRef] [PubMed]

4. Senedese, J.M.; Rinaldi-Neto, F.; Furtado, R.A.; Nicollela, H.D.; de Souza, L.D.R.; Ribeiro, A.B.; Ferreira, L.S.; Magalhães, G.M.; Carlos, I.Z.; da Silva, J.J.M.; et al. Chemopreventive role of *Copaifera reticulata* Ducke oleoresin in colon carcinogenesis. *Biomed. Pharmacother.* **2019**, *111*, 331–337. [CrossRef]

5. Sethi, G.; Shanmugam, M.K.; Ramachandran, L.; Kumar, A.P.; Tergaonkar, V. Multifaceted link between cancer and inflammation. *Biosci. Rep.* **2012**, *32*, 1–15. [CrossRef]

6. Yu, T.; Moh, S.H.; Kim, S.B.; Yang, Y.; Kim, E.; Lee, Y.W.; Cho, C.K.; Kim, K.H.; Yoo, B.C.; Cho, J.Y.; et al. HangAmDan-B, an ethnomedicinal herbal mixture, suppresses inflammatory responses by inhibiting Syk/NF-kappaB and JNK/ATF-2 pathways. *J. Med. Food* **2013**, *16*, 56–65. [CrossRef]

7. Steward, W.P.; Brown, K. Cancer chemoprevention: A rapidly evolving field. *Br. J. Cancer* **2013**, *109*, 1–7. [CrossRef]

8. Dickinson, S.E.; Olson, E.R.; Levenson, C.; Janda, J.; Rusche, J.J.; Alberts, D.S.; Bowden, G.T. A novel chemopreventive mechanism for a traditional medicine: East Indian sandalwood oil induces autophagy and cell death in proliferating keratinocytes. *Arch. Biochem. Biophys.* **2014**, *558*, 143–152. [CrossRef]

9. Mokhtar, M.M.; Shaban, H.M.; Hegazy, M.E.A.F.; Ali, S.S. Evaluating the potential cancer chemopreventive efficacy of two different solvent extracts of *Seriphidium herba-alba* in vitro. *Bull. Fac. Pharm. Cairo Univ.* **2017**, *55*, 195–201. [CrossRef]

10. Kidd, P. Th1/Th2 balance: The hypothesis, its limitations, and implications for health and disease. *Altern. Med. Rev.* **2003**, *8*, 223–246.

11. Gagliani, N.; Huber, S. Basic aspects of T Helper cell differentiation. *Methods Mol. Biol.* **2017**, *1514*, 19–30. [PubMed]

12. Kutukculer, N.; Azarsiz, E.; Aksu, G.; Karaca, N.E. CD4$^+$CD25$^+$Foxp3$^+$ T regulatory cells, Th1 (CCR5, IL-2, IFN-gamma) and Th2 (CCR4, IL-4, IL-13) type chemokine receptors and intracellular cytokines in children with common variable immunodeficiency. *Int. J. Immunopathol. Pharmacol.* **2016**, *29*, 241–251. [CrossRef] [PubMed]

13. Matsuzaki, J.; Tsuji, T.; Imazeki, I.; Ikeda, H.; Nishimura, T. Immunosteroid as a regulator for Th1/Th2 balance: Its possible role in autoimmune diseases. *Autoimmunity* **2005**, *38*, 369–375. [CrossRef] [PubMed]

14. Li, Z.; Peng, A.; Feng, Y.; Zhang, X.; Liu, F.; Chen, C.; Ye, X.; Qu, J.; Jin, C.; Wang, M.; et al. Detection of T lymphocyte subsets and related functional molecules in follicular fluid of patients with polycystic ovary syndrome. *Sci. Rep.* **2019**, *9*, 6040. [CrossRef] [PubMed]

15. Raut, J.S.; Karuppayil, S.M. A status review on the medicinal properties of essential oils. *Ind. Crop. Prod.* **2014**, *62*, 250–264. [CrossRef]

16. Edris, A.E. Pharmaceutical and therapeutic potentials of essential oils and their individual volatile constituents: A review. *Phytother. Res.* **2007**, *21*, 308–323. [CrossRef]

17. Slamenova, D.; Horvathova, E. Cytotoxic, anti-carcinogenic and antioxidant properties of the most frequent plant volatiles. *Neoplasma* **2013**, *60*, 343–354. [CrossRef]

18. Liu, Z.B.; Niu, W.M.; Yang, X.H.; Wang, Y.; Wang, W.G. Study on perfume stimulating olfaction with volatile oil of *Acorus gramineus* for treatment of the Alzheimer's disease rat. *J. Tradit Chin. Med.* **2010**, *30*, 283–287. [CrossRef]

19. Olas, B.; Brys, M. Is it safe to use *Acorus calamus* as a source of promising bioactive compounds in prevention and treatment of cardiovascular diseases? *Chem. Biol. Interact.* **2018**, *281*, 32–36. [CrossRef]

20. Rajput, S.B.; Tonge, M.B.; Karuppayil, S.M. An overview on traditional uses and pharmacological profile of *Acorus calamus* Linn. (Sweet flag) and other Acorus species. *Phytomedicine* **2014**, *21*, 268–276. [CrossRef]

21. Chellian, R.; Pandy, V.; Mohamed, Z. Pharmacology and toxicology of α-and β-asarone: A review of preclinical evidence. *Phytomedicine* **2017**, *32*, 41–58. [CrossRef] [PubMed]

22. Stegmüller, S.; Schrenk, D.; Cartus, A.T. Formation and fate of DNA adducts of alpha-and beta-asarone in rat hepatocytes. *Food Chem. Toxicol.* **2018**, *116*, 138–146. [CrossRef] [PubMed]

23. Xiao, X.; Xu, X.; Li, F.; Xie, G.; Zhang, T. Anti-inflammatory treatment with beta-asarone improves impairments in social interaction and cognition in MK-801 treated mice. *Brain Res. Bull.* **2019**, *150*, 150–159. [CrossRef] [PubMed]

24. Zhao, Z.; He, X.; Han, W.; Chen, X.; Liu, P.; Zhao, X.; Wang, X.; Zhang, L.; Wu, S.; Zheng, X. Genus *Tetradium* L.: A comprehensive review on traditional uses, phytochemistry, and pharmacological activities. *J. Ethnopharmacol.* **2019**, *231*, 337–354. [CrossRef] [PubMed]

25. Pellati, F.; Benvenuti, S.; Yoshizaki, F.; Bertelli, D.; Rossi, M.C. Headspace solid-phase microextraction-gas chromatography–mass spectrometry analysis of the volatile compounds of Evodia species fruits. *J. Chromatogr. A* **2005**, *1087*, 265–273. [CrossRef] [PubMed]

26. Tundis, R.; Loizzo, M.R.; Menichini, F. An overview on chemical aspects and potential health benefits of limonoids and their derivatives. *Crit. Rev. Food Sci. Nutr.* **2014**, *54*, 225–250. [CrossRef]

27. Wang, J.; Zhang, T.; Zhu, L.; Ma, C.; Wang, S. Anti-ulcerogenic effect of Zuojin Pill against ethanol-induced acute gastric lesion in animal models. *J. Ethnopharmacol.* **2015**, *173*, 459–467. [CrossRef]

28. Shen, Y.; Jin, L.; Xiao, P.; Lu, Y.; Bao, J. Total phenolics, flavonoids, antioxidant capacity in rice grain and their relations to grain color, size and weight. *J. Cereal Sci.* **2009**, *49*, 106–111. [CrossRef]

29. Yu, H.; Zheng, L.; Yin, L.; Xu, L.; Qi, Y.; Han, X.; Xu, Y.; Liu, K.; Peng, J. Protective effects of the total saponins from *Dioscorea nipponica* Makino against carbon tetrachloride-induced liver injury in mice through suppression of apoptosis and inflammation. *Int. Immunopharmacol.* **2014**, *19*, 233–244. [CrossRef]

30. Lin, J.Y.; Tang, C.Y. Strawberry, loquat, mulberry, and bitter melon juices exhibit prophylactic effects on LPS-induced inflammation using murine peritoneal macrophages. *Food Chem.* **2008**, *107*, 1587–1596. [CrossRef]

31. Lin, W.C.; Lin, J.Y. Five bitter compounds display different anti-inflammatory effects through modulating cytokine secretion using mouse primary splenocytes in vitro. *J. Agric. Food Chem.* **2011**, *59*, 184–192. [CrossRef] [PubMed]

32. Lin, B.F.; Chiang, B.L.; Lin, J.Y. *Amaranthus spinosus* water extract directly stimulates proliferation of B lymphocytes in vitro. *Int. Immunopharmacol.* **2005**, *5*, 711–722. [CrossRef] [PubMed]

33. Du, Z.; Clery, R.A.; Hammond, C.J. Volatiles from leaves and rhizomes of fragrant *Acorus* spp. (Acoraceae). *Chem. Biodivers.* **2008**, *5*, 887–895. [CrossRef] [PubMed]

34. Sharifi-Rad, J.; Sureda, A.; Tenore, G.C.; Daglia, M.; Sharifi-Rad, M.; Valussi, M.; Tundis, R.; Sharifi-Rad, M.; Loizzo, M.R.; Ademiluyi, A.O.; et al. Biological activities of essential oils: From plant chemoecology to traditional healing systems. *Molecules* **2017**, *22*, 70. [CrossRef]

35. Dima, C.; Dima, S. Essential oils in foods: Extraction, stabilization, and toxicity. *Curr. Opin. Food Sci.* **2015**, *5*, 29–35. [CrossRef]

36. Bhalla, Y.; Gupt, V.K.; Jaitak, V. Anticancer activity of essential oils: A review. *J. Sci. Food Agric.* **2013**, *93*, 3643–3653. [CrossRef]

37. Dong, R.; Gong, Y.; Meng, W.; Yuan, M.; Zhu, H.; Ying, M.; He, Q.; Cao, J.; Yang, B. The involvement of M2 macrophage polarization inhibition in fenretinide-mediated chemopreventive effects on colon cancer. *Cancer Lett.* **2017**, *388*, 43–53. [CrossRef]

38. Sun, K.; Song, X.; Jia, R.; Yin, Z.; Zou, Y.; Li, L.; Yin, L.; He, C.; Liang, X.; Yue, G.; et al. Evaluation of analgesic and anti-Inflammatory activities of water extract of *Galla Chinensis* in vivo models. *Evid. Based Complement. Alternat. Med.* **2018**, 6784032. [CrossRef]

39. Yang, C.; You, L.; Yin, X.; Liu, Y.; Leng, X.; Wang, W.; Sai, N.; Ni, J. Heterophyllin B ameliorates lipopolysaccharide-induced inflammation and oxidative stress in RAW 264.7 macrophages by suppressing the PI3K/Akt pathways. *Molecules* **2018**, *23*, 717. [CrossRef]

 *biomolecules*

Article

Licochalcone D Induces ROS-Dependent Apoptosis in Gefitinib-Sensitive or Resistant Lung Cancer Cells by Targeting EGFR and MET

Ha-Na Oh [1,†], Mee-Hyun Lee [2,3,†], Eunae Kim [4], Ah-Won Kwak [1], Goo Yoon [1], Seung-Sik Cho [1], Kangdong Liu [2,3], Jung-Il Chae [5,*] and Jung-Hyun Shim [1,2,*]

1   Department of Pharmacy, Mokpo National University, Jeonnam 58554, Korea; 17392303@mokpo.ac.kr (H.-N.O.); rhkrdkdnjs12@mokpo.ac.kr (A.-W.K.); gyoon@mokpo.ac.kr (G.Y.); sscho@mokpo.ac.kr (S.-S.C.)
2   The China-US (Henan) Hormel Cancer Institute, Zhengzhou 450008, Henan, China; mhlee@hci-cn.org (M.-H.L.); kangdongliu@126.com (K.L.)
3   Basic Medical College, Zhengzhou University, Zhengzhou 450001, Henan, China
4   College of Pharmacy, Chosun University, Gwangju 61452, Korea; eunaekim@chosun.ac.kr
5   Department of Dental Pharmacology, School of Dentistry, BK21 Plus, Jeonbuk National University, Jeonju 54896, Korea
*   Correspondence: jichae@jbnu.ac.kr (J.-I.C.); s1004jh@gmail.com or s1004jh@mokpo.ac.kr (J.-H.S.); Tel.: +82-63-270-4024 (J.-I.C.); +82-61-450-2684 (J.-H.S.); Fax: +82-63-270-4037 (J.-I.C.); +82-61-450-2689 (J.-H.S.)
†   These authors contributed equally to this work.

Received: 3 January 2020; Accepted: 10 February 2020; Published: 13 February 2020

Abstract: Licochalcone D (LCD), a flavonoid isolated from a Chinese medicinal plant *Glycyrrhiza inflata*, has a variety of pharmacological activities. However, the anti-cancer effects of LCD on non-small cell lung cancer (NSCLC) have not been investigated yet. The amplification of *MET* (hepatocyte growth factor receptor) compensates for the inhibition of epidermal growth factor receptor (EGFR) activity due to tyrosine kinase inhibitor (TKI), leading to TKI resistance. Therefore, EGFR and MET can be attractive targets for lung cancer. We investigated the anti-proliferative and apoptotic effects of LCD in lung cancer cells HCC827 (gefitinib-sensitive) and HCC827GR (gefitinib-resistant) through 3-(4,5-dimethylthiazol-2-yl)-2,5-diphenyltetrazolium bromide (MTT) assay, pull-down/kinase assay, cell cycle analysis, Annexin-V/7-ADD staining, reactive oxygen species (ROS) assay, mitochondrial membrane potential (MMP) assay, multi-caspase assay, and Western blot analysis. The results showed that LCD inhibited phosphorylation and the kinase activity of EGFR and MET. In addition, the predicted pose of LCD was competitively located at the ATP binding site. LCD suppressed lung cancer cells growth by blocking cell cycle progression at the G2/M transition and inducing apoptosis. LCD also induced caspases activation and poly (ADP-ribose) polymerase (PARP) cleavage, thus displaying features of apoptotic signals. These results provide evidence that LCD has anti-tumor effects by inhibiting EGFR and MET activities and inducing ROS-dependent apoptosis in NSCLC, suggesting that LCD has the potential to treat lung cancer.

Keywords: licochalcone D; non-small cell lung cancer; reactive oxygen species; apoptosis

---

1. Introduction

Flavonoids are the most effective and variable biologically active compounds in plants. Licochalcone D (LCD) is an active flavonoid isolated from the Chinese medicinal herb *Glycyrrhiza inflata* [1]. LCD is present in the roots and rhizomes of *G. inflata*. The chemical name of LCD is

(E)-3-(3,4-dihydroxy-2-methoxyphenyl)-1-[4-hydroxy-3-(3-methylbut-2-enyl)phenyl]prop-2-en-1-one. Various pharmacological actions, including anti-oxidant, anti-biotic, anti-ulcer, and anti-carcinogenic effects, have been described for LCD [2]. Previous studies have demonstrated that LCD can induce cell apoptosis and suppress migration and invasion in skin cancer [3]. In addition, LCD can suppress the proliferation in oral cancer cells [4]. However, the mechanism by which LCD exerts its effects on lung cancer has not been fully determined yet.

Non-small cell lung cancer (NSCLC) is the major cause of death from cancer in the world. NSCLC accounts for approximately 85% of lung cancer cases [5]. Chemotherapy remains marginally effective for NSCLC. Chemotherapy can slightly prolong the survival of patients with advanced lung cancer. However, it has clinically significant adverse effects [6]. The current treatment approach includes surgical resection, radiation therapy, and chemotherapy alone or in combination [6]. Despite these therapies, lung cancer is rarely curable, with an overall 5-year survival rate of only 15% [7]. As a result of the low cure rate of NSCLC, it is important to find effective treatment, with a focus on new molecular and targeted therapies.

Epidermal growth factor receptor (EGFR) mutations lead to the outstanding activation of EGFR signaling and carcinogenicity transformation both in vitro and in vivo [8]. Cancers with EGFR mutations (EGFR-mutated cancers) rely on EGFR signaling for growth. They are often sensitive to medical treatment with EGFR tyrosine kinase inhibitors (TKIs) [8,9]. Most patients with lung cancer have tumor-activating EGFR mutations. Treatment with EGFR-TKIs causes tumor reduction; however, the progression of cancer occurs at 6 to 12 months after treatment [10]. Various mechanisms of resistance to EGFR-TKIs (such as erlotinib and gefitinib) have been identified. The comprehension of these mechanisms is critical to evolving treatment strategies in the setting of resistance development. One of the resistance mechanisms is EGFR T790M point mutation within exon 20 [9]. hepatocyte growth factor receptor (MET) and human epidermal growth factor receptor (HER)2 overexpression with an upregulation of parallel signaling pathways have also been reported [9]. A gefitinib-resistant HCC827GR (*MET*-amplified) cell has been generated by exposing these cells to gefitinib for six months [11]. Their results showed that lung cancer cell growth was inhibited by simultaneous treatment with gefitinib and MET inhibitor. Thus, the dual inhibition of EGFR and MET might be a means to overcome lung cancer resistance.

The objective of this study was to investigate whether LCD could inhibit cell proliferation through EGFR and MET dual targets in NSCLC using human gefitinib-sensitive or resistant NSCLC cells. We found that LCD could induce the apoptosis of HCC827 and HCC827GR cells by inhibiting both EGFR and the MET signaling pathway. To confirm whether LCD induced apoptosis, we carried out cell proliferation, cell cycle distribution, reactive oxygen species (ROS) production, mitochondrial membrane potential (MMP) depolarization, and multi-caspase activation assays. The results of this study might shed light on the mechanism involved in the effect of LCD on lung cancer. We expect that LCD for cancer treatment might give improved results.

2. Materials and Methods

2.1. Reagents and Antibodies

LCD was prepared by Professor Goo Yoon according to previous reports [1]. Roswell Park Memorial Institute (RPMI)-1640 medium, phosphate-buffered saline (PBS), fetal bovine serum (FBS), penicillin/streptomycin, and trypsin were purchased from Hyclone (Logan, UT, USA). Dimethyl-sulfoxide (DMSO), 3-(4,5-dimethylthiazol-2-yl)-2,5-diphenyltetrazolium bromide (MTT), and Basal Medium Eagle (BME) were purchased from Sigma Chemical Company (St. Louis, MO, USA). Gefitinib was purchased from Cayman Chemical (Ann Abor, MI, USA). Savolitinib was obtained from Selleckchem (Houston, TX, USA). Primary antibodies against cyclin B1, cdc2, p21, p27, β-actin, Bid, Bcl-xl, Mcl-1, Bad, cytochrome c (cyto c), α-tubulin, COX4, apoptotic protease activating factor-1 (Apaf-1), cleaved poly (ADP-Ribose) polymerase (C-PARP), and ERBB3 (HER3) were obtained from

Santa Cruz Biotechnology (Santa Cruz, CA, USA). Antibodies against phosphorylated (p)-EGFR (Tyr1068), EGFR, p-MET (Tyr1234/1235), MET, p-ERBB3 (Tyr1289), p-AKT (Ser473), and AKT were purchased from Cell Signaling Biotechnology (Beverly, MA, USA).

2.2. Cell Culture

The EGFR mutant (del E746_A750) NSCLC cell line HCC827 was obtained from the American Type Culture Collection (ATCC, Manassas, VA, USA). *MET*-amplified HCC827GR (gefitinib-resistant HCC827) cells were kindly contributed by professor Pasi A. Jänne (Department of Medical Oncology, Dana-Farber Cancer Institute, Boston, MA, USA). HCC827 and HCC827GR cells were cultured in RPMI-1640 medium supplemented with 10% FBS and 100 U/ml penicillin/streptomycin at 37 °C in a humidified atmosphere of 5% CO_2.

2.3. Pull-Down Assay

To confirm the interaction between LCD and EGFR or MET, HCC827 and HCC827GR cell lysates were mixed with Sepharose 4B or LCD-Sepharose 4B beads. The protein extract was incubated with LCD-Sepharose 4B beads or Sepharose 4B beads in reaction buffer [50 mM Tris (pH 7.5), 5 mM EDTA, 150 mM NaCl, 1 mM/L dithiothreitol, 0.01% Nonidet P-40, 2 µg/mL bovine serum albumin, 0.02 mM phenylmethylsulfonyl fluoride, and 1X proteinase inhibitor] at 4 °C for 12 h. The mixture containing beads was washed six times with a washing buffer [50 mM Tris (pH 7.5), 5 mM EDTA, 150 mM NaCl, 1 mM dithiothreitol, 0.01% Nonidet P-40, and 0.02 mM phenylmethylsulfonyl fluoride]. Then, bound proteins were eluted with sodium dodecyl sulfate–polyacrylamide gel electrophoresis (SDS-PAGE) sample buffer and subjected to Western blot analysis.

2.4. Western Blotting

Whole cells were lysed with Radio-Immunoprecipitation Assay (RIPA) buffer (iNtRON Biotechnology, Korea). Protein concentrations were determined using a Bio-Rad DC Protein Assay kit (Bio-Rad, Hercules, CA, USA). An equal amount of protein was loaded on 8%–15% SDS-PAGE gels. After separation, proteins were transferred to polyvinylidene fluoride membranes (Millipore, Bedford, MA, USA), blocked with 3% or 5% skim milk in PBS containing 0.1% Tween-20 (PBST) at room temperature (RT) for 1 h or 2 h, and incubated with each primary antibody against its specific protein at 4 °C overnight. After washing six times, membranes were incubated with secondary antibodies for 2 h. Blots were scanned with an Image Quant LAS 500 (GE Healthcare, Uppsala, Sweden) using Western blotting luminol reagent (Santa Cruz, CA, USA).

2.5. ATP-Competitive Binding Assay

To determine whether LCD might compete with ATP, 100 ng of recombinant active EGFR or MET and ATP were pre-incubated at RT for 2 h. Subsequently, Sepharose 4B beads or LCD conjugated-Sepharose 4B beads were incubated at 4°C overnight and then washed with washing buffer. Bound proteins were subjected to Western blot analysis.

2.6. Kinase Assay

To determine the kinase activity of EGFR or MET in response to ATP, kinase reactions were incubated with EGFR (1.8 ng/µL; #3831), or MET (7 ng/µL; #3361), ATP (5 µM or 10 µM), substrates (0.2 µg/µL), LCD, gefitinib (1 µM), or savolitinib (5 nM) in a kinase reaction buffer containing 40 mM Tris (pH 7.5), 20 mM $MgCl_2$, 0.1 mg/ml BSA, 50 µM dithiothreitol, 2 mM $MnCl_2$, and 100 µM sodium vanadate using kinase enzyme systems (Promega, Madison, WI, USA). Before the plate was incubated at RT (22–25 °C) for 1 h, all kinase reactions were performed in 384-well plates with a volume of 5 µL. For the purpose of depleting the remaining ATP, 5 µL of ADP-Glo reagent (ADP-Glo kinase assay kit; Promega) was added to each well at RT for 40 min. Finally, 10 µL of kinase detection solution was added into each well of the 384-well plate. Luminescence reaction was proceeded and measured with a Centro LB 960 microplate luminometer (Berthold Technologies, Germany) for 0.5 s.

2.7. Molecular Modeling and Simulation

To predict the binding pose of the receptor tyrosine kinase, we performed a molecular docking simulation and a molecular dynamic (MD)s simulation. To perform the docking simulation, three-dimensional (3D) ligand and receptor structures were needed such as input files of the docking software, Vina. Receptor structures were downloaded from the protein database bank (PDB), including EGFR kinase in complex with gefitinib (PDB entry 2ITO) and MET kinase in complex with MK-2461 (PDB entry 3Q6W). LCD such as a ligand was made with Marvin sketch software. To run efficient searching, the assignment of the binding site was important for the docking simulation. Based on the binding interactions of the complex, the important ATP binding site was predicted to be Lys745 and Asp855 in EGFR and Lys1110 and Asp1222 in MET. Motifs of tyrosine kinase commonly included a glycine-rich nucleotide phosphate-binding loop, a G-loop (EGFR: Gly721-Gly724, MET: Gly1087-Gly1090), a hinge (EGFR: Leu792-Pro794, MET: Tyr1159-Lys1161), a Asp-Phe-Gly (DFG) motif (EGFR: Asp855-Gly857, MET: Asp1222-Gly1224), and an A-loop (EGFR: Asp855-Val876, MET: Asp1222-Leu1245). These binding poses were efficiently investigated for all important motifs. According to the result of the docking simulation, the top three binding poses were chosen based on the score of binding affinity. To confirm thermal stability in aqua solvent environment, MD simulation was performed using Gromacs software, which could overcome the lack of a rigid docking simulation. Complexes were solvated by a TIP3P water model and neutralized by adding ions. The Amber14SB force field (ff14SB) and general amber force field (GAFF) were applied for the protein and the ligand, respectively. Collected structures from the docking simulation were run in a physiological condition (310 K and 1 atm). After running for 500 ps in isothermal–isobaric (NPT) condition, canonical (NVT) ensemble was equilibrated for 30 ns. The stable structure was averaged in the converged state of MD simulation.

2.8. MTT Assay

HCC827 and HCC827GR cells were seeded into 96-well plates and incubated for 24 h. Then, these cells were treated with various concentrations of LCD for 24 h or 48 h. After incubation, MTT reagent was added to each well and incubated at 37 °C for 1 h. The culture medium and MTT were removed from each well, and the formazan in each well was dissolved with 100 μL of DMSO. The absorbance was measured at 570 nm using a Multiskan GO spectrophotometer (Thermo Scientific, Vantaa, Finland).

2.9. Anchorage-Independent Cell Growth Assay

Cells were seeded into 0.3% top agar over a layer of 0.6% bottom agar in a 6-well plate at a density of 8000 cells/well. Various concentrations of LCD and DMSO were added to the top and bottom layers containing culture medium (BME, 10% FBS, 2 mM L-glutamine and 5 μg/mL gentamicin). Plates were incubated at 37 °C for two weeks. The number of colonies was counted under a light microscope (Leica Microsystems, Wetzlar, Germany).

2.10. Annexin V/7-Minoactinomycin D (7-AAD) Staining

To evaluate NSCLC cell death with LCD treatment, Annexin V/7-AAD staining was performed using a Muse™ Annexin V and Dead Cell kit (MCH100105, Merck Millipore, Billerica, MA, USA). The HCC827 (1.95×10^5) and HCC827GR (1.8×10^5) cells were seeded onto a 6-well plate and treated with DMSO or LCD at different concentrations for 48 h. Cells were collected and subjected to Annexin V/7-AAD staining using 100 μL of Muse™ Annexin V and Dead Cell reagent according to the manufacturer's protocol. Apoptotic cells were detected with a Muse™ Cell Analyzer (Merck Millipore).

2.11. Cell Cycle Analysis

A Muse™ Cell Cycle kit (MCH100106, Merck Millipore) was used to perform cell cycle analysis. Briefly, HCC827 and HCC827GR cells were collected by centrifugation at 4000 rpm for 5 min at 4 °C, washed three times with 1X PBS, and fixed with 70% cold ethanol at −20 °C for 24 h. These cells were collected by centrifugation at 4000 rpm for 10 min at 4 °C and washed once with PBS. Subsequently, Muse™ Cell Cycle Reagent was added to cell pellet followed by incubation at RT for 30 min in the dark. A Muse™ Cell Analyzer was used to obtain cell cycle data.

2.12. ROS Measurement

Intracellular ROS was measured with a Muse™ Oxidative Stress Kit (MCH100111, Merck Millipore). First, cells were grown in 6-well plates and treated with 5, 10, or 20 μM LCD for 48 h. Cells were washed with 1X assay buffer and incubated with a Muse™ Oxidative Stress Reagent working solution at 37 °C for 30 min. The level of fluorescence was determined with a Muse™ Cell Analyzer.

2.13. MMP Assay

MMP was measured using a Muse™ MitoPotential Kit (MCH100110, Merck Millipore). In brief, cells were exposed to 5, 10, or 20 μM of LCD for 48 h at 37 °C in a CO_2 incubator. Cells were washed with 1× assay buffer, and fluorescence was then measured using Muse™ MitoPotential working solution. After incubation with 7-AAD for 5 min, the MMP was determined with a Muse™ Cell Analyzer.

2.14. Isolation of Cytosol and Mitochondrial Fractionation

Whole-cell extracts were obtained from LCD untreated or treated HCC827 and HCC827GR cells. Cells were resuspended in a plasma membrane extraction buffer containing 250 mM sucrose, 10 mM HEPES (pH 8.0), 10 mM KCl, 1.5 mM $MgCl_2 \cdot 6H_2O$, 1 mM EDTA, 1 mM EGTA, 0.1 mM phenylmethylsulfonyl fluoride, 0.01 mg/mL aprotinin, and 0.01 mg/mL leupeptin. Then, these cells were homogenized using 0.1% of digitonin and centrifuged at 13,000 rpm for 5 min. Supernatants were centrifuged at 13,000 rpm for 30 min to separate the cytosol fraction. The pellet was rinsed with plasma membrane extraction buffer and resuspended with 0.5 % of Triton X-100 in plasma membrane extraction buffer. Lysates were centrifuged at 13,000 rpm for 30 min to obtain supernatants as mitochondria fractions.

2.15. Multi-Caspase Assay

Multi-caspase (caspase-1, -3, -4, -5, -6, -7, -8, and -9) activity was analyzed with a Muse™ Multi-Caspase Kit (MCH100109, Merck Millipore). Briefly, HCC827 (1.95×10^5 cells/well) and HCC827GR (1.8×10^5 cells/well) cells were allowed to adhere for 24 h on 6-well plates. After treatment with LCD for 48 h, cells were harvested and washed with 1X caspase buffer. Then, these cells were incubated with Muse™ Multi-Caspase Reagent working solution at 37 °C for 30 min. After Muse™ Caspase 7-AAD working solution was added, flow cytometry analysis was carried out with a Muse™ Cell Analyzer.

2.16. Statistical Analysis

Statistical significance was evaluated using the software GraphPad Prism statistics (v5, GraphPad Software, USA, RRID: SCR_002798). Differences among multiple groups were tested using one-way or two-way ANOVA followed by Dunnett's post hoc test. All data are expressed as mean ± standard deviation (SD). Differences were considered significant at $p < 0.05$.

3. Results

3.1. LCD Targets EGFR or MET

To understand the direct binding of LCD with EGFR or MET, we performed ex vivo pull-down assays (Sepharose 4B or LCD-Sepharose 4B beads) and in vitro ATP competitive binding assays. We used the gefitinib-sensitive NSCLC cell line HCC827 and gefitinib-resistant NSCLC cell line HCC827GR. As shown in Figure 1B, the pull-down assay and Western blotting results revealed that LCD bound to EGFR or MET in HCC827 and HCC827GR cells. However, there was no interaction between LCD and AKT. The results showed that the interaction between LCD and EGFR or MET was offset in the presence of 10 or 100 μM of ATP (Figure 1C). To further determine the interaction between LCD and EGFR or MET, we conducted kinase assay using gefitinib (EGFR inhibitor) and savolitinib (MET inhibitor) as positive controls. Compared to the untreated control group, EGFR and MET kinase activity were inhibited by treatment with LCD (Figure 1D). The inhibition levels of EGFR and MET kinase activity by LCD were similar to those by positive controls. These results suggest that LCD can suppress the kinase activity of EGFR or MET as an ATP competitive inhibitor. Figure 1E shows the predicted binding poses of LCD in EGFR and MET. In the complex of EGFR (Figure 1E left panel), LCD had two hydrogen bonds formed by the backbone of Met793 in the hinge and the sidechain of Asp855 in the DFG loop. The 4-hydroxy-3-(3-methylbut-2-enyl) phenyl group (A ring) and 3,4-dihydroxy-2-methoxyphenyl group (B ring) lay on the same plane and became jammed between the hydrophobic cores such as Leu718, Val726, and Ala743 of the P-loop and Leu844. In the complex of MET (Figure 1E right panel), the ketone group of LCD formed a hydrogen bond with the backbone of Met1160 in the hinge. The phenol ring (A ring) of LCD was surrounded by a hydrophobic pocket. Tyr1159 of the hinge and Ile1084, Val1092, Ala1108, and Lys1110 of the P-loop were covered similar to a lid. In addition, LCD was deeply supported by the hydrophobic sidechains of Met1160 of the hinge and Leu1140, Met1211, and Ala1221 of the ATP pocket downwards. The binding pose of EGFR was totally similar to that of MET driven by forming hydrogen bonds, and the hydrophobic interaction and exactly LCD was located at the ATP binding region of EGFR and MET. Notably, the stabilization of the complex would be enhanced by the hydrophobic interaction. Therefore, the predicted result was consistent with the experimental data, showing that LCD inhibited EGFR and MET competitively.

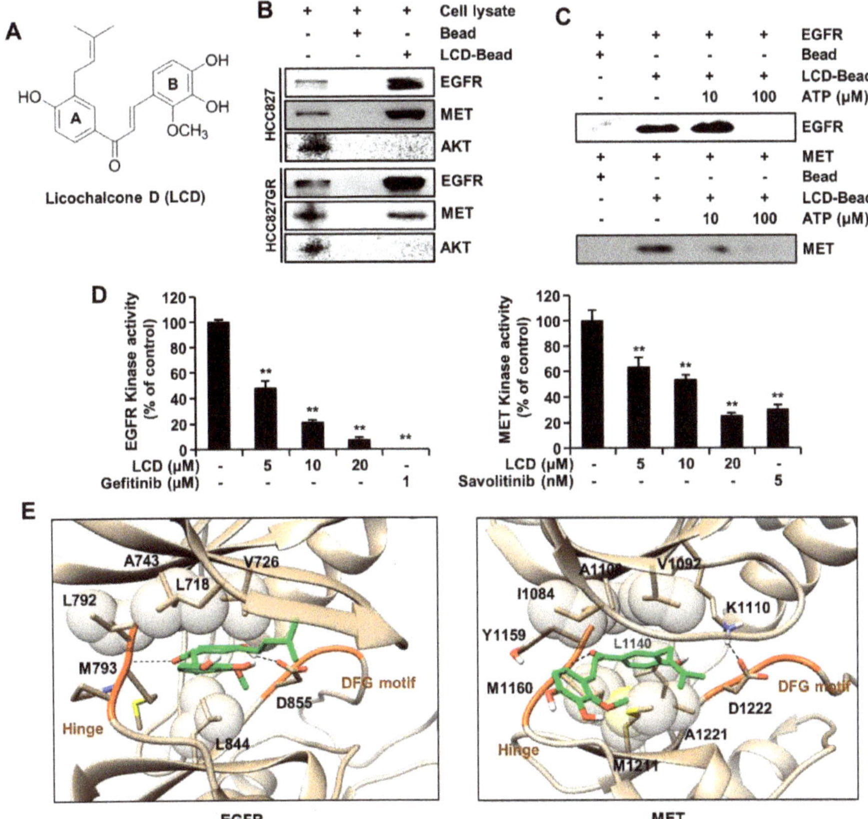

Figure 1. Licochalcone D (LCD) and epidermal growth factor receptor (EGFR) or hepatocyte growth factor receptor (MET) protein interaction. (**A**) Structure of LCD. (**B**) Pull-down assay. Whole cell lysates and Sepharose 4B or LCD-Sepharose 4B beads were incubated together. After washing, proteins bound to bead were released with SDS-PAGE loading buffer (lane 1, control; lane 2, Sepharose 4B beads; and lane 3, LCD-Sepharose 4B beads). (**C**) LCD binds to EGFR or MET competitively with ATP. Active EGFR (100 ng) or MET (100 ng) was incubated with ATP at different concentrations (0, 10, or 100 µM). Proteins were subjected to Western blotting. (**D**) EGFR or MET kinase activity of LCD by ADP-Glo kinase assay: percent inhibition rate shown at 5, 10, and 20 µM of LCD treatment. Data represent the mean value ± SD (n = 3). **, $p < 0.01$ compared with the control. (**E**) Predicted binding poses for EGFR (left) and MET (right). The dash line represented a hydrogen bond. LCD was tightly sandwiched between the hydrophobic sidechains (sphere). LCD was stabilized by hydrogen bonds and the hydrophobic interactions.

3.2. LCD Regulates Cellular Signaling Pathways Associated with EGFR or MET

To elucidate whether LCD could regulate EGFR and MET signaling cascades, Western blot analysis was used to evaluate the effects of LCD on expression and activation levels of proteins in RTK signaling pathways, including EGFR, MET, and their downstream signaling molecules. As displayed in Figure 2A, with an increasing concentration of LCD, the expression levels of p-EGFR, p-MET, and p-AKT showed a tendency to decrease in HCC827 and HCC827GR cells. These results indicate that LCD is involved in the down-regulation of the EGFR and MET signaling pathways.

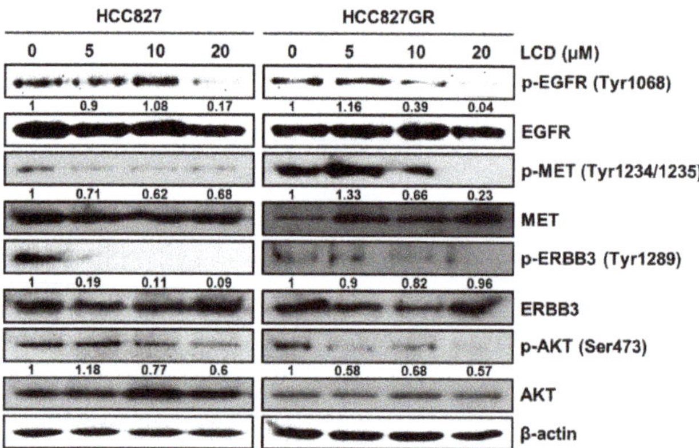

Figure 2. Comparison of EGFR or MET-related protein expression in HCC827 and HCC827GR cells. HCC827 and HCC827GR cells were treated with 5, 10, and 20 µM of LCD for 48 h. Levels of phosphorylated (p)-EGFR (Tyr1068), EGFR, p-MET (Tyr1234/1235), MET, p-ERBB3 (Tyr1289), ERBB3, p-AKT (Ser473), and AKT were analyzed by Western blotting. β-actin was used as a loading control. Band intensities were quantified using Image J.

3.3. LCD Inhibits the Growth of HCC827 and HCC827GR Cells

To analyze the effect of LCD on the viability of lung cancer cells, we used MTT to confirm the change in viability of HCC827 and HCC827GR cells. These two cell lines showed different gefitinib sensitivities. HCC827 cells were more sensitive to gefitinib than HCC827GR cells (Figure 3A). As shown in Figure 3A, LCD significantly decreased the viability of HCC827 cells. The same phenomenon was observed in HCC827GR cells (Figure 3A). The IC_{50} values of LCD for the viability of HCC827 and HCC827GR cells were 17.9 ± 0.97 µM and 19.1 ± 0.5 µM, respectively. To determine whether LCD could inhibit anchorage-independent growth, we carried out soft agar assays to measure the effects of LCD on HCC827 and HCC827GR cells. The number of colonies was decreased by treatment with LCD in a concentration-dependent manner (Figure 3B). These results show that LCD decreases the viability of NSCLC cell lines HCC827 and HCC827GR.

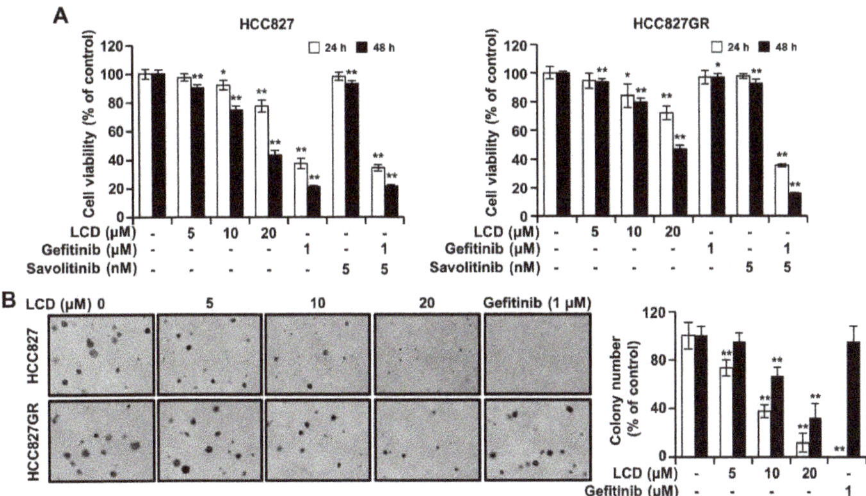

Figure 3. Effects of LCD on cell viability and colony formation. (**A**) HCC827 and HCC827GR cells were incubated in 96-well plates for 48 h in culture medium supplemented with LCD at an increasing concentration from 5 µM to 20 µM, and control cells. Cell viability was measured by 3-(4,5-dimethylthiazol-2-yl)-2,5-diphenyltetrazolium bromide (MTT) assay based on absorbance at 570 nm. Values shown in the graph represent mean ± SD (n = 6). (**B**) Anchorage-independent soft agar assay was performed with HCC827 and HCC827GR cell lines. Data are representative of triplicate experiments. *, $p < 0.05$ and **, $p < 0.01$ compared with the control.

3.4. Cell Cycle Accumulation at G2/M Phase after Treatment with LCD

To assess the cause of the inhibited cell viability by LCD, we determined cell cycle distribution using a flow cytometric assay. The results indicated that the G2/M arrest of the cell cycle was increased in HCC827 cells and HCC827GR cells treated with LCD (5, 10, and 15 µM) for 48 h (Figure 4A). As shown in Figure 4B, the treatment of HCC827 and HCC827GR cells with LCD for 48 h resulted in a prominent increase in the expression of the Sub-G1 population in a dose-dependent manner. To confirm the underlying biochemical mechanisms related to the regulation of G2/M cell cycle arrest, we tested the effect of LCD on protein levels of cyclins and cyclin-dependent kinase (cdk)s during the G2/M cell cycle arrest progression. Treatment with LCD dramatically decreased the expression levels of cyclin B1 and cdc2 but increased p21 and p27 expression in HCC827 and HCC827GR cells (Figure 4C). The increased sub-G1 population of the cell cycle might suggest induced cell apoptosis. Thus, we performed flow cytometry analysis to confirm the level of apoptosis. HCC827 cells treated with LCD at concentrations of 5, 10, and 20 µM showed 16.1 ± 0.6%, 36.2 ± 0.3%, and 45.2 ± 0.6% of total apoptotic cells, respectively (Figure 4D). HCC827GR cells treated with LCD at the same concentrations of 5, 10, and 20 µM resulted in 12.7 ± 1.7%, 26.1 ± 0.7%, and 43.5 ± 0.4% of total apoptotic cells, respectively (Figure 4D). These results suggest that LCD can induce G2/M cell cycle arrest and apoptosis.

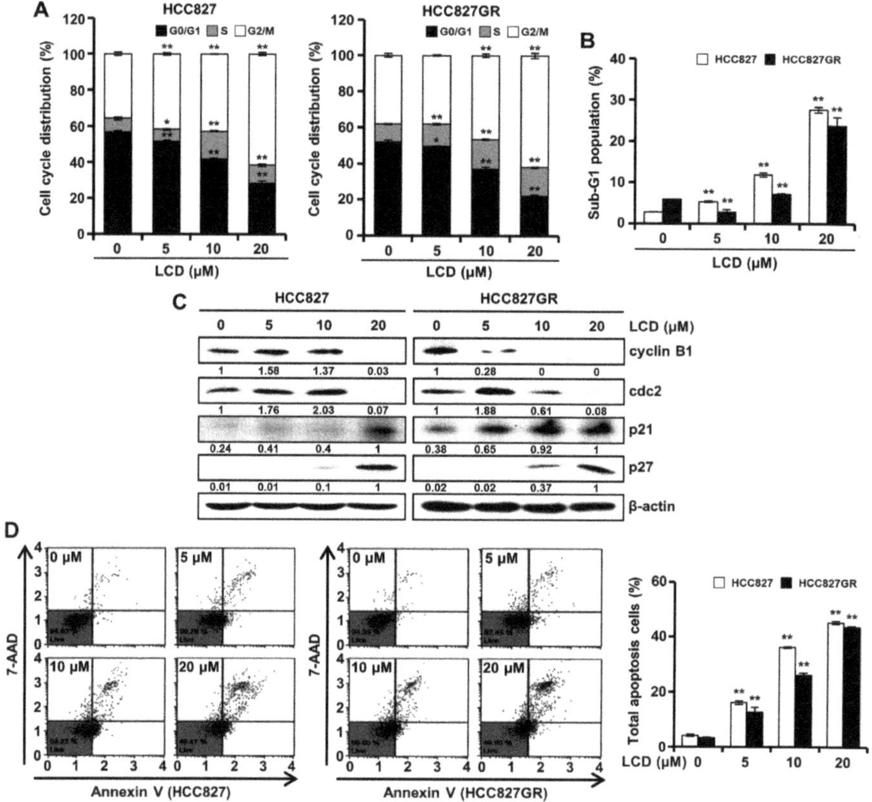

Figure 4. Cell cycle regulation and apoptosis induction of LCD. HCC827 and HCC827GR cells were exposed to LCD for 48 h. (**A,B**) Flow cytometry assay. LCD induced the arrest of non-small cell lung cancer (NSCLC) cells at the G2/M phase of cell cycle. In the G0/G1, S, G2/M, and sub-G1 phases, each value represents the mean ± SD (n = 3). (**C**) Expression levels of cell cycle-related proteins in HCC827 and HCC827GR cells were determined by Western blot. Relative protein levels of cyclin B1, cdc2, p21 and p27 were quantified using Image J software. (**D**) Apoptosis was determined using Annexin V/7-Aminoactinomycin D (7-ADD) staining. Annexin V/7-AAD double-stained cells were detected with a Muse™ Cell Analyzer. United Annexin V/7-ADD reactivity allowed the classification of cells into four groups: early apoptotic cells [Annexin V (+) and 7-AAD (−)], late apoptotic or dead cells [Annexin V (+) and 7-AAD (+)], dead cells [Annexin V (−) and 7-AAD (+)], and live cells [Annexin V (−) and 7-AAD (−)]. Data are presented as the mean ± SD of three independent experiments. *, $p < 0.05$ and **, $p < 0.01$.

3.5. LCD Induces ROS-Dependent Apoptosis

To understand the mechanism of LCD-induced apoptosis, we tested the effects of LCD on intracellular ROS generation. Levels of ROS in HCC827 and HCC827GR cells were measured at 48 h after cells were treated with LCD. ROS values of HCC827 cells treated with LCD at concentrations of 5, 10, and 20 μM were 12.2 ± 1.2%, 21.5 ± 0.7%, and 34.3 ± 0.5%, respectively. ROS values of HCC827GR cells treated with LCD at concentrations of 5, 10, and 20 μM were 9.5 ± 0.3%, 18 ± 0.4%, and 31.3 ± 2.9%, respectively (Figure 5A). To further examine the significance of ROS in LCD-induced apoptosis, HCC827 and HCC827GR cells were pretreated with N-acetylcysteine (NAC), an ROS inhibitor, for 3 h and then treated with 20 μM of LCD for 48 h. NAC treatment resulted in similar ROS fluorescence levels compared to controls. However, ROS produced after treatment with LCD was markedly blocked by co-treatment with NAC (Figure 5B). As shown in Figure 5C, NAC significantly suppressed LCD-induced NSCLC cell growth inhibition. In addition, protein expression levels after co-treatment with LCD and NAC were determined by Western blot analysis using corresponding antibodies. Interestingly, the co-treatment of NAC and LCD did not restore the expression of p-EGFR and p-MET to control levels (Figure 5D). On the other hand, the expression of PARP was increased and the expression of C-PARP was decreased compared to LCD treatment (Figure 5D). These results indicate that LCD can induce ROS-dependent apoptosis in both NSCLC cells.

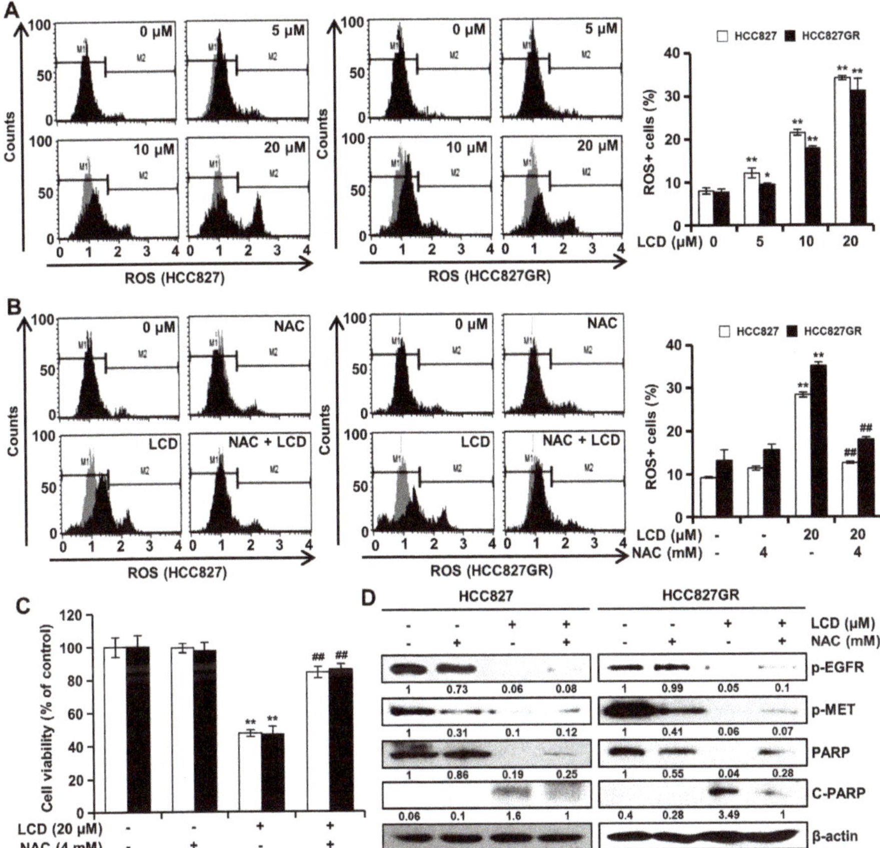

Figure 5. LCD treatment leads to intracellular reactive oxygen species (ROS) generation. (**A**) HCC827 and HCC827GR cells were treated with LCD (5, 10, and 20 μM) for 48 h, and intracellular ROS levels were determined using a Muse™ Cell Analyzer. (**B**) NSCLC cells were pretreated with 4 mM of N-acetylcysteine (NAC; an ROS inhibitor) for 3 h and then treated with 20 μM of LCD for 48 h. Values are expressed as means ± SD (n = 3). (**C**) MTT assay was performed to detect the viability of NSCLC cells treated with increasing concentrations of LCD and NAC. Values shown in the graph represent mean ± SD (n = 6). (**D**) After NAC pretreatment and LCD treatment, whole cell lysates were then subjected to Western blotting with antibodies against phosphorylated (p)-EGFR, p-MET, poly (ADP-ribose) polymerase (PARP), and β-actin. The intensity of the p-EGFR, p-MET, PARP, and C-PARP bands was quantified using Image J. *, $p < 0.05$ and **, $p < 0.01$ compared with the control. ##, $p < 0.01$ significantly different from LCD-treated cells.

3.6. LCD Induces Apoptosis through the Mitochondrial Pathway

MMP is an important index to estimate the permeability of the mitochondrial membrane. To determine whether LCD can induce apoptosis by disrupting the mitochondrial membrane, NSCLC cells were stained with MitoPotential dye, a cationic and lipophilic dye, to monitor MMP. Results showed that the loss of MMP was increased drastically after treatment with LCD (Figure 6A). Western blot analysis was performed to check the levels of proteins associated with the mitochondria. As shown in Figure 6B, Bid was cleaved as the concentration of LCD increased. LCD increased the expression of a pro-apoptotic protein Bad (Figure 6B). In contrast, the expression levels of anti-apoptotic proteins Mcl-1 and Bcl-xl were decreased following LCD treatment (Figure 6B). During apoptotic cell death, early events that occur include mitochondrial depolarization and the loss of cyto c from the mitochondrial intermembrane space. Cyto c was increased in the cytosol fraction but decreased in the mitochondrial fraction of NSCLC cells after LCD treatment (Figure 6B). The expression of Apaf-1 known to form apoptosome with cyto c was also increased by LCD. Furthermore, the expression level of a DNA repair enzyme PARP was reduced by LCD in a dose-dependent manner (Figure 6B). To determine whether LCD-induced apoptosis was associated with caspase activation, levels of multi-caspase in NSCLC cells were assessed with a Muse™ Cell Analyzer. As shown in Figure 6C, LCD induced activities of multi-caspases dose-dependently (25%, 40%, and 46% for HCC827 cells and 28%, 35%, and 48% for HCC827GR cells by LCD at concentrations of 5, 10, and 20 µM, respectively). These results demonstrate that LCD sensitizes NSCLC cells to apoptosis via the mitochondrial pathway and the activation of caspases.

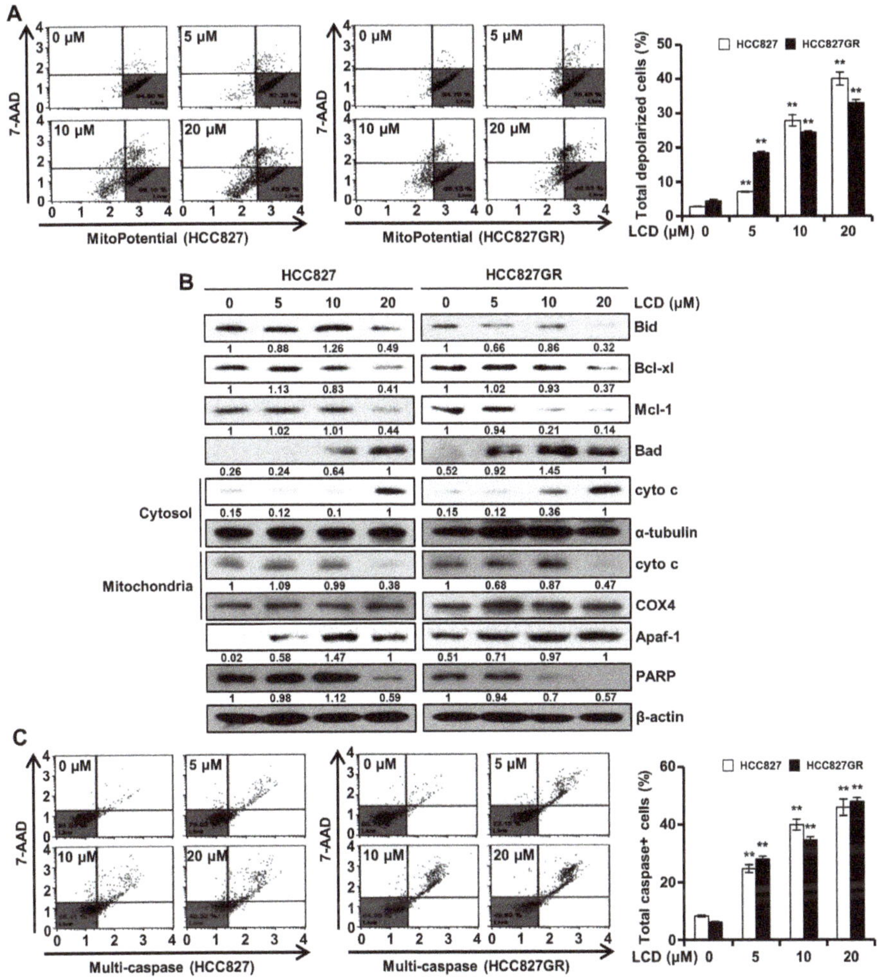

Figure 6. Effect of LCD on mitochondrial membrane potential (MMP) and activities of caspases in NSCLC cells. (**A**) Cells were exposed to LCD (5, 10, and 20 M) or DMSO for 48 h and MMP was analyzed by using a Muse™ Cell Analyzer. The fluorescence from right to left means the depolarization of MMP. Data are expressed as means ± SD of three independent experiments performed in triplicate. (**B**) After treatment of LCD, both cell lysates were subjected to Western blotting. Expression levels of Bid, Bcl-xl, Mcl-1, Bad, cyto c, α-tubulin, COX4, Apaf-1, and PARP were normalized to that of β-actin. Band intensities were quantified using Image J software. (**C**) HCC827 and HCC827GR cells were treated with an increasing concentration of LCD for 48 h. Multi-caspase (caspase-1, -3, -4, -5, -6, -7, -8, and -9) activity was determined using a Muse™ Cell Analyzer. Values are means (caspase-positive) ± SD. **, $p < 0.01$.

4. Discussion

Chemotherapy drugs generally include cisplatin and taxanes in the treatment of lung cancer. Unfortunately, many patients acquire resistance to the drug either intrinsically or after medication [6]. The occurrence of gefitinib resistance is a barrier to have effective clinical therapies for a number of solid tumors, including lung cancer. Various gefitinib resistance mechanisms, including EGFR mutations and *MET* amplification, have been reported in several studies [9,10]. More than 90% of

EGFR mutations occur in exons 19–21, with exon 19 mutations being the most frequent [9]. The deletion of exon 19 and EGFR mutation of L858R of exon 21 have higher sensitivities than the TKI response in those with wild-type EGFR [12]. Conversely, resistance to TKI can be induced through the acquisition of secondary mutations of EGFR (T790M, L747S, D761Y, and T854A) or the activation of other signaling bypass pathways [9]. Resistance obtained after TKI treatment is almost unavoidable, and the success rate of treatment is low. Overcoming these resistances requires a strategy to target molecules related to resistance or in combination with other compounds.

Some clinical studies of EGFR- or MET-targeted inhibitors have been discontinued. However, others have shown encouraging results [10]. The overexpression of EGFR and MET has been reported in NSCLC, which can activate various downstream signaling molecules involved in cell growth and survival [8,13]. Blocking EGFR or MET alone for tumor suppression can lead to cell survival by activating other alternative pathways. Previous reports have shown that treatment with a single inhibitor of EGFR and MET, respectively, has no effect on cell proliferation, although cell survival is suppressed when EGFR and MET inhibitors are combined [11,14]. A dual blockade of EGFR and MET can significantly inhibit the proliferation several carcinoma cells, including lung cancer cells [14,15], head and neck cancer cells [16], and colon cancer cells [17]. Similarly, our data showed that treatment with EGFR and MET alone did not affect cell viability. However, they inhibited cell survival upon combination treatment. LCD inhibited lung cancer cell proliferation through ATP competitive inhibition of EGFR and MET as a single drug. Other studies have reported that inhibitors of EGFR and MET alone could inhibit cell survival, whereas a combination treatment of EGFR and MET inhibitors can significantly inhibit cell survival, showing synergistic effects on the induction of apoptosis [13].

In lung cancer cells, a signaling network exists between the same RTK family: EGFR, MET, and ERBB3. The results of the present study revealed that the levels of phosphorylated proteins of EGFR, MET, and ERBB3 decreased with an increasing concentration of LCD. Since the amplification of *MET* can induce gefitinib resistance by inducing ERBB3-dependent activation [11], it is thought that LCD targeting MET can also inhibit the expression of ERBB3. EGFR and MET can activate and share important downstream molecules involved in biological activities such as cell growth and survival [18]. The activation of AKT in lung cancer cells is involved in imparting resistance to TKI as well as cell growth and proliferation [10]. The reduction of p-AKT caused by the dual targeting of EGFR and MET might be a pathway to overcome resistance by preventing the conversion of compensation pathways. Thus, the dual targeting of EGFR and MET might be a promising therapeutic strategy to overcome lung cancer that is sensitive or resistant to TKI.

One of the meaningful findings in our study was that the mechanism underlying the apoptosis induced by LCD involved the inhibition of EGFR and MET. Apoptotic cell death is induced through two pathways: the extrinsic pathway and the intrinsic pathway. Apoptosis is caused by a variety of extrinsic and intrinsic factors such as ROS, DNA damage, and heat shock [19]. ROS can regulate physiological functions such as cell proliferation and cell cycle progression. However, excessive ROS levels can cause cell damage and lead to apoptotic cell death [19]. The exposure of NSCLC cells to LCD significantly increased ROS production, whereas ROS levels were restored to control levels by NAC. The expression of p-EGFR and p-MET, which were inhibited by LCD treatment, showed no significant difference by NAC. These results indicate that ROS is downstream of EGFR and MET, suggesting that ROS might be closely involved in the cell proliferation inhibition of LCD. ROS are known to affect EGFR signaling. ROS can induce TKI resistance by activating the EGFR signaling pathway [20,21]. High levels of ROS play an opposite role in tumor progression and drug resistance, leading to cell cycle arrest, apoptosis induction, and toxic effects on cancer cells [21]. This indicates that an appropriate level of ROS can mediate drug resistance, whereas excessive ROS production can lead to cell death. LCD can induce the apoptosis of lung cancer cells through the accumulation of ROS. ROS is closely associated with the mitochondrial pathway. In fact, the location of most intracellular ROS production is mitochondria [19]. LCD increased MMP loss and up- or down-regulated mitochondrial-related proteins, leading to apoptotic cell death through mitochondrial (intrinsic) pathways. Apoptosis molecular cascade by LCD led to the release of cyto c into the cytoplasm, the activation of Apaf-1 and caspase, and the cleavage of PARP. It is interesting to note that ROS plays an important role in the apoptosis-inducing mechanism of LCD because PARP, which was reduced by LCD, increased PARP expression by co-treatment with NAC.

ROS is also known to be involved in cell cycle progression, which is regulated by cyclin and cdk [20]. The link between cell cycle and apoptosis has been demonstrated in many studies [22]. It regulates cell proliferation. The cyclin B1/cdc2 complex is involved in the G2/M phase transition, and cdk activity is mediated by cdk inhibitors (CKI) such as p21 and p27 [22,23]. LCD decreased cyclin B1/cdc2 protein expression but increased CKI, causing G2/M cell cycle arrest. LCA and LCB of the same licorice family as LCD can also block G2/M cell cycle progression and induce the apoptosis of lung cancer cells [14,24]. Some agents that can induce apoptosis may also increase cytotoxicity in association with G2/M checkpoint arrest [23].

Importantly, LCD shows anti-tumor activity without causing weight loss in vivo using a xenograft model of oral cancer cells [4]. In addition, LCD (1 µg/ml) shows myocardial protective effects in the cardiac tissues of injured rats [25]. These findings indicate that LCD is low in toxicity. In addition, while common flavonoids generally have low bioavailability in humans, LCs have been shown to be well absorbed through passive diffusion through the Caco-2 cell monolayer, which is a human intestinal cell line [26]. Since intestinal permeability is an important factor influencing the bioavailability of drugs, LCs can be expected to have high bioavailability in the human body. Liquiritin, another component of licorice, enhanced the cell proliferation inhibitory effect of cisplatin in gastric cancer cells resistant to cisplatin, and showed synergistic effects on tumor growth inhibition in vivo [27]. In the future, we will evaluate the anti-tumor efficacy of lung cancer cells with different resistance mechanisms for the combination of anti-cancer drugs and LCD.

5. Conclusions

In summary, the results of this study suggested a molecular mechanism for the anti-cancer effect of LCD in gefitinib-sensitive or gefitinib-resistant lung cancer cells. The anti-tumor efficacy of LCD was shown through the dual inhibition of EGFR and MET activity, suggesting that it was a principal target of LCD. In addition, LCD induced ROS-dependent apoptotic cell death and inhibited the proliferation of lung cancer cells. In conclusion, LCD offers clinical benefit for TKI-sensitive or TKI-resistant lung cancer. It shows potential as an effective anti-cancer compound.

Author Contributions: Data curation, H.-N.O., M.-H.L., E.K., A.-W.K., G.Y., S.-S.C., and K.L.; Funding acquisition, J.-I.C. and J.-H.S.; Supervision, J.-I.C. and J.-H.S.; Writing—original draft, H.-N.O. and M.-H.L.; Writing—review and editing, H.-N.O., M.-H.L., J.-I.C., and J.-H.S. All authors have read and agreed to the published version of the manuscript.

Funding: This research was funded by the Basic Science Research program through the NRF Funded by the Ministry of Education, Science and Technology (2019R1A2C1005899).

Conflicts of Interest: The authors declare no conflict of interest.

References

1. Wang, Z.; Liu, Z.; Cao, Y.; Paudel, S.; Yoon, G.; Cheon, S.H. Short and Efficient Synthesis of Licochalcone B and D Through Acid-Mediated Claisen-Schmidt Condensation. *Bull. Korean Chem. Soc.* **2013**, *34*, 3906–3908. [CrossRef]
2. Furusawa, J.-I.; Funakoshi-Tago, M.; Mashino, T.; Tago, K.; Inoue, H.; Sonoda, Y.; Kasahara, T. Glycyrrhiza inflata-derived chalcones, Licochalcone A, Licochalcone B and Licochalcone D, inhibit phosphorylation of NF-kappaB p65 in LPS signaling pathway. *Int. Immunopharmacol.* **2009**, *9*, 499–507. [CrossRef]
3. Si, L.; Yan, X.; Hao, W.; Ma, X.; Ren, H.; Ren, B.; Li, D.; Dong, Z.; Zheng, Q. Licochalcone D induces apoptosis and inhibits migration and invasion in human melanoma A375 cells. *Oncol. Rep.* **2018**, *39*, 2160–2170.

4.   Seo, J.-H.; Choi, H.-W.; Oh, H.-N.; Lee, M.-H.; Kim, E.; Yoon, G.; Cho, S.-S.; Park, S.-M.; Cho, Y.-S.; Chae, J.-I. Licochalcone D directly targets JAK2 to induced apoptosis in human oral squamous cell carcinoma. *J. Cell Physiol.* **2019**, *234*, 1780–1793. [CrossRef]

5.   Paez, J.G. EGFR Mutations in Lung Cancer: Correlation with Clinical Response to Gefitinib Therapy. *Science* **2004**, *304*, 1497–1500. [CrossRef]

6.   Huang, C.-Y.; Ju, D.-T.; Chang, C.-F.; Reddy, P.M.; Velmurugan, B.K. A review on the effects of current chemotherapy drugs and natural agents in treating non–small cell lung cancer. *Biomedicine* **2017**, *7*, 23. [CrossRef]

7.   Zappa, C.; Mousa, S.A. Non-small cell lung cancer: Current treatment and future advances. *Transl. Lung Cancer Res.* **2016**, *5*, 288–300. [CrossRef]

8.   Zandi, R.; Larsen, A.B.; Andersen, P.; Stockhausen, M.-T.; Poulsen, H.S. Mechanisms for oncogenic activation of the epidermal growth factor receptor. *Cell. Signal.* **2007**, *19*, 2013–2023. [CrossRef]

9.   Huang, L.; Fu, L. Mechanisms of resistance to EGFR tyrosine kinase inhibitors. *Acta Pharm. Sin. B* **2015**, *5*, 390–401. [CrossRef]

10.   Lin, Y.; Wang, X.; Jin, H. EGFR-TKI resistance in NSCLC patients: Mechanisms and strategies. *Am. J. Cancer Res.* **2014**, *4*, 411–435.

11.   Engelman, J.A.; Zejnullahu, K.; Mitsudomi, T.; Song, Y.; Hyland, C.; Park, J.O.; Lindeman, N.; Gale, C.-M.; Zhao, X.; Christensen, J.; et al. MET Amplification Leads to Gefitinib Resistance in Lung Cancer by Activating ERBB3 Signaling. *Science* **2007**, *316*, 1039–1043. [CrossRef]

12.   Lynch, T.J.; Bell, D.W.; Sordella, R.; Gurubhagavatula, S.; Okimoto, R.A.; Brannigan, B.W.; Harris, P.L.; Haserlat, S.M.; Supko, J.G.; Haluska, F.G.; et al. Activating mutations in the epidermal growth factor receptor underlying responsiveness of non-small-cell lung cancer to gefitinib. *N. Engl. J. Med.* **2004**, *350*, 2129–2139. [CrossRef]

13.   Puri, N.; Salgia, R. Synergism of EGFR and c-Met pathways, cross-talk and inhibition, in non-small cell lung cancer. *J. Carcinog.* **2008**, *7*, 9. [CrossRef]

14.   Oh, H.-N.; Lee, M.-H.; Kim, E.; Yoon, G.; Chae, J.-I.; Shim, J.-H. Licochalcone B inhibits growth and induces apoptosis of human non-small-cell lung cancer cells by dual targeting of EGFR and MET. *Phytomedicine* **2019**, *63*, 153014. [CrossRef]

15.   Jung, S.K.; Lee, M.H.; Lim, D.Y.; Lee, S.Y.; Jeong, C.H.; Kim, J.E.; Lim, T.G.; Chen, H.; Bode, A.M.; Lee, H.J.; et al. Butein, a novel dual inhibitor of MET and EGFR, overcomes gefitinib-resistant lung cancer growth. *Mol. Carcinog.* **2015**, *54*, 322–331. [CrossRef]

16.   Xu, H.; Stabile, L.P.; Gubish, C.T.; Gooding, W.E.; Grandis, J.R.; Siegfried, J.M. Dual blockade of EGFR and c-Met abrogates redundant signaling and proliferation in head and neck carcinoma cells. *Clin. Cancer Res.* **2011**, *17*, 4425–4438. [CrossRef]

17.   Qiu, P.; Wang, S.; Liu, M.; Ma, H.; Zeng, X.; Zhang, M.; Xu, L.; Cui, Y.; Xu, H.; Tang, Y.; et al. Norcantharidin Inhibits cell growth by suppressing the expression and phosphorylation of both EGFR and c-Met in human colon cancer cells. *BMC Cancer* **2017**, *17*, 55. [CrossRef]

18.   Guo, A.; Villen, J.; Kornhauser, J.; Lee, K.A.; Stokes, M.P.; Rikova, K.; Possemato, A.; Nardone, J.; Innocenti, G.; Wetzel, R.; et al. Signaling networks assembled by oncogenic EGFR and c-Met. *Proc. Natl. Acad. Sci. USA* **2008**, *105*, 692–697. [CrossRef]

19.   Redza-Dutordoir, M.; Averill-Bates, D.A. Activation of apoptosis signalling pathways by reactive oxygen species. *Biochim. Biophys. Acta Bioenerg.* **2016**, *1863*, 2977–2992. [CrossRef]

20.   Verbon, E.H.; Post, J.A.; Boonstra, J. The influence of reactive oxygen species on cell cycle progression in mammalian cells. *Gene* **2012**, *511*, 1–6. [CrossRef]

21.   Weng, M.-S.; Chang, J.-H.; Hung, W.-Y.; Yang, Y.-C.; Chien, M.-H. The interplay of reactive oxygen species and the epidermal growth factor receptor in tumor progression and drug resistance. *J. Exp. Clin. Cancer Res.* **2018**, *37*, 61. [CrossRef] [PubMed]

22.   Pucci, B.; Kasten, M.; Giordano, A. Cell Cycle and Apoptosis. *Neoplasia* **2000**, *2*, 291–299. [CrossRef] [PubMed]

23.   DiPaola, R.S. To arrest or not to G2-M Cell-cycle arrest: Commentary re: AK Tyagi et al., Silibinin strongly synergizes human prostate carcinoma DU145 cells to doxorubicin-induced growth inhibition, G2-M arrest, and apoptosis. *Clin. Cancer Res.* **2002**, *8*, 3311–3314. [PubMed]

24. Qiu, C.; Zhang, T.; Zhang, W.; Zhou, L.; Yu, B.; Wang, W.; Yang, Z.; Liu, Z.; Zou, P.; Liang, G. Licochalcone A Inhibits the Proliferation of Human Lung Cancer Cell Lines A549 and H460 by Inducing G2/M Cell Cycle Arrest and ER Stress. *Int. J. Mol. Sci.* **2017**, *18*, 1761. [CrossRef]
25. Yuan, X.; Niu, H.-T.; Wang, P.-L.; Lu, J.; Zhao, H.; Liu, S.-H.; Zheng, Q.-S.; Li, C.-G. Cardioprotective Effect of Licochalcone D against Myocardial Ischemia/Reperfusion Injury in Langendorff-Perfused Rat Hearts. *PLoS ONE* **2015**, *10*, e0128375. [CrossRef]
26. Wang, X.-X.; Liu, G.-Y.; Yang, Y.-F.; Wu, X.-W.; Xu, W.; Yang, X.-W. Intestinal Absorption of Triterpenoids and Flavonoids from Glycyrrhizae radix et rhizoma in the Human Caco-2 Monolayer Cell Model. *Molecules* **2017**, *22*, 1627. [CrossRef]
27. Wei, F.; Jiang, X.; Gao, H.-Y.; Gao, S.-H. Liquiritin induces apoptosis and autophagy in cisplatin (DDP)-resistant gastric cancer cells in vitro and xenograft nude mice in vivo. *Int. J. Oncol.* **2017**, *51*, 1383–1394. [CrossRef]

 *biomolecules*

Article

Accumulation of Anthocyanins through Overexpression of AtPAP1 in *Solanum nigrum* Lin. (Black Nightshade)

Saophea Chhon [1], **Jin Jeon** [1], **Joonyup Kim** [2,*] and **Sang Un Park** [1,*]

[1] Department of Crop Science, Chungnam National University, 99 Daehak-ro, Yuseong-gu, Daejeon 34134, Korea; saophea.chhon91@gmail.com (S.C); jeonjin519@cnu.ac.kr (J.J.)

[2] Department of Horticultural Science, Chungnam National University, 99, Daehak-Ro, Yuseong-gu, Daejeon 34134, Korea

* Correspondence: jykim12@cnu.ac.kr (J.K.); supark@cnu.ac.kr (S.U.P.);
Tel.: +82-42-821-5738 (J.K.); +82-42-821-5730 (S.U.P.);
Fax: +82-42-823-1382 (J.K.); +82-42-822-2631 (S.U.P.)

Received: 4 January 2020; Accepted: 10 February 2020; Published: 11 February 2020

Abstract: Black nightshade (*Solanum nigrum*) belongs to the *Solanaceae* family and is used as a medicinal herb with health benefits. It has been reported that the black nightshade plant contains various phytochemicals that are associated with antitumor activities. Here we employed a genetic approach to study the effects of overexpression of *Arabidopsis thaliana* production of anthocyanin pigment 1 (AtPAP1) in black nightshade. Ectopic expression of AtPAP1 resulted in enhanced accumulation of anthocyanin pigments in vegetative and reproductive tissues of the transgenic plants. Analysis of anthocyanin revealed that delphinidin 3-O-rutinoside-5-O-glucoside, delphinidin 3,5-O-diglucoside, delphinidin 3-O-rutinoside, petunidin 3-O-rutinoside (*cis-p*-coumaroyl)-5-O-glucoside, petunidin 3-(feruloyl)-rutinoside-5-glucoside, and malvidin 3-(feruloyl)-rutinoside-5-glucoside are highly induced in the leaves of AtPAP1 overexpression lines. Furthermore, ectopic expression of AtPAP1 evoked expression of early and late biosynthetic genes of the general phenylpropanoid and flavonoid pathways that include phenylalanine ammonia-lyase (*PAL*), cinnamate-4-hydroxylase (*C4H*), 4-coumarate CoA ligase (*4CL*), chalcone isomerase (*CHI*), and quinate hydroxycinnamoyl transferase (*HCT*), which suggests these genes might be transcriptional targets of AtPAP1 in black nightshade. Concomitantly, the total content of anthocyanin in the transgenic black nightshade plants was higher compared to the control plants, which supports phenotypic changes in color. Our data demonstrate that a major anthocyanin biosynthetic regulator, AtPAP1, can induce accumulation of anthocyanins in the heterologous system of black nightshade through the conserved flavonoid biosynthesis pathway in plants.

Keywords: Anthocyanin; *Solanum nigrum* L.; flavonoid biosynthesis; AtPAP1

---

1. Introduction

Black nightshade (*Solanum nigrum*) is widely used as a leafy vegetable, fruit and source of various therapeutic drugs. Consumption of the leaves and fruits as food is widespread in Africa and Southeast Asia. As a whole plant, the black nightshade has been used as a folk medicine in Asia to treat inflammation, edema, and mastitis. Phytochemical screening of the crude extracts from the plant revealed the presence of secondary metabolites such as alkaloids, glycoproteins, flavonoids, polyphenols, and triterpenoids [1,2]. Although a wide range of bioactivities including anti-inflammatory, antioxidant, antinociceptive, antipyretic, antitumor, antiulcerogenic, cancer chemopreventive, hepatoprotective, and immunomodulatory effects have been reported, the roles of anthocyanins associated with these activities for the black nightshade were not fully determined. Anthocyanins are flavonoids linked to the pigmentation of plant flowers and fruit that ranges from red to blue. They also attract pollinators and animals for pollination and seed dispersal and defend plants against various biotic and abiotic stresses [3]. Anthocyanins also play a positive role in the reduction of many chronic diseases that result from oxidative stress [4]. The biosynthesis of anthocyanins is regulated by activities of many enzymes involved in the general phenylpropanoid pathway, which can be divided into two groups [5]. The upstream biosynthesis genes, chalcone synthase (*CHS*), chalcone isomerase (*CHI*), and flavone 3-hydroxylase (*F3H*), are the early biosynthetic genes (EBGs) common to the biosynthesis of all downstream flavonoids. The late biosynthesis genes (LBGs) such as *flavonoid 3′-hydroxylase* (*F3′H*), flavonoid 3′,5′-hydroxylase (*F3′5′H*), dihydroflavonol 4-reductase (*DFR*), anthocyanidin synthase/leucoanthocyanidin dioxygenase (*ANS/LDOX*), and flavonoid 3-O-glucosyltransferase (*UFGT*) are responsible for the biosynthesis of specific classes of flavonoids, including anthocyanins.

Previous studies have demonstrated that gene expression for the EBGs and the content of anthocyanins are not consistently correlated, and vary depending on the plants species [6]. On the contrary, it has been shown that there appears to be a consistent positive correlation between expression levels of the LBGs and anthocyanin levels in several *Solanaceae* plants. Gene expression of the EBGs (e.g., *CHS*, *CHI*, *F3H*), *F3′H*, and flavanol synthase (*FLS*) are shown to be regulated by R2R3 type MYB transcription factors. In regard to the LBGs including *DFR*, *LDOX*, and UDP-glucose: flavonoid 3-O-glucosyl transferase (*UF3GT*), it has been shown that expression of these genes are coordinately regulated by the MYB-bHLH-WD40 (MBW) transcription factor complex, consisting of R2R3 type MYB activators, bHLH activators, and the WD-repeat protein transparent testa glabara 1 (TTG1) [7–9].

Arabidopsis AtPAP1 encodes an MYB75 transcription factor, which is crucial to the biosynthesis of anthocyanins in *Arabidopsis*. Both knockout and knockdown mutants of AtPAP1 display deficiencies in anthocyanin pigmentation in seedlings, while ectopic expression of AtPAP1 overproduces anthocyanins throughout the plant (e.g., roots, flowers) resulting from coordinated up-regulation of anthocyanin biosynthesis genes [10]. Transient transactivation assay in a combination with bioinformatics assay uncovered that transcriptional activation of anthocyanin biosynthesis genes is mediated by AtPAP1 that targets a common *cis*-regulatory element present in the upstream region of the 5′-UTR of biosynthesis genes. The AtPAP1 binds to the 10 bp CCACG-containing PAP1 *cis*-regulatory element in the promoter regions of these anthocyanin biosynthesis-related genes (preferentially LBGs). In addition, light is essential for the transcriptional activation of AtPAP1 [11]. This light-dependent transcriptional activation of AtPAP1 is mediated by the leucine-zipper transcription factor HY5 that directly binds to G- and ACE-boxes in the promoter region of AtPAP1, which acts as a downstream component of phytochrome (PHY), cryptochrome (CRY), and UV-B (UVR8) photoreceptors [12]. Further, this light-dependent anthocyanin biosynthesis is positively regulated by sugar and cytokinin and negatively regulated by ethylene [13] and nitrogen deficiency [14]. In addition, it has been demonstrated that AtPAP1 activity is negatively regulated by repressor protein associated with the R3 type MYB-related protein, MYBL2 [8] and miR156-targeted squamosa lateral organ promoter binding protein-like 9 (SPL9) that destabilizes the MYB–bHLH–WD40 complex [15].

Here we utilized a genetic approach to study the effects of constitutive expression of AtPAP1 in black nightshade. Constitutive expression of AtPAP1 led to enhanced accumulation of anthocyanins in black nightshade. Using liquid chromatography-mass spectrometry (LC/MS/MS) six anthocyanins were identified in the AtPAP1 overexpression lines. Total anthocyanins analysis measured by spectrophotometry revealed that the transgenic plants contained a higher amount of anthocyanins compared with the control plants. In addition, ectopic expression of *Arabidopsis* AtPAP1 up-regulated several genes that are possibly involved in the biosynthesis of anthocyanins in black nightshade. Our results demonstrate that a major biosynthetic regulator can induce enhanced accumulation of anthocyanins through the conserved flavonoid biosynthesis pathways in black nightshade.

2. Materials and Methods

2.1. Identification of Phenylpropanoid Biosynthesis Genes in Black Nightshade

Genes involved in phenylpropanoid biosynthesis in the black nightshade (*S. nigrum*) were identified from RNA-seq data. The sequences of transcripts that correspond to the functional annotation available in the *Solanaceae* database were selected. In addition, the phenylpropanoid biosynthetic genes of *Arabidopsis* obtained from TAIR (https://www.arabidopsis.org/) were used as queries to search for homologous sequences in the *S. nigrum* transcriptome database. The deduced amino acid sequences of the retrieved genes were further analyzed for homology using BLAST in the NCBI GenBank database (https://blast.ncbi.nlm.nih.gov/Blast.cgi). The genes with a maximum identity were used for the further transcriptional study [16,17].

2.2. HPLC Analysis of Anthocyanin Content

Leaf samples were harvested and freeze-dried at −80 °C for 3 days. The freeze-dried samples were pulverized using a mortar and pestle. Each powdered sample (100 mg) was extracted with 2.0 mL of water: formic acid (95:5, v/v) vortexing for 5 min, followed by sonication for 20 min at room temperature. The slurry mixture was centrifuged at 10,000 rpm at 4 °C for 15 min (IEC Clinical Centrifuge at Damon/IEC Division, Needham, MA, USA). The supernatant was filtered through a 13 mm (0.45 µm) PTFE syringe filter (Advantech DISMIC-13HP, Toyo Roshi Kaisha, Ltd., Tokyo, Japan), and 1.5 mL of the extract was used for further HPLC analysis. The HPLC system (1200 series, Agilent Technologies, Palo Alto, CA, USA) used in the study was equipped with a PDA LC detector. Individual anthocyanins within the extract solution were separated on a Synergy 4 µ Polar-RP 80A (250 × 4.6 mm, i.d.) column with a Security Guard AQ C18 (4 × 3 mm, i.d.) both purchased from Phenomenex (Torrance, CA, USA). The detection wavelength and temperature of the column oven were set at 520 nm and 40 °C, respectively. Solvent A consists of water: formic acid (95:5, v/v) and solvent B consisted of acetonitrile: formic acid (95:5, v/v). The flow rate was maintained at 1.0 mL/min and the injection volume was 10 µL. Samples were eluted with gradient condition as follows: started at 0–8 min, 5%–13% B; 8–13 min, 13% B; 13–20 min, 13%–17% B; 20–23 min, 17% B; 23–30 min, 17–20% B; 30–40 min, 20% B; 40–40.1 min, 20-5% B; and 40.1–50 min, 5% B (total 50 min) [18]. Peaks were identified and qualified using Empower 3 software (Waters Corporation, Milford, MA, USA). The samples of T1 and T2 generations of transgenic black nightshade plants were analyzed by LC/MS/MS mass spectrometry. The contents of total anthocyanins were measured by a spectrophotometric device.

2.3. LC/MS/MS Analysis

For the LC-MS/MS of transgenic black nightshade plants was analyzed using a Triple TOF 5600 system composed of a Hybrid Quadrupole-TOF LC/MS/MS Mass Spectrometer (AB Sciex Instruments, Framingham, MA, USA); Model 5035153/M; Serial Number BN24891607; Source Housing DuoSpray Ion Source; Sampler G7129B; Metering 40 µL Analytical Heat; DAD G1315D Serial Number DEAAX09767; Vacuum Gauge (10×10^{-5} Torr) at pressure 3.0 temperature 500.0 °C. The LC-MS conditions were set as follows: using the positive ion mode ([M]+) MS/MS High Resolution; duration of 3 µs; scan range from 200 to 1500 *m/z*; pulser frequency value of 15.392 kHz; sample acquisition duration 65 min; ion tolerance 50.000 mDa.

2.4. Measurement of Total Anthocyanins Content by Spectrophotometry

To determine the total anthocyanin content, 50 mg of the dry weight of the sample in 15 mL tube was extracted using 3 mL of methanol in 1% of HCl (v/v) and incubated at 4 °C to avoid light for 16 h. The extracts were centrifuged at 10,000 rpm at 4 °C for 15 min. The supernatant was transferred to a new 15 mL tube. Then the sample was diluted five times with an extraction buffer for measuring the spectrophotometric absorption at 530 nm and 650 nm. Total anthocyanin content was calculated as previously described [19].

2.5. Plant Growth and Optimal Kanamycin Selection Media

Leaf explants of black nightshade were tested to select the optimal concentration of kanamycin on plant regeneration medium. The various concentrations of kanamycin (10, 20, 30, 50, 100 mg/L) were used in Murashige and Skoog (MS) media supplemented with 3% sucrose and 0.8% agar that contained 2.0 mg/L of 6-benzylaminopurine (BAP) and 0.1 mg/L of (1-naphthalene acetic acid (NAA). Excised leaves within 5 mm × 5 mm size were placed on plant regeneration media and then incubated at room temperature and 25 °C under 16 h light/8 h dark condition. After 1 month approximately, no shoot was induced on medium containing 50 mg/L of kanamycin. Thus, a full range of MS supplemented with 2.0 mg/L of BAP, 0.1 mg/L of NAA, and 50 mg/L of kanamycin was used as transgenic selection media.

2.6. Construction of AtPAP1 Overexpression Vector

The sequence of *AtPAP1* was clone into pDONR 221 vector using BP clonase II (Invitrogen) according to the manufacturer's instructions. The amplified genomic coding region was transformed into TOP10 cells (Invitrogen, Carlsbad, CA, USA) and transferred to the pK7WGF2 destination vector for N-terminal GFP-fusion and C-terminal GFP-fusion (http//www.psb.ugent.be/gateway) using LR clone (Invitrogen). The resultant AtPAP1 overexpression constructs consisted of a 35S promoter, Green fluorescence protein (GFP), the respective gene sequences, and the neomycin phosphotransferase (NPTII) gene as a selectable marker (Figure 1). The pk7WGF2 plasmid was then transferred into *Agrobacterium tumefaciens* GV3101 by electroporation and grew at 28 °C on Luria-Bertani (LB) medium with rifampicin, and spectinomycin at a concentration of 50 g/mL. A pK7WGF2 vector consisted of a 35S promoter GFP-GUS fusion and the neomycin phosphotransferase (*NPTII*) gene as a selection marker was used as a control. Excised leaves of *S. nigrum* from 1-month old seedlings were used as the explant material for co-culture with into *A. tumefaciens* GV3101 harboring GUS, and AtPAP1 overexpression constructs.

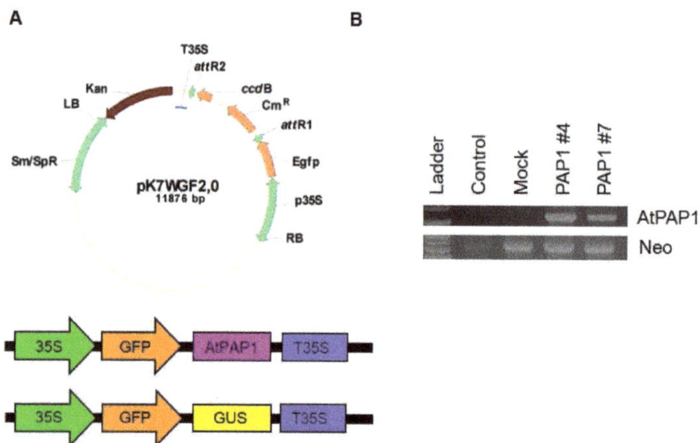

Figure 1. The schematic diagrams of T-DNA region of transformation vectors. (**A**) Plasmids pK7WGF2-GFP-PAP1 and pK7WGF2-GFP-GUS were used to transform *S. nigrum*. (**B**) qRT-PCR analysis of AtPAP1 gene and kanamycin selection marker (Neo) gene in AtPAP1overexpression lines.

2.7. Agrobacterium Preparation

A. tumefaciences GV3101:pK7wgf2: GFP: PAP1 was inoculated from glycerol stock and cultured in a 30 mL of LB (1% tryptone, 0.5% yeast extract, and 1% NaCl, pH 7.0) liquid media supplemented with 50 mg/L of gentamycin and 25 mg/L of rifampicin, following overnight incubation at 28 °C.

2.8. Co-Culture with Agrobacterium

Agrobacterium mid-log phase (O.D A600 = 0.6) cells were harvested by centrifugation at 4,000 rpm at 4 °C for 10 min. *A. tumefaciens* was resuspended in half-strength MS liquid media and cell density was adjusted to an O.D A600 of 1.0 for plant co-culture. The leaf explant from 3 weeks old seedling was wounded by using a scalpel, then immersed in *A. tumefaciens* GV3101 suspended liquid medium for 15–20 min. The leaf explants were then incubated on MS media supplemented with 2.0 mg/L of BAP and 0.1 mg/L of NAA without antibiotics for two days. The infected leaf explant was transferred to the plant regeneration media containing MS supplement with 2.0 mg/L of BAP, 0.1 mg/L of NAA, and antibiotics (50 mg/L of kanamycin + 500/250 mg/L of cefotaxime) for one week. The single shoot emerged from the wounded site was separated from the leaf explant and transferred to the rooting media containing $\frac{1}{2}$ MS media supplemented with antibiotics (50 mg/L of kanamycin + 250 mg/L of cefotaxime). After 2 weeks of rooting, a single whole plant was transplanted to the pot and acclimatized for two weeks before transferring to the growth chamber.

2.9. PCR Analysis of the Transgenic Lines

Plant genomic DNA for polymerase chain reaction (PCR) was extracted by the Genomic DNA mini kit (Geneaid Biotech Ltd., New Taipei City, Taiwan), following the manufacturer's instructions. The leaves of transformation (AtPAP1) lines and control (GUS and GV3101) plants were ground into a fine powder (50–100 mg fresh weight) using a mortar and pestle. Extracted genomic DNA was subjected to PCR analysis using gene-specific primers provided in Table 1.

Table 1. Comparison of phenylpropanoid biosynthetic genes of *S. nigrum* with most orthologous genes.

Genes	Length (amino acid)	Orthologous genes (Accession No.)	Identity (%)
SnPAL	712	*Solanum pennellii* PAL XP_015055648.1	97
		Solanum lycopersicum PAL XP_004249558.1	97
		Capsicum baccatum PAL PHT36452.1	96
SnC4H	505	*Solanum tuberosum* C4H ABC69046	99
		Solanum pennellii C4H XP_015078931.1	99
		Capsicum baccatum C4H PHT46927.1	97
Sn4CL	545	*Solanum tuberosum* 4CL NP_001305568.1	95
		Solanum pennellii 4CL XP_015070224.1	95
		Solanum lycopersicum 4CL NM_001346841.1	95
SnC3H	510	*Solanum tuberosum* C3H XP_006362631.1	94
		Solanum pennellii C3H XP_015084850.1	93
		Solanum lycopersicum C3H XP_004228867.1	93
SnCHS	271	*Nicotiana tabacum* CHS XP_016494186.1	81
		Solanum tuberosum CHS XP_006367318.1	76
		Solanum pennellii CHS XP_015060638.1	75
SnCHI	217	*Solanum brevicaule* CHI APZ86742.1	94
		Solanum tuberosum CHI APZ86744.1	93
		Solanum melongena CHI ANN02871.1	90
SnUGT75C1	473	*Solanum tuberosum* UGT75C1 XP_006358760.1	92
		Solanum pennellii UGT75C1 XP_015088519.1	91
		Lycium barbarum UGT75C1 BAG80544.1	86
SnCOMT	363	*Solanum tuberosum* COMT XP_015164331.1	95
		Nicotiana attenuata COMT OIT03318.1	93
		Solanum pennellii COMT XP_015070697.1	93
SnHCT	435	*Solanum pennellii* HCT XP_015070028.1	97
		Solanum lycopersicum HCT XP_004235891.1	96
		Capsicum annuum HCT NP_001311756.1	96

The PCR conditions consisted of denaturation at 95 °C for 30 s, annealing at 58 °C for 30 s and extension at 72 °C for 30 s, followed by 35 cycles and the final extension at 72 °C for 5 min. The PCR products were analyzed on a 1% agarose gel stained with ethidium bromide and visualized under UV light.

2.10. Transgenic Plant Seed Segregation Assay

The transgenic plants were grown in a growth chamber for four months and the mature seeds of T0 primary transgenic plants were harvested. Twelve AtPAP1 overexpression transgenic seeds (T1) were harvested from seeds germinated on half-strength MS media with 50 mg/L Kanamycin selection marker. Then five shoots of kanamycin-resistant in each line were grown on the pot (T_2) for three months. The transgenic plants were grown in a growth chamber for four months and the mature seeds of T_0 primary transgenic plants were harvested. Following generations (i.e., T1 and T2) screened on 50 mg/L kanamycin media were subjected to phenotype and anthocyanin analysis.

2.11. Total RNA Extraction and cDNA Synthesis

Total RNA was isolated from *S. nigrum* at fruit maturity stage by using RNA Mini Kit (Geneaid) according to the manufacturer's protocol. The quantity and quality of total RNA were examined on a 1.0% agarose gel and using NanoVue Plus Spectrophotometer (GE Healthcare Life Sciences, Pittsburgh, PA, USA), respectively.

2.12. cDNA Synthesis and Quantitative Real-time PCR

The cDNA was synthesized from 1 μg of DNA free total RNA using ReverTra Ace-α Kit with oligo (dT) 20 primer (Toyobo, Osaka, Japan). The resulting cDNA products were used as templates for real time-PCR analysis. For all target genes, primers were designed using Primer Design software (GeneRunner.exe) and using an online program (https://www.genscript.com/ssl-bin/app/primer). Quantitative real-time PCR was performed in a BIO-RAD CFX96 Real-time PCR system (Bio-Rad Laboratories, Hercules, CA, USA) with 2X QuantiTect SYBR Green RT-PCR Master Mix, (QIAGEN, Hilden, Germany) under the condition as follows: reverse transcription at 50 °C for 30 min, PCR initial at 95 °C for 15 min, denaturation at 94 °C for 15 s, annealing at 60 °C for 30 s, extension at 72 °C for 30 s and reaction cycle was repeated for 39 cycles at extension followed by a final extension at 72 °C for 1 min. All reactions were from three biological replicates.

2.13. Statistical Analysis

Statistical analyses were performed using GraphPad InStat software (Graphpad Instat Software, San Diego, CA, USA). One-way ANOVA was applied for comparisons with the Turkey test. All data are expressed as mean ± SEM. The significance threshold was at a level of 5% (* $P < 0.05$; ** $P < 0.01$, *** $P < 0.001$). All the experiments were biological triplicates.

3. Results

3.1. Identification of Phenylpropanoid Biosynthesis Genes in Black Nightshade

The sequences of genes involved in the phenylpropanoid biosynthesis pathway were identified from the RNA-seq data of the black nightshade. The retrieved sequences were subject to a BLAST search to confirm and compare with the orthologs of other *Solanaceae* plants. The genes identified in black nightshade associated with general phenylpropanoid pathways were designated as *SnPAL* (accession MT032183, 712 amino acids), *SnC4H* (accession MT032184, 505 amino acids), *Sn4CL* (accession MT032185, 545 amino acid), *SnC3H* (accession MT032186, 510 amino acids), and *SnCHI* (accession MT032187, 217 amino acids). In addition, a biosynthesis gene involved in the anthocyanin pathway (e.g., anthocyanin-5-*o*-glucosyl transferase) was designated as *SnUGT75C1* (accession MT032188, 473 amino acids), and genes more closely linked to the lignin synthesis (e.g., caffeate *O*-methyltransferase

(*COMT*)) and hydroxycinnamoyl-CoA shikimate (*HCT*)) were designated as *SnCOMT* (accession MT032189, 363 amino acids), and *SnHCT* (accession MT032190, 435 amino acids) (Table 1). To add perspective to the identified sequence of black nightshade, orthologs of several other *Solanaceae* plants that showed most similarity to the black nightshade genes were also compared. Except for *SnCHS*, all the deduced amino acid sequences from various *Solanaceae* plants had sequence identity from 90 to 99 percentile between orthologous genes in *Solanaceae* plants.

3.2. Induction of Anthocyanin by Overexpression of AtPAP1 in Black Nightshade

We used a gain-of-function approach to examine the effect of overexpression of AtPAP1 on black nightshade. To this end, we generated the transgenic black nightshade plants that harbor AtPAP1 cDNA and GUS (control) both driven by CaMV 35S (Figure 1A). The presence of the transgenes was confirmed by PCR analysis of AtPAP1 transgenic plants using Kan primers (Neo). Expected sizes of amplicons (630 bp) were observed in the corresponding *S. nigrum* plant lines (Figure 1B). In addition, the expression of *AtPAP1* transgene was verified by quantitative RT-PCR (Figure 1B).

More than 15 individual transgenic lines were obtained and maintained in a growth chamber (Figure 2A–F), for which the transgene was also confirmed by PCR using Kan primers (Neo) (Figure 2G). Overexpression of *Arabidopsis* AtAPAP1 resulted in enhanced color changes all over the plant including leaves and flowers compared with the control plant (Figure 3A). The appearance of purple color was especially notable in leaves and floral organs such as petals and anthers (Figure 3B,C). Thus, our data demonstrate that overexpression of a single *Arabidopsis* regulatory gene, *AtPAP1*, might have increased the content of anthocyanin in the heterologous system of black nightshade.

Figure 2. Schematic diagram of AtPAP1 overexpression transgenic lines of *S. nigrum* by Agrobacterium-mediated transformation. (**A**) Shoot regeneration for 30 days after AtPAP1 transformation. (**B**) Shoot regeneration after sub-culture for 30 days of AtPAP1 transformation. (**C**) & (**D**) Development of the whole plant. (**E**) Transgenic plant rooting. (**F**) Transgenic plant after transfer to the pot for 40 days. (**G**) Confirmation of transgene using kanamycin (Neo) primer in diverse transgenic lines.

Figure 3. Phenotype comparison of AtPAP1 overexpression transgenic lines of *S. nigrum*. (**A**) Whole plants of AtPAP1 transgenic lines and the control line at fruit development stage in a growth chamber. (**B**) Leaf and (**C**) flower phenotype comparison between AtPAP1 overexpression lines and control (GFP-GUS (Control) and GV3101 (Mock)) lines.

3.3. Expression of Anthocyanin Biosynthesis Genes in the AtPAP1-overexpressed Black Nightshade Plants

To understand the mechanism underlying the increased content of anthocyanin in the transgenic black nightshade, we examined gene expression of anthocyanin biosynthesis that might have been affected by overexpression of AtPAP1. As expected, expression of AtPAP1 transgene was up-regulated in the transgenic black nightshade plants (PAP1 #4 and #7) compared with the control (GFP-GUS) and Mock (GV3101) lines (Figure 4). When we examined the genes identified in this study, associated with the phenylpropanoid biosynthesis pathway, most of the genes were highly induced in the AtPAP1 overexpression lines (Figure 5). Of the genes identified in the current study, genes known to be associated with general phenylpropanoid pathways (EBGs), such as *SnPAL* (accession MT032183), *SnC4H* (accession MT032184), *Sn4CL* (accession MT032185), and *SnCHI* (accession MT032187) were all highly induced in the AtPAP1 overexpressed plants. Expression of other biosynthesis genes associated with anthocyanin (*SnUGT75C1*, accession MT032188) and lignin (*SnHCT*, accession MT032190) was largely unaltered. This suggests that black nightshade genes involved in the general phenylpropanoid biosynthesis pathway may have been regulated by the AtPAP1 transcription factor that rendered the purple color in *AtPAP1*-overexpressing *S. nigrum* plants, which ultimately led to an increase in the levels of anthocyanin in vegetative tissue and flower parts.

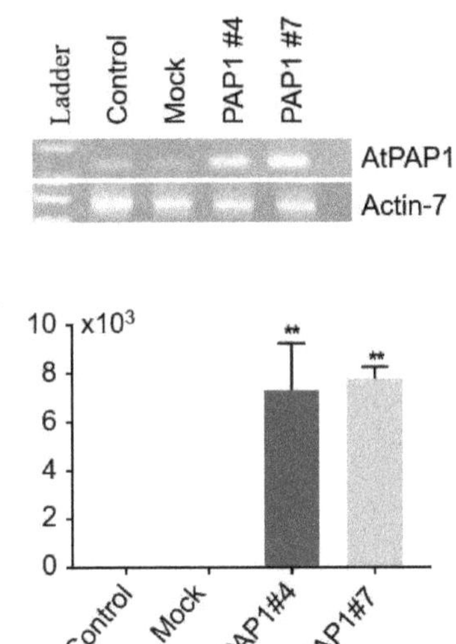

Figure 4. Expression of AtPAP1 gene in AtPAP1 overexpressing *S. nigrum* transgenic lines and control lines. All data are expressed as mean and SEM. The significance threshold was at a level of 5% (* $P <$ 0.05; ** $P < 0.01$).

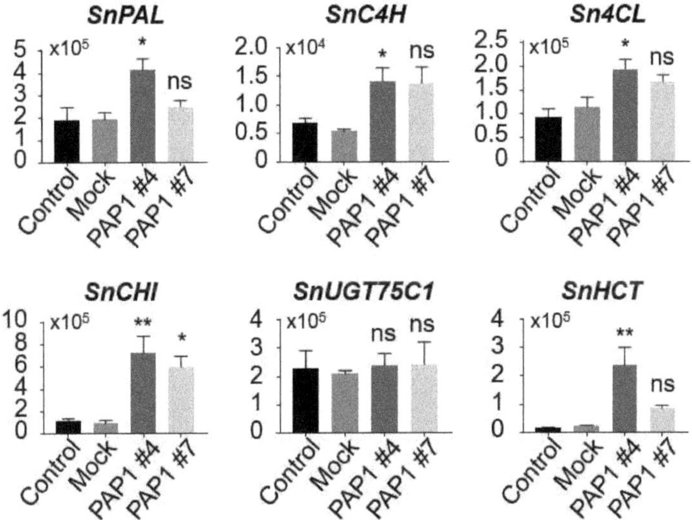

Figure 5. Expression of flavonoid biosynthetic genes in AtPAP1 overexpressing *S. nigrum* transgenic lines and control lines. The AtPAP1 transgenic lines were harvested and proceeded to qRT-PCR analysis. SnPAL, phenylalanine ammonia-lyase; SnC4H, cinnamate-4-hydroxylase; Sn4CL, 4-coumarate: CoA ligase; SnCHI, chalcone isomerase; SnUGT75C1, UDP-glycosyltransferase 75C1; SnHCT, quinate hydroxycinnamoyl transferase. All data are presented as mean and SEM. The significance threshold was at a level of 5% (* $P < 0.05$; ** $P < 0.01$,). All the experiments were biological triplicates.

3.4. Identification of Anthocyanins in the AtPAP1-overexpressed Black Nightshade Plants

To understand the types of anthocyanins that might have accumulated resulting from the overexpression of AtPAP1, we investigated the metabolites that correspond to anthocyanins using

HPLC. From the first and second generation of AtPAP1-overexpressed transgenic black nightshade lines, we have detected several peaks associated with the various anthocyanins in black nightshade transgenic lines, while no detectable peaks have been observed in the control line (Figure 6).

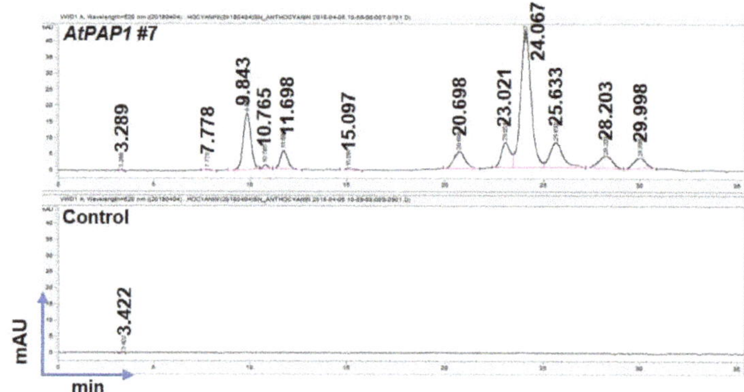

Figure 6. Identification of anthocyanins in AtPAP1 overexpressed *S. nigrum* transgenic lines by HPLC analysis. Above, AtPAP1 overexpression *S. nigrum* line. Detected eight peaks represented unknown anthocyanins at different retention time; Below, *S. nigrum* control line. No peak detected indicated the absence of anthocyanin.

Subsequent analysis of LC-MS/MS mass spectrometry has revealed that each peak of the HPLC data obtained demonstrated the same pattern of the mass ionization in both T0 and T1 transgenic lines, indicating that anthocyanins detected in two generations of transgenic plants were identical. The total ion chromatogram (TIC) was separated into 18 main peaks, from which only six peaks that represent anthocyanins were identified. The *m/z* ratio of each intact anthocyanin and its daughter fragments are listed in Table 2. Of these, three aglycones were determined as delphinidin aglycone (Dd, *m/z* 303), petunidin aglycone (Pt, *m/z* 317) and malvidin aglycone (Mv, *m/z* 331) [20]. In addition, three delphinids, two petunidin and one malvidin were detected in the AtPAP1-overexpressed black nightshade transgenic lines.

Specifically, the mass spectrum of peak 8 with a retention time at 16.41 min that had a molecular ion *m/z* of 773.2136 [M]+ and fragment *m/z* of 627.156 ([M-146]+) identified as delphinidin 3-O-rutinoside-5-O-glucoside with a loss of one molecule of rhamnose [21]. The peak 9 with retention time at 20.47 min that had a molecular ion *m/z* of 627.1557 and fragment *m/z* of 611.1610 was identified to be a 3,5-diglucoside glycone (m/z 303 [Dd]+ known as delphinidin aglycon), and this peak represented delphinidin-3,5-O-diglucoside. The peak 12 with a retention time at 25.42 min with *m/z* 611.1611[M]+ and fragmentation *m/z* of 464.1031 ([M-146]+) corresponds to a delphinidin aglycone that is derived from loss of one molecule of rhamnose with *m/z* 303 [Dp]+, was identified as an anthocyanin derivative (delphinidin 3-0-rutinoside) [21]. Peak 13 with a retention time at 28.46 min that showed a molecular ion *m/z* and fragment *m/z* of 933.2679 and 641.1721 [M-146-146], respectively, appeared as a petunidin aglycone with a loss of two molecules of coumaric acid (*m/z* 317([Pt]+) identified as petunidin-3-O-rutinoside (*cis-p*-coumaroyl)-5-O-glucoside or petunidin-3-O-rutinoside (*trans-p*-coumaroyl)-5-O-glucoside. The peak 15 with retention time at 32.68 min that had a molecular ion and fragment of 963.2803 and 317 [Pt]+ corresponded to a petunidin aglycone, which was tentatively identified as petunidin-3-(feruloyl)-rutinoside-5-glucoside [21–23]. Finally, the peak 17 with a retention time at 36.54 min was identified as malvidin-3-(feruloyl)-rutinoside-5-glucoside by the ion at *m/z* of 977.2945 [M]+ and the fragment at *m/z* of 801.2463 [M-176]+, produced by elimination of one molecule of ferulic acid (*m/z* 331[Mv]+) that is malvidin aglycone (Figure 7) [24].

Table 2. The molecular ionization characteristic of anthocyanin detected in AtPAP1 overexpression *S. nigrum* lines.

Peak No.	RT (min)	[M]+ (m/z)	Fragmentation (*m/z*)	Tentative Identification	Molecular Formula
1	5.4400	465.0432	ND	Unknown	
2	6.2000	ND	ND	Unknown	
3	9.2600	ND	ND	Unknown	
4	10.4600	ND	ND	Unknown	
5	13.1181	ND	ND	Unknown	
6	13.5200	787.2294	ND	Unknown	
7	14.7600	617.0800	ND	Unknown	
8	16.4103	773.2136	627.1565 ([M-146]+)	Delphinidin 3-O-rutinoside-5-O-glucoside	C33H41O21
9	20.9300	627.1557	611.1610 ([3,5-diglucoside]+); 303.0501 [Dd]+	Delphinidin-3,5-O-diglucoside	C27H30O17
10	22.1800	627.1557	303.0501 [Dd]+	Delphinidin-3,5-O-diglucoside	
11	24.2900	741.2243	ND	Unknown	
12	26.5900	611.1611	464.10331 ([M-146]+); 303.0501 [Dd]+	Delphinidin-3-O-rutinoside	C27H31O16
13	29.5200	933.2679	641.1721 ([M-146-146]+); 317.0605 [Pt]+	Petunidin 3-O-(p-coumaroyl) rutinoside-5-O-glucoside	C43H49O23
14	31.5517	933.2191	641.1721 ([M-146-146]+)	Unknown	
15	32.6800	963.2803	316.9172 [Pt]+	Petunidin-3-(feruloyl)-rutinoside-5-glucoside	C44H51O24
16	35.0600	947.2843	493.0000, 301.0000	Unknown	
17	36.9300	977.2945	801.2463 ([M-176]+); 331.2095 [Mv]+	Malvidin-3-(feruloyl)-rutinoside-5-glucoside	C45 H53 O24
18	40.6300	447.2900	331.0000 [Mv]+	Unknown	

* Abbreviations: M, molecular; Dd, delphinidin; Pt, petunidin; Mv, malvidin. Glycone mass and tentative assignment: 331, malvidin; 317, petunidin; 303, delphinidin; 176, glucose/ ferulic acid; 146, rhamnose/coumaric acid; 611, 3,5-diglucoside.

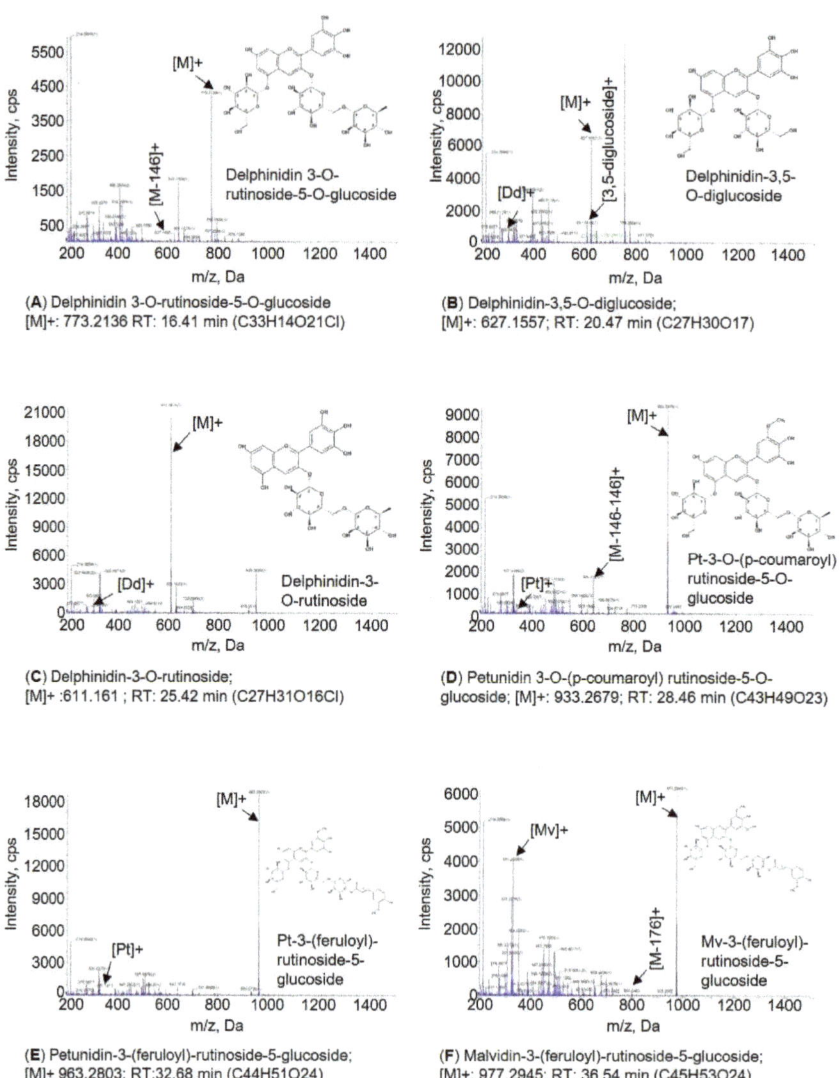

(**A**) Delphinidin 3-O-rutinoside-5-O-glucoside [M]+: 773.2136 RT: 16.41 min (C33H14O21Cl)

(**B**) Delphinidin-3,5-O-diglucoside; [M]+: 627.1557; RT: 20.47 min (C27H30O17)

(**C**) Delphinidin-3-O-rutinoside; [M]+ :611.161 ; RT: 25.42 min (C27H31O16Cl)

(**D**) Petunidin 3-O-(p-coumaroyl) rutinoside-5-O-glucoside; [M]+: 933.2679; RT: 28.46 min (C43H49O23)

(**E**) Petunidin-3-(feruloyl)-rutinoside-5-glucoside; [M]+ 963.2803; RT:32.68 min (C44H51O24)

(**F**) Malvidin-3-(feruloyl)-rutinoside-5-glucoside; [M]+: 977.2945; RT: 36.54 min (C45H53O24)

Figure 7. Mass spectrometric data of six anthocyanins detected in AtPAP1 overexpressed *S. nigrum* transgenic lines. (**A**) Delphinidin 3-O-rutinoside-5-O-glucoside, (**B**) Delphinidin-3,5-O-diglucoside, (**C**) Delphinidin-3-O-rutinoside, (**D**) Petunidin 3-O-(p-coumaroyl) rutinoside-5-O-glucoside, (**E**) Petunidin-3-(feruloyl)-rutinoside-5-glucoside, and (**F**) Malvidin-3-(feruloyl)-rutinoside-5-glucoside.

3.5. Total Anthocyanin Contents induced by Overexpression of AtPAP1 in Black Nightshade

In addition to six anthocyanins identified in the AtPAP1-overexpressed transgenic black nightshade plants, we measured the total anthocyanin content in the leaves of AtPAP1-overexpressed black nightshade transgenic plants using spectrophotometry. Compared with control and mock (GV3101) lines, both first and second generation plants of the AtPAP1-overexpression lines had higher levels of total anthocyanins (Figure 8). Our data collectively demonstrate that ectopic expression of a single regulatory gene, AtPAP1, can induce total anthocyanins in black nightshade plants, likely through regulation of the conserved biosynthesis pathway of anthocyanin in black nightshade plants.

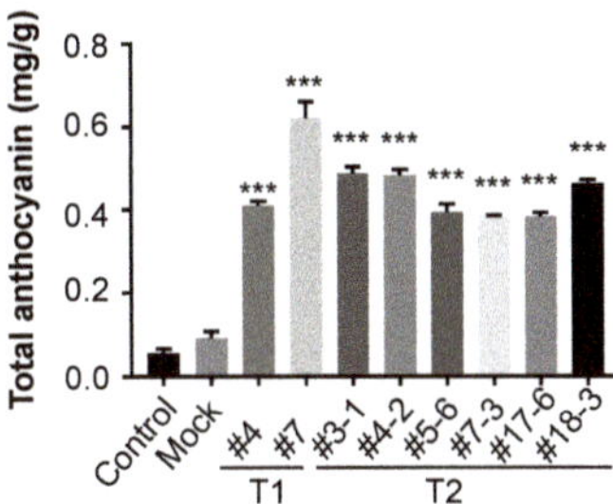

Figure 8. Total anthocyanin contents in AtPAP1 overexpressed *S. nigrum* transgenic lines. Total anthocyanin was measured by spectrophotometry at A530 nm and A650 nm. All data are expressed as mean and SEM. All AtPAP1 overexpressed *S. nigrum* transgenic lines from first (T1) and second generation (T2) were compared with control and mock lines. The significance threshold was at a level of 5% (*** $P < 0.001$). All the experiments were biological triplicates.

4. Discussion

In the current study, we have demonstrated the accumulation of anthocyanins by overexpressing an AtPAP1 in the black nightshade plant. Expression analysis of the genes associated with the biosynthesis of general phenylpropanoid and flavonoid pathways indicates that not all the regulatory genes of anthocyanins may not be transcriptional targets of AtPAP1, and may require unidentified species-specific mechanisms. Nonetheless, a single *Arabidopsis* MYB transcription factor, AtPAP1, was sufficient to enhance the accumulation of anthocyanins in a heterologous system of black nightshade.

In *Arabidopsis*, it has been shown that the biosynthesis of anthocyanins and proanthocyanidins is predominantly regulated by the MBW transcriptional complex that consists of MYB, bHLH, and WD-40 [25]. Alternatively, an *Arabidopsis* MYB transcription factor, AtPAP1, has been shown to be sufficient to promote accumulation of anthocyanin in a heterologous plant system such as dandelion (*Taraxacum brevicorniculatum*) [26]. Novel MYB transcription factors from the American plum (*Prunus americana*) have been also demonstrated to be effective in the accumulation of anthocyanins in tobacco and citrus without requiring any further transcriptional co-regulators except for *PamMYBA3*, which appears to be dependent upon an interacting protein (bHLH) to form a functional transcriptional regulatory unit [27]. Moreover, overexpression of MYB transcription factor, EsMybA1, from *Herba epimedii* (*Epimedium sagittatum)* was sufficient to induce high levels of anthocyanins in the transgenic tobacco and *Arabidopsis*, via up-regulation of major flavonoid biosynthetic pathway genes [28]. These reports indicate that some MYB transcription factors physically associate with the co-regulators such as bHLH, while other MYB transcriptional factors appear to be sufficient to induce the accumulation of anthocyanins in diverse plant species [29].

From our study, ectopic expression of an *Arabidopsis* AtPAP1 alone was sufficient to induce anthocyanin production in black nightshade, resulting in the purple-colored phenotype in both vegetative and reproductive tissues. It is noteworthy that *Gerbera* MYB transcription factor GMYB10, an *Arabidopsis* ortholog of AtPAP1, activates some of EBGs of general phenylpropanoid pathway genes (*PAL, C4H, CHI,* and *F3H*) and accumulate anthocyanins [30]. It would be interesting to know if the genes identified as EBGs of black nightshade are also the transcriptional targets of AtPAP1, which confers the pigmentation of anthocyanins in the transgenic lines.

Although numerous studies about anthocyanins in berries (e.g., cherry, blueberry, blackcurrant) have been reported to date, studies of anthocyanins in black nightshade are still scarce. Nonetheless, a recent study by Wang et al. [20] has shown that eight anthocyanins with four anthocyanin aglycones such as cyanidin (Cy), malvidin (Mv), petunidin (Pt) and delphinidin (Dp) were detected in the peel and flesh of black nightshade fruit [20]. In our study, we have identified six anthocyanins that included three delphinids, two petunidin and one malvidin from leaf tissue of AtPAP1-overexpressed transgenic black nightshade plants. Among the anthocyanins identified in this study, delphinidin 3-O-rutinoside-5-O-glucoside, delphinidin-3,5-O-diglucoside, delphinidin-3-O-rutinoside, and malvidin-3-(feruloyl)-rutinoside-5-glucoside were also detected in the fruit of black nightshade [20]. In addition to the above four anthocyanins, two petunidin, petunidin-3-O-rutinoside (*cis-p*-coumaroyl)-5-O-glucoside and petunidin-3-(feruloyl)-rutinoside-5-glucoside, were identified only in our study. Thus the latter two appear to be leaf-specific anthocyanins that are regulated by AtPAP1 in black nightshade.

In addition to developmental regulation, biosynthesis of anthocyanins is affected by various abiotic factors (e.g., light, sucrose, nitrogen) and phytohormones [5]. Stress tolerance against drought, high salinity, light and cold are often correlated with anthocyanin accumulation in plants [31]. These various abiotic stress conditions are known to induce a distinct set of anthocyanins, indicating that anthocyanins have different biological functions, or that coloration patterns of each anthocyanin may play a unique role in response to a certain stress factor [32]. It would be interesting to examine whether enhancing anthocyanin production in a black nightshade would help mediate stress tolerance against stress conditions. Of interest in this regard was that anthocyanin pigmentation of transgenic black nightshade plants has disappeared when grown in the greenhouse (data not are shown), which suggests that biosynthesis of anthocyanins in black nightshade governed by AtPAP1 is under the control of light and likely temperature conditions. How diverse stress conditions that would affect the accumulation of anthocyanins in black nightshade remain to be investigated.

5. Conclusions

In this present study, we have shown that an *Arabidopsis* MYB transcription factor, AtPAP1, is sufficient to induce the accumulation of anthocyanins in black nightshade that may provide the platform for the basic analysis of phytochemical biosynthetic genes and their final compounds. Furthermore, a black nightshade can be a model system to study AtPAP1-induced anthocyanin in response to various stress conditions. Further identification of detailed mechanisms that are associated with an AtPAP1-mediated and environment-modulated accumulation of anthocyanins remains to be investigated. Taken together, our data lay a foundation to study the molecular mechanisms of biosynthesis of flavonoid and anthocyanin pathways to improve the phytochemical properties in this species.

Author Contributions: S.C. is the first author. J.J. is the contributing author. S.C. and S.U.P. conceived the study. Experimental data were generated and analyzed by S.C., J.J., J.K., S.U.P. All authors reviewed and wrote the manuscript. All authors have read and agreed to the published version of the manuscript.

Funding: This research was supported by the Bio & Medical Technology Development Program of the National Research Foundation (NRF) funded by the Ministry of Science, ICT & Future Planning (2016M3A9A5919548) (S.U.P.) and by the National Research Foundation of Korea (NRF) grant funded by the Korean government (Ministry of Science and ICT) (NRF-2017R1A2B4010356) (J.K.).

Conflicts of Interest: The authors declare no competing financial interest.

References

1. Jain, R.; Sharma, A.; Gupta, S.; Sarethy, I.P.; Gabrani, R. *Solanum nigrum*: Current perspectives on therapeutic properties. *Altern Med. Rev.* **2011**, *16*, 78–85. [PubMed]

2. Sridhar, T.M.; Josthna, P.; Naidu, C.V. In vitro antibacterial activity and phytochemical analysis of *Solanum nigrum* (Linn.)—An important antiulcer medicinal plant. *J. Exp. Sci.* **2011**, *2*, 24–29.

3. Lev-Yadun, S.; Gould, K.S. Role of anthocyanins in plant defence. In *Anthocyanins*; Winefield, C., Davies, K., Gould, K., Eds.; Springer: New York, NY, USA, 2009; pp. 21–48. [CrossRef]

4. Ferrari, D.; Cimino, F.; Fratantonio, D.; Molonia, M.S.; Bashllari, R.; Busa, R.; Saija, A.; Speciale, A. Cyanidin-3-O-glucoside modulates the in vitro inflammatory crosstalk between intestinal epithelial and endothelial cells. *Mediat. Inflamm.* **2017**. [CrossRef] [PubMed]

5. Shi, M.Z.; Xie, D.Y. Biosynthesis and metabolic engineering of anthocyanins in *Arabidopsis thaliana*. *Recent Pat. Biotechnol* **2014**, *8*, 47–60. [CrossRef] [PubMed]

6. Liu, Y.; Tikunov, Y.; Schouten, R.E.; Marcelis, L.F.M.; Visser, R.G.F.; Bovy, A. Anthocyanin biosynthesis and degradation mechanisms in *Solanaceous* vegetables: A review. *Front. Chem.* **2018**, *6*, 52. [CrossRef] [PubMed]

7. Gonzalez, A. Pigment loss in response to the environment: A new role for the WD/bHLH/MYB anthocyanin regulatory complex. *New Phytol.* **2009**, *182*, 1–3. [CrossRef]

8. Matsui, K.; Umemura, Y.; Ohme-Takagi, M. AtMYBL2, a protein with a single MYB domain, acts as a negative regulator of anthocyanin biosynthesis in *Arabidopsis*. *Plant J.* **2008**, *55*, 954–967. [CrossRef]

9. Petroni, K.; Tonelli, C. Recent advances on the regulation of anthocyanin synthesis in reproductive organs. *Plant Sci.* **2011**, *181*, 219–229. [CrossRef]

10. Gonzalez, A.; Zhao, M.; Leavitt, J.M.; Lloyd, A.M. Regulation of the anthocyanin biosynthetic pathway by the TTG1/bHLH/Myb transcriptional complex in *Arabidopsis* seedlings. *Plant J.* **2008**, *53*, 814–827. [CrossRef]

11. Jeong, J.B.; De Lumen, B.O.; Jeong, H.J. Lunasin peptide purified from *Solanum nigrum* L. protects DNA from oxidative damage by suppressing the generation of hydroxyl radical via blocking Fenton reaction. *Cancer Lett.* **2010**, *293*, 58–64. [CrossRef]

12. Shin, D.H.; Choi, M.G.; Lee, H.K.; Cho, M.; Choi, S.B.; Choi, G.; Park, Y.I. Calcium dependent sucrose uptake links sugar signaling to anthocyanin biosynthesis in *Arabidopsis*. *Biochem. Bioph. Res. Co.* **2013**, *430*, 634–639. [CrossRef] [PubMed]

13. Das, P.K.; Shin, D.H.; Choi, S.B.; Yoo, S.D.; Choi, G.; Park, Y.I. Cytokinins enhance sugar-induced anthocyanin biosynthesis in *Arabidopsis*. *Mol. Cells* **2012**, *34*, 93–101. [CrossRef] [PubMed]

14. Rubin, G.; Tohge, T.; Matsuda, F.; Saito, K.; Scheible, W.R. Members of the LBD family of transcription factors repress anthocyanin synthesis and affect additional nitrogen responses in *Arabidopsis*. *Plant Cell* **2009**, *21*, 3567–3584. [CrossRef]

15. Gou, J.Y.; Felippes, F.F.; Liu, C.J.; Weigel, D.; Wang, J.W. Negative regulation of anthocyanin biosynthesis in *Arabidopsis* by a miR156-targeted SPL transcription factor. *Plant Cell* **2011**, *23*, 1512–1522. [CrossRef]

16. Cuong, D.M.; Jeon, J.; Morgan, A.M.A.; Kim, C.; Kim, J.K.; Lee, S.Y.; Park, S.U. Accumulation of charantin and expression of triterpenoid biosynthesis genes in bitter melon (*Momordica charantia*). *J. Agr. Food Chem.* **2017**, *65*, 7240–7249. [CrossRef]

17. Jeon, J.; Bong, S.J.; Park, J.S.; Park, Y.K.; Arasu, M.V.; Al-Dhabi, N.A.; Park, S.U. De novo transcriptome analysis and glucosinolate profiling in watercress (*Nasturtium officinale* R. Br.). *BMC Genomics* **2017**, *18*, 401. [CrossRef]

18. Woo, H.; Kim, J.W.; Han, K.M.; Lee, J.H.; Hwang, I.S.; Lee, J.H.; Kim, J.; Kweon, S.J.; Cho, S.; Chae, K.R.; et al. Simultaneous analysis of 17 diuretics in dietary supplements by HPLC and LC-MS/MS. *Food Addit. Contam. Part A* **2013**, *30*, 209–217. [CrossRef]

19. Lotkowska, M.E.; Tohge, T.; Fernie, A.R.; Xue, G.P.; Balazadeh, S.; Mueller-Roeber, B. The Arabidopsis transcription factor MYB112 promotes anthocyanin formation during salinity and under high light stress. *Plant Physiol.* **2015**, *169*, 1862–1880. [CrossRef]

20. Wang, S.L.; Chu, Z.H.; Ren, M.X.; Jia, R.; Zhao, C.B.; Fei, D.; Su, H.; Fan, X.Q.; Zhang, X.T.; Li, Y.; et al. Identification of anthocyanin composition and functional analysis of an anthocyanin activator in *Solanum nigrum* fruits. *Molecules* **2017**, *22*, 6. [CrossRef]

21. Xu, W.J.; Luo, G.J.; Yu, F.Y.; Jia, Q.X.; Zheng, Y.; Bi, X.Y.; Lei, J.J. Characterization of anthocyanins in the hybrid progenies derived from *Iris dichotoma* and *I. domestica* by HPLC-DAD-ESI/MS analysis. *Phytochemistry* **2018**, *150*, 60–74. [CrossRef]

22. Roldan, M.V.G.; Outchkourov, N.; van Houwelingen, A.; Lammers, M.; de la Fuente, I.R.; Ziklo, N.; Aharoni, A.; Hall, R.D.; Beekwilder, J. An *O*-methyltransferase modifies accumulation of methylated anthocyanins in seedlings of tomato. *Plant J.* **2014**, *80*, 695–708. [CrossRef] [PubMed]

23. Zheng, J.; Ding, C.X.; Wang, L.S.; Li, G.L.; Shi, J.Y.; Li, H.; Wang, H.L.; Suo, Y.R. Anthocyanins composition and antioxidant activity of wild *Lycium ruthenicum* Murr. from Qinghai-Tibet Plateau. *Food Chem.* **2011**, *126*, 859–865. [CrossRef]

24. Su, X.Y.; Xu, J.T.; Rhodes, D.; Shen, Y.T.; Song, W.X.; Katz, B.; Tomich, J.; Wang, W.Q. Identification and quantification of anthocyanins in transgenic purple tomato. *Food Chem.* **2016**, *202*, 184–188. [CrossRef] [PubMed]

25. Xu, W.J.; Dubos, C.; Lepiniec, L. Transcriptional control of flavonoid biosynthesis by MYB-bHLH-WDR complexes. *Trends Plant Sci.* **2015**, *20*, 176–185. [CrossRef] [PubMed]

26. Qiu, J.; Sun, S.Q.; Luo, S.Q.; Zhang, J.C.; Xiao, X.Z.; Zhang, L.Q.; Wang, F.; Liu, S.Z. Arabidopsis AtPAP1 transcription factor induces anthocyanin production in transgenic *Taraxacum brevicorniculatum*. *Plant Cell Rep.* **2014**, *33*, 669–680. [CrossRef] [PubMed]

27. Dasgupta, K.; Thilmony, R.; Stover, E.; Oliveira, M.L.; Thomson, J. Novel R2R3-MYB transcription factors from *Prunus americana* regulate differential patterns of anthocyanin accumulation in tobacco and citrus. *Gm. Crops. Food.* **2017**, *8*, 85–105. [CrossRef]

28. Huang, W.J.; Sun, W.; Lv, H.Y.; Luo, M.; Zeng, S.H.; Pattanaik, S.; Yuan, L.; Wang, Y. A R2R3-MYB transcription factor from *Epimedium sagittatum* regulates the flavonoid biosynthetic pathway. *PLoS ONE* **2013**, *8*, e70778. [CrossRef]

29. Anwar, M.; Wang, G.Q.; Wu, J.C.; Waheed, S.; Allan, A.C.; Zeng, L.H. Ectopic overexpression of a novel R2R3-MYB, *NtMYB2* from Chinese *Narcissus* represses anthocyanin biosynthesis in tobacco. *Molecules* **2018**, *23*, 4. [CrossRef]

30. Laitinen, R.A.E.; Ainasoja, M.; Broholm, S.K.; Teeri, T.H.; Elomaa, P. Identification of target genes for a MYB-type anthocyanin regulator in *Gerbera hybrida*. *J. Exp. Bot.* **2008**, *59*, 3691–3703. [CrossRef]

31. Kovinich, N.; Kayanja, G.; Chanoca, A.; Otegui, M.S.; Grotewold, E. Abiotic stresses induce different localizations of anthocyanins in Arabidopsis. *Plant Signal. Behav.* **2015**, *10*, e1027850. [CrossRef]

32. Kovinich, N.; Kayanja, G.; Chanoca, A.; Riedl, K.; Otegui, M.S.; Grotewold, E. Not all anthocyanins are born equal: Distinct patterns induced by stress in *Arabidopsis*. *Planta* **2014**, *240*, 931–940. [CrossRef] [PubMed]

Article

Asiatic Acid, Extracted from *Centella asiatica* and Induces Apoptosis Pathway through the Phosphorylation p38 Mitogen-Activated Protein Kinase in Cisplatin-Resistant Nasopharyngeal Carcinoma Cells

Yen-Tze Liu [1,2,3,4], Yi-Ching Chuang [4], Yu-Sheng Lo [4], Chia-Chieh Lin [4], Yi-Ting Hsi [4], Ming-Ju Hsieh [1,3,4,5,*] and Mu-Kuan Chen [6,*]

1   Institute of Medicine, Chung Shan Medical University, Taichung 402, Taiwan; 144084@cch.org.tw
2   Department of Family Medicine, Changhua Christian Hospital, Changhua 500, Taiwan
3   Department of Holistic Wellness, Mingdao University, Changhua 52345, Taiwan
4   Oral Cancer Research Center, Changhua Christian Hospital, Changhua 500, Taiwan;
    177267@cch.org.tw (Y.-C.C.); 165304@cch.org.tw (Y.-S.L.); 181327@cch.org.tw (C.-C.L.);
    180082@cch.org.tw (Y.-T.H.)
5   Graduate Institute of Biomedical Sciences, China Medical University, Taichung 404, Taiwan
6   Department of Otorhinolaryngology, Head and Neck Surgery, Changhua Christian Hospital,
    Changhua 500, Taiwan
*   Correspondence: 170780@cch.org.tw (M.-J.H.); 53780@cch.org.tw (M.-K.C.);
    Tel.: +886-4-7238595 (ext. 1018) (M.-K.C.); Fax: +886-4-7232942 (M.-K.C.)

Received: 4 December 2019; Accepted: 23 January 2020; Published: 25 January 2020

Abstract: Nasopharyngeal carcinoma (NPC) is an important issue in Asia because of its unique geographical and ethnic distribution. Cisplatin-based regimens are commonly the first-line used chemotherapy, but resistance and toxicities remain a problem. Therefore, the use of anticancer agents derived from natural products may be a solution. Asiatic acid (AA), extracted from *Centella asiatica*, was found to have anticancer activity in various cancers. The aim of this study is to examine the cytotoxic effect and mediated mechanism of AA in cisplatin-resistant NPC cells. The results shows that AA significantly reduce the cell viability of cisplatin-resistant NPC cell lines (cis NPC-039 and cis NPC-BM) in dose and time dependent manners caused by apoptosis through the both intrinsic and extrinsic apoptotic pathways, including altered mitochondrial membrane potential, activated death receptors, increased Bax expression, and upregulated caspase 3, 8, and 9. The Western blot analysis of AA-treated cell lines reveals that the phosphorylation of MAPK pathway proteins is involved. Further, the results of adding inhibitors of these proteins indicates that the phosphorylation of p38 are the key mediators in AA-induced apoptosis in cisplatin-resistant human NPC cells. This is the first study to demonstrate the AA-induced apoptotic pathway through the phosphorylation p38 in human cisplatin-resistant nasopharyngeal carcinoma. AA is expected to be another therapeutic option for cisplatin-resistant NPC because of the promising anti-cancer effect and fewer toxic properties.

Keywords: Asiatic acid; nasopharyngeal cancer; apoptosis; cisplatin resistance; MAPK pathway

---

1. Introduction

Nasopharyngeal carcinoma (NPC) is an important and widely studied issue in Asia, especially in Eastern Asia and South-Eastern Asia, because of its unique geographical distribution. According to the GLOBOCAN estimates in 2018, the estimated number of new cases is 129,079 worldwide, of which 109,221 are in Asia, accounting for 84.6% [1]. Etiological factors for NPC include Epstein–Barr virus (EBV) infection, human papillomavirus (HPV) infection, genetic susceptibility, and salted fish consumption [2]. Different from other head and neck cancer, the primary and only curative treatment for NPC is radiotherapy instead of surgery for early-stage disease. Chemotherapeutic agents are used as concurrent chemoradiotherapy (CCRT), adjuvant chemotherapy, or induction chemotherapy in locoregionally advanced disease and metastatic disease [3]. Cisplatin-based regimen is commonly the first-line used. Paiar el al. proposed that the overall survival (OS) and disease-free survival by CCRT had statistically significant improvement with five-year OS of about 70% in non-metastatic stage III and IV disease, but the efficacy of chemotherapy varied markedly among different trials [4]. Otherwise, chemoradiotherapy leads to various side effects including hematological acute toxicity, mucositis, xerostomia, loss of taste, dysphagia, skin damage, skull bone damage, sensorineural hearing loss, and osteoradionecrosis. Therefore, new effective treatments with fewer toxicities need to be identified.

Finding anti-cancer agents from natural compound is a burgeoning research filed recent years due to its fewer toxic characteristics. Asiatic acid (AA) is a pentacyclic triterpenoid, one of the subclasses of phytochemicals, and derived from *Centella asiatica* which is a kind of tropical plant used as a diet and medicine [5]. AA has been demonstrated to have antioxidant and anti-inflammatory activity, and to have been involved in numerous molecular mechanisms and cell signaling pathway which related to its cardio-protective, neuro-protective, anti-diabetic, anti-obesity, and anti-cancer effect [5]. Previous studies have reported many mechanisms observed in carcinogenesis and metastatic process. Hsu el al. discovered the anticancer effect of AA in breast cancer through cell cycle arrest and mitochondrial apoptotic pathway [6]. Another research proposed that AA inhibit transforming growth factor-beta1 (TGF-beta1)-induced epithelial-mesenchymal transition in lung cancer [7]. However, the effect of AA on nasopharyngeal cancer cells is still unclear.

As described above, radiotherapy and chemotherapy are effective in early-stage NPC. In contrast, the median overall survival in distant metastases is only about 12 to 15 months after platinum-based chemotherapy [8]. Multidrug combinations can improve the therapeutic effects, but accompanied with significant hematological and mucosal toxicities, even septic deaths [9]. Even patients who initially respond to cisplatin will recur, and it is often refractory to further platinum therapy [10]. Whether it is innate or acquired resistance, to discover effective and less-toxic agents for anti-cancer or chemosensitizers is important. This study aims to examine the cytotoxic effect and mechanisms of AA on cisplatin-resistance NPC cells.

2. Materials and Methods

2.1. Chemicals

Asiatic acid is a triterpenoid isolated from *Centella asiatica* ($\geq$98% purity), purchased from Chemfaces. Molecular Weight: 488.7, Formula: $C_{30}H_{48}O_5$, dissolved in dimethyl sulfoxide (DMSO), and stored at -20 °C. All experiment treatments were consistently less than 0.1% (v/v) DMSO concentration. Other chemicals were obtained from the following companies: MTT (3-(4,5-dimethylthiazol-2-yl)-2,5-diphenyltetrazolium bromide) and DAPI dye (Sigma Aldrich, St Louis, MO, USA), cisplatin (Enzo Life Sciences), specific inhibitors (Santa Cruz Biotechnology, Santa Cruz, CA, USA).

2.2. Cell Culture

The cisplatin-resistant nasopharyngeal carcinoma cell lines (cis NPC-BM and cis NPC-039) were established by human nasopharyngeal carcinoma (NPC) cell lines (NPC-BM and NPC-039), a gift from Dr. Jen-Tsun Lin, Hematology & Oncology, Changhua Christian Hospital, continuous culture with increase concentrations of cisplatin 1–60 nM for 6 months. The culture medium contains 60 nM cisplatin to maintain drug resistance. All cells were cultured in RPMI 1640 medium (with 10% (v/v) fetal bovine serum (FBS), 1 mM glutamine, 1% (v/v) penicillin/streptomycin, 1.5 g/l sodium bicarbonate, and 1 mM sodium pyruvate). All cells were cultured at 37 °C in a 5% CO_2 culture room.

2.3. Cell Cytotoxicity Assay

The effect of Asiatic Acid on cell growth was assayed by the MTT (3-(4,5-dimethylthiazol-2-yl)-2,5-diphenyl tetrazolium bromide) method. Briefly, cells were cultured in 96-well plates (1×10^4/well) and treated with various concentrations of AA (0, 25, 50, 75 µM). After the medium was removed, MTT reagent was added to each well (0.5 mg/mL final concentration) with a further incubation for 4 h at 37 °C in 5% (*v/v*) CO_2. After removing the supernatant, DMSO was added to dissolve the formed blue formazan crystals. The absorbance of the converted dye was measured at 595 nm by the ELISA plate reader. Each condition was performed in triplicate and data were obtained from at least 3 separate experiments.

2.4. Cell Cycle Analysis

Cells were seeded (5×10^5/well) on 6-well plates overnight and incubated with various concentrations of AA for 24 h. After wards, cells were fixed 70% (*v/v*) ice-cold ethanol overnight, and then stained with PI buffer (4 mg/mL PI, 1% (*v/v*) Triton X-100, 0.5 mg/mL RNase A in PBS) for 30 min in the dark at room temperature and then filtered through a 40-mm nylon filter (Falcon, Billerica, MA, USA). The cell cycle distribution was analyzed by flow cytometry.

2.5. DAPI Staining

Observation of morphology by DAPI staining cis NPC-BM and cis NPC-039 cells treated with different concentrations of AA (0, 25, 50, and 75 µM) were incubated for 24 h. The cells were fixed with 4% (*v/v*) formaldehyde for 30 min at room temperature and permeabilized with 0.1% (*v/v*) Triton X-100. Washed twice in PBS, cells were stained with DAPI (50 µg/mL) for 15 min in the dark, then after being washed in PBS, morphological changes were photographed using fluorescence microscopy (Lecia, Bensheim, Germany). Quantitatively, the percentage of apoptotic cells by number of dots was scored on at least 500 cells.

2.6. Annexin V/PI Double Staining Assay

Cells were seeded on 6-well overnight and then treated with different concentrations of AA for 24 h. Cells were harvested and suspended in PBS (2% (*v/v*) BSA), and then incubated with Muse™ Annexin V & Dead Cell reagent (EMD Millipore, Billerica, MA, USA) for 20 min at room temperature in dark. Samples were analyzed by Muse Cell Analyzer flow cytometry (EMD Millipore, Billerica, MA, USA) and by MUSE 1.4 Analysis software (EMD Millipore).

2.7. Mitochondrial Membrane Potential Measurement

Cells were seeded on 6-well plates and incubated with different dose of AA for 24 h. The cells were harvested, washed and suspended in Muse MitoPotential working solution at 37 °C for 20 min. Add 5 µL of 7-AAD and incubate at room temperature for 5 min. Samples were monitored by a Muse Cell Analyzer flow cytometry (EMD Millipore) The data were analyzed by a Muse Cell Analyzer (Millipore).

2.8. Western Blotting Analysis

The cells are washed twice with cold PBS and lysed with RIPA buffer containing protease inhibitor cocktail and phosphatase inhibitor cocktail. The protein quantification of supernatants was determined using BCA protein assay (Pierce) and all samples are separated by sodium dodecyl sulfate polyacrylamide gel electrophoresis (SDS-PAGE) and the separated proteins were transferred electrophoretic ally from the gel to the surface of PVDF membrane (Millipore, Bedford, MA, USA). Membranes were blocked with 5% non-fat milk in TBST for 1 h. The analysis used primary antibodies as described by the manufacturers of the antibodies. Antibodies were purchased from Cell Signaling Technology, Inc. (Danvers, MA, USA). Anti-ERK 1/2 (#4695), Anti-JNK 1/2 (#9252), Anti-p38 (#8690), Anti-Akt (#4685), Anti-phospho-ERK 1/2 (#4370), Anti-phospho-JNK 1/2 (#4668), Anti-phospho-p38

(#4511), Anti-phospho-Akt (#4060), Anti-cleaved caspase 3 (#9664), Anti-cleaved caspase 8(#9496), Anti-cleaved caspase 9 (#9505), Anti-cleaved PARP(#5625), Anti-Bax (#2772), Anti-Bak (#3814), Anti-Fas (#4233), Anti-TNF-R1(#3736), Anti-DCR2 (#8049), Anti-DR5 (#8074) or isotype-control rabbit mAbs (Cell Signaling Technology, Beverly, MA). After the final rinsing with TBST, the membrane was incubated with secondary HRP-linked anti-rabbit IgG or anti-mouse IgG for 1 h. After washing, the immunoblots were visualized using chemiluminescence (ECL) HRP substrate (Millipore).

2.9. Statistical Analysis

Values represent the means ± standard deviation and the experiments were repeated at least three times. Data were analyzed using Student's t-test when two groups were compared. A p value of <0.05 was considered statistically significant. Statistical analysis using one-way analysis of variance (ANOVA). Analysis of more than three groups using Dunnett's test.

3. Results

3.1. Cytotoxicity of Asiatic acid on Cisplatin-Resistance Human NPC Cell Lines

Figure 1a reveals the chemical structure of Asiatic acid, a pentacyclic triterpene, titrated extract from *Centella asiatica*. We assessed the cell viability of two kinds of cisplatin-resistant human NPC cell lines, cis NPC-039 and cis NPC-BM, by MTT assay. These cell lines were treated with an increasing concentration of AA (0, 25, 50, and 75 µM) with or without cisplatin for 24, 48, and 72 h. The results in Figure 1b,c show that the cell viability of both cis NPC-039 and cis NPC-BM was inhibited in a dose- and time-dependent manner regardless of whether with or without cisplatin. Cell viability was significantly reduced in concentration of 50 and 75 µM compared to control (0 µM).

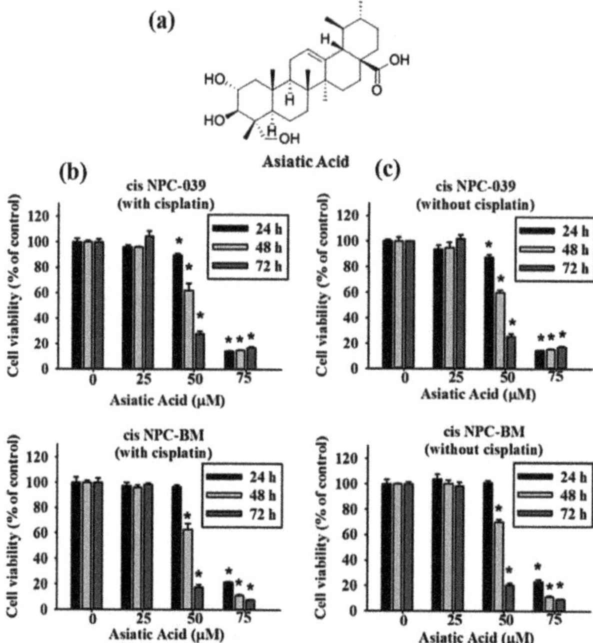

Figure 1. Cytotoxic effects of Asiatic acid on cisplatin-resistant human NPC cell lines (NPC-039 and NPC-BM) (**a**) The chemical structure of Asiatic acid. (**b**) The cell viability significantly decreased in AA-treated cis NPC-BM cell and cis NPC-039 cell with cisplatin and (**c**) without cisplatin in a dose and time dependent manner under MTT assay. * $p < 0.05$ as compared with control group.

3.2. Asiatic Acid Induced Cell Apoptosis and Cell Cycle Arrest in Cisplatin-Resistance Human NPC Cell Lines

PI staining and flow cytometry were applied to determine the cell cycle distribution of cisplatin-resistance human NPC cell lines, cis NPC-039 and cis NPC-BM, and to examine the effect of AA on cell growth suppression. Figure 2 shows that the proportion in the sub-G1 phase increased in a concentration-dependent manner, especially in 75 µM group. Figure 3a illustrates the more condensed nuclei, suggesting the typical characteristic morphology of DNA fragmentation in apoptotic cells, in the higher concentration group of AA-treated NPC cell lines with the fluorescence microscopy after DAPI staining. The difference was statistically significant in 50 and 75 µM groups of both two cell lines (Figure 3b). Furthermore, we performed the Annexin-V and PI double staining and followed by flow cytometric analysis. The results in Figure 3c,d indicated that AA increased apoptotic cells in early stage and late stage of apoptosis, and significantly in 75 µM -treated cell lines.

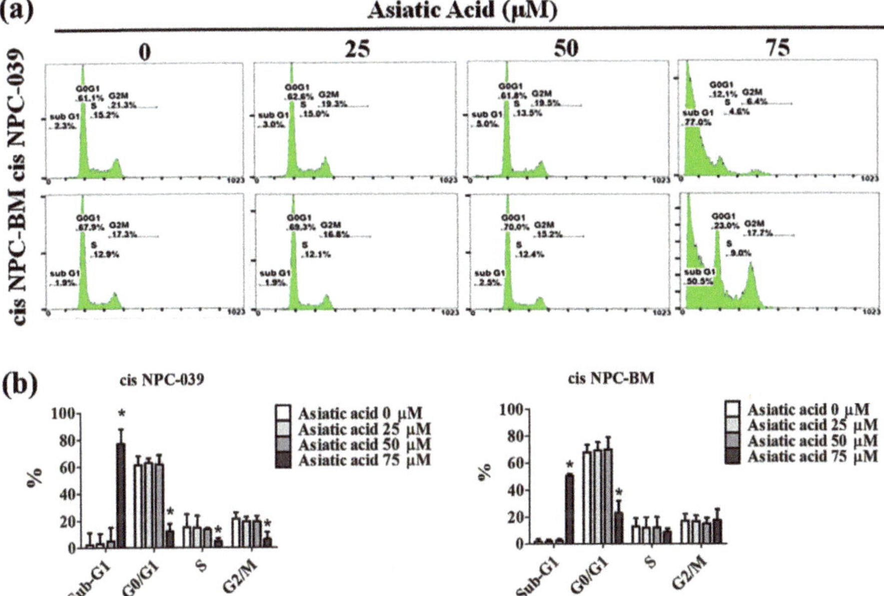

Figure 2. Asiatic acid cause cell cycle arrest in cis NPC-039 and cis NPC-BM cell lines. (**a**) The cisplatin-resistant human NPC cell lines were treated with Asiatic acid (0, 25, 50, and 75 µM). Cell cycle phase distribution (Sub-G1, G0/G1, S, and G2/M) were estimated by flow cytometry. (**b**) The rate of different phase of cell cycle in cis NPC-039 and cis NPC-BM cell lines. * $p < 0.05$ as compared with control group.

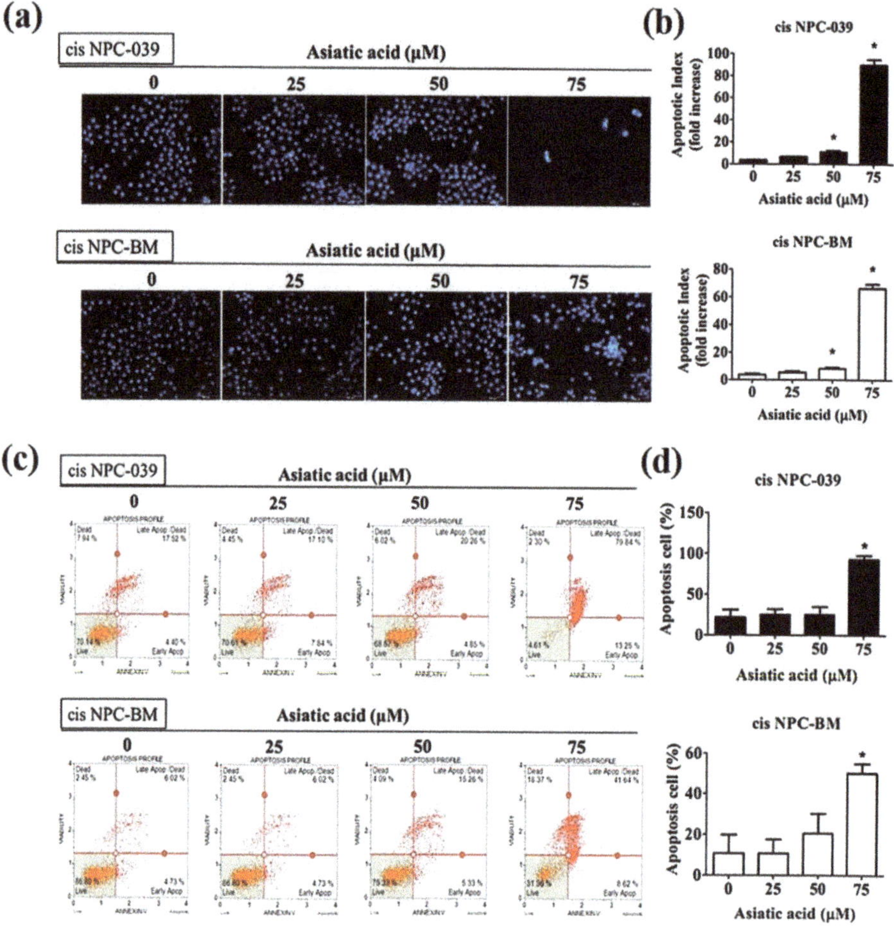

Figure 3. Effect of Asiatic acid on apoptosis induction in cis NPC-039 and cis NPC-BM cell lines. (**a**) Nuclear counterstain level was detected by DAPI staining. Characteristic 'blebbing' morphology of cell nucleus was observed by fluorescence microscopy. (**b**) The DNA condensation fold compared with control. (**c**) and (**d**) Cell apoptosis was analyzed by flow cytometric with Annexin-V/PI Staining. * $p < 0.05$ as compared with control group.

3.3. Asiatic Acid Induced Cell Apoptosis in Cisplatin-Resistance Human NPC Cell Lines through the Mitochondrial Pathway and Death Receptor Related Pathway

The pathway involved in the AA-induced apoptosis in cis NPC-039 and cis NPC-BM cell lines have to be verified. We performed the Muse MitoPotential Kit and Muse Cell Analyzer assays. The results in Figure 4a,b display that the mitochondrial membrane potential was changed and the depolarized cells was significantly increased with AA at 75 µM. On the other hand, the expressions of Fas, DCR2, and DR5 were all increased in AA-treated cisplatin-resistant NPC cell lines by Western blot analysis (Figure 4c,d).

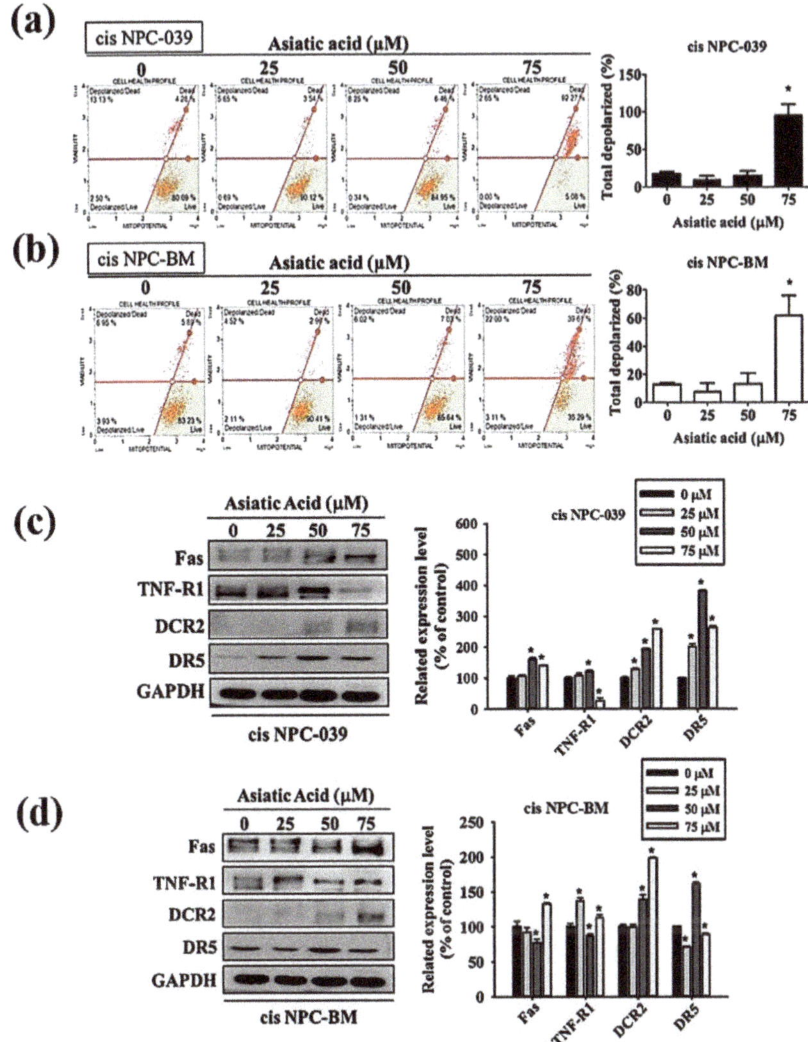

Figure 4. Asiatic acid cause cis NPC-039 and cis NPC-BM cell apoptosis thought the intrinsic pathway and the extrinsic pathway. (**a**) Effect of Asiatic acid on mitochondrial membrane potential was analyzed by flow cytometric (**b**) The related mitochondrial membrane depolarized levels compared with control. (**c**) The expression levels of Fas, TNF-R1, DcR2, and DR5 were detected by Western blot. (**d**) The relative band density was normalized to GAPDH as a densitometer. * $p < 0.05$ versus control group.

3.4. Asiatic Acid Upregulates the Expression of Caspase-3, -8, -9, and Altered the Expression of Bax and Bak in Cisplatin-Resistance Human NPC-039 and NPC-BM Cell Lines

We performed Western blot analysis of the cleaved forms of caspase-3, caspase-8, caspase-9, and Poly (ADP-ribose) polymerase (PARP) in AA-treated NPC cell lines to determine the caspase activation pathway in AA-induced apoptosis. As observed in Figure 5a,b, the expressions of caspase-3, -8, -9, and PARP increase significantly in cis NPC-039 cell lines with AA at 50 μM compared to control group. In cis NPC-BM cell clines, significantly increased expression of caspase-3, -8, and -9 was found in AA treatment of 75 μM for 24 h. In Figure 5c,d, the related expression level of Bax and Bak, two pro-apoptotic proteins, in cis NPC-BM cell line were significantly increased by 183% and 137% respectively. In another cis NPC-039 cell line, 24-h 50 μM and 75 μM AA treatment increased Bax expression, but decreased Bak expression.

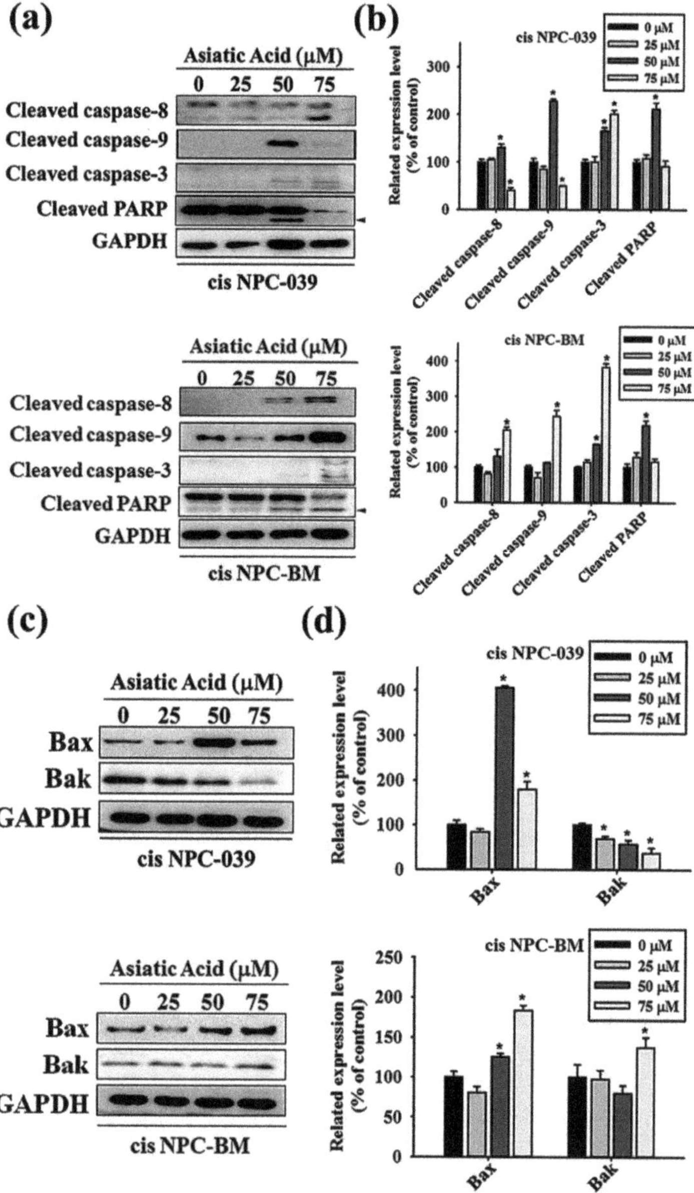

Figure 5. Asiatic acid activates caspase 3, 8, and 9 and upregulated the expression of Bax and Bak. (**a**) The expression levels of cleaved caspase-3, -8, -9 and PARP were detected by Western blot. (**b**) The protein levels were quantified and normalized to GAPDH using a densitometer. (**c**) The expression levels of Bak and Bak were detected by Western blot. (**d**) The protein levels were quantified and normalized to GAPDH using a densitometer. * $p < 0.05$ versus control group.

3.5. Asiatic Acid Induced Cell Apoptosis via p38 Pathway in Cisplatin-Resistance Human NPC-039 and NPC-BM Cell Lines

We applied the Western blot analysis to obtain the level of expression of ERK1/2, AKT, p38, and JNK1/2 to confirm the mechanisms related to AA-mediated apoptosis. The results in Figure 6a,b display the significantly increased expressions of phosphorylation of p38 and JNK1/2 with dose-dependent relationship in cis NPC-039 cell lines, whereas decreased the phosphorylation of AKT in 50- and 75 µM -treated groups and ERK1/2 in 50 µM-treated group. In cis BPC-BM cell lines, AA treatment of 50 and 75 µM for 24 h significantly increased the activation of phosphor-p38 by 298% and 348% respectively and significantly decreased the expression of phosphor-ERK1/2. Decreased expression of phosphor-AKT and increased of phosphor-JNK1/2 were found in 75 µM-treated cell group. To further examine the signal transduction pathways involved in apoptosis, we treated cis NPC-039 and cis NPC-BM cell lines with AA of 50 µM plus ERK1/2 inhibitor (U), AKT inhibitor (LY), p38 inhibitor (SB), and JNK1/2 inhibitor (SP). The related expression level of cleaved caspase 9 was increased when adding ERK1/2 inhibitor (U) and JNK1/2 inhibitor (SP) and decreased when adding AKT inhibitor (LY) and p38 inhibitor (SB) in both cell lines compared to AA-treated group. The expression of cleaved PARP was increased when adding JNK1/2 inhibitor (SP) and decreased when adding AKT inhibitor (LY), and p38 inhibitor (SB) in cis NPC-039. However, the expression of cleaved PARP was increased when adding ERK1/2 inhibitor (U) and decreased when adding AKT inhibitor (LY), p38 inhibitor (SB), and JNK1/2 inhibitor (SP) in cis NPC-BM. Therefore, we might propose that the phosphorylation of p38 plays a considerable role in the pathway of AA-related apoptosis.

4. Discussion

Chemotherapy is one of the mainstay options for the treatment against nasopharyngeal carcinoma, especially in the locoregionally advanced or metastatic stage. Cisplatin is most commonly used, but the therapeutic outcome may be restricted because of its toxicity and resistance. This research attempted to discover the promising natural compound against cisplatin-resistance NPC cells. Our results indicate that the Asiatic acid may induce apoptosis in cisplatin-resistance NPC-039 and NPC-BM cell lines via the p38 and ERK signaling transduction pathway.

It has been proposed that Asiatic acid has various pharmacological activities such as anti-inflammatory and antioxidant, as well as the potent anti-hypertensive, neuro and cardio-protective, antimicrobial, and antitumor activities [5]. Many studies had proved the anti-cancer effects in vitro or in vivo of AA in breast cancer, ovary cancer, colon cancer, gastric cancer, cholangiocarcinoma, glioblastoma, lung cancer, lymphoma, and melanoma [6,11–18]. Our study corresponds with these researches and revealed the anti-cancer effect of AA in both dose and time dependent manners in cisplatin-resistance NPC cell lines (Figure 1b). AA induced cell death by cell cycle arrest and apoptosis in this study (Figures 2 and 3). Previous study also revealed that AA induced cells undergo S-G2/M phase arrest and apoptosis in human breast cancer MCF-7 and MDA-MB-231 cell lines [6]. Siddique et al. suggested that AA induced apoptosis through decreasing the Bcl-2 and increasing the Bax, caspase-3 and -9 in a rat model of colon carcinogenesis [11].

Furthermore, we investigated the intrinsic and extrinsic pathway in apoptosis induced by AA. Figure 4 shows that both the mitochondrial pathway and the death receptor-initiated pathway are involved in AA-induced apoptosis, as well as the upregulated Bax and Bak in Figure 5. The research in 2017 revealed that AA-induced apoptosis in lung cancer was caused by altering the mitochondrial membrane potential and lead to activation of caspase and PARP signaling pathways [14]. Related researches mention that AA cause activation of mitochondrial pathway via regulating B cell CLL/lymphoma 2 (BCL2) family proteins and increasing the ratio of Bax to Bcl2 which lead to cancer cell apoptosis [19,20].

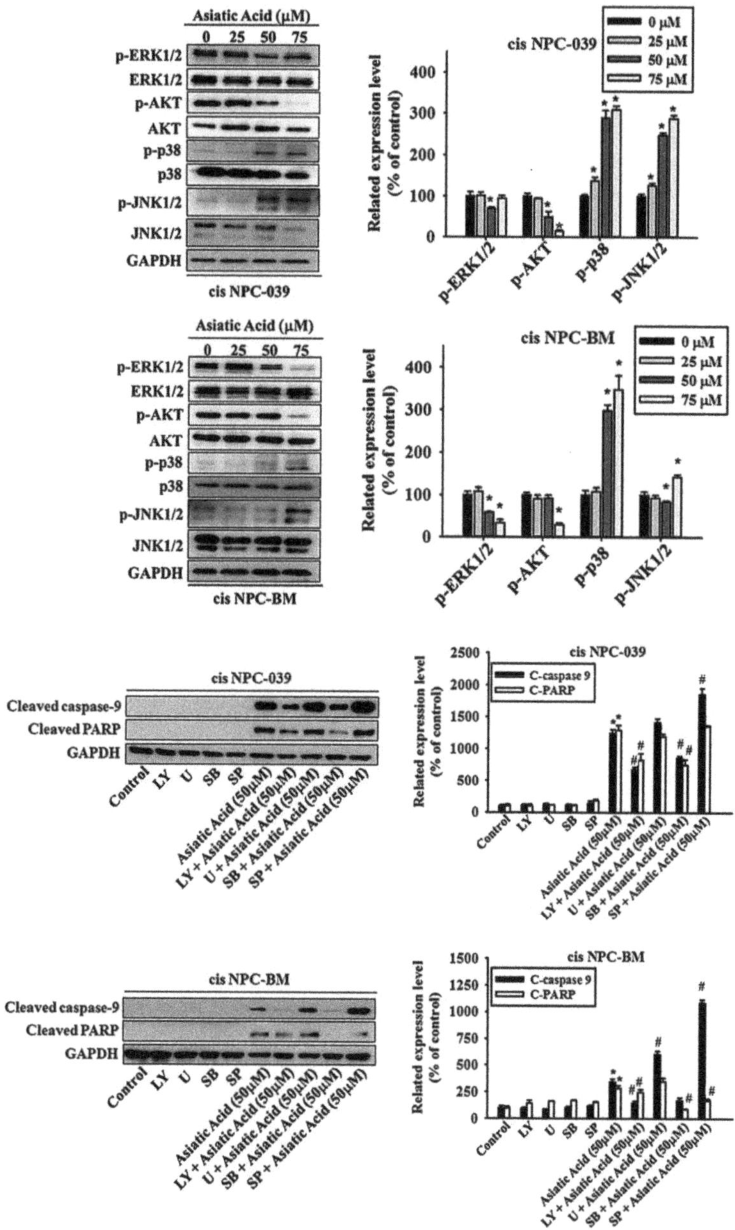

Figure 6. Effect of Asiatic acid on the MAPKs and AKT pathway in cis NPC-039 and cis NPC-BM cell lines (**a**) The expression levels of ERK1/2, AKT, p38, and JNK1/2 were detected by Western blot. (**b**) The protein levels were quantified and normalized to GAPDH using a densitometer. * $p < 0.05$ versus control group. (**c**) After pretreated with ERK1/2, AKT, p38, and JNK1/2 inhibitors separately (U, LY, SB, SP), the expression levels of cleaved caspase-9 and PARP were detected by Western blot. (**d**) The protein levels were quantified and normalized to GAPDH using a densitometer. * $p < 0.05$ versus control group. # $p < 0.05$, compared with Asiatic acid treated only.

Previous studies revealed that the AA-induced apoptosis of human cancer cell is mediated by the mitogen-activated protein kinase (MAPK) pathway with or without involving the activated caspases [5,6,21]. The inhibition of ERK and p38 phosphorylation in AA-treated cancer cell are also noted [15]. Our results in Figures 5 and 6 also indicate that caspase-3, -8, and -9 are upregulated in AA-induced apoptosis and that caspase-9 expression is mediated by the phosphorylation of p38 and ERK1/2 pathway in human NPC cell lines. Other research has indicated the anti-cancer activity of AA by Akt pathway [17,22], but our results didn't show it.

Based on previous research, the resistance of cisplatin comes from three mechanisms: altered DNA repair, cytosolic inactivation of drug, and altered cellular accumulation of drug, and the increased expression of excision repair cross complementation 1 (ERCC1) to alter DNA repair in NPC was related to cisplatin resistance [10]. Liu et al. proposed that the resistance might be sensitized by increasing Bax/Bcl-2 mRNA ratios caused by inhibited NFκB expression and downregulating ERCC1 and thymidine phosphorylase (TP) by inhibiting the phosphorylation of ERK, p38 and PI3K/AKT pathways, which all lead to apoptosis, in human hepatoma HepG2 cells [23]. Another research also revealed the decreased ERCC1 mRNA expression by downregulation of p38 MAPK pathway [21]. Consistent with these studies, our results revealed significantly increased expression of Bax and decreased phosphorylation of ERK in AA-treated cis-NPC cell lines. However, the phosphorylation of p38 was upregulated in our study. Hsieh et al. also proposed that another plant-derived triterpene, Celastrol, also induces cancer cell apoptosis of cisplatin-resistant nasopharyngeal carcinoma though increasing p38 MAPK signaling pathway, as well as ERK1/2 [24]. These inconsistent results of the regulatory signaling pathway and detailed mechanisms require further clarification.

5. Conclusions

In conclusion, this is the first study to demonstrate the AA-induced apoptotic pathway in human cisplatin-resistant nasopharyngeal carcinoma. The mechanism of both intrinsic and extrinsic pathways, including the phosphorylation p38 and Bax upregulation, is involved. Therefore, AA is expected to be another therapeutic option for cisplatin-resistant NPC because of the promising anti-cancer effect and fewer toxic properties.

Data Availability Statement

All the data that supports the findings of this study are available in this article.

Author Contributions: Conceptualization, M.-J.H. and M.-K.C.; methodology, Y.-C.C.; Y.-S.L.; C.-C.L. and Y.-T.H.; software, Y.-T.L.; writing—original draft preparation, Y.-T.L.; writing—review and editing, M.-J.H. and M.-K.C. All authors have read and agreed to the published version of the manuscript.

Funding: This research was funded by grants from National Science Council, Taiwan (MOST 106-2314-B-371-006-MY3) (MOST 106-2314-B-371-005-MY3) and Changhua Christian Hospital (108-CCH-IRP-041, 108-CCH-IRP-002, 108-CCH-ICO-145).

Conflicts of Interest: The authors declare no conflict of interest.

Abbreviations

Nasopharyngeal carcinoma (NPC); Asiatic acid AA); Extracellular signal–regulated kinases (ERKs); c-Jun N-terminal kinases (JNKs); Mitogen-activated protein kinases (MAPK); Propidium iodide (PI)

References

1. Bray, F.; Ferlay, J.; Soerjomataram, I.; Siegel, R.L.; Torre, L.A.; Jemal, A. Global cancer statistics 2018: GLOBOCAN estimates of incidence and mortality worldwide for 36 cancers in 185 countries. *CA: A Cancer J. Clin.* **2018**, *68*, 394–424. [CrossRef] [PubMed]
2. Chua, M.L.K.; Wee, J.T.S.; Hui, E.P.; Chan, A.T.C. Nasopharyngeal carcinoma. *Lancet* **2016**, *387*, 1012–1024. [CrossRef]

3. Chen, Y.-P.; Chan, A.T.C.; Le, Q.-T.; Blanchard, P.; Sun, Y.; Ma, J. Nasopharyngeal carcinoma. *Lancet* **2019**, *394*, 64–80. [CrossRef]
4. Paiar, F.; Di Cataldo, V.; Zei, G.; Pasquetti, E.M.; Cecchini, S.; Meattini, I.; Mangoni, M.; Agresti, B.; Iermano, C.; Bonomo, P.; et al. Role of chemotherapy in nasopharyngeal carcinoma. *Oncol. Rev.* **2012**, *6*, 1. [CrossRef] [PubMed]
5. Meeran, M.F.N.; Goyal, S.N.; Suchal, K.; Sharma, C.; Patil, C.R.; Ojha, S.K. Pharmacological Properties, Molecular Mechanisms, and Pharmaceutical Development of Asiatic Acid: A Pentacyclic Triterpenoid of Therapeutic Promise. *Front. Pharmacol.* **2018**, *9*.
6. Hsu, Y.L.; Kuo, P.L.; Lin, L.T.; Lin, C.C. Asiatic acid, a triterpene, induces apoptosis and cell cycle arrest through activation of extracellular signal-regulated kinase and p38 mitogen-activated protein kinase pathways in human breast cancer cells. *J. Pharmacol. Experiment. Ther.* **2005**, *313*, 333–344. [CrossRef]
7. Cui, Q.; Ren, J.; Zhou, Q.; Yang, Q.; Li, B. Effect of asiatic acid on epithelial-mesenchymal transition of human alveolar epithelium A549 cells induced by TGF-beta1. *Oncol. Lett.* **2019**, *17*, 4285–4292.
8. Loong, H.H.; Ma, B.B.; Chan, A.T. Update on the Management and Therapeutic Monitoring of Advanced Nasopharyngeal Cancer. *Hematol. Clin. North Am.* **2008**, *22*, 1267–1278. [CrossRef]
9. Su, W.-C.; Kao, R.-H.; Tsao, C.-J.; Chen, T.-Y. Chemotherapy with Cisplatin and Continuous Infusion of 5-Fluorouracil and Bleomycin for Recurrent and Metastatic Nasopharyngeal Carcinoma in Taiwan. *Oncol.* **1993**, *50*, 205–208. [CrossRef]
10. Amable, L. Cisplatin resistance and opportunities for precision medicine. *Pharmacol. Res.* **2016**, *106*, 27–36. [CrossRef]
11. Siddique, A.I.; Mani, V.; Renganathan, S.; Ayyanar, R.; Nagappan, A.; Namasivayam, N. Asiatic acid abridges pre-neoplastic lesions, inflammation, cell proliferation and induces apoptosis in a rat model of colon carcinogenesis. *Chem. Interactions* **2017**, *278*, 197–211. [CrossRef] [PubMed]
12. Wang, G.; Jing, Y.; Cao, L.; Gong, C.; Gong, Z.; Cao, X. A novel synthetic Asiatic acid derivative induces apoptosis and inhibits proliferation and mobility of gastric cancer cells by suppressing STAT3 signaling pathway. *OncoTargets Ther.* **2017**, *10*, 55–66. [CrossRef] [PubMed]
13. Thakor, F.K.; Wan, K.-W.; Welsby, P.J.; Welsby, G. Pharmacological effects of asiatic acid in glioblastoma cells under hypoxia. *Mol. Cell. Biochem.* **2017**, *430*, 179–190. [CrossRef] [PubMed]
14. Wu, T.; Geng, J.; Guo, W.; Gao, J.; Zhu, X. Asiatic acid inhibits lung cancer cell growth in vitro and in vivo by destroying mitochondria. *Acta Pharmaceutica Sinica* **2017**, *7*, 65–72. [CrossRef]
15. Wu, Q.; Lv, T.; Chen, Y.; Wen, L.; Zhang, J.; Jiang, X.; Liu, F. Apoptosis of HL-60 human leukemia cells induced by Asiatic acid through modulation of B-cell lymphoma 2 family proteins and the mitogen-activated protein kinase signaling pathway. *Mol. Med. Rep.* **2015**, *12*, 1429–1434. [CrossRef]
16. Lian, G.-Y.; Wang, Q.-M.; Tang, P.M.-K.; Zhou, S.; Huang, X.-R.; Lan, H.-Y. Combination of Asiatic Acid and Naringenin Modulates NK Cell Anti-cancer Immunity by Rebalancing Smad3/Smad7 Signaling. *Mol. Ther.* **2018**, *26*, 2255–2266. [CrossRef]
17. Ren, L.; Cao, Q.-X.; Zhai, F.-R.; Yang, S.-Q.; Zhang, H.-X. Asiatic acid exerts anticancer potential in human ovarian cancer cells via suppression of PI3K/Akt/mTOR signalling. *Pharm. Boil.* **2016**, *54*, 2377–2382. [CrossRef]
18. Sakonsinsiri, C.; Kaewlert, W.; Armartmuntree, N.; Thanan, R.; Pakdeechote, P. Anti-cancer activity of asiatic acid against human cholangiocarcinoma cells through inhibition of proliferation and induction of apoptosis. *Cell. Mol. Boil.* **2018**, *64*, 28–33. [CrossRef]
19. Kim, K.B.; Kim, K.; Bae, S.; Choi, Y.; Cha, H.J.; Kim, S.Y.; Lee, J.H.; Jeon, S.H.; Jung, H.J.; Ahn, K.J.; et al. MicroRNA-1290 promotes asiatic acidinduced apoptosis by decreasing BCL2 protein level in A549 nonsmall cell lung carcinoma cells. *Oncol. Rep.* **2014**, *32*, 1029–1036. [CrossRef]
20. Gonçalves, B.M.; Salvador, J.A.R.; Marín, S.; Cascante, M. Synthesis and anticancer activity of novel fluorinated asiatic acid derivatives. *Eur. J. Med. Chem.* **2016**, *114*, 101–117. [CrossRef]
21. Planchard, D.; Camara-Clayette, V.; Dorvault, N.; Soria, J.-C.; Fouret, P. p38 mitogen-activated protein kinase signaling, ERCC1 expression, and viability of lung cancer cells from never or light smoker patients. *Cancer* **2012**, *118*, 5015–5025. [CrossRef]
22. Hao, Y.; Huang, J.; Ma, Y.; Chen, W.; Fan, Q.; Sun, X.; Shao, M.; Cai, H. Asiatic acid inhibits proliferation, migration and induces apoptosis by regulating Pdcd4 via the PI3K/Akt/mTOR/p70S6K signaling pathway in human colon carcinoma cells. *Oncol. Lett.* **2018**, *15*, 8223–8230. [CrossRef] [PubMed]

23. Liu, C.L.; Lim, Y.P.; Hu, M.L. Fucoxanthin enhances cisplatin-induced cytotoxicity via NFkappaB-mediated pathway and downregulates DNA repair gene expression in human hepatoma HepG2 cells. *Mar. Drugs* **2013**, *11*, 50–66. [CrossRef] [PubMed]
24. Hsieh, M.-J.; Wang, C.-W.; Lin, J.-T.; Chuang, Y.-C.; Hsi, Y.-T.; Lo, Y.-S.; Lin, C.-C.; Chen, M.-K. Celastrol, a plant-derived triterpene, induces cisplatin-resistance nasopharyngeal carcinoma cancer cell apoptosis though ERK1/2 and p38 MAPK signaling pathway. *Phytomedicine* **2018**, *58*, 152805. [CrossRef] [PubMed]

 *biomolecules*

Review

Can Medicinal Plants and Bioactive Compounds Combat Lipid Peroxidation Product 4-HNE-Induced Deleterious Effects?

Fei-Xuan Wang [1,*], Hong-Yan Li [2], Yun-Qian Li [2] and Ling-Dong Kong [3,*]

[1]   National Center of Supervision Inspection on Processed Food & Food Additives Quality, Nanjing Institute of Product Quality Inspection, Nanjing 210019, China
[2]   School of Basic Medical Sciences, Nanjing Medical University, Nanjing 211166, China; hongyanli@njmu.edu.cn (H.-Y.L.); lyq_52245@163.com (Y.-Q.L.)
[3]   School of Life Sciences, Nanjing University, Nanjing 210023, China
*   Correspondence: cyruswangc@gmail.com (F.-X.W.); kongld@nju.edu.cn (L.-D.K.); Tel.: +86-25-86673778 (F.-X.W.)

Received: 1 December 2019; Accepted: 14 January 2020; Published: 16 January 2020

Abstract: The toxic reactive aldehyde 4-hydroxynonenal (4-HNE) belongs to the advanced lipid peroxidation end products. Accumulation of 4-HNE and formation of 4-HNE adducts induced by redox imbalance participate in several cytotoxic processes, which contribute to the pathogenesis and progression of oxidative stress-related human disorders. Medicinal plants and bioactive natural compounds are suggested to be attractive sources of potential agents to mitigate oxidative stress, but little is known about the therapeutic potentials especially on combating 4-HNE-induced deleterious effects. Of note, some investigations clarify the attenuation of medicinal plants and bioactive compounds on 4-HNE-induced disturbances, but strong evidence is needed that these plants and compounds serve as potent agents in the prevention and treatment of disorders driven by 4-HNE. Therefore, this review highlights the pharmacological basis of these medicinal plants and bioactive compounds to combat 4-HNE-induced deleterious effects in oxidative stress-related disorders, such as neurotoxicity and neurological disorder, eye damage, cardiovascular injury, liver injury, and energy metabolism disorder. In addition, this review briefly discusses with special attention to the strategies for developing potential therapies by future applications of these medicinal plants and bioactive compounds, which will help biological and pharmacological scientists to explore the new vistas of medicinal plants in combating 4-HNE-induced deleterious effects.

Keywords: medicinal plants; bioactive compounds; herbs; biological activity; 4-hydroxynonenal; deleterious effects

---

1. Introduction

The toxic reactive aldehyde 4-hydroxynonenal (4-HNE) belongs to the advanced lipid peroxidation end products [1]. As the second toxic messenger of free radicals, 4-HNE modifies specific amino acid residues on proteins to form highly cytotoxic reactive 4-HNE adducts via Michael addition or Schiff base reaction or reacts with DNA via epoxidation under oxidative stress conditions. Oxidative stress is one of the main pathogeneses for human disorders [2]. Accumulation of 4-HNE and formation of 4-HNE adducts are induced by mainly reactive oxygen species (ROS) and other free radicals; participate in several cytotoxic processes such as protein dysfunction, apoptosis, and inflammatory injury; and produce deleterious effects [2–8]. As a result, 4-HNE serves as a biomarker for both lipid peroxidation and oxidative stress [1,2]. Therefore, the inhibition, neutralization, and detoxification of

4-HNE and 4-HNE adducts has become one of the directions of developing potential therapies for oxidative stress-related disorders.

Medicinal plants have been widely used to prevent and treat various kinds of disorders for thousands of years. Their extracts and bioactive compounds have the ability to prevent the cells from oxidative stress injury [9–11]. Some of them have been suggested to be attractive sources of potential agents to mitigate oxidative stress and attenuate 4-HNE-induced toxicity. However, researchers pay little attention to reviewing these medicinal plants and natural bioactive compounds on combating 4-HNE-induced deleterious effects in neurotoxicity and neurological disorder, eye damage, cardiovascular injury, liver injury, and energy metabolism disorder [12–16].

Therefore, this review focuses on the therapeutic effects of medicinal plants in treating 4-HNE-related disorders under oxidative stress. An open-ended, English-restricted search of PubMed/MEDLINE database and Web of Science database has been conducted (up to 31 December 2019) using terms such as 4-HNE-induced or driven/induced by 4-HNE, and medicinal plant/herb/phytochemical/natural compounds compound/constituent. This review aims to highlight the pharmacological basis of medicinal plants and bioactive compounds in the attenuation of 4-HNE-induced deleterious effects, as well as the underlying molecular mechanisms. Additionally, this review briefly discusses the strategies for developing potential therapies in future applications of these medicinal plants and compounds, which will benefit biological and pharmacological scientists to explore the new vistas of medicinal plants in combating 4-HNE-induced deleterious effects.

2. Neuroprotection

Neurotoxicity exacerbated by oxidative stress is strongly associated with 4-HNE toxicity; its further deterioration cleaves nuclear enzyme poly (ADP-ribose) polymerase (PARP) to affect DNA repair and apoptosis in neurons, leading to neurological disorders [1,17]. Polyphenol extract from red wine offers health benefits and can help to avoid neurodegenerative diseases [18]. This extract inhibits 4-HNE-induced cleavage of PARP and protects against apoptosis in neuronal-like catecholaminergic cells (rat pheochromocytoma, PC12 cells) by reducing intracellular ROS [19]. Polyphenolic flavonoids quercetin and myricetin isolated from red wine extract or *Ginkgo biloba* extract have cytoprotective effects on 4-HNE-induced PC12 cell death [19], further demonstrating the benefits in neurodegeneration [20,21]. Other medicinal plant flavonoids luteolin and apigenin have antioxidant, anti-inflammatory, and neuroprotective effects. These two flavonoid compounds also attenuate 4-HNE-induced PARP-1 and caspase-3 activation as well as cell viability in PC12 cells [22].

The accumulation of 4-HNE up-regulates mitogen-activated protein kinase (MAPK) superfamily, especially C-Jun-N-terminal kinase (JNK), which plays a crucial role in oxidative stress-mediated cellular apoptosis [23,24]. Piceatannol, a bioactive stilbene derivative from medicinal plants *Passiflora edulis* or *Gnetum parvifolium*, restores 4-HNE-induced PARP cleavage and apoptosis regulator Bcl-2 expression, and down-regulates the phosphorylation of JNK (p-JNK) in PC12 cell death and nuclear condensation [25]. Recently, maternal supplementation with piceatannol has been demonstrated to have neuroprotection against neonate brain damage and reverse the sensorimotor deficit and cognitive impairment in rats [26]. Additionally, as a traditional medicinal plant in China and Korea, citri reticulatae viride pericarpium also restores 4-HNE-induced inflammatory injury in PC12 cells via inhibiting the activation of p-JNK and nuclear factor kappa-B (NF-κB) [27]. Mitogen-activated protein kinase kinase 4 (MKK4) as an upstream activator of JNK plays a critical role in response to cellular oxidative stress. Cocoa procyanidin fraction and its major antioxidant compound procyanidin B2 are reported to attenuate 4-HNE-induced nuclear condensation and increase sub-G1 fraction (markers of apoptotic cell death) as well as intracellular ROS accumulation in PC12 cells. This fraction and procyanidin B2 also protect against 4-HNE-induced PC12 cell apoptosis by blocking MKK4 activity [28]. Thus, the modulation of JNK and its upstream activator by medicinal plants and bioactive compounds may have potential therapeutic indications for neurodegenerative diseases.

Nuclear factor (erythroid-derived 2)-like-2 factor (Nrf2) is a ubiquitous master transcription factor that binds to the antioxidant response elements (ARE) and then enhances the expression levels of antioxidant enzymes and cytoprotective genes within DNA in response to oxidative stress [29]. Trans-resveratrol, a natural stilbene constituent isolated from medicinal plant grapevinele, is known to activate Nrf2 signaling pathway. Trans-resveratrol counters the cytotoxic response of 4-HNE consistently and reduces ROS generation and lipid peroxidation in PC-12 cells, showing its improvement of antioxidant defense system. It also restores 4-HNE-induced protein expression changes of mitochondria-mediated apoptosis markers (caspase-3, Bax, and B-cell lymphoma-2 (Bcl-2)) as well as oxidative damage, resulting in the alleviation of apoptotic neurodegeneration in PC12 cells [30]. More importantly, sulforaphane (an isothiocyanate mainly derived from crucifers such as broccoli, cabbages and olives) and carnosic acid (a plastidial catecholic diterpene mainly from *Rosmarinus officinalis*) are two phytochemicals Nrf2/ARE activators, which are found to attenuate 4-HNE-induced inhibition of mitochondrial respiration for complex I; the latter has been found to protect complex II in the isolated cortical mitochondria in young adult male CF-1 mice [14]. Furthermore, sulforaphane and carnosic acid decrease the amount of 4-HNE bound to mitochondrial proteins [14]. Microtubule-associated protein 1 light chain 3 alpha (LC3) is a marker of autophagosome, which is required for clearing dead cells. Study has shown that polyphenolic flavonoids luteolin and apigenin mitigate LC3 conversion and ROS production, activate Nrf2 signaling, and inhibit cytotoxicity in 4-HNE-exposed PC12 cells [22]. Therefore, the above phytochemicals have the capability of Nrf2 signaling induction to prevent the mitotoxicity of 4-HNE and provide neuroprotection.

Hyperactivity of NADPH oxidases (NOX) induced by 4-HNE has a crucial role in ROS overproduction and apoptosis; thus, NOX-derived ROS is a central mechanism in the development of neurodegeneration and neuroinflammation [31]. Kaempferol, a natural flavonoid presented in many medicinal plants, has been proved as a potential neuroprotective agent. It suppresses 4-HNE-mediated p-JNK and apoptosis in PC12 cells. Interestingly, kaempferol directly binds p47(phox), a cytosolic subunit of NOX, and effectively blocks 4-HNE-induced NOX activation in PC12 cells, exerting its potential neuroprotection against NOX-mediated neurodegeneration [13]. Scutellarin, a bioactive flavone isolated from *Scutellaria baicalensis*, has anti-inflammatory, anti-oxidative, anti-apoptotic, and anti-neurotoxic properties. More recently, a molecular docking study showed that scutellarin selectively binds to NOX2 with high affinity. Substantially, scutellarin down-regulates NOX2 expression and reduces 4-HNE and ROS levels in astrocytes subjected to ischemia/reperfusion and in rats with focal cerebral ischemia [32]. These results indicate that scutellarin is a neuroprotective flavone against ischemic injury.

Alzheimer's disease (AD) is a progressive and irreversible neurodegenerative disorder. High levels of 4-HNE are detected in the AD brain with the induction of neuronal apoptosis and oxidative stress, indicating that 4-HNE may play a role in AD subjects and/or animal models of age-related neurodegenerative disorders [33,34]. Acetylcholine esterase (AChE), cyclooxygenase-2 (COX-2), and matrix metalloproteinase-8 (MMP-8) are important target proteins implicated in AD pathogenesis. Phytocompounds albiziasaponin-A, iso-orientin, and salvadorin have antioxidant and anticholinesterase activities. The results from molecular docking studies show that there are strong interactions of albiziasaponin-A, iso-orientin, and salvadorin with AChE, COX2, and MMP8, respectively. Hence, these three bioactive compounds reduce serum 4-HNE levels in colchicine-induced AD model of rats, suggesting that albiziasaponin-A, iso-orientin, and salvadorin may be potential neuroprotective agents [35].

Glutathione *S*-transferase (GST) is a critical enzyme that participates in the detoxification of 4-HNE and has potential to prevent degenerative cellular processes [36]. Naringenin, a flavonoid found in grapefruit juice, is reported to improve learning and memory in a rat model of AD [37]. In an intracerebroventricular-streptozotocin-induced AD-type model of rats with cognitive impairment, naringenin increases GST, glutathione peroxidase (GSH-Px), glutathione reductase (GR), superoxide dismutase (SOD), and catalase (CAT) to detoxify 4-HNE in the hippocampus [38]. Collapsin response mediator protein-2 (CRMP-2) is hyperphosphorylated in AD. More recently, using computational tools with structural and dynamic analyses, naringenin and naringenin-7-O-glucuronide have been found to selectively bind CRMP-2 and then reduce p-CRMP-2, being the therapeutic activity of novel CRMP-2 inhibitor [39,40].

In senescence-accelerated mouse prone 8 (SAMP8) model of mice with age-related cognitive decline and AD, rosemary extract as well as spearmint extract (containing carnosic acid and rosmarinic acid) are reported to reduce 4-HNE levels in brain cortex and protein carbonyls in brain hippocampus, resulting in the improvement of learning and memory [41]. More importantly, rosmarinic acid increases antioxidant enzyme activities (such as SOD, CAT, and GSH-Px) and glutathione levels and then reduces 4-HNE in amyloid beta 42-induced echoic memory decline in rats with oxidative stress and cholinergic impairment, possibly contributing to the improvement of neural network dynamics of auditory processes [42]. Dark tea prevents and treats age-related degenerative diseases. L-theanine, 3,3′-azanediylbis (4-hydroxybenzoic acid) and one of 8-C N-ethyl-2-pyrrolidinone substituted flavan-3-ols were recently isolated for in silico characterization of microbial metabolites extracted from dark tea. Oral administration of these three compounds indeed inhibits the formation of 4-HNE and ubiquitinated protein aggregates, regulates Aβ metabolic pathway, increases endogenous antioxidant capacity, and reduces neuronal apoptosis rate in SAMP8 mice, showing the protection of SAMP8 neurons [43].

Parkinson's disease (PD) is an age-related neurodegenerative disorder that severely affects quality of life. Chrysin, a flavone found in *Passiflora caerulea*, *P. incarnata*, and *Oroxylum indicum*, improves antioxidant defense system to protect against oxidative stress by increasing GSH to reduce 4-HNE. Consistently, chrysin improves behavioral and cognitive ability in a 6-hydroxydopamine-induced mouse model of PD [44]. Quercetin also markedly improves the motor balance and coordination in 1-methyl-4-phenyl-1, 2, 3, 6-tetrahydropyridine-induced PD of mice. Moreover, quercetin significantly decreases 4-HNE immunoreactivity in striatum of mouse brains and increases GPx, SOD, AchE, and dopamine contents, showing the antiparkinsonian property [45].

3. Prevention of Eye Damage

The main visual ocular structures are the cornea and conjunctiva, lens, vitreum, and retina. Any structure dysfunction can cause visual problems and even blindness. Overproduction of ROS leads to the increased level of 4-HNE and induces lipid oxidative damage and inflammation in the cornea and conjunctiva [46,47], whereas 4-HNE accumulation causes the death of lens epithelia cells to induce lens opacification and cataract [48] as well as retinal pigment epithelium (RPE) cell death [49,50]. In the ocular surface of mouse model with dry eye disease, antioxidant extracts from *Schizonepeta tenuifolia*, *Angelica dahurica*, *Rehmannia glutinosa*, and *Cassia tora* reduce the number of 4-HNE-positive cells to protect from lipid peroxidation-related membrane damage. These antioxidant extracts also inhibit extracellular ROS production and decrease IL-1β, IL-6, tumor necrosis factor-α (TNF-α), and interferon-γ (IFN-γ) levels of the ocular surface, resulting in the improvement of clinical signs [51].

Aldose reductase (AR, an NADPH-dependent oxidoreductase) hyperactivity in the lens plays an important role in the pathogenesis of oxidative stress-mediated diabetic cataract [52]. Quercetin possesses therapeutic effect in the management and treatment of diabetic cataract. Quercetin and its glycoside derivative rutin, or *G. biloba* extract, remarkably inhibit AR activity, stimulate GSH production, and decrease the levels of 4-HNE, lipid peroxidation malondialdehyde (MDA), and advanced glycation end-products in the lenses of streptozotocin-induced diabetic cataract rats, delaying the progression of lens opacification [53]. Curcumin, a diketone constituent extracted from *Curcuma longa*, is a wonderful natural antioxidant and anti-cataract agent. This bioactive compound enhances the formation of GST isoenzymes to detoxify 4-HNE by aiding in the conjugation of 4-HNE to glutathione, showing its prevention against cataract [48].

Proliferative diabetic retinopathy, a leading cause of blindness, is also promoted by 4-HNE. Berberine, an isoquinoline alkaloid from a medicinal plant rhizoma of *Coptidis chinesis*, has favorable effects on glucose and lipid metabolism in animal experimental and human clinical studies [54]. In confluent human retinal Müller cells exposed to 4-HNE, cell death is partially attenuated by berberine pretreatment. Berberine also restores 4-HNE-induced autophagy in this cell model, showing its potential to prevent diabetic retinopathy [55].

In fact, 4-HNE can destabilize actin and induce cytoskeletal damage, resulting in retinopathy and age-related macular degeneration [56]. Cyanidin-3-glucoside is a major anthocyanin in mulberry or *Lonicera caerulea,* and its supplement has potential to prevent eye diseases. In human adult retinal pigmented epithelial (ARPE-19) cells, cyanidin-3-glucoside protects against 4-HNE-induced cell apoptosis, inflammatory damage, and angiogenesis [12,57]. Quercetin also increases viability and decreases inflammation and cytotoxicity in 4-HNE-exposed ARPE-19 cells [58]. Recently, solid dispersion of quercetin has been reported to decrease retinal pigment epithelium sediments and Bruch's membrane thickness in Nrf2 wild-type (WT) mice with dry age-related macular degeneration. This solid dispersion reduces ROS and MDA contents and increases SOD, GSH-PX, and CAT activities in serum and retinal tissues of Nrf2 WT mice. Solid dispersion of quercetin also up-regulates Nrf2 mRNA expression and enhances its nuclear translocation, as well as Nrf2 target gene hemeoxygenase-1(HO-1) in retinal tissues of Nrf2 WT mice [59]. These observations suggest that quercetin may alleviate oxidative injury to prevent dry age-related macular degeneration by enhancing Nrf2 activation. Medicinal plants marigold or grape seed (containing macular pigments lutein and zeaxanthin) are reported to have anti-oxidative activity and prevent 4-HNE adduct formation, actin solubility, and lipofuscin accumulation as well as age-related cone and rod photoreceptor dysfunction in β5(-/-) mice with cytoskeletal damage in aging RPE cells [56]. Of note, 4-HNE release and oxidative damage are also induced by irradiation exposure, causing retinopathy. Cyanidin-3-glucoside and quercetin decrease 4-HNE release in rod outer segments incubated with all-trans-retinal to generate bisretinoid under irradiation [60], whereas lutein and zeaxanthin isomers have recently been demonstrated to protect against light-induced retinopathy by reducing oxidative and endoplasmic reticulum stress in BALB/cJ mice [61]. Photodegradation of N-retinylidene-N-retinylethanolamine (A2E) is known to release reactive carbonyls. Medicinal plant compounds cyanidin-3-glucoside, quercetin, ferulic acid, and chlorogenic acid diminish cellular ROS and protect GSH from the reaction with photooxidized A2E in A2E accumulated-RPE cells irradiated with short-wavelength light [60].

This is especially important in exploring the possible molecular mechanisms by which bioactive compounds prevent 4-HNE-related retinopathy. NOD-like receptor protein 3 (NLRP3) inflammasome activation in the eyes is associated with the pathogenesis of age-related macular degeneration in RPE cells [62]. Cyanidin-3-glucoside has been found to inhibit NLRP3 inflammasome activation by reducing NLRP3, caspase-1, IL-1β, and IL-18 levels in 4-HNE-exposed ARPE-19 cells. This inhibitory mechanism may be mediated by regulating JNK-c-Jun/activator protein 1 (AP-1) pathway, further demonstrating the potential of cyanidin-3-glucoside to prevent retinal degenerative diseases [63,64]. Quercetin improves cell membrane integrity and mitochondrial function and decreases IL-6, IL-8, and monocyte chemotactic protein 1 (MCP-1) production, presumably by regulating MAPK/NF-κB signaling pathway in ARPE-19 cells stimulated by inflammatory cytokine [65].

4. Protection against Cardiovascular Injury

The normal redox status can be altered by 4-HNE adducts, which cause cardiomyocyte damage and vascular dysfunction. Thus, 4-HNE may play a key role in the progression of cardiovascular diseases [5,66,67]. *Thymbra capitata* is a Mediterranean culinary herb. Its essential oil prevents 4-HNE-induced cell death and avoids mitochondrial membrane potential loss and ROS generation in primary cultures of neonatal rat cardiomyocytes [15]. Olive (*Olea europaea*) leaf, a popular traditional herbal medicine, has a cardioprotective function. The ethanolic and methanolic extracts of *O. europaea* are reported to inhibit 4-HNE-induced apoptosis, ROS production, viability impairment, mitochondrial dysfunction, and pro-apoptotic activation in rat cardiomyocyte cell line (H9c2 cells). These extracts also reduce 4-HNE-induced phosphorylation of stress-activated transcription factors in

H9c2 cells [68]. Furthermore, phenolic compounds oleuropein, hydroxytyrosol, and quercetin derived from olive leaf are found to prevent against 4-HNE-induced cardiomyocyte carbonyl stress and toxicity and eventually regulate the cellular redox status [68]. Other medicinal plant constituents such as epigallocatechin-3-gallate, sulforaphane, and puerarin are reported to protect against cardiomyopathy. These compounds also reduce 4-HNE and lipid peroxidation to prevent oxidative stress in animal models of cardiovascular injury [69–71].

Aldehyde dehydrogenase 2 (ALDH2) mediates detoxification of toxic aldehydes, being necessary and sufficient to confer cardioprotection. Alpha-lipoic acid isolated from spinach and broccoli acts as a free radical scavenger. This bioactive compound activates myocardial ALDH2 and reduces 4-HNE and MDA levels in a Langendorff model of rat ischemia/reperfusion [72]. A recent report shows that α-lipoic acid prominently down-regulates the expression of 4-HNE and NOX subunit p67^{phox} to inhibit oxidative stress in fistula-created deterioration of cardiac function rats [73], further demonstrating the cardioprotective effect. Flavonoid glycoside baicalin, extracted from *Scutellaria baicalensis* root, protects against hypoxia/reoxygenation-induced H9c2 cell injury. Baicalin also enhances ALDH activity to reduce of 4-HNE and MAD levels and increases SOD activity to suppress ROS production in H9c2 cells, showing its cardioprotection against hypoxia/reoxygenation-induced cardiomyocyte injury [74].

Atherosclerotic lesion is associated with the accumulation of reactive aldehydes derived from oxidized lipids. Accumulation of 4-HNE becomes an important risk factor that contributes to the atherogenicity of oxidized low-density lipoprotein (ox-LDL) and the development of atherosclerosis [75]. *G. biloba leaf* extract is reported to reduce ox-LDL and attenuate 4-HNE-induced production of matrix metalloproteinase-1 (MMP-1), probably through suppressing the activation of tyrosine-phosphorylated form of platelet-derived growth factor receptor beta in human coronary smooth muscle cells [76]. Medicinal plant *Opuntia cladodes* powder containing phenolic acid and flavonoids also reduces ox-LDL and 4-HNE toxicity in normal (Apc +/+) and preneoplastic (Apc min/+) immortalized epithelial colon cells [77]. Moreover, *O. cladodes* significantly reduces the formation of atherosclerotic lesions and the accumulation of 4-HNE adducts in the vascular wall of apoE-KO mice [78]. These observations demonstrate that *O. cladodes* may have therapeutic potential in 4-HNE-related cardiovascular diseases.

5. Protection against Liver Injury

There is evidence that the increased level of 4-HNE is correlated with the pathogenesis of liver injury [2,79]. Ginseng is a well-known medicinal plant. Ginseng extracts exert protection against liver injury via multiple pathways. For example, the fine root extract of ginseng with ginsenosides profiles has been found to attenuate 4-HNE-induced DNA damage in HepG2 cells. Methanolic extract of the main root also shows a decrease of DNA damage in this cell model [80]. Water extract of *P. ginseng*, rich in ginsenosides Rg1 and Rb1, significantly reduces total ROS generation and especially down-regulates 4-HNE signals in radiation-induced steatohepatitis of mice [81]. Red ginseng or black ginseng is a type of the repeated steam-processed *P. ginseng*. Red ginseng extract improves chronic alcohol-induced histopathological changes of liver in mice. This extract also inhibits oxidative stress and lipid peroxidation by reducing the formation of 4-HNE and the number of 4-HNE-positive cells and protects hepatocytes from inflammation and necrosis by activating AMP-activated protein kinase (AMPK)/Sirt1 pathway [82]. Recently, black ginseng extract has been reported to maintain the cellular redox status by restoring NOX and GSH level change and alleviate oxidative stress by reducing intracellular ROS production and lipid peroxide 4-HNE signals, resulting in the protection of hydrogen peroxide-induced oxidative damage in AML-12 cells [83]. American ginseng berry, the ripe fruit of *Panax quinquefolius*, contains a variety of protopanaxadiol ginsenosides and protopanaxatriol ginsenosides. This ripe fruit extract inhibits alanine aminotransferase and aspartate transaminase activity and mitigates oxidative stress and lipid peroxidation by reinforcing SOD activity, suppressing GSH depletion, and reducing 4-HNE formation in acetaminophen-induced liver injury of mice [84].

Several other bioactive compounds of medicinal plants also mitigate oxidative stress, improve antioxidant function, and reduce 4-HNE, protecting against liver injury. Tanshinone II-A is a diterpene quinone constituent of *Salvia miltiorrhiza*. This bioactive compound restores 4-HNE-induced hepatocyte damage in normal liver tissue NCTC 1469 cells. Tanshinone II-A also up-regulates peroxisome proliferator-activated receptor α (PPARα) expression and scavenges 4-HNE in this hepatocyte model [85]. Fisetin, one of the most popular polyphenols in fruits and vegetables, exhibits senotherapeutic activity in treating aged mice. It reduces liver oxidative stress damage by decreasing 4-HNE adducts and increasing the ratio of intracellular oxidized glutathione [86,87]. *p*-Coumaric acid, rich in fruits, vegetables, and plants, protects hepatocytes against oxidative stress and lipid peroxidation efficiently by reducing high levels of ROS and formation of 4-HNE protein adducts. Consistently, this active compound attenuates acetaminophen-induced hepatotoxicity, apoptosis, and inflammation in mice by suppressing the translocation of apoptosis-inducing factor and MAPK signaling and decreasing NF-κB activity [88]. Chlorogenic acid is an active compound in *Ligustrum lucidum* fruit that protects hepatocyte mitochondria from oxidative stress mediated by the activation of AMPK in carbon tetrachloride-induced acute liver injury of mice [89]. Water extract of *L. lucidum* inhibits ROS generation and increases intracellular GSH to reduce 4-HNE levels in this animal model, demonstrating that *L. lucidum* inhibits oxidative stress to prevent liver injury via AMPK activation [89].

Of note, high 4-HNE levels cause rapid cell death associated with the depletion of sulfhydryl groups and inhibition of key metabolic enzymes. Aldo-keto reductase family 7 member A2 (AKR7A2) is the most abundant anthracycline metabolizing enzyme, which has been found to protect against 4-HNE toxicity. AKR7A2 over-expression rescues the effect of Nrf2-knockdown on 4-HNE-induced cytotoxicity [90]. The active coumarin compound 7-Hydroxycoumain is presented in many medicinal plants, such as the seeds of *Clausena lansium*, *Parasenecio forrestii*, *Artemisia argyi*, and *Pharbitis purpurea*. This phytochemical acts as an AKR7A2 inducer and protects against 4-HNE-induced HepG2 cell damage via AKR7A2 induction. Thus, the inducible AKR7A2 has provided a new therapeutic target to treat oxidative stress-related chronic liver disease [90]. Additionally, the effects of bioactive compounds on AR activity have been reported. Plant triterpenoid saponins soyasaponin Ba (V), soyasaponin Bb, soyasaponin Bd (sandosaponin A), soyasaponin αg, 3-O-[R-L-rhamnopyranosyl (1→2)-α-D-glucopyranosyl (1→2)-α-D-glucuronopyranosyl]olean-12-en-22-oxo-3α,-24-diol, and soyasaponin βg isolated from methanolic extract of *Phaseolus vulgaris* seeds (Zolfino bean) have the ability to inhibit highly purified human recombinant aldose reductase (hAKR1B1), as well as hAKR1B1-dependent reduction of 4-HNE [91]. This work suggests that the reduction of 4-HNE by metabolic enzymes is an important determinant of the prevention of oxidative stress-driven liver injury.

6. Improvement of Energy Metabolism Disorder

Oxidative stress induced by 4-HNE also causes energy metabolism disorder that leads to tissue injury [92–94]. A harmful effect is exerted on adipocytes by 4-HNE, and 4-HNE also promotes insulin resistance [92,95,96]. In fact, 4-HNE induction and its lipolytic response are observed in insulin resistance patients with obesity and type 2 diabetes mellitus [92]. Insulin signaling pathway plays an important role in mediating energy metabolism. Tocotrienol-rich fraction in olive oil is reported to protect against oxidative stress, inflammation, and apoptosis, prevent the increased levels of 4-HNE and protein carbonyls from hyperglycemia-induced muscle damage, and effectively reduce insulin resistance in type 2 diabetic mice, via regulating the AMPK/Sirt1/peroxisome proliferator-activated receptor γ coactivator 1 (PGC1) pathway [97]. Carnosic acid decreases 4-HNE-mediated free fatty acid release and activates Tyr632 phosphorylation of insulin receptor substrate-1 (p-IRS-1$^{\text{Tyr632}}$) and protein kinase B (p-Akt), inhibits as well as p-IRS-1$^{\text{Ser307}}$ in insulin signaling pathway in 3T3-L1 adipocytes. This rosemary diterpene compound also suppresses 4-HNE-induced phosphorylation of protein kinase A (p-PKA) and hormone-sensitive lipase (HSL) and attenuates 4-HNE-induced down-regulation of p-AMPK and acetyl-CoA carboxylase, demonstrating the alleviation of insulin resistance driven by 4-HNE [16].

The improvement of antioxidant defense system in treating energy metabolism disorders is also important. GST can protect against 4-HNE-induced insulin resistance [36]. Carnosic acid up-regulates GST expression to reduce 4-HNE-conjugated proteins, being consistent with its attenuation of 4-HNE-induced lipolytic response and insulin resistance [16]. *Morus alba* (mulberry), a popular ornamental plant, has several well-documented beneficial effects to prevent and treat metabolic diseases [98]. The root bark extract of *M. alba*, rich in Diels–Alder type adducts cudraflavone B and other flavonoids, markedly protects against lipid peroxidation-induced pancreatic injury by increasing GSH and reducing 4-HNE in rats [99]. Mulberry leaf contains a variety of flavonoids to control obesity [100]. Recent study shows that mulberry leaf aqueous extract significantly reduces plasma levels of 4-HNE adducts and enhances the cellular antioxidant defense system to inhibit oxidative stress via normalizing Nrf2 activation in high fat diet-fed mice [101]. Mulberry fruits, a well-known source of resveratrol, prevent diabetic complications by reducing lipid peroxidation and 4-HNE accumulation in diabetic pregnancy [102]. Consistently, mulberry bioactive compound trans-resveratrol mitigates oxidative stress by reducing lipid peroxidation and inhibiting 4-HNE expression and DNA damage in streptozotocin-induced type 1 diabetes of rats [103]. In addition, the inhibition of NOX plays a key role in the mitigation of oxidative stress and prevention of insulin resistance, steatosis, and muscle damage [104,105]. NOX inhibitor apocynin, isolated from *Picrorhiza kurroa*, has been found to attenuate insulin resistance in skeletal muscle of mice with metabolic syndrome. Another phytochemical, (−)-epicatechin, found in green tea and cocoa, also down-regulates NOX1/NOX4, reduces 4-HNE adducts from ileum cells, and mitigates high fat diet-induced insulin resistance and steatosis in male C57BL/6J mice by preventing oxidative stress [106]. Cyanidin and delphinidin are the most common aglycon forms of anthocyanins and also have NOX inhibitory activity to attenuate high fat-induced steatosis and control the adverse effects by reducing 4-HNE-protein adducts in mice [107].

Another risk factor of obesity and type 2 diabetes is increased following exposure to fine particular matter ($PM_{2.5}$) [108–110]. Research has shown 4-Hydroxytyrosol to be beneficial for metabolic and cardiovascular disorders. It also reduces liver and serum 4-HNE levels and alleviates $PM_{2.5}$-induced, adiposity, and insulin resistance in adult female C57BL/6j mice by mitigating oxidative stress as well as restraining NF-κB activation and gut microbiota [111]. Additionally, there is evidence that 4-hydroxytyrosol significantly reduces 4-HNE expression in red blood cells from hyperlipemic patients and protects red blood cells against 4-HNE toxicity [112].

7. Amelioration of Other Disorders

Medicinal plants and bioactive compounds also combat 4-HNE-induced deleterious effects and ameliorate the clinical signs in other disorders. *Aloe vera* is a medicinal herb that contains numerous bioactive components. Aqueous extract of *A. vera* has an antioxidant capacity on endogenous ROS production and 4-HNE-protein adducts induced by 4-HNE in human cervical cancer (HeLa), human microvascular endothelial cells (HMEC), human keratinocytes (HaCat), and human osteosarcoma (HOS) cell cultures [113]. Triterpene saponin aescin isolated from *Aesculus hippocastanum* restricts lipid peroxidation to reduce cytotoxic 4-HNE levels and keeps more neurons and myelin sheaths alive in the spinal cord injury of rats, which may be mediated by enhancing HO-1 expression and inhibiting NF-κB activation [114,115]. Natural anti-oxidation compounds capsaicin (chili peppers), curcumin (turmeric), and polyphenols as well as *G. biloba* and *Polypodium leucotomos* are reported to inhibit 4-HNE-induced oxidative stress in human melanocytes, resulting in the inhibition of cell apoptosis [116]. Rutin effectively decreases 4-HNE and increases GSH-Px and GSH; thus, it prevents against UV-induced skin fibroblast membrane disruption via the regulation of antioxidant system [117]. Ascorbic acid has a similar effect in reducing 4-HNE levels in human skin fibroblasts exposed to UV radiation and hydrogen peroxide [118].

Additionally, 4-HNE is reported to disrupt gap junction-mediated intercellular communication in the lateral wall structures of the cochlea [119]. Some medicinal plant constituents such as astragaloside IV, baicalein, catalpol, curcumin, kaempferol, luteolin, quercetin, resveratrol, and rosmarinic acid are reported to prevent hair cell death [120]. Astragaloside IV (the major constituent of *Astragalus membranaceus*) significantly down-regulates 4-HNE expression in guinea pig cochlea exposed to impulse noise, being consistent with the attenuation of impulse noise-induced trauma [121,122]. Rosmarinic acid enhances the endogenous antioxidant defense and decreases 4-HNE expression by the regulation of the Nrf2/HO-1 signaling pathway, resulting in the attenuation of noise-induced hearing loss and injury in rat cochlea [123]. Therefore, these natural compounds may be the effective therapy to ameliorate 4-HNE-driven hearing loss.

During food digestion, the absorption of highly toxic 4-HNE from the gastrointestinal tract into the blood system may contribute to the development of multiple oxidative stress-related disorders [124]. Polyphenol-rich beverages such as Japanese Sencha green tea and grape juice promote the stability of polyunsaturated fatty acids to oxidation and inhibit 4-HNE formation during the simulated digestion of linseed oil emulsions in the intestinal phase [125]. Algae (microalgae *Schizochytrium* sp.), the division of lower plants, also has pharmaceutical applications. Algae oil, rich in phenolic compounds β-carotene and tocopherols, is reported to prevent lipid peroxidation and inhibit 4-HNE formation in digestion during both in vitro gastric and duodenal digestions [126]. (−)-epicatechin also restores fat diet-induced increase of intestinal permeability, down-regulation of ileal tight junction proteins, as well as reduction of endotoxemia, being consistent with the improvement of insulin resistance. More importantly, (−)-epicatechin and apocynin are found to suppress NOX1/NOX4 overexpression, 4-HNE adducts and monolayer permeabilization by regulating redox sensitive signals in TNFα-exposed Caco-2 cells, further demonstrating the protection against the high fat diet-induced increased intestinal permeability. These results are consistent with the prevention of steatosis and insulin resistance under the treatment of (−)-epicatechin and apocynin [106]. The edible amaranth plants from around the world have strong antioxidant activity [127]. Clinical study shows that the supplementation of amaranth seed oil in addition to standard anti-helicobacter pylori treatment significantly decreases the accumulation of 4-HNE-histidine adducts in gastric mucosa and increases heart rate variability in duodenal peptic ulcer patients. Therefore, the standard treatment of duodenal peptic ulcer requires additional therapeutic approaches by using amaranth seed oil [128].

Energy metabolism disorders induced by 4-HNE may alter longevity by orchestrating the development of a biological phenotype. Plant fruits rich in high levels of antioxidants can promote longevity and health span. For example, the supplementation with 4% nectarine has been found to extend life and increase fecundity. Nectarine reduces 4-HNE-protein adduct in wild-type females fed with high-fat diet. Moreover, nectarine ameliorates aging-related death and reduces oxidative damage in female sod1-mutant flies [129]. These results demonstrate that the attenuation of 4-HNE-induced energy metabolism disorders by medicinal plants and bioactive compounds may promote longevity.

8. The Strategy for Developing Potential Therapy

As a toxic end-product of lipid peroxidation, 4-HNE is an important mediator in physiological adaptive reaction and signal transport, as well as in the pathogenesis of multiple oxidative stress-related disorders. Medicinal plants and bioactive compounds are found to enhance metabolism, detoxification or clearance of 4-HNE by regulating activities of endogenous enzymes (such as AR, GST, and AKRs), which may provide the effective strategies for combating 4-HNE-driven deleterious effects. Because of differential metabolisms of 4-HNE observed in liver, lung, and brain of rodents [130], further study is warranted to determine whether these medicinal plants and bioactive compounds have the capacity to mediate oxidative, reductive, and conjugative pathways to metabolize 4-HNE and 4-HNE adducts in oxidative stress-triggered injury of these tissues.

The Nrf2/ARE pathway is suggested to mediate the adaptive induction of antioxidant and detoxifying enzymes including GST and AKR1C1, resulting in the metabolism of 4-HNE [131]. The capability of the Nrf2/ARE pathway induction may constitute a pleiotropic cytoprotective defense to prevent 4-HNE toxicity. Ongoing studies will determine the therapeutic effects of Nrf2/ARE activators derived from medicinal plants and bioactive compounds to mitigate 4-HNE cytotoxicity-induced pathophysiology of disorders or diseases.

Of note, PPARα modulates several biological processes that are perturbed in energy homeostasis [132]. A recent report shows that PPARα is unable to block β-catenin transcriptional activity induced by a constitutively active mutant of lipoprotein receptor-related protein 6 (LRP6). In fact, 4-HNE-induced ROS production enhances LRP6 stability; this event is inhibited by PPARα overexpression [133]. It is known that anti-obesity hormone leptin signaling as well as insulin signaling mediate the regulation of PPARα. Insulin resistance is improved by 4-HNE [96], which also selectively inhibits leptin signaling, possibly promoting the pathogenesis of leptin resistance in obesity [134]. Some medicinal plants and natural compounds acting on PPARα, leptin, or insulin signaling may be responsible for the therapeutic effects on 4-HNE-induced energy metabolism disorders [85].

However, several important issues related to 4-HNE-driven molecular events in diseases, or the relevance of medicinal plants and bioactive compounds in redox homeostasis, still need more studies and new comprehensive approaches. In this regard, preclinical studies and clinical intervention trials are required, which should include the use of accurate analytical techniques, such as the determination of 4-HNE and 4-HNE adducts by immunohistochemistry and enzyme-linked immunosorbent assay, as well as matrix-assisted laser desorption/ionization–tandem time of flight (MALDI–TOF/TOF) mass spectrometry and liquid chromatography–tandem mass spectrometry (LC–MS/MS) [135–138].

9. Conclusions and Future Perspectives

As a biomarker of lipid peroxidation and oxidative stress, cytotoxic 4-HNE contributes to the progression of oxidative stress-related disorders. Some medicinal plants and bioactive compounds have been demonstrated to reduce 4-HNE levels, detoxify 4-HNE, and inhibit 4-HNE adduct formation through the up-regulation of the antioxidant enzymes and suppression of 4-HNE-mediated downstream signals. More experiments have proved that these medicinal plants and bioactive compounds combat 4-HNE-induced deleterious effects in oxidative stress-related neurological disorder, eye damage, cardiovascular injury, liver injury, energy metabolism disorder, and other disorders (Table 1). The understanding of the molecular basis for the function of medicinal plants and bioactive compounds would be useful to facilitate the selection of 4-HNE metabolism for future intervention investigation and health claim support and develop new treatment for 4-HNE-induced deleterious effects in these disorders. Moreover, the identification and analysis of 4-HNE and 4-HNE adducts are necessary to further support the association between medicinal plants and oxidative stress-related disorders in clinical trials.

Table 1. Medicinal plants and bioactive compounds in the attenuation of 4-HNE-induced deleterious effects.

Medicinal Plants and Bioactive Compounds	Action/Mechanism	4-HNE-Induced Model	References
Neuroprotection			
Polyphenol extract (red wine)	Inhibition of cleavage of PARP, reduction of ROS, protection against apoptosis	PC12 cells with apoptosis	[19]
Quercetin Myricetin (red wine extract or *G. biloba*)	Cytoprotective effects	PC12 cell death	[19–21]
Luteolin Apigenin (plant flavones)	Attenuation of cell death, caspase-3 and PARP-1 activation, mitigation of LC3 conversion and ROS production, activation of Nrf2 signaling	PC12 cells with cell viability	[22]
Piceatannol (*P. edulis* or *G. parvifolium*)	Cytoprotective effect, restoration of PARP cleavage and Bcl-2 expression, down-regulation of p-JNK	PC12 cell death and nuclear condensation	[25]
Citri reticulatae viride pericarpium	Anti-inflammation	PC12 cells with inflammatory injury	[27]
Cocoa procyanidin fraction Procyanidin B2	Attenuation of nuclear condensation, apoptotic cell death and ROS accumulation, blockade of MKK4 activity	PC12 cell death and nuclear condensation	[28]
Trans-resveratrol (grapevinele)	Countering the cytotoxic response, attenuation of apoptotic neurodegeneration	PC12 cells with cytotoxicity	[30]
Sulforaphane (crucifers such as broccoli, cabbages and olives) Carnosic acid (*R. officinalis*)	Increase of mitochondrial respiration Nrf2/ARE induction preventing against mitotoxic effect	Young adult male CF-1 mice Isolated cortical mitochondria with inhibition of mitochondrial respiration	[14]
Kaempferol (flavonoid in many medicinal plants)	Suppression of apoptosis and p-JNK, inhibition of NOX activation	PC12 cells with apoptosis Neuron-like cells with NOX-mediated neurodegeneration	[13]
Prevention of eye damage			
Berberine (*C. chinesis*)	Restoration of autophagy, inhibition of diabetic retinopathy	Confluent human retinal Müller cells with cell death	[55]
Cyanidin-3-glucoside (plant fruits mulberry or *L. caerulea*)	Reduction of apoptosis ratio, inflammation and angiogenesis	ARPE-19 cells with apoptosis, inflammatory damage and angiogenesis	[12,63]
Cyanidin-3-glucoside	Inhibition of NLRP3 inflammasome activation Regulation of JNK-c-Jun/AP-1 pathway	ARPE-19 cells with inflammation	[64]

Table 1. *Cont.*

Medicinal Plants and Bioactive Compounds	Action/Mechanism	4-HNE-Induced Model	References
Quercetin	Anti-inflammation, improvement of cell membrane integrity and mitochondrial function, decrease of IL-6, IL-8 and MCP-1 production, regulation of MAPK pathway	ARPE-19 cells with cytotoxicity	[58,65]
Protection of cardiovascular injury			
Oil of *T. capitata*	Prevention of cell death, mitochondrial membrane potential loss and ROS generation	Primary cultures of neonatal rat cardiomyocytes with cell death	[15]
Ethanolic and methanolic extracts of olive leaf (*O. europaea*)	Inhibition of apoptosis, ROS production, viability impairment, mitochondrial dysfunction and pro-apoptotic activation, reduction of phosphorylation of stress-activated transcription factors	Rat cardiomyocytes with cell death	[68]
Oleuropein Hydroxytyrosol Quercetin (olive leaf)	Prevention of carbonyl stress and toxicity, regulation of cellular redox status	Rat cardiomyocytes with cell death	[68]
G. biloba leaf extract	Reduction of ox-LDL, attenuation of MMP-1 production, inhibition of the tyrosine-phosphorylated form of platelet-derived growth factor receptor beta activation	Human coronary smooth muscle cells with injury	[76]
O. cladodes powder (containing phenolic acid and flavonoids)	Protection against toxicity	Normal (Apc +/+) and preneoplastic (Apc min/+) immortalized epithelial colon cells with toxicity	[77]
Protection against liver injury			
Fine root extract of ginseng with ginsenosides profiles Methanolic extract of the main root	Inhibitory capacity against DNA damage	HepG2 cells with DNA damage	[80]
Tanshinone II-A (*S. miltiorrhiza*)	Alleviation of hepatocyte damage up-regulation of PPARα, and scavenging 4-HNE	NCTC 1469 cells with damage	[85]
7- Hydroxycoumain (*C. lansium, P. forrestii, A. argyi,* and *P. purpurea*)	Hepatoprotection via AKR7A2 induction	HepG2 cells with cytotoxicity	[90]
Improvement of energy metabolism disorder			
Carnosic acid (*R. officinalis*)	Reduction of free fatty acid release, activation of p^{Tyr632} IRS-1 and p-Akt, p^{Ser307}IRS-1, suppression of the PKA/HSL pathway activation, decrease of p-AMPK and acetyl-CoA carboxylase, alleviation of insulin resistance	3T3-L1 adipocytes with insulin signaling impairment	[16]

Table 1. *Cont.*

Medicinal Plants and Bioactive Compounds	Action/Mechanism	4-HNE-Induced Model	References
Carnosic acid	Attenuation of free fatty acid release, up-regulation of GST, reduction of 4-HNE-conjugated proteins attenuation of the lipolytic response	Human subcutaneous adipocytes with lipolysis	[16]
4-Hydroxytyrosol (olive leaf)	Protection of red blood cells with oxidative damage	Hyperlipemic patients	[112]
Repair of other disorders			
A. vera	Antioxidant capacity for the reduction of ROS and 4-HNE-protein adducts	HeLa, HMEC, HaCat, and HOS cells with over-production of ROS and -HNE-protein adducts	[113]
Capsaicin (chili peppers) Curcumin (turmeric) Polyphenols *G. biloba* extract *P. leucotomos* extract	Inhibition of oxidative stress and cell apoptosis	Human melanocytes with oxidative stress and apoptosis	[116]

Author Contributions: Conceptualization, F.-X.W. and L.-D.K.; software, F.-X.W.; writing—original draft preparation, F.-X.W., H.-Y.L., and Y.-Q.L.; review and editing, F.-X.W. and H.-Y.L.; revising, L.-D.K. and F.-X.W. All authors have read and agreed to the published version of the manuscript.

Funding: This research received no external funding.

Conflicts of Interest: The authors declared no conflict of interest.

Abbreviations

4-HNE	4-hydroxynonenal
ROS	reactive oxygen species
PARP	poly (ADP-ribose) polymerase
PC12 cells	neuronal-like catecholaminergic cells (rat pheochromocytoma)
Nrf-2	nuclear factor (erythroid-derived 2)-like-2 factor
ARE	antioxidant response elements
Bcl-2	B-cell lymphoma-2
LC3	protein 1 light chain 3 alpha
JNK	c-Jun-N-terminal kinase
MAPK	mitogen-activated protein kinases
MKK4	mitogen-activated protein kinase kinase 4
NOX	NADPH oxidase
NF-κB	nuclear factor kappa-B
AD	Alzheimer's disease
AChE	acetylcholine esterase
COX-2	cyclooxygenase-2
MMP-8	matrix metalloproteinase-8
CRMP-2	collapsin response mediator protein-2
GST	glutathione *S*-transferase
GSH-Px	glutathione peroxidase
GR	glutathione reductase
SOD	superoxide dismutase
CAT	catalase
SAMP8	senescence-accelerated mouse prone 8
PD	Parkinson's disease
GSH	glutathione
RPE	retinal pigment epithelium
NLRP3	NOD-like receptor protein 3
IL-1β	interleukin 1β
AP-1	activator protein 1
MCP-1	monocyte chemotactic protein 1
MDA	malondialdehyde
HO-1	hemeoxygenase-1
TNF-α	tumor necrosis factor-α
IFN-γ	interferon-γ
AR	aldose reductase
A2E	N-retinylidene-N-retinylethanolamine
H9c2 cells	rat cardiomyocyte cell line
ALDH2	aldehyde dehydrogenase 2
ox-LDL	oxidized low-density lipoprotein
MMP-1	matrix metalloproteinase-1
PPARα	peroxisome proliferator-activated receptor α
AMPK	AMP-activated protein kinase
AKR7A2	Aldo-Keto reductase family 7 member A2
hAKR1B1	human recombinant aldose reductase

PGC1	peroxisome proliferator-activated receptor γ coactivator 1
IRS-1	insulin receptor substrate-1
Akt	protein kinase B
PKA	protein kinase A
HSL	hormone-sensitive lipase
HeLa	human cervical cancer
HMEC	human microvascular endothelial cells
HaCat	human keratinocytes
HOS	human osteosarcoma
LRP6	lipoprotein receptor-related protein 6
MALDI-TOF/TOF	matrix-assisted laser desorption/ionization-tandem time of flight
LC-MS/MS	liquid chromatography-tandem mass spectrometry

References

1. Dalleau, S.; Baradat, M.; Guéraud, F.; Huc, L. Cell death and diseases related to oxidative stress: 4-hydroxynonenal (HNE) in the balance. *Cell Death Differ.* **2013**, *20*, 1615–1630. [CrossRef] [PubMed]

2. Breitzig, M.; Bhimineni, C.; Lockey, R.; Kolliputi, N. 4-Hydroxy-2-nonenal: A critical target in oxidative stress? *Am. J. Physiol. Cell Physiol.* **2016**, *311*, C537–C543. [CrossRef] [PubMed]

3. Zarkovic, K.; Jakovcevic, A.; Zarkovic, N. Contribution of the HNE-immunohistochemistry to modern pathological concepts of major human diseases. *Free Radic. Biol. Med.* **2017**, *111*, 110–126. [CrossRef] [PubMed]

4. Luczaj, W.; Gegotek, A.; Skrzydlewska, E. Antioxidants and HNE in redox homeostasis. *Free Radic. Biol. Med.* **2017**, *111*, 87–101. [CrossRef]

5. Mali, V.R.; Palaniyandi, S.S. Regulation and therapeutic strategies of 4-hydroxy-2-nonenal metabolism in heart disease. *Free Radic. Res.* **2014**, *48*, 251–263. [CrossRef]

6. Cohen, G.; Riahi, Y.; Sunda, V.; Deplano, S.; Chatgilialoglu, C.; Ferreri, C.; Kaiser, N.; Sasson, S. Signaling properties of 4-hydroxyalkenals formed by lipid peroxidation in diabetes. *Free Radic. Biol. Med.* **2013**, *65*, 978–987. [CrossRef]

7. Csala, M.; Kardon, T.; Legeza, B.; Lizák, B.; Mandl, J.; Margittai, É.; Puskás, F.; Száraz, P.; Szelényi, P.; Bánhegyi, G. On the role of 4-hydroxynonenal in health and disease. *Biochim. Biophys. Acta* **2015**, *1852*, 826–838. [CrossRef]

8. Xiao, M.; Zhong, H.; Xia, L.; Tao, Y.; Yin, H. Pathophysiology of mitochondrial lipid oxidation: Role of 4-hydroxynonenal (4-HNE) and other bioactive lipids in mitochondria. *Free Radic. Biol. Med.* **2017**, *111*, 316–327. [CrossRef]

9. Rashid, S.; Ahmad, M.; Zafar, M.; Anwar, A.; Sultana, S.; Tabassum, S.; Ahmed, S.N. Ethnopharmacological evaluation and antioxidant activity of some important herbs used in traditional medicines. *J. Tradit. Chin. Med.* **2016**, *36*, 689–694. [CrossRef]

10. Tresserra-Rimbau, A.; Lamuela-Raventos, R.M.; Moreno, J.J. Polyphenols, food and pharma. Current knowledge and directions for future research. *Biochem. Pharmacol.* **2018**, *156*, 186–195. [CrossRef]

11. Olszowy, M. What is responsible for antioxidant properties of polyphenolic compounds from plants? *Plant Physiol. Biochem.* **2019**, *144*, 135–143. [CrossRef] [PubMed]

12. Wang, Y.; Qi, W.; Huo, Y.; Song, G.; Sun, H.; Guo, X.; Wang, C. Cyanidin-3-glucoside attenuates 4-hydroxynonenal- and visible light-induced retinal damage in vitro and in vivo. *Food Funct.* **2019**, *10*, 2871–2880. [CrossRef] [PubMed]

13. Jang, Y.J.; Kim, J.; Shim, J.; Kim, J.; Byun, S.; Oak, M.H.; Lee, K.W.; Lee, H.J. Kaempferol attenuates 4-hydroxynonenal-induced apoptosis in PC12 cells by directly inhibiting NADPH oxidase. *J. Pharmacol. Exp. Ther.* **2011**, *337*, 747–754. [CrossRef] [PubMed]

14. Miller, D.M.; Singh, I.N.; Wang, J.A.; Hall, E.D. Administration of the Nrf2-ARE activators sulforaphane and carnosic acid attenuates 4-hydroxy-2-nonenal-induced mitochondrial dysfunction ex vivo. *Free Radic. Biol. Med.* **2013**, *57*, 1–9. [CrossRef]

15. Hortigon-Vinagre, M.P.; Blanco, J.; Ruiz, T.; Henao, F. Thymbra capitata essential oil prevents cell death induced by 4-hydroxy-2-nonenal in neonatal rat cardiac myocytes. *Planta Med.* **2014**, *80*, 1284–1290. [CrossRef]

16. Liu, K.L.; Kuo, W.C.; Lin, C.Y.; Lii, C.K.; Liu, Y.L.; Cheng, Y.H.; Tsai, C.W. Prevention of 4-hydroxynonenal-induced lipolytic activation by carnosic acid is related to the induction of glutathione S-transferase in 3T3-L1 adipocytes. *Free Radic. Biol. Med.* **2018**, *121*, 1–8. [CrossRef]

17. Neely, M.D.; Boutte, A.; Milatovic, D.; Montine, T.J. Mechanisms of 4-hydroxynonenal-induced neuronal microtubule dysfunction. *Brain Res.* **2005**, *1037*, 90–98. [CrossRef]

18. Caruana, M.; Cauchi, R.; Vassallo, N. Putative role of red wine polyphenols against brain pathology in Alzheimer's and Parkinson's disease. *Front. Nutr.* **2016**, *3*, 00031. [CrossRef]

19. Jang, Y.J.; Kang, N.J.; Lee, K.W.; Lee, H.J. Protective effects of red wine flavonols on 4-hydroxynonenal-induced apoptosis in PC12 cells. *Ann. N. Y. Acad. Sci.* **2009**, *1171*, 170–175. [CrossRef]

20. Elumalai, P.; Lakshmi, S. Role of quercetin benefits in neurodegeneration. *Adv. Neurobiol.* **2016**, *12*, 229–245.

21. Joshi, V.; Mishra, R.; Upadhyay, A.; Amanullah, A.; PoluriK, M.; Singh, S.; Kumar, A.; Mishra, A. Polyphenolic flavonoid (Myricetin) upregulated proteasomal degradation mechanisms: Eliminates neurodegenerative proteins aggregation. *J. Cell Physiol.* **2019**, *234*, 20900–20914. [CrossRef] [PubMed]

22. Wu, P.S.; Yen, J.H.; Kou, M.C.; Wu, M.J. Luteolin and apigenin attenuate 4-hydroxy-2-nonenal-mediated cell death through modulation of UPR, Nrf2-ARE and MAPK pathways in PC12 cells. *PLoS ONE* **2015**, *10*, e0130599. [CrossRef] [PubMed]

23. Singhal, S.S.; Singh, S.P.; Singhal, P.; Horne, D.; Singhal, J.; Awasthi, S. Antioxidant role of glutathione S-transferases: 4-Hydroxynonenal, a key molecule in stress-mediated signaling. *Toxicol. Appl. Pharmacol.* **2015**, *289*, 361–370. [CrossRef] [PubMed]

24. Liu, W.; Kato, M.; Akhand, A.A.; Hayakawa, A.; Suzuki, H.; Miyata, T.; Kurokawa, K.; Hotta, Y.; Ishikawa, N.; Nakashima, I. 4-Hydroxynonenal induces a cellular redox status-related activation of the caspase cascade for apoptotic cell death. *J. Cell Sci.* **2000**, *113*, 635–641. [PubMed]

25. Jang, Y.J.; Kim, J.E.; Kang, N.J.; Lee, K.W.; Lee, H.J. Piceatannol attenuates 4-hydroxynonenal-induced apoptosis of PC12 cells by blocking activation of c-Jun N-terminal kinase. *Ann. N. Y. Acad. Sci.* **2009**, *1171*, 176–182. [CrossRef]

26. Dumont, U.; Sanchez, S.; Olivier, B.; Chateil, J.F.; Pellerin, L.; Beauvieux, M.C.; Bouzier-Sore, A.K.; Roumes, H. Maternal consumption of piceatannol: A nutritional neuroprotective strategy against hypoxia-ischemia in rat neonate. *Brain Res.* **2019**, *1717*, 86–94. [CrossRef]

27. Ye, Y.J.; Kim, Y.S.; Kang, M.S. Effects of citri reticulatae viride pericarpium on 4-hydroxynonenal-induced inflammation in PC12 cells. *J. Korean Med. Obes. Res.* **2016**, *16*, 79–84. [CrossRef]

28. Cho, E.S.; Jang, Y.J.; Kang, N.J.; Hwang, M.K.; Kim, Y.T.; Lee, K.W.; Lee, H.J. Cocoa procyanidins attenuate 4-hydroxynonenal-induced apoptosis of PC12 cells by directly inhibiting mitogen-activated protein kinase kinase 4 activity. *Free Radic. Biol. Med.* **2009**, *46*, 1319–1327. [CrossRef]

29. Nguyen, T.; Nioi, P.; Pickett, C.B. The Nrf2-antioxidant response element signaling pathway and its activation by oxidative stress. *J. Biol. Chem.* **2009**, *284*, 13291–13295. [CrossRef] [PubMed]

30. Siddiqui, M.A.; Kashyap, M.P.; Kumar, V.; Al-Khedhairy, A.A.; Musarrat, J.; Pant, A.B. Protective potential of trans-resveratrol against 4-hydroxynonenal induced damage in PC12 cells. *Toxicol. In Vitro* **2010**, *24*, 1592–1598. [CrossRef] [PubMed]

31. Hernandes, M.S.; D'Avila, J.C.; Trevelin, S.C.; Reis, P.A.; Kinjo, E.R.; Lopes, L.R.; Castro-Faria-Neto, H.C.; Cunha, F.Q.; Britto, L.R.; Bozza, F.A. The role of Nox2-derived ROS in the development of cognitive impairment after sepsis. *J. Neuroinflamm.* **2014**, *11*, 36. [CrossRef] [PubMed]

32. Sun, J.B.; Li, Y.; Cai, Y.F.; Huang, Y.; Liu, S.; Yeung, P.K.; Deng, M.Z.; Sun, G.S.; Zilundu, P.L.; Hu, Q.S.; et al. Scutellarin protects oxygen/glucose-deprived astrocytes and reduces focal cerebral ischemic injury. *Neural Regen. Res.* **2018**, *13*, 1396–1407. [PubMed]

33. Di Domenico, F.; Tramutola, A.; Butterfield, D.A. Role of 4-hydroxy-2-nonenal (HNE) in the pathogenesis of alzheimer disease and other selected age-related neurodegenerative disorders. *Free Radic. Biol. Med.* **2017**, *111*, 253–261. [CrossRef] [PubMed]

34. Benedetti, E.; D'Angelo, B.; Cristiano, L.; Di Giacomo, E.; Fanelli, F.; Moreno, S.; Cecconi, F.; Fidoamore, A.; Antonosante, A.; Falcone, R.; et al. Involvement of peroxisome proliferator-activated receptor β/δ (PPAR β/δ) in BDNF signaling during aging and in Alzheimer disease: Possible role of 4-hydroxynonenal (4-HNE). *Cell Cycle* **2014**, *13*, 1335–1344. [CrossRef]

35. Rasool, M.; Malik, A.; Waquar, S.; Tul-Ain, Q.; Jafar, T.H.; Rasool, R.; Kalsoom, A.; Ghafoor, M.A.; Sehgal, S.A.; Gauthaman, K.; et al. In-Silico characterization and in-vivo validation of albiziasaponin-A, iso-orientin, and salvadorin using a rat model of Alzheimer's disease. *Front. Pharmacol.* **2018**, *9*, 730. [CrossRef]

36. Curtis, J.M.; Grimsrud, P.A.; Wright, W.S.; Xu, X.; Foncea, R.E.; Graham, D.W.; Brestoff, J.R.; Wiczer, B.M.; Ilkayeva, O.; Cianflone, K.; et al. Downregulation of adipose glutathione S-transferase A4 leads to increased protein carbonylation, oxidative stress, and mitochondrial dysfunction. *Diabetes* **2010**, *59*, 1132–1142. [CrossRef]

37. Ghofrani, S.; Joghataei, M.T.; Mohseni, S.; Baluchnejadmojarad, T.; Bagheri, M.; Khamse, S.; Roghani, M. Naringenin improves learning and memory in an Alzheimer's disease rat model: Insights into the underlying mechanisms. *Eur. J. Pharmacol.* **2015**, *764*, 195–201. [CrossRef]

38. Khan, M.B.; Khan, M.M.; Khan, A.; Ahmed, M.E.; Ishrat, T.; Tabassum, R.; Vaibhav, K.; Ahmad, A.; Islam, F. Naringenin ameliorates Alzheimer's disease (AD)-type neurodegeneration with cognitive impairment (AD-TNDCI) caused by the intracerebroventricular-streptozotocin in rat model. *Neurochem. Int.* **2012**, *61*, 1081–1093. [CrossRef]

39. Lawal, M.F.; Olotu, F.A.; Agoni, C.; Soliman, M.E. Exploring the C-terminal tail dynamics: Structural and molecular perspectives into the therapeutic activities of novel CRMP-2 inhibitors, naringenin and naringenin-7-O-glucuronide, in the treatment of Alzheimer's disease. *Chem. Biodivers.* **2018**, *15*, e1800437. [CrossRef]

40. Lawal, M.; Olotu, F.A.; Soliman, M.E.S. Across the blood-brain barrier: Neurotherapeutic screening and characterization of naringenin as a novel CRMP-2 inhibitor in the treatment of Alzheimer's disease using bioinformatics and computational tools. *Comput. Biol. Med.* **2018**, *98*, 168–177. [CrossRef]

41. Farr, S.A.; Niehoff, M.L.; Ceddia, M.A.; Herrlinger, K.A.; Lewis, B.J.; Feng, S.L.; Welleford, A.; Butterfield, D.A.; Morley, J.E. Effect of botanical extracts containing carnosic acid or rosmarinic acid on learning and memory in SAMP8 mice. *Physiol. Behav.* **2016**, *165*, 328–338. [CrossRef] [PubMed]

42. Kantar Gok, D.; Hidisoglu, E.; Ocak, G.A.; Er, H.; Acun, A.D.; Yargıcoglu, P. Protective role of rosmarinic acid on amyloid beta 42-induced echoic memory decline: Implication of oxidative stress and cholinergic impairment. *Neurochem. Int.* **2018**, *118*, 1–13. [CrossRef] [PubMed]

43. Cai, S.; Yang, H.; Wen, B.; Zhu, K.; Zheng, X.; Huang, J.; Wang, Y.; Liu, Z.; Tu, P. Inhibition by microbial metabolites of Chinese dark tea of age-related neurodegenerative disorders in senescence-accelerated mouse prone 8 (SAMP8) mice. *Food Funct.* **2018**, *9*, 5455–5462. [CrossRef] [PubMed]

44. Goes, A.T.R.; Jesse, C.R.; Antunes, M.S.; Lobo Ladd, F.V.; Lobo Ladd, A.A.B.; Luchese, C.; Paroul, N.; Boeira, S.P. Protective role of chrysin on 6-hydroxydopamine-induced neurodegeneration a mouse model of Parkinson's disease: Involvement of neuroinflammation and neurotrophins. *Chem. Biol. Interact.* **2018**, *279*, 111–120. [CrossRef] [PubMed]

45. Lv, C.; Hong, T.; Yang, Z.; Zhang, Y.; Wang, L.; Dong, M.; Zhao, J.; Mu, J.; Meng, Y. Effect of quercetin in the 1-methyl-4-phenyl-1, 2, 3, 6-tetrahydropyridine-induced mouse model of Parkinson's disease. *Evid. Based Complement. Alternat. Med.* **2012**, *2012*, 928643. [CrossRef]

46. Deng, R.; Hua, X.; Li, J.; Chi, W.; Zhang, Z.; Lu, F.; Zhang, L.; Pflugfelder, S.C.; Li, D.Q. Oxidative stress markers induced by hyperosmolarity in primary human corneal epithelial cells. *PLoS ONE* **2015**, *10*, e0126561. [CrossRef]

47. Seen, S.; Tong, L. Dry eye disease and oxidative stress. *Acta Ophthalmol.* **2018**, *96*, E412–E420. [CrossRef]

48. Raman, T.; Ramar, M.; Arumugam, M.; Nabavi, S.M.; Varsha, M.K. Cytoprotective mechanism of action of curcumin against cataract. *Pharmacol. Rep.* **2016**, *68*, 561–569. [CrossRef]

49. Chen, J.; Wang, L.; Chen, Y.; Sternberg, P.; Cai, J. Phosphatidylinositol 3 kinase pathway and 4-hydroxy-2-nonenal-induced oxidative injury in the RPE. *Investig. Ophthalmol. Vis. Sci.* **2009**, *50*, 936–942. [CrossRef]

50. Vatsyayan, R.; Chaudhary, P.; Sharma, A.; Sharma, R.; Rao Lelsani, P.C.; Awasthi, S.; Awasthi, Y.C. Role of 4-hydroxynonenal in epidermal growth factor receptor-mediated signaling in retinal pigment epithelial cells. *Exp. Eye Res.* **2011**, *92*, 147–154. [CrossRef]

51. Choi, W.; Lee, J.B.; Cui, L.; Li, Y.; Li, Z.R.; Choi, J.S.; Lee, H.S.; Yoon, K.C. Therapeutic efficacy of topically applied antioxidant medicinal plant extracts in a mouse model of experimental dry eye. *Oxid. Med. Cell Longev.* **2016**, *2016*, 4727415. [CrossRef] [PubMed]

52. Chang, K.C.; Shieh, B.; Petrash, J.M. Role of aldose reductase in diabetes-induced retinal microglia activation. *Chem. Biol. Interact.* **2019**, *302*, 46–52. [CrossRef] [PubMed]

53. Lu, Q.; Hao, M.; Wu, W.Y.; Zhang, N.; Isaac, A.T.; Yin, J.L.; Zhu, X.; Du, L.; Yin, X.X. Antidiabetic cataract effects of GbE, rutin and quercetin are mediated by the inhibition of oxidative stress and polyol pathway. *Acta Biochim. Pol.* **2018**, *65*, 35–41. [CrossRef] [PubMed]

54. Feng, X.; Sureda, A.; Jafari, S.; Memariani, Z.; Tewari, D.; Annunziata, G.; Barrea, L.; Hassan, S.T.S.; Smejkal, K.; Malanik, M.; et al. Berberine in cardiovascular and metabolic diseases: From mechanisms to therapeutics. *Theranostics* **2019**, *9*, 1923–1951. [CrossRef]

55. Yang, S.H.; Wu, M.Y.; Fu, D.X.; Chen, J.P.; Zhang, J.; Wilson, K.; Elliot, M.; Du, M.; Lyo, T. Berberine acts as a novel autophagy blocker to protect human müLler cell from 4-hne induced cell death. *Investig. Ophthalmol. Vis. Sci.* **2012**, *53*, 2019.

56. Yu, C.C.; Nandrot, E.F.; Dun, Y.; Finnemann, S.C. Dietary antioxidants prevent age-related retinal pigment epithelium actin damage and blindness in mice lacking alpha v beta 5 integrin. *Free Radic. Biol. Med.* **2012**, *52*, 660–670. [CrossRef]

57. Xu, D.; Hu, M.J.; Wang, Y.Q.; Cui, Y.L. Antioxidant activities of quercetin and its complexes for medicinal application. *Molecules* **2019**, *24*, 1123. [CrossRef]

58. Hytti, M.; Piippo, N.; Salminen, A.; Honkakoski, P.; Kaarniranta, K.; Kauppinen, A. Quercetin alleviates 4-hydroxynonenal-induced cytotoxicity and inflammation in ARPE-19 cells. *Exp. Eye Res.* **2015**, *132*, 208–215. [CrossRef]

59. Shao, Y.; Yu, H.; Yang, Y.; Li, M.; Hang, L.; Xu, X. A solid dispersion of quercetin shows enhanced Nrf2 activation and protective effects against oxidative injury in a mouse model of dry age-related macular degeneration. *Oxid. Med. Cell. Longev.* **2019**, *2019*, 1479571. [CrossRef]

60. Wang, Y.; Kim, H.J.; Sparrow, J.R. Quercetin and cyanidin-3-glucoside protect against photooxidation and photodegradation of A2E in retinal pigment epithelial cells. *Exp. Eye Res.* **2017**, *160*, 45–55. [CrossRef]

61. Yu, M.Z.; Yan, W.M.; Beight, C. Lutein and zeaxanthin isomers protect against light-induced retinopathy via decreasing oxidative and endoplasmic reticulum stress in BALB/cJ mice. *Nutrients* **2018**, *10*, 842. [CrossRef] [PubMed]

62. Tseng, W.A.; Thein, T.; Kinnunen, K.; Lashkari, K.; Gregory, M.S.; D'Amore, P.A.; Ksander, B.R. NLRP3 inflammasome activation in retinal pigment epithelial cells by lysosomal destabilization: Implications for age-related macular degeneration. *Investig. Ophthalmol. Vis. Sci.* **2013**, *54*, 110–120. [CrossRef] [PubMed]

63. Wang, Y.; Huo, Y.Z.; Zhao, L.; Lu, F.; Wang, O.; Yang, X.; Ji, B.P.; Zhou, F. Cyanidin-3-glucoside and its phenolic acid metabolites attenuate visible light-induced retinal degeneration in vivo via activation of Nrf2/HO-1 pathway and NF-κB suppression. *Mol. Nutr. Food Res.* **2016**, *60*, 1564–1577. [CrossRef] [PubMed]

64. Jin, X.L.; Wang, C.T.; Wu, W.; Liu, T.T.; Ji, B.P.; Zhou, F. Cyanidin-3-glucoside alleviates 4-hydroxyhexenal-induced NLRP3 inflammasome activation via JNK-c-Jun/AP-1 pathway in human retinal pigment epithelial cells. *J. Immunol. Res.* **2018**, *2018*, 5604610. [CrossRef] [PubMed]

65. Cheng, S.C.; Huang, W.C.; JH, S.P.; Wu, Y.H.; Cheng, C.Y. Quercetin inhibits the production of IL-1beta-induced inflammatory cytokines and chemokines in ARPE-19 cells via the MAPK and NF-kappaB signaling pathways. *Int. J. Mol. Sci.* **2019**, *20*, 2957. [CrossRef]

66. Chapple, S.J.; Cheng, X.; Mann, G.E. Effects of 4-hydroxynonenal on vascular endothelial and smooth muscle cell redox signaling and function in health and disease. *Redox Biol.* **2013**, *1*, 319–331. [CrossRef]

67. Phaniendra, A.; Jestadi, D.B.; Periyasamy, L. Free radicals: Properties, sources, targets, and their implication in various diseases. *Indian J. Clin. Biochem.* **2015**, *30*, 11–26. [CrossRef]

68. Bali, E.B.; Ergin, V.; Rackova, L.; Bayraktar, O.; Kucukboyaci, N.; Karasu, C. Olive leaf extracts protect cardiomyocytes against 4-hydroxynonenal-induced toxicity in vitro: Comparison with oleuropein, hydroxytyrosol, and quercetin. *Planta Med.* **2014**, *80*, 984–992. [CrossRef]

69. Othman, A.I.; El-Sawi, M.R.; El-Missiry, M.A.; Abukhalil, M.H. Epigallocatechin-3-gallate protects against diabetic cardiomyopathy through modulating the cardiometabolic risk factors, oxidative stress, inflammation, cell death and fibrosis in streptozotocin-nicotinamide-induced diabetic rats. *Biomed. Pharmacother.* **2017**, *94*, 362–373. [CrossRef]

70. Xin, Y.; Bai, Y.; Jiang, X.; Zhou, S.; Wang, Y.; Wintergerst, K.A.; Cui, T.; Ji, H.; Tan, Y.; Cai, L. Sulforaphane prevents angiotensin II-induced cardiomyopathy by activation of Nrf2 via stimulating the Akt/GSK-3ß/Fyn pathway. *Redox Biol.* **2018**, *15*, 405–417. [CrossRef]

71. Liu, B.; Zhao, C.; Li, H.; Chen, X.; Ding, Y.; Xu, S. Puerarin protects against heart failure induced by pressure overload through mitigation of ferroptosis. *Biochem. Biophys. Res. Commun.* **2018**, *497*, 233–240. [CrossRef] [PubMed]

72. He, L.; Liu, B.; Dai, Z.; Zhang, H.F.; Zhang, Y.S.; Luo, X.J.; Ma, Q.L.; Peng, J. Alpha lipoic acid protects heart against myocardial ischemia-reperfusion injury through a mechanism involving aldehyde dehydrogenase 2 activation. *Eur. J. Pharmacol.* **2012**, *678*, 32–328. [CrossRef] [PubMed]

73. Kurumazuka, D.; Kitada, K.; Tanaka, R.; Mori, T.; Ohkita, M.; Takaoka, M.; Matsumura, Y. α-Lipoic acid exerts a primary prevention for the cardiac dysfunction in aortocaval fistula-created rat hearts. *Heliyon* **2019**, *5*, e02371. [CrossRef] [PubMed]

74. Jiang, W.B.; Zhao, W.; Chen, H.; Wu, Y.Y.; Wang, Y.; Fu, G.S.; Yang, X.J. Baicalin protects H9c2 cardiomyocytes against hypoxia/reoxygenation-induced apoptosis and oxidative stress through activation of mitochondrial aldehyde dehydrogenase 2. *Clin. Exp. Pharmacol. Physiol.* **2018**, *45*, 303–311. [CrossRef] [PubMed]

75. Negre-Salvayre, A.; Garoby-Salom, S.; Swiader, A.; Rouahi, M.; Pucelle, M.; Salvayre, R. Proatherogenic effects of 4-hydroxynonenal. *Free Radic. Biol. Med.* **2017**, *111*, 127–139. [CrossRef]

76. Akiba, S.; Yamaguchi, H.; Kumazawa, S.; Oka, M.; Sato, T. Suppression of oxidized LDL-induced PDGF receptor beta activation by Ginkgo biloba extract reduces MMP-1 production in coronary smooth muscle cells. *J. Atheroscler. Thromb.* **2007**, *14*, 219–225. [CrossRef]

77. Keller, J.; Camaré, C.; Bernis, C.; Astello-García, M.; de la Rosa, A.P.; Rossignol, M.; del Socorro Santos Díaz, M.; Salvayre, R.; Negre-Salvayre, A.; Guéraud, F. Antiatherogenic and antitumoral properties of Opuntia cladodes: Inhibition of low density lipoprotein oxidation by vascular cells, and protection against the cytotoxicity of lipid oxidation product 4-hydroxynonenal in a colorectal cancer cellular model. *J. Physiol. Biochem.* **2015**, *71*, 577–587. [CrossRef]

78. Garoby-Salom, S.; Guéraud, F.; Camaré, C.; de la Rosa, A.P.; Rossignol, M.; Santos Díaz Mdel, S.; Salvayre, R.; Negre-Salvayre, A. Dietary cladode powder from wild type and domesticated Opuntia species reduces atherogenesis in apoE knock-out mice. *J. Physiol. Biochem.* **2016**, *72*, 59–70. [CrossRef]

79. Shearn, C.T.; Reigan, P.; Petersen, D.R. Inhibition of hydrogen peroxide signaling by 4-hydroxynonenal due to differential regulation of Akt1 and Akt2 contributes to decreases in cell survival and proliferation in hepatocellular carcinoma cells. *Free Radic. Biol. Med.* **2012**, *53*, 1–11. [CrossRef]

80. Seo, B.Y.; Choi, M.J.; Kim, J.S.; Park, E. Comparative analysis of ginsenoside profiles: Antioxidant, antiproliferative, and antigenotoxic activities of Ginseng extracts of gine and main roots. *Prev. Nutr. Food Sci.* **2019**, *24*, 128–135. [CrossRef]

81. Kim, H.G.; Jang, S.S.; Lee, J.S.; Kim, H.S.; Son, C.G. Panax ginseng Meyer prevents radiation-induced liver injury via modulation of oxidative stress and apoptosis. *J. Ginseng Res.* **2017**, *41*, 159–168. [CrossRef] [PubMed]

82. Han, J.Y.; Lee, S.; Yang, J.H.; Kim, S.; Sim, J.; Kim, M.G.; Jeong, T.C.; Ku, S.K.; Cho, I.J.; Ki, S.H. Korean Red Ginseng attenuates ethanol-induced steatosis and oxidative stress via AMPK/Sirt1 activation. *J. Ginseng Res.* **2015**, *39*, 105–115. [CrossRef]

83. Choudhry, Q.N.; Kim, J.H.; Cho, H.T.; Heo, W.; Lee, J.J.; Lee, J.H.; Kim, Y.J. Ameliorative effect of black ginseng extract against oxidative stress-induced cellular damages in mouse hepatocytes. *J. Ginseng Res.* **2019**, *43*, 179–185. [CrossRef] [PubMed]

84. Xu, X.Y.; Wang, Z.; Ren, S.; Leng, J.; Hu, J.N.; Liu, Z.; Chen, C.; Li, W. Improved protective effects of American ginseng berry against acetaminophen-induced liver toxicity through TNF-α-mediated caspase-3/-8/-9 signaling pathways. *Phytomedicine* **2018**, *51*, 128–138. [CrossRef] [PubMed]

85. Qian, Q.Y.; Ying, N.; Yang, Z.; Zhou, L.; Liu, Q.S.; Hu, Z.Y.; Fan, C.L.; Li, S.T.; Dou, X.B. Mechanisms of tanshinone II_A in reducing 4-HNE-induced hepatocyte damage by activating PPARα. *Zhongguo Zhong Yao Za Zhi* **2019**, *44*, 1862–1868. (In Chinese)

86. Sundarraj, K.; Raghunath, A.; Perumal, E. A review on the chemotherapeutic potential of fisetin: In vitro evidences. *Biomed. Pharmacother.* **2018**, *97*, 928–940. [CrossRef]

87. Yousefzadeh, M.J.; Zhu, Y.; McGowan, S.J.; Angelini, L.; Fuhrmann-Stroissnigg, H.; Xu, M.; Ling, Y.Y.; Melos, K.I.; Pirtskhalava, T.; Inman, C.L.; et al. Fisetin is a senotherapeutic that extends health and lifespan. *EBioMedicine* **2018**, *36*, 18–28. [CrossRef]

88. Cha, H.; Lee, S.; Lee, J.H.; Park, J.W. Protective effects of p-coumaric acid against acetaminophen-induced hepatotoxicity in mice. *Food Chem. Toxicol.* **2018**, *121*, 131–139. [CrossRef]

89. Seo, H.L.; Baek, S.Y.; Lee, E.H.; Lee, J.H.; Lee, S.G.; Kim, K.Y.; Jang, M.H.; Park, M.H.; Kim, J.H.; Kim, K.J.; et al. Liqustri lucidi Fructus inhibits hepatic injury and functions as an antioxidant by activation of AMP-activated protein kinase in vivo and in vitro. *Chem. Biol. Interact.* **2017**, *262*, 57–68. [CrossRef]

90. Li, D.; Gu, Z.; Zhang, J.; Ma, S. Protective effect of inducible aldo-keto reductases on 4-hydroxynonenal-induced hepatotoxicity. *Chem. Biol. Interact.* **2019**, *304*, 124–130. [CrossRef]

91. Balestri, F.; De Leo, M.; Sorce, C.; Cappiello, M.; Quattrini, L.; Moschini, R.; Pineschi, C.; Braca, A.; La Motta, C.; Da Settimo, F.; et al. Soyasaponins from Zolfino bean as aldose reductase differential inhibitors. *J. Enzyme Inhib. Med. Chem.* **2019**, *34*, 350–360. [CrossRef] [PubMed]

92. Elrayess, M.A.; Almuraikhy, S.; Kafienah, W.; Al Menhali, A.; Al-Khelaifi, F.; Bashah, M.; Zarkovic, K.; Zarkovic, N.; Waeg, G.; Alsayrafi, M.; et al. 4-Hydroxynonenal causes impairment of human subcutaneous adipogenesis and induction of adipocyte insulin resistance. *Free Radic. Biol. Med.* **2017**, *104*, 129–137. [CrossRef] [PubMed]

93. Pillon, N.J.; Croze, M.L.; Vella, R.E.; Soulere, L.; Lagarde, M.; Soulage, C.O. The lipid peroxidation by-product 4-hydroxy-2-nonenal (4-HNE) induces insulin resistance in skeletal muscle through both carbonyl and oxidative stress. *Endocrinology* **2012**, *153*, 2099–2111. [CrossRef] [PubMed]

94. Wang, Z.; Dou, X.; Gu, D.; Shen, C.; Yao, T.; Nguyen, V.; Braunschweig, C.; Song, Z. 4-Hydroxynonenal differentially regulates adiponectin gene expression and secretion via activating PPARgamma and accelerating ubiquitin-proteasome degradation. *Mol. Cell Endocrinol.* **2012**, *349*, 222–231. [CrossRef]

95. Killion, E.A.; Reeves, A.R.; El Azzouny, M.A.; Yan, Q.W.; Surujon, D.; Griffin, J.D.; Bowman, T.A.; Wang, C.; Matthan, N.R.; Klett, E.L.; et al. A role for long-chain acyl-CoA synthetase-4 (ACSL4) in diet-induced phospholipid remodeling and obesity-associated adipocyte dysfunction. *Mol. Metab.* **2018**, *9*, 43–56. [CrossRef]

96. Guo, L.; Zhang, X.M.; Zhang, Y.B.; Huang, X.; Chi, M.H. Association of 4-hydroxynonenal with classical adipokines and insulin resistance in a Chinese non-diabetic obese population. *Nutr. Hosp.* **2017**, *34*, 363–368. [CrossRef]

97. Lee, H.; Lim, Y. Tocotrienol-rich fraction supplementation reduces hyperglycemia-induced skeletal muscle damage through regulation of insulin signaling and oxidative stress in type 2 diabetic mice. *J. Nutr. Biochem.* **2018**, *57*, 77–85. [CrossRef]

98. Mahboubi, M. Morus alba (mulberry), a natural potent compound in management of obesity. *Pharmacol. Res.* **2019**, *146*, 104341. [CrossRef]

99. Kavitha, Y.; Geetha, A. Anti-inflammatory and preventive activity of white mulberry root bark extract in an experimental model of pancreatitis. *J. Tradit. Complement. Med.* **2018**, *8*, 497–505. [CrossRef]

100. Adisakwattana, S.; Intrawangso, J.; Hemrid, A.; Chanathong, B.; Makynen, K. Extracts of edible plants inhibit pancreatic lipase, cholesterol esterase and cholesterol micellization, and bind bile acids. *Food Technol. Biotechnol.* **2012**, *50*, 11–16.

101. Ann, J.Y.; Eo, H.; Lim, Y. Mulberry leaves (Morus alba L.) ameliorate obesity-induced hepatic lipogenesis, fibrosis, and oxidative stress in high-fat diet-fed mice. *Genes Nutr.* **2015**, *10*, 46. [CrossRef] [PubMed]

102. Singh, C.K.; Kumar, A.; Lavoie, H.A.; Dipette, D.J.; Singh, U.S. Diabetic complications in pregnancy: Is resveratrol a solution? *Exp. Biol. Med.* **2013**, *238*, 482–490. [CrossRef] [PubMed]

103. Al-Hussaini, H.; Kilarkaje, N. Trans-resveratrol mitigates type 1 diabetes-induced oxidative DNA damage and accumulation of advanced glycation end products in glomeruli and tubules of rat kidneys. *Toxicol. Appl. Pharmacol.* **2018**, *339*, 97–109. [CrossRef]

104. Sukumar, P.; Viswambharan, H.; Imrie, H.; Cubbon, R.M.; Yuldasheva, N.; Gage, M.; Galloway, S.; Skromna, A.; Kandavelu, P.; Santos, C.X.; et al. Nox2 NADPH oxidase has a critical role in insulin resistance-related endothelial cell dysfunction. *Diabetes* **2013**, *62*, 2130–2134. [CrossRef] [PubMed]

105. Mahmoud, A.M.; Ali, M.M.; Miranda, E.R.; Mey, J.T.; Blackburn, B.K.; Haus, J.M.; Phillips, S.A. Nox2 contributes to hyperinsulinemia-induced redox imbalance and impaired vascular function. *Redox Biol.* **2017**, *13*, 288–300. [CrossRef] [PubMed]

106. Cremonini, E.; Wang, Z.; Bettaieb, A.; Adamo, A.M.; Daveri, E.; Mills, D.A.; Kalanetra, K.M.; Haj, F.G.; Karakas, S.; Oteiza, P.I. (−)-Epicatechin protects the intestinal barrier from high fat diet-induced permeabilization: Implications for steatosis and insulin resistance. *Redox Biol.* **2018**, *14*, 588–599. [CrossRef] [PubMed]

107. Daveri, E.; Cremonini, E.; Mastaloudis, A.; Hester, S.N.; Wood, S.M.; Waterhouse, A.L.; Anderson, M.; Fraga, C.G.; Oteiza, P.I. Cyanidin and delphinidin modulate inflammation and altered redox signaling improving insulin resistance in high fat-fed mice. *Redox Biol.* **2018**, *18*, 16–24. [CrossRef] [PubMed]

108. Yang, B.Y.; Guo, Y.; Markevych, I.; Qian, Z.M.; Bloom, M.S.; Heinrich, J.; Dharmage, S.C.; Rolling, C.A.; Jordan, S.S.; Komppula, M.; et al. Association of long-term exposure to ambient air pollutants with risk factors for cardiovascular disease in China. *JAMA Netw. Open* **2019**, *2*, e190318. [CrossRef]

109. De Bont, J.; Casas, M.; Barrera-Gomez, J.; Cirach, M.; Rivas, I.; Valvi, D.; Alvarez, M.; Dadvand, P.; Sunyer, J.; Vrijheid, M. Ambient air pollution and overweight and obesity in school-aged children in Barcelona, Spain. *Environ. Int.* **2019**, *125*, 58–64. [CrossRef]

110. Lucht, S.; Hennig, F.; Moebus, S.; Fuhrer-Sakel, D.; Herder, C.; Jockel, K.H.; Hoffmann, B.; Heinz, G. Air pollution and diabetes-related biomarkers in non-diabetic adults: A pathway to impaired glucose metabolism? *Environ. Int.* **2019**, *124*, 370–392. [CrossRef]

111. Wang, N.; Ma, Y.; Liu, Z.; Liu, L.; Yang, K.; Wei, Y.; Liu, Y.; Chen, X.; Sun, X. and Wen, D. Hydroxytyrosol prevents PM2.5-induced adiposity and insulin resistance by restraining oxidative stress related NF-kappaB pathway and modulation of gut microbiota in a murine model. *Free Radic. Biol. Med.* **2019**, *141*, 393–407. [CrossRef]

112. Gallo, G.; Bruno, R.; Taranto, A.; Martino, G. Are polyunsaturated fatty acid metabolites, the protective effect of 4-hydroxytyrosol on human red blood cell membranes and oxidative damage (4-hydroxyalkenals) compatible in hypertriglyceridemic patients? *Pharmacogn. Mag.* **2017**, *13*, S561–S566. [CrossRef] [PubMed]

113. Cesar, V.; Jozić, I.; Begović, L.; Vuković, T.; Mlinarić, S.; Lepeduš, H.; Borović Šunjić, S.; Žarković, N. Cell-type-specific modulation of hydrogen peroxide cytotoxicity and 4-hydroxynonenal binding to human cellular proteins in vitro by antioxidant Aloe vera extract. *Antioxidants* **2018**, *7*, 125. [CrossRef]

114. Cheng, P.; Kuang, F.; Ju, G. Aescin reduces oxidative stress and provides neuroprotection in experimental traumatic spinal cord injury. *Free Radic. Biol. Med.* **2016**, *99*, 405–417. [CrossRef] [PubMed]

115. Sirtori, C.R. Aescin: Pharmacology, pharmacokinetics and therapeutic profile. *Pharmacol. Res.* **2001**, *44*, 183–193. [CrossRef] [PubMed]

116. Li, Q.; Lin, F.Q.; Wang, S.Q.; Hong, W.S.; Xu, A. Protective capacity of five kinds of plant extracts on 4HNE-induced melanocyte. *Chin. Arch. Tradit. Chin. Med.* **2015**, *33*, 1610–1613.

117. Gegotek, A.; Bielawska, K.; Biernacki, M.; Dobrzynska, I.; Skrzydlewska, E. Time-dependent effect of rutin on skin fibroblasts membrane disruption following UV radiation. *Redox Biol.* **2017**, *12*, 733–744. [CrossRef]

118. Gegotek, A.; Bielawska, K.; Biernacki, M.; Zareba, I.; Surazynski, A.; Skrzydlewska, E. Comparison of protective effect of ascorbic acid on redox and endocannabinoid systems interactions in in vitro cultured human skin fibroblasts exposed to UV radiation and hydrogen peroxide. *Arch. Dermatol. Res.* **2017**, *309*, 285–303. [CrossRef]

119. Yamaguchi, T.; Yoneyama, M.; Hinoi, E.; Ogita, K. Involvement of calpain in 4-hydroxynonenal-induced disruption of gap junction-mediated intercellular communication among fibrocytes in primary cultures derived from the cochlear spiral ligament. *J. Pharmacol. Sci.* **2015**, *129*, 127–134. [CrossRef]

120. Castaneda, R.; Natarajan, S.; Jeong, S.Y.; Hong, B.N.; Kang, T.H. Traditional oriental medicine for sensorineural hearing loss: Can ethnopharmacology contribute to potential drug discovery? *J. Ethnopharmacol.* **2018**, *231*, 409–428. [CrossRef]

121. Xiong, M.; He, Q.L.; Lai, H.W.; Wang, J. Astragaloside IV inhibits apoptotic cell death in the guinea pig cochlea exposed to impulse noise. *Acta Otolaryngol.* **2012**, *132*, 467–474. [CrossRef] [PubMed]

122. Xiong, M.; Lai, H.; He, Q.; Wang, J. Astragaloside IV attenuates impulse noise-induced trauma in guinea pig. *Acta Otolaryngol.* **2011**, *131*, 809–816. [CrossRef] [PubMed]

123. Fetoni, A.R.; Paciello, F.; Rolesi, R.; Eramo, S.L.M.; Mancuso, C.; Troiani, D.; Paludetti, G. Rosmarinic acid up-regulates the noise-activated Nrf2/HO-1 pathway and protects against noise-induced injury in rat cochlea. *Free Radic. Biol. Med.* **2015**, *85*, 269–281. [CrossRef] [PubMed]

124. Kanner, J.; Selhub, J.; Shpaizer, A.; Rabkin, B.; Shacham, I.; Tirosh, O. Redox homeostasis in stomach medium by foods: The Postprandial Oxidative Stress Index (POSI) for balancing nutrition and human health. *Redox Biol.* **2017**, *12*, 929–936. [CrossRef]

125. Lamothe, S.; Guérette, C.; Dion, F.; Sabik, H.; Britten, M. Antioxidant activity of milk and polyphenol-rich beverages during simulated gastrointestinal digestion of linseed oil emulsions. *Food Res. Int.* **2019**, *122*, 149–156. [CrossRef]

126. Tullberg, C.; Vegarud, G.; Undeland, I. Oxidation of marine oils during in vitro gastrointestinal digestion with human digestive fluids – Role of oil origin, added tocopherols and lipolytic activity. *Food Chem.* **2019**, *270*, 527–537. [CrossRef]

127. Pasko, P.; Barton, H.; Zagrodzki, P.; Izewska, A.; Krosniak, M.; Gawlik, M.; Gawlik, M.; Gorinstein, S. Effect of amaranth seeds in diet on oxidative status in plasma and selected tissues of high fructose-fed rats. *Plant. Foods Hum. Nutr.* **2010**, *65*, 146–151. [CrossRef]

128. Cherkas, A.; Zarkovic, K.; Cipak Gasparovic, A.; Jaganjac, M.; Milkovic, L.; Abrahamovych, O.; Yatskevych, O.; Waeg, G.; Yelisyeyeva, O.; Zarkovic, N. Amaranth oil reduces accumulation of 4-hydroxynonenal-histidine adducts in gastric mucosa and improves heart rate variability in duodenal peptic ulcer patients undergoing Helicobacter pylori eradication. *Free Radic. Res.* **2018**, *52*, 135–149. [CrossRef]

129. Boyd, O.; Weng, P.; Sun, X.P.; Alberico, T.; Laslo, M.; Obenland, D.M.; Kern, B.; Zou, S. Nectarine promotes longevity in Drosophila melanogaster. *Free Radic. Biol. Med.* **2011**, *50*, 1669–1678. [CrossRef]

130. Zheng, R.J.; Dragomir, A.C.; Mishin, V.; Richardson, J.R.; Heck, D.E.; Laskin, D.L.; Laskin, J.D. Differential metabolism of 4-hydroxynonenal in liver, lung and brain of mice and rats. *Toxicol. Appl. Pharm.* **2014**, *279*, 43–52. [CrossRef]

131. Huang, Y.; Li, W.; Kong, A.N. Anti-oxidative stress regulator NF-E2-related factor 2 mediates the adaptive induction of antioxidant and detoxifying enzymes by lipid peroxidation metabolite 4-hydroxynonenal. *Cell Biosci.* **2012**, *2*, 40. [CrossRef] [PubMed]

132. Gross, B.; Pawlak, M.; Lefebvre, P.; Staels, B. PPARs in obesity-induced T2DM, dyslipidaemia and NAFLD. *Nat. Rev. Endocrinol.* **2017**, *13*, 36–49. [CrossRef] [PubMed]

133. Cheng, R.; Ding, L.X.; He, X.M.; Takahashi, Y.; Ma, J.X. Interaction of PPAR alpha with the canonic Wnt pathway in the regulation of renal fibrosis. *Diabetes* **2016**, *65*, 3730–3743. [CrossRef] [PubMed]

134. Hosoi, T.; Kuwamura, A.; Thon, M.; Tsuchio, K.; Abd El-Hafeez, A.A.; Ozawa, K. Possible involvement of 4-hydroxy-2-nonenal in the pathogenesis of leptin resistance in obesity. *Am. J. Physiol. Cell Physiol.* **2019**, *316*, C641–C648. [CrossRef]

135. Aslebagh, R.; Pfeffer, B.A.; Fliesler, S.J.; Darie, C.C. Mass spectrometry-based proteomics of oxidative stress: Identification of 4-hydroxy-2-nonenal (HNE) adducts of amino acids using lysozyme and bovine serum albumin as model proteins. *Electrophoresis* **2016**, *37*, 2615–2623. [CrossRef]

136. Mendez, D.; Hernaez, M.L.; Diez, A.; Puyet, A.; Bautista, J.M. Combined proteomic approaches for the identification of specific amino acid residues modified by 4-hydroxy-2-nonenal under physiological conditions. *J. Protome. Res.* **2010**, *9*, 5770–5781. [CrossRef]

137. Delosiere, M.; Sante-Lhoutellier, V.; Chantelauze, C.; Durand, D.; Thomas, A.; Joly, C.; Pujos-Guillot, E.; Remond, D.; Comte, B.; Gladine, C.; et al. Quantification of 4-hydroxy-2-nonenal-protein adducts in the in vivo gastric digesta of mini-pigs using a GC-MS/MS method with accuracy profile validation. *Food Funct.* **2016**, *7*, 3497–3504. [CrossRef]

138. Colzani, M.; Criscuolo, A.; De Maddis, D.; Garzon, D.; Yeum, K.J.; Vistoli, G.; Carini, M.; Aldini, G. A novel high resolution MS approach for the screening of 4-hydroxy-trans-2-nonenal sequestering agents. *J. Pharmaceut. Biomed.* **2014**, *91*, 108–118. [CrossRef]

 *biomolecules*

Review

Mechanistic Pathways and Molecular Targets of Plant-Derived Anticancer *ent*-Kaurane Diterpenes

Md. Shahid Sarwar [1,2], **Yi-Xuan Xia** [1], **Zheng-Ming Liang** [1], **Siu Wai Tsang** [1] and **Hong-Jie Zhang** [1,*] 

[1]   School of Chinese Medicine, Hong Kong Baptist University, Kowloon Tong, Hong Kong, China; 15484327@life.hkbu.edu.hk or sarwar@nstu.edu.bd (M.S.S.); xiayixuan@hkbu.edu.hk (Y.-X.X.); chem@outlook.com (Z.-M.L.); tsang@hkbu.edu.hk (S.W.T.)

[2]   Department of Pharmacy, Noakhali Science and Technology University, Sonapur, Noakhali 3814, Bangladesh

*   Correspondence: zhanghj@hkbu.edu.hk; Tel.: +852-3411-2956

Received: 18 December 2019; Accepted: 14 January 2020; Published: 16 January 2020

Abstract: Since the first discovery in 1961, more than 1300 *ent*-kaurane diterpenoids have been isolated and identified from different plant sources, mainly the genus *Isodon*. Chemically, they consist of a perhydrophenanthrene subunit and a cyclopentane ring. A large number of reports describe the anticancer potential and mechanism of action of *ent*-kaurane compounds in a series of cancer cell lines. Oridonin is one of the prime anticancer *ent*-kaurane diterpenoids that is currently in a phase-I clinical trial in China. In this review, we have extensively summarized the anticancer activities of *ent*-kaurane diterpenoids according to their plant sources, mechanistic pathways, and biological targets. Literature analysis found that anticancer effect of *ent*-kauranes are mainly mediated through regulation of apoptosis, cell cycle arrest, autophagy, and metastasis. Induction of apoptosis is associated with modulation of BCL-2, BAX, PARP, cytochrome c, and cleaved caspase-3, -8, and -9, while cell cycle arrest is controlled by cyclin D1, c-Myc, p21, p53, and CDK-2 and -4. The most common metastatic target proteins of *ent*-kauranes are MMP-2, MMP-9, VEGF, and VEGFR whereas LC-II and mTOR are key regulators to induce autophagy.

Keywords: *isodon* genus; *ent*-kaurane diterpenoids; cancer; natural compounds; pathways

---

1. Introduction

Cancer is a heterogeneous disease which caused 9.6 million deaths globally in 2018 and the incidence of cancer is projected to increase to 13.1 million deaths by 2030 [1,2]. In addition, a large number of cancer patients have forms of the disease that are resistant to the mainstay anticancer drugs. Currently used anticancer drugs are mainly obtained from natural sources (plants, microorganisms, and marine organisms) or from synthetic reactions [3]. Unfortunately, most of these drugs also kill normal cells generating severe side effects [4,5]. These factors indicate the urgent need for the development of new therapeutics and treatment strategies.

Medicinal plants are an important source of drug entities. Over the past few decades, plants became an indispensable source of anticancer agents which are commonly used to treat different types of cancers in clinical settings. The World Health Organization (WHO) estimates that 80% of the total world's population residing in rural areas rely on plant sources as a part of their primary health care system. In addition, more than one-quarter of the prescribed drugs in developed countries are prepared directly or indirectly from plants [6]. In developing countries, 75% of the population use plant-derived drugs for the treatment of various illnesses [7]. Since the early 1900s, thousands of medicinal plants have been thoroughly investigated to screen cytotoxic compounds, yielding numerous anticancer drugs and lead compounds [5]. The structure of the cytotoxic plant metabolite podophyllotoxin was first described in 1932 [8], this was followed by the discovery of vinca alkaloids (vinblastine and vincristine) in 1958 [9], and later, the identification of paclitaxel in 1971 [10]. These plant-derived molecules have revolutionized cancer treatment, yet more lead structures continue to be needed.

ent-Kaurane diterpenoids have recently attracted the attention of scientists and doctors due to their promising anticancer effect and safety profile [11]. They are subclass of tetracyclic diterpenes which consist of a perhydrophenantrene unit (A, B, and C rings) and cyclopentane unit (D ring) connected by a bridge of two carbons between C-8 and C-13 (Figure 1). Diterpenes containing a kaurane scaffold with configurational inversion at all chiral centers are known as *ent*-kaurane diterpenes. In this case, the prefix "*ent*-" is added before the name of this class of compounds to indicate that they are enantiomers of kaurene diterpenes. Moreover, International Union of Pure and Applied Chemistry (IUPAC) has established structured-guidelines for the stereochemistry, nomenclature, and numbering of the *ent*-kaurane skeleton. With the exception of those compounds containing an alkene functionality between C-9 and C-11, most of *ent*-kauranes show characteristic negative values for the specific optical rotation [12]. This class of compounds are mainly available in *Isodon* genus which mainly consists of a group of flowering plants. Other plant species that also contain *ent*-kaurane compounds belong to different families such as Asteraceae, Lamiaceae, Euphorbiaceae, Jungermanniaceae, and Chrysobalanaceae [13]. Accumulating evidences described the anticancer activity and underlying mechanism of *ent*-kaurane compounds against numerous cancers such as lung, colon, breast, prostate, liver, and gastric cancer both in vitro and in vivo. [14,15]. In this article, we discuss the anticancer *ent*-kaurane diterpenoids reported in the literature along with their source and the biological targets regulated by this type of compound.

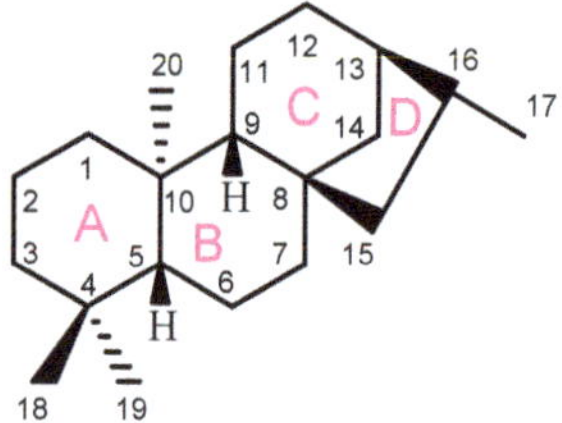

Figure 1. General structure and numbering system of *ent*-kaurane diterpenes.

Methodology

Literature search was performed through the commonly used databases such as PubMed, MEDLINE, Scopus, Embase, Google Scholar, ScienceDirect, and Wiley Online. The articles which described the anticancer activity, mechanism of action and protein targets of *ent*-kaurane diterpenes were included in the review whilst those papers which mentioned only the isolation and characterization without further exploration of their molecular pathways were excluded. The molecular structures of diterpenoids were generated using the ChemDraw software. The chemical formulae, molecular weight, and related information of the diterpenoids were comprehensively matched with SciFinder®, a well-known portal for chemical compounds.

2. Anticancer *ent*-Kauranes and Their Targets

2.1. Oridonin

Oridonin ($C_{20}H_{28}O_6$, MW: 364.44, Figure 2), was first isolated from *Isodon rubescens* in the 1960s and one of its derivatives L-alanine-(14-oridonin) ester trifluoroacetate (HAO472) (Figure 2) has progressed to phase-I clinical trials by Hengrui Medicine Co. Ltd., Lianyungang, China [16]. Numerous studies reported the anticancer activities of oridonin against diverse human cell lines such as HCT116 (colorectal carcinoma, IC_{50}: 32.6 µM) [17], OCI-AML3 (acute myeloid leukemia, IC_{50}: 3.27 µM) [18], and BxPC-3 (pancreatic carcinoma, IC_{50}: 53.0 µM) [19] after 24 h treatment; KYSE-150 (esophageal carcinoma,

IC$_{50}$: 28.7 μM), EC9706 (esophageal carcinoma, IC$_{50}$: 34.4 μM), KYS-30 cells (esophageal carcinoma, IC$_{50}$: 32.3 μM) [20], Jurkat (T-cell leukemia, IC$_{50}$: 0.73 μM) [21], MG-63 (osteosarcoma, IC$_{50}$: 10.9 μM), HOS (osteosarcoma, IC$_{50}$: 11.9 μM), Saos-2 (osteosarcoma, IC$_{50}$: 17.3 μM), U-2OS cells (osteosarcoma, IC$_{50}$: 17.7 μM) [22], EC109 (esophageal carcinoma, IC$_{50}$: 19.7 μM), EC9706 (esophageal carcinoma, IC$_{50}$: 31.3 μM), EC1 (esophageal carcinoma, IC$_{50}$: 25.8 μM) [23], HUVECs (human umbilical vein endothelial cells, IC$_{50}$: 420 μM) [24], CNE1 (nasopharyngeal carcinoma, IC$_{50}$: 3.66 μM), and CNE2 (nasopharyngeal carcinoma, IC$_{50}$: 5.93 μM) cells [25] after 48 h treatment; and EC109 (esophageal carcinoma, IC$_{50}$: 67.1 μM), SHG-44 (glioblastoma, IC$_{50}$: 53.5 μM), MCF-7 (breast carcinomas, IC$_{50}$: 72.1 μM) [26], and SGC-7901 (gastric carcinoma, IC$_{50}$: 22.7 μM) [27] after 72 h treatment. Kadioglu et al. (2018) tested the cytotoxicity of oridonin in spectrum of drug-resistant cancer cell lines including CCRF-CEM (leukemia, IC$_{50}$: 1.65 μM), CEM/ADR5000 (leukemia, IC$_{50}$: 8.53 μM), MDA-MB231 (breast carcinoma, IC$_{50}$: 6.06 μM), MDA-MB231/BCRP (breast carcinoma, IC$_{50}$: 9.74 μM), HCT116 (p53$^{+/+}$) (IC$_{50}$: 18.0 μM), HCT-116 (p53$^{-/-}$) (IC$_{50}$: 34.7 μM), U87MG (glioblastoma, IC$_{50}$: 17.4 μM), HepG2 (liver carcinoma, IC$_{50}$: 25.7 μM), and AML12 (liver normal cells, IC$_{50}$: >109 μM) cells, but did not mention the incubation time with oridonin in these cytotoxic test experiments [28].

A wide variety of sources have reported the anticancer mechanism of oridonin in multiple cancer cell lines [17,24,27,29–32]. According to the study of Yao et al. (2017) oridonin induced autophagy by decreasing the protein levels of glucose transporter 1 (GLUT1) and monocarboxylate transporter 1 (MCT1) in SW480 human colorectal cancer cells, in the BALB/c xenograft model. Autophagy induction by oridonin was correlated to increased expression of light chain-I (LC-I) and LC-II, while decreasing the phosphorylation of adenosine monophosphate-activated protein kinase (p-AMPK) [17]. In human umbilical vein endothelial cells (HUVECs), oridonin suppressed the proliferation, tube formation, migration, and invasion, but induced apoptosis. Oridonin was also found to block the angiogenesis of zebrafish by decreasing the mRNA expression of vascular endothelial growth factor A (VEGFA), vascular endothelial growth factor receptor 2 (VEGFR2) and VEGFR3, while reducing the expression of metastatic proteins claudin-1, -4, and -7. Antimetastatic activity of oridonin was further confirmed in a xenograft zebrafish model tested at a dose of 8 mg/kg that showed significant effect ($p < 0.01$) in comparison to the vehicle control group [24]. Gao et al. (2016) reported that oridonin induced apoptosis by downmodulating the expression level of B-cell lymphoma 2 (BCL-2), but upmodulating the expression of BAX, thus reducing the BCL-2/BAX ratio in human gastric cancer SGC-7901 cells. Moreover, oridonin treatment activated caspase-3 by promoting the release of cytochrome c from mitochondria to the cytosol [27]. In another study, oridonin potentiated the anticancer activity of lentinan, a polysaccharide isolated from shiitake mushrooms, by upregulating the expression levels of caspase-3, -8, -9, BAX, p53, and p21 while downregulating the expression of BCL-2, B-cell lymphoma extra-large (BCL-X$_L$) and epidermal growth factor (EGF) in SMMC-7721 human hepatoma cells [29].

In prostate cancer cells (PC-3 and DU-145), oridonin increased the expression levels of p53, p21, caspase-3, -9, and poly (ADP-ribose) polymerase (PARP), while it decreased cyclin-dependent kinase 1 (CDK1) levels [30]. Moreover, it inhibited the expression of phosphoinositide 3-kinase (PI3K) and blocked phosphorylation of protein kinase B (p-Akt). Sun et al. (2018) reported synergistic anticancer activity of oridonin in combination with lentinan by decreasing the expression of BCL-2 and nuclear factor kappa B (NF-κB) and increasing the expression of caspase-3, -9, p53, p21, NF-κB inhibitor-α (IκB-α) by increasing transcription of mRNA and translation of the mRNA into their respective protein products in human hepatoblastoma HepG2 cells [31]. In 4T1 human breast cancer cells, oridonin inhibited cellular proliferation, migration, and invasion via a negative modulation of notch1-4. Furthermore, the administration of oridonin (5 mg/kg) significantly ($p < 0.01$) reduced the weight (84%) and volume (72%) of 4T1 tumors compared with the negative control group in a xenograft nude mouse study [32].

2.2. Eriocalyxin B

Eriocalyxin B ($C_{20}H_{24}O_5$, MW: 344.41, Figure 2) was isolated from the plant *Isodon eriocalyx* var. laxiflora [33]. The cytotoxicity of eriocalyxin B was tested in panel of human cell lines such as SMMC-7721 (hepatocarcinoma, IC_{50}: 0.76 μM) [33], MCF-7 (IC_{50}: 0.75 μM), MDA-MB-231 (IC_{50}: 0.47 μM) [34] after 48 h treatment; PANC1 (pancreatic carcinoma, IC_{50}: 1.79 μM), CAPAN1 (pancreatic carcinoma, IC_{50}: 0.86 μM), CAPAN2 (pancreatic carcinoma, IC_{50}: 0.73 μM), SW1990 (pancreatic carcinoma, IC_{50}: 1.40 μM), WRL68 (normal human liver cells, IC_{50}: >3.58 μM), PBMC (human peripheral blood mononuclear cells, IC_{50}: >5.83 μM) [35], SU-DHL-4 (lymphoma, IC_{50}: 1.00 μM), Namalwa (lymphoma, IC_{50}: 1.50 μM), Raji (lymphoma, IC_{50}: 2.00 μM), Jurkat (lymphoma, IC_{50}: 2.00 μM), U266 (lymphoma, IC_{50}: 5.60 μM) and HUT78 (lymphoma, IC_{50}: 2.50 μM) cells [36] after 72 h treatment.

The molecular pathways of eriocalyxin B against different cancers have been intensively investigated [33,34,37–39]. In hepatocellular carcinoma SMMC-7721 cells, eriocalyxin B induced apoptosis by interfering with the binding of NF-κB with the response elements via targeting cysteine 62 moiety of p50 [33]. In a recent study, eriocalyxin B was reported to induce autophagy upon an upregulation of LC3B-II and beclin-1 in MCF-7 and MDA-MB-231 human breast cancer cells; however, the expression of p62 was downregulated. In addition to autophagy, eriocalyxin B also induced apoptosis by activating caspase-3 and PARP while decreasing BCL-2. In fact, the phosphorylation levels of Akt, mammalian target of rapamycin (mTOR), and p70S6K were decreased in a dose- and time-dependent manner post eriocalyxin B treatment. Therefore, the mechanism of action of eriocalyxin B is suggested to be associated with the Akt/mTOR/p70S6K signaling pathway [34].

Earlier studies demonstrated that eriocalyxin B induced G1 phase cell cycle arrest by decreasing the expression of cyclin D1, CDK4, and phosphorylated retinoblastoma (p-Rb). When applied to HUVECs at 50 and 100 nM, eriocalyxin B notably inhibited the VEGF-induced cell proliferation, tube formation, cell migration, and invasion [37]. Further to its modulation on the VEGF cascade, eriocalyxin B suppressed the VEGF-induced phosphorylation of VEGFR-2 via an interaction with various ATP-binding sites, thus leading to the repression of several VEGFR-2 downstream molecules such as VEGFR-1, focal adhesion kinase (FAK), Src, $pSer^{473}$-Akt, extracellular signal-regulated kinase (ERK1/2), and $pThr^{180}/Tyr^{182}$-mitogen-activated protein kinase (p38-MAPK). When administered in vivo, eriocalyxin B restrained the formation of new blood vessels, vascularization, and growth of the 4T1 breast tumor xenografts. According to the study of Lu et al. (2016), eriocalyxin B inhibited the cell proliferation, migration, invasion, and angiogenesis of human colon cancer cells SW1116 [38]. Mechanistic studies found that it inhibited the phosphorylation of janus kinase 2 (JAK2) and signal transducer and activator of transcription-3 (STAT3) as well decreased the expression of VEGF, VEGFR-2, matrix metallopeptidase-2 (MMP-2), MMP-9, and proliferating cell nuclear antigen (PCNA). Yu et al., (2015) reported that eriocalyxin B inhibited both constitutive- and interleukin-6 (IL-6)-induced phosphorylation of STAT-3 in A549 lung cancer cells without affecting the upstream kinases such as JAK1, JAK2, and tyrosine kinase 2 (TYK2) [39].

2.3. Excisanin A

Excisanin A ($C_{20}H_{30}O_5$, MW: 350.455, Figure 2) was isolated from *Isodon macrocalyxin* which exhibited cytotoxic potential in breast cancer MDA-MB-231 (IC_{50}: 22.4 μM), SKBR3 (IC_{50}: 27.3 μM) [40] and MDA-MB-453 (IC_{50}: 10.3 μM) cells as well as in hepatoma Hep3B (IC_{50}: 6.45 μM) cells [41] after 72 h treatment. At 10–40 μM concentration, excisanin A suppressed the expression of MMP-2, MMP-9, integrin β1, β-catenin, and reduced the phosphorylation of FAK and Src in breast cancer (MDA-MB-231 and SKBR3) cells. Moreover, the phosphorylation of PI3K, Akt and glycogen synthase kinase 3β (GSK3β) was also reduced after treatment with excisanin A [40]. In addition to its effect on breast cancer metastasis, excisanin A inhibited the proliferation of human hepatocellular carcinoma cell line Hep3B and breast cancer cell line MDA-MB-453 cells by inducing apoptosis. It significantly ($p < 0.01$) reduced the tumor weight (46.4%) as well as induced tumor cell apoptosis in Hep3B xenograft mice at a dose of

20 mg/kg/day. Mechanistic studies showed that excisanin A reduced the expression of p-GSK-3α/β, Thr308-Akt, Ser473-Akt, p-mTOR, and p-FKHR expression in both Hep3B and MDA-MB-453 cells [41]. Zhang et al. (2013) reported dose- and time-dependent induction of autophagy by increasing the expression of LC3-II and decreasing the expression of p62 in nasopharyngeal carcinoma (NPC) cell lines CNE1 and CNE2 cells after excisanin A treatment. Moreover, excisanin A also upregulated p-JNK (c-jun N-terminal kinase), p-c-Jun and sestrin 2 expression in NPC cells [42].

2.4. Ponicidin

Ponicidin ($C_{20}H_{26}O_6$, MW: 362.42, Figure 2) was isolated from *Isodon rubescens* and *I. japonicas*, and showed cytotoxic activity against the human carcinoma cell lines K562 (leukemia), Bcap37 (breast), GC823 (gastric), BIU87 (bladder), HeLa (cervical) and PC-3 (prostate) cells [43]. Further studies also described the cytotoxicity of ponicidin in HeLa (IC_{50}: 23.1 μM) [44], A549 (IC_{50}: 38.0 μM) and GLC-82 (lung, IC_{50}: 32.0 μM) cancer cell lines after 24 h; A549 cells (IC_{50}: 31.0 μM) and GLC-82 (IC_{50}: 26.0 μM) after 48 h; A549 cells (IC_{50}: 15.0 μM) and GLC-82 cells (IC_{50}: 13.0 μM) [45] after 72 h. It suppressed the growth of gastric carcinoma MKN28 cells in dose- and time-dependent manner by downregulating BCL-2, p-JAK2, and p-STAT3 expression, while upregulating BAX and cleaved caspase-3 expression [43]. An earlier study reported induction of apoptosis and disruption of mitochondrial membrane potential in ponicidin-treated lung cancer A549 and GLC-82 cells by the upregulating expression of cleaved caspase-3, -8, -9, and BAX, and downregulating the expression of BCL-2 and survivin [45]. Du et al., (2015) reported that ponicidin induced apoptosis and cell cycle arrest in colon cancer cells HCT-116 by increasing caspase-3, BAX, and p-p38 expression, while decreasing BCL-2, p-ERK, and p-Akt expression [46].

2.5. Pharicins A and B

Pharicins A ($C_{24}H_{34}O_7$, MW: 434.53, Figure 2) and B ($C_{24}H_{34}O_8$, MW: 450.53) were isolated from *Isodon pharicus*. The only structural difference between these two compounds is the position of hydroxyl group which is located at C-1 in pharicin A and at C-12 in pharicin B. Pharicin A induced mitotic arrest in leukemia cell line Jurkat and solid tumor-derived cell line raji by inhibiting auto phosphorylation activity of BubR1 [47]. Investigation of the effects of pharicin B in several acute myeloid leukemia (AML) cell lines and primary leukemia cells from AML patients such as NB-4, U937 and THP-1 cells showed cytotoxicity with IC_{50} values around 3.5 μM after 48 h of treatment. Further studies found that it can quickly stabilize the retinoic acid receptor-α (RAR-α) protein, even in the presence of all-trans-retinoic acid (ATRA) which generally induce the loss of RAR-α protein. Moreover, pharicin B also increased ATRA-dependent transcriptional activity of RAR-α protein in NB4 cells, a type of promyelocytic leukemia–RAR-α–positive APL cell line [48].

2.6. Jaridonin

Jaridonin ($C_{22}H_{32}O_5$, MW: 376.49, Figure 2) was isolated from the Chinese herb *Isodon rubescens* which exhibited cytotoxicity in esophageal squamous cancer EC109 (IC_{50}: 12.0 μM), EC9706 (IC_{50}: 11.2 μM), and EC1 (IC_{50}: 4.60 μM) cells [23] after 48 h as well as in human glioma cell line SHG-44 (IC_{50}: 14.7 μM) and breast cancer MCF-7 (IC_{50}: 16.7 μM) cell [26] after 72 h. A mechanistic study found that it induced apoptosis and G2/M phase cell cycle arrest in esophageal cancer EC109, EC9706 and EC1 cells. Jaridonin treatment caused remarkable reduction of mitochondrial membrane potential, release of cytochrome c into the cytosol, and increased expression of caspase-3 and -9, leading to activation of the mitochondria-mediated apoptosis. Moreover, jaridonin also increased the production of reactive oxygen species (ROS) and upregulated p53, p21waf1/Cip1, and BAX expression [23]. Another study confirmed the induction of G2/M phase arrest in esophageal squamous cancer EC9706 cells [49]. However, cell cycle arrest was related with increased phosphorylation of ataxia-telangiectasia mutated (ATM) (Ser1981) protein kinase, cell division control 2 (Cdc2) (Tyr15), Cdc25C, and H2A histone family member X (H2A.X)(Ser139) as well as increased expression of check point kinase 1(Chk1) and Chk2.

2.7. Jungermannenones A and B

Jungermannenones A ($C_{20}H_{28}O_2$, MW: 300.44, Figure 2) and B ($C_{20}H_{28}O_3$, MW: 316.44, Figure 2) were isolated from *Jungermannia fauriana*, and exhibited cancer cell killing activity in multiple cell lines [50]. The IC_{50} values of jungermannenone A were determined in PC3 (IC_{50}: 1.34 µM), DU145 (prostate carcinoma, IC_{50}: 5.01 µM), LNCaP (prostate carcinoma, IC_{50}: 2.78 µM), A549 (IC_{50}: 8.64 µM), MCF-7 (IC_{50}: 18.3 µM), HepG2 (IC_{50}: 5.29 µM) and RWPE1 (normal prostate epithelial, IC_{50}: 5.09 µM) cell lines, while the IC_{50} values of jungermannenone B were measured in PC3 (4.93 µM), DU145 (5.50 µM), LNCaP (3.18 µM), A549 (5.26 µM), MCF-7 (14.2 µM), HepG2 (6.02 µM), and RWPE1 (18.2 µM). Both compounds induced apoptosis by stimulating ROS accumulation and inducing cell cycle arrest. Western blot analysis showed reduced expression of c-myc, cyclin D1, cyclin E, CDK4, and p-Cdc2, while increased expression of p21 and p-ERK. Furthermore, both jungermannenones A and B induced DNA damage by reducing the expression of DNA repair proteins Ku70/Ku80 and RDA51 [50].

2.8. Effusanin E

Effusanin E ($C_{20}H_{28}O_6$, MW: 364.44, Figure 2) was isolated from *Rabdosia serra* (*Rabdosia* is a synonym of *Isodon*), and is cytotoxic to nasopharyngeal carcinoma cell CNE2 (IC_{50}: ~60 µM) but non-toxic up to 500 µM in normal nasopharyngeal epithelial cell line after 48 h. It inhibited cell proliferation and induced apoptosis in CNE2 cells by increasing the expression of cleaved PARP, caspase-3 and -9 proteins, as well as decreasing nuclear translocation of p65 NF-κB [51]. Moreover, the binding ability of NF-κB to the promoter region of cyclooxygenase-2 (COX-2) abolished after treating the cells with effusanin E, therefore, inhibiting the expression and promoter activity of COX-2. In vivo studies also found significant reduction of tumor growth ($p < 0.05$) at 30 mg/kg without any sign of toxicity in comparison to negative control group (DMSO) as well as decreased expression of p50 NF-κB and COX-2 in tumor tissue.

2.9. Longikaurin A

Longikaurin A ($C_{20}H_{28}O_5$, MW: 348.44, Figure 2) was described from *Isodon ternifolius*, and displayed cytotoxic potential in human hepatocarcinoma SMMC-7721 (IC_{50}: ~1.8 µM), HepG2 (IC_{50}: ~2 µM), BEL7402 (IC_{50}: ~6 µM), Huh7 (IC_{50}: ~6 µM), and LO2 (IC_{50}: ~9 µM) cells [52], and the nasopharyngeal carcinoma CNE1 (IC_{50}: 1.26 µM) and CNE2 (IC_{50}: 1.52 µM) cells [25] with 48 h incubation. This *ent*-kaurane compound induced apoptosis and G2/M phase cell cycle arrest in SMMC-7721 and HepG2 cells. In vivo studies found longikaurin A treatment (6 mg/kg) significantly suppressed tumor development ($p < 0.01$) in a SMMC-7721 xenograft mouse model study and the antitumor effect was comparable to the positive control 5-FU (fluorouracil) (10 mg/kg). A mechanistic study showed reduced expression of S-phase kinase-associated protein 2 (Skp2) after longikaurin A treatment, which was correlated with increased expression of p21 and p-Cdc2 (Try15) and decreased expression of cyclin B1 and Cdc2 proteins [52]. It also induced ROS production and phosphorylation of JNK.

In nasopharyngeal carcinoma (NPC) cell lines S18 and S26, longikaurin A exerted anticancer activity and suppressed the stemness of cells. Western blot analysis showed downregulation of c-myc and fibronectin in NPC cells [53]. At a low concentration (0.2 µM), longikaurin A induced S phase arrest while at a higher concentration (3.12 µM) it induced apoptosis in human NPC cell lines CNE1 and CNE2 cells [25]. Mechanistic studies found upregulation of cleaved caspase-3, cleaved PARP, and BAX, while downregulation of BCL-xL, p-Akt, and p-GSK-3β. In CNE2 xenograft model, longikaurin A significantly suppressed tumor growth without affecting the body weights of the mice [53].

2.10. Glaucocalyxins A and B

Glaucocalyxin A ($C_{20}H_{28}O_4$, MW: 332.44, Figure 2) was isolated from *Isodon japonica*, and demonstrated strong anticancer effects in the human cell lines of HL-60 (leukemia, IC_{50}: 6.15 µM) cells [54] after 24 h; Focus (hepatocarcinoma, IC_{50}: 2.70 μM), SMMC-7721 (IC_{50}: 5.58 μM), HepG2 (IC_{50}: 8.22 μM), SK-HEP1 (hepatocarcinoma, IC_{50}: 2.87 μM) [55], HOS (IC_{50}: 7.02 μM), Saos-2 (IC_{50}: 7.32 μM), U-2OS (IC_{50}: 8.36 μM), MG-63 (IC_{50}: 5.30 μM) [56], and UMUC3 (bladder carcinoma, IC_{50}: 9.77 μM) cells [57] after 48 h; and MCF-7 (IC_{50}: 1.00 μM) and Hs578T (breast carcinoma, IC_{50}: 4.00 μM) [58] after 72 h.

Glaucocalyxin A dose- and concentration-dependently inhibited the growth of the liver cancer Focus and SMMC7721 cells. Mechanistic studies found induction of G2/M phase cell cycle arrest as well as increased expression of cleaved caspase-3 and PARP [55]. In MCF-7 and Hs578T breast cancer cells, glaucocalyxin A induced apoptosis, and G2/M phase arrest by increasing the expression of intrinsic apoptotic markers cleaved caspase-3, BAX, and p53, while decreasing BCL-2 expression. It also upregulated the expression of extrinsic apoptotic markers such as Fas and Fas ligand (FasL) at the mRNA and protein level. Moreover, the expression of p-ERK and p-JNK increased in a dose-dependent manner after glaucocalyxin A treatment [58]. Furthermore, it concentration-dependently suppressed the proliferation and induced apoptosis in human brain glioblastoma U87MG cells by activating caspase-3, while downregulating p-Akt, p-Bad, and X-linked inhibitor of apoptosis protein (XIAP) [59]. In another study, glaucocalyxin A induced a dose-dependent apoptosis in HL-60 cells by increasing the expression of caspase-3, -9, and BAX, while decreasing the expression of BCL-2. It caused loss of mitochondrial membrane potential and release of cytochrome c from mitochondria into the cytosol as well as elevated intracellular ROS generation [60].

Glaucocalyxin B ($C_{22}H_{30}O_5$, MW: 374.48, Figure 2), which differs from glaucocalyxin A by an acetylation of the hydroxyl group at C14, exerted anticancer activity in leukemia cell line HL-60 (IC_{50}: 5.86 μM) after 24 h [54], gastric cancer cell line SGC-7901 (IC_{50}: 13.4 μM) [61] after 60 h, and cervical cancer cell lines HeLa (IC_{50}: 4.61 μM) and SiHa (IC_{50}: 3.11 μM) after 72 h [62]. The anticancer activity in HeLa and SiHa cells was related with induction of apoptosis and autophagy by increasing the expression of phosphatase and tensin homolog (PTEN) protein and cleaved PARP, as well as increasing the cleavage of LC3 II/I protein. Moreover, glaucocalyxin B also reduced the expression of p-Akt in HeLa and SiHa cells [62].

2.11. Lasiodin

Lasiodin ($C_{22}H_{30}O_7$, MW: 406.48, Figure 2) was isolated from *Isodon serra*, and inhibited the proliferation and migration of human nasopharyngeal carcinoma cells CNE1 (IC_{50}: ~6 μM) and CNE2 (IC_{50}: ~5 μM) at 24 h incubation [63]. Mechanistic studies showed that lasiodin-induced apoptosis was related to increased expression of apoptotic protease activating factor 1 (Apaf-1), release of cytochrome c, and cleavages of PARP, caspase-3 and -9. Moreover, the expression of p-Akt, p-ERK1/2, p-p38, and p-JNK was also reduced after 24 h treatment with lasiodin. However, lasiodin-mediated inhibition of cell proliferation was blocked in the cells treated with Akt or MAPK inhibitors. Lasiodin treatment also downregulated the expression of COX-2, revoked NF-κB binding to the COX-2 promoter and stimulated the nuclear translocation of NF-κB. Therefore, lasiodin inhibited the proliferation of CNE1 and CNE2 cells by activating Apaf-1/caspase-dependent apoptotic pathways and suppressing Akt/MAPK and COX-2/NF-κB signaling cascades.

2.12. Adenanthin

Adenanthin ($C_{26}H_{34}O_9$, MW: 490.55, Figure 2) was isolated from the leaves of *Isodon adenanthus*, and showed antiproliferative effects against hepatocellular carcinoma cells such as HepG2 (IC_{50}: 2.31 μM), Bel-7402 (IC_{50}: 6.67 μM), and SMMC-7721 (IC_{50}: 8.13 μM) while exhibiting less toxicity in the two normal hepatic cell lines QSG-7701 (IC_{50}: 19.6 μM) and HL-7702 cells (IC_{50}: 20.4 μM) at 48 h exposure [64]. At 72 h treatment, the IC_{50} values of adenanthin in esophageal carcinoma cell line EC109, glioma cell line SHG-44, and breast cancer cell line MCF-7 were 6.50, 4.80, and 7.60 μM, respectively [26]. According to the study of Hou et al. (2014), adenanthin killed malignant liver cells by stimulating ROS generation and targeting peroxiredoxin (Prx) I and II proteins which are considered essential for survival of HCC cells. In-vivo studies with SMMC-7721 cells xenograft mice showed significantly reduced tumor size at 10 mg/kg without any notable side effects. However, the body weight of the mice decreased after treatment with 20 mg/kg of adenanthin [64]. In a different study, the same research group also found that adenanthin stimulated acute promyelocytic leukemia (APL) cell differentiation by targeting Prx I and Prx II as well as blocking their peroxidase activities. Moreover, the level H_2O_2 increased after adenanthin treatment, which led to activation of p-ERK, p-c-Jun, and increased transcription of CCAAT-enhancer-binding protein β (C/EBPβ) [65].

2.13. Kaurenic Acid

Kaurenic acid or kaurenoic acid (KA) ($C_{21}H_{32}O_2$, MW: 316.49, Figure 2), chemically known as *ent*-kaur-16-en-19-oic acid, was isolated from *Espeletia semiglobulata*, and exhibited antimelanoma effects with an IC_{50} value of 0.79 μM in B16F1 cells. An in vivo study found that KA (160 mg/kg) markedly ($p < 0.001$) reduced the tumor sizes (49.51%) compared with the vehicle control group (87.5%) in a C57BL/6 mice model. RT-PCR analysis showed reduced expression of BCL-xL at the mRNA level after KA treatment in C57BL/6 mice [66]. In another study, KA selectively reduced the cell viability of two breast cancer cells MCF7 (proficient P53) and SKBR3 (mutated p53), while KA treatment was resistant to HB4A cell line [67]. At 70 μM, KA caused 40% and 25% cell death of MCF7 and SKBR3 cells suggesting that p53 protein may play a vital role in anticancer activity of KA. A recent study reported significant genotoxicity, apoptosis, and cell cycle arrest in gastric cancer cells after KA treatment [68]. At 10 μg/mL concentration, KA downregulated the expression of c-myc, CCND1, BCL-2, and caspase-3, while it upregulated ATM, Chk2, and TP53 expression.

2.14. Weisiensin B

Weisiensin B ($C_{20}H_{28}O_5$, MW: 348.44, Figure 2) was isolated from the traditional Chinese herb *Isodon weisiensis* and inhibited the growth and proliferation of human hepatoma cell lines BEL-7402 and HepG2, ovarian cancer HO-8910 cells, and gastric cancer SGC-7901 cells [69]. The IC_{50} values of the compound were 10.0, 3.24, 32, and 4.34 μM in BEL-7402, HepG2, HO-8910 cells and SGC-7901 cells, respectively, after 48 h. DNA fragmentation assay and Hoechst 33,258 staining showed that weisiensin B significantly induced apoptosis. Flow cytometry analysis with propidium iodide (PI) staining revealed induction of G2/M phase arrest after weisiensin B treatment. Another study reported weisinensis B-mediated induction of apoptosis, G2/M phase arrest, and significant ROS generation in human chronic myeloid leukemia K562 cells [70].

2.15. Inflexinol

Inflexinol ($C_{24}H_{34}O_8$, MW: 450.53, Figure 2) was isolated from *Isodon excisus*, which is native to China, Korea, and Japan. It inhibited the growth of the colon cancer cell lines SW620 (IC_{50}: 29.0 μM), HCT116 (p53$^{+/+}$) (IC_{50}: 30.0 μM) and HCT116 (p53$^{-/-}$) (IC_{50}: 34.0 μM) but did not show toxicity in the normal colon CCD-112 CoN cells up to 40 μM [71]. Antiproliferative effects of inflexinol were mediated through induction of apoptosis via downregulation of antiapoptotic markers BCL-2, XIAP, and cIAP1/2, with simultaneous upregulation of cleaved caspase-3, -9, and PARP. Inflexinol treatment also significantly increased the ratio of BAX/BCL-2 and suppressed the expression of cyclin D1 and BCL-2. Moreover, inhibitory effects of inflexinol on the growth of colon cancer cells were related to inactivation of NF-κB by modification of a cysteine residue in the p50 subunit. An in vivo study in a SW620 xenograft model showed dose-dependent reduction of tumor volumes (64.5% to 47.8%) and weights (75.9% to 58%) at 12 and 36 mg/kg, respectively, compared with the control group. Moreover, it also exhibited inhibition of DNA binding activity of NF-κB in tumor tissue and suppressed nuclear translocation of p65 and p50 as well as inhibited the phosphorylation of IκB in the cytosol [71].

2.16. Xerophilusin B

Xerophilusin B ($C_{20}H_{26}O_5$, MW: 346.42, Figure 2) was isolated from *Isodon xerophilus* and showed dose-dependent growth inhibition of esophageal squamous cell carcinoma KYSE-140 (IC_{50}: 2.80 µM), KYSE-150 (IC_{50}: 1.20 µM), KYSE-450 (IC_{50}: 1.70 µM), and KYSE-510 (IC_{50}: 2.60 µM) cells when incubated for 72 h. Further studies found that xerophilusin B induced apoptosis and G2/M phase cell cycle arrest in KYSE-150 and KYSE-450 cells [72]. Treatment with xerophilusin B increased the release of mitochondrial cytochrome c and upregulated the expression of cleaved caspase-3 and -9, while downregulated caspase-7 and PARP levels. Moreover, the ratio of BCL-2/BAX decreased after xerophilusin B treatment. An in vivo study in BALB/c nude mice showed significant inhibition of tumor growth at 15 mg/kg without any major adverse effects.

2.17. Henryin

Henryin ($C_{22}H_{32}O_6$, MW: 392.49, Figure 2) was isolated from *Isodon rubescens*, and inhibited the proliferation of colorectal cancer SW480 (IC_{50}: 0.27 µM), HT-29 (IC_{50}: 0.77 µM), and HCT-116 cells (IC_{50}: 0.90 µM), and lung cancer A549 cells (IC_{50}: 2.47 µM), but is slightly less toxic to normal colon cells CCD-841-CoN (IC_{50}: 2.98 µM) and normal bronchus cells BEAS-2B (IC_{50}: 3.55 µM). The inhibitory effects of henryin were correlated to reduced expression of cyclin D1 and c-myc. Moreover, it inhibited the binding of β-catenin to transcription factor 4 (TCF-4) [73].

2.18. 11α, 12α-epoxyleukamenin E

11α, 12α-epoxyleukamenin E (EPLE) ($C_{22}H_{30}O_6$, MW: 390.48, Figure 2) was isolated from *Salvia cavaleriei*, and exhibited cytotoxic activity in colorectal cancer cell lines HCT-116 and SW480 by downregulating the markers of the Wnt-signaling pathway such as c-myc, axin2 and survivin as well as inhibiting β-catenin transcriptional activity [74]. It induced apoptosis by decreasing the expression of BCL-2 and Bcl-xL, while increasing the expression of Bim and caspase-3. Moreover, a combination of EPLE and 5-fluorouracil produced synergistic anticancer effects in colon cancer cells. An in vivo study in the xenograft model also showed significant reduction of tumor size after EPLE treatment [74].

2.19. DEK

Ent-kaur-2-one-16β,17-dihydroxy-acetone-ketal (DEK) ($C_{23}H_{36}O_3$, MW: 360.54, Figure 2) was isolated from the leaves of *Rubus corchorifolius*, and exhibited cytotoxic activity (IC_{50} = 40.0 µM) in HCT-116 human colon cancer cells after 72 h treatment but was non-toxic up to 100 µM in human colonic myofibroblasts CCD-18Co cells [75]. Mechanistic studies found that DEK induced apoptosis by increasing the expression of cleaved caspase-3, -9, PARP, p53, BAX, and p21Cip1/Waf1, while decreasing the expression of cell cycle markers such as cyclin D1, CDK2, and CDK4. Moreover, the expression of two carcinogenic proteins including epidermal growth factor receptor (EGFR) and COX-2 were reduced and the activation of Akt was inhibited after DEK treatment.

2.20. JDA-202

JDA-202 ($C_{21}H_{30}O_5$, MW: 362.47, Figure 2) was isolated from *Isodon rubescens*, and exhibited anticancer effects in esophageal cancer EC109 (IC_{50}: 8.60 µM) and EC9706 (IC_{50}: 9.40 µM) cells, yet was considerably non-toxic in normal cell lines KYSE-450 (IC_{50}: 26.2 µM) and HET-1A (IC_{50}: 36.1 µM) at 24 h drug incubation. Antiproliferative activity was related to direct binding of JDA-202 to the antioxidant protein Prx I, inhibition of its activity and expression, as well as induction of hydrogen peroxide (H_2O_2)-related cell death in esophageal cancer cell lines. Moreover, JDA-202 treatment also led to the increased phosphorylation of JNK, p38, and ERK, which were suppressed by ROS scavenger *N*-acetylcysteine (NAC) and H_2O_2 scavenger catalase. Intravenous administration of JDA-202 at 20 mg/kg/day in Balb/c nude mice significantly inhibited the tumor growth without loss of body weight and organ toxicities [26].

2.21. Pterisolic Acid G

Pterisolic acid G (PAG) ($C_{20}H_{28}O_6$, MW: 364.44, Figure 2), isolated from *Pteris semipinnata*, dose- and time-dependently inhibited the growth of human colorectal cancer cell line HCT-116 with IC_{50} values 20.4, 16.2, and 4.07 µM for 24, 48, and 72 h, respectively. Mechanistic studies found that PAG reduced the expression of disheveled segment polarity protein 2 (Dvl-2), GSK-3β, β-catenin, cyclin D1, and c-myc in HCT-116 cells. Moreover, it induced apoptosis by increasing the expression of p53, puma, cleaved PARP, and cleaved caspase-3, as well as decreasing the expression of p-p65, BCL-2, and BCL-xL [76].

2.22. Rabdoternin B and Maoecrystal I

Rabdoternin B ($C_{21}H_{30}O_7$, MW: 394.46, Figure 2) and maoecrystal I ($C_{22}H_{30}O_8$, MW: 422.47, Figure 2) were isolated from the aerial parts of *Isodon rosthornii* [77] and the leaves of *I. xerophilus* [78], respectively. After 48 h, rabdoternin B showed cytotoxicity in colon cancer SW480 (IC_{50}: 23.2 µM), HT-29 (IC_{50}: 36.3 µM), and HCT-116 (IC_{50}: 20.7 µM) cells, but no toxicity in normal colon CCD-841-CoN (IC_{50}: >40 µM) cells. The IC_{50} values of maoecrystal I were 16.2, 11.4, 26.2, and >40 µM in SW480, HT-29, HCT-116, and CCD-841-CoN cells, respectively. Both compounds induced G2/M phase arrest in SW480 cells. Further research with maoecrystal I found that its anticancer activity was related with downregulation of Wnt-signaling target genes such as c-myc, cyclin D1, surviving, and axin2 [79].

2.23. CHKA

Huang et al. (2016) isolated 3α-cinnamoyloxy-9β-hydroxy-*ent*-kaura-16-en-19-oic acid (CHKA) ($C_{29}H_{36}O_5$, MW: 464.6, Figure 2) from the ethanol extract of *Wedelia chinensis*. CHKA from petroleum ether fraction showed most potent anti-angiogenic activity in zerbrafish model. Moreover, CHKA treatment also considerably blocked a series of VEGF-stimulated angiogenesis events such as proliferation, invasion, and tube formation of endothelial cells. Additionally, it also diminished the activity of VEGFR-2 tyrosine kinase and inhibited several downstream targets including p-VEGFR-2, p-mTOR, p-Akt, and p-ERK in HUVECs. In a mouse model, CHKA significantly blocked sprouts formation of the aortic ring, and inhibited subsequent formation of vessels [80].

2.24. CrT1

Ent-18-acetoxy-7β-hydroxy kaur-15-oxo-16-ene (CrT1) ($C_{22}H_{32}O_4$, MW: 360.49, Figure 2) was isolated from the traditional Vietnamese medicinal plant *Croton tonkinensis*, and exhibited dose- and time-dependent antiproliferative activity in various cancer cell lines with IC_{50} values ranging between 8.40 and 31.2 µM but no toxicity in fibroblast NIH-3T3 cells (IC_{50} not reported) [81]. CrT1 induced apoptosis and G1 cell cycle arrest in human hepatocellular carcinoma SK-HEP1 cells by the activation of caspase-3, -7, -8, -9, and PARP. Moreover, the expression of p53 and BAX increased, while the expression of BCL-2 decreased after CrT1 treatment. CrT1 also increased the cytoplasmic translocation of cytochrome c, upregulated the expression of p-AMPK, while downregulating the expression of p-mTOR and p-p70S6K.

2.25. Ent-16β-17α-dihydroxykaurane

Ent-16β-17α-dihydroxykaurane (DHK) ($C_{20}H_{34}O_2$, MW: 306.49, Figure 2) was isolated from the bark of *Croton malambo*, and exerted pro-apoptotic effects in human breast cancer cell line MCF-7 (IC_{50}: 40.8 µM) after 72 h by downregulating BCL-2 expression at both the mRNA and protein levels, as well as decreasing human telomerase reverse transcriptase (hTERT) expression at the mRNA level [82]. The same research group later found dissociation of activator protein 2 alpha (Ap2α)–Rb activating complex in MCF-7 cells after DHK treatment, which affects the binding ability of the complex to BCL-2 gene promoter. This process leads to downregulation of BCL-2 as well as upregulation of E2F transcription factor 1 (E2F1) and its target pro-apoptotic gene puma [83].

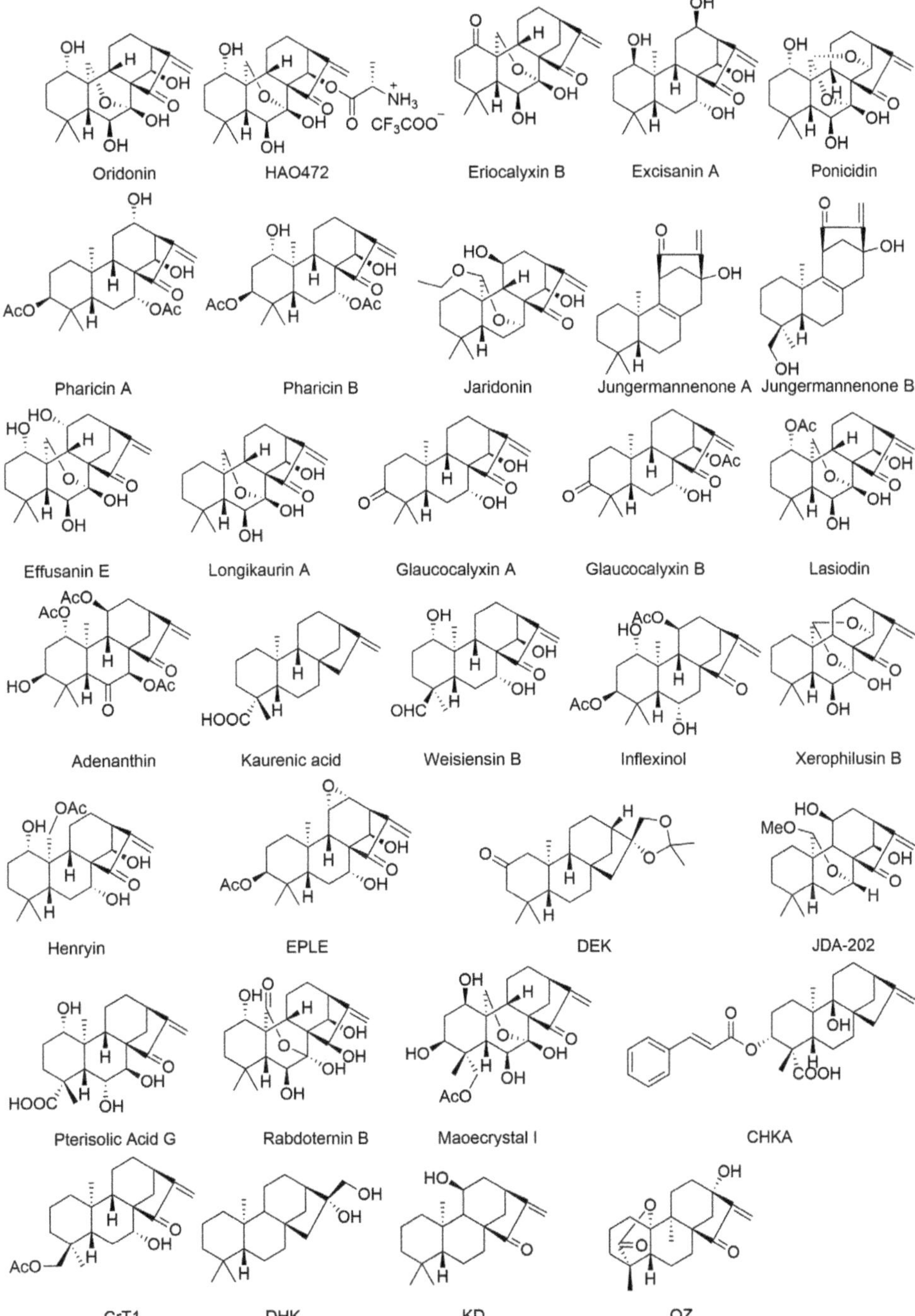

Figure 2. Structures of plant-derived anticancer *ent*-kaurane diterpenoids.

2.26. Ent-11α-hydroxy-16-kauren-15-one

Ent-11α-hydroxy-16-kauren-15-one (KD) ($C_{20}H_{30}O_2$, MW: 302.46, Figure 2) was isolated from the Japanese liverwort *Jungermannia truncata*, and induced apoptosis in human promyelocytic leukemia HL-60 (IC_{50}: 0.56 µM) cells through activation of caspase-8. Treatment with KD also resulted in activation of caspase-9 but caspase-9-specific inhibitor could not diminish KD-induced apoptosis. Moreover, KD treatment resulted in time-dependent cleavage of caspase-8-substrate Bid as well as proteolytic processing of procaspase-8, suggesting association of caspase-8-dependent pathway in KD-induced apoptosis [84].

2.27. Hydroxy-15-oxo-zoapatlin

Hydroxy-15-oxo-zoapatlin (OZ) ($C_{20}H_{26}O_4$, MW: 330.42, Figure 2) was isolated from plants in the genus *Parinari* and has been thoroughly investigated for its anticancer activity. Evaluation of this molecule for its proapoptotic activity in acute lymphoblastic leukemia Molt4 (IC_{50}: 5.00 µM) cells showed that OZ induced the externalization of hypodiploidia and phosphatidylserine which are considered two hallmarks of apoptosis. Furthermore, OZ treatment also increased the expression of PARP and caspase-3 in Molt4 cells [85].

The most commonly reported anticancer pathways of the *ent*-kaurane diterpenoids with their biological targets from different pathways have been summarized in Table S1 (Supplementary Materials) and Figure 3.

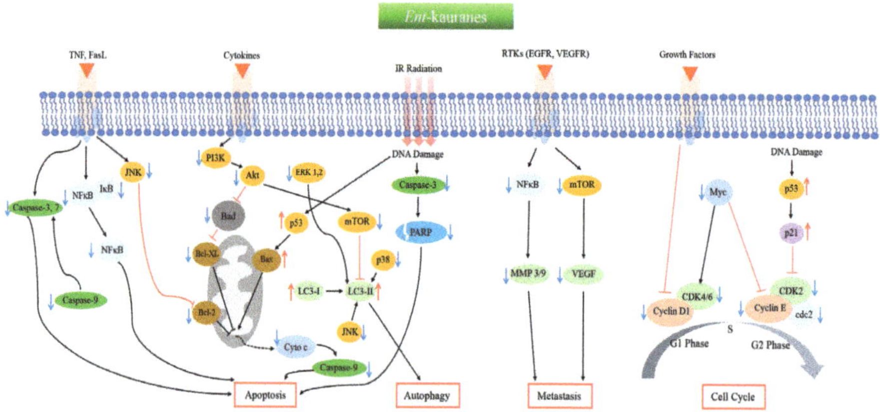

Figure 3. The common biological targets and mechanistic pathways of anticancer *ent*-kaurane diterpenoids.

3. Conclusions

Nature has blessed humanity with a wide variety of drugs for the treatment of cancer, and over the last few decades a large number of natural products have been widely investigated for their anticancer potential. In recent years, *ent*-kaurane diterpenoids have drawn great attention from the scientific community to evaluate their cancer inhibiting effects. Despite the identification of mechanistic pathways of a large number of compounds in both in vitro and in vivo models, there still remains much to understand about their mechanism of action, direct drug binding activity, and most importantly, to design and/or conduct preclinical and clinical research. We hope this article will provide a new platform for encouraging thorough investigation of *ent*-kaurane diterpenoids for the development of new anticancer drug candidates to treat the growing number of cancer patients. Although we have discussed the anticancer activity and mechanistic pathways of several *ent*-kaurane diterpenes, there remain many compounds with similar structures that we have not included that are certainly worthy of further exploration as anticancer agents. Additional work on the structure–activity relationship (SAR) of the *ent*-kaurane diterpenoids with their various molecular targets could also shed light as a way towards bringing some of these potent anticancer molecules into the clinic.

Supplementary Materials: The following are available online at http://www.mdpi.com/2218-273X/10/1/144/s1, Table S1: Plant-derived anticancer *ent*-kaurane diterpenoids with their biological targets.

Author Contributions: Conceptualization, M.S.S. and H.-J.Z.; Writing—original draft preparation, M.S.S.; Review and editing, S.W.T.; Revising, M.S.S., Y.-X.X., and Z.-M.L.; Supervision, H.-J.Z. All authors read and approved the final version of the manuscript.

Funding: This work is a part of PhD study supported by Hong Kong PhD Fellowship Scheme (HKPFS). This work was funded by the Research Grants Council of the Hong Kong Special Administrative Region, China (Project no. HKBU 12102219), and Hong Kong Baptist University, Research Committee, Initiation Grant—Faculty Niche Research Areas (RC-IG-FNRA/17-18/12).

Acknowledgments: The authors offer heartfelt thanks to Stephen Deyrup, Associate Professor, Department of Chemistry and Biochemistry, Siena College, USA, for his generous effort to edit the manuscript. The authors are also grateful to Yu Zhu, School of Chinese Medicine, Hong Kong Baptist University, for checking and revising the chemical structures of the compounds.

Conflicts of Interest: The authors declare no conflict of interest.

References

1. World Health Organization (WHO). Cancer. Available online: https://www.who.int/news-room/fact-sheets/detail/cancer (accessed on 31 December 2019).
2. Soerjomataram, I.; Lortet-Tieulent, J.; Parkin, D.M.; Ferlay, J.; Mathers, C.; Forman, D.; Bray, F. Global burden of cancer in 2008: A systematic analysis of disability-adjusted life-years in 12 world regions. *Lancet* **2012**, *380*, 1840–1850. [CrossRef]
3. Buccioni, M.; Dal Ben, D.; Lambertucci, C.; Maggi, F.; Papa, F.; Thomas, A.; Santinelli, C.; Marucci, G. Antiproliferative evaluation of isofuranodiene on breast and prostate cancer cell lines. *Sci. World J.* **2014**, *2014*, 264829. [CrossRef] [PubMed]
4. DeVita, V.T.; Chu, E. A history of cancer chemotherapy. *Cancer Res.* **2008**, *68*, 8643–8653. [CrossRef] [PubMed]
5. Cragg, G.M.; Grothaus, P.G.; Newman, D.J. Impact of natural products on developing new anti-cancer agents. *Chem Rev.* **2009**, *7*, 3012–3043. [CrossRef]
6. World Health Organisation. Summary of WHO guidelines for the assessment of herbal medicines. *Herb. Gram.* **1993**, *28*, 13–14.
7. Nyirenda, K.K.; Saka, J.D.K.; Naidoo, D.; Maharaj, V.J.; Muller, C.J.F. Antidiabetic, anti-oxidant and antimicrobial activities of *Fadogia ancylantha* extracts from Malawi. *J. Ethnopharmacol.* **2012**, *143*, 372–376. [CrossRef]
8. Jones, G.B. History of Anticancer Drugs. In *eLS*; John Wiley & Sons, Ltd.: Chichester, UK, 2014; pp. 1–4.
9. Sottomayor, M.; RosBarcelo, A. The vinca alkaloids: From biosynthesis and accumulation in plant cells, to uptake, activity and metabolism in animal cells. *Stud. Nat. Prod. Chem.* **2006**, *33*, 813–857.
10. Barbuti, A.M.; Chen, Z.S. Paclitaxel through the ages of anticancer therapy: Exploring its role in chemoresistance and radiation therapy. *Cancers* **2015**, *7*, 2360–2371. [CrossRef]
11. Wang, L.; Li, D.; Wang, C.; Zhang, Y.; Xu, J. Recent progress in the development of natural ent-kaurane diterpenoids with anti-tumor activity. *Mini Rev. Med. Chem.* **2011**, *11*, 910–919. [CrossRef]
12. García, P.A.; de Oliveira, A.B.; Batista, R. Occurrence, biological activities and synthesis of kaurane diterpenes and their glycosides. *Molecules* **2007**, *12*, 455–483. [CrossRef]
13. Li, J.L.; Chen, Q.Q.; Jin, Q.P.; Gao, J.; Zhao, P.J.; Lu, S.; Zeng, Y. IeCPS2 is potentially involved in the biosynthesis of pharmacologically active Isodon diterpenoids rather than gibberellin. *Phytochemistry* **2012**, *76*, 32–39. [CrossRef] [PubMed]
14. Xia, Y.; Feng, M.; Wang, E.; Chen, L.; Wang, J.; Hou, R.; Zhao, Y. An ent-Kaurane diterpenoid isolated from *Rabdosia excisa* suppresses Bcr-Abl protein expression in vitro and in vivo and induces apoptosis of CML cells. *Chem. Biodivers.* **2019**, *16*, e1900443. [CrossRef] [PubMed]

15. Zhu, J.; Chen, Y.; Chen, Z.; Wei, J.; Zhang, H.; Ding, L. Leukamenin E, an ent-kaurane diterpenoid, is a novel and potential keratin intermediate filament inhibitor. *Eur. J. Pharmacol.* **2019**, *846*, 86–99. [CrossRef] [PubMed]

16. Ding, Y.; Ding, C.; Ye, N.; Liu, Z.; Wold, E.A.; Chen, H.; Wild, C.; Shen, Q.; Zhou, J. Discovery and development of natural product oridonin-inspired anticancer agents. *Eur. J. Med. Chem.* **2016**, *122*, 102–117. [CrossRef] [PubMed]

17. Yao, Z.; Xie, F.; Li, M.; Liang, Z.; Xu, W.; Yang, J.; Liu, C.; Li, H.; Zhou, H.; Qu, L.H. Oridonin induces autophagy via inhibition of glucose metabolism in p53-mutated colorectal cancer cells. *Cell Death Dis.* **2017**, *8*, e2633. [CrossRef]

18. Li, F.F.; Yi, S.; Wen, L.; He, J.; Yang, L.J.; Zhao, J.; Zhang, B.P.; Cui, G.H.; Chen, Y. Oridonin induces NPM mutant protein translocation and apoptosis in NPM1c+ acute myeloid leukemia cells in vitro. *Acta Pharmacol. Sin.* **2014**, *35*, 806–813. [CrossRef]

19. Chen, R.Y.; Xu, B.; Chen, S.F.; Chen, S.S.; Zhang, T.; Ren, J.; Xu, J. Effect of oridonin-mediated hallmark changes on inflammatory pathways in human pancreatic cancer (BxPC-3) cells. *World J. Gastroenterol.* **2014**, *20*, 14895–14903. [CrossRef]

20. Jiang, J.H.; Pi, J.; Jin, H.; Cai, J.Y. Oridonin-induced mitochondria-dependent apoptosis in esophageal cancer cells by inhibiting PI3K/AKT/mTOR and Ras/Raf pathways. *J. Cell. Biochem.* **2019**, *120*, 3736–3746. [CrossRef]

21. Vasaturo, M.; Cotugno, R.; Fiengo, L.; Vinegoni, C.; Dal Piaz, F.; De Tommasi, N. The anti-tumor diterpene oridonin is a direct inhibitor of Nucleolin in cancer cells. *Sci. Rep.* **2018**, *8*, 16735. [CrossRef]

22. Lu, Y.; Sun, Y.; Zhu, J.; Yu, L.; Jiang, X.; Zhang, J.; Dong, X.; Ma, B.; Zhang, Q. Oridonin exerts anticancer effect on osteosarcoma by activating PPAR-γ and inhibiting Nrf2 pathway. *Cell Death Dis.* **2018**, *9*, 15. [CrossRef]

23. Ma, Y.C.; Ke, Y.; Zi, X.; Zhao, W.; Shi, X.J.; Liu, H.M. Jaridonin, a novel ent-kaurene diterpenoid from *Isodon rubescens*, inducing apoptosis via production of reactive oxygen species in esophageal cancer cells. *Curr. Cancer Drug Targets* **2013**, *13*, 611–624. [CrossRef] [PubMed]

24. Tian, L.; Xie, K.; Sheng, D.; Wan, X.; Zhu, G. Antiangiogenic effects of oridonin. *BMC Complement. Altern. Med.* **2017**, *17*, 192. [CrossRef] [PubMed]

25. Zou, Q.F.; Du, J.K.; Zhang, H.; Wang, H.B.; Hu, Z.D.; Chen, S.P.; Du, Y.; Li, M.Z.; Xie, D.; Zou, J.; et al. Anti-tumour activity of longikaurin A (LK-A), a novel natural diterpenoid, in nasopharyngeal carcinoma. *J. Transl. Med.* **2013**, *11*, 200. [CrossRef] [PubMed]

26. Shi, X.J.; Ding, L.; Zhou, W.; Ji, Y.; Wang, J.; Wang, H.; Ma, Y.; Jiang, G.; Tang, K.; Ke, Y.; et al. Pro-apoptotic effects of JDA-202, a novel natural diterpenoid, on esophageal cancer through targeting peroxiredoxin I. *Antioxid. Redox Signal.* **2017**, *27*, 73–92. [CrossRef]

27. Gao, S.; Tan, H.; Zhu, N.; Gao, H.; Lv, C.; Gang, J.; Ji, Y. Oridonin induces apoptosis through the mitochondrial pathway in human gastric cancer SGC-7901 cells. *Int. J. Oncol.* **2016**, *48*, 2453–2460. [CrossRef]

28. Kadioglu, O.; Saeed, M.; Kuete, V.; Greten, H.J.; Efferth, T. Oridonin Targets Multiple Drug-Resistant Tumor Cells as Determined by in Silico and in Vitro Analyses. *Front. Pharmacol.* **2018**, *9*, 355. [CrossRef]

29. Xu, T.; Jin, F.; Wu, K.; Ye, Z.; Li, N. Oridonin enhances in vitro anticancer effects of lentinan in SMMC-7721 human hepatoma cells through apoptotic genes. *Exp. Ther. Med.* **2017**, *14*, 5129–5134. [CrossRef]

30. Lu, J.; Chen, X.; Qu, S.; Yao, B.; Xu, Y.; Wu, J.; Jin, Y.; Ma, C. Oridonin induces G2/M cell cycle arrest and apoptosis via the PI3K/Akt signaling pathway in hormone-independent prostate cancer cells. *Oncol. Lett.* **2017**, *13*, 2838–2846. [CrossRef]

31. Sun, Z.; Han, Q.; Duan, L.; Yuan, Q.; Wang, H. Oridonin increases anticancer effects of lentinan in HepG2 human hepatoblastoma cells. *Oncol. Lett.* **2018**, *15*, 1999–2005. [CrossRef]

32. Xia, S.; Zhang, X.; Li, C.; Guan, H. Oridonin inhibits breast cancer growth and metastasis through blocking the Notch signaling. *Saudi Pharm. J.* **2017**, *25*, 638–643. [CrossRef]

33. Kong, L.M.; Deng, X.; Zuo, Z.L.; Sun, H.D.; Zhao, Q.S.; Li, Y. Identification and validation of p50 as the cellular target of eriocalyxin B. *Oncotarget* **2014**, *5*, 11354–11364. [CrossRef]

34. Zhou, X.; Yue, G.G.L.; Chan, A.M.L.; Tsui, S.K.W.; Fung, K.P.; Sun, H.; Pu, J.; Lau, C.B.S. Eriocalyxin B, a novel autophagy inducer, exerts anti-tumor activity through the suppression of Akt/mTOR/p70S6K signaling pathway in breast cancer. *Biochem. Pharmacol.* **2017**, *142*, 58–70. [CrossRef] [PubMed]

35. Li, L.; Yue, G.G.; Lau, C.B.; Sun, H.; Fung, K.P.; Leung, P.C.; Han, Q.; Leung, P.S. Eriocalyxin B induces apoptosis and cell cycle arrest in pancreatic adenocarcinoma cells through caspase- and p53-dependent pathways. *Toxicol. Appl. Pharmacol.* **2012**, *262*, 80–90. [CrossRef] [PubMed]

36. Zhang, Y.W.; Jiang, X.X.; Chen, Q.S.; Shi, W.Y.; Wang, L.; Sun, H.D.; Shen, Z.X.; Chen, Z.; Chen, S.J.; Zhao, W.L. Eriocalyxin B induces apoptosis in lymphoma cells through multiple cellular signaling pathways. *Exp. Hematol.* **2010**, *38*, 191–201. [CrossRef] [PubMed]

37. Zhou, X.; Yue, G.G.L.; Liu, M.; Zuo, Z.; Lee, J.K.M.; Li, M.; Tsui, S.K.W.; Fung, K.P.; Sun, H.D.; Pu, J.; et al. Eriocalyxin B, a natural diterpenoid, inhibited VEGF-induced angiogenesis and diminished angiogenesis-dependent breast tumor growth by suppressing VEGFR-2 signaling. *Oncotarget* **2016**, *7*, 82820–82835. [PubMed]

38. Lu, Y.M.; Chen, W.; Zhu, J.S.; Chen, W.X.; Chen, N.W. Eriocalyxin B blocks human SW1116 colon cancer cell proliferation, migration, invasion, cell cycle progression and angiogenesis via the JAK2/STAT3 signaling pathway. *Mol. Med. Rep.* **2016**, *13*, 2235–2240. [CrossRef]

39. Yu, X.; He, L.; Cao, P.; Yu, Q. Eriocalyxin B inhibits STAT3 signaling by covalently targeting STAT3 and blocking phosphorylation and activation of STAT3. *PLoS ONE* **2015**, *10*, e0128406. [CrossRef]

40. Qin, J.; Tang, J.; Jiao, L.; Ji, J.; Chen, W.D.; Feng, G.K.; Gao, Y.H.; Zhu, X.F.; Deng, R. A diterpenoid compound, excisanin A, inhibits the invasive behavior of breast cancer cells by modulating the integrin β1/FAK/PI3K/AKT/β-catenin signaling. *Life Sci.* **2013**, *93*, 655–663. [CrossRef]

41. Deng, R.; Tang, J.; Xia, L.P.; Li, D.D.; Zhou, W.J.; Wang, L.L.; Feng, G.K.; Zeng, Y.X.; Gao, Y.H.; Zhu, X.F. ExcisaninA, a diterpenoid compound purified from Isodon MacrocalyxinD, induces tumor cells apoptosis and suppresses tumor growth through inhibition of PKB/AKT kinase activity and blockade of its signal pathway. *Mol. Cancer Ther.* **2009**, *8*, 873–882. [CrossRef]

42. Zhang, X.Y.; Wu, X.Q.; Deng, R.; Sun, T.; Feng, G.K.; Zhu, X.F. Upregulation of sestrin 2 expression via JNK pathway activation contributes to autophagy induction in cancer cells. *Cell. Signal.* **2013**, *25*, 150–158. [CrossRef]

43. Liu, Y.F.; Lu, Y.M.; Qu, G.Q.; Liu, Y.; Chen, W.X.; Liao, X.H.; Kong, W.M. Ponicidin induces apoptosis via JAK2 and STAT3 signaling pathways in gastric carcinoma. *Int. J. Mol. Sci.* **2015**, *16*, 1576–1589. [CrossRef] [PubMed]

44. Liu, Y.; Fan, Y.; Chen, Y.; Hou, Y. Ponicidin induces apoptosis of human cervical cancer HeLa cell line through the PI3K/Akt and MAPK signaling pathways. *Int. J. Clin. Exp. Med.* **2016**, *9*, 23214–23221.

45. Liu, J.J.; Huang, R.W.; Lin, D.J.; Peng, J.; Zhang, M.; Pan, X.; Hou, M.; Wu, X.Y.; Lin, Q.; Chen, F. Ponicidin, an *ent*-kaurane diterpenoid derived from a constituent of the herbal supplement PC-SPES, *Rabdosia rubescens*, induces apoptosis by activation of caspase-3 and mitochondrial events in lung cancer cells in vitro. *Cancer Investig.* **2006**, *24*, 136–148. [CrossRef]

46. Du, J.; Chen, C.; Sun, Y.; Zheng, L.; Wang, W. Ponicidin suppresses HT29 cell growth via the induction of G1 cell cycle arrest and apoptosis. *Mol. Med. Rep.* **2015**, *12*, 5816–5820. [CrossRef] [PubMed]

47. Xu, H.Z.; Huang, Y.; Wu, Y.L.; Zhao, Y.; Xiao, W.L.; Lin, Q.S.; Sun, H.D.; Dai, W.; Chen, G.Q. Pharicin A, a novel natural ent-kaurene diterpenoid, induces mitotic arrest and mitotic catastrophe of cancer cells by interfering with BubR1 function. *Cell Cycle* **2010**, *9*, 2897–2907. [CrossRef] [PubMed]

48. Gu, Z.M.; Wu, Y.L.; Zhou, M.Y.; Liu, C.X.; Xu, H.Z.; Yan, H.; Zhao, Y.; Huang, Y.; Sun, H.D.; Chen, G.Q. Pharicin B stabilizes retinoic acid receptor-α and presents synergistic differentiation induction with ATRA in myeloid leukemic cells. *Blood* **2010**, *116*, 5289–5297. [CrossRef] [PubMed]

49. Ma, Y.C.; Su, N.; Shi, X.J.; Zhao, W.; Ke, Y.; Zi, X.; Zhao, N.M.; Qin, Y.H.; Zhao, H.W.; Liu, H.M. Jaridonin-induced G2/M phase arrest in human esophageal cancer cells is caused by reactive oxygen species-dependent Cdc2-tyr15 phosphorylation via ATM-Chk1/2-Cdc25C pathway. *Toxicol. Appl. Pharmacol.* **2015**, *282*, 227–236. [CrossRef]

50. Guo, Y.X.; Lin, Z.M.; Wang, M.J.; Dong, Y.W.; Niu, H.M.; Young, C.Y.F.; Lou, H.X.; Yuan, H.Q. Jungermannenone A and B induce ROS-and cell cycle-dependent apoptosis in prostate cancer cells in vitro. *Acta Pharmacol. Sin.* **2016**, *37*, 814–824. [CrossRef]

51. Zhuang, M.; Zhao, M.; Qiu, H.; Shi, D.; Wang, J.; Tian, Y.; Lin, L.; Deng, W. Effusanin E suppresses nasopharyngeal carcinoma cell growth by inhibiting NF-κB and COX-2 signaling. *PLoS ONE* **2014**, *9*, e109951. [CrossRef]

52. Liao, Y.J.; Bai, H.Y.; Li, Z.H.; Zou, J.; Chen, J.W.; Zheng, F.; Zhang, J.X.; Mai, S.J.; Zeng, M.S.; Sun, H.D.; et al. Longikaurin A, a natural *ent-kaurane*, induces G2/M phase arrest via downregulation of Skp2 and apoptosis induction through ROS/JNK/c-Jun pathway in hepatocellular carcinoma cells. *Cell Death Dis.* **2014**, *5*, e1137. [CrossRef]

53. Yuan, Y.; Du, Y.; Hu, X.Y.; Liu, M.Y.; Du, J.K.; Liu, X.M.; Yu, H.E.; Wang, T.Z.; Pu, J.X.; Zhong, Q.; et al. Longikaurin A, a natural *ent-kaurane*, suppresses stemness in nasopharyngeal carcinoma cells. *Oncol. Lett.* **2017**, *13*, 1672–1680. [CrossRef] [PubMed]

54. Yang, W.H.; Zhang, Z.; Sima, Y.H.; Zhang, J.; Wang, J.W. Glaucocalyxin A and B-induced cell death is related to GSH perturbation in human leukemia HL-60 cells. *Anticancer Agents Med. Chem.* **2013**, *13*, 1280–1290. [CrossRef] [PubMed]

55. Tang, L.; Jin, X.; Hu, X.; Hu, X.; Liu, Z.; Yu, L. Glaucocalyxin A inhibits the growth of liver cancer Focus and SMMC-7721 cells. *Oncol. Lett.* **2016**, *11*, 1173–1178. [CrossRef] [PubMed]

56. Zhu, J.; Sun, Y.; Lu, Y.; Jiang, X.; Ma, B.; Yu, L.; Zhang, J.; Dong, X.; Zhang, Q. Glaucocalyxin A exerts anticancer effect on osteosarcoma by inhibiting GLI1 nuclear translocation via regulating PI3K/Akt pathway. *Cell Death Dis.* **2018**, *9*, 708. [CrossRef]

57. Lin, W.; Xie, J.; Xu, N.; Huang, L.; Xu, A.; Li, H.; Li, C.; Gao, Y.; Watanabe, M.; Liu, C.; et al. Glaucocalyxin A induces G2/M cell cycle arrest and apoptosis through the PI3K/Akt pathway in human bladder cancer cells. *Int. J. Biol. Sci.* **2018**, *14*, 418–426. [CrossRef]

58. Li, M.; Jiang, X.; Gu, Z.; Zhang, Z. Glaucocalyxin A activates FasL and induces apoptosis through activation of the JNK pathway in human breast cancer cells. *Asian Pac. J. Cancer Prev.* **2013**, *14*, 5805–5810. [CrossRef]

59. Xiao, X.; Cao, W.; Jiang, X.; Zhang, W.; Zhang, Y.; Liu, B.; Cheng, J.; Huang, H.; Huo, J.; Zhang, X. Glaucocalyxin A, a negative Akt regulator, specifically induces apoptosis in human brain glioblastoma U87MG cells. *Acta Biochim. Biophys. Sin.* **2013**, *45*, 946–952. [CrossRef]

60. Gao, L.W.; Zhang, J.; Yang, W.H.; Wang, B.; Wang, J.W. Glaucocalyxin A induces apoptosis in human leukemia HL-60 cells through mitochondria-mediated death pathway. *Toxicol. Vitr.* **2011**, *25*, 51–63. [CrossRef]

61. Ur Rahman, M.S.; Zhang, L.; Wu, L.; Xie, Y.; Li, C.; Cao, J. Sensitization of gastric cancer cells to alkylating agents by glaucocalyxin B via cell cycle arrest and enhanced cell death. *Drug Des. Devel Ther.* **2017**, *11*, 2431–2441. [CrossRef]

62. Pan, Y.; Bai, J.; Shen, F.; Sun, L.; He, Q.; Su, B. Glaucocalyxin B induces apoptosis and autophagy in human cervical cancer cells. *Mol. Med. Rep.* **2016**, *14*, 1751–1755. [CrossRef]

63. Lin, L.; Deng, W.; Tian, Y.; Chen, W.; Wang, J.; Fu, L.; Shi, D.; Zhao, M.; Luo, W. Lasiodin inhibits proliferation of human nasopharyngeal carcinoma cells by simultaneous modulation of the Apaf-1/Caspase, AKT/MAPK and COX-2/NF-κB signaling pathways. *PLoS ONE* **2014**, *9*, e97799. [CrossRef] [PubMed]

64. Hou, J.K.; Huang, Y.; He, W.; Yan, Z.W.; Fan, L.; Liu, M.H.; Xiao, W.L.; Sun, H.D.; Chen, G.Q. Adenanthin targets peroxiredoxin I/II to kill hepatocellular carcinoma cells. *Cell Death Dis.* **2014**, *5*, e1400. [CrossRef] [PubMed]

65. Liu, C.X.; Yin, Q.Q.; Zhou, H.C.; Wu, Y.L.; Pu, J.X.; Xia, L.; Liu, W.; Huang, X.; Jiang, T.; Wu, M.X.; et al. Adenanthin targets peroxiredoxin I and II to induce differentiation of leukemic cells. *Nat. Chem. Biol.* **2012**, *8*, 486–493. [CrossRef]

66. Sosa-Sequera, M.C.; Chiurillo, M.; Moscoso, J.; Dolinar, J.; Suarez, O.; Neira, N.; Mendoza, H.; Rivero-Paris, M. Kaurenic acid: Evaluation of the in vivo and in vitro antitumor activity on murine melanoma. *Indian J. Pharmacol.* **2011**, *43*, 683–688. [PubMed]

67. Peria, F.M.; Tiezzi, D.G.; Tirapelli, D.P.; Neto, F.S.; Tirapelli, C.R.; Ambrosio, S.; Oliveira, H.F.; Tirapelli, L. Kaurenoic acid antitumor activity in breast cancer cells. *J. Clin. Oncol.* **2010**, *28*, e13641. [CrossRef]

68. dos Santos Cardoso, P.C.; da Rocha, C.A.; Leal, M.F.; de Oliveira Bahia, M.; Alcântara, D.D.; dos Santos, R.A.; dos Santos Gonçalves, N.; Ambrósio, S.R.; Cavalcanti, B.C.; Moreira-Nunes, C.A.; et al. Effect of diterpenoid kaurenoic acid on genotoxicity and cell cycle progression in gastric cancer cell lines. *Biomed. Pharmacother.* **2017**, *89*, 772–780. [CrossRef] [PubMed]

69. Ding, L.; Zhang, S.D.; Yang, D.J.; Liu, B.; Zhou, Q.Y.; Yang, H. Cytotoxicity and apoptosis induction of weisiensin B isolated from *Rabdosia weisiensis* C.Y. Wu in human hepatoma cells. *J. Asian Nat. Prod. Res.* **2008**, *10*, 1055–1062. [CrossRef]

70. Liu, G.A.; Chang, J.C.; Feng, X.L.; Ding, L. Apoptosis induced by weisiensin B isolated from Rabdosia weisiensis C.Y. Wu in K562. *Pharmazie* **2015**, *70*, 263–268.

71. Ban, J.O.; Oh, J.H.; Hwang, B.Y.; Moon, D.C.; Jeong, H.S.; Lee, S.; Kim, S.; Lee, H.; Kim, K.B.; Han, S.B.; et al. Inflexinol inhibits colon cancer cell growth through inhibition of nuclear factor-kappaB activity via direct interaction with p50. *Mol. Cancer Ther.* **2009**, *8*, 1613–1624. [CrossRef]

72. Yao, R.; Chen, Z.; Zhou, C.; Luo, M.; Shi, X.; Li, J.; Gao, Y.; Zhou, F.; Pu, J.; Sun, H.; et al. Xerophilusin b induces cell cycle arrest and apoptosis in esophageal squamous cell carcinoma cells and does not cause toxicity in nude mice. *J. Nat. Prod.* **2015**, *78*, 10–16. [CrossRef]

73. Li, X.; Pu, J.; Jiang, S.; Su, J.; Kong, L.; Mao, B.; Sun, H.; Li, Y. Henryin, an *ent-kaurane* diterpenoid, inhibits Wnt signaling through interference with β-Catenin/TCF4 interaction in colorectal cancer cells. *PLoS ONE* **2013**, *8*, e68525. [CrossRef] [PubMed]

74. Ye, Q.; Yao, G.; Zhang, M.; Guo, G.; Hu, Y.; Jiang, J.; Cheng, L.; Shi, J.; Li, H.; Zhang, Y.; et al. A novel *ent-kaurane* diterpenoid executes antitumor function in colorectal cancer cells by inhibiting Wnt/β-catenin signaling. *Carcinogenesis* **2014**, *36*, 318–326. [CrossRef] [PubMed]

75. Chen, X.; Wu, X.; Ouyang, W.; Gu, M.; Gao, Z.; Song, M.; Chen, Y.; Lin, Y.; Cao, Y.; Xiao, H. Novel *ent*-kaurane diterpenoid from *Rubus corchorifolius* L. f. inhibits human colon cancer cell growth via inducing cell cycle arrest and apoptosis. *J. Agric. Food Chem.* **2017**, *65*, 1566–1573. [CrossRef] [PubMed]

76. Qiu, S.; Wu, X.; Liao, H.; Zeng, X.; Zhang, S.; Lu, X.; He, X.; Zhang, X.; Ye, W.; Wu, H.; et al. Pteisolic acid G, a novel ent-kaurane diterpenoid, inhibits viability and induces apoptosis in human colorectal carcinoma cells. *Oncol. Lett.* **2017**, *14*, 5540–5548. [CrossRef]

77. Zhan, R.; Li, X.N.; Du, X.; Wang, W.G.; Dong, K.; Su, J.; Li, Y.; Pu, J.X.; Sun, H.D. Bioactive *ent*-kaurane diterpenoids from *Isodon rosthornii*. *J. Nat. Prod.* **2013**, *76*, 1267–1277. [CrossRef]

78. Li, L.M.; Weng, Z.Y.; Huang, S.X.; Pu, J.X.; Li, S.H.; Huang, H.; Yang, B.B.; Han, Y.; Xiao, W.L.; Li, M.N.; et al. Cytotoxic ent-kauranoids from the medicinal plant Isodon xerophilus. *J. Nat. Prod.* **2007**, *70*, 1295–1301. [CrossRef]

79. Zhang, J.; Kong, L.M.; Zhan, R.; Ye, Z.N.; Pu, J.X.; Sun, H.D.; Li, Y. Two natural ent-kauranoids as novel Wnt signaling inhibitors. *Nat. Prod. Bioprospect.* **2014**, *4*, 135–140. [CrossRef]

80. Huang, W.; Liang, Y.; Wang, J.; Li, G.; Wang, G.; Li, Y.; Chung, H.Y. Anti-angiogenic activity and mechanism of kaurane diterpenoids from *Wedelia chinensis*. *Phytomedicine* **2016**, *23*, 283–292. [CrossRef]

81. Sul, Y.H.; Lee, M.S.; Cha, E.Y.; Thuong, P.T.; Khoi, N.M.; Song, I.S. An *ent-kaurane* diterpenoid from *Croton tonkinensis* induces apoptosis by regulating AMP-activated protein kinase in SK-HEP1 human hepatocellular carcinoma cells. *Biol. Pharm. Bull.* **2013**, *36*, 158–164. [CrossRef]

82. Morales, A.; Pérez, P.; Compagnone, R.M.R.; Suarez, A.I.; Arvelo, F.; Ramírez, J.L.; Galindo-Castro, I. Cytotoxic and proapoptotic activity of ent-16β-17α-dihydroxykaurane on human mammary carcinoma cell line MCF-7. *Cancer Lett.* **2005**, *218*, 109–116. [CrossRef]

83. Morales, A.; Alvarez, A.; Francisco, F.; Suárez, A.I.; Compagnone, R.S.; Galindo-Castro, I. The natural diterpene ent-16β-17α-dihydroxykaurane downregulates Bcl-2 by disruption of the Ap-2α/Rb transcription activating complex and induces E2F1 upregulation in MCF-7 cells. *Apoptosis* **2011**, *16*, 1245–1252. [CrossRef] [PubMed]

84. Kondoh, M.; Suzuki, I.; Sato, M.; Nagashima, F.; Simizu, S.; Harada, M.; Fujii, M.; Osada, H.; Asakawa, Y.; Watanabe, Y. Kaurene diterpene induces apoptosis in human leukemia cells partly through a caspase-8-dependent pathway. *J. Pharmacol. Exp. Ther.* **2004**, *311*, 115–122. [CrossRef] [PubMed]

85. Dal Piaz, F.; Nigro, P.; Braca, A.; De Tommasi, N.; Belisario, M.A. 13-Hydroxy-15-oxo-zoapatlin, an *ent-kaurane* diterpene, induces apoptosis in human leukemia cells, affecting thiol-mediated redox regulation. *Free Radic. Biol. Med.* **2007**, *43*, 1409–1422. [CrossRef] [PubMed]

 *biomolecules*

Review

Neuroprotective Effects of Quercetin in Alzheimer's Disease

Haroon Khan [1,*], Hammad Ullah [1], Michael Aschner [2], Wai San Cheang [3] and Esra Küpeli Akkol [4]

1 Department of Pharmacy, Abdul Wali Khan University, Mardan 23200, Pakistan; hamm.swabian@gmail.com
2 Department of Molecular Pharmacology, Albert Einstein College of Medicine, Forchheimer 209, 1300 Morris Park Avenue, Bronx, NY 10461, USA; michael.aschner@einstein.yu.edu
3 Institute of Chinese Medical Sciences, State Key Laboratory of Quality Research in Chinese Medicine, University of Macau, Macau, China; annacheang@um.edu.mo
4 Department of Pharmacognosy, Faculty of Pharmacy Gazi University, 06330 Etiler/Ankara, Turkey; esrak@gazi.edu.tr
* Correspondence: hkdr2006@gmail.com or haroonkhn@awkum.edu.pk; Tel.: +92-332-9123171

Received: 14 November 2019; Accepted: 22 December 2019; Published: 30 December 2019

Abstract: Quercetin is a flavonoid with notable pharmacological effects and promising therapeutic potential. It is widely distributed among plants and found commonly in daily diets predominantly in fruits and vegetables. Neuroprotection by quercetin has been reported in several in vitro studies. It has been shown to protect neurons from oxidative damage while reducing lipid peroxidation. In addition to its antioxidant properties, it inhibits the fibril formation of amyloid-β proteins, counteracting cell lyses and inflammatory cascade pathways. In this review, we provide a synopsis of the recent literature exploring the relationship between quercetin and cognitive performance in Alzheimer's disease and its potential as a lead compound in clinical applications.

Keywords: quercetin; polyphenols; Alzheimer's disease; mechanistic insights; clinical directions

---

1. Introduction

Alzheimer's disease (AD) contributes to 60–80% of total dementia cases, and it mostly affects elder people (65 years of age or older) [1]. The pathogenesis of AD is typically associated with the accumulation of amyloid-β (Aβ) aggregates and the hyperphosphorylation of tau proteins, leading to neurofibrillary tangles (NFTs) and synaptic dysfunction [2–4]. Around 35.6 million people worldwide are estimated to be affected with AD, with a prevalence rate of 4.6 million new cases each year. The prevalence rate of AD increases with age: the rate doubles every 5 years from 60 years of age [5,6].

Early studies led to the cholinergic deficit hypothesis of AD, which states that deficiency in acetylcholine is the main cause of the disease. In the pursuit of drugs that are able to restore acetylcholine levels, the first acetyl cholinesterase inhibitors were developed, including tacrine. Since then, other drugs in the same class have been pursued, namely donepezil, rivastigmine, and galantamine. Current AD therapy consists of cholinesterase inhibitors and N-methyl-D-aspartate (NMDA) antagonists, including memantine. Acetyl cholinesterase inhibitors prevent the hydrolysis of acetylcholine, and memantine modulates NMDA receptor activity, causing a reduction in excitatory glutamate signals. However, these drugs offer little palliative effects, and they also have numerous undesirable safety profiles with a number of adverse side effects [7,8]. Acetyl cholinesterase inhibitors are associated with gastrointestinal side effects such as nausea, diarrhea, and abdominal pain, as well as urinary incontinence, insomnia, and nightmares. The use of tacrine has been limited because of its poor bioavailability and reported hepatotoxicity. Memantine is clinically less effective compared to acetyl cholinesterase inhibitors. Additionally, these drugs are not targeting the root cause of the disorder [9].

Phytochemicals are best known to reduce the risk of chronic diseases, such as cardiovascular diseases, hypertension, diabetes, and cancers [10–12]. Flavonoids are the most diverse group of phytochemicals and are widely distributed in higher plants with outstanding therapeutic potential [13–16]. Flavonoids are further divided into six classes on the basis of their chemical skeleton: flavanols, flavanones, flavones, flavonols, isoflavonoids, and anthocyanidins [17]. While targeting multiple targets, they have been proven beneficial in the prevention of neurodegenerative disorders and may delay the process of neurodegeneration [18]. Flavonoids are extensively studied for their antioxidant and anti-inflammatory activities, both of which are important in triggering the pathogenesis of AD [19]. Studies suggested that flavonoids are capable of crossing the blood–brain barrier (BBB), which makes them potential agents in preventing neurodegenerative disorders; however, different flavonoid subclasses differ in their ability to cross the BBB [20,21]. In the case of AD, their efficacy is attributed to the reduction of Aβ toxicity and decreasing oxidative stress [22,23]. Nevertheless, anti-AD effects of certain flavonoids, such as myricetin, rutin, fisetin, catechins, quercetin, kaempferol, and apigenin have been reported [24–27].

Quercetin is one of the most potent antioxidants of plant origin and is one of the predominant flavonoids found more commonly in edible plants [28]. It belongs to the flavonols class of flavonoids, representing a major class of polyphenols. The dietary intake of total flavonoids is estimated to be 200–350 mg/day, and the intake of quercetin is 10–16 mg/day. The recommended dosage of quercetin aglycone as a dietary supplement is 1 g/day [29–31]. It exhibits numerous beneficial effects on human health, acting as anti-carcinogenic, anti-inflammatory, anti-infective, and psychostimulant agent. It also inhibits lipid peroxidation and platelet aggregation, and it stimulates mitochondrial biogenesis [32]. Several studies have reported on the neuroprotective effects of quercetin, both in vitro and in vivo models of neurodegenerative disorders, such as cognitive impairment [33], ischemia, traumatic injury [34], Parkinson's disease (PD) [35], and Huntington's disease (HD) [36]. The aim of the present review is to provide a summary of the recent literature exploring the relationship between quercetin and cognitive performance. Our primary focus is on the chemical basis and pharmacology of quercetin and its anti-AD mechanisms.

2. Chemistry of Quercetin

A flavonoid is structurally diphenyl propane containing 15 carbon atoms in its structure (Figure 1A). It contains a close heterocyclic pyran ring in addition to two benzene rings. The term 4-oxo-flavonoid (Figure 1B) is often used to describe flavonoid containing a carbonyl group on position C-4 of ring C. Substitution and oxidation in the heterocyclic pyran ring classifies flavonoids into subclasses, namely flavones, flavonols, flavanones, flavononol, isoflavones, and flavan-3-ols. Substitution in the benzene rings of flavonoid structures leads to differences in individual compounds within specific classes [37,38]. Quercetin belongs to the flavone class of flavonoids having a chemical formula of $C_{15}H_{10}O_7$. Its IUPAC name is 3,3′,4′,5,7-pentahydroxyflavanone or 3,3′,4′,5,7-pentahydroxy-2-phenylchromen-4-one (Figure 2). Quercetin contains an OH group at positions 3, 5, 7, 3′, and 4′. Quercetin is an aglycone, (lacking an attached sugar molecule). Attaching a glycosyl group (glucose, rhamnose, or rutinose) most commonly at position 3 replacing the OH group leads to the formation of quercetin glycoside. Quercetin is insoluble or sparingly soluble in water, while it is quite soluble in alcohol and lipids. A glycosyl group increases its water solubility, and thus the quercetin glycoside is soluble in water [32,39]. In addition to antioxidant activities, multiple OH groups in the structure of quercetin may also lead to its photodegradation. Dall'Acqua et al. (2012) reported that OH groups at positions 3, 3′, and 4′ are mainly involve in photolability, while OH groups at positions 5 and 7 do not play a crucial role in the photo-oxidative mechanism [40].

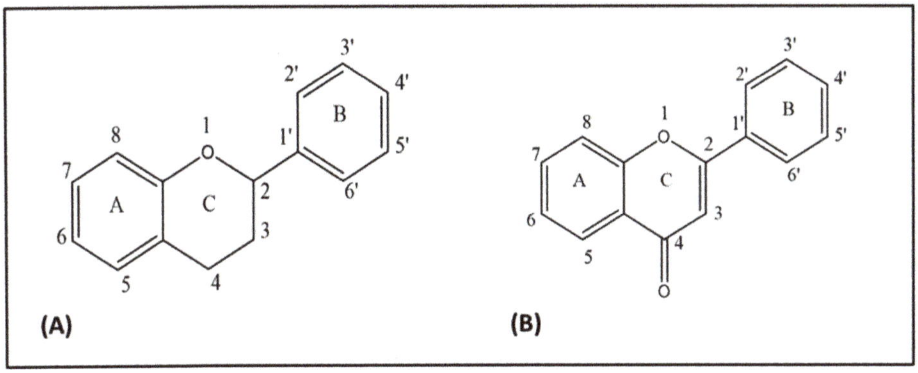

Figure 1. (**A**) Flavan nucleus, (**B**) 4-oxo-flavonoid nucleus.

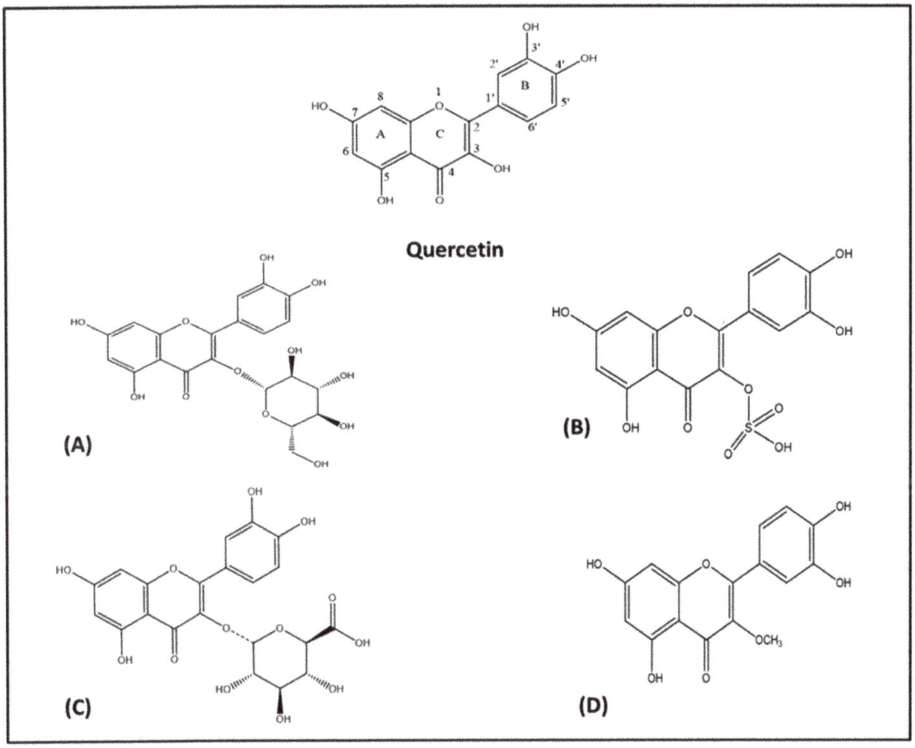

Figure 2. Chemical skeleton of quercetin and derivatives. (**A**) Quercetin glucoside, (**B**) quercetin-3-*O*-sulfate, (**C**) quercetin-3-*O*-glucuronide, and (**D**) 3-*O*-methyl quercetin.

Several flavonoids have anti-inflammatory activity, among which flavanones such as naringenin have weak activity, while flavonols including quercetin, kaempferol, and myricetin have strong anti-inflammatory activity. Flavonols are more active against phospholipase A-2, given the presence of the C-ring-2,3 double bond. Furthermore, the glycosylation of quercetin, such as rutin, reduces the anti-inflammatory activity of quercetin [41]. Glycosylation at position 3 of quercetin leads to a decrease in the free radical neutralizing activity. Glycosylation also decreases the acetyl cholinesterase inhibition activity of quercetin [42]. The methylation of quercetin at positions 4′ and 7 results in improved anticancer activities. The replacement of OH groups with O-methylated moieties also increases the metabolic stability of a compound [43]. The presence of a double bond between C2 and C3 as well as the OH group at ring B is essential for the thrombin inhibition activity of quercetin. Substituting the OH group with an OCH_3 group in ring B and ring C reduces the thrombin inhibition activity, whereas replacement of the OH groups in ring A with an OCH_3 group improves the thrombin inhibition activity of quercetin [44].

Quercetin may act as a pro-oxidant phytochemical by generating reactive oxygen species (ROS) and reactive electrophilic quinine type metabolites because of the presence of catechol moiety in the B ring, C-ring-2, 3 double bond, and free OH group at the C-3 position [45]. Furthermore, in vitro studies have confirmed that the esters-based precursors of quercetin increase the bioavailability of quercetin [46]. In considering the general structure of flavonoids, OH group substitution at positions C5, C7, and C3, the substitution of OCH_3 or H at C3', and OH or OCH_3 substitution at position C4' improve their neuroprotective activities as confirmed by in vitro studies, utilizing neuronal cell cultures [47].

3. Sources

The name quercetin has been used since 1857. It is derived from Latin *Quercetum*, after *Quercus* (oak) [48]. It is a flavonoid that is most commonly occurring in higher plants and in the glycosidal form, such as rutin (quercetin-3-*O*-rutinoside), isoquercetin (quercetin-3-*O*-glucoside), and hyperin (quercetin-3-*O*-galactoside). It can also be isolated in free-form from leaf surfaces, fruits, or bud extracts. Plant families rich in quercetin are Compositae, Passiflorae, Rhamnaceae, and Solanaceae [49]. Onions, asparagus, red leaf lettuce, apples, capers, and berries contain relatively high concentrations of quercetin [29,50]. The botanical sources of quercetin have been summarized in Table 1.

Table 1. Botanical sources of quercetin.

S.No.	Botanical Name	Family	Common Name	Active Parts	References
01	*Punica granatum*	*Lythraceae*	*Pomegranate*	Fruits	[51]
02	*Ruta graveolens*	*Rutaceae*	*Rue*	Leaves	[51]
03	*Camellia sinensis*	*Theaceae*	*Green tea*	Leaves	[52,53]
04	*Allium cepa*	*Amaryllidaceae*	*Red onion*	Fruits	[54,55]
05	*Mangifera indica*	*Anacardiaceae*	*Mango*	Fruits	[56]
06	*Moringa oleifera*	*Moringaceae*	*Drumstick tree*	Leaves	[57,58]
07	*Cydonia oblonga*	*Rosaceae*	*Quince*	Fruits and leaves	[59]
08	Solidago *canadensis* L.	*Compositae/ Asteraceae*	*Goldenrod*	Flowering parts	[60]
09	*Vaccinium angustifolium* and *Vaccinium corymbosum*	*Ericaceae*	*Blueberries*	Fruits	[61,62]
10	*Phaleria macrocarpa*	*Thymelaceae*	*Mahkotadewa*	Seeds	[63]
11	*Lepidium latifolium*	*Brassicaceae*	*Papperweed*	Roots and leaves	[64,65]
12	*Achras sapota (Manilkara zapota)*	*Sapotaceae*	*Sapodilla*	Fruits	[66]
13	*Cichorium intybus*	*Compositae/ Asteraceae*	*Chicory*	Leaves	[67]
14	*Solanum lycopersicum*	*Solanaceae*	*Tomato*	Fruits	[68]
15	*Malus domestica*	*Rosaceae*	*Apple*	Fruits	[69]
16	*Vitis vinifera*	*Vitaceae*	*Grapevines*	Fruits	[70,71]
17	*Rhamnus alaternus*	*Rhamnaceae*	*Buckthorn*	Bark	[72]
18	*Passiflora incarnate*	*Passifloraceae*	*Passion flower*	Leaves	[73]
19	*Morus alba*	*Moraceae*	*White mulberry or Tut*	Leaves	[74,75]
20	*Ginkgo biloba*	*Ginkgoaceae*	*Maidenhair tree*	Leaves	[76,77]
21	*Hypericum perforatum*	*Hypericaceae*	*St. John's wort or hypericum*	Aerial parts	[78]
22	Achillea millefolium L.	*Compositae/ Asteraceae*	*Yarrow*	Flowering tops	[79]

4. Pharmacokinetic Parameters of Quercetin

Studies have revealed that the low bioavailability of quercetin limits its use for therapeutic purposes, although it has a wide range of pharmacological properties. Low solubility, poor absorption, and rapid metabolism are major aspects of the low bioavailability of quercetin [80]. Ader et al. (2000) have explored the bioavailability and metabolism of quercetin in pigs [81]. Quercetin was administered to each animal in a single intravenous dose of 0.4 mg/kg body weight and a single oral dose of 50 mg/kg body weight one week later. Blood samples were collected to analyze the pharmacokinetic parameters of quercetin. The apparent bioavailability was 0.54 ± 0.19% when considering free quercetin alone, 8.6 ± 3.8% including conjugated quercetin, and 17.0 ± 7.1% including quercetin metabolites (kaempferol, tamarixetin, and isorhamnetin). The authors also reported that the conjugation of quercetin to glucuronic acid and sulfuric acid preferentially occurs in the intestinal wall.

Moon et al. (2008) have studied the pharmacokinetics of quercetin in human beings [82]. Quercetin was administered to subjects in doses of 500 mg three times daily, and the plasma and urine samples were collected from subjects to analyze the concentration of quercetin aglycone and metabolites. The average peak plasma concentration reported after the administration of quercetin at 500 mg three times daily was 463 ng/mL at 3.5 h. Re-entry peaks on plasma concentration versus time curves showed enterohepatic recirculation. The oral clearance of quercetin was reported to be high (3.5×10^4 L/h) with an average half life of 3.5 h. The urinary recovery of quercetin aglycone and conjugated metabolites were 0.05% to 3.6% and 0.08% to 2.6%, respectively. Previously, Day et al. (2001) while studying the identification of quercetin metabolites in plasma have reported that quercetin is found exclusively as a glucuronated or sulfate conjugate in plasma after oral administration [83].

Moreover, quercetin aglycone is less effective to cross the BBB, as it is a substrate for efflux transporter p-glycoprotein. On the other hand, quercetin conjugates are also predicted to less effectively cross the BBB because of high polarity [84,85]. However, the literature supported that some glucuronides are effectively crossing the BBB because of aided active transport [86]. Kawei et al. (2008) revealed that at sites of inflammation, the intracellular deconjugation of quercetin 3-*O*-glucuronide by β-glucuronidase increases, and this phenomena is more relevant inside the central nervous system (CNS) in the case of neurodegenerative disorders, rather than being of relevance in the penetration of the BBB [87].

5. Pathophysiological Mechanisms of Alzheimer's Disease

AD is mainly characterized by the accumulation of extracellular amyloid plaques, which are known as senile plaques. The amyloid plaques are associated with neuroinflammatory changes as well as neurofibrillary tangles, affecting processes that are essential in maintaining neuronal health: communication, metabolism, and repair. Aβ is produced by the sequential cleavage of amyloid β precursor protein by β and γ secretases [88,89]. The amyloid precursor protein (APP) exists in three different forms: the shortest isoform, APP695 mostly expressed in neurons, whereas two other isoforms APP751 and APP770 are expressed in glial cells, such as astrocytes. The physiological role of APP is to stimulate cell growth and proliferation. Aβ plaques lead to neuronal loss in several brain regions, including the entorhinal cortex, hippocampus, neocortex, amygdale, and subcortical areas [90,91]. Two different enzymatic processes of APP occur in the cells: the non-amyloidogenic and the amyloidogenic pathways. In the non-amyloidogenic processing pathway, the APP is cleaved by α and γ-secretases, which yields long secreted forms of APP (sAPPα) and C-terminal fragments. In the amyloidogenic processing pathway, APP is cleaved by β and γ-secretases, which yields long secreted forms of APP (sAPPβ), C-terminal fragments, and Aβs. The Aβ fragment is chemically sticky-forming plaques, and blocks cell-to-cell signaling via synapses, in turn leading to neuronal cell death [92,93].

Neurofibrillary tangles (NFTs) are produced by the hyperphosphorylation of microtubule-associated proteins, which are known as tau [94], representing another key feature of AD. The excessive phosphorylation of tau proteins decreases their binding ability to microtubules, resulting in the formation of NFTs [95]. Both the density and neuroanatomic localization of NFTs are important determinants in the pathogenesis of AD. NFTs in neocortical regions are commonly associated with severe cognition impairment [96]. Aβ induces the phosphorylation of tau proteins, thus triggering the formation of NFTs and causing neuronal cell death by increasing oxidative stress, altering calcium homeostasis and excitotoxicity [90]. Imbalance between Aβ production and clearance leads to the formation of Aβ aggregates, especially in late onset AD. The impairment of Aβ clearance most commonly occurs in late onset AD. Heparan sulfate proteoglycans (HSPGs) are most critical in the pathogenesis of AD by affecting Aβ metabolism and decreasing Aβ clearance. HSPGs bind to Aβ and accelerate its aggregation, mediating the neurotoxicity and neuroinflammation induced by Aβ. Under normal conditions, HSPGs perform various important physiological functions in development, growth factor signaling, cell proliferation, adhesion, migration, and homeostasis [97].

Oxidative stress commonly occurs in AD, contributing to neurodegeneration. ROS generation triggers Aβ aggregation. Oxidative stress is inherent to mild cognitive impairment (MCI) in the progression of AD. Patients with MCI usually develop cognitive impairment with a minimal impairment of instrumental activities of daily life, and it can be the first cognitive expression of AD. Alteration in the phosphorylation of proteins, such as heme oxygenase-1 and biliverdin reductase A, has been noted to occur, affecting the signaling of the most critical antioxidant pathways [98–100]. This leads to mitochondrial damage, which is concomitant with the decreased activity of mitochondrial energy-related proteins, including pyruvate dehydrogenase complex and alpha-ketoglutarate dehydrogenase. Defective mitochondria, in turn, trigger the generation of high levels of reactive oxygen species (ROS), for which antioxidant defenses may be deficient [101–103]. As such, the process proceeds unabated, as in a brush fire.

Neuroinflammation is a hallmark in the pathogenesis of AD. The innate immune cells of the brain are fast to respond to systemic events, mostly in aged and diseased brains. Misfolded and aggregated proteins binding to pattern recognition receptors on microglia and astroglia trigger an immune response, releasing a number of inflammatory cytokines and chemokines [104,105]. In contrast to oxidative stress, which is short lived, chronic inflammation is long lasting, resulting in the sustained release of inflammatory mediators.

Microglia surrounding senile plaques is commonly activated in AD, resulting in the upregulation of human leukocyte antigen-DR (HLA-DR). In addition, the generation of inflammatory mediators may lead to microglial activation and neurotoxicity secondary to the CD14-dependent process. Despite having phagocytotic activity, microglia fail to phagocytose Aβ due to the presence of inflammatory cytokines and various extracellular matrix proteins [106]. Receptor of advanced glycation end products (RAGE) activation leads to neurodegeneration as it triggers an increase in inflammatory mediators and oxidative stress. RAGE activation also leads to activation of downstream regulatory pathways such as the NF-kB, STAT, and JNK pathways [107].

Acetylcholinergic, glutamatergic, and serotonergic neurons are mostly affected in AD. Early histological changes involve the loss of cholinergic neurotransmission in the cerebral cortex. The latter, along with the hippocampus, receive the highest cholinergic input, and the loss of cholinergic neurons in these regions results in cognitive deficits and memory impairment. Choline acetyltransferase activity is greatly reduced in individuals with AD. It has also been shown that acetyl cholinesterase (AChE) accelerates the aggregation of Aβ [108]. The degeneration of serotonergic neurons in the raphe nucleus and noradrenergic neurons in the locus coeruleus mediates on cognitive symptoms associated with AD [109]. Figure 3 illustrates the pathogenesis of Alzheimer's disease.

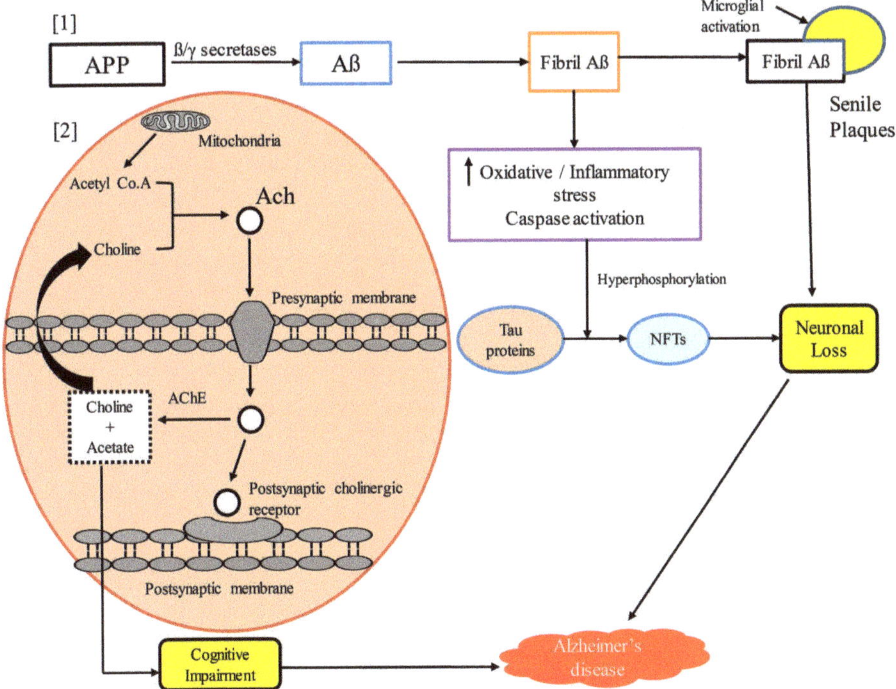

Figure 3. Schematic presentation of the pathogenesis of Alzheimer's disease. [1] Amyloid precursor protein (APP) is hydrolyzed by β and γ secretases to form β-amyloid (Aβ), which aggregates to form fibril Aβs. Fibril Aβs upregulate oxidative stress, the inflammatory cascade, and caspase activation, which results into the hyperphosphorylation of the Tau protein to form neurofibrillary tangles(NFTs), and the ultimate result is neuronal cells loss. Extensive fibrils along with activated microglia accumulated to form senile plaques, which lead to neuronal and synaptic loss. [2] Upstream regulating acetyl cholinesterase (AChE) enzyme promotes acetylcholine (Ach) degradation, resulting in neurotransmitter deficit, which leads to cognitive impairment. Amyloid precursor protein (APP), amyloid beta proteins (Aβ), neurofibrillary tangles (NFTs), acetylcholine (Ach), acetyl cholinesterase (AChE).

6. Neuroprotective Efficacy of Quercetin

The neuroprotective effects of quercetin have been extensively studied. At low micromolar concentrations, it antagonizes cell toxicity by oxidative stress in neurons. It suppresses neuroinflammatory processes by downregulating pro-inflammatory cytokines, such as NF-kB and iNOS, while stimulating neuronal regeneration. After absorption, quercetin metabolites are glucuronated, methylated, or sulfated, and all have been shown to afford neuroprotection. The neuroprotective efficacy of quercetin has been studied in both in vitro and in vivo models [50,110,111]. A study using drug screening in *Caenorhabditis elegans* nematodes with neuronal expression of human exon-1 huntingtin (128Q) and mutant Htt striatal cells derived from knock-in HD mice, concluded that isoquercetin improved motor functions in acute spinal cord injury, reduced α-synuclein fibrillization, reduced hippocampal neuronal cell death, improved synaptic plasticity, and reversed histopathological hallmarks of AD [112]. Quercetin also protects against mitochondrial dysfunction and progressive dopaminergic neurodegeneration by activating PKD1-Akt cell survival signaling axis Cell Culture and MitoPark transgenic mouse models of Parkinson's disease [113].

Quercetin has shown therapeutic efficacy, improving learning, memory, and cognitive functions in AD [114]. Khan et al. (2009) and Shimmyo et al. (2008) concluded that quercetin administration resulted in the inhibition of AChE and secretase enzymes using in vitro models, thus preventing the degradation of acetylcholine, and decreasing Aβ production, respectively [115,116]. Sabogal-Guáqueta et al. (2015) have been reported that quercetin administration reverses extracellular β-amyloidosis and decreases tauopathies, astrogliosis, and microgliosis in the hippocampus and amygdale, thus protecting cognitive and emotional function in age triple transgenic Alzheimer's disease model mice [117]. Wand et al. (2014) studied the effects of the long-term administration of quercetin on cognition and mitochondrial dysfunction in a mouse model of Alzheimer's disease. They noted that quercetin ameliorates mitochondrial dysfunction by restoring mitochondrial membrane potential, decreases ROS production, and restores ATP synthesis. It also increased the expression of AMP-activated protein kinase (AMPK), which is a key cell regulator of energy metabolism. Activated AMPK can decrease ROS generation by inhibiting NADPH oxidase activity or by increasing the antioxidant activity of enzymes such as superoxide dismutase-2 and uncoupling protein-2. The activation of AMPK also decreased Aβ deposition, regulating APP processing and promoting Aβ clearance. These mechanisms likely account for some of the therapeutic efficacy of quercetin on cognition and the attenuation of Aβ-induced neurotoxicity [118]. Quercetin and rutin have also been reported to function as memory enhancers in scopolamine-induced memory impairment in zebrafish, thus possibly enhancing cholinergic neurotransmission [119].

7. Anti-Alzheimer's Disease Mechanisms of Quercetin

7.1. Inhibition of AβAggregation and Tau Phosphorylation

The aggregation of Aβ is a key hallmark of AD [120]. Quercetin interferes with the formation of neurotoxic oligomeric Aβ species and displays fibril destabilizing effects on preformed fibrillar Aβ, reversing Aβ-induced neurotoxicity [110]. The structure of efficient polyphenolic inhibitors of Aβ contains two aromatic rings with two to six atom linkers. The aromatic rings contain a minimum number of three hydroxyl groups, which play an important role in fibril inhibition through hydrophobic interaction between the aromatic rings with β-sheet structures, forming hydrogen bonds. The phenolic hydroxyls increase the electron density in the aromatic rings, which may increase the binding of quercetin with the aromatic amino acids of the peptide beta-sheet structures. Quercetin possesses these structural requirements containing hydrophobic moieties and thus arrests fibril formation. The more hydroxyl groups present in the structure of the molecule, the higher its anti-amyloidogenic activity [121,122]. It is also suggested that the catechol structure may be auto-oxidized to form o-quinone on ring B, which then forms an O-quinone-Aβ42 adduct by targeting Lys residues at positions 16 and 28 of Aβ42. This phenomenon explains why quercetin has higher Aβ aggregation inhibitory activities compared to kaempferol, morin, and datiscetin [123].

Quercetin is also reported from in vitro and in silico studies to inhibit beta-secretase-1 (BACE-1) enzyme activity through the formation of hydrogen bonds. The OH group at position C-3 has a significant role in BACE-1 inhibition [116]. It has been documented from in vitro and molecular docking studies performed by Paris et al. (2011) that NF-kB regulates the production of Aβ by regulation of the β cleavage of APP, and that the quercetin-induced inhibition of NF-kB affects the regulation of BACE-1 expression [124]. Tauopathy commonly begins in the hippocampus, affecting hippocampal-dependent cognitive tasks followed by progression to other brain areas. Quercetin has been shown to decrease the phosphorylation of tau proteins and to inhibit the formation of NFTs in age triple transgenic Alzheimer's disease model mice [117]. Kinases and protein phosphatases such as protein phosphatase 2A (PP2A) play a role in regulating the hyperphosphorylation of tau proteins. The hyperphosphorylation of tau proteins is mostly due to the imbalance between phosphorylation and dephosphorylation mechanisms. Thus, the inhibition of PP2A may lead to the hyperphosphorylation of tau proteins. Quercetin reverses the hyperphosphorylation of tau proteins via MAPKs and PI3K/Akt/GSK3β signaling pathways in HT22 cells (a cell line from mouse hippocampal neurons) [125].

7.2. Acetylcholinesterase Inhibition

The inhibition of AChE is one of the therapies most commonly pursued in the treatment of mild to moderate AD. AChE is an enzyme that is responsible for the degradation of acetylcholine, and its inhibition results in increased acetylcholine levels, thus improving the cognitive symptoms of AD. In vitro studies have shown that quercetin is a competitive inhibitor of AChE and butyrylcholinesterase (BChE). It inhibits both enzymes in a concentration-dependent manner [126]. Quercetin inhibits AChE secondary to hydrophobic interactions and strong hydrogen bonding with the enzyme, reducing the hydrolysis of ACh, thus increasing ACh levels in the synaptic cleft, as reported by Abdalla et al. (2013) using cadmium-exposed rats as a model of study [127]. Studies have linked the presence/absence of OH groups on the phenyl rings of the test compound to the inhibition of AChE and BChE, given that the OH groups form hydrogen bonds with amino acid residues at the active site of the enzyme. This phenomenon may explain the inhibitory efficacy of quercetin for both of these enzymes. Moreover, quercetin exhibits greater potency for the inhibition of AChE and BChE than its glycosidal form, rutin. Previous studies have posited the presence of a sugar moiety in the molecule is essential for its enzymatic inhibition [115,128,129]. Jung et al. (2007) studied the acetylcholinesterase inhibiting the potential of flavonoids, including quercetin isolated from *Agrimonia pilosa* and they reported the IC50 value of quercetin to be 19.8 [130].

7.3. Attenuation of Oxidative Stress

Oxidative stress is caused by the accumulation of ROS in cells, and is an important factor in various neurodegenerative disorders. It is the most common mechanism of age-related degenerative processes. Mitochondria are the primary site for ROS production, and mitochondrial dysfunction leads to the overproduction of ROS followed by ATP depletion and ultimately cell death [50,131]. The major source of ROS is the superoxide anion radicals generated by the electron transport chain during oxidative phosphorylation. Superoxide dismutase (SOD) converts superoxide ions to hydrogen peroxide, which can be detoxified by catalase and glutathione peroxidases [132]. It has been reported that Aβ triggers oxidative stress, leading to lipid peroxidation and protein oxidation, which results in damaged mitochondria and the dysfunction of key enzymes associated with various pathways, including glucose metabolism [133].

Among the phytochemicals, quercetin is a potent antioxidant. It has been shown to effectively reduce the concentrations of superoxide anion free radicals, and its antioxidant potential makes it a versatile choice in the management of various disorders, including AD [49]. Previous studies have shown that quercetin has direct radical scavenging action. The presence of two pharmacophores in its structure is responsible for its antioxidant activities: one is a catechol group in the B ring, and the other is the OH group at position C-3. Quercetin also modulates the cell's own antioxidant pathways, by inducing Nrf-2-ARE and paraoxonase 2 (PON2), which is an antioxidant enzyme. Nuclear factor (erythroid-derived 2)-like 2 (Nrf-2) is an important regulator of cellular defenses against oxidative stress. Heme oxygenase-1, glutamate cysteine ligase, glutathione S-transferase, glutathione peroxidase, SOD, catalase, sulfiredoxin, and thioredoxin are enzymes downstream of Nrf-2-ARE. Thus, the activation of the Nrf-2-ARE pathway likely modulates the formation and degradation of misfolded protein aggregates in AD [50,64,134].

7.4. Attenuation of Neuroinflammation by Quercetin

Reducing the neuroinflammatory events in microglia might afford a beneficial strategy for the prevention of the progression of inflammatory-mediated neurodegenerative disorders. Quercetin has been reported to have anti-inflammatory actions, and is a suitable candidate among phytochemicals for future studies on its efficacy to reverse neuroinflammation [135,136]. Quercetin has already been shown to inhibit neuroinflammation by reducing nitric oxide production, iNOS gene expression in microglia, the production of inflammatory cytokines such as tumor necrosis factor-α (TNF-α), IFN-γ, interleukin-1β (IL-1β), IL-6, IL-12, and COX-2 in activated macrophages, as well as a reduction in cytokine expression as reported from in vivo studies. Quercetin also downregulates JNK/Jun phosphorylation and inhibits TNF-α production in mice, thus protecting neurons against LPS-induced inflammation [137]. Figures 4 and 5 have been summarizing anti-Alzheimer's targets and mechanistic insights of quercetin, respectively.

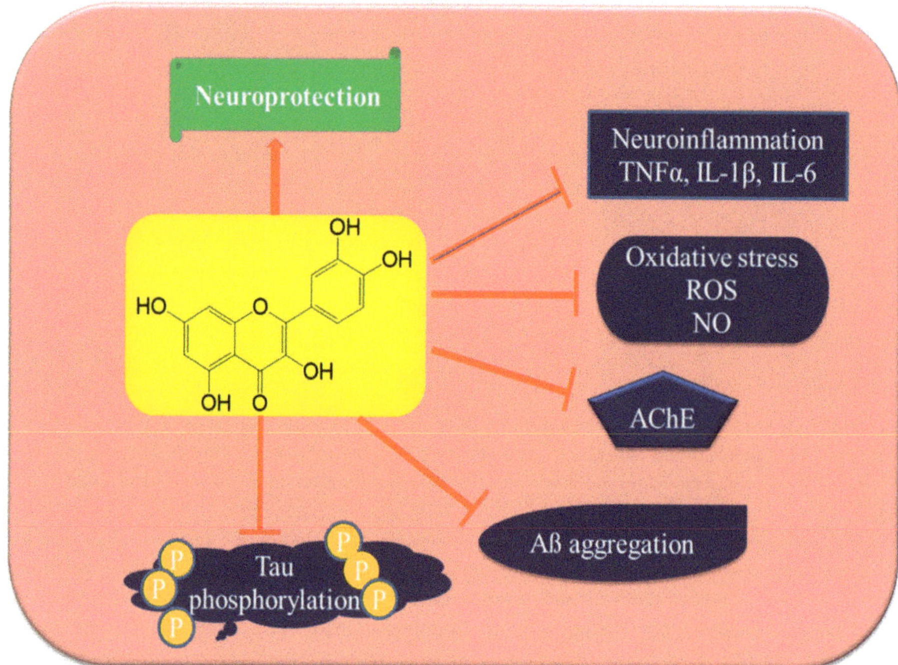

Figure 4. Anti-Alzheimer's disease targets of quercetin. Quercetin may utilize several mechanistic targets for neuroprotection in Alzheimer's disease such as the downstream regulation of oxidative stress and neuroinflammation, which leads to the direct protection of neurons, inhibiting AChE enzymes and resulting in increasing acetylcholine levels and reducing Tau phosphorylation and Aβ aggregation. Tumor necrosis factor-α (TNFα), interleukin-1β (IL-1β), interleukin-6 (IL-6), reactive oxygen species (ROS), nitric oxide (NO), acetyl cholinesterase (AChE), and amyloid beta protein (Aβ).

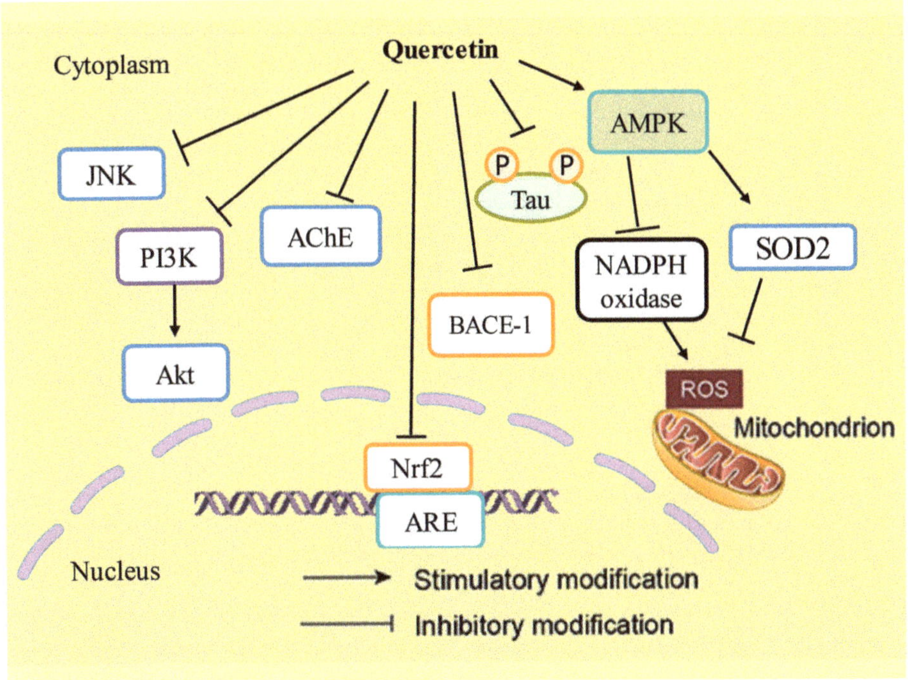

Figure 5. Mechanistic insights of quercetin in Alzheimer's disease. Quercetin has inhibitory effects on JNK, PI3K/Akt pathways, acetyl cholinesterase (AChE), nuclear factor (erythroid-derived 2)-like 2 (Nrf-2), beta-secretase-1 (BACE-1) enzyme activity, and the hyperphosphorylation of tau proteins. On the other hand, it stimulates the expression of AMP-activated protein kinase (AMPK), which thereby decreases reactive oxygen species (ROS) generation by inhibiting NADPH oxidase activity or by increasing the antioxidant activity of enzymes such as superoxide dismutase-2 (SOD2).

8. Anti-Alzheimer's Disease Potential: In Vitro Studies on the Efficacy of Quercetin

Quercetin protects neurons from oxidative damage, while reducing lipid peroxidation. Furthermore, given its antioxidant properties, quercetin inhibits the fibril formation of Aβ proteins, counteracting cell lyses and inflammatory cascade pathways [7,138,139]. In vitro studies on quercetin have shown that its antioxidant activities are concentration-dependent. It acts as an antioxidant at lower concentrations (neuronal cells treated with 5 µM and 10 µM), but it possesses toxic effects at higher doses (neuronal cells treated with 20 µM and 40 µM) [140]. It also inhibits iNOS and regulates the expression of COX-2 in various models, reflecting upon its anti-inflammatory activities. Most of the absorbed quercetin is present as metabolites, including glucuronidated, methylated, and sulfated metabolites. All exert neuroprotective effects, but limited testing of these metabolites has been carried out to date [117]. It has also been reported that quercetin decreases APP maturation, thus altering Aβ synthesis and aggregation [141]. The neuroprotective effects of quercetin glycosides have also been reported, which are characterized by antagonizing changes in gene expression, such as Park2, Park5, Park7, Casp3, and Casp7 genes [142].

9. Anti-Alzheimer's Disease Potential: In Vivo Studies on the Efficacy of Quercetin

The protective effects of a diet rich in polyphenols have been reported in several pathologic conditions, such as cardiovascular diseases, metabolic disorders, infections, cancers, and neurodegenerative disorders. It has been reported in several studies that quercetin exerts neuroprotective effects when administered in vivo. It protects neuronal cells from oxidative stress induced by various chemicals and prevents hippocampal apoptosis [50,143]. Quercetin improves memory, learning, and cognitive functions, and all these effects have been shown to be associated with its antioxidant properties [144]. In vivo studies using mice as an animal model have supported that quercetin increases spatial memory tasks, and decreases β-amyloidosis, tauopathies, astrogliosis, and microgliosis by increasing AMPK activity and decreasing mitochondrial dysfunction [117,118].

Keddy et al. (2012) investigated the neuroprotective and anti-inflammatory effects of the flavonoids-enriched fraction containing quercetin and its glucosides in a mouse model of hypoxic-ischemic brain injury. The study had concluded that the repeated administration of the flavonoid-enriched fraction prior to an experimental stroke produced by hypoxic-ischemia prevents the neuronal loss in the striatum and dorsal hippocampus. Due to the low bioavailability of quercetin and its glucosides, it required injection through the intraperitonial or intravenous route to achieve neuroprotective effects [145]. Tota (2010) et al. studied the effects of quercetin on cerebral blood flow and memory impairment in mice and linked the ability of quercetin to increase cerebral blood flow and energy metabolism to its memory-enhancing effects [146]. After oral administration in humans, quercetin is extensively metabolized during its absorption from the gut, affecting its bioavailability. Clinical efficacy trials/studies of quercetin have yet to be carried out, which is likely given its low BBB penetrability [147]. Furthermore, the metabolites of quercetin have long half-lives in vivo, and repeated dosing may lead to plasma accumulation [145]. These are important factors, which will need to be considered in the design of quercetin analogs for clinical studies.

10. Conclusions

Quercetin is a flavonoid with notable pharmacological effects and promising therapeutic potential. It is widely distributed among plants and found commonly in our daily diet, such as in fruits and vegetables. It has beneficial properties against general mechanisms of AD etiology in a variety of in vitro and in vivo models. It protects neuronal cells by attenuating oxidative stress and neuroinflammation. The anti-Alzheimer's disease properties of quercetin include the inhibition of Aβ aggregation and tau phosphorylation. It restores acetylcholine levels through the inhibition of hydrolysis of acetylcholine by AChE enzyme. Although showing neuroprotective efficacy in several in vitro and animal models, in vivo studies have reported that it is extensively metabolized upon absorption from the gut, affecting its bioavailability. It also has low BBB penetrability, thus limiting its efficacy in combating neurodegenerative disorders. Therefore, future clinical trials of quercetin and its analogs as neuroprotective agents must improve its bioavailability, developing related molecules with greater gut and brain penetrability, which will likely improve clinical efficacy.

Funding: MA was supported in part by grants from the National Institutes of Environmental Health Sciences R01 ES10563, R01 ES07331 and R01 ES020852 (MA).

Conflicts of Interest: The authors declare that they have no conflict of interest.

References

1. DeTure, M.A.; Dickson, D.W. The neuropathological diagnosis of Alzheimer's disease. *Mol. Neurodegener.* **2019**, *14*, 1–18. [CrossRef] [PubMed]
2. Parent, M.J.; Zimmer, E.R.; Shin, M.; Kang, M.S.; Fonov, V.S.; Mathieu, A.; Aliaga, A.; Kostikov, A.; Do Carmo, S.; Dea, D. Multimodal imaging in rat model recapitulates Alzheimer's Disease biomarkers abnormalities. *J. Neurosci.* **2017**, *37*, 12263–12271. [CrossRef]
3. Kommaddi, R.P.; Das, D.; Karunakaran, S.; Nanguneri, S.; Bapat, D.; Ray, A.; Shaw, E.; Bennett, D.A.; Nair, D.; Ravindranath, V. Aβ mediates F-actin disassembly in dendritic spines leading to cognitive deficits in Alzheimer's disease. *J. Neurosci.* **2018**, *38*, 1085–1099. [CrossRef]
4. Wallace, R.A.; Dalton, A.J. What can we learn from study of Alzheimer's disease in patients with Down syndrome for early-onset Alzheimer's disease in the general population? *Alzheimer's Res. Ther.* **2011**, *3*, 13. [CrossRef]

5. Hollingworth, P.; Harold, D.; Jones, L.; Owen, M.J.; Williams, J. Alzheimer's disease genetics: Current knowledge and future challenges. *Int. J. Geriatr. Psychiatry* **2011**, *26*, 793–802. [CrossRef]

6. Mayeux, R.; Stern, Y. Epidemiology of Alzheimer disease. *Cold Spring Harb. Perspect. Med.* **2012**, *2*, a006239. [CrossRef]

7. Aliev, G.; Obrenovich, M.E.; Reddy, V.P.; Shenk, J.C.; Moreira, P.I.; Nunomura, A.; Zhu, X.; Smith, M.A.; Perry, G. Antioxidant therapy in Alzheimer's disease: Theory and practice. *Mini Rev. Med. Chem.* **2008**, *8*, 1395–1406. [CrossRef]

8. Rountree, S.D.; Chan, W.; Pavlik, V.N.; Darby, E.J.; Siddiqui, S.; Doody, R.S. Persistent treatment with cholinesterase inhibitors and/or memantine slows clinical progression of Alzheimer disease. *Alzheimers Res.* **2009**, *1*, 7. [CrossRef] [PubMed]

9. Kim, J.; Lee, H.J.; Lee, K.W. Naturally occurring phytochemicals for the prevention of Alzheimer's disease. *J. Neurochem.* **2010**, *112*, 1415–1430. [CrossRef] [PubMed]

10. Farooq, U.; Khan, A.; Naz, S.; Rauf, A.; Khan, H.; Khan, A.; Ullah, I.; Bukhari, S.M. Sedative and antinociceptive activities of two new sesquiterpenes isolated from *Ricinus communis*. *Chin. J. Nat. Med.* **2018**, *16*, 225–230. [CrossRef]

11. Jawad, M.; Khan, H.; Pervez, S.; Bawazeer, S.S.; Abu-Izneid, T.; Saeed, M.; Kamal, M.A. Pharmacological validation of the anxiolytic, muscle relaxant and sedative like activities of *Capsicum annuum* in animal model. *Bangladesh J. Pharmacol.* **2017**, *12*, 439–447. [CrossRef]

12. Karim, N.; Khan, I.; Khan, H.; Ayub, B.; Abdel-Halim, H.; Gavande, N. Anxiolytic potential of natural flavonoids. *Sm J. Steroids Horm.* **2018**, *1*, 1001.

13. Khan, H.; Marya; Amin, S.; Kamal, M.A.; Patel, S. Flavonoids as acetylcholinesterase inhibitors: Current therapeutic standing and future prospects. *Biomed. Pharmacother.* **2018**, *101*, 860–870. [CrossRef] [PubMed]

14. Nabavi, S.F.; Khan, H.; D'Onofrio, G.; Šamec, D.; Shirooie, S.; Dehpour, A.R.; Argüelles, S.; Habtemariam, S.; Sobarzo-Sanchez, E. Apigenin as neuroprotective agent: Of mice and men. *Pharmacol. Res.* **2018**, *128*, 359–365. [CrossRef] [PubMed]

15. Rengasamy, K.R.R.; Khan, H.; Gowrishankar, S.; Lagoa, R.J.L.; Mahomoodally, F.M.; Khan, Z.; Suroowan, S.; Tewari, D.; Zengin, G.; Hassan, S.T.S.; et al. The role of flavonoids in autoimmune diseases: Therapeutic updates. *Pharmacol. Ther.* **2019**, *194*, 107–131. [CrossRef]

16. Nabavi, S.F.; Atanasov, A.G.; Khan, H.; Barreca, D.; Trombetta, D.; Testai, L.; Sureda, A.; Tejada, S.; Vacca, R.A.; Pittalà, V.; et al. Targeting ubiquitin-proteasome pathway by natural, in particular polyphenols, anticancer agents: Lessons learned from clinical trials. *Cancer Lett.* **2018**, *434*, 101–113. [CrossRef]

17. Khan, H.; Ullah, H.; Martorell, M.; Valdes, S.E.; Belwal, T.; Tejada, S.; Sureda, A.; Kamal, M.A. Flavonoids nanoparticles in cancer: Treatment, prevention and clinical prospects. *Semin. Cancer Biol.* **2019**, 1–13. [CrossRef]

18. Solanki, I.; Parihar, P.; Mansuri, M.L.; Parihar, M.S. Flavonoid-based therapies in the early management of neurodegenerative diseases. *Adv. Nutr. Int. Rev. J.* **2015**, *6*, 64–72. [CrossRef]

19. Devore, E.E.; Kang, J.H.; Breteler, M.; Grodstein, F. Dietary intakes of berries and flavonoids in relation to cognitive decline. *Ann. Neurol.* **2012**, *72*, 135–143. [CrossRef]

20. Gao, X.; Cassidy, A.; Schwarzschild, M.; Rimm, E.; Ascherio, A. Habitual intake of dietary flavonoids and risk of Parkinson disease. *Neurology* **2012**, *78*, 1138–1145. [CrossRef]

21. Ullah, H.; Khan, H. Anti-Parkinson potential of silymarin: Mechanistic insight and therapeutic standing. *Front. Pharmacol.* **2018**, *9*, 422. [CrossRef]

22. Baptista, F.I.; Henriques, A.G.; Silva, A.M.; Wiltfang, J.; da Cruz e Silva, O.A. Flavonoids as therapeutic compounds targeting key proteins involved in Alzheimer's disease. *Acs Chem. Neurosci.* **2014**, *5*, 83–92. [CrossRef] [PubMed]

23. Deng, Y.-H.; Wang, N.-N.; Zou, Z.-X.; Zhang, L.; Xu, K.-P.; Chen, A.F.; Cao, D.-S.; Tan, G.-S. Multi-Target Screening and Experimental Validation of Natural Products from Selaginella Plants against Alzheimer's Disease. *Front. Pharmacol.* **2017**, *8*, 539. [CrossRef] [PubMed]

24. Hamaguchi, T.; Ono, K.; Murase, A.; Yamada, M. Phenolic compounds prevent Alzheimer's pathology through different effects on the amyloid-β aggregation pathway. *Am. J. Pathol.* **2009**, *175*, 2557–2565. [CrossRef] [PubMed]

25. Ayaz, M.; Ullah, F.; Sadiq, A.; Kim, M.O.; Ali, T. Natural Products-Based Drugs: Potential Therapeutics against Alzheimer's Disease and other Neurological Disorders. *Front. Pharmacol.* **2019**, *10*, 1417. [CrossRef]

26. Maher, P. The potential of flavonoids for the treatment of neurodegenerative diseases. *Int. J. Mol. Sci.* **2019**, *20*, 3056. [CrossRef] [PubMed]

27. Kouhestani, S.; Jafari, A.; Babaei, P. Kaempferol attenuates cognitive deficit via regulating oxidative stress and neuroinflammation in an ovariectomized rat model of sporadic dementia. *Neural Regen. Res.* **2018**, *13*, 1827.

28. Brüll, V.; Burak, C.; Stoffel-Wagner, B.; Wolffram, S.; Nickenig, G.; Müller, C.; Langguth, P.; Alteheld, B.; Fimmers, R.; Naaf, S. Effects of a quercetin-rich onion skin extract on 24 h ambulatory blood pressure and endothelial function in overweight-to-obese patients with (pre-) hypertension: A randomised double-blinded placebo-controlled cross-over trial. *Br. J. Nutr.* **2015**, *114*, 1263–1277. [CrossRef]

29. Kim, D.H.; Khan, H.; Ullah, H.; Hassan, S.T.; Šmejkal, K.; Efferth, T.; Mahamoodally, M.F.; Xu, S.; Habtemariam, S.; Filosa, R. MicroRNA targeting by quercetin in cancer treatment and chemoprotection. *Pharmacol. Res.* **2019**, 104346. [CrossRef]

30. Kawabata, K.; Mukai, R.; Ishisaka, A. Quercetin and related polyphenols: New insights and implications for their bioactivity and bioavailability. *Food Funct.* **2015**, *6*, 1399–1417. [CrossRef]

31. Harwood, M.; Danielewska-Nikiel, B.; Borzelleca, J.; Flamm, G.; Williams, G.; Lines, T. A critical review of the data related to the safety of quercetin and lack of evidence of in vivo toxicity, including lack of genotoxic/carcinogenic properties. *Food Chem. Toxicol.* **2007**, *45*, 2179–2205. [CrossRef] [PubMed]

32. Li, Y.; Yao, J.; Han, C.; Yang, J.; Chaudhry, M.T.; Wang, S.; Liu, H.; Yin, Y. Quercetin, Inflammation and Immunity. *Nutrients* **2016**, *8*, 167. [CrossRef]

33. Rishitha, N.; Muthuraman, A. Therapeutic evaluation of solid lipid nanoparticle of quercetin in pentylenetetrazole induced cognitive impairment of zebrafish. *Life Sci.* **2018**, *199*, 80–87. [CrossRef] [PubMed]

34. Li, X.; Wang, H.; Gao, Y.; Li, L.; Tang, C.; Wen, G.; Zhou, Y.; Zhou, M.; Mao, L.; Fan, Y. Protective effects of quercetin on mitochondrial biogenesis in experimental traumatic brain injury via the Nrf2 signaling pathway. *PLoS ONE* **2016**, *11*, e0164237. [CrossRef] [PubMed]

35. El-Horany, H.E.; El-latif, R.N.A.; ElBatsh, M.M.; Emam, M.N. Ameliorative effect of quercetin on neurochemical and behavioral deficits in rotenone rat model of Parkinson's disease: Modulating autophagy (quercetin on experimental Parkinson's disease). *J. Biochem. Mol. Toxicol.* **2016**, *30*, 360–369. [CrossRef] [PubMed]

36. Sandhir, R.; Mehrotra, A. Quercetin supplementation is effective in improving mitochondrial dysfunctions induced by 3-nitropropionic acid: Implications in Huntington's disease. *Biochim. Et Biophys. Acta* **2013**, *1832*, 421–430. [CrossRef] [PubMed]

37. Tsao, R.; McCallum, J. Chemistry of flavonoids. In *Fruit and Vegetable Phytochemicals*; De la Rosa, L.A., Alvarez-Parrilla, E., Gonzalez-Aguilar, G.A., Eds.; Wiley-Blackwell: Ames, IA, USA, 2010; pp. 131–153.

38. Aherne, S.A.; O'Brien, N.M. Dietary flavonols: Chemistry, food content, and metabolism. *Nutrition* **2002**, *18*, 75–81. [CrossRef]

39. Kumar, S.; Sharma, H.; Yadav, K. Quercetin and metabolic syndrome. *EJPMR* **2016**, *3*, 701–709.

40. Dall'Acqua, S.; Miolo, G.; Innocenti, G.; Caffieri, S. The photodegradation of quercetin: Relation to oxidation. *Molecules* **2012**, *17*, 8898–8907. [CrossRef]

41. Kim, H.P.; Son, K.H.; Chang, H.W.; Kang, S.S. Anti-inflammatory plant flavonoids and cellular action mechanisms. *J. Pharmacol. Sci.* **2004**, *96*, 229–245. [CrossRef] [PubMed]

42. Ganeshpurkar, A.; Saluja, A.K. The pharmacological potential of rutin. *Saudi Pharm. J.* **2017**, *25*, 149–164. [CrossRef] [PubMed]

43. Shi, Z.H.; Li, N.G.; Tang, Y.P.; Shi, Q.P.; Tang, H.; Li, W.; Zhang, X.; Fu, H.A.; Duan, J.A. Biological Evaluation and SAR Analysis of O-Methylated Analogs of Quercetin as Inhibitors of Cancer Cell Proliferation. *Drug Dev. Res.* **2014**, *75*, 455–462. [CrossRef] [PubMed]

44. Shi, Z.-H.; Li, N.-G.; Tang, Y.-P.; Yang, J.-P.; Duan, J.-A. Metabolism-based synthesis, biologic evaluation and SARs analysis of O-methylated analogs of quercetin as thrombin inhibitors. *Eur. J. Med. Chem.* **2012**, *54*, 210–222. [CrossRef] [PubMed]

45. Gliszczyńska-Świgło, A.; van der Woude, H.; de Haan, L.; Tyrakowska, B.; Aarts, J.M.; Rietjens, I.M. The role of quinone reductase (NQO1) and quinone chemistry in quercetin cytotoxicity. *Toxicol. Vitr.* **2003**, *17*, 423–431. [CrossRef]

46. Biasutto, L.; Marotta, E.; De Marchi, U.; Zoratti, M.; Paradisi, C. Ester-based precursors to increase the bioavailability of quercetin. *J. Med. Chem.* **2007**, *50*, 241–253. [CrossRef]

47. Echeverry, C.; Arredondo, F.; Abin-Carriquiry, J.A.; Midiwo, J.O.; Ochieng, C.; Kerubo, L.; Dajas, F. Pretreatment with natural flavones and neuronal cell survival after oxidative stress: A structure–activity relationship study. *J. Agric. Food Chem.* **2010**, *58*, 2111–2115. [CrossRef]

48. Jaimand, K.; Rezaee, M.B.; Behrad, Z.; Najafy-Ashtiany, A. Comparison of extraction and measurement of quercetin from stigma, style, sepals, petals and stamen of Crocus sativus by HPLC in combination with heat and ultrasonic. *J. Med. Plants By-Prod.* **2012**, *1*, 167–170.

49. Alok, S.; Jain, S.K.; Verma, A.; Kumar, M.; Mahor, A.; Sabharwal, M. Herbal antioxidant in clinical practice: A review. *Asian Pac. J. Trop. Biomed.* **2014**, *4*, 78–84. [CrossRef]

50. Costa, L.G.; Garrick, J.M.; Roquè, P.J.; Pellacani, C. Mechanisms of neuroprotection by quercetin: Counteracting oxidative stress and more. *Oxidative Med. Cell. Longev.* **2016**, *2016*, 1–10. [CrossRef]

51. Choudhary, M.; Kumar, V.; Malhotra, H.; Singh, S. Medicinal plants with potential anti-arthritic activity. *J. Intercult. Ethnopharmacol.* **2015**, *4*, 147. [CrossRef]

52. Zhou, T.-S.; Zhou, R.; Yu, Y.-B.; Xiao, Y.; Li, D.-H.; Xiao, B.; Yu, O.; Yang, Y.-J. Cloning and characterization of a flavonoid 3′-hydroxylase gene from tea plant (Camellia sinensis). *Int. J. Mol. Sci.* **2016**, *17*, 261. [CrossRef]

53. Thakur, D.; Das, S.; Sabhapondit, S.; Tamuly, P.; Deka, D. Antimicrobial Activities of Tocklai Vegetative Tea Clones. *Indian J. Microbiol.* **2011**, *51*, 450–455. [CrossRef]

54. Henagan, T.; Cefalu, W.; Ribnicky, D.; Noland, R.; Dunville, K.; Campbell, W.; Stewart, L.; Forney, L.; Gettys, T.; Chang, J. In vivo effects of dietary quercetin and quercetin-rich red onion extract on skeletal muscle mitochondria, metabolism, and insulin sensitivity. *Genes Nutr.* **2015**, *10*, 2. [CrossRef]

55. Oliveira, T.T.; Campos, K.M.; Cerqueira-Lima, A.T.; Carneiro, T.C.B.; da Silva Velozo, E.; Melo, I.C.A.R.; Figueiredo, E.A.; de Jesus Oliveira, E.; de Vasconcelos, D.F.S.A.; Pontes-de-Carvalho, L.C. Potential therapeutic effect of Allium cepa L. and quercetin in a murine model of Blomia tropicalis induced asthma. *Daru J. Pharm. Sci.* **2015**, *23*, 18. [CrossRef] [PubMed]

56. Gondi, M.; Rao, U.P. Ethanol extract of mango (Mangifera indica L.) peel inhibits α-amylase and α-glucosidase activities, and ameliorates diabetes related biochemical parameters in streptozotocin (STZ)-induced diabetic rats. *J. Food Sci. Technol.* **2015**, *52*, 7883–7893. [CrossRef] [PubMed]

57. Leone, A.; Spada, A.; Battezzati, A.; Schiraldi, A.; Aristil, J.; Bertoli, S. Cultivation, genetic, ethnopharmacology, phytochemistry and pharmacology of Moringa oleifera leaves: An overview. *Int. J. Mol. Sci.* **2015**, *16*, 12791–12835. [CrossRef] [PubMed]

58. Saini, R.K.; Sivanesan, I.; Keum, Y.-S. Phytochemicals of Moringa oleifera: A review of their nutritional, therapeutic and industrial significance. *3 Biotech.* **2016**, *6*, 203. [CrossRef]

59. Oliveira, A.P.; Costa, R.M.; Magalhães, A.S.; Pereira, J.A.; Carvalho, M.; Valentão, P.; Andrade, P.B.; Silva, B.M. Targeted metabolites and biological activities of Cydonia oblonga Miller leaves. *Food Res. Int.* **2012**, *46*, 496–504. [CrossRef]

60. Apati, P.; Szentmihályi, K.; Balázs, A.; Baumann, D.; Hamburger, M.; Kristó, T.S.; Szőke, É.; Kéry, Á. HPLC analysis of the flavonoids in pharmaceutical preparations from Canadian goldenrod (Solidago canadensis). *Chromatographia* **2002**, *56*, S65–S68. [CrossRef]

61. Huang, W.-y.; Zhang, H.-c.; Liu, W.-x.; Li, C.-y. Survey of antioxidant capacity and phenolic composition of blueberry, blackberry, and strawberry in Nanjing. *J. Zhejiang Univ. Sci. B* **2012**, *13*, 94–102. [CrossRef]

62. Roopchand, D.E.; Kuhn, P.; Rojo, L.E.; Lila, M.A.; Raskin, I. Blueberry polyphenol-enriched soybean flour reduces hyperglycemia, body weight gain and serum cholesterol in mice. *Pharmacol. Res.* **2013**, *68*, 59–67. [CrossRef] [PubMed]

63. Altaf, R.; Asmawi, M.Z.B.; Dewa, A.; Sadikun, A.; Umar, M.I. Phytochemistry and medicinal properties of Phaleria macrocarpa (Scheff.) Boerl. extracts. *Pharmacogn. Rev.* **2013**, *7*, 73. [CrossRef] [PubMed]

64. Kaur, T.; Hussain, K.; Koul, S.; Vishwakarma, R.; Vyas, D. Evaluation of nutritional and antioxidant status of Lepidium latifolium Linn.: A novel phytofood from Ladakh. *PLoS ONE* **2013**, *8*, e69112. [CrossRef] [PubMed]

65. Gupta, S.M.; Singh, S.; Pandey, P.; Grover, A.; Ahmed, Z. Semi-quantitative analysis of transcript accumulation in response to drought stress by Lepidium latifolium seedlings. *Plant. Signal. Behav.* **2013**, *8*, e25388. [CrossRef] [PubMed]

66. Srivastava, M.; Hegde, M.; Chiruvella, K.K.; Koroth, J.; Bhattacharya, S.; Choudhary, B.; Raghavan, S.C. Sapodilla plum (Achras sapota) induces apoptosis in cancer cell lines and inhibits tumor progression in mice. *Sci. Rep.* **2014**, *4*, 6147. [CrossRef] [PubMed]

67. Street, R.A.; Sidana, J.; Prinsloo, G. Cichorium intybus: Traditional uses, phytochemistry, pharmacology, and toxicology. *Evid. Based Complement. Altern. Med.* **2013**, *2013*. [CrossRef]

68. Bovy, A.; Schijlen, E.; Hall, R.D. Metabolic engineering of flavonoids in tomato (Solanum lycopersicum): The potential for metabolomics. *Metabolomics* **2007**, *3*, 399. [CrossRef] [PubMed]

69. Newcomb, R.D.; Crowhurst, R.N.; Gleave, A.P.; Rikkerink, E.H.; Allan, A.C.; Beuning, L.L.; Bowen, J.H.; Gera, E.; Jamieson, K.R.; Janssen, B.J. Analyses of expressed sequence tags from apple. *Plant. Physiol.* **2006**, *141*, 147–166. [CrossRef] [PubMed]

70. Manela, N.; Oliva, M.; Ovadia, R.; Sikron-Persi, N.; Ayenew, B.; Fait, A.; Galili, G.; Perl, A.; Weiss, D.; Oren-Shamir, M. Phenylalanine and tyrosine levels are rate-limiting factors in production of health promoting metabolites in Vitis vinifera cv. Gamay Red cell suspension. *Front. Plant Sci.* **2015**, *6*, 538. [CrossRef]

71. Esatbeyoglu, T.; Ewald, P.; Yasui, Y.; Yokokawa, H.; Wagner, A.E.; Matsugo, S.; Winterhalter, P.; Rimbach, G. Chemical characterization, free radical scavenging, and cellular antioxidant and anti-inflammatory properties of a stilbenoid-rich root extract of Vitis vinifera. *Oxidative Med. Cell. Longev.* **2016**, *2016*, 8591286. [CrossRef]

72. Boussahel, S.; Speciale, A.; Dahamna, S.; Amar, Y.; Bonaccorsi, I.; Cacciola, F.; Cimino, F.; Donato, P.; Ferlazzo, G.; Harzallah, D. Flavonoid profile, antioxidant and cytotoxic activity of different extracts from Algerian Rhamnus alaternus L. bark. *Pharmacogn. Mag.* **2015**, *11*, S102. [PubMed]

73. Aman, U.; Subhan, F.; Shahid, M.; Akbar, S.; Ahmad, N.; Ali, G.; Fawad, K.; Sewell, R.D. *Passiflora incarnata* attenuation of neuropathic allodynia and vulvodynia apropos GABA-ergic and opioidergic antinociceptive and behavioural mechanisms. *BMC Complement. Altern. Med.* **2016**, *16*, 77. [CrossRef] [PubMed]

74. Khan, H.; Saeed, M.; Khan, M.A.; Muhammad, N.; Ghaffar, R. Isolation of long-chain esters from the rhizome of *Polygonatum verticillatum* by potent tyrosinase inhibition. *Med. Chem. Res.* **2013**, *22*, 2088–2092. [CrossRef]

75. Król, E.; Jeszka-Skowron, M.; Krejpcio, Z.; Flaczyk, E.; Wójciak, R.W. The effects of supplementary Mulberry leaf (Morus alba) extracts on the trace element status (Fe, Zn and Cu) in relation to diabetes management and antioxidant indices in diabetic rats. *Biol. Trace Elem. Res.* **2016**, *174*, 158–165. [CrossRef] [PubMed]

76. Shepperd, W.D. Ginkgo biloba L.: Ginkgo. In *The Woody Plant Seed Manual*; Agricultural Handbook No. 727; Bonner, F.T., Karrfalt, R.P., Eds.; Department of Agriculture, Forest Service: Washington, DC, USA, 2008; pp. 559–561.

77. Chan, P.-C.; Xia, Q.; Fu, P.P. Ginkgo biloba leave extract: Biological, medicinal, and toxicological effects. *J. Environ. Sci. Health C Environ. Carcinog. Ecotoxicol. Rev.* **2007**, *25*, 211–244. [CrossRef] [PubMed]

78. Oliveira, A.I.; Pinho, C.; Sarmento, B.; Dias, A.C. Neuroprotective activity of Hypericum perforatum and its major components. *Front. Plant Sci.* **2016**, *7*, 1004. [CrossRef]

79. Benedek, B.; Kopp, B. Achillea millefolium L. sl revisited: Recent findings confirm the traditional use. *Wien. Med. Wochenschr.* **2007**, *157*, 312–314. [CrossRef]

80. Cai, X.; Fang, Z.; Dou, J.; Yu, A.; Zhai, G. Bioavailability of quercetin: Problems and promises. *Curr. Med. Chem.* **2013**, *20*, 2572–2582. [CrossRef]

81. Ader, P.; Wessmann, A.; Wolffram, S. Bioavailability and metabolism of the flavonol quercetin in the pig. *Free Radic. Biol. Med.* **2000**, *28*, 1056–1067. [CrossRef]

82. Moon, Y.J.; Wang, L.; DiCenzo, R.; Morris, M.E. Quercetin pharmacokinetics in humans. *Biopharm. Drug Dispos.* **2008**, *29*, 205–217. [CrossRef]

83. Day, A.J.; Mellon, F.; Barron, D.; Sarrazin, G.; Morgan, M.R.; Williamson, G. Human metabolism of dietary flavonoids: Identification of plasma metabolites of quercetin. *Free Radic. Res.* **2001**, *35*, 941–952. [CrossRef] [PubMed]

84. Youdim, K.A.; Qaiser, M.Z.; Begley, D.J.; Rice-Evans, C.A.; Abbott, N.J. Flavonoid permeability across an in situ model of the blood–brain barrier. *Free Radic. Biol. Med.* **2004**, *36*, 592–604. [CrossRef] [PubMed]

85. Youdim, K.A.; Dobbie, M.S.; Kuhnle, G.; Proteggente, A.R.; Abbott, N.J.; Rice-Evans, C. Interaction between flavonoids and the blood–brain barrier: In vitro studies. *J. Neurochem.* **2003**, *85*, 180–192. [CrossRef] [PubMed]

86. Kroemer, H.K.; Klotz, U. Glucuronidation of drugs. *Clin. Pharmacokinet.* **1992**, *23*, 292–310. [CrossRef]

87. Kawai, Y.; Nishikawa, T.; Shiba, Y.; Saito, S.; Murota, K.; Shibata, N.; Kobayashi, M.; Kanayama, M.; Uchida, K.; Terao, J. Macrophage as a target of quercetin glucuronides in human atherosclerotic arteries implication in the anti-atherosclerotic mechanism of dietary flavonoids. *J. Biol. Chem.* **2008**, *283*, 9424–9434. [CrossRef]

88. Moore, B.D.; Chakrabarty, P.; Levites, Y.; Kukar, T.L.; Baine, A.-M.; Moroni, T.; Ladd, T.B.; Das, P.; Dickson, D.W.; Golde, T.E. Overlapping profiles of Aβ peptides in the Alzheimer's disease and pathological aging brains. *Alzheimer's Res. Ther.* **2012**, *4*, 18. [CrossRef]

89. Swomley, A.M.; Förster, S.; Keeney, J.T.; Triplett, J.; Zhang, Z.; Sultana, R.; Butterfield, D.A. Abeta, oxidative stress in Alzheimer disease: Evidence based on proteomics studies. *Biochim. Et Biophys. Acta* **2014**, *1842*, 1248–1257. [CrossRef]

90. Revett, T.J.; Baker, G.B.; Jhamandas, J.; Kar, S. Glutamate system, amyloid β peptides and tau protein: Functional interrelationships and relevance to Alzheimer disease pathology. *J. Psychiatry Neurosci.* **2013**, *38*, 6. [CrossRef]

91. Dawkins, E.; Small, D.H. Insights into the physiological function of the β-amyloid precursor protein: Beyond Alzheimer's disease. *J. Neurochem.* **2014**, *129*, 756–769. [CrossRef]

92. Teich, A.F.; Arancio, O. Is the amyloid hypothesis of Alzheimer's disease therapeutically relevant? *Biochem. J.* **2012**, *446*, 165–177. [CrossRef]

93. Chow, V.W.; Mattson, M.P.; Wong, P.C.; Gleichmann, M. An overview of APP processing enzymes and products. *Neuromolecular Med.* **2010**, *12*, 1–12. [CrossRef] [PubMed]

94. Šimić, G.; Babić Leko, M.; Wray, S.; Harrington, C.; Delalle, I.; Jovanov-Milošević, N.; Bažadona, D.; Buée, L.; De Silva, R.; Di Giovanni, G. Tau protein hyperphosphorylation and aggregation in Alzheimer's disease and other tauopathies, and possible neuroprotective strategies. *Biomolecules* **2016**, *6*, 6. [CrossRef] [PubMed]

95. Piedrahita, D.; Hernández, I.; López-Tobón, A.; Fedorov, D.; Obara, B.; Manjunath, B.; Boudreau, R.L.; Davidson, B.; LaFerla, F.; Gallego-Gómez, J.C. Silencing of CDK5 reduces neurofibrillary tangles in transgenic alzheimer's mice. *J. Neurosci.* **2010**, *30*, 13966–13976. [CrossRef]

96. Nelson, P.T.; Alafuzoff, I.; Bigio, E.H.; Bouras, C.; Braak, H.; Cairns, N.J.; Castellani, R.J.; Crain, B.J.; Davies, P.; Tredici, K.D. Correlation of Alzheimer disease neuropathologic changes with cognitive status: A review of the literature. *J. Neuropathol. Exp. Neurol.* **2012**, *71*, 362–381. [CrossRef] [PubMed]

97. Liu, C.-C.; Zhao, N.; Yamaguchi, Y.; Cirrito, J.R.; Kanekiyo, T.; Holtzman, D.M.; Bu, G. Neuronal heparan sulfates promote amyloid pathology by modulating brain amyloid-β clearance and aggregation in Alzheimer's disease. *Sci. Transl. Med.* **2016**, *8*, ra44–ra332. [CrossRef] [PubMed]

98. Butterfield, D.A.; Swomley, A.M.; Sultana, R. Amyloid β-peptide (1–42)-induced oxidative stress in Alzheimer disease: Importance in disease pathogenesis and progression. *Antioxid. Redox Signal.* **2013**, *19*, 823–835. [CrossRef]

99. Butterfield, D.A.; Di Domenico, F.; Barone, E. Elevated risk of type 2 diabetes for development of Alzheimer disease: A key role for oxidative stress in brain. *Biochim. Et Biophys. Acta* **2014**, *1842*, 1693–1706. [CrossRef]

100. Petersen, R.C.; Lopez, O.; Armstrong, M.J.; Getchius, T.S.; Ganguli, M.; Gloss, D.; Gronseth, G.S.; Marson, D.; Pringsheim, T.; Day, G.S. Practice guideline update summary: Mild cognitive impairment: Report of the Guideline Development, Dissemination, and Implementation Subcommittee of the American Academy of Neurology. *Neurology* **2018**, *90*, 126–135. [CrossRef]

101. García-Escudero, V.; Martín-Maestro, P.; Perry, G.; Avila, J. Deconstructing mitochondrial dysfunction in Alzheimer disease. *Oxidative Med. Cell. Longev.* **2013**, *2013*. [CrossRef]

102. Hardas, S.S.; Sultana, R.; Clark, A.M.; Beckett, T.L.; Szweda, L.I.; Murphy, M.P.; Butterfield, D.A. Oxidative modification of lipoic acid by HNE in Alzheimer disease brain. *Redox Biol.* **2013**, *1*, 80–85. [CrossRef]

103. Yan, M.H.; Wang, X.; Zhu, X. Mitochondrial defects and oxidative stress in Alzheimer disease and Parkinson disease. *Free Radic. Biol. Med.* **2013**, *62*, 90–101. [CrossRef]

104. Heneka, M.T.; Carson, M.J.; El Khoury, J.; Landreth, G.E.; Brosseron, F.; Feinstein, D.L.; Jacobs, A.H.; Wyss-Coray, T.; Vitorica, J.; Ransohoff, R.M. Neuroinflammation in Alzheimer's disease. *Lancet Neurol.* **2015**, *14*, 388–405. [CrossRef]

105. De Strooper, B.; Karran, E. The cellular phase of Alzheimer's disease. *Cell* **2016**, *164*, 603–615. [CrossRef] [PubMed]

106. Frank-Cannon, T.C.; Alto, L.T.; McAlpine, F.E.; Tansey, M.G. Does neuroinflammation fan the flame in neurodegenerative diseases? *Mol. Neurodegener.* **2009**, *4*, 47. [CrossRef] [PubMed]

107. Ray, R.; Juranek, J.K.; Rai, V. RAGE axis in neuroinflammation, neurodegeneration and its emerging role in the pathogenesis of amyotrophic lateral sclerosis. *Neurosci. Biobehav. Rev.* **2016**, *62*, 48–55. [CrossRef] [PubMed]

108. Ahmed, T.; Zahid, S.; Mahboob, A.; Mehpara Farhat, S. Cholinergic system and post-translational modifications: An insight on the role in Alzheimer's disease. *Curr. Neuropharmacol.* **2017**, *15*, 480–494. [CrossRef] [PubMed]

109. Savonenko, A.V.; Melnikova, T.; Hiatt, A.; Li, T.; Worley, P.F.; Troncoso, J.C.; Wong, P.C.; Price, D.L. Alzheimer's therapeutics: Translation of preclinical science to clinical drug development. *Neuropsychopharmacology* **2012**, *37*, 261–277. [CrossRef]

110. Caruana, M.; Cauchi, R.; Vassallo, N. Putative role of red wine polyphenols against brain pathology in Alzheimer's and Parkinson's disease. *Front. Nutr.* **2016**, *3*, 31. [CrossRef]

111. Jantan, I.; Ahmad, W.; Bukhari, S.N.A. Plant-derived immunomodulators: An insight on their preclinical evaluation and clinical trials. *Front. Plant Sci.* **2015**, *6*, 655. [CrossRef]

112. Figueira, I.; Menezes, R.; Macedo, D.; Costa, I.; Nunes dos Santos, C. Polyphenols beyond barriers: A glimpse into the brain. *Curr. Neuropharmacol.* **2017**, *15*, 562–594. [CrossRef]

113. Ay, M.; Luo, J.; Langley, M.; Jin, H.; Anantharam, V.; Kanthasamy, A.; Kanthasamy, A.G. Molecular mechanisms underlying protective effects of quercetin against mitochondrial dysfunction and progressive dopaminergic neurodegeneration in cell culture and MitoPark transgenic mouse models of Parkinson's Disease. *J. Neurochem.* **2017**, *141*, 766–782. [CrossRef] [PubMed]

114. David, A.V.A.; Arulmoli, R.; Parasuraman, S. Overviews of biological importance of quercetin: A bioactive flavonoid. *Pharmacogn. Rev.* **2016**, *10*, 84.

115. Khan, M.T.H.; Orhan, I.; Şenol, F.; Kartal, M.; Şener, B.; Dvorská, M.; Šmejkal, K.; Šlapetová, T. Cholinesterase inhibitory activities of some flavonoid derivatives and chosen xanthone and their molecular docking studies. *Chem. Biol. Interact.* **2009**, *181*, 383–389. [CrossRef] [PubMed]

116. Shimmyo, Y.; Kihara, T.; Akaike, A.; Niidome, T.; Sugimoto, H. Flavonols and flavones as BACE-1 inhibitors: Structure–activity relationship in cell-free, cell-based and in silico studies reveal novel pharmacophore features. *Biochim. Et Biophys. Acta* **2008**, *1780*, 819–825. [CrossRef]

117. Sabogal-Guáqueta, A.M.; Munoz-Manco, J.I.; Ramírez-Pineda, J.R.; Lamprea-Rodriguez, M.; Osorio, E.; Cardona-Gómez, G.P. The flavonoid quercetin ameliorates Alzheimer's disease pathology and protects cognitive and emotional function in aged triple transgenic Alzheimer's disease model mice. *Neuropharmacology* **2015**, *93*, 134–145. [CrossRef]

118. Wang, D.-M.; Li, S.-Q.; Wu, W.-L.; Zhu, X.-Y.; Wang, Y.; Yuan, H.-Y. Effects of long-term treatment with quercetin on cognition and mitochondrial function in a mouse model of Alzheimer's disease. *Neurochem. Res.* **2014**, *39*, 1533–1543. [CrossRef]

119. Richetti, S.; Blank, M.; Capiotti, K.; Piato, A.; Bogo, M.; Vianna, M.; Bonan, C. Quercetin and rutin prevent scopolamine-induced memory impairment in zebrafish. *Behav. Brain Res.* **2011**, *217*, 10–15. [CrossRef]

120. Regitz, C.; Marie Dußling, L.; Wenzel, U. Amyloid-beta (Aβ1–42)-induced paralysis in Caenorhabditis elegans is inhibited by the polyphenol quercetin through activation of protein degradation pathways. *Mol. Nutr. Food Res.* **2014**, *58*, 1931–1940. [CrossRef]

121. Jiménez-Aliaga, K.; Bermejo-Bescós, P.; Benedí, J.; Martín-Aragón, S. Quercetin and rutin exhibit antiamyloidogenic and fibril-disaggregating effects in vitro and potent antioxidant activity in APPswe cells. *Life Sci.* **2011**, *89*, 939–945. [CrossRef]

122. Porat, Y.; Abramowitz, A.; Gazit, E. Inhibition of amyloid fibril formation by polyphenols: Structural similarity and aromatic interactions as a common inhibition mechanism. *Chem. Biol. Drug Des.* **2006**, *67*, 27–37. [CrossRef]

123. Sato, M.; Murakami, K.; Uno, M.; Nakagawa, Y.; Katayama, S.; Akagi, K.-i.; Masuda, Y.; Takegoshi, K.; Irie, K. Site-specific inhibitory mechanism for amyloid β42 aggregation by catechol-type flavonoids targeting the Lys residues. *J. Biol. Chem.* **2013**, *288*, 23212–23224. [CrossRef] [PubMed]

124. Paris, D.; Mathura, V.; Ait-Ghezala, G.; Beaulieu-Abdelahad, D.; Patel, N.; Bachmeier, C.; Mullan, M. Flavonoids lower Alzheimer's Aβ production via an NFκB dependent mechanism. *Bioinformation* **2011**, *6*, 229. [CrossRef] [PubMed]

125. Jiang, W.; Luo, T.; Li, S.; Zhou, Y.; Shen, X.-Y.; He, F.; Xu, J.; Wang, H.-Q. Quercetin protects against okadaic acid-induced injury via MAPK and PI3K/Akt/GSK3β signaling pathways in HT22 hippocampal neurons. *PLoS ONE* **2016**, *11*, e0152371. [CrossRef] [PubMed]

126. Abdalla, F.H.; Schmatz, R.; Cardoso, A.M.; Carvalho, F.B.; Baldissarelli, J.; de Oliveira, J.S.; Rosa, M.M.; Nunes, M.A.G.; Rubin, M.A.; da Cruz, I.B. Quercetin protects the impairment of memory and anxiogenic-like behavior in rats exposed to cadmium: Possible involvement of the acetylcholinesterase and Na+, K+-ATPase activities. *Physiol. Behav.* **2014**, *135*, 152–167. [CrossRef]

127. Abdalla, F.H.; Cardoso, A.M.; Pereira, L.B.; Schmatz, R.; Gonçalves, J.F.; Stefanello, N.; Fiorenza, A.M.; Gutierres, J.M.; da Silva Serres, J.D.; Zanini, D. Neuroprotective effect of quercetin in ectoenzymes and acetylcholinesterase activities in cerebral cortex synaptosomes of cadmium-exposed rats. *Mol. Cell. Biochem.* **2013**, *381*, 1–8. [CrossRef]

128. Ademosun, A.O.; Oboh, G.; Bello, F.; Ayeni, P.O. Antioxidative properties and effect of quercetin and its glycosylated form (Rutin) on acetylcholinesterase and butyrylcholinesterase activities. *J. Evid. Based Complement. Altern. Med.* **2016**, *21*, NP11–NP17. [CrossRef]

129. Baldissarelli, J.; Santi, A.; Schmatz, R.; Abdalla, F.H.; Cardoso, A.M.; Martins, C.C.; Dias, G.R.M.; Calgaroto, N.S.; Pelinson, L.P.; Reichert, K.P. Hypothyroidism enhanced ectonucleotidases and acetylcholinesterase activities in rat synaptosomes can be prevented by the naturally occurring polyphenol quercetin. *Cell. Mol. Neurobiol.* **2017**, *37*, 53–63. [CrossRef]

130. Jung, M.; Park, M. Acetylcholinesterase inhibition by flavonoids from Agrimonia pilosa. *Molecules* **2007**, *12*, 2130–2139. [CrossRef]

131. Kennedy, M.A.; Moffat, T.C.; Gable, K.; Ganesan, S.; Niewola-Staszkowska, K.; Johnston, A.; Nislow, C.; Giaever, G.; Harris, L.J.; Loewith, R. A signaling lipid associated with Alzheimer's disease promotes mitochondrial dysfunction. *Sci. Rep.* **2016**, *6*, 19332. [CrossRef]

132. Lee, J.; Jo, D.-G.; Park, D.; Chung, H.Y.; Mattson, M.P. Adaptive cellular stress pathways as therapeutic targets of dietary phytochemicals: Focus on the nervous system. *Pharmacol. Rev.* **2014**, *66*, 815–868. [CrossRef]

133. Butterfield, D.A. The 2013 discovery award from the society for free radical biology and medicine: Selected discoveries from the Butterfield Laboratory of Oxidative Stress and its sequelae in brain in cognitive disorders exemplified by Alzheimer disease and chemotherapy induced cognitive impairment. *Free Radic. Biol. Med.* **2014**, *157*. [CrossRef]

134. Lakhanpal, P.; Rai, D.K. Quercetin: A versatile flavonoid. *Internet J. Med. Update* **2007**, *2*, 22–37. [CrossRef]

135. Testa, G.; Gamba, P.; Badilli, U.; Gargiulo, S.; Maina, M.; Guina, T.; Calfapietra, S.; Biasi, F.; Cavalli, R.; Poli, G. Loading into nanoparticles improves quercetin's efficacy in preventing neuroinflammation induced by oxysterols. *PLoS ONE* **2014**, *9*, e96795. [CrossRef] [PubMed]

136. Kim, B.-W.; Koppula, S.; Park, S.-Y.; Hwang, J.-W.; Park, P.-J.; Lim, J.-H.; Choi, D.-K. Attenuation of inflammatory-mediated neurotoxicity by Saururus chinensis extract in LPS-induced BV-2 microglia cells via regulation of NF-κB signaling and anti-oxidant properties. *BMC Complement. Altern. Med.* **2014**, *14*, 502. [CrossRef] [PubMed]

137. Qureshi, A.A.; Tan, X.; Reis, J.C.; Badr, M.Z.; Papasian, C.J.; Morrison, D.C.; Qureshi, N. Inhibition of nitric oxide in LPS-stimulated macrophages of young and senescent mice by δ-tocotrienol and quercetin. *Lipids Health Dis.* **2011**, *10*, 239. [CrossRef] [PubMed]

138. Davis, J.M.; Murphy, E.A.; Carmichael, M.D. Effects of the dietary flavonoid quercetin upon performance and health. *Curr. Sports Med. Rep.* **2009**, *8*, 206–213. [CrossRef]

139. Belo, D.; André, C.; Lucho, A.P.d.B.; Vinadé, L.; Rocha, L.; Seibert França, H.; Marangoni, S.; Rodrigues-Simioni, L. In vitro antiophidian mechanisms of Hypericum brasiliense choisy standardized extract: Quercetin-dependent neuroprotection. *Biomed Res. Int.* **2013**, *2013*, 943520.

140. Ansari, M.A.; Abdul, H.M.; Joshi, G.; Opii, W.O.; Butterfield, D.A. Protective effect of quercetin in primary neurons against Aβ (1–42): Relevance to Alzheimer's disease. *J. Nutr. Biochem.* **2009**, *20*, 269–275. [CrossRef]

141. West, S.; Bhugra, P. Emerging drug targets for Aβ and tau in Alzheimer's disease: A systematic review. *Br. J. Clin. Pharmacol.* **2015**, *80*, 221–234. [CrossRef]

142. Magalingam, K.B.; Radhakrishnan, A.; Ramdas, P.; Haleagrahara, N. Quercetin glycosides induced neuroprotection by changes in the gene expression in a cellular model of Parkinson's disease. *J. Mol. Neurosci.* **2015**, *55*, 609–617. [CrossRef]

143. Dong, Y.-s.; Wang, J.-l.; Feng, D.-y.; Qin, H.-z.; Wen, H.; Yin, Z.-m.; Gao, G.-d.; Li, C. Protective effect of quercetin against oxidative stress and brain edema in an experimental rat model of subarachnoid hemorrhage. *Int. J. Med. Sci.* **2014**, *11*, 282. [CrossRef]

144. Singh, S.K.; Srivastav, S.; Yadav, A.K.; Srikrishna, S.; Perry, G. Overview of Alzheimer's disease and some therapeutic approaches targeting Aβ by using several synthetic and herbal compounds. *Oxidative Med. Cell. Longev.* **2016**, *2016*, 7361613. [CrossRef] [PubMed]

145. Keddy, P.G.; Dunlop, K.; Warford, J.; Samson, M.L.; Jones, Q.R.; Rupasinghe, H.V.; Robertson, G.S. Neuroprotective and anti-inflammatory effects of the flavonoid-enriched fraction AF4 in a mouse model of hypoxic-ischemic brain injury. *PLoS ONE* **2012**, *7*, e51324. [CrossRef] [PubMed]

146. Tota, S.; Awasthi, H.; Kamat, P.K.; Nath, C.; Hanif, K. Protective effect of quercetin against intracerebral streptozotocin induced reduction in cerebral blood flow and impairment of memory in mice. *Behav. Brain Res.* **2010**, *209*, 73–79. [CrossRef] [PubMed]

147. Ossola, B.; Kääriäinen, T.M.; Männistö, P.T. The multiple faces of quercetin in neuroprotection. *Expert Opin. Drug Saf.* **2009**, *8*, 397–409. [CrossRef]

 *biomolecules*

Review

Anticancer Plants: A Review of the Active Phytochemicals, Applications in Animal Models, and Regulatory Aspects

Tariq Khan [1,*] , Muhammad Ali [2,*] , Ajmal Khan [3], Parveen Nisar [2], Sohail Ahmad Jan [4] , Shakeeb Afridi [2] and Zabta Khan Shinwari [2,5]

1   Department of Biotechnology, University of Malakand, Chakdara 18800, Pakistan
2   Department of Biotechnology, Quaid-i-Azam University, Islamabad 45320, Pakistan;
    parveennisar@yahoo.com (P.N.); shakeebafridi@outlook.com (S.A.); shinwari2008@gmail.com (Z.K.S.)
3   Department of Zoology, University of Buner, Sowari 17290, Pakistan; ajmal.sci@gmail.com
4   Department of Biotechnology, Hazara University, Mansehra 21120, Pakistan; sjan.parc@gmail.com
5   National Council for Tibb, Islamabad, Pakistan
*   Correspondence: tariqkhan@uom.edu.pk (T.K.); alibiotech01@gmail.com (M.A.);
    Tel.: +92-51-90644133 (M.A.)

Received: 15 November 2019; Accepted: 25 December 2019; Published: 27 December 2019

Abstract: The rising burden of cancer worldwide calls for an alternative treatment solution. Herbal medicine provides a very feasible alternative to western medicine against cancer. This article reviews the selected plant species with active phytochemicals, the animal models used for these studies, and their regulatory aspects. This study is based on a meticulous literature review conducted through the search of relevant keywords in databases, Web of Science, Scopus, PubMed, and Google Scholar. Twenty plants were selected based on defined selection criteria for their potent anticancer compounds. The detailed analysis of the research studies revealed that plants play an indispensable role in fighting different cancers such as breast, stomach, oral, colon, lung, hepatic, cervical, and blood cancer cell lines. The in vitro studies showed cancer cell inhibition through DNA damage and activation of apoptosis-inducing enzymes by the secondary metabolites in the plant extracts. Studies that reported in vivo activities of these plants showed remarkable results in the inhibition of cancer in animal models. Further studies should be performed on exploring more plants, their active compounds, and the mechanism of anticancer actions for use as standard herbal medicine.

Keywords: cancer; apoptosis; herbs; cell lines; in vivo

---

1. Introduction

The burden of cancer rose to 18.1 million new cases and 9.6 million deaths in 2018. With 36 different types, cancer mainly affects men in the form of colorectal, liver, lung, prostate, and stomach cancer and women in the form of breast, cervix, colorectal, lung, and thyroid cancer [1]. Treating cancer has become a whole new area of research. There are conventional as well as very modern techniques applied against cancers. A variety of techniques i.e., chemotherapy, radiation therapy, or surgery are used for treating cancer. However, all of them have some disadvantages [2]. The use of conventional chemicals bears side effects and toxicities [3]. But as the problem persists, new approaches are needed for the control of diseases, especially, because of the failure of conventional chemotherapeutic approaches. Therefore, there is a need for new strategies for the prevention and cure of cancer to control the death rate because of this disease.

Herbal medicine has become a very safe, non-toxic, and easily available source of cancer-treating compounds. Herbs are believed to neutralize the effects of diseases in a body because of various characteristics they possess [4]. For instance, among the many anticancer medicinal plants, *Phaleria macrocarpa* (local name: *Mahkota dewa*) and *Fagonia indica* (local name: *Dhamasa*) have been used traditionally for the anticancer properties of their active ingredients [5,6]. Metabolites extracted from the plant material are used to induce apoptosis in cancer cells. Gallic acid as the active component was purified from the fruit extract of *P. macrocarpa* and has demonstrated a role in the induction of apoptosis in lung cancer, leukemia, and colon adenocarcinoma cell lines [7,8]. It is a polyhydroxy phenolic compound and a natural antioxidant that can be obtained from a variety of natural products i.e., grapes, strawberries, bananas, green tea, and vegetables [9]. It also plays a critical role in preventing malignancy transformation and the development of cancer [10]. Similarly, other compounds such as vinca alkaloids, podophyllotoxin, and camptothecin obtained from various plants are used for the treatment of cancer.

With the advancement in the industrial sector and industrial medicine, the use of herbs was forgotten for a long period of time [11]. Hurdles regarding natural compounds are reduced because of the advent of new techniques and interest has been developed in the use of such natural ingredients in the pharmaceutical industry [12,13]. It has been estimated by the world health organization that 80% of the world is using traditional treatment methods [14]. Understanding of the effects or actions of herbs on various targets comes with the help of modern biomolecular science which recognizes some important properties i.e., anticancer, anti-inflammatory, and anti-virus. With the increasing understanding of the effects of such herbal medicine, their effects against different types of cancers have also been identified. For instance, hepatocellular carcinomas (HCC) are considered as the fifth most common malignancy in the world with increasing incidence [15,16]. Many studies have been performed on the treatment and prevention of using herbal medicine against HCC in which it is shown that all phases of HCC such as initiation, promotion, and progression could be affected by components of herbs [17,18].

However, as far as herbal compounds are considered as drugs, it is erroneously believed that they have no issues in terms of safety and side effects. There are hundreds of species of plants that are toxic to health. In the same way, there are many compounds in otherwise friendly plants that cause cytotoxicity. Based upon testing it has been proved that even anticancer plants result in cytotoxic effects [19].

Herbs are regulated under the "dietary supplement health and education act" as a dietary supplement in the United States of America. This review highlights the mechanism of some very important anticancer plants, the research related to their mechanism of action, their active ingredients, and the guidelines in place for their regulations.

2. Sources and Methodology

The most relevant literature was retrieved through a meticulous search on the electronic databases, Web of Science, Scopus, PubMed, and Google Scholar. The keywords and phrases used during the search were "Medicinal plants," "Anticancer activity," "Anticancer herbs," "Anticancer plants," "Mechanism of action," "Animal models," "in vitro activity," and "in vivo activity." The number of relevant articles finalized after extraction and analysis through the combination of the above keywords/phrases and the inclusion criteria was 200. The inclusion was based on two sets of criteria. According to the first set, i.e., "general criteria," articles selected for this manuscript had (i) reported the traditional anticancer activity of plants and their parts, (ii) reported the anticancer role of extract or pure compounds from plants.

The second set of criteria was used for selecting specific anticancer plants whose phytochemicals are discussed in detail. For this purpose, twenty plants were selected for which recent articles were available that (a) studied in vitro and in vivo anticancer activities of herbal products, (b) reported the anticancer/antitumor activity of active compounds from the plants, and (c) assessed the in vivo anticancer activity of the herbal anticancer products.

All the data were extracted in a table and the mechanisms of action were explained in respective subheadings and demonstrated through different figures.

3. Selected Plants and Their Anticancer Activity

Research so far has tested the anticancer activity of a plethora of plants and plant-based compounds. Some of these plants and their compounds prove to be very effective against one or more types of cancers. Based on their activities, the following plants are selected for the in vitro and in vivo anticancer activities of their compounds. The rest of the important plants shortlisted for their activities are presented in Table 1 along with their activities.

3.1. Artemisia annua

The genus Artemisia, widespread in Europe, Asia, North America, and South Africa has approximately 400 species worldwide [20]. Plants of the genus were used for centuries in classical medicine [21]. *Artemisia annua* is an annual short-day plant that belongs to family Asteraceae, having a brownish rigid stem. *A. annua* is known as sweet wormwood (Chinese: qīnghāo) and "dona" in the Urdu language in India and Pakistan [8]. *A. annua* was used by old Chinese for the preparation of anti-malarial drugs known as artemisinin (Figure 1). Having a unique ability of environmental adaptation it consistently resists insects and pathogens [22].

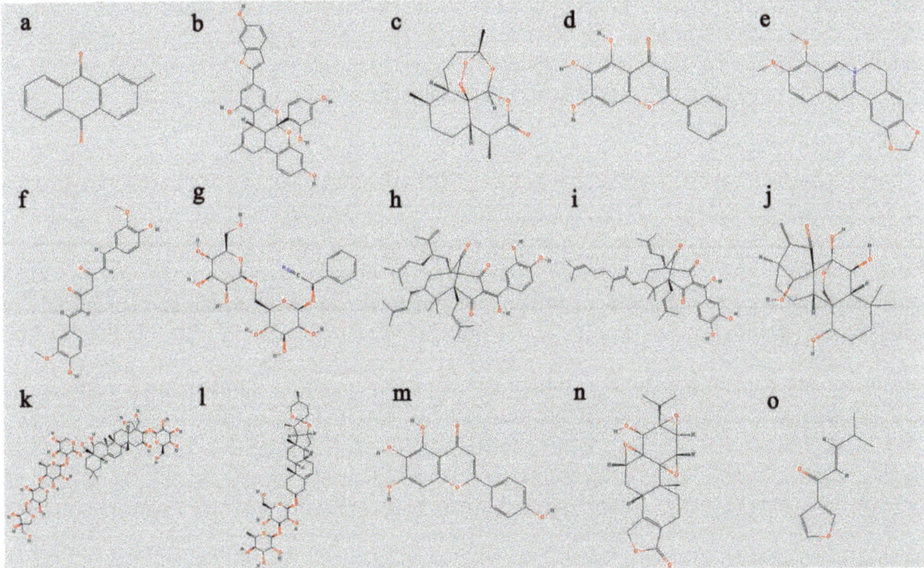

Figure 1. Structural representation of important anticancer secondary metabolites from plants. The structures are adapted from NCBI cited as National Center for Biotechnology Information. PubChem Database. (**a**) 2-Methylanthraquinone, Compound identification number (CID) = 6773; (**b**) albanol A, CID = 44567218; (**c**) artemisinin, CID = 68827; (**d**) baicalein, CID = 5281605; (**e**) berberine, CID = 2353; (**f**) curcumin, CID = 969516; (**g**) D-amygdalin, CID = 656516; (**h**) garcinol, CID = 5281560; (**i**) oblongifolin A CID = 53364454; (**j**) oridonin, CID = 5321010; (**k**) platycodin D, CID = 162859; (**l**) polyphyllin C, CID = 44429637; (**m**) scutellarein, CID = 5281697, and (**n**) triptolide, CID = 107985. (**o**) isoegomaketone, CID = 5318556; https://pubchem.ncbi.nlm.nih.gov/compound/ (accessed on 18 July 2019).

A. annua also synthesize scopoletin and 1,8-cineole compounds. Similarly, semi-synthetic derivatives of artemisinin are also generated such as arteether, artemether, and artesunate. Artesunate has been studied to be a very effective anticancer compound. Efferth [23] studied the effect of artesunate on 55 different cancer cell lines including leukemia, melanoma, lung cancer, colon cancer, renal cancer, ovarian cancer, and tumors of the central nervous system. They suggested that artesunate was most effective against leukemia and colon cancers. Furthermore, it was observed through these studies that the artesunate was more active than the drugs used for such cancers.

The stem and leaves *A. annua* were subject to extraction with the help of 80% ethanol and water. Several quantitative phenolic compounds from *A. annua* were identified using high-performance liquid chromatography (HPLC). The extracts were tested against HeLa and AGS cell lines. The cell growth inhibition activity of stem extracts was lower compared to leaf extracts. The ethanolic extracts of leaves lead to growth inhibitions (57.24% and 67.07%) in HeLa and AGS cells, respectively at a concentration of 500 mg/mL. HPLC analysis showed that the amount of phenolic acids was lower in stem extract than in leaves extract of *A. annua*. It was concluded from the data that the antioxidant and anticancer capacity was the result of phenolic compounds as well as unidentified compounds within *A. annua* [24].

3.2. Coptis chinensis

Coptis chinensis, the Chinese goldthread, is a herb used as a traditional medicine in China thus officially enlisted in the Chinese pharmacopeia [25]. It is widely known for its traditional use against various diseases like diarrhea, dysentery, acute febrile, and supportive infections. The organic extract of *C. chinensis* possesses anti-inflammatory and anti-oxidant properties [26,27]. *C. chinensis* extract has wide use in the treatment of cholera, dysentery, diabetes, blood and lung cancer because of its strong antibacterial activity [28]. *Coptis* genus contains the most important and active components, such as an alkaloid i.e., berberine (Figure 1). Berberines alkaloids are used frequently as criteria in the quality control of *Rhizoma coptidis* (Huang Lian) products and lead to the apoptosis of human leukemia HL-60 cells by down regulating nucleophosmin/B23 and telomerase activity.

3.3. Curcuma longa

Curcuma longa (Turmeric) belongs to the ginger family Zingiberaceae. It is a rhizomatous herbaceous perennial plant [29]. It is naturally found in Southeast Asia and the Indian subcontinent. These plants are annually collected for their rhizomes and are then propagated from some of those rhizomes [30]. *C. longa* possesses a broad range of pharmacological activities including anti- HIC (human immunodeficiency virus), anti-inflammatory, antioxidant effects, nematocidal and anti-bacterial activities.

Curcumin, the main component of *C. longa*, plays an important role in the therapeutic activities of *C. longa* [31]. Curcumin shows anticancer and anti-inflammatory activities as reported by many different studies. Cyclooxygenase (COX)-2 plays a vital role in the formation of colon cancer. In a study conducted by Goel et al. [32], the HT-29 colon cancer cells of humans were treated with different concentrations of curcumin to study the effect of curcumin on the expression of COX-2. The cell growth of HT-29 cells was inhibited by curcumin in a concentration- and time-dependent manner. Curcumin affected COX-2 by inhibiting its mRNA and protein expression, but no such inhibitory effect was found against COX-1. From this data, it can be suggested that the in vitro growth of HT-29 cells is significantly affected by a non-toxic concentration of curcumin. Curcumin may thus play an important role in the prevention of colon cancer. Furthermore, the anticancer effects of curcumin on human breast cancer cell lines (MCF-7) were assessed through lactate dehydrogenase and 3-(4,5-dimethyl-2-thiazolyl)-2, 5-diphenyl-2H-tetrazolium bromide assays to assess cytotoxicity and cell viability, respectively. The results showed that curcumin induced cytotoxicity and inhibited cells in a time- and concentration-dependent manner. This was observed through increased caspase 3/9 activity and induction of apoptosis. The results also indicated that curcumin downregulated miR-21 the expression of miR-21 in MCF-7 cells by upregulating the PTEN/Akt signaling pathway [33].

3.4. Fagonia indica

Fagonia indica, locally known as "dhamasa" is a flowering plant and belongs to the family of caltrop, Zygophyllaceae [34]. Members of Fagonia genus are known for their use as traditional medicine and are found effective in the treatment of many skin problems [35]. Traditionally, it was also used as a medicine for curing cancer as well as ailments resulting from poisons [5]. Amino acids and proteins [36], flavonoids [37], alkaloids [38], saponins [39], and terpenoid [40] are the phytochemicals found in the Fagonia species. *F. indica* is found to have liver protective [41] and antioxidant properties as well [42].

The aqueous extracts of *F. indica* have been found very effective against different types of cancer specifically breast cancers. For instance, Waheed et al. [43] performed bioactivity-guided fractionation to isolate the active and potent fraction of the *F. indica* extract. The activity was assessed against three cancer cell lines: MCF-7 estrogen-dependent breast cancer, MDA-MB-468 estrogen-independent breast cancer, and Caco-2 colon cancer cells (Figure 2). The results through different pieces of evidence such as the activity of pan-caspase inhibitor Z-VAD-fmk, caspase-3 cleavage, and DNA ladder assays suggested that apoptosis was stimulated in MDA-MB-468 and Caco-2 cells. Furthermore, a new steroidal saponin glycoside caused necrosis through cell lysis in MCF-7 cells. Similarly, Lam et al. [44] also demonstrated significant activity against breast cancer cells line MCF-7 through an aqueous extract of *F. indica*.

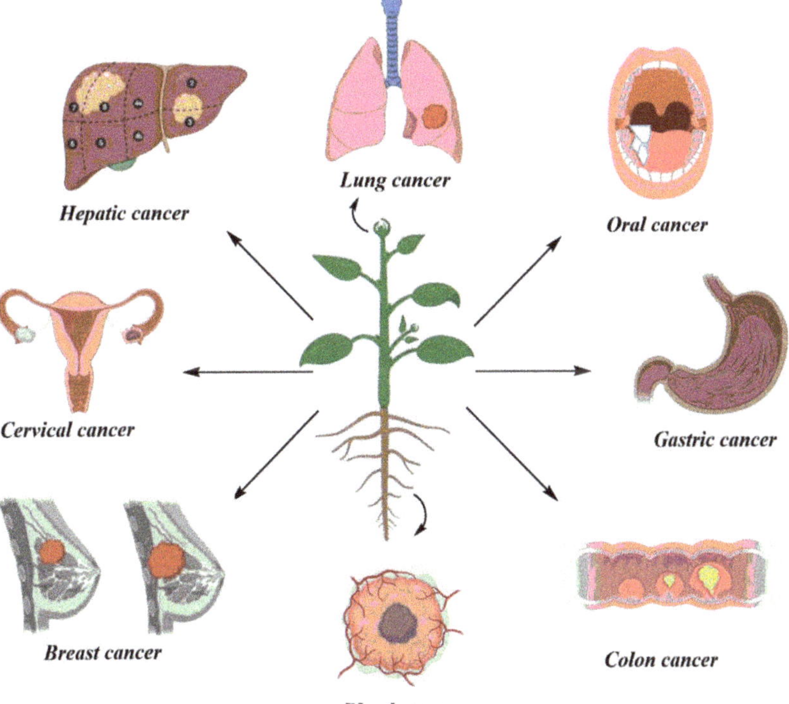

Figure 2. Illustration of activity of plants against several types of cancers. The icons were taken from Biorender illustrator and constructed through ChemBiodraw v14.0.

3.5. Garcinia oblongifolia

Garcinia oblongifolia (Lingnan Garcinia) belongs to the family of Clusiaceae and has a wide range of pharmaceutical activities. The important metabolites of the *G. oblongifolia* species; polyisoprenylated benzophenones and xanthones have anticancer, antioxidant, antifungal, apoptotic, and anti-pathogenic properties [45,46]. In vitro study showed that the bark of *G. oblongifolia* contains important secondary metabolites including oblongifolin A–G, oblongixanthones A–C along with other important compounds. These metabolites showed maximum apoptotic activities in HeLa-C3 cell lines and cytotoxic properties in the cervical cancer cells [47,48]. Li et al. [49] isolated about 40 different compounds from fruit, leaves, branches, and other parts of *G. oblongifolia*. They noted very high cytotoxic activities of these metabolites in the tested MCF-7 breast cancer cell line. However, they found the higher anti-cytotoxic activity of branch as compared to other plant parts. A small vacuole body formation was found at a low bark concentration of 0.250 g/mL. The vacuole size was increased at high concentrations of 500 g/mL and 1000 g/mL. The leaf part showed mild vacuole formation at a high concentration of 500 g/mL. Similarly, Feng, Huang, Gao, Xu and Luo [48] tested the pro-apoptotic activities of twenty different isolated compounds from *G. oblongifolia* in cervical cancer HeLa cells. Among all tested compounds the oblongifolins F and G, xanthone, nigrolineaxanthone T, and garcicowin B gave high pro-apoptotic properties at 10 µM concentration.

3.6. Garcinia indica

Garcinia indica, commonly known as kokum, is also an important medicinal plant that belongs to the Garcinia genus. The garcinol of *G. indica* shows positive activities in the experimental HT-29 and HCT-116 colon cancer cells along with normal immortalized intestinal cells (IEC-6 and INT-407). In another study, the fruit extract of *G. indica* was used for the isolation of garcinol. The garcinol at IC_{50} values (3.2–21.4 µM) for 72 h treatment shows strong inhibitory properties in all intestinal cells. The anticancer properties were higher in the cancer cells as compared to normal immortalized cells [50].

Similarly, Liao et al. [51] also observed a high tumor-inhibiting activity of *G. indica* in a human colorectal cancer cell line (HT-29). The garcinol at 10 µM concentration retarded the cell invasion activities several folds. The fruit extracts of *G. indica* has been shown very effective in the activation of caspase-3/CPP32 and the breakdown poly (ADP-ribose) polymerase (PARP) protein to inhibit leukemia in humans in the HL-60 cells [52]. These results indicated that garcinol (IC_{50} = 9.42 µM) shows strong growth inhibitory effects against human leukemia HL-60 cells.

3.7. Hedyotis diffusa

Hedyotis diffusa (Chinese: sheshecao) is a member of the family Rubiaceae. It is spread over the northeast regions of Asia. *H. diffusa* has been commonly used to cure inflammatory diseases i.e., urethritis, bronchitis, and appendicitis [53,54].

Because of the recent advances in pharmacological practices, this herb received importance for having antitumor properties and showed effective results in treating cancers of the liver, colon, lungs, brain, and pancreas [55]. *H. diffusa* contains important bioactive derivatives of polysaccharides, triterpenes, and anthraquinones [56,57].

Methyl anthraquinonesare, one of the bioactive compounds in *H. diffusa*, is responsible for apoptosis of many cancers. It shows apoptosis and inhibitory effect on the MCF-7 cell line of breast cancer via activation of the caspase-4/Ca^{2+}/calpain pathway when applied in a concentration of 18.62 µM for 24 h. It was observed that the S phase of the cell cycle and the percentage of the apoptotic cells were markedly increased when methyl anthraquinone was applied to MCF-7 cells [58]. Similarly, a concentrated extract of *H. diffusa* cause an inhibitory effect on the cervical cancer proliferation and induces apoptosis of Hela cells. Studies on the effect of *H. diffusa* ethanolic extracts on anti-colorectal cancer showed that these extracts cause an inhibitory effect on the Ct-26 cells by applying different concentrations (0.06 mg/mL, 0.08 mg/mL, 0.10 mg/mL and 0.12 mg/mL) with the rate of 35.46% to 71.84% [59].

3.8. Loranthus parasiticus and Scurrulus parasitica

Loranthus parasiticus, also known as Sang Ji Sheng (in Chinese), is a member of the Loranthaceae family and is widely distributed in the Southwestern regions of China. *L. parasiticus* is a semiparasitic plant, historically used as traditional folk medicine in China and Japan [60]. Since they are parasitic in nature, their biological activities including phytoconstituents are highly dependent on the host trees [61,62]. In fact, certain studies indicated the reverse effects of *L. parasiticus* on a host tree in decreasing sugar and chlorophyll content [63]. In vitro and in vivo studies have been conducted to evaluate the possible anticancer and antitumor potential of *L. parasiticus*. Cytotoxicity analysis of aqueous extracts of *L. parasiticus* has shown positive activity against ovarian cancer cell lines; SKOV3, CAOV3, and OVCAR-3 [64].

A comparative study conducted by Xiao et al. [65] on flavonoids extracted in 80% ethanol from *Scurrulus parasitica* harvested from different hosts showed a good anticancer potential on acute myeloid leukemia cell line HL-60. Flavonoids from *S. parasitica* on *N. indicum* induced apoptosis and inhibited cell proliferation with an IC_{50} value of 0.60 mg/L on HL-60 cells and arrested cell cycle at G_0–G_1 phase. *S. parasitica* parasitizing on *Morus alba* also showed an IC_{50} value of 2.49 mg/L. Similarly, Xiao et al. [66] isolated a polysaccharide in aqueous and ethanol extract from the leaves of *S. parasitica* and conducted proliferation inhibition assay on S180, K562, HL-60 cell lines. They observed inhibition in sarcoma S180 growth in mice with a 54% tumor inhibition rate Loranthus parasiticus on the optimal dose of 100 mg kg^{-1} d^{-1}. *S. parasitica* downregulate expression of CyclinD1, Bcl-2, and Ki-67 protein, and upregulate the expression of Bax protein which helps in the inhibition of cancer cell line and apoptosis of cancer cell in vivo.

3.9. Morus alba

M. alba, commonly called white mulberry, is native to China, Japan, India and is cultivated throughout the world where silkworm is raised. Their leaves are the main source of food for silkworms. Extracts from *M. alba* are traditionally used to cure cough, edema, insomnia, bronchitis, asthma, nose bleeding, wound healing, eye infections, and diabetes [67]. *M. alba* contains many pharmaceutically important compounds like kuwanol, hydroxymoricin, moranoline, morusin, calystegin, albafuran, and albanol. The leaves of *M. alba* contain some active compounds such as quercetin, rutin, apigenin, 1-deoxynojirimycin [67].

A study by Chon et al. [68] on methanolic extract of *M. alba* leaves showed anti-proliferative effects on different human cell lines like pulmonary carcinoma (Calu-6), colon carcinoma (HCT-116) and breast adenocarcinoma (MCF-7). These results showed a strong link to the concentrations of the investigated extracts. Anti-proliferative activity in the methanolic extract of *M. alba* leaves was observed on the human cell line of gastric carcinoma (SNU-601) in a concentration of 1000 mg/mL.

In another study on albanol A, isolated from *M. alba* root extract, Kikuchi et al. [69] showed apoptosis-inducing, cytotoxic activity (IC_{50} = 1.7 µM) in HL-60 cell line. It induced topoisomerase II (IC_{50} = 22.8 µM), clearly reduced the levels of pro-caspases 3, 8, and 9. Furthermore, the Bax/Bcl-2 ratio also increased and induced HL-60 apoptotic cell death through stimulation of the death receptor. The results of a study conducted on *M. alba* leaves in different extracts on human cell line hepatoma (HepG2) showed that methanolic leaf extract also showed inhibition (IC_{50} = 33.1 µg/mL) of HepG2 cell. It was concluded that *M. alba* leaves extract contains different phenolic compounds in different solvents which showed an anti-proliferative effect on the HepG2 cell line through the arrest of the cell cycle in G2/M phase. This was achieved with $p27^{Kip1}$ protein expression, activated caspases to induced cell apoptosis and inhibited topoisomerase IIα activity [70].

Furthermore, lectin was isolated from the leaves of *M. alba* which showed anti-proliferative activity on the human breast cell line (MCF-7) at a concentration of 8.5 µg/mL. This compound also showed cell cycle arrest and cytotoxicity in a human colorectal cell line (HCT-15) with inhibiting concentration of 16 µg/mL. The mechanism of inhibition of cancer cell lines was linked to the induction of apoptosis through activation and release of caspase-3 [71].

Methanolic extracts of *M. alba* showed significant anti-proliferative activity on the HepG2 cell line. These extracts showed significant inhibition of cells reducing cleared viable cell count. The cell growth was inhibited by suppressing the activity of NF-kB gene expression and biochemical marker modulation [72].

Further, Qin et al. [73] isolated 15 compounds from the root bark of *M. alba* which contain six diels-alder adducts and nine prenylated flavanones. The study observed that two new compounds, soroceal B and sanggenol Q showed cytotoxic activity. One of the isolated compounds showed selective cytotoxic activity in cells (HL-60 and AGS) at inhibiting concentrations of 3.4 μM.

3.10. Paris polyphylla

Paris polyphylla (called "Love Apple") belongs to family Liliaceae and contains 24 species throughout the world [74]. *P. polyphylla* is mostly used by Indian and Chinese traditional medicine system for having potential anticancer properties. *P. polyphylla* consists of important secondary metabolites such as polyphyllin D, formosanin C, β-ecdysterone, dioscin, daucosterol heptasaccharide, oligosaccharides, octasaccharide, protogracillin, trigofoenoside A, yunnanosides G-J, padelaoside B, pinnatasterone, and other saponins [75]. Steroidal saponins are the main active components because of its structural diversity and bio-activities such as antitumor, immune-stimulator, analgesic, and hemostatic properties [76–81].

Aqueous and ethanol extracts of *P. polyphylla* showed potential antitumor activity against human liver carcinoma (HepG2 and SMMC-7721) cell line, human gastric (BGC-823) cell line, human colon adenocarcinoma (LoVo and SW-116) cell line, and human esophagus adenocarcinoma (CaEs-17) cell lines. Ethanolic extract showed a strong inhibitory effect with IC_{50} values ranging from 10 μg/mL to 30 μg/mL [82]. Extract of *P. polyphylla* also showed an antitumor effect in esophageal cancer ECA109 cells by increasing the connexin26 mRNA and protein expression. Studies reported that *P. polyphylla* extracts increased the *Bad* genes expression and decreased the expression of *Bcl-2* genes, inhibiting the growth of ECA109 cells by proliferation and inducing cell apoptosis [83].

3.11. Perilla frutescens

Perilla frutescens, commonly called perilla or Korean perilla or Beefsteak plant, is widely distributed in Vietnam, China, Japan, and most Asian regions belong to the Labiatae family [84,85]. Economically, one of the most significant crops, cultivation of *P. frutescens* in China and some other Asian countries is more than 2000 years old [86,87]. Stem, seed, and leaf parts of *P. frutescens* have been used to treat poisoning, cold, bloating, and headache [74,88]). Multiple in vivo and in vitro studies have been conducted to evaluate the anticancer and antitumor potential of *P. frutescens*. Leaf extract of *P. frutescens* showed the highest anticancer activity in HepG2 cells through cell proliferation inhibition and upregulation of apoptosis-related gene expression [89]. Other studies revealed that ethanolic leaf extract of *P. frutescens* promoted apoptotic induction and tumorigenesis through death-mediated receptors and scavenging the reactive oxygen species (ROS) [90,91]

Similarly, Lin, Kuo, Wang, Cheng, Huang and Chen [89] used leaf extract of *P. frutescens* to evaluate the proliferation and apoptosis in HepG2 cells. Anti-proliferation activity was observed in HepG2 cells treated with *P. frutescens* leaf extract at a concentration of 105 μg/mL. The study reported that significant apoptosis was observed through flow cytometry. Furthermore, microarray results showed apoptosis-related gene expression in a time-dependent manner. The activity of *P. frutescens* leaf extract was compared to the activity of rosmarinic acid which showed less effective results in apoptosis-related gene expression and apoptosis induction in HepG2 cells.

The essential oil component "isoegomaketone" isolated from *P. frutescens* also induced apoptosis in human colon cancer (DLD1) cells. A study by Cho et al. [92] reported the inhibition of cell growth by isoegomaketone when treated for over 24 h, cleaved caspase-3, 8, and 9 in a time-dependent and dose-dependent manner. Isoegomaketone treatment triggered PARP cleavage, translocation of the protein Bax, cleaved Bid protein and the release of cytochrome *c* to the cytoplasm from mitochondria, induced apoptosis and translocation of apoptosis-inducing factor from mitochondria into the nucleus. It was suggested that isoegomaketone from *P. frutescens* induced apoptosis in DLD1 cells via both caspase-dependent and caspase-independent pathways.

3.12. Platycodon grandiflorus

Platycodon grandifloras, commonly known as balloon flower, or Chinese bellflower, belongs to the family Campanulaceae, which is distributed through Northeast Asia. The rhizomes of *P. grandiflorus* are very effective and are used as a traditional medicine in China, North Korea, and Japan for treatment of different diseases like cough, sore throat, phlegm, and other ailments [93]. *P. grandiflorus* contains many biologically active compounds which include saponins, flavonoids, anthocyanins, phenolics, and polysaccharide. These compounds have significant immune-stimulatory [94], anti-inflammatory [95], hepatoprotective [96], and antitumor activities. The antitumor activity of *P. grandifloras* was shown in a dose-dependent manner by reducing PKC enhancement of matrix metallopeptidases (MmP-9 and MmP-2), which caused the death of HT-80 cells [97]. Yu and Kim [98] isolated platycodin D from the root of *P. grandiflorus* and treated MCF-7 cells with a concentration of 5–100 μM which reduce cell viability and proliferation in a dose-dependent and time-dependent manner as compared with controlled cells. Induction of anticancer activity was observed because of caspase 8 and 9 activation and PARP cleavage. In addition, platycodin D upregulate cellular levels of protein Bax and Bcl-2 and downregulate the activation of caspase-9. Further, it also induces proteolytic activation of Bid (a protein of the proapoptotic Bcl-2 family).

Platycodin D is a triterpene saponin isolated from the roots of *P. grandiflorus* shows cytotoxic effects on the human leukemia cells. It inhibited telomerase activity and showed a cytotoxic effect in a dose-dependent manner with a concentration of 10–20 μM. This was shown to be achieved through downregulating the expression of human telomerase reverse transcriptase (hTERT). The results of the study by Kim et al. [99] showed that suppression of telomerase activity and cytotoxic effect on leukemia cells by platycodin D was through post-translational and transcriptional inhibition of hTERT. Platycodin D induced apoptosis and cell death of human leukemia (U937) cell by inducing the production of ROS through *Egr-1* gene activation and as a result, decreased in mitochondrial membrane potential, activating caspase-3 and PARP cleavage [100].

The extract of *P. grandiflorus* also showed antitumor effects on ovarian (SKOV3) cancer cells. Results showed that induced apoptosis occurs through the downregulation of Bcl-2 expression, upregulation of Bax expression, activation of caspase (3, 8, 9), and mitochondrial cytochrome *c* released to the cytosol [101].

3.13. Prunus armeniaca

Prunus armeniaca (Armenian plum) belongs to an important plant family Rosacea. Various parts of the plant are used as the major source of some important antioxidant substances and are commonly used against cancer and some other cardiovascular diseases [102]. The fruit part of *P. armeniaca* contains various important secondary metabolites like β-carotene, flavonoids, organic acids, thiamine, minerals, and oils [103]. The seeds of *P. armeniaca* contains plenty of cyanogenic glycosides, used against different types of cancers [104]. Amygdalin is one of the important glycosides of *P. armeniaca*, used for the treatment of prostate cancer [105]. Gomaa [106] reported the antioxidant and anticancer activities of *P. armeniaca* at different combinations of methanolic and ethanolic extracts with water. The study concluded that kernels of *P. armeniaca* have better antioxidant capacity compared to almonds that belong to the same genus. The study showed the highest antioxidant activity (67%), total lycopene content (4.70 mg/mL), and total phenolic content (3.290 mg/g of dry extract) in *P. armeniaca* kernels. Furthermore, *P. armeniaca* kernels were also found active in the inhibition of certain cancers when tested on the colon (HCT-116), human breast (MCF-7), and HepG2 cell lines in different concentrations and combinations. The highest cytotoxicity was observed from methanolic extract (IC$_{50}$ = 10.1 μg) against HepG2 cells. According to Madrau et al. [107], the fruit part contains a large amount of phenolic compound that shows strong antioxidant and immune-stimulant properties. Similarly, the study conducted by Liu et al. [108] found low anti-scavenging activity against DPPH of dry seed extract of apricot. Vardi et al. [109] reported the strong protective activity of apricot against intestinal oxidative damage in the rat model experiment.

3.14. Rabdosiae rubescens

Rabdosiae rubescens (Chinese: Dong Ling Cao) is a Chinese medicinal herb that belongs to the family Lamiaceae. It possesses multiple biological activities like antibacterial, anti-inflammatory, anti-parasitic, and anticancer [110]. *R. rubescens* contain important chemical compounds including monoterpenes, sesquiterpene, diterpene, and terpenoids. Oridonin, a tetracyclic terpenoid, is the main active compound in *R. rubescens* [111]. Oridonin gained its attention because of the remarkable properties of growth inhibition and the induction of apoptosis in cancer cells. In vitro and in vivo studies showed the induction of apoptosis in a variety of cancer cells by oridonin as in hepatocellular carcinoma, breast, gastric, skin, colorectal, gallbladder, and pancreatic cancers [112].

Bao et al. [113] isolated oridonin from *R. rubescens* which showed a potent anticancer potential in gallbladder both in vitro and in vivo. SGC996 and NOZ cells treated with oridonin results in inhibition of colony formation and growth of tumor cells in S-phase and induced apoptosis. Wang, et al. [114] showed the molecular mechanism of oridonin as an anticancer compound in HepG2 cancer cells. Oridonin in a concentration of 41.77 μM inhibited the growth of HepG2 cells while in a concentration of 44 μM, it induced G2/cell cycle arrest and apoptosis when applied for 24 h. The expression of nine different proteins was observed through proteomic analysis. Eight out of the nine proteins are already reported to be involved in anticancer activity. Oridonin up-regulate STRAP, Hsp70.1, Sti1, TCTP, and PPase, downregulate hnRNP-E1 which are significantly involved in the G2/M cell cycle arrest and apoptosis. The upregulation of HP1 beta and GlyRS by oridonin are responsible for the inhibitory effects on tyrosine kinase and telomerase.

Wang et al. [115] also reported that oridonin from *R. rubescens* is a potent cytotoxic agent that inhibited the growth and migration of highly metastatic human breast cancer cell lines (MCF-7 and MDA-MB-231). Oridonin significantly inhibited the growth of human breast cancer cells in a dose- and time-dependent manner, by arresting the cell cycle in the G2/M phase and accumulated cells to SUB-G1 phase. Results from the study also showed that oridonin triggered apoptosis by reducing Bcl-2/Bax ratio, NF-κB (p65), caspase-8, phospho-mTOR, IKKβ, IKKα and increasing the expression level of PPARγ, PARP, and Fas cleaving in a time-dependent manner. Oridonin significantly suppressed MDA-MB-231 cell invasion and migration through decreasing MmPs expression and regulation of integrin β1/FAK pathway and inhibited the growth and apoptosis via DNA damage and activated the extrinsic and intrinsic apoptotic pathway in breast cancer cells.

3.15. Scutellaria baicalensis

Scutellaria baicalensis is one of the important medicinal plants species of family Lamiaceae. It is commonly known as Baikal skullcap or Chinese skullcap and is found in different regions of the world including East Asia, Europe, and the Russian Federation. Its root part is known as *Scutellariae radix* and used as traditional Chinese medicine for the treatment of hepatitis, respiratory, and gastrointestinal diseases [116]. The root parts have maximum flavonoid content having multiple pharmacological properties [117].

About 60 different flavonoids have been identified in *S. baicalensis* which showed maximum antioxidant activities [116,118]. The four flavones metabolites also showed antimutagenic properties [119]. Woźniak et al. [120] studied the antioxidant potentials of four flavones: baicalein, baicalin, wogonin, and their glucuronides compounds and wogonoside. These flavones have different antioxidant capacity depending on the chemical structure and mechanisms of activity. However, among these, baicalein showed maximum antioxidant activities while the wogonin provided high protection to the linoleic acid from oxidation but did not show any antioxidant activity [121].

The *S. baicalensis* extract is useful against a wide range of cancer cells like brain tumor cells [122], prostate cancer cells [123], and head and neck squamous cell carcinoma (HNSCC) cell lines [124]. The aqueous extracts of roots led to programmed cell death, and thus inhibit the growth and development of apoptosis. They suppressed the growth of lymphoma and myeloma cell lines via disturbing normal expression level of *Bcl* and *c-myc* genes, while increased the expression level of cyclin-dependent kinase inhibitor p27 (KIP1) [121]. Hongwei et al. [125] used three different concentrations (80, 120, and 160 µmol/L) of baicalin for 2 days against BGC-823 and MGC-803 gastric cancer cells. The results of 3-(4,5-dimethyl-2-thiazolyl)-2,5-diphenyl-2-H-tetrazolium bromide assay showed a lower rate at different doses. Flow cytometric analysis showed that baicalin induces apoptosis in a dose-dependent way. Also, it was found that baicalin increase the gene expression of caspase-3, caspase-9, and other B cell lymphoma (Bcl-2)-associated X protein but lowers the expression of the *Bcl-2* gene.

A recent study reported the angiogenic properties of baicalin by using chick embryo chorioallantoic membrane [126]. The aqueous extract of baicalin showed both inhibitory and angiogenic activities while the baicalein showed only anti-proliferative property. Quantitative PCR results showed the expression of 84 different angiogenesis-related genes at different doses of baicalin and baicalein. The low dose significantly increases angiogenic genes expression while the high concentration showed the opposite activity by decreasing angiogenesis and increasing the lethality. The low to high levels of baicalein decrease the level of gene expression of multiple angiogenic genes and decreased cell proliferation.

Sato et al. [127] studied the anticancer potential of the root extract of *S. baicalensis* by using human oral squamous cell carcinoma (OSCC) cell line. The findings of the study showed 100 µg/mL root extract retard the monolayer and anchorage-independent growth rate many times but do not affect the cell adhering ability. Downregulated expression of cyclin-dependent kinase 4 cyclin D1, G1 phase arrest, and PARP cleavage was observed with the application of the root extract of *S. baicalensis*.

3.16. Scutellaria barbata

Scutellaria barbata, the barbed skullcap is a key medicinal plant species of family Lamiaceae, used to treat inflammatory and cancer diseases [128]. It is rich in important secondary metabolites like alkaloids, flavones, steroids, and polysaccharides [129,130]. In vitro studies showed positive activities against a vast range of cancers i.e., colon cancer, lung cancer, hepatoma, and skin cancer [128].

The apigenin and luteolin isolated from *S. barbata* gave cytotoxic activity against both human breast cancer cell line MDA-MB-231 and non-transformed breast cell line (MCF10A) [131]. Similarly, scutellarein was found to possess strong anti-breast cancer activity demonstrated in MDA-MB-468 cell lines [132]. Scutellarein increased the concentration of mitochondrial superoxide and peroxide while decreasing the level of glycolysis, retarding the growth of cancer cells by lowering ATP synthesis. Other secondary metabolites of *S. barbata* showed cytotoxic properties by leading ROX and DNA damage [131]. The other important alkaloid of *S. barbata*, scutebarbatine A (SBT-A) also resulted in high antitumor and apoptosis activities in experimental A549 cells.

3.17. Tripterygium wilfordii

Tripterygium wilfordii of the family Celastraceae is also known as "Thunder God Vine," and is native to Korea, China, and Japan. It is commonly used for the treatment of multiple diseases such as rheumatoid arthritis, systemic lupus erythematosus, nephritis, asthma, and cancers [133–135]. *T. wilfordii* produces important bioactive compound triptolide which is used as an immunosuppressive and anti-proliferative agent [134]. It has a five-membered unsaturated lactone ring and is used against different breast cancer cells by activation of pro-apoptotic compounds by modulating several signaling pathways [136].

In vitro studies showed anti-proliferative and pro-apoptotic activities against tumor cell lines [137–139]. He et al. [140] studied the in vitro properties of triptolide of *T. wilfordii* activity by using human umbilical vein endothelial cells (HUVECs). The IC$_{50}$ value for HUVECs proliferation was 45 nM. The results of semi-quantitative RT-PCR and Western blot analysis in human umbilical vein endothelial cells showed that triptolide cause downregulation of proangiogenic Tie2 and VEGFR-2 expression after 24 h at 50 nM concentration. At high concentration (100 nM), the VEGFR-2 mRNA expression was completely blocked. But on the other hand, the knockdown of Tie2 decreases the inhibitory activities of triptolide on endothelial network formation. The overall results showed that anticancer properties of triptolide are directly correlated with the blockage of two endothelial receptor-mediated signaling pathways. Triptolide showed more antagonistic activities against the proliferation of HUVECs as compared to normal cells like skin keratinocytes HaCaT cells and other liver cells L-02 [141].

Celastrol is another important compound of *T. wilfordii*, having cytotoxic activity against a broad range of cancer cells [142]. It demonstrates its anticancer activities by blocking the NF-κB via targeting IκB kinase and TAK1-induced NF-κB activation [143,144]. Other studies showed that triptolide of *T. wilfordii* has anticancer properties in many model systems such as experiments have also demonstrated triptolide's therapeutic efficacy in several model systems including neuroblastoma in nude mice model [145], other xenografts of human melanoma, breast cancer, bladder cancer, gastric carcinoma [146].

3.18. Tussilago farfara

Tussilago farfara (commonly called coltsfoot) is one of the important medicinal plants, grown in Europe and various regions of western and central Asia, commonly used against cancer. It possesses a high quantity of flavonoids and other phenolic compounds and some trace elements (Zn, Mg, and Se). The presence of these substances plays a key role in the anticancer activities of this plant. Maximum scavenging activity was recorded in water extract as compared to ethanol extract. It shows a 20.9% antioxidant activity. Further, this plant showed maximum antioxidant activity both using DPPH and yeast model [147]. The quercetin-glycosides isolated from the flower bud of *T. farfara* shows the highest antioxidant activity [148]. Lee et al. [149] reported the (TF)-induced cytotoxic and apoptotic activities of the flower part of *T. farfara* in human colon cancer cell line (HT-29) by using a methanolic extract. Fatykhova et al. [150] showed the genotoxic activity of *T. farfara* herb juice against known genotoxic compounds like nalidixic acid in SOS chromotest and furacilin in Rec assay. Their findings showed that dilution of the herb juice gave maximum antimutagenic properties in SOS chromotest as compared to furacilin in Rec assay.

Lee et al. [151] reported the activity of *T. farfara* as the TNF-related apoptosis-inducing ligand (TRAIL)-induced apoptosis via MKK7/JNK activation by inhibition of mitogen protein kinase-TOR signaling pathway regulator-like protein (MKK7-TIPRL) in human hepatocellular carcinoma cells. The *T. farfara* extract decreases 50% interaction between MKK7-TIPRL. HPLC data further verified the presence of many important phenolic compounds that decrease MKK7-TIPRL interaction and also examined the activation of MKK7/JNK.

3.19. Wedelia chinensis

Wedelia chinensis (Chinese: Peng qi ju), indigenous to India, South-East Asia, and China, is one of the important anticancer plants belonging to family Asteraceae which is rich in many important secondary metabolites like phenol, flavonoids, and tannin [152].

The essential oils of *W. chinensis* give a positive effect on lung cancer during the in vitro study. The GC-MS analysis recorded the presence of two important compounds carvacrol and trans-caryophyllene. High anti-scavenging activities were found at different levels of dose. The study of B16F-10 melanoma metastatic cell line showed that the concentrations of some important antioxidant enzymes (including catalase, superoxide dismutase, and glutathione peroxidase) increased many folds in the treatment groups. Similarly, the amount of glutathione also increased while the concentrations of other compounds such as lipid peroxidation and nitric oxide were decreased. The histopathology studies further verified that these essential oils show negative effects on cancer development [153].

4. In Vivo Studies of Anticancer Herbal Medicine: An Overview

The herbal medicines are tested both in vitro and in vivo. The anticancer activities of the various medicinal plants have been tested in vivo using different animal models (Figure 3). There are many studies available on in vivo experiments of the many different anticancer plants in mice models. For instance, dihydroartemisinin was reported to inhibit tumor tissue, increase the level of interferon-gamma (IFN-γ), and decrease interleukin 4 (IL-4) in tumor-bearing mice [154]. Similarly, artesunate, a derivative of artemisinin is also reported to be a promising drug against angiogenic Kaposi's sarcoma [155], growth inhibition of A549 and H1299 lung tumors by 100 mg/kg dose [156], the suppression of human prostate cancer xenograft [157] and the inhibition of leukemia growth in mice [158].

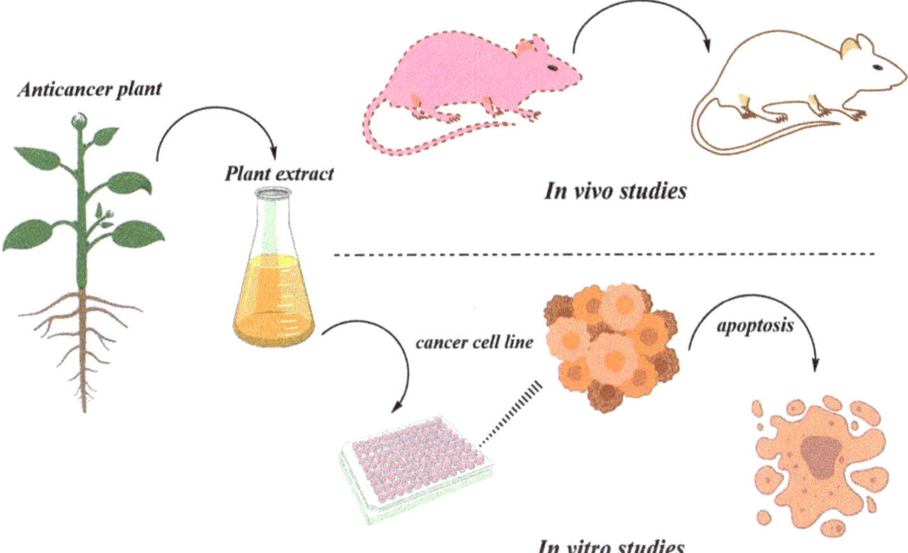

Figure 3. A depiction of general strategies applied for assaying extracts/phytochemicals from important medicinal plants for their anticancer activity both in vitro and in vivo.

Irradiation of C57BL/6 mice combined with a dose of 2 mg/kg twice a week was proved effective against lung carcinoma [159]. The effectiveness of berberine was enhanced when it was used in combination with other agents. Coptisine, another alkaloid of *Coptidis rhizoma* is proved to have anticancer effects when used in concentrations of 150 mg/kg against BALB/c nude mice by suppressing tumor growth and reducing cancer metastasis. The inhibition of the RAS-ERK pathway was suggested as the mechanism for this activity [160]. Another study was also performed on the nude mice on the HepG2 cells by applying the aqueous extract of *H. diffusa* which inhibits proliferation of cells in a dose-dependent manner, also delay S phase and arrest cells in G_0/G_1 phase [161].

Similarly, a high anticancer activity of SBT-A was found in transplanted tumor nude mice. Yang et al. [162] reported the anticancer activity of the polysaccharides isolated from *S. barbata* by 95-D Xenograft model. The results showed that polysaccharides give strong anti-proliferative activities against a 95-D cell line. It also lowered the expression of phospho-c-Met and other signaling elements like phospho-Erk and phospho-Akt. In vivo study also gave maximum antitumor activity by using a 95-D subcutaneous xenograft model. After one daily intraperitoneal injection for 3 weeks, the tumor growth was significantly decreased (47.72 % and 13.6%) at 100 and 200 mg/kg treatments. The ex vivo studies also showed that polysaccharides of *S. barbata* inhibit the phosphorylation of c-Met signaling pathway.

Furthermore, Li et al. [163] isolated a steroidal saponin from *P. polyphylla* which inhibited tumor growth in Lewis bearing-C57BL/6 mice and induced apoptosis in A549 cells. Results showed that steroidal saponin in concentration of 2.5, 5.0, and 7.5 mg/kg showed significant inhibition rate of $26.49 \pm 17.30\%$, $40.32 \pm 18.91\%$, and $54.94 \pm 16.48\%$, remarkably increased thymus and sleep indices, decreased inflammatory cytokines (TNF-α, IL-8, and IL-10). This in turn inhibited the tumor growth in C57BL/6 mice by reduced volume and weight of tumor. Nuclear changes, DNA condensation, chromatin fragmentation, and apoptosis are induced in A549 cells with a concentration of 0.25, 0.50, and 0.75 mg/mL steroidal saponin. Tumor growth inhibited by steroidal saponin was associated with decreased ROS, inflammatory response, and induction of apoptosis.

Furthermore, Wanga et al. [164] reported the effect of isoegomaketone from *P. frutescens* on Huh-7 hepatoma cell carcinoma and tumor-xenograft nude mice. Results showed that isoegomaketone inhibited cells and decreased tumor weight and volume. Isoegomaketone in the concentration of 10 nM/L decreased pAkt without affecting Akt. Hepatoma cell carcinoma tumor growth was suppressed by isoegomaketone from *P. frutescens* through PI3K/Akt signaling pathway blocking. *R. coptidis* is also showed anticancer activity in rats as suggested by the inhibition of cyclooxygenase 2 activity. The number of aberrant crypt foci in the rat colon was decreased by 54% after the administration of *R. coptidis* extracts [165].

Manjamalai and Grace [166] reported the apoptosis along with lowering angiogenesis and lung metastasis activities of the essential oils of *W. chinensis* by using B16F-10 melanoma cell line in C57BL/6 mice. The mice were injected with B16F-10 melanoma cells through the tail vein and treated with different doses of essential oil. A 50-μg essential oil concentration showed maximum cytotoxic activities with 65.17% lethality within 24 h. The numbers of apoptotic cells increased many times in experimental samples as compared to the control group. They also recorded high levels of important proteins like p53 and caspase-3 in essential oil-treated samples compared to other non-treated samples. They recommended this plant for the treatment and control of cancer.

In vivo activities of oridonin from *R. rubescens* showed a potent anticancer potential in the gallbladder [113]. When injected intra-peritoneally with a concentration of 5, 10, 15 mg/kg for 3 weeks to athymic nude mice, oridonin significantly inhibited NOZ xenografts growth. Oridonin also inhibited NF-κB nuclear translocation, increased Bax/Bcl-2 ratio, activated caspase-3, caspase-9, and PARP-1 which showed that the mitochondrial pathway is concerned with apoptosis mediated by oridonin.

Studies have reported anticancer activities of two artemisinin dimer, dimer-hydrazone (dimer-Sal) and dimer-alcohol (dimer-OH) and one monomer dihydroartemisinin (DHA) compared to the control against MTLn3 breast tumors in rats. Results of the study reported that dimer-Sal, dimer-OH, and DHA significantly suppressed tumors in rats compared to the control group. It was also observed that the dimers were more potent as compared to the monomers [167].

It is also reported that artemisinin is responsible for preventing breast cancer in rats treated with a single oral dose (50 mg/kg) of 7,12-dimethalbenz anthracene (DMBA) which is known for rapidly inhibiting the multiple breast tumors. After the feeding DMBA with 0–2% artemisinin to the target group and plain food in powdered form to the control group, both groups of experimental rats were monitored for breast tumors for 40 weeks. Oral artemisinin significantly reduced the development of breast tumors (57%) as compared to control fed (96%). The research indicates that artemisinin might be a potent cancer chemoprevention agent, having lesser side effects [168]. Similarly, oral administration of curcumin to rats reduced the level of Gp A72 (glycoprotein) by 73% hence lowering paw inflammation [169].

Tanaka et al. [170] observed activities of fruit extracts of *G. indica* in the azoxymethane (AOM)-induced colonic aberrant crypt foci in male model rats (F344). They found lower proliferating cell nuclear antigen index and high concentrations of glutathione S-transferase and quinone reductase. They also observed the maximum chemo-preventive activities of garcinol.

Besides mice and rats, there are many other animal models employed for studying anticancer activities. Zebrafish models are also employed for technical advantages including the ease of advanced genetic studies, expression of tumor in any organ and the striking resemblance to human malignancies [171]. Zhu et al. [172] used furanodiene which is a terpenoid isolated from *Rhizoma curcumae*, for their anticancer effects in zebrafish models. They observed that furanodiene showed anticancer effects in a pancreatic cell line (JF 305) and human breast cancer cells (MCF-7) transplanted into zebrafish. Furanodiene showed effective results through ROS production, anti-angiogenesis, apoptosis induction, and DNA strand breaks.

Similarly, the artemisinin type compound can have anticancer activities against different types of tumors including leukemia, carcinomas of breast, kidneys, lungs, and ovaries, lymphoma, melanoma, and brain tumors [23,173,174]. Currently, reports of in vivo activities of *A. annua* are accumulating. One study reported anticancer activities of *A. annua* against four animal models aged 10 including a male cat with malignant fibrosarcoma, a male dog with malignant mesenchymal neoplasia, a female dog with breast cancer and another male dog with a malignant fibrosarcoma. The animals were treated with various doses of 150 mg/day (3 capsules), 450 mg/day (2 capsules), 450 mg/day (3 capsules), and 450 mg/day (2 capsules). All the animals showed complete reduction with no tumor relapse [175].

5. Regulatory Aspects of Herbal Anticancer Drugs

It is generally established that the drugs including the anticancer compounds require phase III clinical research trials for marketing permissions. The Food and Drug Administration (FDA) and European Medicines Agency (EMA) guidelines require at least one controlled trial in Phase III with statistically significant results for the green signal to market them [176]. Except for exceptional circumstances, all the drugs need to go through all the phases of trials according to the guidelines of international agencies such as the FDA and EMA. However, it has been observed that pharmaceutical companies deviate from the standard protocol and start testing new compounds on human subjects earlier than the defined timeline. The reason for such practices is to accelerate the approval of these compounds under the pressure of investors [176]. This means that the drug is presented for approval with insufficient data on its quality, safety, and efficacy.

Although plant-based compounds have shown be less toxic compared to conventional synthetic compounds, there is growing evidence on the side effects of the unregulated use of these plants against different diseases. The problem is that there is insufficient data available regarding the quality, safety, and efficacy of herbal drugs. *F. indica*, for instance, has shown potent activity against breast cancer when tested in the MDA-MB-231 cell line [44]. *F. indica* is used traditionally to treat many disorders and people have even started the use of its herbal tea against breast cancer. However, the question remains that there are only a few reports available on the anticancer activity of the plant. Globally, the process of oncology drug development and marketing is regulated through the involvement of experts and an advisory process mediated by regulatory authorities [177].

There are several regulatory framework models available for prescribing such drugs but there is a need for harmony among regulating agencies and improvement in the regulation process. For instance, the FDA has recently adopted the questions and answers guidelines of the International Council for Harmonization on the nonclinical evaluation of drugs intended to treat cancer. These guidelines include 41 questions and answers which provide additional information about anticancer drug development and are aimed at bringing harmonization in the process of anticancer drug development [178]. It is, however, suggested that regulatory authorities, while bringing harmony with other agencies working for regulating anticancer herbal compounds, should increase the focus on combining information from traditional knowledge about that drug and the scientific studies on it [179].

Moreover, it is evident that plants of the same species grown in different areas vary in their profile of medicinal compounds [180]. This calls for the need to focus on the production of uniform and high-quality plants with a uniform metabolite profile that once tested is declared safe or unsafe once and for all. This might be achieved through the help of in vitro growth and biotechnological and genetic studies on these anticancer plants [181,182].

6. Modern Trends in Traditional Medicine Informatics and Opportunities for Anticancer Plant Products

With the advancement of information technology and bioinformatics, there is an increasing trend to build resources and databases that report herbal formulations, active components of the herb, and related information. There are several efforts like Chinese Medicine Integrated Database (TCMID) [183], Collective Molecular Activities of Useful Plants (CMAUP) [184], SymMap [185], encyclopedia of traditional Chinese medicine (ETCM) [186] etc. In addition, several researchers have developed strategies for in silico pharmacokinetic properties of molecules/drugs [187–191]. Such approaches are also applicable to phytochemicals and plant-based active drug components for their virtual screening, possible mode of action, and advanced drug discovery [192–195]. Several plant-based anticancer compounds have been evaluated using in silico and systems pharmacology tools [196–201]. The current study encourages further studies on anticancer active ingredients (of plant origin) for their in silico screening and pharmacokinetic activities. Considering the fact that plant-based drug formulations usually consists of several phytochemicals or even more than one plants. The major challenge on this direction would be to predict the role of phytochemicals other than active compounds and are present in the traditional medicine.

7. Conclusions

This detailed analysis of different plants showed that medicinal herbs promise a huge anticancer potential. This article comprehensively highlights the mechanism of antitumor action of some of the important plants. This is generally done through regulating signaling pathways. Many studies have reported inhibition of enzymes that stops tumor growth. These studies are mainly performed in human cell lines. It is highlighted that these plants play an important anticancer role through their different classes of secondary metabolites (Table 1). However, the study of these plants should not limit the study of a plethora of anticancer plants some of which are still unexplored. Studies are needed to highlight the mechanism of anticancer action of many already explored and many unexplored plants.

Table 1. Some of the important anticancer medicinal plants, their active components, and in vitro and in vivo activity.

S.No.	Plant Name	Common Name	Parts Used	Extract Used (Aqueous/Methanolic etc.)	Active Components Used	Dose Concentration	Cancer Cell Line Applied To	Animal Models Applied To	References
1	Allium sativum	Garlic	Leaves	Aqueous extracts	Allicin, flavonoids, and phenolic components	20 mg/kg/0.2 mL	Wehi-164 tumor cells	Balb/c mice	[202]
2	Alpinia galangal	Lengkuas, greater galangal, and blue ginger	Rhizomes	Ethyl acetate extract	Chrysin	1.3 mg/kg	Murine daltons lymphoma ascite (dla) and human lung cancer (a549) cells	Balb/c mice	[203]
3	Alstonia scholaris	Blackboard or devil's tree	Stem bark	Ethyl Alcohol extract	–	210 mg/kg	Hela cells lines	Swiss Albino mice	[204]
4	Andrographis paniculata	Creat or green chireta	Aerial parts	Methanolic extract	Diterpenes	10 µg/mL	Cancer cell lines sw620 and a498	Swiss Albino mice	[205]
5	Angelica archangelica	Garden angelica, wild celery, and Norwegian angelica	Root and rhizome	Ethanolic extract	Angelicin	500mg/kg	Mcf7 and 4t1 cell lines	Female balb/c mice	[206]
6	Aralia elata	Chinese angelica-tree, Japanese angelica-tree, and Korean angelica-tree	Leaves	Ethanol extract	–	300 mg/kg	Mcf-7 cells	Tumor bearing-nude mice	[207]
7	Artemisia annua	Sweet wormwood, sweet annie, and sweet sagewort	–	–	Artemisinin	0.02%	Breast cancer	Rats	[168]
8	Asclepia scurassavica	Tropical milkweed	Shade dried leaves	Ethyl acetate and methanolextract	B-sitosterol	10–20 mg/kg b.w.	Human colo 320 dm and monkey vero cell lines	Male wistar rats	[208]
9	Astragalus membranaceus	Mongolian milkvetch	–	–	Polysaccharide	400 mg/kg	Liver cancer	H22 hepatocarcinoma transplanted balb/c mice	[209]
10	Copaifera multijuga	Hayne oil, Copaiba	Trunk of the tree	Oil resin	Clerodane Diterpenes	2 g/Kg	B16f10 melanoma cells	Male C57/black mice	[210]
11	Coptidis rhizoma	Huanglian, Copaiba, and Copaibera	–	–	Berberine	200 µM and 400 µM	Human hepatic carcinoma cell lines HepG2 and mhcc97-l.	–	[211]
12	Curcuma longa	Turmeric	–	–	Curcumin	75 µM	Ht-29 colon cancer cells of human	–	[32]

Table 1. *Cont.*

S.No.	Plant Name	Common Name	Parts Used	Extract Used (Aqueous/Methanolic etc.)	Active Components Used	Dose Concentration	Cancer Cell Line Applied To	Animal Models Applied To	References
13	*Elephantopus scaber*	Elephant's Foot	–	Dimethyl sulfoxide extract	Deoxyelephantopin (doe)	25mg/kg	Murine ehrlich ascites carcinoma (eac)	Male swiss albino mice	[212]
14	*Fagonia schweinfurthii*	bush candle	Whole plant	Ethanolic extract	Carbon tetrachloride (ccl4)	200 µg/mL	HepG2 cell line	Male albino rats	[213]
15	*Garcinia indica*	Kokum	Fruits	Ethanol extract	Garcinol	<1 µM	Ht-29 and hct-116 colon cancer cells	–	[50]
16	*Garcinia oblongifolia*	Lingnan garcinia	Branch	Methanol extract	Xanthone	1000 µg/mL	Mcf-7 breast cancer cell line	–	[49]
17	*Garcinia preussii*	–	Fruits and leaves	Meohextract	Benzophenones		Du145, hela, ht-29, and a431 cell lines	–	[214]
18	*Hedyotis diffusa*	Snake-needle grass	–	–			Hela cells	Nude mice xenograft	[215]
19	*Hedyotis* spp.	–	*Aerial parts, stem and leaves*	Methanol extract	–	20 µM	Cem-ss cell line	–	[54]
20	*Kaempferia parviflora*	Black ginger	Rhizomes	Ethanolic extract	–	1 mg/mL	Ovarian cancer cell line, skov3	–	[216]
21	*Litchi chinensis*	litchi or lychee	Fruit pericarp	Ethanolic extract	Polyphenolic compounds	0.3 mg/mL	Human smmc-7721 hepatocellular carcinoma cell Line	Murine hepatoma bearing-mice	[217]
22	*Menyanthes trifoliata*	Bogbean, Buckbean, and Marsh Trefoil	Aerial part and root	Aqueous methanol extract	Polyphenolic compounds	1.5 mg/mL	Grade iv glioma cells	–	[218]
23	*Morus alba*	white mulberry	Root	N-hexane and methanolextracts.	Albanol a	30 µM	HL-60 (human leukemia) and Crl1579 (human melanoma) cell lines	–	[69]
24	*Morus nigra*	Black mulberry or blackberry	Aerial parts	dimethyl sulfoxide extract	Phenolic compounds especially Ascorbic acid and chlorogenic acid	1000 µg/mL	human prostate adenocarcinoma (PC-3)	–	[219]
25	*Nitraria retusa*	Salt tree or Nitre bush	Leaves	Chloroform extract	B-sitosterol and palmitic acid	50 mg/Kg b.w	B16-f10 cells lines	Balb/c mice	[220]

Table 1. *Cont.*

S.No.	Plant Name	Common Name	Parts Used	Extract Used (Aqueous/Methanolic etc.)	Active Components Used	Dose Concentration	Cancer Cell Line Applied To	Animal Models Applied To	References
26	*Paeonia lactiflora*	Chinese Peony	Root	Aqueous extract	–	15 mg/mL	Human hepatoma cell lines (HepG2 and hep3b)	–	[221]
27	*Paris polyphylla*	Herb Paris	Rhizomes	Methanol extract	Steroidal saponins	7.5 mg/kg	A549 cell line	Tumor-bearing c57bl/6 mice	[163]
28	*Perilla frutescens*	Beafsteak plant	Leaves	Meoh extract	Isoegomaketone	10nmol/l	Huh-7 hepatoma cell carcinoma	Tumor-xenograft nude mice	[164]
29	*Perilla frutescens*	Beafsteak plant	Leaf	–	Rosmarinic acid	105 µg/mL	Human hepatoma (HepG2) cells	–	[89]
30	*Platycodon grandiflorus*	balloon-flower	Root	Platycodin d was dissolved in Phosphate-buffered saline	Platycodin D	8 µg/mL	Human breast cancer cell line, mcf-7	–	[98]
31	*Pleurotus pulmonarius*	Indian Oyster, Italian Oyster, Phoenix Mushroom, or the Lung Oyster	Edible part	Aqueous extract	–	20 mg/kg	Huh7 liver cancer cells	Nude mice	[222]
32	*Rabdosia rubescens*	Bing Ling Cao, BlushredRabdosia, and Isodonrubescens	–	–	Oridonin	30 µmol/L	Human gallbladder cancer cell lines sgc996 and noz	Athymic nude mice	[113]
33	*Rhodamnia rubescens*	Scrub stringybark, brush turpentine, or brown malletwood	–	–	Tetracycline diterpenoidoridonin	50 µM	Human breast (mcf-7 and mda-mb-231) cancer cells	–	[115]
34	*Scutellaria barbata*	Barbed Skullcap	–	–	Polysaccharides	40 µg/mL	95-d cell line	Xenograft model	[162]
35	*Scutellaria baicalensis*	Baikal skullcap	Root	Aqueous extract	Baicalin	100 µg/mL	Human oral squamous cell carcinoma (oscc) cell line	–	[127]
36	*Tripterygium wilfordii*	Thunder god vine	–	–	Triptolide	250 nmol/L	Neuroblastoma cell lines (n2a and sknsh)	Neuroblastoma (nude mice model)	[145]
37	*Tussilago farfara*	Coltsfoot	Flower buds	Methanol extract	Quercetin-glycosides		Ht-29 human colon cancer cells	–	[149]
38	*Wedelia chinensis*	Chinese Wedelia	Leaves	Essential oils	Carvocrol and trans-caryophyllene		B16f-10 melanoma metastatic cell line	C57bl/6 mice	[153]
39	*Zuojin wan*		–	Aqueous extract	Palmatine, berberine, epiberberine, and coptisine	10 mg/mL	S180 tumor cells	Chinese kunming (km) mice	[223]

Author Contributions: Conceptualization, T.K., and M.A.; methodology, T.K. and M.A.; software, T.K.; investigation, T.K.; resources, S.A.J. and A.K.; data curation, P.N.; writing—original draft preparation, T.K., P.N., S.A. and A.K.; writing—review and editing, T.K. and M.A.; visualization, T.K.; supervision, Z.K.S. and M.A.; project administration, M.A. All authors have read and agreed to the published version of the manuscript.

Funding: This research received no external funding.

Acknowledgments: The authors acknowledge the efforts of all researchers working in the area of traditional medicine against cancer. The authors apologize to all scientists whose work has not been cited because of space limitation or unknowingly due to the great amount of literature in the field of anticancer phytochemicals.

Conflicts of Interest: The authors declare no conflict of interest.

References

1. Bray, F.; Ferlay, J.; Soerjomataram, I.; Siegel, R.L.; Torre, L.A.; Jemal, A. Global cancer statistics 2018: GLOBOCAN estimates of incidence and mortality worldwide for 36 cancers in 185 countries. *CA Cancer J. Clin.* **2018**, *68*, 394–424. [CrossRef] [PubMed]

2. Karpuz, M.; Silindir-Gunay, M.; Ozer, A.Y. Current and Future Approaches for Effective Cancer Imaging and Treatment. *Cancer Biother. Radiopharm.* **2018**, *33*, 39–51. [CrossRef] [PubMed]

3. Nobili, S.; Lippi, D.; Witort, E.; Donnini, M.; Bausi, L.; Mini, E.; Capaccioli, S. Natural compounds for cancer treatment and prevention. *Pharmacol. Res.* **2009**, *59*, 365–378. [CrossRef] [PubMed]

4. Cheng, H. *Advanced Textbook on Traditional Chinese Medicine and Pharmacology*; New World Press: Beijing, China, 1995.

5. Shehab, N.G.; Mahdy, A.; Khan, S.A.; Noureddin, S.M. Chemical constituents and biological activities of *Fagonia indica* Burm F. *Res. J. Med. Plant.* **2011**, *5*, 531–546. [CrossRef]

6. Faried, A.; Kurnia, D.; Faried, L.; Usman, N.; Miyazaki, T.; Kato, H.; Kuwano, H. Anticancer effects of gallic acid isolated from Indonesian herbal medicine, *Phaleria macrocarpa* (Scheff.) Boerl, on human cancer cell lines. *Int. J. Oncol.* **2007**, *30*, 605–613. [CrossRef]

7. Sohi, K.K.; Mittal, N.; Hundal, M.K.; Khanduja, K.L. Gallic acid, an antioxidant, exhibits antiapoptotic potential in normal human lymphocytes: A Bcl-2 independent mechanism. *J. Nutr. Sci. Vitaminol. (Tokyo)* **2003**, *49*, 221–227. [CrossRef]

8. Inoue, M.; Suzuki, R.; Koide, T.; Sakaguchi, N.; Ogihara, Y.; Yabu, Y. Antioxidant, gallic acid, induces apoptosis in HL-60RG cells. *Biochem. Biophys. Res. Commun.* **1994**, *204*, 898–904. [CrossRef]

9. Sun, J.; Chu, Y.-F.; Wu, X.; Liu, R.H. Antioxidant and antiproliferative activities of common fruits. *J. Agric. Food Chem.* **2002**, *50*, 7449–7454. [CrossRef]

10. Taraphdar, A.K.; Roy, M.; Bhattacharya, R. Natural products as inducers of apoptosis: Implication for cancer therapy and prevention. *Curr. Sci.* **2001**, 1387–1396.

11. Pal, S.K.; Shukla, Y. Herbal medicine: Current status and the future. *Asian Pac. J. Cancer Prev.* **2003**, *4*, 281–288.

12. Koehn, F.E.; Carter, G.T. The evolving role of natural products in drug discovery. *Nat. Rev. Drug Discov.* **2005**, *4*, 206–220. [CrossRef] [PubMed]

13. Saklani, A.; Kutty, S.K. Plant-derived compounds in clinical trials. *Drug Discov. Today* **2008**, *13*, 161–171. [CrossRef] [PubMed]

14. Wang, C.-Z.; Calway, T.; Yuan, C.-S. Herbal medicines as adjuvants for cancer therapeutics. *Am. J. Chin. Med.* **2012**, *40*, 657–669. [CrossRef] [PubMed]

15. Fattovich, G.; Stroffolini, T.; Zagni, I.; Donato, F. Hepatocellular carcinoma in cirrhosis: Incidence and risk factors. *Gastroenterology* **2004**, *127*, S35–S50. [CrossRef]

16. Llovet, J.M. Updated treatment approach to hepatocellular carcinoma. *J. Gastroenterol.* **2005**, *40*, 225–235. [CrossRef]

17. Ruan, W.-J.; Lai, M.-D.; Zhou, J.-G. Anticancer effects of Chinese herbal medicine, science or myth? *J. Zhejiang Univ. Sci. B* **2006**, *7*, 1006–1014. [CrossRef]

18. Trease, G.; Evans, W. *Textbook of Pharmacognosy*; Balliere: London, UK, 1983; pp. 57–59.

19. Ghorani-Azam, A.; Sepahi, S.; Riahi-Zanjani, B.; Ghamsari, A.A.; Mohajeri, S.A.; Balali-Mood, M. Plant toxins and acute medicinal plant poisoning in children: A systematic literature review. *J. Res. Med. Sci.* **2018**, *23*, 26.

20. Abad, M.J.; Bedoya, L.M.; Bermejo, P. Essential Oils from the Asteraceae Family Active against Multidrug-Resistant Bacteria. In *Fighting Multidrug Resistance with Herbal Extracts, Essential Oils and Their Components*; Rai, M.K., Kon, K.V., Eds.; Academic Press: San Diego, CA, USA, 2013; pp. 205–221.

21. Tan, R.X.; Zheng, W.; Tang, H. Biologically active substances from the genus *Artemisia*. *Planta Med.* **1998**, *64*, 295–302. [CrossRef]

22. Lu, H.; Zou, W.X.; Meng, J.C.; Hu, J.; Tan, R.X. New bioactive metabolites produced by *Colletotrichum* sp., an endophytic fungus in *Artemisia annua*. *Plant. Sci.* **2000**, *151*, 67–73. [CrossRef]

23. Efferth, T. Mechanistic perspectives for 1, 2, 4-trioxanes in anti-cancer therapy. *Drug Resist. Updat.* **2005**, *8*, 85–97. [CrossRef]

24. Ryu, J.-H.; Lee, S.-J.; Kim, M.-J.; Shin, J.-H.; Kang, S.-K.; Cho, K.-M.; Sung, N.-J. Antioxidant and anticancer activities of *Artemisia annua* L. and determination of functional compounds. *J. Korean Soc. Food Sci. Nutr.* **2011**, *40*, 509–516. [CrossRef]

25. Xie, W.; Gu, D.; Li, J.; Cui, K.; Zhang, Y. Effects and action mechanisms of berberine and *Rhizoma coptidis* on gut microbes and obesity in high-fat diet-fed C57BL/6J mice. *PLoS ONE* **2011**, *6*, e24520. [CrossRef] [PubMed]

26. Yin, J.; Zhang, H.; Ye, J. Traditional Chinese medicine in treatment of metabolic syndrome. *Endocr. Metab. Immune Disord. Drug Targets* **2008**, *8*, 99–111. [CrossRef] [PubMed]

27. Hu, Y.; Davies, G.E. Berberine inhibits adipogenesis in high-fat diet-induced obesity mice. *Fitoterapia* **2010**, *81*, 358–366. [CrossRef] [PubMed]

28. Tang, J.; Feng, Y.; Tsao, S.; Wang, N.; Curtain, R.; Wang, Y. Berberine and *Coptidis rhizoma* as novel antineoplastic agents: A review of traditional use and biomedical investigations. *J. Ethnopharmacol.* **2009**, *126*, 5–17. [CrossRef]

29. Ammon, H.P.; Wahl, M.A. Pharmacology of *Curcuma longa*. *Planta Med.* **1991**, *57*, 1–7. [CrossRef]

30. Liu, F.; Ng, T. Antioxidative and free radical scavenging activities of selected medicinal herbs. *Life Sci.* **2000**, *66*, 725–735. [CrossRef]

31. Schinella, G.; Tournier, H.; Prieto, J.; De Buschiazzo, P.M.; Ríos, J. Antioxidant activity of anti-inflammatory plant extracts. *Life Sci.* **2002**, *70*, 1023–1033. [CrossRef]

32. Goel, A.; Boland, C.R.; Chauhan, D.P. Specific inhibition of cyclooxygenase-2 (COX-2) expression by dietary curcumin in HT-29 human colon cancer cells. *Cancer Lett.* **2001**, *172*, 111–118. [CrossRef]

33. Wang, X.; Hang, Y.; Liu, J.; Hou, Y.; Wang, N.; Wang, M. Anticancer effect of curcumin inhibits cell growth through miR-21/PTEN/Akt pathway in breast cancer cell. *Oncol. Lett.* **2017**, *13*, 4825–4831. [CrossRef]

34. Beier, B.A. A revision of the desert shrub *Fagonia* (Zygophyllaceae). *Syst. Biodivers.* **2005**, *3*, 221–263. [CrossRef]

35. Akhtar, N.; Begum, S. Ethnopharmacological important plants of Jalala, district Mardan, Pakistan. *Pak. J. Pl. Sci.* **2009**, *15*, 95–100.

36. Sharrma, S.; Gupta, V.; Sharma, G. Phytopharmacology of *Fagonia indica* (L): A review. *J. Nat. Cons.* **2010**, *1*, 143–147.

37. Ibrahim, L.F.; Kawashty, S.A.; El-Hagrassy, A.M.; Nassar, M.I.; Mabry, T.J. A new kaempferol triglycoside from *Fagonia taeckholmiana*: Cytotoxic activity of its extracts. *Carbohydr. Res.* **2008**, *343*, 155–158. [CrossRef] [PubMed]

38. Sharawy, S.M.; Alshammari, A.M. Checklist of poisonous plants and animals in Aja Mountain, Ha'il Region, Saudi Arabia. *Aust. J. Basic Appl. Sci.* **2009**, *3*, 2217–2225.

39. Shaker, K.H.; Bernhardt, M.; Elgamal, M.H.A.; Seifert, K. Triterpenoid saponins from *Fagonia indica*. *Phytochemistry* **1999**, *51*, 1049–1053. [CrossRef]

40. Perrone, A.; Masullo, M.; Bassarello, C.; Hamed, A.I.; Belisario, M.A.; Pizza, C.; Piacente, S. Sulfated triterpene derivatives from *Fagonia arabica*. *J. Nat. Prod.* **2007**, *70*, 584–588. [CrossRef]

41. Bagban, I.; Roy, S.; Chaudhary, A.; Das, S.; Gohil, K.; Bhandari, K. Hepatoprotective activity of the methanolic extract of *Fagonia indica* Burm in carbon tetra chloride induced hepatotoxicity in albino rats. *Asian Pac. J. Trop. Biomed.* **2012**, *2*, S1457–S1460. [CrossRef]

42. Eman, A.A. Morphological, phytochemical and biological screening on three Egyptian species of *Fagonia*. *Acad Arena* **2011**, *3*, 18–27.

43. Waheed, A.; Barker, J.; Barton, S.J.; Owen, C.P.; Ahmed, S.; Carew, M.A. A novel steroidal saponin glycoside from *Fagonia indica* induces cell-selective apoptosis or necrosis in cancer cells. *Eur. J. Pharm. Sci.* **2012**, *47*, 464–473. [CrossRef]

44. Lam, M.; Carmichael, A.R.; Griffiths, H.R. An aqueous extract of *Fagonia cretica* induces DNA damage, cell cycle arrest and apoptosis in breast cancer cells via FOXO3a and p53 expression. *PLoS ONE* **2012**, *7*, e40152. [CrossRef] [PubMed]

45. Wu, S.-B.; Long, C.; Kennelly, E.J. Structural diversity and bioactivities of natural benzophenones. *Nat. Prod. Rep.* **2014**, *31*, 1158–1174. [CrossRef] [PubMed]

46. Hemshekhar, M.; Sunitha, K.; Santhosh, M.S.; Devaraja, S.; Kemparaju, K.; Vishwanath, B.; Niranjana, S.; Girish, K. An overview on genus *Garcinia*: Phytochemical and therapeutical aspects. *Phytochem. Rev.* **2011**, *10*, 325–351. [CrossRef]

47. Huang, S.-X.; Feng, C.; Zhou, Y.; Xu, G.; Han, Q.-B.; Qiao, C.-F.; Chang, D.C.; Luo, K.Q.; Xu, H.-X. Bioassay-guided isolation of xanthones and polycyclic prenylated acylphloroglucinols from *Garcinia oblongifolia*. *J. Nat. Prod.* **2008**, *72*, 130–135. [CrossRef] [PubMed]

48. Feng, C.; Huang, S.-X.; Gao, X.-M.; Xu, H.-X.; Luo, K.Q. Characterization of Proapoptotic Compounds from the Bark of *Garcinia oblongifolia*. *J. Nat. Prod.* **2014**, *77*, 1111–1116. [CrossRef] [PubMed]

49. Li, P.; AnandhiSenthilkumar, H.; Wu, S.-B.; Liu, B.; Guo, Z.-Y.; Fata, J.E.; Kennelly, E.J.; Long, C.-L. Comparative UPLC-QTOF-MS-based metabolomics and bioactivities analyses of *Garcinia oblongifolia*. *J. Chromatogr. B* **2016**, *1011*, 179–195. [CrossRef]

50. Hong, J.; Kwon, S.J.; Sang, S.; Ju, J.; Zhou, J.-N.; Ho, C.-T.; Huang, M.-T.; Yang, C.S. Effects of garcinol and its derivatives on intestinal cell growth: Inhibitory effects and autoxidation-dependent growth-stimulatory effects. *Free Radic. Biol. Med.* **2007**, *42*, 1211–1221. [CrossRef]

51. Liao, C.H.; Sang, S.; Ho, C.T.; Lin, J.K. Garcinol modulates tyrosine phosphorylation of FAK and subsequently induces apoptosis through down-regulation of Src, ERK, and Akt survival signaling in human colon cancer cells. *J. Cell. Biochem.* **2005**, *96*, 155–169. [CrossRef]

52. Pan, M.-H.; Chang, W.-L.; Lin-Shiau, S.-Y.; Ho, C.-T.; Lin, J.-K. Induction of apoptosis by garcinol and curcumin through cytochrome c release and activation of caspases in human leukemia HL-60 cells. *J. Agric. Food Chem.* **2001**, *49*, 1464–1474. [CrossRef]

53. Lin, C.-C.; Ng, L.-T.; Yang, J.-J.; Hsu, Y.-F. Anti-inflammatory and hepatoprotective activity of peh-hue-juwa-chi-cao in male rats. *Am. J. Chin. Med.* **2002**, *30*, 225–234. [CrossRef]

54. Ahmad, R.; Ali, A.M.; Israf, D.A.; Ismail, N.H.; Shaari, K.; Lajis, N.H. Antioxidant, radical-scavenging, anti-inflammatory, cytotoxic and antibacterial activities of methanolic extracts of some *Hedyotis* species. *Life Sci.* **2005**, *76*, 1953–1964. [CrossRef] [PubMed]

55. Fang, Y.; Zhang, Y.; Chen, M.; Zheng, H.; Zhang, K. The active component of *Hedyotis diffusa* Willd. *Chin. Tradit. Plant. Med.* **2004**, *26*, 577–579.

56. Ahmad, R.; Shaari, K.; Lajis, N.H.; Hamzah, A.S.; Ismail, N.H.; Kitajima, M. Anthraquinones from *Hedyotis capitellata*. *Phytochemistry* **2005**, *66*, 1141–1147. [CrossRef] [PubMed]

57. Li, C.; Xue, X.; Zhou, D.; Zhang, F.; Xu, Q.; Ren, L.; Liang, X. Analysis of iridoid glucosides in *Hedyotis diffusa* by high-performance liquid chromatography/electrospray ionization tandem mass spectrometry. *J. Pharm. Biomed. Anal.* **2008**, *48*, 205–211. [CrossRef] [PubMed]

58. Liu, Z.; Liu, M.; Liu, M.; Li, J. Methylanthraquinone from *Hedyotis diffusa* WILLD induces Ca^{2+}-mediated apoptosis in human breast cancer cells. *Toxicol. Vitr.* **2010**, *24*, 142–147. [CrossRef]

59. Wu, Z.; Jin, C.; Li, J.; Chen, X.; Yao, Q.; Zhu, Q. Inhibition of Colon Cancer Cells by Ethanol Extract of *Oldenlandia diffusa*. *J. Kunming Med. Univ.* **2013**, *10*, 31–34.

60. Vidal-Russell, R.; Nickrent, D.L. Evolutionary relationships in the showy mistletoe family (Loranthaceae). *Am. J. Bot.* **2008**, *95*, 1015–1029. [CrossRef]

61. Fukunaga, T.; Kajikawa, I.; Nishiya, K.; Takeya, K.; Itokawa, H. Studies on the constituents of the *Japanese mistletoe, Viscum album* L. var. coloratum Ohwi grown on different host trees. *Chem. Pharm. Bull.* **1989**, *37*, 1300–1303. [CrossRef]

62. Osadebe, P.; Okide, G.; Akabogu, I. Study on anti-diabetic activities of crude methanolic extracts of *Loranthus micranthus* (Linn.) sourced from five different host trees. *J. Ethnopharmacol.* **2004**, *95*, 133–138. [CrossRef]

63. Qin, Y.; Mo, T.; Liu, X.; Pan, Z.; Ye, M. Effects of Loranthus parasiticus on Nutrition Metabolism of Osmanthus fragrans and Cinnamomum burmannii. *Acta Agric. Boreali-Occident. Sin.* **2010**, *4*, 34.

64. Powell, C.B.; Fung, P.; Jackson, J.; Dall'Era, J.; Lewkowicz, D.; Cohen, I.; Smith-McCune, K. Aqueous extract of herba *Scutellaria barbatae*, a Chinese herb used for ovarian cancer, induces apoptosis of ovarian cancer cell lines. *Gynecol. Oncol.* **2003**, *91*, 332–340. [CrossRef] [PubMed]

65. Xiao, Y.-J.; Chen, Y.-Z.; Chen, B.-H.; Chen, J.-H.; Lin, Z.-X.; Fan, Y.-L. Study on cytotoxic activities on human leukemia cell line HL-60 by flavonoids extracts of *Scurrula parasitica* from four different host trees. *China J. Chin. Mater. Med.* **2008**, *33*, 427–432.

66. Xiao, Y.; Fan, Y.; Chen, B.; Zhang, Q.; Zeng, H. Polysaccharides from *Scurrula parasitica* L. inhibit sarcoma S180 growth in mice. *China J. Chin. Mater. Med.* **2010**, *35*, 381–384.

67. Devi, B.; Sharma, N.; Kumar, D.; Jeet, K. *Morus alba* Linn: A phytopharmacological review. *Int. J. Pharm. Pharm. Sci.* **2013**, *5*, 14–18.

68. Chon, S.-U.; Kim, Y.-M.; Park, Y.-J.; Heo, B.-G.; Park, Y.-S.; Gorinstein, S. Antioxidant and antiproliferative effects of methanol extracts from raw and fermented parts of mulberry plant (*Morus alba* L.). *Eur. Food Res. Technol.* **2009**, *230*, 231–237. [CrossRef]

69. Kikuchi, T.; Nihei, M.; Nagai, H.; Fukushi, H.; Tabata, K.; Suzuki, T.; Akihisa, T. Albanol A from the root bark of *Morus alba* L. induces apoptotic cell death in HL60 human leukemia cell line. *Chem. Pharm. Bull.* **2010**, *58*, 568–571. [CrossRef]

70. Naowaratwattana, W.; De-Eknamkul, W.; De Mejia, E.G. Phenolic-containing organic extracts of mulberry (*Morus alba* L.) leaves inhibit HepG2 hepatoma cells through G2/M phase arrest, induction of apoptosis, and inhibition of topoisomerase IIα activity. *J. Med. Food* **2010**, *13*, 1045–1056. [CrossRef]

71. Deepa, M.; Priya, S. Purification and characterization of a novel anti-proliferative lectin from *Morus alba* L. leaves. *Protein Pept. Lett.* **2012**, *19*, 839–845. [CrossRef]

72. Fathy, S.A.; Singab, A.N.B.; Agwa, S.A.; El Hamid, D.M.A.; Zahra, F.A.; El Moneim, S.M.A. The antiproliferative effect of mulberry (*Morus alba* L.) plant on hepatocarcinoma cell line HepG2. *Egypt. J. Med. Hum. Genet.* **2013**, *14*, 375–382. [CrossRef]

73. Qin, J.; Fan, M.; He, J.; Wu, X.-D.; Peng, L.-Y.; Su, J.; Cheng, X.; Li, Y.; Kong, L.-M.; Li, R.-T. New cytotoxic and anti-inflammatory compounds isolated from *Morus alba* L. *Nat. Prod. Res.* **2015**, *29*, 1711–1718. [CrossRef]

74. Huang, X.; Gao, W.; Man, S.; Zhao, Z. Advances in studies on saponins in plants of Paris L. and their biosynthetic approach. *Chin. Tradit. Herb. Drugs* **1994**.

75. Negi, J.S.; Bisht, V.K.; Bhandari, A.K.; Bhatt, V.P.; Singh, P.; Singh, N. *Paris polyphylla*: Chemical and biological prospectives. *Anti-Cancer Agents Med. Chem.* **2014**, *14*, 833–839. [CrossRef] [PubMed]

76. Zhang, X.-F.; Cui, Y.; Huang, J.-J.; Zhang, Y.-Z.; Nie, Z.; Wang, L.-F.; Yan, B.-Z.; Tang, Y.-L.; Liu, Y. Immuno-stimulating properties of diosgenyl saponins isolated from *Paris polyphylla*. *Bioorg. Med. Chem. Lett.* **2007**, *17*, 2408–2413. [CrossRef] [PubMed]

77. Zhang, T.; Liu, H.; Liu, X.-T.; Chen, X.-Q.; Wang, Q. Steroidal saponins from the rhizomes of *Paris delavayi*. *Steroids* **2009**, *74*, 809–813. [CrossRef] [PubMed]

78. Deng, D.; Lauren, D.R.; Cooney, J.M.; Jensen, D.J.; Wurms, K.V.; Upritchard, J.E.; Cannon, R.D.; Wang, M.Z.; Li, M.Z. Antifungal saponins from *Paris polyphylla* Smith. *Planta Med.* **2008**, *74*, 1397–1402. [CrossRef] [PubMed]

79. Zhao, Y.; Kang, L.-P.; Liu, Y.-X.; Liang, Y.-G.; Tan, D.-W.; Yu, Z.-Y.; Cong, Y.-W.; Ma, B.-P. Steroidal saponins from the rhizome of *Paris polyphylla* and their cytotoxic activities. *Planta Med.* **2009**, *75*, 356. [CrossRef]

80. Fu, Y.L.; Yu, Z.Y.; Tang, X.M.; Zhao, Y.; Yuan, X.L.; Wang, S.; Ma, B.P.; Cong, Y.W. Pennogenin glycosides with a spirostanol structure are strong platelet agonists: Structural requirement for activity and mode of platelet agonist synergism. *J. Thromb. Haemost.* **2008**, *6*, 524–533. [CrossRef]

81. Guo, L.; Su, J.; Deng, B.; Yu, Z.; Kang, L.; Zhao, Z.; Shan, Y.; Chen, J.; Ma, B.; Cong, Y. Active pharmaceutical ingredients and mechanisms underlying phasic myometrial contractions stimulated with the saponin extract from *Paris polyphylla* Sm. var. yunnanensis used for abnormal uterine bleeding. *Hum. Reprod.* **2008**, *23*, 964–971. [CrossRef]

82. Sun, J.; Liu, B.R.; Hu, W.J.; Yu, L.X.; Qian, X.P. In vitro anticancer activity of aqueous extracts and ethanol extracts of fifteen taditional Chinese medicines on human digestive tumor cell lines. *Phytother. Res.* **2007**, *21*, 1102–1104. [CrossRef]

83. Li, F.-R.; Jiao, P.; Yao, S.-T.; Sang, H.; Qin, S.-C.; Zhang, W.; Zhang, Y.-B.; Gao, L.-L. *Paris polyphylla* Smith extract induces apoptosis and activates cancer suppressor gene connexin26 expression. *Asian Pac. J. Cancer Prev.* **2012**, *13*, 205–209. [CrossRef]

84. Heci, Y. Valuable ingredients from herb perilla: A mini review. *Innov. Food Technol.* **2001**, *29*, 32–33.

85. Asif, M.; Kumar, A. Nutritional and functional characterization of *Perilla frutescens* seed oil and evaluation of its effect on gastrointestinal motility. *Malay. J. Pharm. Sci.* **2010**, *8*, 1–12.

86. Lee, J.K.; Ohnishi, O. Geographic differentiation of morphological characters among *Perilla* crops and their weedy types in East Asia. *Breed. Sci.* **2001**, *51*, 247–255. [CrossRef]

87. Lee, J.K.; Ohnishi, O. Genetic relationships among cultivated types of *Perilla frutescens* and their weedy types in East Asia revealed by AFLP markers. *Genet. Resour. Crop. Evol.* **2003**, *50*, 65–74. [CrossRef]

88. Mao, Q.-Q.; Huang, Z.; Zhong, X.-M.; Feng, C.-R.; Pan, A.-J.; Li, Z.-Y.; Ip, S.-P.; Che, C.-T. Effects of SYJN, a Chinese herbal formula, on chronic unpredictable stress-induced changes in behavior and brain BDNF in rats. *J. Ethnopharmacol.* **2010**, *128*, 336–341. [CrossRef]

89. Lin, C.-S.; Kuo, C.-L.; Wang, J.-P.; Cheng, J.-S.; Huang, Z.-W.; Chen, C.-F. Growth inhibitory and apoptosis inducing effect of *Perilla frutescens* extract on human hepatoma HepG2 cells. *J. Ethnopharmacol.* **2007**, *112*, 557–567. [CrossRef]

90. Osakabe, N.; Yasuda, A.; Natsume, M.; Yoshikawa, T. Rosmarinic acid inhibits epidermal inflammatory responses: Anticarcinogenic effect of *Perilla frutescens* extract in the murine two-stage skin model. *Carcinogenesis* **2004**, *25*, 549–557. [CrossRef]

91. Kwak, C.S.; Yeo, E.J.; Moon, S.C.; Kim, Y.W.; Ahn, H.J.; Park, S.C. *Perilla* leaf, *Perilla frutescens*, induces apoptosis and G1 phase arrest in human leukemia HL-60 cells through the combinations of death receptor-mediated, mitochondrial, and endoplasmic reticulum stress-induced pathways. *J. Med. Food* **2009**, *12*, 508–517. [CrossRef]

92. Cho, B.O.; Jin, C.H.; Park, Y.D.; Ryu, H.W.; Byun, M.W.; Seo, K.I.; Jeong, I.Y. Isoegomaketone induces apoptosis through caspase-dependent and caspase-independent pathways in human DLD1 cells. *Biosci. Biotechnol. Biochem.* **2011**, *75*, 1306–1311. [CrossRef]

93. Zhang, L.; Wang, Y.; Yang, D.; Zhang, C.; Zhang, N.; Li, M.; Liu, Y. *Platycodon grandiflorus*—An ethnopharmacological, phytochemical and pharmacological review. *J. Ethnopharmacol.* **2015**, *164*, 147–161. [CrossRef]

94. Han, S.B.; Park, S.H.; Lee, K.H.; Lee, C.W.; Lee, S.H.; Kim, H.C.; Kim, Y.S.; Lee, H.S.; Kim, H.M. Polysaccharide isolated from the radix of *Platycodon grandiflorum* selectively activates B cells and macrophages but not T cells. *Int. Immunopharmacol.* **2001**, *1*, 1969–1978. [CrossRef]

95. Xie, Y.; Pan, H.; Sun, H.; Li, D. A promising balanced Th1 and Th2 directing immunological adjuvant, saponins from the root of *Platycodon grandiflorum*. *Vaccine* **2008**, *26*, 3937–3945. [CrossRef] [PubMed]

96. Khanal, T.; Choi, J.H.; Hwang, Y.P.; Chung, Y.C.; Jeong, H.G. Saponins isolated from the root of *Platycodon grandiflorum* protect against acute ethanol-induced hepatotoxicity in mice. *Food Chem. Toxicol.* **2009**, *47*, 530–535. [CrossRef] [PubMed]

97. Lee, K.J.; Hwang, S.J.; Choi, J.H.; Jeong, H.G. Saponins derived from the roots of *Platycodon grandiflorum* inhibit HT-1080 cell invasion and MMPs activities: Regulation of NF-κB activation via ROS signal pathway. *Cancer Lett.* **2008**, *268*, 233–243. [CrossRef] [PubMed]

98. Yu, J.S.; Kim, A.K. Platycodin D induces apoptosis in MCF-7 human breast cancer cells. *J. Med. Food* **2010**, *13*, 298–305. [CrossRef]

99. Kim, M.-O.; Moon, D.-O.; Choi, Y.H.; Shin, D.Y.; Kang, H.S.; Choi, B.T.; Lee, J.-D.; Li, W.; Kim, G.-Y. Platycodin D induces apoptosis and decreases telomerase activity in human leukemia cells. *Cancer Lett.* **2008**, *261*, 98–107. [CrossRef]

100. Shin, D.Y.; Kim, G.Y.; Li, W.; Choi, B.T.; Kim, N.D.; Kang, H.S.; Choi, Y.H. Implication of intracellular ROS formation, caspase-3 activation and Egr-1 induction in platycodon D-induced apoptosis of U937 human leukemia cells. *Biomed. Pharmacother.* **2009**, *63*, 86–94. [CrossRef]

101. Hu, Q.; Pan, R.; Wang, L.; Peng, B.; Tang, J.; Liu, X. *Platycodon grandiflorum* induces apoptosis in SKOV3 human ovarian cancer cells through mitochondrial-dependent pathway. *Am. J. Chin. Med.* **2010**, *38*, 373–386. [CrossRef]

102. Yiğit, D.; Yiğit, N.; Mavi, A. Antioxidant and antimicrobial activities of bitter and sweet apricot (*Prunus armeniaca* L.) kernels. *Braz. J. Med. Biol. Res.* **2009**, *42*, 346–352. [CrossRef]

103. Hacıseferoğulları, H.; Gezer, I.; Özcan, M.M.; MuratAsma, B. Post-harvest chemical and physical–mechanical properties of some apricot varieties cultivated in Turkey. *J. Food Eng.* **2007**, *79*, 364–373. [CrossRef]

104. Yan, J.; Tong, S.; Li, J.; Lou, J. Preparative isolation and purification of amygdalin from *Prunus armeniaca* L. with high recovery by high-speed countercurrent chromatography. *J. Liq. Chromatogr. Relat. Technol.* **2006**, *29*, 1271–1279. [CrossRef]

105. Akcicek, E.; Otles, S.; Esiyok, D. Cancer and its prevention by some horticultural and field crops in Turkey. *Asian Pac. J. Cancer Prev* **2005**, *6*, 224–230. [PubMed]

106. Gomaa, E.Z. In vitro antioxidant, antimicrobial, and antitumor activities of bitter almond and sweet apricot (*Prunus armeniaca* L.) kernels. *Food Sci. Biotechnol.* **2013**, *22*, 455–463. [CrossRef]

107. Madrau, M.A.; Piscopo, A.; Sanguinetti, A.M.; Del Caro, A.; Poiana, M.; Romeo, F.V.; Piga, A. Effect of drying temperature on polyphenolic content and antioxidant activity of apricots. *Eur. Food Res. Technol.* **2009**, *228*, 441. [CrossRef]

108. Liu, H.; Chen, F.; Yang, H.; Yao, Y.; Gong, X.; Xin, Y.; Ding, C. Effect of calcium treatment on nanostructure of chelate-soluble pectin and physicochemical and textural properties of apricot fruits. *Food Res. Int.* **2009**, *42*, 1131–1140. [CrossRef]

109. Vardi, N.; Parlakpinar, H.; Ozturk, F.; Ates, B.; Gul, M.; Cetin, A.; Erdogan, A.; Otlu, A. Potent protective effect of apricot and β-carotene on methotrexate-induced intestinal oxidative damage in rats. *Food Chem. Toxicol.* **2008**, *46*, 3015–3022. [CrossRef]

110. Sun, H.-D.; Huang, S.-X.; Han, Q.-B. Diterpenoids from Isodon species and their biological activities. *Nat. Prod. Rep.* **2006**, *23*, 673–698. [CrossRef]

111. Liu, H.-M.; Yan, X.; Kiuchi, F.; Liu, Z. A new diterpene glycoside from *Rabdosia rubescens*. *Chem. Pharm. Bull.* **2000**, *48*, 148–149. [CrossRef]

112. Zhang, Y.; Liang, Y.; He, C. Anticancer activities and mechanisms of heat-clearing and detoxicating traditional Chinese herbal medicine. *Chin. Med.* **2017**, *12*, 20. [CrossRef]

113. Bao, R.; Shu, Y.; Wu, X.; Weng, H.; Ding, Q.; Cao, Y.; Li, M.; Mu, J.; Wu, W.; Ding, Q. Oridonin induces apoptosis and cell cycle arrest of gallbladder cancer cells via the mitochondrial pathway. *BMC Cancer* **2014**, *14*, 217. [CrossRef]

114. Wang, H.; Ye, Y.; Pan, S.-Y.; Zhu, G.-Y.; Li, Y.-W.; Fong, D.W.; Yu, Z.-L. Proteomic identification of proteins involved in the anticancer activities of oridonin in HepG2 cells. *Phytomedicine* **2011**, *18*, 163–169. [CrossRef] [PubMed]

115. Wang, S.; Zhong, Z.; Wan, J.; Tan, W.; Wu, G.; Chen, M.; Wang, Y. Oridonin induces apoptosis, inhibits migration and invasion on highly-metastatic human breast cancer cells. *Am. J. Chin. Med.* **2013**, *41*, 177–196. [CrossRef] [PubMed]

116. Shang, X.; He, X.; He, X.; Li, M.; Zhang, R.; Fan, P.; Zhang, Q.; Jia, Z. The genus *Scutellaria* an ethnopharmacological and phytochemical review. *J. Ethnopharmacol.* **2010**, *128*, 279–313. [CrossRef] [PubMed]

117. Gasiorowski, K.; Lamer-Zarawska, E.; Leszek, J.; Parvathaneni, K.; Yendluri, B.B.; Błach-Olszewska, Z.; Aliev, G. Flavones from root of *Scutellaria baicalensis* Georgi: Drugs of the future in neurodegeneration? *CNS Neurol. Disord. Drug Targets* **2011**, *10*, 184–191. [CrossRef]

118. Matkowski, A.; Jamiolkowska-Kozlowska, W.; Nawrot, I. Chinese Medicinal Herbs as Source of Antioxidant Compounds—Where Tradition Meets the Future. *Curr. Med. Chem.* **2013**, *20*, 984–1004. [PubMed]

119. Wozniak, D.; Lamer-Zarawska, E.; Matkowski, A. Antimutagenic and antiradical properties of flavones from the roots of *Scutellaria baicalensis* Georgi. *Food/Nahrung* **2004**, *48*, 9–12. [CrossRef]

120. Woźniak, D.; Dryś, A.; Matkowski, A. Antiradical and antioxidant activity of flavones from *Scutellariae baicalensis* radix. *Nat. Prod. Res.* **2015**, *29*, 1567–1570. [CrossRef]

121. Kumagai, T.; Müller, C.I.; Desmond, J.C.; Imai, Y.; Heber, D.; Koeffler, H.P. *Scutellaria baicalensis*, a herbal medicine: Anti-proliferative and apoptotic activity against acute lymphocytic leukemia, lymphoma and myeloma cell lines. *Leuk. Res.* **2007**, *31*, 523–530. [CrossRef]

122. Scheck, A.C.; Perry, K.; Hank, N.C.; Clark, W.D. Anticancer activity of extracts derived from the mature roots of *Scutellaria baicalensis* on human malignant brain tumor cells. *BMC Complement. Altern. Med.* **2006**, *6*, 27. [CrossRef]

123. Ye, F.; Jiang, S.; Volshonok, H.; Wu, J.; Zhang, D.Y. Molecular mechanism of anti-prostate cancer activity of *Scutellaria baicalensis* extract. *Nutr. Cancer* **2007**, *57*, 100–110. [CrossRef]

124. Zhang, D.Y.; Wu, J.; Ye, F.; Xue, L.; Jiang, S.; Yi, J.; Zhang, W.; Wei, H.; Sung, M.; Wang, W. Inhibition of cancer cell proliferation and prostaglandin E2 synthesis by *Scutellaria baicalensis*. *Cancer Res.* **2003**, *63*, 4037–4043. [PubMed]

125. Wang, H.; Li, H.; Luo, J.; Gu, J.; Wang, H.; Wu, H.; Xu, Y. Baicalin extracted from Huang qin (*Radix Scutellariae Baicalensis*) induces apoptosis in gastric cancer cells by regulating B cell lymphoma (Bcl-2)/Bcl-2-associated X protein and activating caspase-3 and caspase-9. *J. Tradit. Chin. Med.* **2017**, *37*, 229–235. [PubMed]

126. Zhu, N.; Wang, S.; Lawless, J.; He, J.; Zheng, Z. Dose Dependent Dual Effect of Baicalin and Herb Huang Qin Extract on Angiogenesis. *PLoS ONE* **2016**, *11*, e0167125. [CrossRef] [PubMed]

127. Sato, D.; Kondo, S.; Yazawa, K.; Mukudai, Y.; Li, C.; Kamatani, T.; Katsuta, H.; Yoshihama, Y.; Shirota, T.; Shintani, S. The potential anticancer activity of extracts derived from the roots of *Scutellaria baicalensis* on human oral squamous cell carcinoma cells. *Mol. Clin. Oncol.* **2013**, *1*, 105–111. [CrossRef]

128. Suh, S.J.; Yoon, J.W.; Lee, T.K.; Jin, U.H.; Kim, S.L.; Kim, M.S.; Kwon, D.Y.; Lee, Y.C.; Kim, C.H. Chemoprevention of *Scutellaria bardata* on human cancer cells and tumorigenesis in skin cancer. *Phytother. Res.* **2007**, *21*, 135–141. [CrossRef]

129. Qu, G.-W.; Yue, X.-D.; Li, G.-S.; Yu, Q.-Y.; Dai, S.-J. Two new cytotoxic ent-clerodane diterpenoids from *Scutellaria barbata*. *J. Asian Nat. Prod. Res.* **2010**, *12*, 859–864. [CrossRef]

130. Dai, Z.-J.; Gao, J.; Li, Z.-F.; Ji, Z.-Z.; Kang, H.-F.; Guan, H.-T.; Diao, Y.; Wang, B.-F.; Wang, X.-J. In Vitro and In Vivo Antitumor Activity of Scutellaria barbate Extract on Murine Liver Cancer. *Molecules* **2011**, *16*, 4389–4400. [CrossRef]

131. Chen, V.; Staub, R.E.; Baggett, S.; Chimmani, R.; Tagliaferri, M.; Cohen, I.; Shtivelman, E. Identification and Analysis of the Active Phytochemicals from the Anti-Cancer Botanical Extract Bezielle. *PLoS ONE* **2012**, *7*, e30107. [CrossRef]

132. Androutsopoulos, V.P.; Ruparelia, K.; Arroo, R.R.; Tsatsakis, A.M.; Spandidos, D.A. CYP1-mediated antiproliferative activity of dietary flavonoids in MDA-MB-468 breast cancer cells. *Toxicology* **2009**, *264*, 162–170. [CrossRef]

133. Qiu, D.; Kao, P.N. Immunosuppressive and anti-inflammatory mechanisms of triptolide, the principal active diterpenoid from the Chinese medicinal herb *Tripterygium wilfordii* Hook. f. *Drugs R D* **2003**, *4*, 1–18. [CrossRef]

134. Brinker, A.M.; Ma, J.; Lipsky, P.E.; Raskin, I.J.P. Medicinal chemistry and pharmacology of genus *Tripterygium* (Celastraceae). *Phytochemistry* **2007**, *68*, 732–766. [CrossRef] [PubMed]

135. Ziaei, S.; Halaby, R. Immunosuppressive, anti-inflammatory and anti-cancer properties of triptolide: A mini review. *Avicenna J. Phytomed.* **2016**, *6*, 149. [PubMed]

136. Sarkar, S.; Paul, S. Triptolide Mediated Amelioration of Breast Cancer via Modulation of Molecular Pathways. *Pharmacogn. J.* **2017**, *9*, 838–845. [CrossRef]

137. Kiviharju, T.M.; Lecane, P.S.; Sellers, R.G.; Peehl, D.M. Antiproliferative and proapoptotic activities of triptolide (PG490), a natural product entering clinical trials, on primary cultures of human prostatic epithelial cells. *Clin. Cancer Res.* **2002**, *8*, 2666–2674.

138. Chang, W.-T.; Kang, J.J.; Lee, K.-Y.; Wei, K.; Anderson, E.; Gotmare, S.; Ross, J.A.; Rosen, G.D. Triptolide and chemotherapy cooperate in tumor cell apoptosis A role for the p53 pathway. *J. Biol. Chem.* **2001**, *276*, 2221–2227. [CrossRef]

139. Lou, Y.J.; Jin, J. Triptolide down-regulates bcr-abl expression and induces apoptosis in chronic myelogenous leukemia cells. *Leuk. Lymphoma* **2004**, *45*, 373–376. [CrossRef]

140. He, M.-F.; Huang, Y.-H.; Wu, L.-W.; Ge, W.; Shaw, P.-C.; But, P.P.-H. Triptolide functions as a potent angiogenesis inhibitor. *Int. J. Cancer* **2010**, *126*, 266–278. [CrossRef]

141. Yao, J.; Jiang, Z.; Duan, W.; Huang, J.; Zhang, L.; Hu, L.; He, L.; Li, F.; Xiao, Y.; Shu, B.; et al. Involvement of mitochondrial pathway in triptolide-induced cytotoxicity in human normal liver L-02 cells. *Boil. Pharm. Bull.* **2008**, *31*, 592–597. [CrossRef]

142. Yang, H.; Chen, D.; Cui, Q.C.; Yuan, X.; Dou, Q.P. Celastrol, a triterpene extracted from the Chinese "Thunder of God Vine," is a potent proteasome inhibitor and suppresses human prostate cancer growth in nude mice. *Cancer Res.* **2006**, *66*, 4758–4765. [CrossRef]

143. Sethi, G.; Ahn, K.S.; Pandey, M.K.; Aggarwal, B.B. Celastrol, a novel triterpene, potentiates TNF-induced apoptosis and suppresses invasion of tumor cells by inhibiting NF-κB–regulated gene products and TAK1-mediated NF-κB activation. *Blood* **2007**, *109*, 2727–2735. [CrossRef]

144. Lee, J.-H.; Koo, T.H.; Yoon, H.; Jung, H.S.; Jin, H.Z.; Lee, K.; Hong, Y.-S.; Lee, J.J. Inhibition of NF-κB activation through targeting IκB kinase by celastrol, a quinone methide triterpenoid. *Biochem. Pharmacol.* **2006**, *72*, 1311–1321. [CrossRef] [PubMed]

145. Antonoff, M.B.; Chugh, R.; Borja-Cacho, D.; Dudeja, V.; Clawson, K.A.; Skube, S.J.; Sorenson, B.S.; Saltzman, D.A.; Vickers, S.M.; Saluja, A.K. Triptolide therapy for neuroblastoma decreases cell viability in vitro and inhibits tumor growth in vivo. *Surgery* **2009**, *146*, 282–290. [CrossRef] [PubMed]

146. Yang, S.; Chen, J.; Guo, Z.; Xu, X.-M.; Wang, L.; Pei, X.-F.; Yang, J.; Underhill, C.B.; Zhang, L. Triptolide inhibits the growth and metastasis of solid tumors1. *Mol. Cancer Ther.* **2003**, *2*, 65–72. [PubMed]

147. Ravipati, A.S.; Zhang, L.; Koyyalamudi, S.R.; Jeong, S.C.; Reddy, N.; Bartlett, J.; Smith, P.T.; Shanmugam, K.; Münch, G.; Wu, M.J.; et al. Antioxidant and anti-inflammatory activities of selected Chinese medicinal plants and their relation with antioxidant content. *BMC Complement. Altern. Med.* **2012**, *12*, 173. [CrossRef] [PubMed]

148. Kim, M.-R.; Lee, J.Y.; Lee, H.-H.; Aryal, D.K.; Kim, Y.G.; Kim, S.K.; Woo, E.-R.; Kang, K.W. Antioxidative effects of quercetin-glycosides isolated from the flower buds of *Tussilago farfara* L. *Food Chem. Toxicol.* **2006**, *44*, 1299–1307. [CrossRef]

149. Lee, M.-R.; Cha, M.-R.; Jo, K.-J.; Yoon, M.-Y.; Park, H.-R. Cytotoxic and apoptotic activities of *Tussilago farfara* extract in HT-29 human colon cancer cells. *Food Sci. Biotechnol.* **2008**, *17*, 308–312.

150. Fatykhova, D.G.; Karamova, N.S.; Abdrahimova, Y.R.; Ilinskaya, O.N. Evaluation of antigenotoxic effects of juices of plants *Chelidonium majus* L., *Plantago major* L. *Tussilago farfara* L. *Ecol. Genet.* **2010**, *8*, 56–65. [CrossRef]

151. Lee, H.-J.; Cho, H.-S.; Jun, S.; Lee, J.-J.; Yoon, J.; Lee, J.-H.; Song, H.; Choi, S.; Kim, S.; Saloura, V.; et al. Tussilago farfara L. augments TRAIL-induced apoptosis through MKK7/JNK activation by inhibition of MKK7 TIPRL in human hepatocellular carcinoma cells. *Oncol. Rep.* **2014**, *32*, 1117–1123. [CrossRef]

152. Nair, N.C.; Sheela, D. Quantification of secondary metabolites and anti-oxidant potential of selected members of the tribe Heliantheae. *J. Pharmacogn. Phytochem.* **2016**, *5*, 163–166.

153. Manjamalai, A.; Grace, V. Antioxidant activity of essential oils from *Wedelia chinensis* (Osbeck) in vitro and in vivo lung cancer bearing C57BL/6 mice. *Asian Pac. J. Cancer Prev.* **2012**, *13*, 3065–3071. [CrossRef]

154. Noori, S.; Hassan, Z.M. Dihydroartemisinin shift the immune response towards Th1, inhibit the tumor growth in vitro and in vivo. *Cell. Immunol.* **2011**, *271*, 67–72. [CrossRef] [PubMed]

155. Dell'Eva, R.; Pfeffer, U.; Vené, R.; Anfosso, L.; Forlani, A.; Albini, A.; Efferth, T. Inhibition of angiogenesis in vivo and growth of Kaposi's sarcoma xenograft tumors by the anti-malarial artesunate. *Biochem. Pharmacol.* **2004**, *68*, 2359–2366. [CrossRef] [PubMed]

156. Katiyar, S.K.; Meeran, S.M.; Katiyar, N.; Akhtar, S. p53 cooperates berberine-induced growth inhibition and apoptosis of non-small cell human lung cancer cells in vitro and tumor xenograft growth in vivo. *Mol. Carcinog.* **2009**, *48*, 24–37. [CrossRef] [PubMed]

157. Choi, M.S.; Oh, J.H.; Kim, S.M.; Jung, H.Y.; Yoo, H.S.; Lee, Y.M.; Moon, D.C.; Han, S.B.; Hong, J.T. Berberine inhibits p53-dependent cell growth through induction of apoptosis of prostate cancer cells. *Int. J. Oncol.* **2009**, *34*, 1221–1230.

158. Harikumar, K.B.; Kuttan, G.; Kuttan, R. Inhibition of progression of erythroleukemia induced by Friend virus in BALB/c mice by natural products—Berberine, Curcumin and Picroliv. *J. Exp. Ther. Oncol.* **2008**, *7*, 275–284.

159. Peng, P.-L.; Kuo, W.-H.; Tseng, H.-C.; Chou, F.-P. Synergistic Tumor-Killing Effect of Radiation and Berberine Combined Treatment in Lung Cancer: The Contribution of Autophagic Cell Death. *Int. J. Radiat. Oncol.* **2008**, *70*, 529–542. [CrossRef]

160. Huang, T.; Xiao, Y.; Yi, L.; Li, L.; Wang, M.; Tian, C.; Ma, H.; He, K.; Wang, Y.; Han, B.; et al. Coptisine from *Rhizoma coptidis* Suppresses HCT-116 Cells-related Tumor Growth in vitro and in vivo. *Sci. Rep.* **2017**, *7*, 38524. [CrossRef]

161. Chen, X.-Z.; Cao, Z.-Y.; Chen, T.-S.; Zhang, Y.-Q.; Liu, Z.-Z.; Su, Y.-T.; Liao, L.-M.; Du, J. Water extract of *Hedyotis diffusa* Willd suppresses proliferation of human HepG2 cells and potentiates the anticancer efficacy of low-dose 5-fluorouracil by inhibiting the CDK2-E2F1 pathway. *Oncol. Rep.* **2012**, *28*, 742–748. [CrossRef]

162. Yang, X.; Yang, Y.; Tang, S.; Tang, H.; Yang, G.; Xu, Q.; Wu, J. Anti-tumor effect of polysaccharides from *Scutellaria barbata* D. Don on the 95-D xenograft model via inhibition of the C-met pathway. *J. Pharmacol. Sci.* **2014**, *125*, 255–263. [CrossRef] [PubMed]

163. Li, Y.; Gu, J.-F.; Zou, X.; Wu, J.; Zhang, M.-H.; Jiang, J.; Qin, D.; Zhou, J.-Y.; Liu, B.-X.-Z.; Zhu, Y.-T. The anti-lung cancer activities of steroidal saponins of *P. polyphylla* Smith var. chinensis (Franch.) Hara through enhanced immunostimulation in experimental Lewis tumor-bearing C57BL/6 mice and induction of apoptosis in the A549 cell line. *Molecules* **2013**, *18*, 12916–12936. [CrossRef] [PubMed]

164. Wanga, Y.; Huangb, X.; Hanc, J.; Zhenga, W.; Maa, W. Extract of *Perilla frutescens* inhibits tumor proliferation of HCC via PI3K/AKT signal pathway. *Afr. J. Tradit. Complementary Altern. Med.* **2013**, *10*, 251–257. [CrossRef]

165. Fukutake, M.; Yokota, S.; Kawamura, H.; Iizuka, A.; Amagaya, S.; Fukuda, K.; Komatsu, Y. Inhibitory Effect of *Coptidis rhizoma* and Scutellariae Radix on Azoxymethane-Induced Aberrant Crypt Foci Formation in Rat Colon. *Biol. Pharm. Bull.* **1998**, *21*, 814–817. [CrossRef] [PubMed]

166. Manjamalai, A.; Grace, B. Chemotherapeutic effect of essential oil of *Wedelia chinensis* (Osbeck) on inducing apoptosis, suppressing angiogenesis and lung metastasis in C57BL/6 mice model. *J. Cancer Sci. Ther.* **2013**, *5*, 271–281. [CrossRef]

167. Singh, N.P.; Lai, H.C.; Park, J.S.; Gerhardt, T.E.; Kim, B.J.; Wang, S.; Sasaki, T. Effects of artemisinin dimers on rat breast cancer cells in vitro and in vivo. *Anticancer Res.* **2011**, *31*, 4111–4114. [PubMed]

168. Lai, H.; Singh, N.P. Oral artemisinin prevents and delays the development of 7, 12-dimethylbenz [a] anthracene (DMBA)-induced breast cancer in the rat. *Cancer Lett.* **2006**, *231*, 43–48. [CrossRef] [PubMed]

169. Joe, B.; Rao, U.J.; Lokesh, B.R. Presence of an acidic glycoprotein in the serum of arthritic rats: Modulation by capsaicin and curcumin. *Mol. Cell. Biochem.* **1997**, *169*, 125–134. [CrossRef]

170. Tanaka, T.; Kohno, H.; Shimada, R.; Kagami, S.; Yamaguchi, F.; Kataoka, S.; Ariga, T.; Murakami, A.; Koshimizu, K.; Ohigashi, H. Prevention of colonic aberrant crypt foci by dietary feeding of garcinol in male F344 rats. *Carcinogenesis* **2000**, *21*, 1183–1189. [CrossRef]

171. Liu, S.; Leach, S.D. Zebrafish models for cancer. *Annu. Rev. Pathol.* **2011**, *6*, 71–93. [CrossRef]

172. Zhu, X.-Y.; Guo, D.-W.; Lao, Q.-C.; Xu, Y.-Q.; Meng, Z.-K.; Xia, B.; Yang, H.; Li, C.-Q.; Li, P. Sensitization and synergistic anti-cancer effects of Furanodiene identified in zebrafish models. *Sci. Rep.* **2019**, *9*, 4541. [CrossRef]

173. Efferth, T.; Dunstan, H.; Sauerbrey, A.; Miyachi, H.; Chitambar, C.R. The anti-malarial artesunate is also active against cancer. *Int. J. Oncol.* **2001**, *18*, 767–773. [CrossRef]

174. Efferth, T. Molecular pharmacology and pharmacogenomics of artemisinin and its derivatives in cancer cells. *Curr. Drug Targets* **2006**, *7*, 407–421. [CrossRef] [PubMed]

175. Breuer, E.; Efferth, T. Treatment of Iron-Loaded Veterinary Sarcoma by Artemisia annua. *Nat. Prod. Bioprospecting* **2014**, *4*, 113–118. [CrossRef] [PubMed]

176. Apolone, G.; Joppi, R.; Garattini, S. Ten years of marketing approvals of anticancer drugs in Europe: Regulatory policy and guidance documents need to find a balance between different pressures. *Br. J. Cancer* **2005**, *93*, 504. [CrossRef] [PubMed]

177. Farrell, A.; Papadouli, I.; Hori, A.; Harczy, M.; Harrison, B.; Asakura, W.; Marty, M.; Dagher, R.; Pazdur, R. The advisory process for anticancer drug regulation: A global perspective. *Ann. Oncol.* **2005**, *17*, 889–896. [CrossRef] [PubMed]

178. Mezher, M. *FDA Adopts ICH Guideline on Nonclinical Evaluation for Anticancer Drugs*; Regulatory Affairs Professional Society: Rockville, MD, USA, 2018.

179. Calixto, J. Efficacy, safety, quality control, marketing and regulatory guidelines for herbal medicines (phytotherapeutic agents). *Braz. J. Med. Biol. Res.* **2000**, *33*, 179–189. [CrossRef] [PubMed]

180. Silva, T.C.D.; Silva, J.M.D.; Ramos, M.A. What Factors Guide the Selection of Medicinal Plants in a Local Pharmacopoeia? A Case Study in a Rural Community from a Historically Transformed Atlantic Forest Landscape. *Evid.-Based Complement. Altern. Med.* **2018**, *2018*, 2519212. [CrossRef]

181. Khan, T.; Abbasi, B.H.; Khan, M.A.; Shinwari, Z.K. Differential Effects of Thidiazuron on Production of Anticancer Phenolic Compounds in Callus Cultures of *Fagonia indica*. *Appl. Biochem. Biotechnol.* **2016**, *179*, 46–58. [CrossRef]

182. Khan, T.; Abbasi, B.H.; Khan, M.A.; Azeem, M. Production of biomass and useful compounds through elicitation in adventitious root cultures of *Fagonia indica*. *Ind Crop. Prod.* **2017**, *108*, 451–457. [CrossRef]

183. Huang, L.; Xie, D.; Yu, Y.; Liu, H.; Shi, Y.; Shi, T.; Wen, C. TCMID 2.0: A comprehensive resource for TCM. *Nucleic Acids Res.* **2018**, *46*, D1117–D1120. [CrossRef]

184. Zeng, X.; Zhang, P.; Wang, Y.; Qin, C.; Chen, S.; He, W.; Tao, L.; Tan, Y.; Gao, D.; Wang, B.; et al. CMAUP: A database of collective molecular activities of useful plants. *Nucleic Acids Res.* **2018**, *47*, D1118–D1127. [CrossRef]

185. Wu, Y.; Zhang, F.; Yang, K.; Fang, S.; Bu, D.; Li, H.; Sun, L.; Hu, H.; Gao, K.; Wang, W.; et al. SymMap: An integrative database of traditional Chinese medicine enhanced by symptom mapping. *Nucleic Acids Res.* **2018**, *47*, D1110–D1117. [CrossRef] [PubMed]

186. Xu, H.-Y.; Zhang, Y.-Q.; Liu, Z.-M.; Chen, T.; Lv, C.-Y.; Tang, S.-H.; Zhang, X.-B.; Zhang, W.; Li, Z.-Y.; Zhou, R.-R.; et al. ETCM: An encyclopaedia of traditional Chinese medicine. *Nucleic Acids Res.* **2018**, *47*, D976–D982. [CrossRef] [PubMed]

187. Jia, C.-Y.; Li, J.-Y.; Hao, G.-F.; Yang, G.-F. A drug-likeness toolbox facilitates ADMET study in drug discovery. *Drug Discov. Today* **2019**. [CrossRef] [PubMed]

188. Keefe, L.J.; Stoll, V.S. Accelerating pharmaceutical structure-guided drug design: A successful model. *Drug Discov. Today* **2019**, *24*, 377–381. [CrossRef]

189. Ferreira, L.L.G.; Andricopulo, A.D. ADMET modeling approaches in drug discovery. *Drug Discov. Today* **2019**, *24*, 1157–1165. [CrossRef]

190. Bergström, F.; Lindmark, B. Accelerated drug discovery by rapid candidate drug identification. *Drug Discov. Today* **2019**, *24*, 1237–1241. [CrossRef]

191. Fang, J.; Liu, C.; Wang, Q.; Lin, P.; Cheng, F. In silico polypharmacology of natural products. *Brief. Bioinform.* **2018**, *19*, 1153–1171. [CrossRef]

192. Yang, B.; Mao, J.; Gao, B.; Lu, X. Computer-Assisted Drug Virtual Screening Based on the Natural Product Databases. *Curr. Pharm. Biotechnol.* **2019**, *20*, 293–301. [CrossRef]

193. Qu, Y.; Zhang, Z.; Lu, Y.; Zheng, D.; Wei, Y. Network Pharmacology Reveals the Molecular Mechanism of Cuyuxunxi Prescription in Promoting Wound Healing in Patients with Anal Fistula. *Evidence-Based Complement. Altern. Med.* **2019**, *2019*, 3865121-9. [CrossRef]

194. Lv, Y.; Hou, X.; Zhang, Q.; Li, R.; Xu, L.; Chen, Y.; Tian, Y.; Sun, R.; Zhang, Z.; Xu, F. Untargeted Metabolomics Study of the In vitro Anti-Hepatoma Effect of Saikosaponin d in Combination with NRP-1 Knockdown. *Molecules* **2019**, *24*, 1423. [CrossRef]

195. Wang, Y.; Jafari, M.; Tang, Y.; Tang, J. Predicting Meridian in Chinese traditional medicine using machine learning approaches. *PLoS Comput. Boil.* **2019**, *15*, e1007249. [CrossRef] [PubMed]

196. Gaur, R.; Yadav, D.K.; Kumar, S.; Darokar, M.P.; Khan, F.; Bhakuni, R.S. Molecular modeling based synthesis and evaluation of in vitro anticancer activity of indolyl chalcones. *Curr. Top. Med. Chem.* **2015**, *15*, 1003–1012. [CrossRef] [PubMed]

197. Tabana, Y.M.; Hassan, L.E.; Ahamed, M.B.; Dahham, S.S.; Iqbal, M.A.; Saeed, M.A.; Khan, M.S.; Sandai, D.; Majid, A.S.; Oon, C.E.; et al. Scopoletin, an active principle of tree tobacco (*Nicotiana glauca*) inhibits human tumor vascularization in xenograft models and modulates ERK1, VEGF-A, and FGF-2 in computer model. *Microvasc Res.* **2016**, *107*, 17–33. [CrossRef] [PubMed]

198. Sharma, P.; Prakash, O.; Shukla, A.; Rajpurohit, C.S.; Vasudev, P.G.; Luqman, S.; Srivastava, S.K.; Pant, A.B.; Khan, F. Structure-Activity Relationship Studies on Holy Basil (*Ocimum sanctum* L.) Based Flavonoid Orientin and its Analogue for Cytotoxic Activity in Liver Cancer Cell Line HepG2. *Comb. Chem. High Throughput Screen.* **2016**, *19*, 656–666. [CrossRef] [PubMed]

199. Ntie-Kang, F.; Simoben, C.V.; Karaman, B.; Ngwa, V.F.; Judson, P.N.; Sippl, W.; Mbaze, L.M. Pharmacophore modeling and in silico toxicity assessment of potential anticancer agents from African medicinal plants. *Drug Des. Dev. Ther.* **2016**, *10*, 2137–2154. [CrossRef]

200. Li, Y.; Wang, J.; Lin, F.; Yang, Y.; Chen, S.S. A Methodology for Cancer Therapeutics by Systems Pharmacology-Based Analysis: A Case Study on Breast Cancer-Related Traditional Chinese Medicines. *PLoS ONE* **2017**, *12*, e0169363. [CrossRef]

201. Sharma, P.; Shukla, A.; Kalani, K.; Dubey, V.; Luqman, S.; Srivastava, S.K.; Khan, F. In-silico & In-vitro Identification of Structure-Activity Relationship Pattern of Serpentine & Gallic Acid Targeting PI3Kgamma as Potential Anticancer Target. *Curr. Cancer Drug Targets* **2017**, *17*, 722–734. [CrossRef]

202. Shirzad, H.; Taji, F.; Rafieian-Kopaei, M. Correlation between antioxidant activity of garlic extracts and WEHI-164 fibrosarcoma tumor growth in BALB/c mice. *J. Med. Food* **2011**, *14*, 969–974. [CrossRef]

203. Lakshmi, S.; Suresh, S.; Rahul, B.; Saikant, R.; Maya, V.; Gopi, M.; Padmaja, G.; Remani, P. In vitro and in vivo studies of 5, 7-dihydroxy flavones isolated from *Alpinia galanga* (L.) against human lung cancer and ascetic lymphoma. *Med. Chem. Res.* **2019**, *28*, 39–51. [CrossRef]

204. Jagetia, G.C.; Baliga, M.S. Evaluation of anticancer activity of the alkaloid fraction of *Alstonia scholaris* (Sapthaparna) in vitro and in vivo. *Phytother. Res. Int. J. Devoted Pharmacol. Toxicol. Eval. Nat. Prod. Deriv.* **2006**, *20*, 103–109.

205. Kumar, R.A.; Sridevi, K.; Kumar, N.V.; Nanduri, S.; Rajagopal, S. Anticancer and immunostimulatory compounds from *Andrographis paniculata*. *J. Ethnopharmacol.* **2004**, *92*, 291–295. [CrossRef] [PubMed]

206. Oliveira, C.R.; Spindola, D.G.; Garcia, D.M.; Erustes, A.; Bechara, A.; Palmeira-dos-Santos, C.; Smaili, S.S.; Pereira, G.J.; Hinsberger, A.; Viriato, E.P. Medicinal properties of *Angelica archangelica* root extract: Cytotoxicity in breast cancer cells and its protective effects against in vivo tumor development. *J. Integr. Med.* **2019**, *17*, 132–140. [CrossRef] [PubMed]

207. Li, F.; Wang, W.; Xiao, H. The evaluation of anti-breast cancer activity and safety pharmacology of the ethanol extract of *Aralia elata* Seem. leaves. *Drug Chem. Toxicol.* **2019**, 1–10. [CrossRef] [PubMed]

208. Baskar, A.A.; Ignacimuthu, S.; Paulraj, G.M.; Al Numair, K.S. Chemopreventive potential of β-sitosterol in experimental colon cancer model-an in vitro and in vivo study. *BMC Complement. Altern. Med.* **2010**, *10*, 24. [CrossRef]

209. Yang, B.; Xiao, B.; Sun, T. Antitumor and immunomodulatory activity of *Astragalus membranaceus* polysaccharides in H22 tumor-bearing mice. *Int. J. Biol. Macromol.* **2013**, *62*, 287–290. [CrossRef]

210. Lima, S.R.; Junior, V.F.V.; Christo, H.B.; Pinto, A.C.; Fernandes, P.D. In vivo and in vitro studies on the anticancer activity of *Copaifera multijuga* Hayne and its fractions. *Phytother. Res.* **2003**, *17*, 1048–1053. [CrossRef]

211. Wang, N.; Feng, Y.; Zhu, M.; Tsang, C.M.; Man, K.; Tong, Y.; Tsao, S.W. Berberine induces autophagic cell death and mitochondrial apoptosis in liver cancer cells: The cellular mechanism. *J. Cell. Biochem.* **2010**, *111*, 1426–1436. [CrossRef]

212. Kabeer, F.A.; Rajalekshmi, D.S.; Nair, M.S.; Prathapan, R. In vitro and in vivo antitumor activity of deoxyelephantopin from a potential medicinal plant *Elephantopus scaber* against Ehrlich ascites carcinoma. *Biocatal. Agric. Biotechnol.* **2019**, *19*, 101106. [CrossRef]

213. Pareek, A.; Godavarthi, A.; Issarani, R.; Nagori, B.P. Antioxidant and hepatoprotective activity of *Fagonia schweinfurthii* (Hadidi) Hadidi extract in carbon tetrachloride induced hepatotoxicity in HepG2 cell line and rats. *J. Ethnopharmacol.* **2013**, *150*, 973–981. [CrossRef]

214. Biloa Messi, B.; Ho, R.; Meli Lannang, A.; Cressend, D.; Perron, K.; Nkengfack, A.E.; Carrupt, P.-A.; Hostettmann, K.; Cuendet, M. Isolation and biological activity of compounds from *Garcinia preussii*. *Pharm. Biol.* **2014**, *52*, 706–711. [CrossRef]

215. Li, J.; Sun, J.; Song, J. Experimental research on effect of *Hedyotis diffusa* Willd on blood metastasis in H22 mice. *Lishizhen Med. Mater. Med. Res.* **2012**, *23*, 2434–2435.

216. Paramee, S.; Sookkhee, S.; Sakonwasun, C.; Takuathung, M.N.; Mungkornasawakul, P.; Nimlamool, W.; Potikanond, S. Anti-cancer effects of *Kaempferia parviflora* on ovarian cancer SKOV3 cells. *BMC Complement. Altern. Med.* **2018**, *18*, 178. [CrossRef] [PubMed]

217. Wang, X.; Yuan, S.; Wang, J.; Lin, P.; Liu, G.; Lu, Y.; Zhang, J.; Wang, W.; Wei, Y. Anticancer activity of litchi fruit pericarp extract against human breast cancer in vitro and in vivo. *Toxicol. Appl. Pharmacol.* **2006**, *215*, 168–178. [CrossRef] [PubMed]

218. Kowalczyk, T.; Sitarek, P.; Skała, E.; Toma, M.; Wielanek, M.; Pytel, D.; Wieczfińska, J.; Szemraj, J.; Śliwiński, T. Induction of apoptosis by in vitro and in vivo plant extracts derived from *Menyanthes trifoliata* L. in human cancer cells. *Cytotechnology* **2019**, *71*, 165–180. [CrossRef] [PubMed]

219. Turan, I.; Demir, S.; Kilinc, K.; Burnaz, N.A.; Yaman, S.O.; Akbulut, K.; Mentese, A.; Aliyazicioglu, Y.; Deger, O. Antiproliferative and apoptotic effect of *Morus nigra* extract on human prostate cancer cells. *Saudi Pharm. J.* **2017**, *25*, 241–248. [CrossRef]

220. Boubaker, J.; Ben Toumia, I.; Sassi, A.; Bzouich-Mokded, I.; Ghoul Mazgar, S.; Sioud, F.; Bedoui, A.; Safta Skhiri, S.; Ghedira, K.; Chekir-Ghedira, L. Antitumoral potency by immunomodulation of chloroform extract from leaves of *Nitraria retusa*, Tunisian medicinal plant, via its major compounds β-sitosterol and palmitic acid in BALB/c mice bearing induced tumor. *Nutr. Cancer* **2018**, *70*, 650–662. [CrossRef]

221. Lee, S.M.Y.; Li, M.L.Y.; Tse, Y.C.; Leung, S.C.L.; Lee, M.M.S.; Tsui, S.K.W.; Fung, K.P.; Lee, C.Y.; Waye, M.M.Y. Paeoniae Radix, a Chinese herbal extract, inhibit hepatoma cells growth by inducing apoptosis in a p53 independent pathway. *Life Sci.* **2002**, *71*, 2267–2277. [CrossRef]

222. Xu, W.W.; Li, B.; Lai, E.T.C.; Chen, L.; Huang, J.J.H.; Cheung, A.L.M.; Cheung, P.C.K. Water extract from *Pleurotus pulmonarius* with antioxidant activity exerts in vivo chemoprophylaxis and chemosensitization for liver cancer. *Nutr. Cancer* **2014**, *66*, 989–998. [CrossRef]

223. Wang, X.-N.; Xu, L.-N.; Peng, J.-Y.; Liu, K.-X.; Zhang, L.-H.; Zhang, Y.-K. In vivo inhibition of S180 tumors by the synergistic effect of the Chinese medicinal herbs *Coptis chinensis* and *Evodia rutaecarpa*. *Planta Med.* **2009**, *75*, 1215–1220. [CrossRef]

Article

Bioactivities of *Geranium wallichianum* Leaf Extracts Conjugated with Zinc Oxide Nanoparticles

Banzeer Ahsan Abbasi [1,†], **Javed Iqbal** [1,*,†], **Riaz Ahmad** [2], **Layiq Zia** [3], **Sobia Kanwal** [4], **Tariq Mahmood** [1], **Canran Wang** [5] and **Jen-Tsung Chen** [6,*]

[1] Department of Plant Sciences, Quaid-i-Azam University, Islamabad 45320, Pakistan; benazirahsanabbasi786@gmail.com (B.A.A.); tmahmood.qau@gmail.com (T.M.)

[2] College of Life Sciences, Shaanxi Normal University, Xi'an 710119, China; riaz17qau@gmail.com

[3] Superconductivity and Magnetism Laboratory, Department of Physics Quaid-i-Azam University, Islamabad 45320, Pakistan; layiqxia@gmail.com

[4] Department of Zoology, University of Gujrat, Sub-Campus Rawalpindi, Punjab 46300, Pakistan; sobiakanwal16@gmail.com

[5] Department of Biochemistry and Molecular Cell Biology, Shanghai Key Laboratory for Tumor Microenvironment and Inflammation, Shanghai Jiao Tong University School of Medicine, Shanghai 200025, China; canranw@gmail.com

[6] Department of Life Sciences, National University of Kaohsiung, Kaohsiung 811, Taiwan

* Correspondence: javed89qau@gmail.com (J.I.); jentsung@nuk.edu.tw (J.-T.C.)

† Equally contributed.

Received: 26 October 2019; Accepted: 17 December 2019; Published: 26 December 2019

Abstract: This study attempts to obtain and test the bioactivities of leaf extracts from a medicinal plant, *Geranium wallichianum* (GW), when conjugated with zinc oxide nanoparticles (ZnONPs). The integrity of leaf extract-conjugated ZnONPs (GW-ZnONPs) was confirmed using various techniques, including Ultraviolet–visible spectroscopy, X-Ray Diffraction, Fourier Transform Infrared Spectroscopy, energy-dispersive spectra (EDS), scanning electron microscopy, transmission electron microscopy, and Raman spectroscopy. The size of ZnONPs was approximately 18 nm, which was determined by TEM analysis. Additionally, the energy-dispersive spectra (EDS) revealed that NPs have zinc in its pure form. Bioactivities of GW-ZnONPs including antimicrobial potentials, cytotoxicity, antioxidative capacities, inhibition potentials against α-amylase, and protein kinases, as well as biocompatibility were intensively tested and confirmed. Altogether, the results revealed that GW-ZnONPs are non-toxic, biocompatible, and have considerable potential in biological applications.

Keywords: Zinc oxide nanoparticles; anticancer; antileishmanial; antimicrobial; biocompatibility

---

1. Introduction

Nanotechnology is one of the important thriving and rapidly developing interdisciplinary sciences involving a combination of knowledge from materials science, physics, chemistry, and biology, etc. The term "nano" is a Greek word which means "extremely small or dwarf" in size range of about one to one-hundred nanometers. NPs possess unique and fascinating magnetic, electrical, optical, and chemical properties with small size, different shapes, and surface effects compared to bulk materials [1–3]. NPs possess breakthrough applications in different fields such as food, agriculture, medicine, cosmetics, energy, environment, and many more [4,5]. Among the different metal nanoparticles (MNPs), zinc oxide nanoparticles (ZnONPs) has received due attention due to their multifunctional and tunable nature. ZnONPs possess 3.37 eV direct band gap with high excitation energy (60 meV), which make it perfect to be utilized in UV photodetector, transistors, and semiconductor diodes [6–8]. In addition, ZnONPs are used in the field of bio-imaging, drug delivery, mineral based sunscreens, lotions, and ointments; biomedicine, especially in the fields of anticancer and antibacterial fields, which are involved with their potent ability to trigger excess reactive oxygen species (ROS) production, release zinc ions, and induce cell apoptosis [9].

Numerous physical and chemical routes have been developed to synthesize ZnONPs with disparate morphologies and sizes [10–13]. These physical synthesis routes face numerous problems; they need high energy, costly instruments, and require high temperature and pressure [14]. Meanwhile, chemical approach involves synthesis of ZnONPs using several reducing agents, costly metal salts, toxic solvents, and reductants. These physical and chemical methods are not only expensive at industrial level production, but also possess potential environmental and biological hazards [15–18]. In contrast, biological/biogenic synthesis of NPs is an emerging area and economically feasible option in the field of "green chemistry" [19–22]. It is considered to be simple, safer, greener, easily scaled up, non-toxic, eco-friendly, energy-efficient, cost-effective, and performed at room temperature and pressure in the absence of non-hazardous solvents and reductants [19–21]. Biological fabrication of NPs can be accomplished using different bacteria, algae, diatoms, and medicinal plants [22,23]. The major disadvantage associated with microbial sources is maintaining contamination free environment, high isolation cost and their maintenance in culture media. Therefore, plants (phytofabrication) promise to be an excellent route in the formation of NPs due to simplicity, replacing chemicals to lessen or even remove materials that are harmful to human health and the ecosystem [24]. The use of a plant-extract in the synthesis of NPs has recently gained significant popularity [20,25]. The phytochemicals available in plants act as a reducing agents, leading to a synthesis of capped NPs, and thus reducing costs and chemical reagents [25,26]. Further, phytochemicals and aqueous environment replace many harmful organic/inorganic solvents and chemical compounds [26]. Currently, plant extracts of various plant parts have been used in the biofabrication of NPs [25]. Plant extracts possess potential biological activities in biomedical applications due to the presence of different phytomolecules: alkaloid, flavonoid, phenolic, terpenes, amino acids, and vitamins may function as strong reducing and stabilizing agents, reducing cost and eliminating the use of toxic chemicals agents [27–30].

Numerous investigations have focused on the green synthesis of ZnONPs and have reported different biological activities [31–34]. Considering the importance of green synthesis, in the present study, ZnONPs have been synthesized using GW leaves extract. The plant possesses many therapeutic application in the treatment of rheumatism, leucorrhoeas, arthritis, gonorrhea, heart, and liver related problems. Among the various plants used in the formation of green NPs, GW was selected due to large amount of bioactive compounds; ursolic acid, herniarin, stigmasterol, β-sitosterol, herniarin, etc. which can help in the reduction, stabilization, and capping of metal ions [35].

The aim of the current research study was to biosynthesize ZnONPs (Zinc nitrate hexahydrate) using natural GW leaves extract without utilizing any surfactants and chemical solvents. The reaction condition and synthesis protocol have been discussed comprehensively. To the best of our knowledge, this is the first research study reporting green fabrication of ZnONPs using GW leaves extract. Furthermore, ZnONPs were extensively studied using various characterizations techniques, followed by investigations of its diverse bio-potentials using number of activities and assays.

2. Experimental

2.1. Plant Sampling and Extract Preparation

Geranium wallichianum (Geraniaceae) was collected in its flowering stage from Nathiagali Mountains (34°04′ N 73°23′ E), Khyber Pakhtunkhwa, Pakistan. The soil is typically moist, loamy, and very shallow. The weather remains cool, pleasant and foggy in summers, cold and chilly in winters with heavy snow [35]. The leaves of GW were thoroughly washed with deionized water, shade dried for 10–15 days so that water content gets removed entirely. The leaves were crushed to powder and preserved in dry and airtight container. The plant extract was prepared by mixing 10 g of GW fresh leaves powder with 250 mL distilled water. The mixture was heated for 2 h at 80 °C under continuous stirring. The resulting solution was cool down at room temperature and filtered three times utilizing Whatman filter papers. The resulting filtered extract was preserved at 4 °C for future use in the biofabrication of ZnONPs.

2.2. Synthesis of ZnONPs

Synthesis of ZnONPs was successfully performed by reducing $Zn(NO_3)_2 \cdot 6H_2O$ using GW leaf extract. In detail, 50 mL filtered GW leaves extract was taken and mixed with 3 gm $Zn(NO_3)_2 \cdot 6H_2O$ salt, heated at 60 °C, and continuously stirred at 500 rpm for 2 h. The obtained solution was centrifuged at 12,000 rpm/30 min. The pellet containing ZnONPs was carefully washed 3–4 times with double distilled water. The obtained powder assumed as ZnONPs was placed in an oven at ~100 °C for 3 h. Further, ZnONPs were calcined in a furnace to obtain crystalline ZnONPs. The calcined ZnONPs were kept in a cool, dry, and dark place, and their characterizations were performed. Figure 1 shows a study layout of ZnONPs from synthesis to characterization and biological applications.

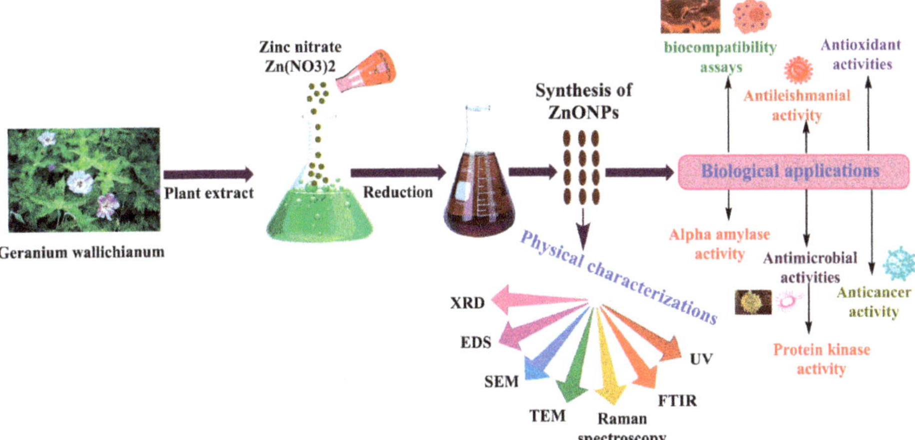

Figure 1. The process of green synthesis of leaf extract-conjugated zinc oxide nanoparticles.

2.3. Characterization of ZnONPs

The biogenic ZnONPs were characterized morphologically, physically, and chemically using various analytical techniques. The bio reduction of zinc ions to ZnONPs was determined by measuring the absorption spectra of reaction solutions via UV-4000 UV-Vis spectrophotometer (Germany) between 200–700 nm. The morphology of the ZnONPs was analyzed through SEM (EM (NOVA FEISEM-450 applied with EDX detectors). TEM analyses was performed to study size and shape of ZnONPs. The average particle size and particle size distribution of ZnONPs were studied via DLS system, using Malvern Zetasizer Nano (Malvern instrument). FTIR (Alpha, Bruker, Germany) analysis was performed between 500–4500 cm^{-1} to study diverse types of functional groups which are responsible in reduction and effective stabilizations for ZnONPs using various modes of vibrations. The crystalline structure of biogenic ZnONPs was determined by PANalytical XRD (Netherland) and crystal size was calculated. The Raman spectroscopy analyses was performed for ZnONPs to study their vibrational properties. The EDS analyses were done to detect the elemental composition of ZnONPs.

2.4. Bio-Potentials of Green GW-ZnONPs

2.4.1. Metabolic Activity of GW-ZnONPs against HepG2 Cells

The MTT cytotoxicity assay was performed according to the previously described method to assess the survival (viability) of the HepG2 (liver cancer cells) by assessing the mitochondrial activity inside the cells after treatment with ZnONPs. For this purpose, HepG2 cells were cultured in DMEM media provided with 10% FBS, 1% Pen-Strep. Cells were seeded in 96-well plates and placed in 5% CO_2 incubator at 37 °C for cell attachment. To assess the cytotoxicity potentials of green ZnONPs, HepG2 cells were treated with different doses of zinc oxide nanoparticles (7.8125–1000 µg/mL) for 48 h and MTT assay was performed. The ZnONPs treated cells were placed in DMEM medium and placed in an incubator for 24 h. After incubation, 100 µL MTT solutions was loaded and set aside for ~2 h. The MTT dye reacted with the oxidoreductase enzyme present in the mitochondria and was then converted into crystals of formazan. This process occurred only in viable cells. Further, the formazan crystals were dissolved with dimethyl sulfoxide (DMSO) and its absorbance was measured using a plate reader at 570 nm. Untreated cells were considered as control, while % inhibition of HepG2 cell lines treated with the synthesized ZnONPs nanoparticles was calculated using the formula below.

$$\%\text{inhibition} = \frac{1 - \text{OD of sample}}{\text{OD of control}} \times 100 \tag{1}$$

2.4.2. Antileishmanial Potential of ZnONPs

The cytotoxicity potentials of ZnONPs were further evaluated using *Leishmania tropica* "KWH23 strain" (both amastigotes and promastigotes culture) utilizing cytotoxicity assay [30,36]. The MI99 media was used added with 10% FBS to culture leishmanial parasites. The leishmanial parasites were treated with various concentrations of ZnONPs (1–200 µg/mL) to evaluate their antileishmanial potentials. During the experiment, Amphotericin-B was considered positive and DMSO as negative control. The Leishmanial parasites in 96 well plates were treated with various concentrations of ZnONPs and kept in 5% CO_2 incubator for 72 h at 24 °C, and absorbance was measured at 540 nm. After treatment, all living leishmanial parasites were counted under a microscope and IC_{50} values were recorded. Median lethal concentration (IC_{50}) was calculated using GraphPad software, while percent inhibition was calculated using the following formula:

$$\%\text{inhibition} = \frac{1 - \text{sample absorbance}}{\text{absorbance of control}} \times 100 \tag{2}$$

2.4.3. Alpha Amylase (AA) Inhibition Potential

The AA inhibition potential of biogenic ZnONPs was determined using a previously described method [36]. The reaction mixture for the activity was made by mixing 25 µL of AA enzyme, 10 µL ZnONPs, 40 µL starch solutions, and 15 µL FBS. The reaction solution with all component was incubated at 50 °C for 30 min by adding 1 M HCL (20 µL) and iodine solutions (90 µL). The acarbose was utilized as positive and distilled water was utilized as negative control during experiment. Median lethal concentration (IC_{50}) was calculated using GraphPad software while percent inhibition was calculated using the following formula:

$$\%\text{inhibition} = \frac{\text{sample absorbance} - \text{absorbance of negative control}}{\text{absorbance of blank} - \text{absorbance of negative control}} \times 100 \tag{3}$$

2.4.4. Protein Kinase (PK) Inhibition Potential

The ZnONPs were also evaluated for its PK inhibition potential using *Streptomyces* 85E strain according to a previously published method [37]. The PK inhibition activity was conducted in a sterile environment. The SP4 minimal media was utilized to prepare even lawns of *Streptomyces*. 10 µL of ZnONPs loaded on filter discs were placed on the petri plates to determine their PK inhibition potential. The surfactin was taken as positive and DMSO as negative control. To target the growth of *Streptomyces* 85E strain, incubation was performed at 30 °C for 72 h. After 24 h, clear and bald zones appeared around the discs, which showed spores inhibition and mycelia development. Finally, zone of inhibition (ZOI) were measured.

2.4.5. Antifungal Assays of ZnONPs

Fungicidal potentials of biogenic ZnONPs were investigated against different fungal strains via disc diffusion method. Before fungicidal assay was performed, fungal spores were sub-cultured in Sabouraud Dextrose liquid media and kept in an incubator for 24 h at 37 °C. The liquid cultures of different fungal strains were adjusted to OD of 0.5. Further, SDA solid media was made for culturing fungal strains and poured into petri plates. Filter discs laden with different concentrations of ZnONPs were kept on media plates. The Amphotericin-B was taken as positive and DMSO as negative controls. After loading test samples and both positive and negative controls, fungal plates were kept in an incubator for ~48 h at 28 °C to observe the ZOI. The fungal strains were treated with various concentrations of ZnONPs ranging from 31.25–1000 µg/mL, and their MIC value was recorded to determine their antifungal potential.

2.4.6. Antibacterial Activity of ZnONPs

The bacterial inhibition potential of GW-ZnONPs were evaluated using various bacterial strains through discs-diffusion method. Before the activity was conducted, bacterial strains were sub-cultured overnight in nutrients broth media and incubated at 37 °C for 24 h. To determine the antibacterial potency of ZnONPs, an overnight culture of bacterial strains was spread on pre-prepared agar media and allowed to dry for 5 min. Subsequently, filter discs laden with different concentration of ZnONPs (31.25–1000 µg/mL) were dried and kept on surface of plates. The plates were kept in incubator at 37 °C for 24 h and observed for ZOI. The standard antibiotic oxytetracycline was taken as positive and 5% DMSO as negative control. Further, MIC values of GW-ZnONPs were determined by calculating ZOI.

2.4.7. Antioxidant Capacities

The radical scavenging potential of ZnONPs was determined using spectrophotometric procedure. The working solution was prepared by mixing DPPH (2.4 mg) into 25 mL of methanol as free radicals. Before the activity was started, various concentrations (1–200 µg/mL) of ZnONPs were prepared and evaluated for their antioxidant potential. The ascorbic acid (AA) was taken as positive and DMSO as negative control. The 200 µL of reaction mixtures was comprised of 180 µL of reagent solution and 20 µL ZnONPs sample. The reaction mixture was then kept for 2 h under dark conditions and absorbance of reaction mixture was measured at 517 nm. The scavenging potential of GW-ZnONPs on free radicals are presented as follows:

$$\% \text{ DPPH scavenging} = 1 - \left(\frac{\text{Absorbance of sample}}{\text{Absorbance of control}}\right) \times 100 \tag{4}$$

The antioxidant potential of ZnONPs was further studied by total antioxidant capacity (TAC) using previously described phosphomollybdenum method [38]. The absorbance was recorded at 695 nm and results were indicated as microgram equivalent of AA per/mg of test samples. AA was used as positive control and DMSO as negative control. Furthermore, total reducing power (TRP) of the asynthesized ZnONPs were studied using Potassium-ferricyanide procedure [39]. AA was taken as positive and DMSO was taken as negative controls. The absorbance of mixture solutions was recorded at 630 nm. The reducing power of asynthesized ZnONPs was measured as AA equivalents per milligrams (AAE/mg).

2.4.8. Biocompatibility of ZnONPs with Human Macrophages

The biocompatible nature of green ZnONPs were investigated using human macrophages via previously used method [40]. The macrophages were cultured in RPMI media supplemented with FBS (10%), Herpes (25 mM), antibiotics (Pen-Strep. Further, macrophages were seeded and cultured in 96-well plates, kept in 5% CO_2 incubator for 24 h for cell attachment. The macrophages were exposed to various concentrations of biogenic ZnONPs (1–200 µg/mL) for 24 h. The absorbance was measured and % inhibition was calculated using the equation below.

$$\% \text{ inhibition } = \frac{1 - \text{ Absorbance of sample}}{\text{Absorbance of control}} \times 100 \tag{5}$$

2.4.9. Biocompatibility of ZnONPs with Human RBCs

The biocompatible nature of asynthesized ZnONPs was further confirmed using human RBCs through previously described method [41]. For hemolytic assay, 1 mL of fresh human RBCs was taken and stored in EDTA falcon to avoid blood coagulation. Further, centrifugation was performed for human RBCs at 12,000 rpm for 10 min. The erythrocytes suspension was made by adding 200 µL erythrocytes into 9.8 mL of PBS (pH 7.2). The 100 µL erythrocytes suspensions was treated with various doses of ZnONPs and incubated at 35 °C for 1 h. Further, centrifugation was performed at 12,000 rpm and supernatants was removed. Further, the supernatant was shifted into 96-well-plate and hemoglobin release was studied at 540 nm. During experiment, Triton X-100 was utilized as positive and DMSO as negative control. The results are calculated as % hemolysis produced by different concentration of ZnONPs and can be calculated employing the formula below:

$$\% \text{ hemolysis } = \frac{\text{Sample abs} - \text{Negative control abs}}{\text{Positive control abs} - \text{Negative control abs}} \times 100 \tag{6}$$

3. Results and Discussion

3.1. Biosynthesis of ZnONPs

In this present study, ZnONPs were rapidly fabricated using GW leaves extract as a bioreductant and stabilizing agents. Different characterization techniques were performed to determine the formation of zinc oxide nanoparticles. The progress of synthesis of ZnONPs was detected by color change after the addition of precursor salt to plant extract at 60 °C. The color change in solutions (reddish black) signal the biosynthesis of ZnONPs. This color change in the solution is due to surface plasmon resonance (SPR) [42]. To confirm the stable nature of asynthesized ZnONPs, 1 mg/mL solutions of nanoparticles was prepared and sonicated for ~40 min. The turbid colloidal suspension was allowed to remain stable for 48 h, and SPR was observed with varied time interval. The wavelength scale was fixed between 200–700 nm and solution was scanned between this range. The UV spectra showed absorption peak at 398 nm, indicating that colloidal suspension remained stable for 48 h. The reduction in absorption peak was noticed after 60 h which showed settlement of NPs at the bottom. The UV-Vis spectrum for green ZnONPs are presented in Figure 2A,B. The XRD configurations of thermally annealed green GW-ZnONPs are presented in Figure 2C. The XRD analysis has confirmed the crystalline nature of ZnONPs. The resulting Bragg peaks were in accordance with single and pure phase hexagonal zincite with JCPD card no: 00–036–1451. The XRD spectra showed different distinct diffraction peak with 2∅ value of 32.44, 35.13, 37.34, 47.21, 57.51, 62.58, 65.76 and 67.61, which are corresponding to (100), (002), (101), (102), (110), (103), (112) and (201) Bragg's reflections. XRD spectrum obtained indicates the absence of impurities. The presence of these peaks are due to leaves extract containing organic compounds, thus play significant role in reduction of zinc ions and stabilization of resultant ZnONPs [42]. The average size calculated as ~18 nm. The XRD configuration for ZnONPs are consistent with earlier studies using green synthesis protocol [43,44].

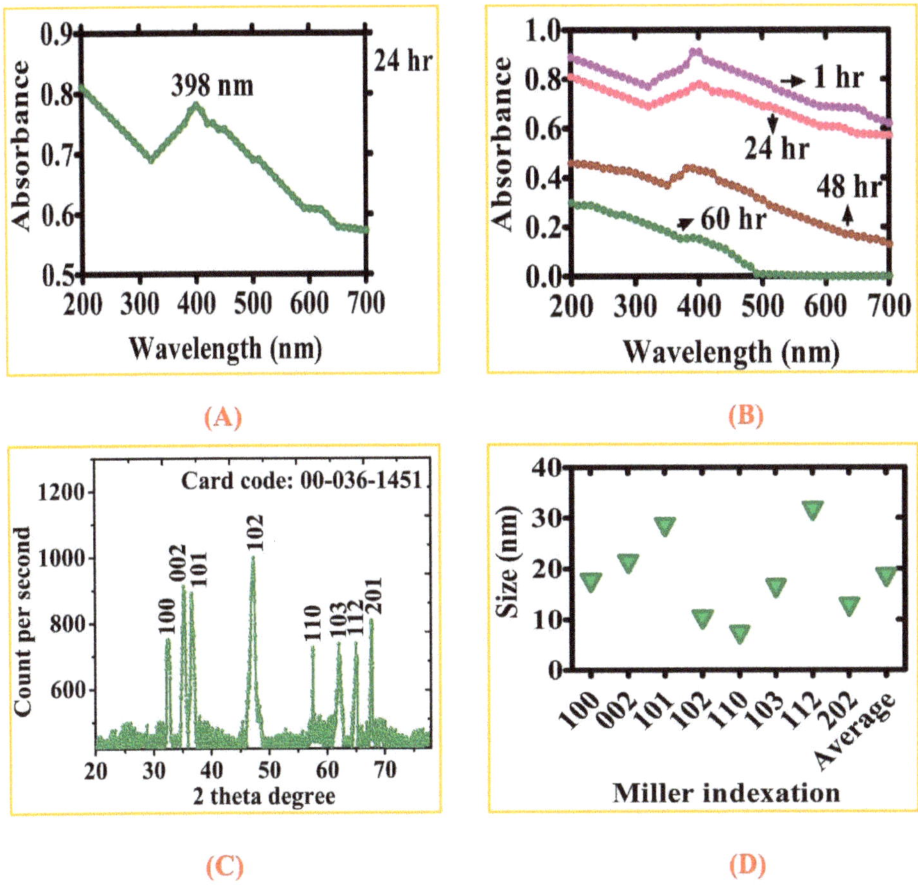

Figure 2. UV and XRD spectra analysis for biogenic ZnONPs (**A**) UV visible spectra (**B**) Stability of biosynthesized ZnONPs (**C**) XRD spectra of *Geranium wallichianum* mediated ZnONPs (**D**) Size calculation via Scherer approximation.

The Raman and FTIR analyses were performed to determine the vibrational properties of ZnONPs. The distinct major modes of Raman spectra are located at 92.5 cm^{-1} (E_2L), 180.77 cm^{-1} (2TAM), 313.23 cm^{-1} (2E2M), 403.04 cm^{-1} (E1TO), 565.12 cm^{-1} (E2H + E2L). Our results of Raman spectra of GW mediated ZnONPs are in agreement with previous studies using *S. thea* [44]. The Raman spectroscopy results are shown in Figure 3A. FTIR analysis was done to recognize the major functional group and their possible role in synthesis and stabilization of ZnONPs. The spectrum of leaves extract mediated zinc oxide nanoparticles is presented in Figure 3B. The peak centered at 3273.31 cm^{-1}, 1395.87 cm^{-1}, 1262.46 cm^{-1}, 1060.47 cm^{-1} indicate O-H, C-C, -CH and C = C stretching. The other peaks located at 1565.48 cm^{-1} correspond to C−C. Peaks at 530 cm^{-1} signify Zn−O bond vibrations from ZnONPs. There was a shift in the peaks of biogenic ZnONPs which suggests that different functional groups of GW leaf extract are involved in synthesis of ZnONPs and prevent agglomeration [45]. The surface morphology of the biogenic ZnONPs was explored using SEM. The SEM images of biogenic ZnONPs are shown in Figure 4A–C. The smaller the size of nanoparticles, the larger the surface area and stronger the activity will be. Thus, ZnONPs have shown significant biological applications. Further, an insight into the morphology and size detail of ZnONPs was revealed by TEM. TEM image in (Figure 4D) shows that ZnONPs are hexagonal in shape with average size of ~18 nm which are consistent with calculation from XRD. Smaller size NPs with larger surface area have shown numerous applications especially in the field of medicine, chemo, and electrochemistry [46,47]. EDX spectra analysis revealed the surface chemical composition of biogenic ZnONPs. The EDX results have shown that all the ionic zinc was resulted into synthesis of ZnONPs leaving no ionic zinc peak. The EDX peaks showed that both Zn and oxygen exist in test samples while no other elements were observed in the EDX spectrum. The absence of other elements confirms the purity of biosynthesized ZnONPs. The EDX spectrum is shown in Figure 5A. The EDX pattern clearly shows that reduction of Zn salt with GW leaves extract yielding in crystalline ZnONPs. The size distributions, PDI and ζ-potentials (ZP) of thermally annealed ZnONPs were detected by DLS analyses. The results indicated larger particles aggregate of 98.26 nm. The ZP and PDI of ZnONPs were −8.53 mV and 0.232 (Figure 5B,C). The details about zeta size and ZP are given in Table 1. Our DLS results are similar to the previous report of ZnONPs using *Ixora coccinea* [48]. DLS is performed to confirm the size of NPs in colloid suspension in the range of nano and submicron, and its ZP measurement is based on particles movement under electric field.

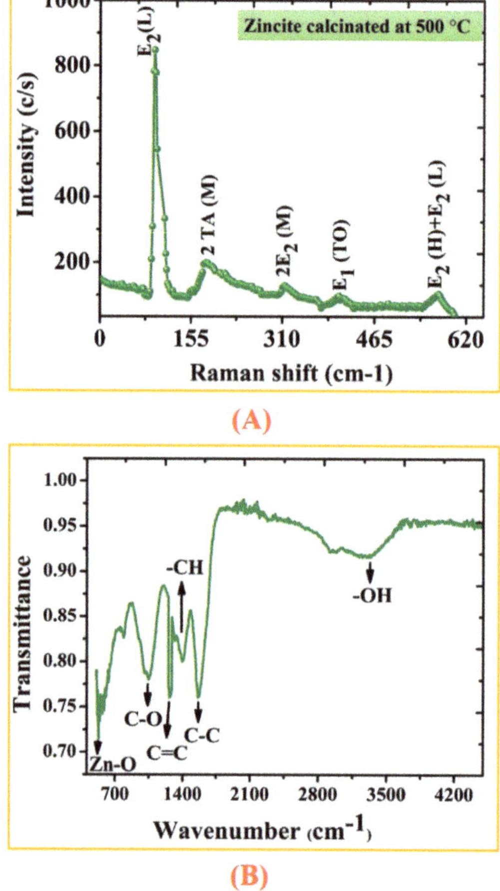

Figure 3. (**A**) Raman spectra of the ZnONPs biosynthesized using zinc nitrate hexahydrate as precursor (**B**) FTIR spectra of ZnONPs.

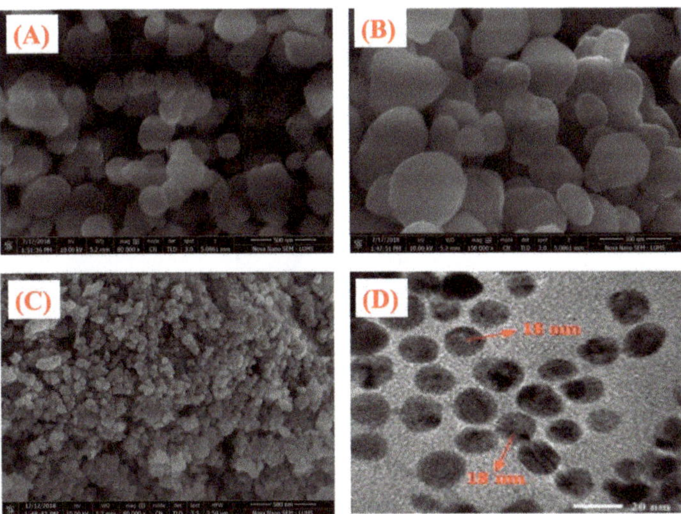

Figure 4. SEM and TEM images of *Geranium wallichianum* mediated ZnONPs using zinc nitrate as a precursor (**A–C**) HR-SEM images (**D**) TEM image.

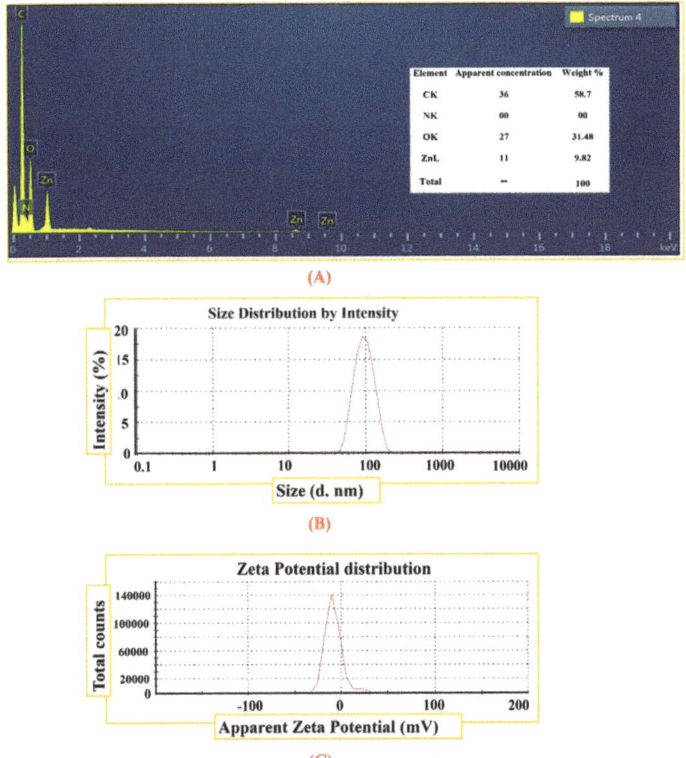

Figure 5. (**A**) Elemental composition using EDX (**B**) Size distribution of *Geranium wallichianum* mediated ZnONPs (**C**) Zeta potential measurement of ZnONPs.

Table 1. Zeta potential measurements of the ZnONPs.

Zeta Size (d. nm) and Potential (mV)	
Zeta size	98.26 (d. nm)
Z-Average	98.09 (d. nm)
PdI	0.232
Intercept	0.943
Zeta potential	−8.53 mV
Zeta deviation	9.16 mV
Conductivity	0.00275 mS/cm
Result quality	Good

3.2. Bio-Potentials of Biogenic GW-ZnONPs

3.2.1. Metabolic Activity of GW-ZnONPs against HepG2 Cells

Cancer is a fatal disease, a major cause of deaths around the globe, and is continuously increasing in cause of death by an estimated ~21 million by the year 2030 [30,49,50]. Among the numerous types of cancers, liver cancer is presently the second deadliest cancer in males and sixth in female causes (~745,517) deaths. The different risk factors related are viral infection, extensive alcohol use, and toxin exposures (aflatoxin) [51]. The cytotoxicity potential of the synthesized zinc oxide nanoparticles against liver cancer cells (HepG2) was evaluated using MTT cytotoxicity assay. The key results obtained by MTT cytotoxicity assay in HepG2 cells treated with various doses of ZnONPs ranging from 7.8125–1000 µg/mL for 48 h are summarized in Figure 6A. Our results of ZnONPs have determined strong reduction in the metabolic activity of HepG2 cancer cells. The metabolic activity was reducing continuously with increase in ZnONPs concentrations. The highest inhibition potential (~71% mortality) was achieved at 1000 µg/mL and cytotoxicity potency was decreasing with a decrease in concentration. The reduction in metabolic activity has shown that ZnONPs might have potential anticancer activity. The IC_{50} value recorded for GW mediated ZnONPs against HepG2 cell lines was 39.26 µg/mL. The cytotoxic effects induced by GW-NPs at lower concentrations could be due to the plant components attached to the ZnONPs. The results obtained from this study are also very well supported with various evidences for the cytotoxic effect of green ZnONPs using *Rhamnus virgata* leaf extract against the liver cancer HepG2 cell line in vitro [52,53].

3.2.2. Antileishmanial Potential of ZnONPs

Leishmania is a tropical disease with an excessive epidemiological diversity caused by at least 20 *Leishmania* species and is transferred by the bite of female sandflies [54]. Leishmanial parasites have wide distribution range in ~100 countries around the world. The drugs present in commercial market for the treatment of leishmaniasis are generally toxic, less potent and expensive. As for example, Antimonials designed as a potential candidate for leishmaniasis therapy has lost its potential against Leishmanial parasites and has developed resistance against it. Thus, pharmaceutical industries are working hard in this direction to develop novel drugs which will be more effective, less toxic, and cost effective for the treatment. Significant research studies have been performed to design MNPs for the treatment of Leishmania. Various NPs have been used in diverse studies to determine cytotoxic potentials against *L.* parasites [34]. However, the biosynthesized ZnONPs are poorly studied for the treatment of *L. tropica*.

In the present report, antileishmanial potentials of ZnONPs was evaluated using *L. tropica* (KMH23) Figure 6B. The *L. tropica* parasites were exposed to various doses (1–200 µg/mL) of ZnONPs for 72 h. The antileishmanial potentials of ZnONPs were increasing with an increase in concentrations of ZnONPs and determined significant antileishmanial potentials against *L. tropica* promastigotes with IC_{50}: 15.60 µg/mL. Similarly, ZnONPs also shown significant potential against *L. tropica* amastigotes with IC_{50}: 34.5 µg/mL which are consistent with earlier studies of biosynthesized ZnONPs [34,53].

The dose-dependent nature and low IC_{50} value signify that these ZnONPs can be utilized in drug delivery for the treatment of leishmaniasis.

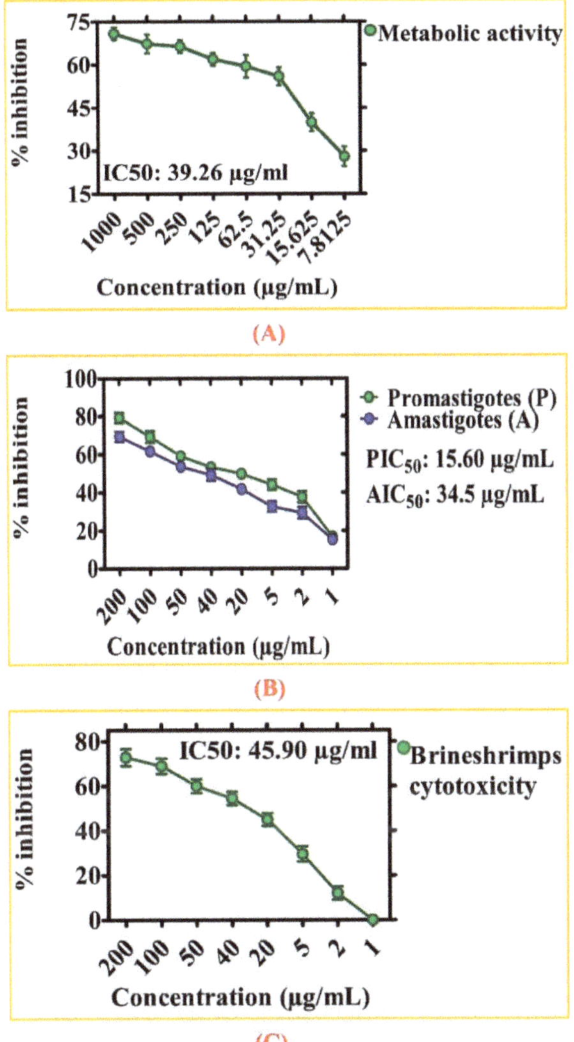

Figure 6. Cytotoxicity assays. The data in all figures represents the mean of three replicates (**A**) Cytotoxicity activities of *Geranium wallichianum* mediated ZnONPs against HepG2 cell line (**B**) Antileishmanial activities of ZnONPs (**C**) Cytotoxicity against brine shrimps.

3.2.3. Antibacterial and Antifungal Activities

The biogenic zinc oxide nanoparticles have also shown significant antibacterial potential against various bacterial strains (BS). For this purpose, different bacterial strains were treated with various doses of ZnONPs ranging from 31.25–1000 µg/mL Figure 7A. The activity was done against different gram positive (*S. aureus* and *B. subtilis*) and gram negative BS (*P. aeruginosa*, *K. pneumoniae*, *E. coli*). Most BS were found susceptible where ZnONPs have shown potential results by inhibiting them. *B. subtilis* was reported to be the most susceptible strain with MIC score of 7.8 µg/mL while *K. pneumoniae* was found to be the least susceptible strain with an MIC value of 125 µg/mL. Oxytetracycline was taken as positive control while treating BS to different doses of ZnONPs. No single dose has shown stronger potential than positive control. Overall, our ZnONPs have shown potential antibacterial activities against different BS which are in line with the previous reports of biosynthesized nanoparticles [28,55]. The increased antibacterial potential of ZnONPs is due to the bioactive functional groups attached on the surface of NPs. In a nutshell, ZnONPs concluded concentration dependent response against different strains of bacteria. Some other studies have discussed the antibacterial potentials of ZnONPs and shown that ROS generation is the core mechanisms that give antimicrobial potentials to NPs. Further, membrane damage (membrane protein damage) due to NPs absorption on surface result in bacterial cell damaging. Similarly, surface defect in the symmetry of NPs is responsible for bacteria inhibition and causes injury to cells [56]. Besides, different functional groups attached from GW leaves extract result in capped ZnONPs, which play an important role in the bacterial inhibition.

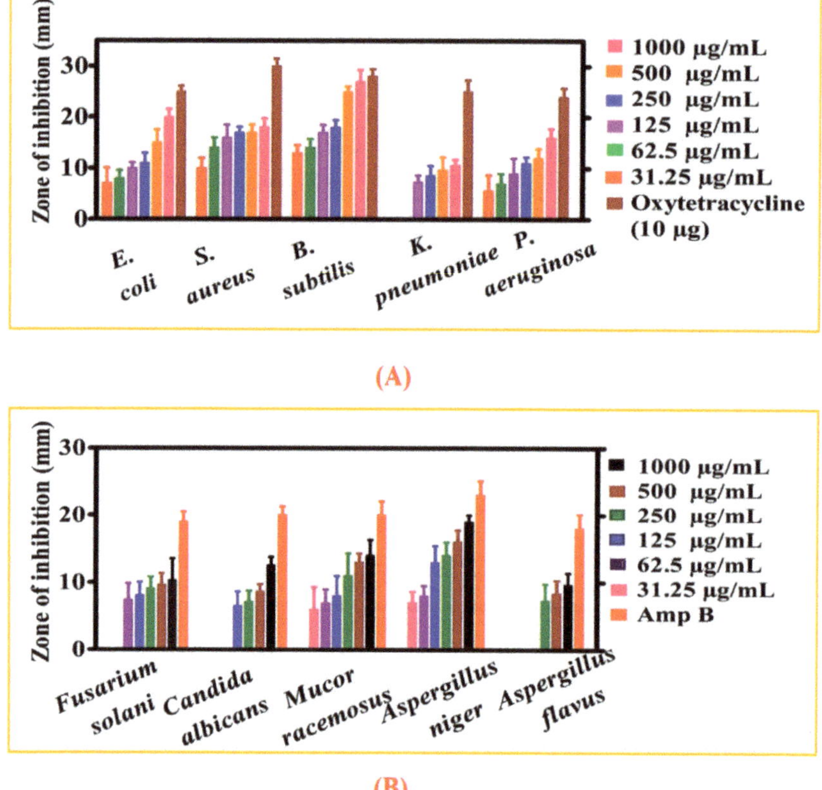

Figure 7. Antibacterial and antifungal assays. The data in all figures represents the mean of three replicates (**A**) Antibacterial potential of biogenic ZnONPs (**B**) Antifungal potential of biogenic ZnONPs.

Furthermore, fungicidal potential of ZnONPs was evaluated using various fungal strains (FS). Before activity was performed different concentration of ZnONPs were prepared ranging from 31.25–1000 µg/mL. The drug amp-B was employed as positive control to compare inhibition potential of ZnONPs. Various FS were used; *Fusarium solani, M. racemosus, A. niger, A. flavus* and *C. albicans* (Figure 7B). Substantial work has been performed on antibacterial potentials of ZnONPs while limited fungicidal potentials have been reported for biogenic ZnONPs. Our current study for the first reported the fungicidal activity of GW-ZnONPs. For this purpose, different FS were treated with various concentrations of ZnONPs (31.25–1000 µg/mL) to determine their antifungal potentials. Generally, a dose dependent inhibition response was reported for ZnONPs where *A. flavus* was the least susceptible fungal strain (MIC: 250 µg/mL while *A. niger* and *M. racemosus* were the most susceptible strain with MIC: 31.25 µg/mL. Among the various FS, *M. racemosus*, *A. niger* and *F. solani* were inhibited at all doses. Besides, ROS, previous studies claim that interaction of ZnONPs with fungal hypea and spores leads to inhibition of their growth. Significant concentration mediated fungicidal assays are reported in previous studies using different fungal strains [34] and are consistent with our present GW-ZnONPs study. MIC values for various bacteria and fungus strain are presented in Table 2.

Table 2. MICs values of different bacterial and fungal strains.

Antibacterial Activity	
Bacterial Strain	MIC (µg/mL)
Gram Positive	
B. subtilis (ATCC: 6633)	7.8
S. aureus (ATCC: 25923)	15.625
Gram Negative	
P. aeruginosa (ATCC: 9721)	31.25
E. coli (ATCC:15224)	15.625
K. pneumonia (ATCC: 4617)	125
Antifungal Activity	
Fungal Strain	MIC (µg/mL)
Aspergillus flavus (FCBP: 0064)	250
Aspergillus niger (FCBP: 0918)	31.25
Candida albicans (FCBP: 478)	125
Fusarium solani (FCBP: 0291)	62.5
Mucor racemosus (FCBP: 0300)	31.25

3.2.4. Enzyme Inhibition Potentials of ZnONPs

Figure 8A shows protein kinase (PK) enzyme inhibition potential of ZnONPs. These enzymes play significant role in phosphorylation of important amino acids; serine-threonine, tyrosine residues, regulate important processes inside cells like metabolism, apoptosis, proliferations, and differentiation. Deregulated phosphorylation of amino acids can result into genetic abnormalities and result in the development of cancer. Therefore, any substance with potentials to inhibit PK enzymes are of great attention in the field of cancer research [57]. PK phosphorylation has played an important role in the formation of hyphae in *Streptomyces* fungal strain and same mechanism is used to evaluate the PK inhibition potentials and is utilized to study medicinal compounds for determining PK inhibition [5]. The PK inhibition activity was done via disc diffusion method using different doses of ZnONPs ranging from 31.25–1000 µg/mL. The surfactin was taken as a positive control. Different ZOI were observed at different doses of ZnONPs, and the highest ZOI was measured as 15 mm at 1000 µg/mL, which shows important protein kinase inhibition potency of ZnONPs. A concentration-dependent activity is reported for ZnONPs. No single dose has shown stronger potential than the positive control. Therefore, a potential signal transduction inhibitor is identified in the form of nanoscaled ZnO that can be further exploited for anti-infective and anticancer properties. Because of the protein kinase inhibition property, one can pre-deduce that biosynthesized ZnONPs may play an important role in cancer therapeutics. Our results of PK inhibition are in line with the previous findings [53].

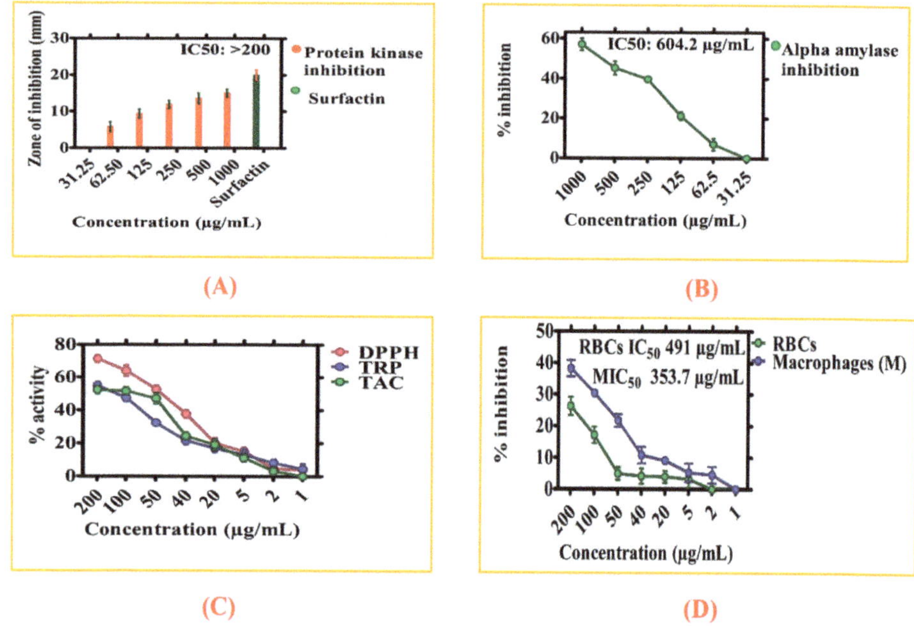

Figure 8. (**A**) Inhibition potential of *Geranium wallichianum* mediated ZnONPs against protein kinase (**B**) Inhibition potential against alpha amylase (**C**) Antioxidant potential of ZnONPs (**D**) Biocompatibility potential of ZnONPs against human RBCs and macrophages.

Besides PK inhibition assay, alpha amylase (AA) inhibition potency of ZnONPs was determined using various concentrations (31.25–1000 µg/mL) of ZnONPs. The AA play significant role by converting carbohydrates into glucose [57], therefore, blocking or by inhibiting the activity of alpha amylase can prevent the level of glucose formation; thus, it may provide new insights into a nano level treatment of diabetes [58]. In our study, biogenic ZnONPs were explored for their AA inhibition activity and have determined significant potential by the inhibition of AA. The highest inhibition rate was observed 57% at 1000 µg/mL, while AA inhibition potentials was slowly decreasing with a decrease in concentration of ZnONPs. Figure 8B shows the biological inhibition potential of ZnONPs against AA. The results of AA activity are in agreement with the previous findings [44,53].

3.2.5. Antioxidant Activities of ZnONPs

Figure 8C indicates antioxidant potentials of ZnONPs. The antioxidant activities were evaluated in concentration ranging from 1–200 µg/mL. Maximum score for TAC of ZnONPs in terms of AA equivalent/milligrams was reported as 52.43% at 200 µg/mL. TAC assay is mainly used to evaluate the scavenging effect of tested chemicals towards reactive oxygen species (ROS). In the present study, water extract of GW leaves extract was utilized in the reduction and stabilizations of metal ions. It can be concluded from our antioxidant activities, that several phenolic compounds available in the GW leaves extract scavenge ROS which are attached on the surface of ZnONPs.

To further explore the antioxidants species coated on the surface of biogenic nanoparticles, TRP assay was determined. This activity was done to study the reductones that play an important role in the antioxidant potential by providing H-atoms and causing damage to free radical chains [59]. The biosynthesized ZnONPs showed potential antioxidant activity. The reducing power of ZnONPs was decreasing with decrease in concentrations of ZnONPs. The maximum TRP (55%) was observed at its highest concentration of 200 µg/mL. Significant DPPH radicals scavenging activity (71.36%) was observed for ZnONPs at 200 µg/mL. From data presented in Figure 8C, it can be concluded that numerous antioxidants compound may be responsible in reduction and stabilization of ZnONPs via GW leaf extracts. Our antioxidants result of GW-mediated ZnONPs are consistent with the previous studies of ZnONPs via *S. thea* and *F. indica* [34,44]. The variations and disagreement compare to other studies may be due to various important factors like experiment condition, method of nanoparticles fabrication, plant, plant part used, and nanoparticle size, etc.

3.2.6. Biocompatibility Potential Assays

The biocompatibility and toxicological effect of zinc oxide nanoparticles were determined using human macrophages and RBCs. Biological substance with hemolytic activity of greater than 5% are known as hemolytic, between 2–5% are slightly hemolytic, while less than 2% is non-hemolytic [60]. If a given nanoparticle is hemolytic, it will rupture red blood cells, which further result in hemoglobin release. To confirm the bio-safe nature, hemolysis assay was conducted using human RBCs. The RBCs were exposed to various concentrations of ZnONPs in a concentration ranging from 200–1 μg/mL. The data obtained have shown that the synthesized are non-hemolytic at lower concentration (2 μg/mL), slightly hemolytic at 5 to 50 μg/mL, while hemolytic at concentrations of >50 μg/mL. These results are in agreement to previous reports of *S. thea* mediated ZnONPs [44]. The IC_{50} value of ZnONPs against human red blood cells was recorded is 491 μg/mL. Our research study confirmed that biosynthesized ZnONPs are non-hemolytic and are considered biocompatible in low concentrations.

The biocompatibility assay was further confirmed by using human macrophages. For this purpose, human macrophages were seeded in 96-well plate and were cultured in RPMI media for 24 h for cells attachment. Further, cells were treated with various concentrations of ZnONPs (200–1 μg/mL). MTT cell viability assay was done to perform biocompatible nature of ZnONPs. The macrophage responded to ZnONPs treatment in a dose dependent manner. The results indicated that ZnONPs at 200 μg/mL inhibit growth of the macrophages by ~32% which determine the bio-safe behavior of biogenic ZnONPs. Normally, macrophages have established mechanisms to deal with ROS produce from external source. According to research studies, ROS are non-toxic to both RBCs and macrophages at a lower concentration unless concentration increases beyond that limit which will be considered toxic for RBCs and macrophages [61]. The IC_{50} value for ZnONPs was calculated >353.7 μg/mL. The results of biocompatibility assays of ZnONPs are presented in Figure 8D.

4. Conclusions and Future Perspectives

In summary, a simple, safe, ecofriendly, and one-step process was used for the biofabrication of ZnONPs utilizing GW *leaves* extract without utilizing any chemical reagents or surfactants. Different characterization techniques were performed. TEM analysis showed that ZnONPs was of ~18 nm with a hexagonal shape. Furthermore, different in vitro biological activities of ZnONPs were performed. The bio-potentials of ZnONPs were investigated against different pathogenic microbial strains and confirmed that ZnONPs have shown significant antimicrobial potential. The ZnONPs displayed strong anticancer and antileishmaniasis activities. Furthermore, moderate antioxidant and enzymes inhibition activities have been investigated. The biosafe nature of ZnONPs was confirmed using human RBCs and macrophages. Based on the above findings, we can say that green synthesis is the way forward and new frontier for designing nanomedicine and can be utilized in different theranostic applications for the treatment of different diseases. In addition, different in vivo studies are encouraged on toxicity aspects in different animal models, and once their biocompatibility and bio-safe nature is confirmed, only then can these NPs can be utilized in clinical applications. Further studies are encouraged on the mechanistic and synthesis aspects of the ZnONPs by using different medicinal plant materials.

Author Contributions: Conceptualization, B.A.A., J.I., R.A., L.Z., S.K., C.W., T.M., and J.-T.C.; writing—original draft preparation, B.A.A., J.I., and R.A.; writing—review and editing, L.Z., S.K., C.W., T.M., and J.-T.C. All authors have read and agreed to the published version of the manuscript.

Funding: This research received no external funding.

Conflicts of Interest: The authors declare no conflict of interest.

References

1. Munir, A.; Haq, T.U.; Hussain, I.; Qurashi, A.; Ullah, U.; Iqbal, M.J.; Hussain, I. Ultrasmall Co@Co(OH)$_2$ Nanoclusters Embedded in N-Enriched Mesoporous Carbon Network as Efficient Electrocatalysts for Durable Water Oxidation. *ChemSusChem* **2019**, *12*, 5117–5125. [CrossRef]
2. Mayedwa, N.; Mongwaketsi, N.; Khamlich, S.; Kaviyarasu, K.; Matinise, N.; Maaza, M. Green synthesis of nickel oxide, palladium and palladium oxide synthesized via *Aspalathus linearis* natural extracts: Physical properties & mechanism of formation. *Appl. Surf. Sci.* **2018**, *446*, 266–272.
3. Yoon, W.J.; Jung, K.Y.; Liu, J.; Duraisamy, T.; Revur, R.; Teixeira, F.L.; Berger, P.R. Lasmon-enhanced optical absorption and photocurrent in organic bulk heterojunction photovoltaic devices using self-assembled layer of silver nanoparticles. *Sol. Energy Mater. Sol. Cells* **2010**, *94*, 128–132. [CrossRef]
4. Iqbal, J.; Abbasi, B.A.; Ahmad, R.; Mahmood, T.; Ali, B.; Khalil, A.T.; Kanwal, S.; Shah, S.A.; Alam, M.M.; Badshah, H.; et al. Nanomedicines for developing cancer nanotherapeutics: From benchtop to bedside and beyond. *Appl. Microbiol. Biotechnol.* **2018**, *102*, 9449–9470. [CrossRef] [PubMed]
5. Abbasi, B.A.; Iqbal, J.; Mahmood, T.; Ahmad, R.; Kanwal, S.; Afridi, S. Plant-mediated synthesis of nickel oxide nanoparticles (NiO) via *Geranium wallichianum*: Characterization and different biological applications. *Mater. Res. Exp.* **2019**, *6*, 0850a7. [CrossRef]
6. Thema, F.T.; Manikandan, E.; Dhlamini, M.S.; Maaza, M. Green synthesis of ZnO nanoparticles via *Agathosma betulina* natural extract. *Mater. Lett.* **2015**, *15*, 124–127. [CrossRef]
7. Kalusniak, S.; Sadofev, S.; Puls, J.; Henneberger, F. ZnCdO/ZnO—A new heterosystem for green-wavelength semiconductor lasing. *Laser Photonics Rev.* **2009**, *3*, 233–242. [CrossRef]
8. Yang, Y.; Jin, P.; Zhang, X.; Ravichandran, N.; Ying, H.; Yu, C.; Wu, M. New epigallocatechin gallate (EGCG) nanocomplexes co-assembled with 3-mercapto-1-hexanol and β-lactoglobulin for improvement of antitumor activity. *J. Biomed. Nanotechnol.* **2017**, *13*, 805–814. [CrossRef]
9. Mirzaei, H.; Darroudi, M. Zinc oxide nanoparticles: Biological synthesis and biomedical applications. *Ceram. Int.* **2017**, *43*, 907–914. [CrossRef]
10. Kawakami, M.; Hartanto, A.B.; Nakata, Y.; Okada, T. Synthesis of ZnO nanorods by nanoparticle assisted pulsed-laser deposition. *Jpn. J. Appl. Phys.* **2003**, *42*, 1–33. [CrossRef]
11. Liu, X.; Wu, X.; Cao, H.; Chang, R.P. Growth mechanism and properties of ZnO nanorods synthesized by plasma-enhanced chemical vapor deposition. *J. Appl. Phys.* **2004**, *95*, 3141–3147. [CrossRef]
12. Zhang, X.H.; Liu, Y.C.; Wang, X.H.; Chen, S.J.; Wang, G.R.; Zhang, J.Y.; Fan, X.W. Structural properties and photoluminescence of ZnO nanowalls prepared by two-step growth with oxygen-plasma-assisted molecular beam epitaxy. *J. Phys. Condens. Matter* **2005**, *17*, 3035. [CrossRef]
13. Gondal, M.A.; Drmosh, Q.A.; Yamani, Z.H.; Saleh, T.A. Synthesis of ZnO2 nanoparticles by laser ablation in liquid and their annealing transformation into ZnO nanoparticles. *Appl. Surf. Sci.* **2009**, *256*, 298–304. [CrossRef]
14. Guzmán, M.G.; Dille, J.; Godet, S. Synthesis of silver nanoparticles by chemical reduction method and their antibacterial activity. *Int. J. Chem. Biomol. Eng.* **2009**, *2*, 104–111.
15. Tahir, M.N.; Natalio, F.; Cambaz, M.A.; Panthöfer, M.; Branscheid, R.; Kolb, U.; Tremel, W. Controlled synthesis of linear and branched Au@ZnO hybrid nanocrystals and their photocatalytic properties. *Nanoscale* **2013**, *5*, 9944–9949. [CrossRef]
16. Polavarapu, L.; Liz-Marzán, L.M. Growth and galvanic replacement of silver nanocubes in organic media. *Nanoscale* **2013**, *5*, 4355–4361. [CrossRef]
17. Yadav, R.S.; Mishra, P.; Pandey, A.C. Growth mechanism and optical property of ZnO nanoparticles synthesized by sonochemical method. *Ultrason. Sonochem.* **2008**, *15*, 863. [CrossRef]
18. Barhoum, A.; Van Assche, G.; Rahier, H.; Fleisch, M.; Bals, S.; Delplancked, M.P.; Bahnemann, D. Sol-gel hot injection synthesis of ZnO nanoparticles into a porous silica matrix and reaction mechanism. *Mater. Des.* **2017**, *119*, 270–276. [CrossRef]
19. Iravani, S.; Korbekandi, H.; Mirmohammadi, S.V.; Zolfaghari, B. Synthesis of silver nanoparticles: Chemical, physical and biological methods. *Res. Pharm. Sci.* **2014**, *9*, 385.

20. Duran, N.; Seabra, A.B. Biogenic synthesized Ag/Au nanoparticles: Production, characterization, and applications. *Curr. Nanosci.* **2018**, *14*, 82–94. [CrossRef]

21. Seabra, A.B.; Duran, N. Nanotoxicology of metal oxide nanoparticles. *Metals* **2015**, *5*, 934–975. [CrossRef]

22. Singh, P.; Kim, Y.J.; Zhang, D.; Yang, D.C. Biological synthesis of nanoparticles from plants and microorganisms. *Trends. Biotechnol.* **2016**, *34*, 588–599. [CrossRef] [PubMed]

23. Durán, N.; Marcato, P.D.; Durán, M.; Yadav, A.; Gade, A.; Rai, M. Mechanistic aspects in the biogenic synthesis of extracellular metal nanoparticles by peptides, bacteria, fungi, and plants. *Appl. Microbiol. Biotechnol.* **2011**, *90*, 1609–1624. [CrossRef] [PubMed]

24. KS, S.; Vellora Thekkae Padil, V.; Senan, C.; Pilankatta, R.; George, B.; Wacławek, S.; Černík, M. (Green Synthesis of High Temperature Stable Anatase Titanium Dioxide Nanoparticles Using Gum Kondagogu: Characterization and Solar Driven Photocatalytic Degradation of Organic Dye. *Nanomaterials* **2018**, *8*, 1002.

25. Iqbal, J.; Abbasi, B.A.; Batool, R.; Khalil, A.T.; Hameed, S.; Kanwal, S.; Mahmood, T. Biogenic synthesis of green and cost effective cobalt oxide nanoparticles using Geranium wallichianum leaves extract and evaluation of in vitro antioxidant, antimicrobial, cytotoxic and enzyme inhibition properties. *Mater. Res. Express* **2019**, *6*, 115407. [CrossRef]

26. Hameed, S.; Shah, S.A.; Iqbal, J.; Numan, M.; Muhammad, W.; Junaid, M.; Umer, F. Cannabis sativa Mediated Synthesis of Gold Nanoparticles and its Biomedical Properties. *Bioinspired Biomim. Nanobiomater.* **2019**, 1–8. [CrossRef]

27. Mohamed, H.E.A.; Afridi, S.; Khalil, A.T.; Zia, D.; Iqbal, J.; Ullah, I.; Maaza, M. Biosynthesis of silver nanoparticles from Hyphaene thebaica fruits and their in vitro pharmacognostic potential. *Mater. Res. Express* **2019**, *6*, 1050c9. [CrossRef]

28. Abbasi, B.A.; Iqbal, J.; Mahmood, T.; Qyyum, A.; Kanwal, S. Biofabrication of iron oxide nanoparticles by leaf extract of *Rhamnus virgata*: Characterization and evaluation of cytotoxic, antimicrobial and antioxidant potentials. *Appl. Organomet. Chem.* **2019**, *33*, e4947. [CrossRef]

29. Iqbal, J.; Abbasi, B.A.; Ahmad, R.; Mahmood, T.; Kanwal, S.; Ali, B.; Badshah, H. Ursolic acid a promising candidate in the therapeutics of breast cancer: Current status and future implications. *Biomed. Pharmacother.* **2018**, *108*, 752–756. [CrossRef]

30. Abbasi, B.A.; Iqbal, J.; Mahmood, T.; Khalil, A.T.; Ali, B.; Kanwal, S.; Ahmad, R. Role of dietary phytochemicals in modulation of miRNA expression: Natural swords combating breast cancer. *Asian. Pac. J. Trop. Med.* **2018**, *11*, 501–509.

31. Bhuyan, T.; Mishra, K.; Khanuja, M.; Prasad, R.; Varma, A. Biosynthesis of zinc oxide nanoparticles from *Azadirachta indica* for antibacterial and photocatalytic applications. *Mater. Sci. Semicond. Process.* **2015**, *32*, 55–61. [CrossRef]

32. Hameed, S.; Iqbal, J.; Ali, M.; Khalil, A.T.; Abbasi, B.A.; Numan, M.; Shinwari, Z.K. Green synthesis of zinc nanoparticles through plant extracts: Establishing a novel era in cancer theranostics. *Mater. Res. Express* **2019**, *6*, 102005. [CrossRef]

33. Matinise, N.; Fuku, X.G.; Kaviyarasu, K.; Mayedwa, N.; Maaza, M. ZnO nanoparticles via *Moringa oleifera* green synthesis: Physical properties & mechanism of formation. *Appl. Surf. Sci.* **2017**, *406*, 339–347.

34. Ismail, M.; Ibrar, M.; Iqbal, Z.; Hussain, J.; Hussain, H.; Ahmed, M.; Choudhary, M.I. Chemical constituents and antioxidant activity of *Geranium wallichianum*. *Records. Natl. Prod.* **2009**, *3*, 193–197.

35. Ellis, S.; Taylor, D.M.; Masood, K.R. Soil formation and erosion in the Murree Hills, northeast Pakistan. *Catena* **1994**, *22*, 69–78. [CrossRef]

36. Iqbal, J.; Abbasi, B.A.; Mahmood, T.; Hameed, S.; Munir, A.; Kanwal, S. Green synthesis and characterizations of Nickel oxide nanoparticles using leaf extract of *Rhamnus virgata* and their potential biological applications. *Appl. Organometal. Chem.* **2019**, e4950. [CrossRef]

37. Fatima, H.; Khan, K.; Zia, M.; Ur-Rehman, T.; Mirza, B.; Haq, I.U. Extraction optimization of medicinally important metabolites from *Datura innoxia* Mill.: An in vitro biological and phytochemical investigation. *BMC Complement. Altern. Med.* **2015**, *15*, 376–394. [CrossRef]

38. Satpathy, S.; Patra, A.; Ahirwar, B.; Delwar Hussain, M. Antioxidant and anticancer activities of green synthesized silver nanoparticles using aqueous extract of tubers of *Pueraria tuberosa*. *Artif. Cells Nanomed. Biotechnol.* **2018**, *46*, 71–85. [CrossRef]

39. Baqi, A.; Tareen, R.B.; Mengal, A.; Khan, N.; Behlil, F.; Achakzai, A.K.K.; Faheem, M. Determination of antioxidants in two medicinally important plants, *Haloxylon griffithii* and *Convolvulus leiocalycinus* of Balochistan. *Pure Appl. Biol.* **2018**, *7*, 296–308. [CrossRef]

40. De Almeida, M.C.; Silva, A.C.; Barral, A.; Barral Netto, M. A simple method for human peripheral blood monocyte isolation. *Memorias do Instituto Oswaldo Cruz* **2000**, *95*, 221–223. [CrossRef]

41. Malagoli, D. A full-length protocol to test hemolytic activity of palytoxin on human erythrocytes. *Invertebrate. Surviv. J.* **2007**, *4*, 92–94.

42. Ibrahim, H.M. Green synthesis and characterization of silver nanoparticles using banana peel extract and their antimicrobial activity against representative microorganisms. *J. Radic. Res. Appl. Sci.* **2015**, *8*, 265. [CrossRef]

43. Zak, A.K.; Razali, R.; Majid, W.A.; Darroudi, M. Synthesis and characterization of a narrow size distribution of zinc oxide nanoparticles. *Int. J. Nanomed.* **2011**, *6*, 1399–1403.

44. Khalil, A.T.; Ovais, M.; Ullah, I.; Ali, M.; Shinwari, Z.K.; Hassan, D.; Maaza, M. *Sageretia thea* (Osbeck.) modulated biosynthesis of NiO nanoparticles and their in vitro pharmacognostic, antioxidant and cytotoxic potential. *Artif. Cells Nanomed. Biotechnol.* **2018**, *46*, 838–852. [CrossRef]

45. Suresh, J.; Pradheesh, G.; Alexramani, V.; Sundrarajan, M.; Hong, S.I. Green synthesis and characterization of zinc oxide nanoparticle using insulin plant (*Costus pictus* D. Don) and investigation of its antimicrobial as well as anticancer activities. *Adv. Nat. Sci. Nanosci. Nanotechnol.* **2018**, *9*, 015008. [CrossRef]

46. Tahir, K.; Nazir, S.; Ahmad, A.; Li, B.; Khan, A.U.; Khan, Z.U.H.; Rahman, A.U. Facile and green synthesis of phytochemicals capped platinum nanoparticles and in vitro their superior antibacterial activity. *J. Photochem. Photobiol. B Biol.* **2017**, *166*, 246. [CrossRef]

47. Khan, F.U.; Chen, Y.; Khan, N.U.; Ahmad, A.; Tahir, K.; Khan, Z.U.; Wan, P. Visible light inactivation of *E. coli*, Cytotoxicity and ROS determination of biochemically capped gold nanoparticles. *Microb. Pathog.* **2017**, *107*, 419–424. [CrossRef]

48. Yedurkar, S.; Maurya, C.; Mahanwar, P. Biosynthesis of zinc oxide nanoparticles using ixora coccinea leaf extract—A green approach. *Open J. Synth. Theory Appl.* **2016**, *5*, 1–14. [CrossRef]

49. Iqbal, J.; Abbasi, B.A.; Ahmad, R.; Batool, R.; Mahmood, T.; Ali, B.; Bashir, S. Potential phytochemicals in the fight against skin cancer: Current landscape and future perspectives. *Biomed. Pharmacother.* **2019**, *109*, 1381–1393. [CrossRef]

50. Iqbal, J.; Abbasi, B.A.; Batool, R.; Mahmood, T.; Ali, B.; Khalil, A.T.; Ahmad, R. Potential phytocompounds for developing breast cancer therapeutics: nature's healing touch. *Eur. J. Pharmacol.* **2018**, *827*, 125–148. [CrossRef]

51. Daher, S.; Massarwa, M.; Benson, A.A.; Khoury, T. Current and future treatment of hepatocellular carcinoma: An updated comprehensive review. *J. Clin. Trans. Hepatol.* **2018**, *6*, 69–78. [CrossRef] [PubMed]

52. Hassan, H.F.H.; Mansour, A.M.; Abo-Youssef, A.M.H.; Elsadek, B.E.; Messiha, B.A.S. Zinc oxide nanoparticles as a novel anticancer approach; in vitro and in vivo evidence. *Clin. Exp. Pharmacol. Physiol.* **2017**, *44*, 235–243. [CrossRef] [PubMed]

53. Iqbal, J.; Abbasi, B.A.; Mahmood, T.; Kanwal, S.; Ahmad, R.; Ashraf, M. Plant-extract mediated green approach for the synthesis of ZnONPs: Characterization and evaluation of cytotoxic, antimicrobial and antioxidant potentials. *J. Mol. Struct.* **2019**, *1189*, 315–327. [CrossRef]

54. Kaye, P.; Scott, P. Leishmaniasis: Complexity at the host–pathogen interface. *Nat. Rev. Microbiol.* **2011**, *9*, 604–611. [CrossRef] [PubMed]

55. Safawo, T.; Sandeep, B.V.; Pola, S.; Tadesse, A. Synthesis and characterization of zinc oxide nanoparticles using tuber extract of anchote (*Coccinia abyssinica* (Lam.) Cong.) for antimicrobial and antioxidant activity assessment. *Open Nano* **2018**, *3*, 56–63. [CrossRef]

56. Li, Y.; Zhang, W.; Niu, J.; Chen, Y. Mechanism of photogenerated reactive oxygen species and correlation with the antibacterial properties of engineered metal-oxide nanoparticles. *ACS Nano* **2012**, *6*, 5164–5173. [CrossRef]

57. Yao, G.; Sebisubi, F.M.; Voo, L.Y.C.; Ho, C.C.; Tan, G.T.; Chang, L.C. Citrinin derivatives from the soil filamentous fungus Penicillium sp. H9318. *J. Braz. Chem. Soc.* **2011**, *22*, 1125–1129. [CrossRef]

58. Oyedemi, S.O.; Oyedemi, B.O.; Ijeh, I.I.; Ohanyerem, P.E.; Coopoosamy, R.M.; Aiyegoro, O.A. Alpha-amylase inhibition and antioxidative capacity of some antidiabetic plants used by the traditional healers in Southeastern Nigeria. *Sci. World J.* **2017**, *2017*, 3592491. [CrossRef]

59. Ul-Haq, I.; Ullah, N.; Bibi, G.; Kanwal, S.; Ahmad, M.S.; Mirza, B. Antioxidant and cytotoxic activities and phytochemical analysis of *Euphorbia wallichii* root extract and its fractions. *Iran. J. Pharm. Res.* **2012**, *11*, 241–249.

60. Dobrovolskaia, M.A.; Clogston, J.D.; Neun, B.W.; Hall, J.B.; Patri, A.K.; McNeil, S.E. Method for analysis of nanoparticle hemolytic properties in vitro. *Nano Lett.* **2008**, *8*, 2180–2187. [CrossRef]

61. Prach, M.; Stone, V.; Proudfoot, L. Zinc oxide nanoparticles and monocytes: Impact of size, charge and solubility on activation status. *Toxicol. Appl. Pharmacol.* **2013**, *266*, 19–26. [CrossRef] [PubMed]

 *biomolecules*

Article

Combined Use of Deep Eutectic Solvents, Macroporous Resins, and Preparative Liquid Chromatography for the Isolation and Purification of Flavonoids and 20-Hydroxyecdysone from *Chenopodium quinoa* Willd

Jia Zeng [1,2,†], Xianchao Shang [1,2,†], Peng Zhang [1], Hongwei Wang [3,*] , Yanlong Gu [4,*] and Jia-Neng Tan [1,*]

[1] Tobacco Research Institute, Chinese Academy of Agricultural Sciences, Qingdao 266101, China; zj253700112@126.com (J.Z.); sxc3220341@163.com (X.S.); Zhangpeng2008@caas.cn (P.Z.)

[2] Graduate School of Chinese Academy of Agricultural Sciences, Beijing 100081, China

[3] State Key Laboratory Cultivation Base, Shandong Provincial Key Laboratory of Ophthalmology, Shandong Eye Institute, Shandong First Medical University & Shandong Academy of Medical Sciences, Qingdao 266071, China

[4] Key Laboratory of Energy Conversion and Storage, Ministry of Education, School of Chemistry and Chemical Engineering, Huazhong University of Science and Technology, Wuhan 430074, China

* Correspondence: whw20051256@163.com (H.W.); klgyl@hust.edu.cn (Y.G.); tanjianeng@caas.cn (J.-N.T.)

† These authors contributed equally to this work.

Received: 16 September 2019; Accepted: 21 November 2019; Published: 25 November 2019

Abstract: Deep eutectic solvents (DESs) were used in combination with macroporous resins to isolate and purify flavonoids and 20-hydroxyecdysone from *Chenopodium quinoa* Willd by preparative high-performance liquid chromatography (HPLC). The extraction performances of six DESs and the adsorption/desorption performances of five resins (AB-8, D101, HPD 400, HPD 600, and NKA-9) were investigated using the total flavonoid and 20-hydroxyecdysone extraction yields as the evaluation criteria, and the best-performing DES (choline chloride/urea, DES-6) and macroporous resin (D101) were further employed for phytochemical extraction and DES removal, respectively. The purified extract was subjected to preparative HPLC, and the five collected fractions were purified in a successive round of preparative HPLC to isolate three flavonoids and 20-hydroxyecdysone, which were identified by spectroscopic techniques. The use of a DES in this study significantly facilitated the preparative-scale isolation and purification of polar phytochemicals from complex plant systems.

Keywords: DESs; preparative-scale purification; flavonoids; 20-hydroxyecdysone; *Chenopodium quinoa*

---

1. Introduction

Deep eutectic solvents (DESs) exhibit the advantages of negligible volatility, adjustable viscosity, high solubility, preparation simplicity, eco-friendliness, and biodegradability, and have therefore attracted much attention [1–6]. Typically, DESs are homogeneous solutions of quaternary ammonium salts, which act as hydrogen-bond acceptors (HBAs) and engage in strong intermolecular hydrogen-bonding interactions with different hydrogen-bond donors. Among the diverse HBAs, cheap, biodegradable, and non-toxic choline chloride (ChCl) is most commonly used, [7] and ChCl-based DESs have been widely applied as reaction media in the fields of organic synthesis, biomass refinery, polymerization, and materials science [8–13]. Moreover, DESs can be used to extract phytochemicals, such as flavonoids, phenolic acids, κ-carrageenan, anthocyanins, and saponins, from various types of natural sources [14–19]. DES extraction technologies, however, have all been restricted to the analysis of active constituents, which could only be detected by HPLC, HPLC-MS/MS, or some other analysis instruments using standard chemicals. Until now, there have been almost no studies on the preparative-scale separation and purification of monomer compounds that are not available on-hand, particularly for plant samples, based on DES extraction. Thus, introducing DESs into the area of phytochemistry will provide an interesting approach for researchers.

Chenopodium quinoa Willd. (quinoa) is one of the oldest cultivated plants in the Andean region and is well known as a functional food and nutraceutical source rich in essential amino acids, unsaturated fats, phytoecdysteroids, flavonoids, phenolic acids, betalains, and saponins, which can allegedly reduce the risk of cardiovascular disease, cancer, osteoporosis, and diabetes [20–31]. For example, flavonoids are commonly used to treat and prevent diabetes and obesity [32,33], while 20-hydroxyecdysone is believed to exhibit antioxidant, antidiabetic, anti-obesity, antihypertensive, anticancer, and anti-inflammatory properties [34,35]. Therefore, the isolation and characterization of phytochemicals from *C. quinoa* is of high practical significance.

The extraction of phytochemicals from *C. quinoa* has gained significant attention [28–30]. For example, Dini et al. isolated and characterized phenolic constituents (e.g., kaempferol, quercetin (QE), and vanillic acid glycosides) from quinoa seeds [29] by partitioning the methanolic seed extract between *n*-butanol and water. The *n*-butanol phase was subjected to defatting with chloroform (CHCl$_3$), and the obtained residue was loaded onto a Sephadex LH-20 column and subjected to droplet counter-current chromatography assisted by thin-layer chromatography to isolate several individual glycosides. In a report by Ho et al., where a series of ecdysteroids were isolated from *C. quinoa* [30], a suspension of the ethanolic (95 vol% aqueous ethanol) extract of ground seeds in water was sequentially treated with hexane, ethyl acetate, and *n*-butanol. Ecdysteroids, including 20-hydroxyecdysone, were separated from the *n*-butanol fraction by the combined use of Diaion HP-20 gel, silica gel, and RP-18 reverse-phase column chromatography. Notably, the above-mentioned organic solvent-based purification procedures are not efficient due to low extraction efficiencies and long extraction times. Therefore, the development of more efficient alternatives for the practical extraction of phytochemicals from *C. quinoa* is required.

Herein, we achieved the separation and purification of flavonoids and 20-hydroxyecdysone from *C. quinoa* by using DESs as extracting solvents in conjunction with the use of macroporous resins, employing preparative high-performance liquid chromatography (HPLC) (Figure 1). The extraction performances of six DESs were investigated and compared with some selected organic solvents. The best DES was thereafter employed in some downstream experiments. A mixture consisting of the DES and quinoa extract was separated using a macroporous resin. Then, the extract was subjected to preparative HPLC for the isolation and purification of monomer compounds, which were further identified by spectroscopic techniques based on comparison with previously reported data.

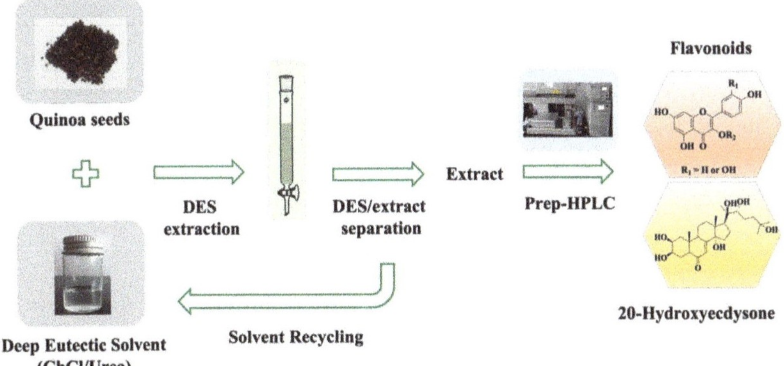

Figure 1. Schematic for routine isolation and purification of flavonoids and 20-hydroxyecdysone using the developed technique.

2. Materials and Methods

2.1. Chemicals, Reagents, and Instruments

ChCl, aminoethyl alcohol, D-glucose, glycerol, oxalic acid, lactic acid, urea, methanol-d_4, QE, 20-hydroxyecdysone, sodium nitrite, and aluminium trichloride ($AlCl_3$) were purchased from Aladdin Industrial Co., Ltd. (Shanghai, China). HPLC-grade methanol and acetonitrile (ACN) were purchased from Shanghai Macklin Biochemical Co., Ltd. (Shanghai, China). Macroporous resins (D101, HPD 600, NKA-9, HPD 400, and AB-8) and analytical-grade methanol were purchased from Sinopharm Chemical Reagent Co., Ltd. (Shanghai, China). Deionized water was obtained using a Unique-R20 purification system (Xiamen RSJ Scientific Instruments Co., Ltd., Xiamen, China).

Nuclear magnetic resonance (NMR) spectra were recorded on an Agilent DD2 spectrometer (500 MHz for ^{1}H and 125 MHz for ^{13}C NMR; Agilent, CA, USA). High-resolution electrospray ionization–tandem mass spectrometry (HRESI-MS) was performed using an LTQ Orbitrap XL spectrometer (Thermo Scientific, MA, USA).

2.2. Determination of Total Flavonoid (TF) and 20-Hydroxyecdysone Contents

TF content was determined using a method obtained by the modification of a previously reported method [36]. A 96-well plate was sequentially charged with the concentrate (50 μL), 0.066 M sodium nitrite solution (100 μL), and 10% $AlCl_3$ solution (*w/v*, 15 μL) and incubated at room temperature for 6 min. Thereafter, 0.5 M aqueous sodium hydroxide (NaOH, 100 μL) was added to terminate the reaction, and the absorbance of the reaction mixture at 510 nm was measured by a microplate reader using QE as a standard. TF content was expressed as milligrams of QE per gram of dry quinoa material.

20-Hydroxyecdysone content was quantified by ultra-performance liquid chromatography–triple quadrupole tandem mass spectrometry (UPLC-QqQ-MS/MS) using an Acquity UPLC system (Waters Corp., MA, USA) coupled with a TSQ Quantum triple quadrupole tandem mass spectrometer (Thermo Scientific, CA, USA). The mobile phase, comprising ACN (eluent A) and 0.1 vol% aqueous formic acid (eluent B), was supplied at a flow rate of 0.3 mL/min. The following gradient program was used: 0–2 min, 5–10 vol% ACN; 2–4 min, 10–20 vol% ACN; 4–6 min, 20–30 vol% ACN; 6–8 min, 30 vol% ACN; 8–10 min, 30–5 vol% ACN. The injection volume and column temperature were 10 μL and 25 °C, respectively.

Low-resolution mass spectrometric detection was carried out in the selected reaction-monitoring mode using an electrospray ionization (ESI) interface. The optimized parameters corresponded to a vaporizer temperature of 450 °C, corona discharge voltage of 4.0 kV, capillary temperature of 225 °C, sheath nitrogen gas pressure of 17 psi, auxiliary nitrogen gas pressure of 5 psi, and collision gas pressure of 1.5 mTorr. The Xcalibur software was used to control the LC-MS system and analyze the collected data. The mass spectrometric analysis of 20-hydroxyecdysone was initially performed over the full scan range in a negative-ion mode, and a pseudomolecular ion at m/z 479.176 (t_R = 4.65 min) as well as four product fragment ions at m/z 319.1 (collision energy = 27 eV), 301.1 (collision energy = 33 eV), 159.0 (collision energy = 27 eV), and 83.0 (collision energy = 37 eV) were detected. The fragmentation patterns of 20-hydroxyecdysone were compared with those previously reported in the literature [37].

2.3. Preparation and Selection of DESs

The ChCl/urea DES was prepared using a modification of a previously reported method [12] (Table 1). Briefly, a glass vial was charged with ChCl and urea, and the obtained mixture was stirred at 280 rpm at 80 °C for 2 h. The obtained ChCl/urea DES (clear stable liquid) was mixed with water (7/3, v/v), and the solution was used for further experiments. Other DESs were prepared in a similar way.

Ultrasound-assisted extraction was used to extract TF and 20-hydroxyecdysone. The extraction capacities of six DESs were evaluated using the extraction yields (E_y) of TF and 20-hydroxyecdysone. First, 100 mg of quinoa seed powder and 1.0 mL of DES were added to a 10 mL glass tube in sequence. The mixture was placed on a vortex meter, stirred for 2 min, and further ultrasonicated at 50 °C for 30 min (25 kHz, 200 W). After the tube cooled down to the room temperature, the mixture was centrifuged at 3000 rpm for 10 min. The supernatant (1 mL) was diluted with methanol (4 mL), and the mixture was passed through a 0.22 μm filter for subsequent analyses. Each extraction was performed in triplicate, and E_y was calculated as follows:

$$E_y = (C_0 \times V_0)/M_0, \tag{1}$$

where C_0 is the concentration of TF or 20-hydroxyecdysone found in the DES, V_0 is the volume of the diluted liquid, and M_0 is the mass of the sample.

Table 1. Compositions and physicochemical properties of deep eutectic solvents (DESs) used in this study.

DESs	Composite of DES	Molar Ratio	Viscosity [a] Pa·s	Conductivity μS/cm	Density g/cm^3
DES-1	ChCl/aminoethyl alcohol	1:6	0.037	3330	1.058
DES-2	ChCl/D-glucose	1:1	- [c]	98	1.304
DES-3	ChCl/glycerol	1:2	0.227	1750	1.171
DES-4	ChCl/oxalic acid [b]	1:1	0.152	1043	1.230
DES-5	ChCl/lactic acid	1:1	0.270	3670	1.133
DES-6	ChCl/urea	1:2	0.658	1880	1.184

[a] Determined at 30 °C. [b] Dehydrate. [c] Not available.

2.4. Response Surface Methodology (RSM)

The optimal values for three variables in the selected DES, including water content (a), extraction temperature (b), and solid (quinoa seed)–liquid (DES) ratio (c) at three levels (−1, 0, and +1) were determined using RSM based on a Box–Behnken design (BBD). Table S1 (Supplementary Materials) shows the investigated variables and their values in the three-level BBD. The experiment was designed to reveal the effect of each parameter on TF and 20-hydroxyecdysone contents as well as the interactions between the three main factors. The entire study was comprised of 17 separate experiments including 5 center points, each of which was conducted in triplicate.

Statistical comparisons were made using a single factor analysis of variance (ANOVA); p-values < 0.05 were considered significant. Data were processed using Design-Expert 8.5 statistical software and the social package for statistical studies (SPSS 19.0).

2.5. Adsorption/Desorption Capacities of the Five Employed Resins

The adsorption/desorption capacities of the five kinds of macroporous resins were investigated using the adsorption/desorption yields of TF and 20-hydroxyecdysone. The DES extract solution was added to a 200 mL flask containing 1.0 g of the pretreated resins. The mixture was shaken at 120 rpm at room temperature for 24 h to reach adsorption equilibrium, and the resins were washed by deionized water and then desorbed with 50 mL of methanol in the flask, which was continually shaken at 120 rpm at room temperature for 24 h. The adsorption/desorption yields of resins were calculated according to the following Equations (2) and (3):

$$D_a = (C_0 - C_a)/C_0, \tag{2}$$

$$D_d = C_d \times V_d/(C_0 - C_a) \times V_0, \tag{3}$$

where D_a is the adsorption yield at adsorption equilibrium; D_d is the desorption yield after adsorption equilibrium; C_0, C_a, and C_d represent the content of TF or 20-hydroxyecdysone at initial equilibrium, absorption equilibrium, and desorption equilibrium, respectively; V_0 and V_d are the initial sample volume and desorption solution volume (mL), respectively.

2.6. Small- and Preparative-Scale Sample Extraction

Quinoa seeds, desaponified by soaking in distilled water for 24 h without the presence of foam, were purchased from Xinjing Quinoa Cultivation and Promotion Co., Ltd. (Shanxi, China) [38]. For small-scale extraction, the seeds were ground and 1.0 g of the obtained powder was mixed with aqueous DES (10 mL). The resulting mixture was sonicated (100 W) at 50 °C for 30 min in an ultrasonic bath (Scientz SB25-12D, Ningbo Scientz Biotechnology Co., Ltd, Ningbo, China) and then centrifuged at 5000 rpm for 10 min. Small-scale extraction was used in the selection of DESs and macroporous resins as well as the RSM experiments. For preparative-scale extraction, powdered seeds (1.0 kg) were suspended in 10 L of aqueous DES. The extraction process was similar to that of small-scale extraction. The collected supernatants were loaded onto a macroporous resin column (60 mm × 1200 mm), which was sequentially flushed with deionized water (5000 mL) at a flow rate of 15 mL/min (to remove the DES) and methanol (2000 mL) at a flow rate of 20 mL/min under reduced pressure. The methanolic extract was concentrated, and the concentrate was stored at 4 °C for further use.

2.7. Analytical and Preparative HPLC

For the first separation, analytical HPLC of quinoa extract was performed on a Waters 2489 chromatography system (Waters, MA, USA) equipped with a C18 separation column (5 μm, 4.6 × 150 mm) using ACN and water for gradient elution at a flow rate of 1.0 mL/min. UV detection was performed at 210 nm. The gradient program was as follows: 0–5 min, 8 vol% ACN; 5–7.5 min, 8–15 vol% ACN; 7.5–10 min, 15–30 vol% ACN; 10–15 min, 30–50 vol% ACN; 15–17.5 min, 50–60 vol% ACN; 17.5–30 min, 60 vol% ACN. The first-round preparative HPLC of quinoa extract was performed on a DAC100 preparative HPLC system (Beijing, China) equipped with a C18 separation column (5 μm, 100 × 250 mm) at a flow rate of 300 mL/min, while the other chromatographic conditions were identical to those used for analytical HPLC. The obtained eluent was separated into five (I–V) fractions.

In the second round of separation, fractions I–V were subjected to analytical and preparative HPLC performed on the Waters 2489 chromatography system equipped with 4.6 × 150 mm (5 μm) and 10 × 250 mm (5 μm) C18 columns, respectively. The flow rates of analytical and preparative separations were set to 1.0 and 6.0 mL/min, respectively. The mobile phase composition and UV detection parameters were identical to those used for the first round of separation. The gradient program was as follows: 0–5 min, 5 vol% ACN; 5–10 min, 5–10 vol% ACN; 10–15 min, 10 vol% ACN; 15–20 min, 10–15 vol% ACN; 20–25 min, 15 vol% ACN; 25–30 min, 15–30 vol% ACN; 30–35 min, 30 vol% ACN; 35–40 min, 30–50 vol% ACN; 40–50 min, 50 vol% ACN.

3. Results and Discussion

3.1. Preparation and Optimization of DES Systems

The non-toxicity, biodegradability, and low cost of ChCl make it suitable for combination with a series of hydrogen-bond donors (HBDs) to afford biocompatible and renewable DESs. Herein, six DESs were synthesized by combining ChCl with HBDs belonging to the classes of inexpensive and easily available fine chemicals (aminoethyl alcohol and urea), polyalcohols (glycerol), organic acids (oxalic and lactic acids), and sugars (D-glucose). The composition and physicochemical properties (density, conductivity, and viscosity) of the six DESs are listed in Table 1.

DES extraction performance was evaluated by considering the TF and 20-hydroxyecdysone contents of the corresponding extracts. Figure 2A compares the obtained TF contents, revealing that the value obtained for DES-6 (3.93 ± 0.15 mg/g) exceeded those obtained for DES-4 (3.45 ± 0.09 mg/g), DES-3 (3.31 ± 0.10 mg quercetin equivalents (QE)/g), DES-5 (2.78 ± 0.17 mg QE/g), DES-1 (2.52 ± 0.06 mg QE/g), and DES-2 (2.25 ± 0.13 mg QE/g). Dini et al. [39] determined the TF contents before and after cooking in sweet and bitter quinoa seeds. The samples were extracted in methanol–water (80:20 *v/v*). It was found that the bitter quinoa seeds processed higher TF content before cooking (1.39 ± 0.35 mg catechin equivalents (CE)/1.0 g) than the sweet seeds (0.81 ± 0.10 mg CE/1.0 g). After cooking, TF contents of bitter and sweet quinoa seeds decreased significantly to 0.63 ± 0.15 mg CE/g and 0.18 ± 0.07 mg CE/g, respectively. In consideration of the similar molecular weights of QE and CE, the TF contents obtained by Dini et al. were lower than those in our study. Figure 2B shows that similar 20-hydroxyecdysone contents were obtained for DES-3 (551.50 ± 4.51 mg/g), DES-5 (546.33 ± 7.60 mg/g), and DES-2 (504.69 ± 12.05 mg/g), while higher values were obtained for DES-6 (606.76 ± 7.98 mg/g), and significantly lower values were obtained for DES-1 (384.80 ± 12.16 mg/g) and DES-4 (107.65 ± 2.88 mg/g). Lafont et al. [30] determined the 20-hydroxyecdysone contents of quinoa seeds obtained from four producing countries (Bolivia, Chile, Ecuador, and Peru). Seed powder was extracted with 20 mL methanol–water (35:65, *v/v*) overnight with magnetic stirring. The obtained 20-hydroxyecdysone contents (0.32–0.42 mg/g) were significantly lower than those obtained after extraction by DES-2, DES-3, DES-5, and DES-6. In addition, the 20-hydroxyecdysone content of the DES-6 extract was also much higher than the values previously reported by Raskin et al. (0.18–0.49 mg/g) [28]. It is possible that the weakly alkaline HBD (urea) possessed stronger hydrogen-bond interactions with phenolic hydroxyls and alcoholic hydroxyls in the flavonoids or alcoholic hydroxyls in 20-hydroxyecdysone than the neutral (D-glucose and glycerol), strong alkaline (aminoethyl alcohol), and weakly acidic (oxalic acid and lactic acid) HBDs, contributing to the enhanced extraction yields. Based on the above results, DES-6 (ChCl/urea) was used in subsequent experiments.

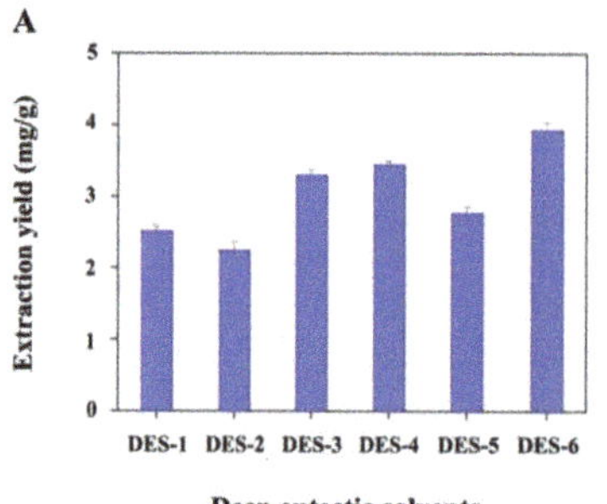

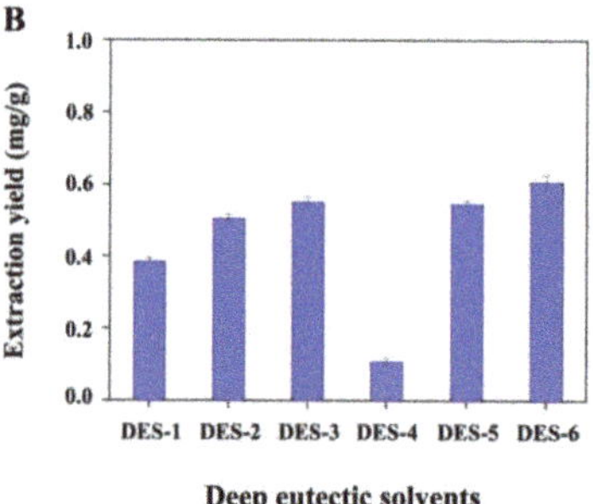

Figure 2. (**A**) Total flavonoid (TF) and (**B**) 20-hydroxyecdysone extraction efficiencies of the employed DESs. TF extraction efficiency was expressed as milligrams QE equivalents per gram of quinoa sample.

3.2. BBD Experimental Design

Three values, including water content in the DES (variable a), extraction temperature (variable b), and solid–liquid ratio (variable c), were used as independent variables to efficiently optimize the extraction yields of TF and 20-hydroxyecdysone from *C. quinoa* in the RSM experiments. The extraction yields were considered as related responses in order to evaluate the efficiency of the extraction procedure. The experiments were performed in a random order to avoid systematic error. The results, including the coded variables and related responses, are presented in Table 2. A second-order polynomial equation was applied to express the proposed model after multiple regression analysis of the experimental data. The regression model equations for the responses and variables in terms of the coded levels are as follows:

$$Y_{TF} = 3.94 + 0.48a + 0.017b - 0.25c - 0.097ab - 0.028ac + 0.20bc - 0.86a^2 - 0.41b^2 - 0.34c^2, \quad (4)$$

$$Y_{20\text{-Hydroxyecdysone}} = 0.50 + 0.033a - 0.017b - 0.025c - 0.037ab - 0.020ac - 0.046bc - 0.068a^2 - 0.026b^2 - 0.12c^2, \quad (5)$$

where a is the DES water content, b is the extraction temperature, and c is the solid–liquid ratio.

Variances of the ANOVA regression model equations of TF and 20-hydroxyecdysone are shown in Tables S2 and S3 (Supplementary Materials). The coefficients (R^2) of the variables of response were 0.9553 and 0.9501 for TF and 20-hydroxyecdysone, respectively. The F-values for the lack-of-fit model were all non-significant, which supported our assumption that the models were sufficient to accurately represent the experimental data.

A response surface plot of the model was used for graphically interpreting the significant effects of interactions among the three variables on the contents of TF and 20-hydroxyecdysone (Figure 3). Results showed that the extraction yields of TF and 20-hydroxyecdysone were apparently related to the main variable. In the model, the concentration of water (%) in the DES solution (variable a) and solid–liquid ratio (variable c) exhibited statistically significant effects ($p < 0.05$) on the extraction yields of TF and 20-hydroxyecdysone. In contrast, extraction temperature (variable b) showed a non-significant effect ($p > 0.05$) on the contents.

Table 2. Response surface optimization experiments using DES-6 extraction of investigated variables.

No.	Coded Values			Variables			Extraction Yields (mg/g)	
	a	b	c	a (%)	b (°C)	c (mg/mL)	TF	20-Hydroxyecdysone
1	1	0	1	50	50	125	2.90	0.29
2	0	1	1	30	60	125	3.20	0.25
3	0	0	0	30	50	100	4.49	0.51
4	−1	−1	0	10	40	100	2.08	0.33
5	0	0	0	30	50	100	4.29	0.53
6	1	1	0	50	60	100	3.07	0.34
7	0	0	0	30	50	100	4.13	0.50
8	1	0	−1	50	50	75	3.20	0.34
9	0	−1	1	30	40	125	2.72	0.38
10	0	1	−1	30	60	75	3.27	0.42
11	1	−1	0	50	40	100	2.78	0.47
12	−1	0	−1	10	50	75	2.53	0.30
13	−1	1	0	10	60	100	2.25	0.34
14	0	0	0	30	50	100	4.28	0.51
15	0	0	0	30	50	100	3.70	0.47
16	0	−1	−1	30	40	75	3.40	0.32
17	−1	0	1	10	50	125	2.03	0.29

Based on RSM results, TF was extracted from *C. quinoa* using the optimum conditions (DES water content, 32.4%; extraction temperature, 50.7 °C; and solid–liquid ratio, 97.2 mg/mL). The optimal extraction conditions for 20-hydroxyecdysone extraction from *C. quinoa*, as well as the related maximal response values (Table 3), were established by applying a simple procedure (DES water content, 32.6%; extraction temperature, 47.8 °C; and solid–liquid ratio, 99.7 mg/mL). A verification experiment performed using these conditions gave optimal extraction yields of 4.23 and 0.51 mg/g for TF and 20-hydroxyecdysone, respectively. In order to simplify the operation, the experimental conditions with a DES water content of 30.0%, an extraction temperature of 50.0 °C, and a solid–liquid ratio of 100.0 mg/mL were selected for DES extraction (10 L). Under this condition, the extraction yields of TF and 20-hydroxyecdysone were 4.11 and 0.50 mg/g, respectively, which were close to the predicted values.

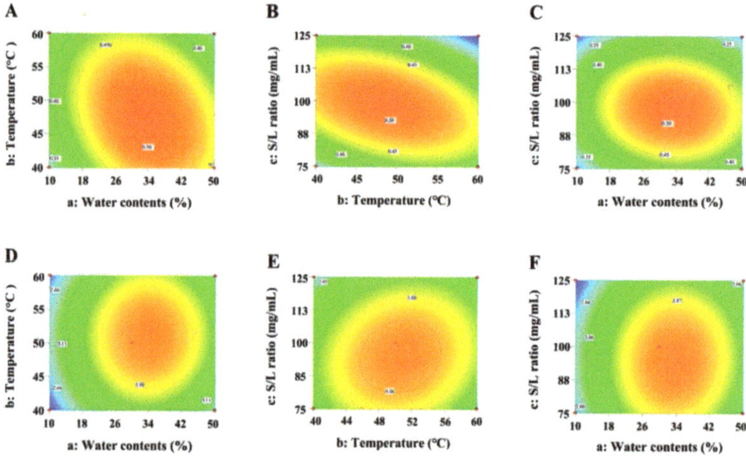

Figure 3. Response surface plots of the models for (**A, B**, and **C**) TF and (**D, E**, and **F**) 20-hydroxyecdysone contents.

Table 3. Optimal conditions of the variables that maximize the response values using response surface methodology.

	Optimal Variable Conditions			Optimum (mg/g)
	a (%)	b (°C)	c (mg/mL)	
TF	32.4	50.7	97.2	4.23
20-Hydroxyecdysone	32.6	47.8	99.7	0.51

3.3. Macroporous Resin Selection

Macroporous resins exhibit the advantages of easy recyclability, low cost, high adsorption/desorption capacity, and suitability for large-scale production, and have therefore been widely used for the enrichment and purification of phytochemicals extracted from complex plant systems. Here, we used TF and 20-hydroxyecdysone contents to evaluate the performances of five macroporous resins (AB-8, D101, HPD 400, HPD 600, and NKA-9) for removing DESs from extracts prior to preparative HPLC (Figure 4A,B). The D101 resin had the highest TF adsorption capacity (81.2%), which was approximately two-fold higher than that of the HPD 400 resin (38.9%) (Figure 4A). The adsorption capacities of AB-8 (73.4%) and HPD 400 (69.3%) were similar to each other and higher than that of NKA-9 (63.6%). The TF desorption capacities of all the tested resins exceeded 70%. In particular, D101 (89.7%) and AB-8 (79.2%) had similar desorption capacities that exceeded those of HPD 400 (75.4%), HPD 600 (73.4%), and NKA-9 (70.2%). Moreover, D101 exhibited the highest 20-hydroxyecdysone adsorption/desorption capacities among the five tested resins (Figure 4B). Therefore, D101 was selected for implementing the next studies.

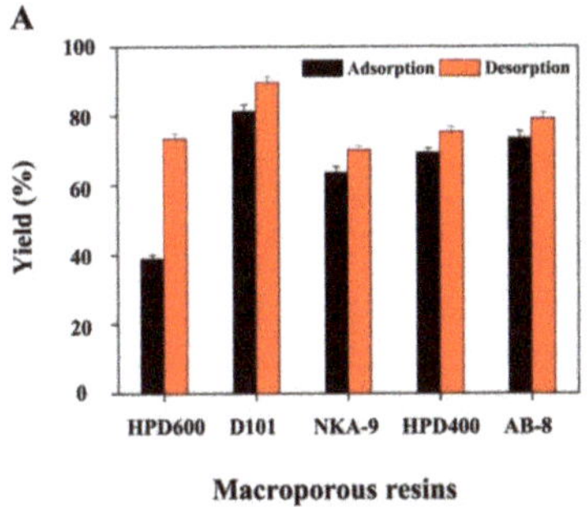
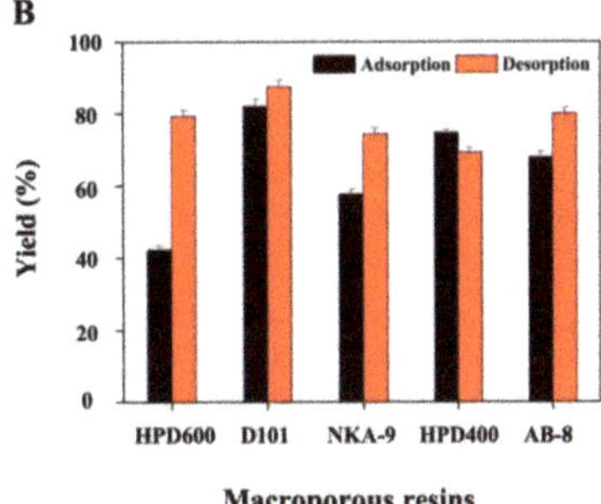

Figure 4. (**A**) TF and (**B**) 20-hydroxyecdysone adsorption/desorption capacities of the five employed resins.

3.4. Preparative-Scale Isolation and Purification of Monomer Compounds

Eluate concentrates obtained at the DES removal stage were sequentially subjected to analytical and preparative HPLC, and the corresponding chromatograms (Figure 5) showed that the resolution of preparative HPLC was similar to that of analytical HPLC. Fractions I (242 mg), II (253 mg), III (206 mg), IV (988 mg), and V (265 mg) separated from 9.2 g of the concentrate were further evaluated by analytical HPLC. The corresponding chromatograms, shown in Figure 6A, revealed that the compounds obtained after the first separation round were not sufficiently pure. Therefore, a second round of preparative HPLC (Figure 6B) was performed using a C18 column with a smaller inner diameter (5 μm, 10 × 250 mm), yielding compound **1** (48.7 mg), compound **2** (47.9 mg), compound **3** (39.9 mg), and compound **4** (230.6 mg) from fractions I–IV, respectively. No compound was isolated from fraction V.

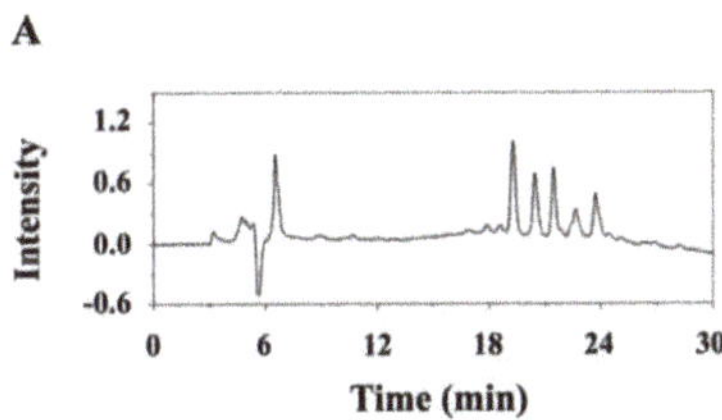
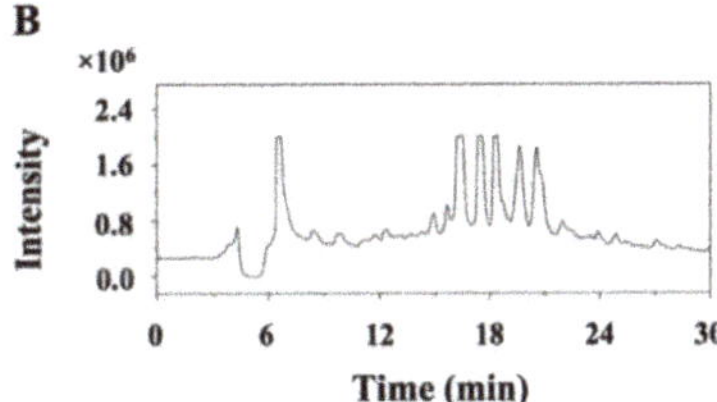

Figure 5. Analytical (**A**) and preparative-scale (**B**) HPLC-UV chromatograms of quinoa extracts.

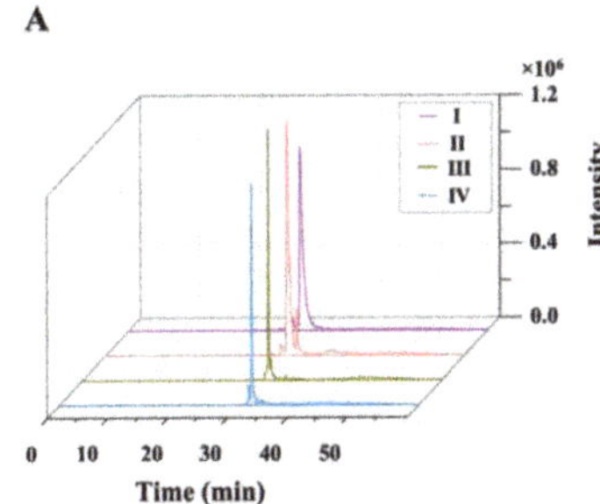
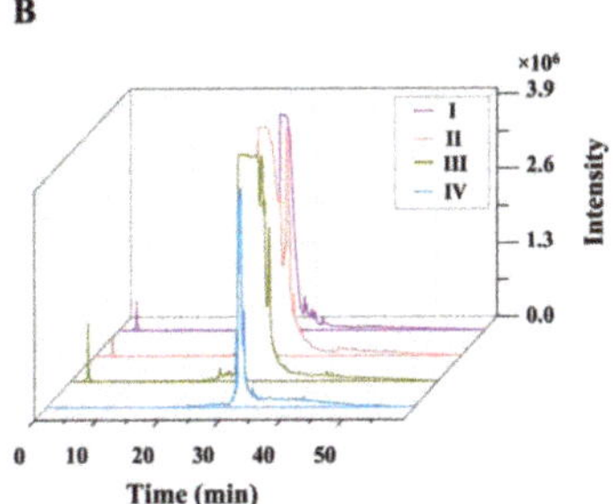

Figure 6. Analytical (**A**) and preparative-scale (**B**) HPLC-UV chromatograms of fractions I–IV.

3.5. Identification of Isolated Compounds

Compounds **1–4** were identified based on their HRESI-MS, ^{1}H NMR, ^{13}C NMR, and 2D NMR spectra (Table 4).

Compound **1**. HRESI-MS: found *m/z* 755.2048 [M-H]$^-$, calcd. for $C_{33}H_{40}O_{20}$ *m/z* 756.2113. ^{1}H NMR (500 MHz, CD$_3$OD): δ (ppm) 7.69 (1H, d, *J* = 2.2 Hz), 7.56 (1H, dd, *J* = 8.5, 2.2 Hz), 6.87 (1H, d, *J* = 8.5 Hz), 6.36 (1H, d, *J* = 2.1 Hz), 6.19 (1H, d, *J* = 2.2 Hz), 5.66 (1H, d, *J* = 7.8 Hz), 5.59 (1H, d, *J* = 7.7 Hz), 5.21 (1H, d, *J* = 1.4 Hz), 4.54 (1H, d, *J* = 1.5 Hz), 1.18 (3H, d, *J* = 6.5 Hz), and 0.95 (3H, d, *J* = 6.5 Hz). ^{13}C NMR (125 MHz, CD$_3$OD): δ (ppm) 178.1, 164.3, 161.7, 156.9, 156.9, 148.2, 144.4, 133.1, 121.9, 121.6, 115.9, 114.7, 104.5, 100.4, 99.6, 98.3, 98.3, 93.2, 76.0, 74.2, 73.8, 72.6, 72.5, 71.0, 70.9, 70.8, 70.6, 69.4, 68.4, 68.3, 65.3, 16.6, 16.0. A comparison with previously reported data [29] allowed compound **1** to be identified as quercetin-3-*O*-(2,6-di-α-L-rhamnopyranosyl)-β-D-galactopyranoside (Figures S4–S6).

Compound **2**. HRESI-MS: found *m/z* 741.1875 [M-H]$^-$, calcd. for $C_{32}H_{38}O_{20}$ *m/z* 742.1956. ^{1}H NMR (500 MHz, CD$_3$OD): δ (ppm) 7.71 (1H, d, *J* = 2.1 Hz), 7.62 (1H, d, *J* = 8.4, 2.1 Hz), 6.86 (1H, d, *J* = 8.4 Hz), 6.37 (1H, d, *J* = 1.9 Hz), 6.17 (1H, d, *J* = 1.9 Hz), 5.40 (1H, d, *J* = 7.9 Hz), 4.52 (1H, d, *J* = 1.5 Hz), 5.46 (1H, d, *J* = 1.2 Hz), 4.06 (1H, d, *J* = 1.2 Hz), 3.99 (1H, dd, *J* = 7.9, 9.4 Hz), 3.79 (1H, m), 3.78 (1H, d, *J* = 3.4 Hz), 3.71 (1H, m), 3.70 (1H, dd, *J* = 12.6, 4.7 Hz), 3.61 (2H, s), 3.57 (1H, dd, *J* = 3.4, 1.5 Hz), 3.55 (1H, d, *J* = 3.5 Hz), 3.51 (1H, dq, *J* = 9.4, 6.2 Hz), 3.42 (1H, dd, *J* = 12.6, 6.8 Hz), 3.27 (1H, t, *J* = 9.4 Hz), 1.18 (3H, d, *J* = 6.2 Hz). ^{13}C NMR (125 MHz, CD$_3$OD): δ (ppm) 178.0, 164.3, 161.3, 157.0, 156.9, 148.3, 144.4, 133.5, 121.8, 121.8, 115.9, 114.7, 109.5, 104.3, 100.6, 100.3, 98.3, 93.2, 79.4, 76.6, 74.6, 74.2, 73.4, 73.3, 72.6, 71.8, 70.8, 69.3, 68.3, 65.5, 64.8, 16.5. The ^{1}H and ^{13}C NMR spectra of compound **2** were similar to those of quercetin-3-*O*-β-D-apiofuranosyl-(1′′′→2′′)-*O*-[α-L-rhamnopyranosyl-(1′′′′→6′′)]-β-D-galactopyranoside-3′,4′-dimethyl ether reported by Dini et al. [29] and compound **2** was concluded to be the demethoxylated version of the above species, namely quercetin-3-*O*-β-D-apiofuranosyl-(1′′′→2′′)-*O*-[α-L-rhamnopyranosyl-(1′′′′→6′′)]-β-D-galactopyranoside. As the NMR data of compound **2** have not been frequently reported, this compound was fully characterized by ^{1}H, ^{13}C, heteronuclear singular quantum correlation (HSQC), heteronuclear multiple bond correlation (HMBC), and ^{1}H-^{1}H correlated spectroscopy (COSY) NMR techniques, and the obtained results are shown in Table S4 (Figures S7–S14).

Compound **3**. HRESI-MS: found *m/z* 739.2099 [M-H]$^-$, calcd. for $C_{33}H_{40}O_{19}$ *m/z* 740.2164. ^{1}H NMR (500 MHz, CD$_3$OD): δ (ppm) 8.06 (2H, d, *J* = 8.9 Hz), 6.89 (2H, d, *J* = 8.9 Hz), 6.37 (1H, d, *J* = 1.7 Hz), 6.17 (1H, d, *J* = 1.5 Hz), 5.60 (1H, d, *J* = 7.6 Hz), 5.21 (1 H, d, *J* = 1.5 Hz), 4.52 (1H, d, *J* = 1.5 Hz), 1.17 (3H, d, *J* = 6.5 Hz) and 0.98 (3H, d, *J* = 6.5 Hz). ^{13}C NMR (125 MHz, CD$_3$OD): δ (ppm) 178.1, 164.3, 161.7, 159.9, 157.2, 157.0, 133.0, 130.8, 130.8, 121.6, 114.8, 114.8, 104.6, 101.1, 100.4, 99.4, 98.4, 93.3, 76.1, 74.3, 73.8, 72.6, 72.5, 71.0, 70.9, 70.9, 70.7, 69.3, 68.4, 68.3, 65.7, 16.5, 16.1. Comparison with previously reported data [29] allowed compound **3** to be identified as kaempferol-3-*O*-(2,6-di-α-L-rhamnopyranosyl)-β-D-galactopyranoside (Figures S15–S17).

Compound **4**. HRESI-MS: found *m/z* 479.3021 [M-H]$^-$, calcd. for $C_{27}H_{44}O_7$ *m/z* 480.3087. ^{1}H NMR (500 MHz, CD$_3$OD): δ (ppm) 5.80 (1H, d, *J* = 2.8 Hz), 3.95 (1H, br s), 3.83 (1H, br s), 1.20 (3H, s), 1.19 (3H, s), 1.18 (3H, s), 0.96 (3H, s), and 0.89 (3H, s). ^{13}C NMR (125 MHz, CD$_3$OD): δ (ppm) 205.0, 166.6, 120.7, 83.8, 77.0, 76.5, 69.9, 67.3, 67.1, 50.4, 49.1, 41.0, 37.8, 35.9, 33.7, 31.4, 31.1, 30.4, 28.3, 27.5, 25.9, 23.0, 20.1, 20.1, 19.6, 16.6. Apart from the missing carbon signal at 47.0 ppm, which was believed to overlap with the solvent peak, the spectral data were in accordance with those reported for 20-hydroxyecdysone [30]. Thus, compound **4** was identified as 20-hydroxyecdysone (Figures S18–S20).

Table 4. Structures of the isolated compounds.

Numbers	Compound Names	R_1	R_2
1	Quercetin-3-*O*-(2,6-di-α-L-rhamnopyranosyl)-β-D-galactopyranoside	OH	3-*O*-(2,6-di-α-L-rhamnopyranosyl)-β-D-galactopyranoside
2	Quercetin-3-*O*-β-D-apiofuranosyl-(1‴→2″)-*O*-[α-L-rhamnopyranosyl-(1⁗→6″)]-β-D-galactopyranoside	OH	3-*O*-β-D-apiofuranosyl-(1‴→2″)-*O*-[α-L-rhamnopyranosyl-(1⁗→6″)]-β-D-galactopyranoside
3	Kaempferol-3-*O*-(2,6-di-α-L-rhamnopyranosyl)-β-D-galactopyranoside	H	3-*O*-(2,6-di-α-L-rhamnopyranosyl)-β-D-galactopyranoside
4	20-Hydroxyecdysone		

4. Conclusions

A technique combining DES-assisted extraction, macroporous resin-facilitated DES removal, and preparative HPLC was developed for the preparative-scale isolation and purification of flavonoids and 20-hydroxyecdysone from *C. quinoa* Willd. Specifically, three flavonoids and 20-hydroxyecdysone were successfully separated and purified, and their structures were identified by HRESI-MS, 1D NMR, and 2D NMR. No saponins were isolated, as the employed *C. quinoa* sample had been subjected to desaponification. DESs were concluded to be appropriate for the extraction of polar compounds, such as glycosides, or other OH-group-rich compounds. The developed method is expected to be well suited for the preparative-scale separation and purification of minor phytochemicals of interest from complex plant systems. Further work will focus on the application of our method to the purification of triterpenoid saponins.

Supplementary Materials: The following are available online at http://www.mdpi.com/2218-273X/9/12/776/s1.

Author Contributions: J.-N.T., H.W. and Y.G. conceived the experiments and revised the manuscript. J.Z., X.S., and P.Z. collected plant materials. J.Z., and X.S. performed the experiments and analyzed the data. J.-N.T., H.W., and Y.G. wrote the manuscript. All authors have read and approved the final version of the manuscript.

Funding: This research was funded by the Fundamental Research Funds for Central Non-Profit Scientific Institution (1610232017010), Agricultural Science and Technology Innovation Program (ASTIP-TRIC05), Shandong Provincial Natural Science Foundation, China (ZR2016BQ43, ZR2019BB061), and National Natural Science Foundation of China (21703284).

Conflicts of Interest: The authors declare no conflict of interest.

References

1. Smith, E.L.; Abbott, A.P.; Ryder, K.S. Deep eutectic solvents (DESs) and their applications. *Chem. Rev.* **2014**, *114*, 11060–11082. [CrossRef] [PubMed]
2. Clarke, C.J.; Tu, W.C.; Levers, O.; Brohl, A.; Hallett, J.P. Green and sustainable solvents in chemical processes. *Chem. Rev.* **2018**, *118*, 747–800. [CrossRef] [PubMed]
3. Ma, C.; Laaksonen, A.; Liu, C.; Lu, X.; Ji, X. The peculiar effect of water on ionic liquids and deep eutectic solvents. *Chem. Soc. Rev.* **2018**, *47*, 8685–8720. [CrossRef] [PubMed]
4. Yang, Q.W.; Zhang, Z.Q.; Sun, X.G.; Hu, Y.S.; Xing, H.B.; Dai, S. Ionic liquids and derived materials for lithium and sodium batteries. *Chem. Soc. Rev.* **2018**, *47*, 2020–2064. [CrossRef]
5. García, J.I.; García-Marín, H.; Pires, E. Glycerol based solvents: synthesis, properties and applications. *Green Chem.* **2013**, *16*, 1007–1033. [CrossRef]
6. Carolin, R.; Burkhard, K. Low melting mixtures in organic synthesis – an alternative to ionic liquids? *Green Chem.* **2012**, *14*, 2969–2982.

7.   Zhao, B.Y.; Xu, P.; Yang, F.X.; Wu, H.; Zong, M.H.; Lou, W.Y. Biocompatible deep eutectic solvents based on choline chloride: Characterization and application to the extraction of rutin from sophora japonica. *ACS Sustain. Chem. Eng.* **2015**, *3*, 2746–2755. [CrossRef]

8.   Xu, J.; Huang, W.; Bai, R.; Queneau, Y.; Jérôme, F.; Gu, Y. Utilization of bio-based glycolaldehyde aqueous solution in organic synthesis: application to the synthesis of 2,3-dihydrofurans. *Green Chem.* **2019**, *21*, 2061–2069. [CrossRef]

9.   Wu, C.; Xiao, H.J.; Wang, S.W.; Tang, M.S.; Tang, Z.L.; Xia, W.; Li, W.F.; Cao, Z.; He, W.M. Natural Deep Eutectic Solvent-Catalyzed Selenocyanation of Activated Alkynes via an Intermolecular H-Bonding Activation Process. *ACS Sustain. Chem. Eng.* **2019**, *7*, 2169–2175. [CrossRef]

10.   Wu, C.; Hong, L.; Shu, H.; Zhou, Q.H.; Wang, Y.; Su, N.; Jiang, S.; Cao, Z.; He, W.M. Practical Approach for Clean Preparation of Z-β-Thiocyanate Alkenyl Esters. *ACS Sustain. Chem. Eng.* **2019**, *7*, 8798–8803. [CrossRef]

11.   Lu, L.H.; Wang, Z.; Xia, W.; Cheng, P.; Zhang, B.; Cao, Z.; He, W.M. Sustainable routes for quantitative green selenocyanation of activated alkynes. *Chin. Chem. Lett.* **2019**, *30*, 1237–1240. [CrossRef]

12.   Chen, J.; Xie, F.W.; Li, X.X.; Chen, L. Ionic liquids for the preparation of biopolymer materials for drug/gene delivery: a review. *Green Chem.* **2018**, *20*, 4169–4200. [CrossRef]

13.   Ma, Z.; Yu, J.; Dai, S. Preparation of inorganic materials using ionic liquids. *Adv. Mater.* **2010**, *22*, 261–285. [CrossRef] [PubMed]

14.   Cunha, S.C.; Fernandes, J.O. Extraction techniques with deep eutectic solvents. *Trends Anal. Chem.* **2018**, *105*, 225–239. [CrossRef]

15.   Duan, L.; Dou, L.L.; Guo, L.; Li, P.; Liu, E.H. Comprehensive evaluation of deep eutectic solvents in extraction of bioactive natural products. *ACS Sustain. Chem. Eng.* **2016**, *4*, 2405–2411. [CrossRef]

16.   Dai, Y.; Rozema, E.; Verpoorte, R.; Choi, Y.H. Application of natural deep eutectic solvents to the extraction of anthocyanins from Catharanthus roseus with high extractability and stability replacing conventional organic solvents. *J. Chromatogr. A* **2016**, *1434*, 50–56. [CrossRef]

17.   Van Osch, D.J.G.P.; Dietz, C.H.J.T.; van Spronsen, J.; Kroon, M.C.; Gallucci, F.; van Sint Annaland, M.; Tuinier, R. A Search for natural hydrophobic deep eutectic solvents based on natural components. *ACS Sustain. Chem. Eng.* **2019**, *7*, 2933–2942. [CrossRef]

18.   Nam, M.W.; Zhao, J.; Lee, M.S.; Jeong, J.H.; Lee, J. Enhanced extraction of bioactive natural products using tailor-made deep eutectic solvents: application to flavonoid extraction from Flos sophorae. *Green Chem.* **2015**, *17*, 1718–1727. [CrossRef]

19.   Liu, Y.; Friesen, J.B.; McAlpine, J.B.; Lankin, D.C.; Chen, S.-N.; Pauli, G.F. Natural deep eutectic solvents: properties, applications, and perspectives. *J. Nat. Prod.* **2018**, *81*, 679–690. [CrossRef]

20.   Pereira, E.; Encina-Zelada, C.; Barros, L.; Gonzales-Barron, U.; Cadavez, V.; Ferreira, I.C.F.R. Chemical and nutritional characterization of Chenopodium quinoa Willd. (quinoa) grains: A good alternative to nutritious food. *Food Chem.* **2019**, *280*, 110–114. [CrossRef]

21.   Chen, Y.S.; Aluwi, N.A.; Saunders, S.R.; Ganjyal, G.M.; Medina-Meza, I.G. Metabolic fingerprinting unveils quinoa oil as a source of bioactive phytochemicals. *Food Chem.* **2019**, *286*, 592–599. [CrossRef] [PubMed]

22.   Ruales, J.; Nair, B.M. Saponins, phytic acid, tannins and protease inhibitors in quinoa (Chenopodium quinoa Willd.) seeds. *Food Chem.* **1993**, *48*, 137–143. [CrossRef]

23.   Kuljanabhagavad, T.; Thongphasuk, P.; Chamulitrat, W.; Wink, M. Triterpene saponins from Chenopodium quinoa Willd. *Phytochemistry* **2008**, *69*, 1919–1926. [CrossRef] [PubMed]

24.   Zhu, N.Q.; Sheng, S.Q.; Sang, S.M.; Jhoo, J.W.; Bai, N.S.; Karwe, M.V.; Rosen, R.T.; Ho, C.T. Triterpene saponins from debittered quinoa. *J. Agric. Food Chem.* **2002**, *50*, 865–867. [CrossRef] [PubMed]

25.   Escribano, J.; Cabanes, J.; Jimenez-Atienzar, M.; Ibanez-Tremolada, M.; Gomez-Pando, L.R.; Garcia-Carmona, F.; Gandia-Herrero, F. Characterization of betalains, saponins and antioxidant power in differently colored quinoa (Chenopodium quinoa) varieties. *Food Chem.* **2017**, *234*, 285–294. [CrossRef] [PubMed]

26.   Gómez-Caravaca, A.M.; Iafelice, G.; Lavini, A.; Pulvento, C.; Caboni, M.F.; Marconi, E. Phenolic compounds and saponins in quinoa samples (Chenopodium quinoa Willd.) grown under different saline and nonsaline irrigation regimens. *J. Agric. Food Chem.* **2012**, *60*, 4620–4627. [CrossRef] [PubMed]

27.   Tang, Y.; Li, X.; Chen, P.X.; Zhang, B.; Hernandez, M.; Zhang, H.; Marcone, M.F.; Liu, R.; Tsao, R. Characterisation of fatty acid, carotenoid, tocopherol/tocotrienol compositions and antioxidant activities in seeds of three Chenopodium quinoa Willd. genotypes. *Food Chem.* **2015**, *174*, 502–508. [CrossRef]

28. Graf, B.L.; Rojo, L.E.; Delatorre-Herrera, J.; Poulev, A.; Calfio, C.; Raskin, I. Phytoecdysteroids and flavonoid glycosides among Chilean and commercial sources of Chenopodium quinoa: variation and correlation to physico-chemical characteristics. *J. Sci. Food Agric.* **2016**, *96*, 633–643. [CrossRef]

29. Dini, I.; Carlo Tenore, G.; Dini, A. Phenolic constituents of Kancolla seeds. *Food Chem.* **2004**, *84*, 163–168. [CrossRef]

30. Kumpun, S.; Maria, A.; Crouzet, S.; Evrard-Todeschi, N.; Girault, J.-P.; Lafont, R. Ecdysteroids from Chenopodium quinoa Willd., an ancient Andean crop of high nutritional value. *Food Chem.* **2011**, *125*, 1226–1234. [CrossRef]

31. Mota, C.; Santos, M.; Mauro, R.; Samman, N.; Matos, A.S.; Torres, D.; Castanheira, I. Protein content and amino acids profile of pseudocereals. *Food Chem.* **2016**, *193*, 55–61. [CrossRef] [PubMed]

32. Graf, B.L.; Poulev, A.; Kuhn, P.; Grace, M.H.; Lila, M.A.; Raskin, I. Quinoa seeds leach phytoecdysteroids and other compounds with anti-diabetic properties. *Food Chem.* **2014**, *163*, 178–185. [CrossRef] [PubMed]

33. Noratto, G.D.; Murphy, K.; Chew, B.P. Quinoa intake reduces plasma and liver cholesterol, lessens obesity-associated inflammation, and helps to prevent hepatic steatosis in obese db/db mouse. *Food Chem.* **2019**, *287*, 107–114. [CrossRef] [PubMed]

34. Van den Driessche, J.J.; Plat, J.; Mensink, R.P. Effects of superfoods on risk factors of metabolic syndrome: a systematic review of human intervention trials. *Food Funct.* **2018**, *9*, 1944–1966. [CrossRef]

35. Mamadalieva, N.Z.; Böhmdorfer, S.; Zengin, G.; Bacher, M.; Potthast, A.; Akramov, D.K.; Janibekov, A.; Rosenau, T. Phytochemical and biological activities of Silene viridiflora extractives. Development and validation of a HPTLC method for quantification of 20-hydroxyecdysone. *Ind. Crop. Prod.* **2019**, *129*, 542–548. [CrossRef]

36. Rodrigues, F.; Palmeira-de-Oliveira, A.; das Neves, J.; Sarmento, B.; Amaral, M.H.; Oliveira, M.B. Coffee silverskin: a possible valuable cosmetic ingredient. *Pharm. Biol.* **2015**, *53*, 386–394. [CrossRef]

37. Wang, H.; Tan, J.; Shang, X.; Zheng, X.; Liu, X.; Wang, J.; Hou, X.; Du, Y. Porous organic cage incorporated monoliths for solid-phase extraction coupled with liquid chromatography-mass spectrometry for identification of ecdysteroids from Chenopodium quinoa Willd. *J. Chromatogr. A* **2019**, *1583*, 55–62. [CrossRef]

38. Caussette, M.; Kershawb, J.L.; Sheltod, D.R. Survey of enzyme activities in desaponified quinoa Chenopodium quinoa Willd. *Food Chem.* **1997**, *60*, 587–592. [CrossRef]

39. Dini, I.; Tenore, G.C.; Dini, A. Antioxidant compound contents and antioxidant activity before and after cooking in sweet and bitter Chenopodium quinoa seeds. *LWT-Food Sci. Technol.* **2010**, *43*, 447–451. [CrossRef]

Availability of Isolated Substances: Not available.

 *biomolecules*

Article

Clerodane Diterpene Ameliorates Inflammatory Bowel Disease and Potentiates Cell Apoptosis of Colorectal Cancer

Jia-Huei Zheng [1], Shian-Ren Lin [1], Feng-Jen Tseng [1,2], May-Jywan Tsai [3], Sheng-I Lue [1,4], Yi-Chen Chia [5], Mindar Woon [6], Yaw-Syan Fu [7,8] and Ching-Feng Weng [7,8,*]

[1]   Department of Life Science and Institute of Biotechnology, National Dong Hwa University, Hualien 97401, Taiwan; jiahueizheng@gmail.com (J.-H.Z.); d9813003@gms.ndhu.edu.tw (S.-R.L.); fengjentseng@yeah.net (F.-J.T.); m655003@kmu.edu.tw (S.-I.L.)
[2]   Department of Orthopedics, Hualien Armed Force General Hospital, Hualien 97144, Taiwan
[3]   Department of Neurosurgery, Neurological Institute, Taipei Veterans General Hospital, Taipei City 11217, Taiwan; mjtsai2@vghtpe.gov.tw
[4]   Department of Physiology & Master's Program, Kaohsiung Medical University, Kaohsiung 80708, Taiwan
[5]   Department of Food Science & Technology, Tajen University, Pingtung 90741, Taiwan; ycchia@tajen.edu.tw
[6]   Department of Radiation Oncology, Yeezen Hospital, Taoyuan 32645, Taiwan; woonmd@yahoo.com
[7]   Department of Biomedical Science and Environmental Biology, Kaohsiung Medical University, Kaohsiung 80708, Taiwan; m805004@kmu.edu.tw
[8]   Institute of Respiratory Disease, Department of Basic Medical Science, Xiamen Medical College, Xiamen 361023, China
*   Correspondence: cfweng@gms.ndhu.edu.tw or cfweng-cfweng@hotmail.com; Tel.: +886-3-8903609

Received: 25 October 2019; Accepted: 20 November 2019; Published: 21 November 2019

Abstract: Inflammatory bowel disease (IBD) is general term for ulcerative colitis and Crohn's disease, which is chronic intestinal and colorectal inflammation caused by microbial infiltration or immunocyte attack. IBD is not curable, and is highly susceptible to develop into colorectal cancer. Finding agents to alleviate these symptoms, as well as any progression of IBD, is a critical effort. This study evaluates the anti-inflammation and anti-tumor activity of 16-hydroxycleroda-3,13-dien-15,16-olide (HCD) in in vivo and in vitro assays. The result of an IBD mouse model induced using intraperitoneal chemical azoxymethane (AOM)/dextran sodium sulfate (DSS) injection showed that intraperitoneal HCD adminstration could ameliorate the inflammatory symptoms of IBD mice. In the in vitro assay, cytotoxic characteristics and retained signaling pathways of HCD treatment were analyzed by MTT assay, cell cycle analysis, and Western blotting. From cell viability determination, the IC_{50} of HCD in Caco-2 was significantly lower in 2.30 µM at 48 h when compared to 5-fluorouracil (5-FU) (66.79 µM). By cell cycle and Western blotting analysis, the cell death characteristics of HCD treatment in Caco-2 exhibited the involvement of extrinsic and intrinsic pathways in cell death, for which intrinsic apoptosis was predominantly activated via the reduction in growth factor signaling. These potential treatments against colon cancer demonstrate that HCD could provide a promising adjuvant as an alternative medicine in combating colorectal cancer and IBD.

Keywords: colorectal cancer; diterpenes; inflammatory bowel diseases; *Polyalthia longifolia*; herbal medicine

1. Introduction

Colorectal cancer (CRC) has a significant health impact worldwide, and is a common cancer type in the United States. In addition, CRC is the third leading cause of cancer deaths for new cases, and second for estimated deaths in both genders in Taiwan [1]. In the majority of cases, it is a consequence of the progressive accumulation of genetic and epigenetic alterations that leads the transformation and progression of normal colorectal mucosa to adenoma and eventually carcinoma, progressing CRC [1]. When analyzed, the incidence of CRC includes 72% colon cancer and 28% rectum cancer [2]. Surgery is mostly the first choice for all stages of CRC treatment, while for stage IV of CRC, recurrent CRC, or liver cancer metastasis, chemotherapy is the main alternative strategy to treat CRC [3]. Clinically, chemotherapeutic drugs, including 5-fluorouracil (5-FU), oxaliplatin, and capecitabine, are commonly used to treat CRC; however, these typically contribute to several adverse effects, such as fatigue, nausea, bone marrow toxicity, immunosuppression, and easy bleeding. Moreover, chemotherapy is a conventional treatment for late-stage and recurrent colorectal cancer, bit the cancer cells frequently become drug-resistant after treatment [3], and the unpredictability of adverse or side effects ordinarily restricts the administration of an accurate dose. As incidences of adverse results are observed, a new discovery of more efficacious and less toxic agents against CRC is of great urgency.

Noticeably, the risk factors of CRC are complicated, and include excessive alcohol use, obesity, hereditary conditions, and long-standing inflammatory bowel disease (IBD) [3]. IBD is comprised of two conditions: ulcerative colitis (UC; ICD-10-CM code K51.90) and Crohn's disease (CD; ICD-10-CM code K50.90), which are caused by uncontrolled gastrointestinal (GI) inflammation or bacterial infection that subsequently result in fever, edema, intestinal fibrosis, and ulcers [4]. Global prevalence rates for IBD exceed 0.3%, which is mostly attributed to human-development-index countries—e.g., the United States, Canada, Germany, Norway, and Australia—where incidence rates are twice as high as those of Asian and African countries [5]. Moreover, the incidence rates of IBD where it occurs in the industrialized countries of Asia, Africa, and South America were between 4% (Taiwan) and 11.1% (Brazil) from the 1990s to 2010 [5]. Therefore, IBD becomes a burden in Western and industrialized societies.

Within an inflamed GI tract, large numbers of immunocytes, including macrophages, T_H cells, neutrophils, and natural killer (NK) cells, are attracted and secrete various forms of cytokines, such as tumor necrosis factor alpha (TNF-α), interferon-γ (IFNγ), and interleukin (IL)-6. When the immunocytes are accumulated, cytokine secretions lead to reductions in incidences of inflammation, and finally potentiate carcinogenesis [6]. Furthermore, one study has illustrated the connection between IBD and the Wnt/β-catenin signaling pathways, which might trigger colon cancer progression and incidences of sporadic colon cancer [7]. Interestingly, vitamin D deficiency, genetic susceptibility, disturbance of the microbiome, and psychological status have recently been confirmed as risk factors for IBD [8,9]. Previous evidence has shown that chronic inflammation may cause GI tumorigenesis, which is familiar and lethal worldwide. Even the known risk factors of IBD that have been broadly explored, as well as the causes, complications, and medications of IBD, including CRC, are still unraveling [10]. Nonetheless, the treatment of IBD remains unsolved and needs to be explored. According to the inquiry of new drug discovery, 46% of new drugs are of natural origin, e.g., derivatives, precursors, or mimetic molecules [11]. This fact exemplifies the continuing research and development of new drugs from natural products, especially medicinal plants, to meet patient demand, as the prospect of exploring new compounds and alternative treatments gains an immediate importance.

Polyalthia longifolia var. pendula Linn. (family Annonaceae) is an ornamental tree originally distributed in India, Sri Lanka, and Pakistan, which also contains numerous biological functions, as presented in the literature [12]. By exploring bioactive components, a clerodane diterpene 16-hydroxycleroda-3,13-dien-15,16-olide (HCD; PubChem ID 124820) has been extensively identified [13]. In previous reports, HCD has been shown to have numerous medicinal values as an anti-inflammation, anti-cancer, anti-fungal, anti-diabetic, and anti-bacterial agent [14]. In our previous studies, HCD performed as an executor to induce autophagy in glioma cells and oral squamous cell carcinoma cells, which consequently triggered cancer cell death [15,16]. Moreover, HCD can induce anoikis and reduce cell proliferation via the regulation of both intracellular growth and focal adhesion signaling in renal carcinoma cells [17,18]. In addition to acting as an anti-tumor agent, HCD could also play a supplementary role in the cytotoxicity of tamoxifen-treated breast cancer via the modulation of the Bax/Bcl-2 ratio, which is directly expressed at cells undergoing apoptosis [19]. Recently, our studies have demonstrated the therapeutic potential of HCD against various types of cancers [19]. Nevertheless, the therapeutic potency of HCD in treating GI inflammation, e.g., IBD and colorectal cancer, has not been clarified. The aim of this study was to determine the dampening effect of HCD on IBD treatment and anticancer activity. In this work, two platforms containing an azoxymethane (AOM)/dextran sodium sulfate (DSS)-induced colitis IBD model (*in vivo*) and colorectal carcinoma cells Caco-2 (*in vitro*) were employed to evaluate the treated efficacy of HCD. Furthermore, the detailed mechanisms of HCD on anti-colorectal cancer were also investigated.

2. Experimental Section

2.1. Study Design

The main focus of this study was to evaluate the potential of HCD against IBD and colon cancer. The experimental design was divided into two parts: IBD induction and colon cancer with HCD treatment. In the IBD model, mice were induced by azoxymethane (AOM)/dextran sodium sulfate (DSS). HCD was intraperitoneally (i.p.) injected, and subsequently GI inflammation was observed. In the colon cancer part, Caco-2 cells were used as the sole platform for the determination of cytotoxicity and cell death characteristics, and the underlying mechanisms of colon cancer cytotoxicity in HCD-treated cells, including cell cycle signaling, growth factor signaling, and inflammatory signaling, were under investigation.

2.2. Chemicals

HCD was kindly provided by Professor Yi-Chen Chia (Department of Food Science and Technology, Tajen University, Taiwan). Isolation and identification of HCD has been described in the previous literature [20]. The reagents and mediums for cell culture were purchased from Thermo-Fisher (Waltham, MA, United States). General chemicals were obtained from Sigma-Aldrich (Merck KGaA, Darmstadt, Germany). The antibodies used in this study are listed in Table 1; these were purchased from Genetex International (Hsinchu, Taiwan), Cell Signaling Technology (Danvers, MA, United States), and Merck Millipore (Burlington, MA, United States).

Table 1. Antibodies used in this study.

Protein Name	Molecular Weight (KDa)	Host	Manufacture	Dilution Factor
iNOS	131	Rabbit	Genetex	1:1000
PARP	119/85	+	Cell Signaling	+
PI3K	85	+	Genetex	+
β-catenin	85	+	+	+
COX-2	69	+	+	+
NF-κB (p65)	65	+	+	+
Akt	60	+	Cell Signaling	+
caspase-8	57/10	+	+	+
p53	53-55	+	+	+
NF-kB (p50)	50	+	Genetex	+
caspase-9	47/35-37	Mouse	Cell Signaling	+
WNT11	39	Rabbit	Genetex	+
cyclin D-	36	+	Cell Signaling	+
PCNA	36	Mouse	+	+
caspase-3	35/17-19	Rabbit	+	+
GADPH	34	+	Genetex	+
p27	27	+	Cell Signaling	+
Bcl-2	26	+	+	+
Bad	23	+	+	+
p21	21	+	Genetex	+

Table 1. *Cont.*

Protein Name	Molecular Weight (KDa)	Host	Manufacture	Dilution Factor
Bax	20	+	Cell Signaling	+
Anti-mouse HRP-conjugated 2nd Ab		Goat	Merck Millipore	1:5000
Anti-mouse HRP-conjugated 2nd Ab		+	+	+

+: Same as above.

2.3. *In Vivo Test*

2.3.1. Animal Source and Care

Eight to ten-week-old C57BL/6 male mice were purchased from BioLASCO (Taipei, Taiwan) and kept in controlled environmental conditions (22 ± 2 °C, 55% $\pm$ 10% humidity, 12/12 h light/dark cycle). The animals were fed a commercial diet and water ad libitum. Mice experimental protocols were used according to the "Guide for the Care and Use of Laboratory Animals" of National Dong-Hwa University, approved by the National Dong-Hwa University Animal Ethics Committee (approval number 001/2016).

2.3.2. Inflammatory Bowel Disease Model Induction and HCD Treatment

The in vivo IBD model was induced by the injection of AOM and DSS, following a previous study with slight modification [21]. At day 0, mice were weighed and i.p. injected with 10 mg/kg B. wt. of AOM. The next day, the mice were freely supplied with 2% DSS solution for an additional seven days, and regular water for a further seven days. This induction cycle was repeated once. The induction of IBD was verified by checking the histological change of the colon after mice were sacrificed at day 35.

Once IBD induction was confirmed, 15 mg/kg B. wt. of 5-FU, as well as 1.6 and 6.4 mg/kg B. wt. of HCD were i.p. injected once every three days until day 65. Mice were sacrificed by CO_2 anesthesia, and their colons were collected for slicing in the literature [16].

2.4. *In Vitro Test*

2.4.1. Cell Culture

Human colorectal carcinoma cell lines Caco-2 and HT-29 were obtained from the American Type Culture Collection (ATCC, Manassas, MA, United States). Caco-2 and HT-29 cells were cultured with Dulbecco's modified Eagle medium (DMEM) or RPMI-1640, with 20% fetal bovine serum (FBS) and 1% penicillin/streptomycin (PS) supplementation, respectively. The environmental conditions were 37 °C and 5% CO_2, and the cultured medium was renewed once every two days. Once cells reached 80% confluence, cells were detached by 0.25% trypsin/EDTA for further experiment. All experiments were carried out within 20 passes, with concern for uniformity and reproducibility.

2.4.2. Cytotoxicity Assay

The cytotoxic effect of HCD was measured by MTT (3-(4, 5-dimethylthiazol-2-yl)-2, 5-diphenyltetrazolium bromide; MDBio Inc., Taipei City, Taiwan) assay, as previously described in the literature [16]. Briefly, 7×10^3 cells per well of two cells were inoculated in 96-well plates before incubating with 0.5, 2.0, 5.0, and 10.0 μM of HCD and 1, 10, 50, and 100 μM of 5-FU (a positive control) for 36 and 48 h, respectively. The optical density (OD) at 570 nm was measured after incubation with MTT solution for 4 h, and was solubilized in DMSO. Cytotoxicity was presented by cell viability, which was the ratio of OD570 between treatment and untreated control (0 μM).

2.4.3. Cell Cycle Analysis

The protocol of cell cycle analysis followed a previous study, with slight modification [16]. In brief, 7×10^4 cells per well of Caco-2 cells were seeded into 12-well plates. Cells were incubated with

0.5, 2.0, and 5.0 μM of HCD for 36 and 48 h, respectively. Treated cells were fixed with 70% freeze ethanol and stained with propidium iodide (PI) at 37 °C for 1 h. The fluorescent intensity of PI within cells was detected by a Cytomics™ FC 500 flow cytometer (Beckman-Coulter, Brea, CA, United States). Data from 10^4 cells in each sample were collected, and the different cell cycles were analyzed.

2.4.4. Western Blotting

A total of 2.5×10^5 cells/well of Caco-2 cells were seeded into a six-well plate and incubated until 80% confluence. Cells were treated with 0.5, 1.0, and 2.0 μM of HCD for 24 and 36 h, respectively. After incubation, cells were homogenized, and the desired protein levels were analyzed according to the protocol described in a previous study [16]. The chemiluminometric intensity of each protein was normalized with GAPDH's chemiluminometric intensity. The protein level change was represented by the ratio of normalized chemiluminometric intensity between treated and untreated groups.

2.5. Statistical Analysis

Data were expressed as mean ± SD from at least three independent experiments. The results were analyzed by one-way analysis of variance (ANOVA) with the Dunnett test. The significant difference ($p < 0.05$) was labelled "*" on the histogram produced by GraphPad Prism Ver 7.0 (GraphPad Software, La Jolla, CA, United States). The IC_{50} of the dose-dependent cytotoxicity was determined using non-linear regression embedded in GraphPad Prism, and the chosen model was the highest R^2 value.

3. Results

3.1. Histological Change of Intestine Tissue after AOM/DSS Induction and HCD Treatment

To generate the IBD mouse model, mice were chemically induced by AOM/DSS. After AOM/DSS induction, enlarged lymph nodes, lymphocyte infiltration, irregular and shorter villi, and thicker muscle mucous and muscle layers were observed in the intestines of mice, which consequently confirmed that mice were successfully induced with IBD after the AOM/DSS given (Figures 1B and 2B), compared to the control without induction (Figures 1A and 2A). In the next experiment, IBD-induced mice were employed to evaluate the amelioration efficacy of 5-FU and HCD on the histopathological signs of IBD. The tissue section showed that the lymphocytes were less or not infiltrated into the lamina propria layer after treatment with 5-FU and HCD (Figure 1C–E). The arranged villi in neat rows were found in an induced group as a positive control (AOM/DSS alone), and this feature was not observed in the 5-FU and HCD-treated groups. Additionally, the lymph nodes were reduced after treating with 5-FU and HCD (Figure 2C–E). These histological changes elicited that HCD could reduce IBD symptoms. The next experiments were performed to evaluate the efficacy of HCD on colorectal cancer cells.

3.2. Cytotoxicity Effects of HCD and 5-Fluorouracil on Colorectal Cancer Cells

To check cytotoxicity of HCD against colorectal cancer cells, cells were treated with various concentrations of HCD and 5-FU (conventional chemotherapeutic agent, as a positive control), respectively. When compared to the untreated control (0 μM), cell viability in HCD-treated groups was significantly decreased (Figure 3A). The IC_{50} values of HCD in Caco-2 cells were 4.10 μM (36 h) and 2.32 μM (48 h), which were lower in 5-FU (100 μM for 36 h; 66.79 μM for 48 h) (Figure 3B). To further validate the potential of HCD for colon cancer treatment, another colon cancer cell, HT-29, was treated with various concentrations of HCD. The results showed that a dose-dependent decrease of cell viability was also observed in HCD-treated HT-29 cells. The IC_{50} values of HCD against HT-29 were 10.18 μM (36 h) and 1.39 μM (48 h), and were higher than those of Caco-2 (Figure 3C). According to these results, we confirmed that the cytotoxicity of HCD in colorectal cancer cells (Caco-2 and HT-29) had a higher potential than 5-FU. Therefore, the subsequent experiments were focused on the investigation of underlying mechanisms in HCD against Caco-2 cells.

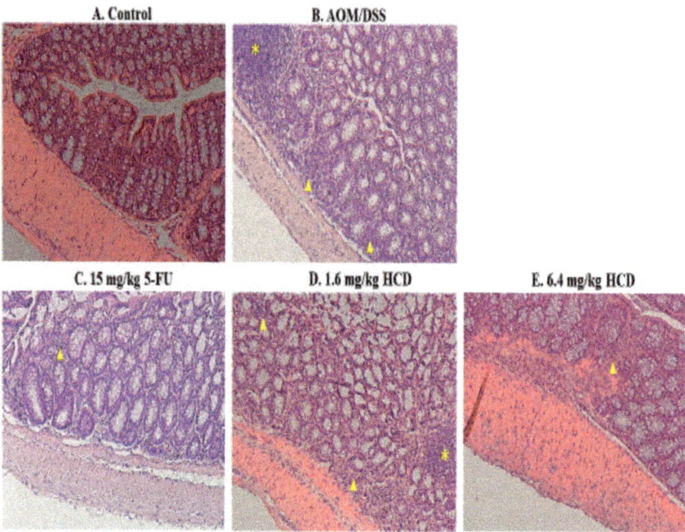

Figure 1. Histological appearances of the longitudinal section in the intestine of mice. Mice treated with (**A**) control and (**B**) azoxymethane (AOM)/dextran sodium sulfate (DSS) induction, as well as AOM/DSS induction followed by (**C**) 15 mg/kg B.wt of 5-fluorouracil (5-FU), (**D**) 1.6 mg/kg B.wt, and (**E**) 6.4 mg/kg B.wt of 16-hydroxycleroda-3,13-dien-15,16-olide (HCD) treatment ($n = 5$ in each group), were sacrificed, and the longitudinal section of tissues was stained with hematoxylin and eosin. The yellow asterisk and triangles indicate lymph node and lymphocyte infiltration, respectively, in the intestine. Magnification = ×100.

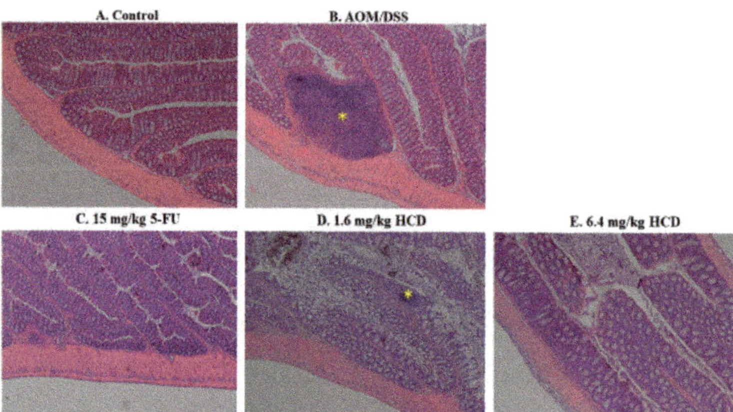

Figure 2. Histological changes of intestinal lymph nodes in HCD-treated mice with inflammatory bowel disease (IBD). The intestinal lymph nodes of (**A**) the control, (**B**) mice with AOM/DSS induction, and mice with AOM/DSS induction followed by (**C**) 15 mg/kg B.wt of 5-FU, (**D**) 1.6 mg/kg B.wt, and (**E**) 6.4 mg/kg B.wt of HCD treatment ($n = 5$ in each group) were stained and photographed. The yellow asterisk indicates the lymph node in the intestine. Magnification = ×40.

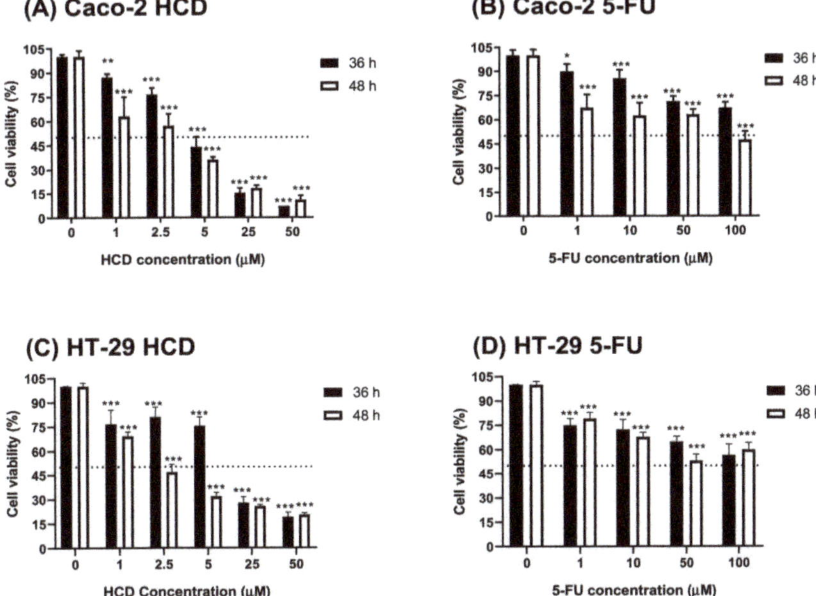

Figure 3. Cell viability of Caco-2 and HT-29 cells after HCD and 5-FU treatments. HCD and 5-FU were used to treat with (**A,B**) Caco-2 and (**C,D**) HT-29 cells for 36 and 48 h, respectively. The cell viability was determined by MTT assay. In addition, the cell viability was calculated according the comparison of control group (0 μM). Data ($n = 3$) were represented as mean ± SD. * $p < 0.05$; ** $p < 0.01$; *** $p <$ 0.001, as compared with the untreated control (0 μM).

3.3. Characteristics of HCD-Induced Cell Death

To identify features of HCD-induced cell death, Caco-2 cells were treated with various concentrations of HCD, and intracellular DNA content was checked using flow cytometry coupling with PI-staining. After 36 h and 48 h of HCD treatment, the sub-G_1 and G_0/G_1 cell cycle exhibited significant change in a dose-dependent fashion (Figure 4A), which could be caused by increasing the sub-G_1 ratio, referred to as the apoptotic population. Interestingly, the significant decrease of the G_2/M phase was found only at 5 μM HCD (Figure 4B). Concurrently, analyzing apoptotic markers, the increase of cleaved caspase-3, -8, -9, and PARP specified that apoptotic cell death appeared at 24 h and 36 h of treatments, respectively (Figure 5). These results clearly illustrate that HCD-caused cell death in Caco-2 cells was dominate in apoptosis. Moreover, by observation of the dynamics of apoptotic markers, intrinsic apoptotic inducer caspase-3 expression was enhanced at the first 24 h, and then cleaved at the following 12 h (Figure 5A). Conversely, Bcl-2, the apoptotic inhibitor, had no significant change until 36 h of treatment (Figure 5B). Additionally, the extrinsic apoptotic markers, cleaved caspase-8 and -9, also had significantly higher expression at either 24 h or 36 h of treatment (Figure 5). The data confirmed that both intrinsic and extrinsic apoptotic signaling pathways were involved in HCD-induced apoptosis, and intrinsic apoptosis might be prior to the extrinsic event. The following experiments would be carried out to determine whether the dynamic change of growth signaling pathway is affected by HCD.

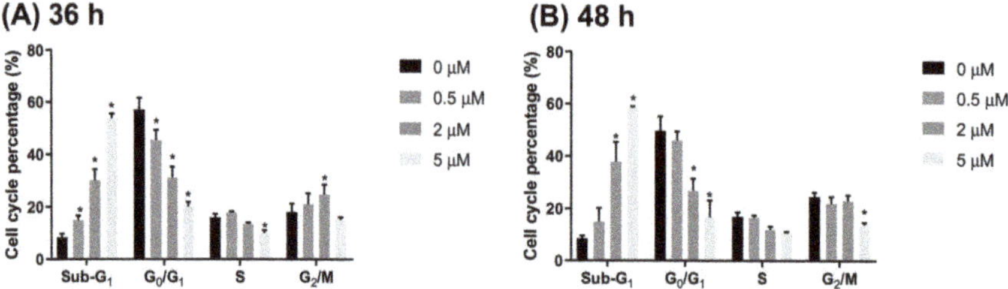

Figure 4. Cell cycle change after HCD treatment. Caco-2 cells were stained by propidium iodide after (**A**) 36 h and (**B**) 48 h of HCD treatment, and further analyzed by fluorescent intensity by flow cytometry. Data ($n = 3$) were represented as mean ± SD. * $p < 0.05$; ** $p < 0.01$; *** $p < 0.001$, as compared with the untreated control (0 μM).

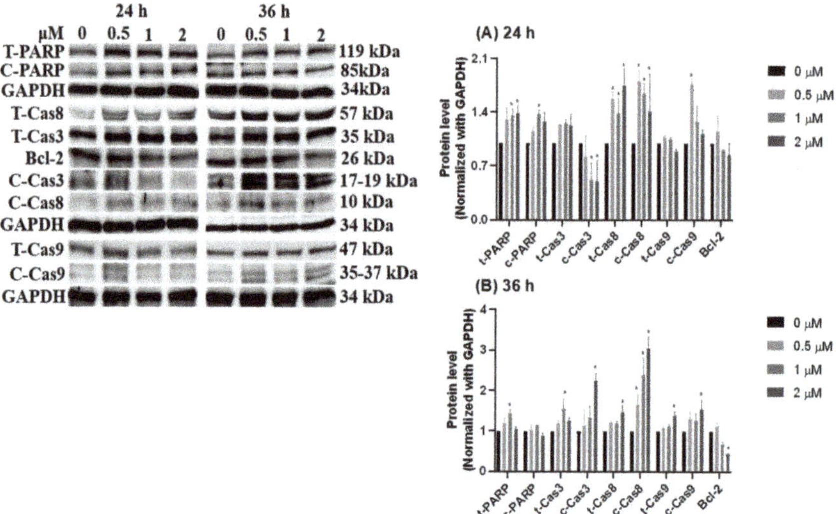

Figure 5. Dynamic change of apoptotic markers after HCD treatment. The protein levels of PARP, caspase-3, caspase-8, and caspase-9 were measured after (**A**) 24 h and (**B**) 36 h of HCD treatment. The protein levels were represented as the ratio of GAPDH-normalized chemiluminometric intensity between the untreated control (0 μM) and treatment. Data ($n = 3$) were represented as mean ± SD. * $p < 0.05$ as compared with the untreated control (0 μM).

3.4. Growth Signal Reduced by HCD Triggered Caco-2 Cell Apoptosis

Notably, the intrinsic apoptotic pathway is regulated by the balance of growth signal and anti-growth signal. The subsequent experiments were applied to measure the signals of PI3K/Akt for growth, p53/p21 for anti-growth, and cyclin D1/PCNA for cell division after HCD treatments. Western blot data indicated that the protein levels of Akt and cyclin D1 were significantly decreased in 2 μM of HCD treatment after the first 24 h, and a significant decrease of the cyclin D1 level at 1 μM of HCD treatment was observed (Figure 6A). Subsequently, the protein levels of p53 and p21 were up-regulated and PCNA was down-regulated at 36 h (Figure 6B). This result hinted that HCD-induced apoptosis could be the reducing result of growth signaling via Akt mediation. The downregulation of growth signaling caused the elevation of p53/p21 protein expression, and consequently turned down cyclin D1 expression, which was the key for overcoming the G1/S checkpoint. However, upstream of Akt, the protein levels of PI3K were not associated with the mediation of Akt (Figure 6), which meant that Akt might be reduced by other signaling pathways. The next section was performed to test the characteristics of the inflammatory pathway within HCD-induced apoptosis.

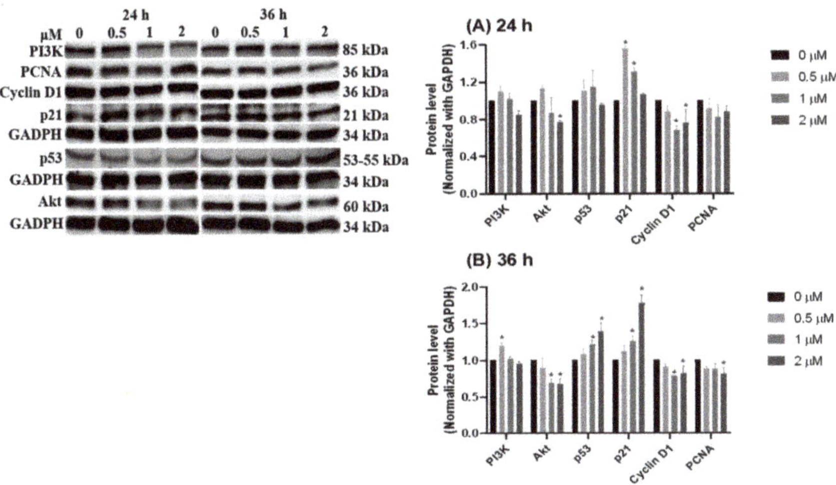

Figure 6. Dynamics of cell growth related signaling pathway after HCD treatment. Caco-2 cells were treated with HCD and the changes in PI3K, Akt, p53, p21, cyclin D1, and PCNA were measured at (**A**) 24 h and (**B**) 36 h. The protein levels were represented as the ratio of GAPDH-normalized chemiluminometric intensity between the untreated control (0 μM) and treatment. Data (*n* = 3) were represented as mean ± SD. * $p < 0.05$ as compared with the untreated control (0 μM).

3.5. Inflammation-Suppressing Effect of HCD in Caco-2 Cells

From the literature, inflammation has been proven as a promoting cause of colorectal cancer [22], and inflammatory signaling could crosstalk with Wnt/β-catenin, as well as being involved in colorectal cancer growth [23]. Thus, the subsequent experiment examined the regulation of inflammatory-related proteins, including COX-2, NF-κB, and β-catenin. At 24 h of HCD treatment, the p50 subunit of NF-κB showed a significantly decreasing manner (Figure 7A). The decrease of p50 was diminished, whereas β-catenin was up-regulated at 36 h (Figure 7B). This result refutes that NF-κB and β-catenin are involved in HCD-mediated apoptosis; however, the underlying mechanisms were not fully interpreted. Of note, inflammatory signaling was proven by the significant characteristics of IBD pathological progression, and then the subsequent experiment will be employed to evaluate whether oral gavage HCD may ameliorate IBD symptom in an AOM/DSS- induced mouse model by the mediation of inflammatory signals.

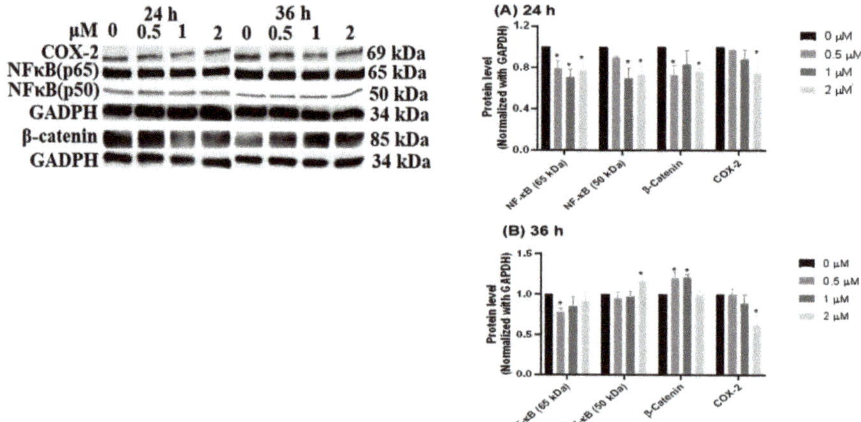

Figure 7. Inflammation-related protein changes after HCD treatments. NF-κB, β-catenin, and Cox-2 were determined by Western blotting at (**A**) 24 h and (**B**) 36 h HCD-treated Caco-2 cells. The chemiluminescent intensity of each protein was normalized with GAPDH and represented as the protein levels. Data (n = 3) were represented as mean ± SD. * p < 0.05 as compared with the untreated control (0 μM).

4. Discussion

In chemical-induced IBD and colitis-associated cancer research, this study broadened the application of HCD in medical use, and could provide a new approach for IBD treatments. These experimental results conferred the anticancer effect of HCD against colon cancer, which led to intrinsic and extrinsic regulation for apoptotic cell death by down-regulating Akt-mediated growth signaling. Moreover, the anti-inflammation fashion of HCD in colon cancer might be one reason for the down-regulation of Akt, and is postulated for the IBD curing effect in vivo.

Usually, the conventional therapies for IBD can be grouped into the two following approaches: anti-inflammation, *e.g.*, corticosteroids, mesalazine, and cyclosporine; and anti-microbial, including ornidazole and rifaximin [24]. Ordinarily, these drugs are useful for treating mild to moderate IBD. However, side effects, such as drug resistance to antibiotics and opportunistic infection, are still of concern [25]. Likewise, about one-third of chronic IBD patients have failed responses to corticosteroid medication, which are valid for acute IBD [25]. Hence, natural components are potentially considered to be the new strategies or approaches for curing IBD. Five types of phenylpropanoids and four types of lignan glycosides—phytochemicals originated from a warm-season perennial legume, *Lespedeza cuneata*—were synthesized, and these compounds could ameliorate UC [26]. Curcumin, the primary active compound of turmeric, alleviates CD and UC by inhibiting NF-κB activity [27]. Macrophage infiltration into the intestines could also be impaired by α-eleostearic acid isolated from *Momordica charantia* [28]. After AOM/DSS induction, polypus and enlargement lymph nodes were found in the colon by histopathological examinations in our study (data not shown). Additionally, the observations of irregular villi arrangement and lymphocyte infiltration into the lamina propia layer in this study are typical characteristics of IBD. The IBD mice treated with HCD had neat rows of villi and lymph nodes that were not enlarged or infiltrated into lamina propia layer, which indicates that HCD could relieve the symptoms of IBD.

In indigenous medication, *P. longifolia* is an antipyretic drug used in past decades [12]. Under modern chemical and medical research, various antimicrobial and anti-inflammatory compounds are purified and identified, such as 16α-hydroxy-cleroda-3,13(14)-Z-diene-15,16-olide, (3S,4R)-3,4,5-trihydroxy pentanoic acid-1,4-lactone and 16-Oxo-cleroda-3,13E-dien-15-oic acid [12]. HCD has been demonstrated to alleviate lipopolysaccharide-induced microglia inflammation via reducing iNOS, COX-2, and NF-κB gene expression [13]. In addition, HCD repressed COX-2 protein expression in Caco-2 and further ameliorated inflammation in AOM/DSS-induced mice. These results imply that HCD represents a novel and potential clinical approach for IBD treatment and CRC chemotherapy.

In our previous study, HCD was demonstrated to be a non-toxic agent to normal cells [16]. When compared to 5-FU, the cytotoxicity of HCD against colorectal cancer was higher than 5-FU (4.10 vs. 100 μM in IC_{50} at 36 h treatment), suggesting that HCD had higher efficacy and more potent when applied to colon cancer treatments. In the cell cycle analysis of Caco-2 cells treated with HCD, the ratio of the sub-G_1 phase was significantly increased, and this increase was associated with an increase of HCD concentrations. This result was accompanied by the analysis of pro-apoptotic markers, such as PARP; caspase-3, -8, and -9; and Bcl-2. During observations from 24 h to 36 h of HCD treatment, the protein levels of pro-apoptotic molecules changed, and these signaling transductions could be precisely determined by caspase-8, caspase-3, Bcl-2, and PARP. Again, the caspase-3 expression level was increased in the first 24 h, which also indicated that intrinsic apoptotic pathways were activated at this time. These results imply that both intrinsic and extrinsic apoptotic signaling pathways were simulanteously activated in Caco-2 cells. This was the first evidence that HCD could induce the apoptosis of cancer cells.

As we know, extrinsic and intrinsic apoptotic signaling pathways exert themselves with different signaling molecules [29]. In general, an extrinsic apoptotic signaling pathway is started from the activation of death receptor (TRAIL receptor or TNF receptor) and terminated at pro-caspase-3 cleavage via caspase-8 activation [29,30]. Different from extrinsic apoptosis, intrinsic apoptotic signaling is mediated by Bcl-2 to cause the loss of mitochondrial membrane potential or growth signaling depletion, which leads to cytochrome c release, pro-caspase-9 cleavage, and consequently, caspase-3 activation [29,31,32]. Therefore, by observing the altered levels of cleavage caspase-8, caspase-9, and Bcl-2, the type of chemical-induced apoptosis could be putatively addressed. Previous studies of HCD-induced cancer cell death were focused on autophagic cell death and intrinsic apoptosis [15,16,33,34]. To the best of our knowledge, this study is the first to show the involvement of HCD-induced extrinsic apoptosis in colorectal cancer. Moreover, the underlined targets of HCD in extrinsic apoptosis still need to be further explored, because this work only focused on the alteration of caspase-8 proteins.

HCD-mediated intrinsic apoptosis was found by down-regulating β-catenin/NF-κB/Akt and activating p53/p21 expression. Previously, HCD had potentiated apoptosis via blocking the PI3K/Akt signaling pathway, promoting Aurora B degradation, and modifying histone-modifying enzymes in leukemia cells [33,34]. Remarkably, the cytotoxicity (cell death) of oral squamous cell carcinoma (OSCC) and glioma cells were also demonstrated in the treatment of HCD through an autophagic manner via Western blotting analysis, without an increase in sub-G_1 [15,16]. Moreover, HCD activated the autophagy in lung cancer (A549) via reducing the protein level of mTOR, PI3K/p85, Akt, and Beclin 1, and suppressed apoptosis by lessening cleaved-PARP formation (Chiu et al., 2019, unpublished data). Thereby, one possibility for cell death is that this difference in apoptosis or autophagy was regardless of different types of cancer cells (adenoma, carcinoma, glioma, or neuroblastoma) and p53 (wild or mutant type). The cell death features of HCD induced in Caco-2 cells and other cancer cells imply that the critical point of the apoptosis/autophagy switch needs to be clarified.

The known roles of p53 are in cell cycle regulation, apoptosis induction, DNA repair activation, and aerobic respiratory improvement [35]. When measuring the apoptosis-related protein levels, the p53 protein level showed an increase at 36 h after HCD treatment. In intestine and colon tissues under IBD, p53 was overexpressed by TNF-α induction, and subsequently triggered cell apoptosis [36]. Interestingly, the induction of TNF-α was only observed in mutant p53, but not in wild-type p53 [37]. Over 50% of colitis-induced colorectal cancer and colon neoplasia could be found with the TP53 mutation. This mutation is believed to be the first step of colitis-associated carcinogenesis when compared with sporadic CRC [38]. Of note, p53 in Caco-2 is an aberrant type [39]. HCD caused the alteration of p53 protein expression in CRC cells, which indicated that HCD might affect p53 protein levels in IBD tissues. Therefore, the effects of HCD on intestinal epithelial cells with wild-type p53 might differ from similar Caco-2 cells.

In addition, as an apoptosis-inducing feature, HCD also possesses inhibiting activity for Wnt/β-catenin in an anti-inflammation manner. Wnt/β-catenin dysregulation has been reported as a key factor for CRC initiation [40]. The Wnt/β-catenin signaling pathway acts as a central regulator in intestine homeostasis and epithelial stem cell proliferation. The Wnt ligand binds to the Frizzled/LRP receptor, and activates a signal cascade to subsequently result in the stabilization of β-catenin. Stable β-catenin can translocate into the nucleus and initiate gene expression of MYC and its downstream target, CCND1 [41]. Impaired activation of the Wnt/β-catenin signaling pathway could cause uncontrolled cell proliferation and finally, elicit colon cells carcinogenesis. Therefore, the Wnt/β-catenin signaling pathway would become the target of CRC prevention, prognosis, and diagnosis. Numerous studies have sought new compounds or herbal medicine for treating CRC. Fermented culture broth of *Antrodia camphorate*, hydnocarpin (a natural lignan), and bark extract of *Mesua ferrea* have been shown to inhibit activity of Wnt/β-catenin in colon cancer cells [42–44]. In this study, the reduction of β-catenin levels after HCD treatment showed the inhibited manner of the Wnt/β-catenin signaling pathway, which indicated a possibility of reducing colon cancer initiation.

5. Conclusions

This study illustrated anti-colorectal cancer activity from HCD by modifying intrinsic growth signaling and inflammatory modulators, which subsequently triggered both intrinsic and extrinsic signals to induce cell apoptosis. Furthermore, the inflammatory symptoms of AOM/DSS-induced enteritis in the in vivo mouse model were also ameliorated by HCD treatments. This is the first evidence showing the medicinal efficacy of HCD on IBD mice and colon cancer, suggesting that HCD could provide an alternative and complementary regimen for anti-colon cancer and IBD treatments.

Author Contributions: Conceptualization, C.-F.W. and Y.-C.C.; methodology, Y.-S.F., S.-R.L., S.-I.L., and C.-F.W.; software, M.-J.T. and F.-J.T.; validation, S.-R.L., F.-J.T., and M.W.; formal analysis, S.-I.L.; investigation, J.-H.Z.; resources, C.-F.W.; data curation, M.-J.T. and Y.-C.C.; writing—original draft preparation, J.-H.Z. and S.-R.L.; writing—review and editing, C.-F.W.; project administration, C.-F.W.; funding acquisition, C.-F.W.

Funding: This study was supported by Ministry of Science and Technology (Grant number 104-2320-B-259-001-MY3)

Conflicts of Interest: The authors declare no conflict of interest.

References

1. Chiang, T.Y.; Wang, C.H.; Lin, Y.F.; You, J.F.; Chen, J.S.; Chen, S.C. Colorectal cancer in Taiwan: A case-control retrospective analysis of the impact of a case management programme on refusal and discontinuation of treatment. *J. Adv. Nurs.* **2018**, *74*, 395–406. [CrossRef]
2. Coppede, F. The role of epigenetics in colorectal cancer. *Expert Rev. Gastroenterol. Hepatol.* **2014**, *8*, 935–948. [CrossRef]
3. Board, P. Colon Cancer Treatment (PDQ(R)): Health Professional Version. In *PDQ Cancer Information Summaries*; PDQ Adult Treatment Editorial Board—National Cancer Institute: Bethesda, MD, USA, 2002.
4. Zippi, M.; Pica, R.; De Nitto, D.; Paoluzi, P. Biological therapy for dermatological manifestations of inflammatory bowel disease. *World J. Clin. Cases* **2013**, *1*, 74–78. [CrossRef]
5. Ng, S.C.; Shi, H.Y.; Hamidi, N.; Underwood, F.E.; Tang, W.; Benchimol, E.I.; Panaccione, R.; Ghosh, S.; Wu, J.C.Y.; Chan, F.K.L.; et al. Worldwide incidence and prevalence of inflammatory bowel disease in the 21st century: a systematic review of population-based studies. *Lancet* **2017**, *390*, 2769–2778. [CrossRef]
6. Neurath, M.F. Cytokines in inflammatory bowel disease. *Nat. Rev. Immunol.* **2014**, *14*, 329–342. [CrossRef] [PubMed]

7.   Robles, A.I.; Traverso, G.; Zhang, M.; Roberts, N.J.; Khan, M.A.; Joseph, C.; Lauwers, G.Y.; Selaru, F.M.; Popoli, M.; Pittman, M.E.; et al.   Whole-Exome Sequencing Analyses of Inflammatory Bowel Disease-Associated Colorectal Cancers. *Gastroenterology* **2016**, *150*, 931–943. [CrossRef] [PubMed]

8.   Gracie, D.J.; Irvine, A.J.; Sood, R.; Mikocka-Walus, A.; Hamlin, P.J.; Ford, A.C. Effect of psychological therapy on disease activity, psychological comorbidity, and quality of life in inflammatory bowel disease: a systematic review and meta-analysis. *Lancet Gastroenterol. Hepatol.* **2017**, *2*, 189–199. [CrossRef]

9.   Van der Sloot, K.W.J.; Amini, M.; Peters, V.; Dijkstra, G.; Alizadeh, B.Z. Inflammatory Bowel Diseases: Review of Known Environmental Protective and Risk Factors Involved. *Inflamm. Bowel Dis.* **2017**, *23*, 1499–1509. [CrossRef]

10.   Gecse, K.B.; Vermeire, S. Differential diagnosis of inflammatory bowel disease: imitations and complications. *Lancet Gastroenterol. Hepatol.* **2018**, *3*, 644–653. [CrossRef]

11.   Newman, D.J.; Cragg, G.M. Natural Products as Sources of New Drugs from 1981 to 2014. *J. Nat. Prod.* **2016**, *79*, 629–661. [CrossRef]

12.   Katkar, K.V.; Suthar, A.C.; Chauhan, V.S. The chemistry, pharmacologic, and therapeutic applications of *Polyalthia longifolia*. *Pharm. Rev.* **2010**, *4*, 62–68. [CrossRef]

13.   Shih, Y.T.; Hsu, Y.Y.; Chang, F.R.; Wu, Y.C.; Lo, Y.C. 6-Hydroxycleroda-3,13-dien-15,16-olide protects neuronal cells from lipopolysaccharide-induced neurotoxicity through the inhibition of microglia-mediated inflammation. *Planta Med.* **2010**, *76*, 120–127. [CrossRef] [PubMed]

14.   Yao, L.J.; Jalil, J.; Attiq, A.; Hui, C.C.; Zakaria, N.A. The medicinal uses, toxicities and anti-inflammatory activity of Polyalthia species (Annonaceae). *J. Ethnopharmacol.* **2019**, *229*, 303–325. [CrossRef] [PubMed]

15.   Thiyagarajan, V.; Sivalingam, K.S.; Viswanadha, V.P.; Weng, C.F. 16-hydroxy-cleroda-3,13-dien-16,15-olide induced glioma cell autophagy via ROS generation and activation of p38 MAPK and ERK-1/2. *Env. Toxicol. Pharm.* **2016**, *45*, 202–211. [CrossRef]

16.   Cheng, M.F.; Lin, S.R.; Tseng, F.J.; Huang, Y.C.; Tsai, M.J.; Fu, Y.S.; Weng, C.F. The autophagic inhibition oral squamous cell carcinoma cancer growth of 16-hydroxy-cleroda-3,14-dine-15,16-olide. *Oncotarget* **2017**, *8*, 78379–78396. [CrossRef]

17.   Chen, Y.C.; Huang, B.M.; Lee, W.C.; Chen, Y.C. 16-Hydroxycleroda-3,13-dien-15,16-olide induces anoikis in human renal cell carcinoma cells: involvement of focal adhesion disassembly and signaling. *Onco Targets* **2018**, *11*, 7679–7690. [CrossRef]

18.   Liu, C.; Lee, W.C.; Huang, B.M.; Chia, Y.C.; Chen, Y.C.; Chen, Y.C. 16-Hydroxycleroda-3, 13-dien-15, 16-olide inhibits the proliferation and induces mitochondrial-dependent apoptosis through Akt, mTOR, and MEK-ERK pathways in human renal carcinoma cells. *Phytomedicine* **2017**, *36*, 95–107. [CrossRef]

19.   Velmurugan, B.K.; Wang, P.C.; Weng, C.F. 16-Hydroxycleroda-3,13-dien-15,16-olide and N-Methyl-Actinodaphine Potentiate Tamoxifen-Induced Cell Death in Breast Cancer. *Molecules* **2018**, *23*, 1966. [CrossRef]

20.   Chen, C.Y.; Chang, F.R.; Shih, Y.C.; Hsieh, T.J.; Chia, Y.C.; Tseng, H.Y.; Chen, H.C.; Chen, S.J.; Hsu, M.C.; Wu, Y.C. Cytotoxic constituents of *Polyalthia longifolia* var. pendula. *J. Nat. Prod.* **2000**, *63*, 1475–1478. [CrossRef]

21.   Parang, B.; Barrett, C.W.; Williams, C.S. AOM/DSS Model of Colitis-Associated Cancer. In *Gastrointestinal Physiology and Diseases*; Ivanov, A., Ed.; Humana Press: New York, NY, USA, 2016; Volume 1422, pp. 297–307.

22.   Janakiram, N.B.; Rao, C.V. The Role of Inflammation in Colon Cancer. In *Inflammation and Cancer*; Aggarwal, B., Sung, B., Gupta, S., Eds.; Springer: Basel, Switzerland, 2014; Volume 816, pp. 25–52.

23.   Ma, B.; Hottiger, M.O. Crosstalk between Wnt/beta-Catenin and NF-kappaB Signaling Pathway during Inflammation. *Front. Immunol.* **2016**, *7*, 378. [CrossRef]

24.   Wheat, C.L.; Ko, C.W.; Clark-Snustad, K.; Grembowski, D.; Thornton, T.A.; Devine, B. Inflammatory Bowel Disease (IBD) pharmacotherapy and the risk of serious infection: a systematic review and network meta-analysis. *Bmc Gastroenterol.* **2017**, *17*, 52. [CrossRef] [PubMed]

25.   Triantafillidis, J.K.; Merikas, E.; Georgopoulos, F. Current and emerging drugs for the treatment of inflammatory bowel disease. *Drug Des. Devel.* **2011**, *5*, 185–210. [CrossRef] [PubMed]

26.   Zhou, J.; Li, C.J.; Yang, J.Z.; Ma, J.; Wu, L.Q.; Wang, W.J.; Zhang, D.M. Phenylpropanoid and lignan glycosides from the aerial parts of Lespedeza cuneata. *Phytochemistry* **2016**, *121*, 58–64. [CrossRef] [PubMed]

27. Vecchi Brumatti, L.; Marcuzzi, A.; Tricarico, P.M.; Zanin, V.; Girardelli, M.; Bianco, A.M. Curcumin and inflammatory bowel disease: potential and limits of innovative treatments. *Molecules* **2014**, *19*, 21127–21153. [CrossRef]
28. Lewis, S.N.; Brannan, L.; Guri, A.J.; Lu, P.; Hontecillas, R.; Bassaganya-Riera, J.; Bevan, D.R. Dietary alpha-eleostearic acid ameliorates experimental inflammatory bowel disease in mice by activating peroxisome proliferator-activated receptor-gamma. *PLoS ONE* **2011**, *6*, e24031. [CrossRef]
29. Galluzzi, L.; Vitale, I.; Aaronson, S.A.; Abrams, J.M.; Adam, D.; Agostinis, P.; Alnemri, E.S.; Altucci, L.; Amelio, I.; Andrews, D.W.; et al. Molecular mechanisms of cell death: recommendations of the Nomenclature Committee on Cell Death 2018. *Cell Death Differ.* **2018**, *25*, 486–541. [CrossRef]
30. Tummers, B.; Green, D.R. Caspase-8: regulating life and death. *Immunol. Rev.* **2017**, *277*, 76–89. [CrossRef]
31. Li, P.; Zhou, L.; Zhao, T.; Liu, X.; Zhang, P.; Liu, Y.; Zheng, X.; Li, Q. Caspase-9: structure, mechanisms and clinical application. *Oncotarget* **2017**, *8*, 23996–24008. [CrossRef]
32. Czabotar, P.E.; Lessene, G.; Strasser, A.; Adams, J.M. Control of apoptosis by the BCL-2 protein family: implications for physiology and therapy. *Nat. Rev. Mol. Cell Biol.* **2014**, *15*, 49–63. [CrossRef]
33. Lin, Y.H.; Lee, C.C.; Chan, W.L.; Chang, W.H.; Wu, Y.C.; Chang, J.G. 16-Hydroxycleroda-3,13-dien-15,16-olide deregulates PI3K and Aurora B activities that involve in cancer cell apoptosis. *Toxicology* **2011**, *285*, 72–80. [CrossRef]
34. Lin, Y.H.; Lee, C.C.; Chang, F.R.; Chang, W.H.; Wu, Y.C.; Chang, J.G. 16-hydroxycleroda-3,13-dien-15,16-olide regulates the expression of histone-modifying enzymes PRC2 complex and induces apoptosis in CML K562 cells. *Life Sci.* **2011**, *89*, 886–895. [CrossRef]
35. Liu, J.; Zhang, C.; Feng, Z. Tumor suppressor p53 and its gain-of-function mutants in cancer. *Acta Biochim. Biophys. Sin. (Shanghai)* **2014**, *46*, 170–179. [CrossRef] [PubMed]
36. Goretsky, T.; Dirisina, R.; Sinh, P.; Mittal, N.; Managlia, E.; Williams, D.B.; Posca, D.; Ryu, H.; Katzman, R.B.; Barrett, T.A. p53 mediates TNF-induced epithelial cell apoptosis in IBD. *Am. J. Pathol.* **2012**, *181*, 1306–1315. [CrossRef] [PubMed]
37. Pastor, D.M.; Irby, R.B.; Poritz, L.S. Tumor necrosis factor alpha induces p53 up-regulated modulator of apoptosis expression in colorectal cancer cell lines. *Dis. Colon Rectum* **2010**, *53*, 257–263. [CrossRef] [PubMed]
38. Yashiro, M. Molecular Alterations of Colorectal Cancer with Inflammatory Bowel Disease. *Dig. Dis. Sci.* **2015**, *60*, 2251–2263. [CrossRef] [PubMed]
39. Ahmed, D.; Eide, P.W.; Eilertsen, I.A.; Danielsen, S.A.; Eknaes, M.; Hektoen, M.; Lind, G.E.; Lothe, R.A. Epigenetic and genetic features of 24 colon cancer cell lines. *Oncogenesis* **2013**, *2*, e71. [CrossRef] [PubMed]
40. Triki, M.; Lapierre, M.; Cavailles, V.; Mokdad-Gargouri, R. Expression and role of nuclear receptor coregulators in colorectal cancer. *World J. Gastroenterol.* **2017**, *23*, 4480–4490. [CrossRef] [PubMed]
41. Koch, S. Extrinsic control of Wnt signaling in the intestine. *Differentiation* **2017**, *97*, 1–8. [CrossRef]
42. Asif, M.; Shafaei, A.; Abdul Majid, A.S.; Ezzat, M.O.; Dahham, S.S.; Ahamed, M.B.K.; Oon, C.E.; Abdul Majid, A.M.S. Mesua ferrea stem bark extract induces apoptosis and inhibits metastasis in human colorectal carcinoma HCT 116 cells, through modulation of multiple cell signalling pathways. *Chin. J. Nat. Med.* **2017**, *15*, 505–514. [CrossRef]
43. Lee, M.A.; Kim, W.K.; Park, H.J.; Kang, S.S.; Lee, S.K. Anti-proliferative activity of hydnocarpin, a natural lignan, is associated with the suppression of Wnt/beta-catenin signaling pathway in colon cancer cells. *Bioorg. Med. Chem. Lett.* **2013**, *23*, 5511–5514. [CrossRef]
44. Hseu, Y.C.; Chao, Y.H.; Lin, K.Y.; Way, T.D.; Lin, H.Y.; Thiyagarajan, V.; Yang, H.L. Antrodia camphorata inhibits metastasis and epithelial-to-mesenchymal transition via the modulation of claudin-1 and Wnt/beta-catenin signaling pathways in human colon cancer cells. *J. Ethnopharmacol.* **2017**, *208*, 72–83. [CrossRef] [PubMed]

 *biomolecules*

Review

Recent Trends in the Application of Chromatographic Techniques in the Analysis of Luteolin and Its Derivatives

Aleksandra Maria Juszczak [1], Marijana Zovko-Končić [2] and Michał Tomczyk [1,*]

1   Department of Pharmacognosy, Faculty of Pharmacy, Medical University of Białystok, ul. Mickiewicza 2a, 15-230 Białystok, Poland; aleksandra.juszczak@umb.edu.pl
2   Department of Pharmacognosy, University of Zagreb, Faculty of Pharmacy and Biochemistry, A. Kovačića 1, 10000 Zagreb, Croatia; mzovko@pharma.hr
*   Correspondence: michal.tomczyk@umb.edu.pl; Tel.: +48-85-748-5694

Received: 1 October 2019; Accepted: 8 November 2019; Published: 12 November 2019

Abstract: Luteolin is a flavonoid often found in various medicinal plants that exhibits multiple biological effects such as antioxidant, anti-inflammatory and immunomodulatory activity. Commercially available medicinal plants and their preparations containing luteolin are often used in the treatment of hypertension, inflammatory diseases, and even cancer. However, to establish the quality of such preparations, appropriate analytical methods should be used. Therefore, the present paper provides the first comprehensive review of the current analytical methods that were developed and validated for the quantitative determination of luteolin and its *C*- and *O*-derivatives including orientin, isoorientin, luteolin 7-*O*-glucoside and others. It provides a systematic overview of chromatographic analytical techniques including thin layer chromatography (TLC), high performance thin layer chromatography (HPTLC), liquid chromatography (LC), high performance liquid chromatography (HPLC), gas chromatography (GC) and counter-current chromatography (CCC), as well as the conditions used in the determination of luteolin and its derivatives in plant material.

Keywords: luteolin; hyphenated techniques; chromatography

---

1. Introduction

Luteolin (Figure 1) is a yellow dye commonly found in fresh plants. It is a flavonoid of the flavone type that is distributed widely throughout the plant kingdom. Similar to other derivatives of 2-phenylbenzo-γ-pyrone, its basic skeleton has a characteristic C_6-C_3-C_6 system, containing two benzene rings and a bridge with a C2-C3 double carbon bond and an attached oxygen atom [1–4]. Structure-activity studies have demonstrated that the pharmacological effects of luteolin and other flavonoids are strongly related to the presence of hydroxyl groups at the C5, C7, C3′ and C4′ carbons as well as to the presence of the double bond in the C2-C3 position [3,5]. The presence of the -OH group at position C3′ distinguishes luteolin from apigenin, and the lack of this group at the C3 carbon is an element that places luteolin in the flavone group [6].

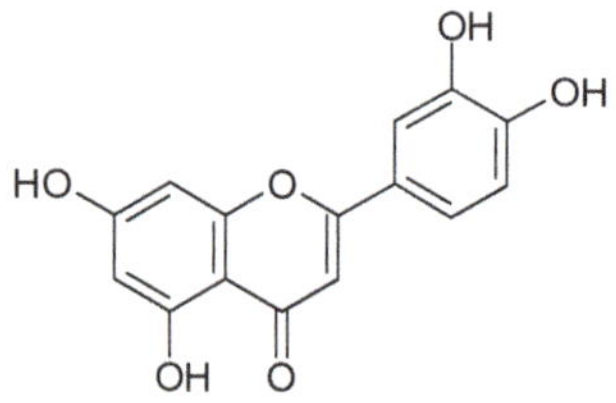

Figure 1. Chemical structure of luteolin.

Luteolin exhibits multiple biological effects such as antioxidant, anti-inflammatory and immunomodulatory activity. Plants rich in luteolin are often used in traditional medicine for treatment of various diseases such as hypertension, inflammatory disorders, and even cancer [7].

Because luteolin bears four hydroxyl groups (at the C5, C7, C3′ and C4′ positions), many derivatives of luteolin can be created. Various types of functional groups and/or sugar molecules can be attached to those positions, creating a huge number of different but structurally similar molecules. The most common are methyl derivatives, as well as *C*- and -*O*-glycosides [8,9].

Orientin, an 8-*C*-glucoside derivative of luteolin, displays an array of health-related biological properties, such as antioxidant, anti-ageing, antiviral, antibacterial, anti-inflammatory, vasodilatation, cardioprotective, radiation protective, neuroprotective, antidepressant-like, anti-adipogenesis, and antinociceptive effects. It may be found in different medicinal plants such as *Ocimum sanctum* (holy basil), *Phyllostachys nigra* (bamboo leaves), *Passiflora* sp. (passion flower), *Linum usitatissimum* (flax), *Euterpe oleracea* (Acai palm) and many others [10]. Another luteolin derivative, isoorientin (luteolin-6-*C*-glucoside) acts as an antioxidant, photoprotective [11], skin lightening [12], hepatoprotective [13] and anti-inflammatory agent [14]. *O*-glucosides of luteolin also display biological activities. For example, luteolin 7-*O*-glucoside alleviates skin lesions in murine models of atopic dermatitis [15] and protects cells against apoptosis induced by hypoxia/reoxygenation [16].

An increasing number of herbal preparations on the market contain luteolin and its derivatives, either as single-ingredient products or in mixtures with other phytochemicals, e.g., in form of medicinal plants extracts. To establish the quality of such products, it is important to use appropriate analytical methods. However, there is a lack of quality reviews of the available methods for quantification of luteolin derivatives, and the information on their comparison is lacking. Therefore, the aim of this article is to systematize knowledge and information in the field of chromatographic analytical techniques used for quantification of luteolin and its derivatives. The presented review is the first description of this type and provides a systematic overview of chromatographic analytical techniques including thin layer chromatography (TLC), high performance thin layer chromatography (HPTLC), liquid chromatography (LC), high performance liquid chromatography (HPLC), gas chromatography (GC) and counter-current chromatography (CCC), as well as the conditions used to assess luteolin and its derivatives.

2. Chromatographic Techniques for the Analysis of Luteolin Derivatives

Chromatography occupies a leading position among other instrumental methods in the analysis of chemical compounds. As a physicochemical method of separation and analysis of mixtures of chemical compounds, it allows detection and identification as well as quantitative determination of the test substance with high accuracy. The coupling of chromatography with other methods of analysis contributes to a more accurate detection and expansion of analytical capabilities, especially for complex mixtures of organic compounds [2,17,18].

Chromatographic techniques are based on the interaction of the mixture components with the mobile and stationary phases of the chromatographic system. This results in the division of the mixture components between the two phases. In addition, the interaction of the mobile and stationary phases is also important in the separation process [17,18]. According to the aggregation state of the mobile phase, chromatographic techniques are divided into gas, liquid and supercritical chromatography. Another criterion for classification of chromatographic techniques is the type of stationary phase. If the stationary phase is a liquid, the chromatography technique is referred to as partition chromatography. In the case of a solid, the technique is referred to as adsorption chromatography. Another example of the classification of chromatographic methods is their division depending on the chromatographic processing method. This classification allows distinguishing column chromatography and planar techniques, which include TLC and paper chromatography [17,19]. Many different chromatographic techniques are used in the analysis of luteolin derivatives. These include TLC, HPTLC, LC, HPLC, GC and CCC [19].

2.1. Thin Layer Chromatography in the Analysis of Luteolin Derivatives

Thin layer chromatography (TLC) is a rather simple but relatively popular method used in the analysis of flavonoids since 1960 [20]. It is a variation of LC that is carried out on a plane and is therefore referred to as planar chromatography. Despite the dynamic development of other chromatography techniques, TLC has not lost its importance in phytochemical analysis [18]. Its values are still recognized as is the basic tool in the qualitative analysis of natural products, and as such, it is still recommended by most modern Pharmacopoeias. In addition, the TLC technique is currently being improved, and the scope of its application is widening, while the results are becoming comparable to those obtained by GC or HPLC [20–22]. The stages of chromatographic analysis consist of placing the sample on the stationary phase and developing the chromatogram, followed by its visualization. In the final step, qualitative and/or quantitative determinations of the tested components are made [18,22,23]. General guidelines for flavonoid separation on TLC plates are presented in Table 1.

Table 1. Recommended combinations of solvents / adsorbents for identification. of different flavonoid types by thin layer chromatography (TLC).

Flavonoid Type	Adsorbent Type/Mobile Phase		
	Cellulose	Polyamide	Silica gel
Polar flavonoid aglycones, e.g., flavones	BuOH:AcOH:H_2O (3:1:1 *v/v/v*) [a] CHCl$_3$:AcOH:H_2O (30:15:2 *v/v/v*) [b]	MeOH:AcOH:H_2O (18:1:1 *v/v/v*)	To:Py:FA (36:9:5 *v/v/v*)
Non-polar flavonoid aglycones, e.g., methylated flavones	10–30% AcOH	—	CHCl$_3$:MEOH (15:1 to 3:1 *v/v*)
Flavonoid glycosides	BuOH:AcOH:H_2O (3:1:1 *v/v/v*) [a] BuOH:AcOH:H_2O (4:1:5 *v/v/v*) [a]	H_2O:MeOH:MEK:methyl acetylacetone (13:3:3:1 *v/v/v/v*)	EtOAc:Py:H_2O:MeOH (80:20:10:5 *v/v/v/v*) (especially flavone C-glycosides)

Abbreviations: AcOH, acetic acid; BuOH, butanol; CHCl3, chloroform; EtOAc, ethyl acetate; FA, formic acid; MeOH, methanol; MEK, methyl ethyl ketone; Py, pyridine; To, toluene. [a] The mobile phase is thoroughly mixed in the separating funnel and the upper phase is used. [b] The mobile phase is thoroughly mixed in the separating funnel and the excess water is discarded.

The main advantage of TLC is that it is a relatively simple and inexpensive technique that allows for rapid qualitative and quantitative analysis of the tested compounds. Samples analyzed with this method usually do not require pre-treatment, such as purification or concentration. In addition, several dozens of samples can be analyzed simultaneously on one plate. A large amount of the diluted sample can be applied to the stationary phase because the solvent evaporates during the application to the plate. Furthermore, due to the evaporation of the solvent phase after the development of the chromatogram, the detection method does not depend on the type of mobile phase used for separation. In TLC, it is possible not only to compare the analyzed components with the standards, but also to differentiate between substances bearing specific functional groups using the appropriate reagents for detection [18].

Thin layer chromatography and column chromatography (CC) are interchangeable techniques that may be combined, which significantly reduces costs and analysis time [22]. To achieve this, the same adsorbents are ideally used for both TLC and CC [18,22,24]. Nevertheless, other solvent systems may also be used. Elution in CC may be carried out in one mobile phase, or its composition may be changed during chromatography (mobile phase gradient), thereby increasing the elution force. In this case, the TLC mobile phase should be changed accordingly [18,24].

The type and quality of the stationary phase greatly affect the separation of mixture components. Thus, the selection of an appropriate adsorbent is very important. However, most TLC analyses are carried out in a normal phase system where hydrophilic (polar) adsorbents are used [18,23,24]. Reversed phase systems with lipophilic (non-polar) stationary phases are common and have little significance in the analysis of flavonoids [20]. Currently, the most commonly used stationary phase for the analysis of flavonoids is silica gel [20]. However, the use polyamide coated chromatography plates, in both normal and reversed phase systems, is not uncommon [24].

Detection of flavonoids on TLC plates is most often conducted under ultraviolet (UV) light at wavelengths of 254 or 366 nm. Luteolin derivatives also display fluorescence, which can be enhanced using the appropriate derivatization reagents, e.g., with the so-called NP/PEG reagent. The most frequent procedure consists of spraying the plate with 1% methanolic diphenylboric acid-β-ethylamino ester (natural product reagents, NP), followed by 5% ethanolic polyethylene glycol 4000 (PEG) solution [20]. A densitometer may also be used for the qualitative analysis of the substance. The analysis is performed by comparing the retardation factor (Rf) and absorption spectrum of the test substance and the standard. Analytes can also be identified by extracting the separated substances from the plate. Then, the analysis is carried out using Fourier transform infrared mass spectrometry (MS), UV spectrometry, Raman spectrometry or other techniques [23]. Even though TLC separation of luteolin derivatives (Table 2) can be performed in different types of stationary phases such as a polyamide phase [26,28], it is most frequently performed on silica gel plates that are often coated with a fluorescent indicator [27,29] (F_{254} plates) for preliminary detection. Such an approach has been used in case of analysis of luteolin 3'-O-glucoside, luteolin 6,8-C-dihexoside and luteolin 7-O-rutinoside in *Phlomis* sp. [27]. However, subsequent analysis with NP/PEG is the standard procedure for TLC analysis of luteolin [26–29], and it is almost always performed regardless of additional types of detection such as detection of flavonoids in *Ligustrum vulgare* with aniline phthalate [26]. Typically, the mobile phase for luteolin derivative separation consists of a mixture of aprotic organic solvents such as ethyl acetate (EtOAc) [26,28] or acetone (Ace) [29] and H_2O with a significant amount of formic (FA) and/or acetic acid (AcOH) to avoid tailing of the separated zones [26–29].

Thin layer chromatography is often used as a complementary method to other chromatographic techniques. For example, analysis of the butanol (BuOH) fraction of the methanol (MeOH) extract from the leaves of the common privet (*Ligustrum vulgare*) conducted by Mučaji et al. [26] allowed isolation of two luteolin derivatives from the plant. TLC was carried out, among other techniques, on polyamide plates, and the optimal mobile phase was found. TLC was used for analysis of individual fractions obtained in column chromatography with or without acid hydrolysis of compounds. MS detection and nuclear magnetic resonance (NMR) spectra were also used [26]. Furthermore, TLC was used together with high performance liquid chromatography combined with mass spectrometry and pulsed amperometric detection (HPLC-PAD-MS) for analysis of luteolin and other phenolic compounds in *Leontopodium alpinum*. In addition to NP/PEG, UV, infrared (IR) and NMR analyses were used for identification of these compounds [29].

Compared to other chromatographic techniques, HPTLC results in reduced time and costs of analysis and provides much greater efficiency of separation. It is suitable even for the analysis of crude extracts similar to TLC, and a relatively small amount of solvent is used to analyze several samples, making this method environmentally friendly [60]. In the analysis of luteolin derivatives (Table 3), HPTLC silica gel 60 is almost exclusively used as the stationary phase [45,47,49,51], while HPTLC NH_2 plates are rarely used, e.g., for separation of flavonoids in some Lamiaceae species such as *Mentha piperita* [53] and *Thymus* sp. [55]. In addition to NP/PEG (e.g., [52,57]), other detection systems may be employed for visualization of luteolin derivatives, such as bis-diazotized sulfanilamide [53] or aqueous solutions of Al^{3+} ions for flavonoids in *M. piperita* [53], honey [49] or *Thymus* sp. [55]. Similar to TLC, mixtures of organic solvents, H_2O, and FA are most often used as the mobile phase [44,45,47–49,51,52].

High performance thin layer chromatography may also be used as a complimentary method to other chromatographic techniques. Chelyn et al. [44] used the HPTLC technique in the analysis of the ethanol (EtOH) extract of *Clinacanthus nutans* leaves. The analysis revealed the presence of, among others, luteolin 8-*C*-glucoside (orientin) and luteolin 6-*C*-glucoside (isoorientin) in the raw material. Their detection was conducted by comparing R_f coefficients using derivatization reagents and 366 nm UV light. In this work, the characteristic fluorescent bands after derivatization provided important clues for the identification of the major flavone present in the samples, while high performance liquid chromatography combined with ultraviolet spectrometry or a diode array detector (HPLC-UV/DAD) technique was employed for the simultaneous detection and quantification of these compounds [44].

Table 2. Thin layer chromatography in the analysis of luteolin derivatives.

Luteolin Derivative	Stationary Phase	Mobile Phase	Detection	Analyzed Species	Ref.
Luteolin	silica gel 60 F_{254} silica gel 60 RP-18 F_{254}S	Hx:EtOAc:AcOH (31:14:5 $v/v/v$); To:DI:AcOH (90:25:4 $v/v/v$) FA:H_2O:MeOH (5.5:47.25:47.25 $v/v/v$)	FBS; UV, 254, 366 nm	*Artemisia annua*	[25]
Luteolin 7-rutoside Luteolin 7-rhamnoside	silica gel cellulose polyamide	EtOAc:FA:AcOH:H_2O (100:11:11:23 $v/v/v/v$) 30% AcOH $CHCl_3$:MeOH:MEK:AcAc (9:4:2:1 $v/v/v/v$); To:MeOH:MEK:BuOH (300:150:150:3 $v/v/v/v$)	NP/PEG, aniline phthalate; UV, 254,366 nm MS, NMR	*Ligustrum vulgare*	[26]
Luteolin 3′-glucoside Luteolin 6,8-dihexoside Luteolin 7-rutinoside	silica gel 60 F_{254}	MeOH:H_2O (15:5 v/v); $CHCl_3$:MeOH (15:5 v/v); 15% AcOH	NP/PEG; UV, 366 nm	*Phlomis persica* *Ph. elliptica*	[27]
Luteolin 7-glucoside	silica gel polyamide plates	EtOAc:FA:H_2O (18:1:1 $v/v/v$) EtOAc:FA:AcOH:H_2O (100:10:10:13 $v/v/v/v$)	NP/PEG; UV, 366 nm	*Carduus acanthoides*	[28]
Luteolin Luteolin 7,4′-diglucoside Luteolin 6-hydroxy-7-glucoside Luteolin 4′-glucoside Luteolin 3′-glucoside Luteolin 7-glucoside	silica gel 60 F_{254}	EtOAc:Ace:FA (8:1:1 $v/v/v$); EtOAc:H_2O:AcOH:FA (10:3:1:1 $v/v/v/v$)	NP; UV, IR, NMR	*Leontopodium alpium*	[29]
Luteolin Luteolin 7-glucoside	silica gel 60 F_{254}	To:Et_2O:AcOH (60:40:10 v/v/v); EtOAc:AcOH:FA:H_2O (100:11:11:26 v/v/v/v)	NP/PEG; UV, 366 nm	*Matricaria recutita, Achillea millefolium,* *Thymus vulgaris, Salvia officinalis*	[30]

Abbreviations: AcAc, acetylacetone; Ace, acetone; AcOH, acetic acid; BuOH, butanol; $CHCl_3$, chloroform; DI, 1,4–dioxane; Et_2O, diethyl ether; EtOAc, ethyl acetate; FA, formic acid; FBS, Fast Blue B Salt; Hx, hexane; MeOH, methanol; MEK, methyl ethyl ketone; NP, 1% methanolic diphenylboric acid-β-ethylamino ester - natural product reagents; PEG, 5% ethanolic polyethylene glycol 4000; To, toluene; UV, ultraviolet spectroscopy; MS, mass spectrometry; NMR, nuclear magnetic resonance; IR, infrared.

Table 3. High performance thin layer chromatography in the analysis of luteolin derivatives.

Luteolin Derivative	Stationary Phase	Mobile Phase	Detection	Analyzed Species	Ref.
Luteolin	HPTLC silica gel 60 F_{254}	Hx:EtOAc:AcOH (31:14:5 *v/v/v*); To:DI:AcOH (90:25:4 *v/v/v*)	FBS; UV, 254, 366 nm	*Artemisia annua*	[25]
	HPTLC diol F_{254}S	$CHCl_3$:Hx:EtOAc (34:4 *v/v/v*)	2% $AlCl_3$; UV, 366 nm	*Oxytropis glabra*	[31]
	HPTLC silica gel 60 RP-18W	BuOH:MeOH:H_2O (300:50:50 *v/v/v*); 15% AcOH	NPR/PEG; UV, 366 nm		
	HPTLC silica gel 60 F_{254}	EtOAc:MeOH:FA:H_2O (20:3:1:2 *v/v/v/v*)	MeOH:H_2SO_4 (95:5 *v/v*); UV, 254, 366 nm	*Asparagus racemosus, Withania somnifera, Vitex negundo, Plumbago zylenica, Butea monosperma, Tephrosia purpurea*	[32]
	HPTLC silica gel 60 F_{254}	To:EtOAc:FA (10:9:1 *v/v/v*)	UV, 254 nm	*Cardiospermum halicacabum*	[33]
	HPTLC silica gel 60 F_{254}	DCM:MeOH (70:30 *v/v*)	UV, 254 nm; NMR	*Satureja montana*	[34]
	HPTLC silica gel G_{60} F_{254}	To:EtOAc:FA (6:4:1 *v/v/v*)	UV, 349 nm	*Hygrophila spinosa*	[35]
	HPTLC silica gel 60 F_{254}	EtOAc:MeOH:H_2O:AcOH (3:1:1:1 *v/v/v/v*)	UV, 254, 366 nm	*Foeniculum vulgare, Cuminum cyminum, Apium graveolens, Petroselinum crispum, Anethum graveolens, Ammi majus*	[36]
	HPTLC silica gel 60 F_{254}	To:EtOAc:FA (3:3:0.8 *v/v/v*)	MeOH:H_2SO_4 (90:10 *v/v*); UV, 254 nm	*Saraca asoca*	[37]
	HPTLC silica gel 60 F_{254}	To:EtOAc:FA (6:4:0.3 *v/v/v*)	NP/PEG; UV, 366 nm	*Premna mucronata*	[38]
	HPTLC silica gel G_{60} F_{254}	To:EtOAc:FA (10:9:1 *v/v/v*)	UV, 254, 366 nm	*Anisochilus carnosus*	[39]

Table 3. *Cont.*

Luteolin Derivative	Stationary Phase	Mobile Phase	Detection	Analyzed Species	Ref.
Luteolin	HPTLC silica gel 60	nHx:EtOAc:FA (30:20:1.5 *v/v/v*)	NP/PEG; UV, 349 nm	*Satureja hortensis*	[40]
	HPTLC silica gel 60 F_{254}	nHx:EtOAc:AcOH (5:3:1 *v/v/v*)	UV	Propolis	[41]
	HPTLC silica gel 60 GF_{254}	To:EtOAc:FA:MeOH (3:3:0.8:0.2 *v/v/v/v*)	UV, 254, 365nm	*Eclipta alba*	[42]
	HPTLC silica gel 60 F_{254}	To:EtOAc:FA (5:3:1 *v/v/v*)	NP; UV, 366, 254 nm	*Vitis vinifera*	[43]
Luteolin 6-glucoside Luteolin 8-glucoside	HPTLC silica gel 60 F_{254}	EtOAc:FA:AcOH:H$_2$O (100:11:11:27 *v/v/v/v*)	NP/PEG; UV, 366 nm	*Clinacanthus nutans*	[44]
	HPTLC silica gel 60 F_{254}	EtOAc:FA:H$_2$O (82:9:9 *v/v/v*)	NP/PEG; UV, 366 nm	*Passiflora alata,* *P. edulis*	[45]
	Nano-DUASIL silica gel 60	THF:To:FA:H$_2$O (16:8:2:1 *v/v/v/v*)	UV, 350 nm	*Phyllostachys pubescens*	[46]
Luteolin glucoside	HPTLC silica gel 60 F_{254}	EtOAc:FA:AcOH:H$_2$O (100:11:11:26 *v/v/v/v*)	NP/PEG; UV, 366 nm	*Equisetum arvense*	[47]
	HPTLC silica gel 60 F_{254}	EtOAc:AcOH:FA:H$_2$O (10: 1.1:1.1:2.6 *v/v/v/v*)	UV, 254, 366 nm	*Aerva javanica*	[48]
Luteolin	NP-HPTLC silica gel	EtOAc:FA:AcOH:H$_2$O (100:11:11:27 *v/v/v/v*) To:EtOAc:AcOH (50:45:5 *v/v/v*)	H$_2$O solution of 4% Al$_2$(SO$_4$)$_3$; UV, 365 nm	*Apis mellifera*, honey	[49]
Luteolin 7-glucoside	HPTLC silica gel 60 F_{254}	To:EtFo:FA (6:4:1 *v/v/v*)	NP; UV, 254, 366 nm	*Potentilla grandiflora,* *P. recta,* *P. anserina,* *P. fruticose,* *P. rupestris,* *P. thuringiaca*	[50]
	modified HPTLC silica gel 60 F_{254} with CN, NH$_2$ HPTLC diol F_{254}	To:EtFo:FA (7:5:1 *v/v/v*) EtOAc:MEK:DIPE:FA (3:10:4:1 *v/v/v/v*)			
Luteolin 7-glucoside	HPTLC silica gel 60	EtOAC:DCM:AcOH:FA:H$_2$O (100:25:10:10:11 *v/v*)	NP; UV, 366 nm	*Lavandula stoechas*	[51]
	HPTLC silica gel 60 F_{254}	EtOAc:AcOH:FA:H$_2$O (100:11:11:26 *v/v/v/v*)	NP/PEG; UV, 254, 366 nm	*Stachys sylvatica,* *S. recta*	[52]

Table 3. *Cont.*

Luteolin Derivative	Stationary Phase	Mobile Phase	Detection	Analyzed Species	Ref.
Luteolin 7-rutinoside	HPTLC NH$_2$ HPTLC RP-18W	Ace:AcOH (85:15 *v/v*) H$_2$O:MeOH (60:40 *v/v*)	MeOH:AlCl$_3$ (98:2 *v/v*), *bis*-diazodized sulfanilamide; UV-VIS, 365 nm IR, MS, NMR	*Mentha piperita*	[53]
Luteolin 7-glucuronide Luteolin 7-glucoside	HPTLC silica gel 60 HPTLC NH$_2$	Et:Ace:H2O:FA (55:25:10:10 *v/v/v/v*) Ace:AcOH (85:15 *v/v*)	NP/PEG; UV, 365 nm	*Mentha piperita,* *Melissa officinalis,* *Salvia officinalis*	[54]
Luteolin 7-rutinoside Luteolin 7-glucoside Luteolin 7-glucuronide	HPTLC NH$_2$ HPTLC silica gel 60	Ace:AcOH (85:15 *v/v*) Ace:FA (85:15 *v/v*) DIPE:Ace:H$_2$O:FA (50:30:10:10 *v/v/v/v*)	MeOH:AlCl$_3$ (98:2 *v/v*), NP/PEG; UV, 365 nm	*Thymus vulgaris,* *Th. serpyllum,* *Majorana hortensis,* *Mentha piperita*	[55] [56]
Luteolin Luteolin acetyl hexuronide Luteolin 7-rutinoside Luteolin acetyl hexoside Luteolin hexuronide	HPTLC silica gel 60 F$_{254}$	pre-develop: CHCl$_3$:MeOH (1:1 *v/v*); *n*Hx:EtOAc:FA (20:19:1 *v/v/v*)	NP: white light, 254, 366, 330 nm; PEG: 254, 297, 340, 366, 430 nm; paraffin–*n*Hx: 254, 320, 360, 366, 400 nm; ESI-MS/MS	*Rosmarinus officinalis* *Salvia officinalis*	[57]
Luteolin 3′,7-diglucoside Luteolin 4′-glucoside Luteolin 7-glucoside Luteolin 8-glucoside Luteolin 6-glucoside	HPTLC silica gel 60 F$_{254}$	EtOAc:MeOH:AcOH:FA:H$_2$O (30:1:2:1:3 *v/v/v/v/v*)	NP; UV, 366 nm	*Colocasia esculenta*	[58]
Luteolin 8-glucoside	HPTLC RP C$_{18}$ HPTLC silica gel F$_{254}$	FA:MeOH:H$_2$O (0.5:6.65:2.85 *v/v/v*) To:EtOAc:FA (6:5:1 *v/v/v*)	ESI-MS/MS, 254, 366 nm	*Cyclanthera pedata*	[59]

Abbreviations: Ace, acetone; AcOH, acetic acid; AlCl$_3$, aluminium trichloride; Al$_2$(SO$_4$)$_3$, aluminium sulfate; BuOH, butanol; CHCl$_3$, chloroform; DCM, dichloromethane; DI, 1,4-dioxane; DIPE, diisopropyl ether; ESI-MS/MS, electrospray ionization combined with tandem mass spectrometry; EtFo, ethyl formate; EtOAc, ethyl acetate; Et, ether; FA, formic acid; H$_2$SO$_4$, sulfuric acid; Hx, hexane; MeOH, methanol; MEK, methyl ethyl ketone; *n*Hx, *n*-hexane; NP, 1% methanolic diphenylboric acid-β-ethylamino ester - natural product reagents; neurotransmitters; PEG, 5% ethanolic polyethylene glycol 4000; THF, tetrahydrofuran; To, toluene; UV-VIS, ultraviolet-visible spectroscopy; ESI-MS, electrospray ionization combined with mass spectrometry.

2.2. High Performance Liquid Chromatography in the Analysis of Luteolin Derivatives

Among the many chromatographic methods, adsorptive chromatography, in which the mobile phase is liquid and the stationary phase is solid, is of the greatest practical importance. For example, this method has a much wider application than GC because it allows the analysis of substances in the form of liquids and soluble solids. Furthermore, it is also suitable for analysis of thermolabile substances because it is usually performed at low temperatures, and such samples are not degraded [61]. Due to the long analysis time, high mobile phase consumption and low efficiency, traditional CC is currently used mainly for preparative purposes. However, improved methods, such as HPLC and especially liquid chromatography combined with mass spectrometry (LC-MS), are increasingly used for analysis of natural compounds including luteolin derivatives [62].

High performance liquid chromatography has been performed since 1960. It is a dynamically developing method with a wide range of uses and has been proven to be very useful in phytochemical analysis. The principle of operation consists of pumping the mobile phase from the tank (or tanks) through the stationary phase-filled column. Eluents are previously filtered and degassed. Some systems are additionally equipped with thermostats that regulate the temperature of the column. If the chromatographic system is properly selected and applied, then the individual components are separated and detected. Strengthened signals are transmitted to the computer, where the obtained data are registered and properly processed [19]. The separation is based on competition of the molecules of the eluent and the substance being analyzed for the space on the adsorbent surface in the column [63,64].

When choosing a column, one should be guided by the size of the sample, time of analysis and expected effect of the separation. Although columns with different diameters are available, those with a diameter of 4.6 mm are by far the most common. However, due to better detection of the components separated in smaller diameter columns, these columns are increasingly being used for separation and analysis [19,63,65]. In addition, improved separation can be achieved using a column filled with smaller particles. In addition to 5 μm particles, which are most common, particles of 3 μm in diameter or even smaller can be used. For example (Table 4), this approach was chosen when separating luteolin 2″-*O*-feruloylhexosyl-6-*C*-hexoside and luteolin 6-*C*-glucoside from *Arenaria montana* [66,67] and isoorientin in *Achillea millefolium* [68,69], as well as separating luteolin 6-*C*-hexosyl-8-*C*-pentoside, luteolin 2″-*O*-deoxyhexosyl-6-*C*-glucoside and luteolin 6-*C*-glucoside in *Cymbopogon citratus* [70,71]. Particles less than 2 μm in diameter were used for ultra-performance liquid chromatography (UPLC) analysis of luteolin derivatives and other flavonoids in *Lactuca sativa* [72,73] and date palm (*Phoenix dactylifera*) [74].

Stationary phases with different polarity may be used in HPLC. In the normal phase system, polar column fillings are used, and most often the filling is silica gel. However, silica gel can adsorb water, which leads to the loss of the original separating properties of the column and thus to impaired reproducibility of the obtained results [19,65]. Therefore, the gel is often modified with the aim of enabling better separation of mixture components. This is performed mainly by bonding alkyl chains (or alkyl chains bearing other functional groups) to functional groups on the gel surface. Silica gels optimized in this way are referred to as the associated phase [2,65]. In the so-called reversed phase (RP) system, non-polar associated phases are used. Such systems are especially useful in the analysis of insoluble or poorly water-soluble compounds, as well as in the analysis of polar compounds, provided that a mobile phase with high water content is used. Typically, an octadecylsilane (ODS) phase, composed of 18 carbon atoms (RP-18), is employed. Such fillers can have different properties depending on the silica gel type and/or production method [19,65]. According to the available literature, the analysis of luteolin derivatives (Table 4) was performed exclusively with RP systems, and the RP-18 system was the most frequently used system, as exemplified in the separation of six luteolin derivatives in *Securigera securidaca* [75] or as many as ten derivatives in *Capsicum annuum* [76]. Stationary phases composed of 8 carbon atoms (RP-8) were rarely used. Examples include the separation of flavonoids in *Coriandrum sativum* [77] and *Achillea millefolium* [68,69].

The selection of an appropriate mobile phase is extremely important for chromatographic separation. The type of analyte, mixture composition, stationary phase and detector used should be taken into account. Mixtures of up to three components are most commonly used. In the reverse system, mixtures of MeOH/H$_2$O or acetonitrile (ACN)/H$_2$O are routinely used. As the amount of organic solvent increases, the retention time for non-polar substances decreases, while the addition of H$_2$O extends their retention times. As a rule of thumb, systems containing ACN/H$_2$O eluents are more efficient than those containing MeOH/H$_2$O [19]. In the normal phase system, however, the base solvent is a non-polar eluent, and its polarity is appropriately modified by the addition of another solvent of higher polarity, e.g., chloroform (CHCl$_3$) [65]. According to the available literature, virtually all mobile phases used for analysis of luteolin derivatives consisted either of ACN/H$_2$O, e.g., [26,66,67,78] or, less often, MeOH/H$_2$O mixtures [72,79–81]. Additionally, a modifier such as FA, e.g., [79,82–84], AcOH, e.g., [44,78,85–87], or trifluoroacetic acid (TFA) [88] was added to avoid peak tailing [89]. Analyses performed without acidic modifiers are rare [90].

In the chromatography process, the elution can be carried out in two ways. The first method is isocratic elution, which involves running with the same composition of mobile phase during the analysis. This means that a constant elution force is maintained throughout the entire separation period. In this case, even a slight change in the composition of the mobile phase may affect the results of the analysis. Isocratic elution is only rarely used for analysis of flavonoids. In the case of luteolin derivatives, this method was used only for analysis of luteolin 7-rhamnosyl- (1-6)-galactoside in *Filago germanica* [91], luteolin 7-rhamnosyl(1-6)galactoside in *Galactites elegans* [90], and various luteolin derivatives in *Apium graveolens* [80]. If the composition of the eluent changes during the division of the mixture, then the gradient elution can be described. The gradient can be linear or specifically programmed. With a change in the composition of the eluent, its elution force increases. For this reason, this method is used particularly for the separation of mixtures composed of substances of different polarity [2,64]. The vast majority of papers describing the separation of luteolin derivatives in plant mixtures use a gradient approach. Examples include the separation of luteolin 8-C-glucoside, luteolin 6-C-glucoside, luteolin 4′-O-glucoside and luteolin 7-O-glucoside from *Chrysanthemum trifurcatum* [92], thirteen luteolin derivatives from *Cymbopogon citratus* [70,71] and many others.

By far, the most common method for detection of luteolin derivatives is using diode array detectors (DADs) (also called photodiode array detectors, PDA), which were used in as many as 46 instances (e.g., for the analysis of luteolin derivatives and other flavonoids in *Cymbopogon citratus* [79], *Dianthus versicolor* [88], *Clinacanthus nutans* [44], *Thymus alternans* and others [93]). More often than not, however, DADs were combined with other detectors for additional structure determination or confirmation, with or without prior isolation. Examples of DADs combined with other detection methods for analysis of luteolin derivatives in plants include the use of tandem mass spectrometry (MS/MS) and NMR for *Ligustrum vulgare* [26], diode array detectors combined with electrospray ionization and mass spectrometry (DAD-ESI-MS) for *Phoenix dactylifera* [94] and *Nerium indicum* [95], electrospray ionization combined with tandem mass spectrometry (ESI-MS/MS) for *Achillea moschata* [96], electrospray ionization combined with time of flight mass spectrometry (ESI-TOF-MS) and electrospray ionization combined with ion trap and tandem mass spectrometry (ESI-IT-MS/MS) for *Aloysia citrodora* and many others.

For example, the HPLC-PAD-MS technique was used in the analysis of aerial flowering parts of the Edelweiss alpine region (Leontopodium alpinum). This method allowed the basic separation of almost all components of the L. alpinum extract prepared by exhaustive dichloromethane (DCM) followed by MeOH extraction. In total, 14 compounds have been isolated from the extract, including several luteolin derivatives. The authors used a gradient as the mobile phase as follows: H$_2$O:0.9% FA:0.1% AcOH:1.5% BuOH (A) and ACN:30% MeOH:0.9% FA:0.1% AcOH (B) and MeOH (C). The structure of the isolated components was additionally confirmed using NMR spectroscopy [29].

Table 4. High performance liquid chromatography/ultra performance liquid chromatography in the analysis of luteolin derivatives.

Luteolin Derivative	Stationary Phase/Column	Mobile Phase	Conditions	Detection	Analyzed Species	Ref.
Luteolin	ODS	0.2% PA:H_2O (A), MeOH (B)	injection volume: 20 µL; flow rate: 1.0 mL/min	DAD, 360 nm	Honey	[97]
	ODS C_{18}	0.2% PA:H_2O (58:41 *v/v*)	injection volume: 50 µL; flow rate: 1.0 mL/min	UV, 350 nm	*Chrysanthemum morifolium*	[98]
	Kinetex C_{18}	ACN (A), 0.05% TFA:H_2O (B)	flow rate: 0.8 mL/min	UV, 254 nm	*Enhalus acoroides*	[99]
	ODS Hypersil C_{18}	MeOH:ACN (1.25:1 *v/v*) (A), 0.5% AcOH:H_2O (B)	flow rate: 0.8 mL/min	MS	*Perilla frutescens*	[100]
	Zorbax SB-C_{18}	PA:H_2O pH 4.0 (A), ACN (B)	injection volume: 50 µL; flow rate: 0.6 ml/min; 25 °C	DAD, 330 nm	*Vernonia condensata*	[101]
	ODS C_{18}	MeOH (A), 0.05% TFA:H_2O (B)	injection volume: 20 µL; flow rate: 1.0 mL/min	UV, 280 nm	*Corchorus olitorius*	[102]
	RP	0.5% PA:H_2O (A), ACN (B)	injection volume: 20 µL; flow rate: 0.8 mL/min; at 35 °C	Triple-TOF-MS	*Veronicastrum latifolium*	[103]
	RP C_{18}	1% FA:H_2O (A), 40% solvent A:ACN (B)	flow rate: 0.5 mL/min; 25 °C	DAD-ESI-MS/MS	*Olea europaea*, olive oil	[104]
	Discovery HS C_{18}	2% AcOH (A), ACN (B)	injection volume: 20 µL; flow rate: 0.8 mL/min	DAD 280, 320 nm	*Agastache foeniculum*	[105]
	Ultrasphere 5 C_{18}	H_2O (A), ACN (B)	injection volume: 10 µL; flow rate: 1 mL/min; 25 °C	ESI-MS; 350 nm	*Rosa rugosa*	[106]
Luteolin 6-glucoside	Vydac RP C_{18}	0.05% TFA:H_2O (A), 0.038% TFA:ACN (*v/v*) (B)	injection volume: 10 µL flow rate: 1 mL/min; 36 °C	UV, 342 nm	*Ficaria verna*	[107]
Luteolin 8-glucoside	RP C_{18}	0.5% FA:H_2O (A), ACN (B)	injection volume: 20 µL; flow rate: 0.5 mL/min; 23 °C	DAD, 254, 340 nm	*Jatropha gossypiifolia, J. mollissima*	[108]
Luteolin 7-glucoside	Hypersil gold C_{18}	0.1% FA:H_2O (A), 0.1% FA:MeOH (B)	injection volume: 10 µL; flow rate: 0.35 mL/min; 38 °C	MS	*Rhoeo discolor*	[109]
	Thermo C_{18}	0.3% FA:H_2O (A), ACN (B)	injection volume:10 µL; flow rate: 1.0 mL/min; 30 °C	DAD-Q-Orbitrap-MS	*Epimedium brevicornum, Anacyclus pyrethrum, Lycium barbarum, Cuscuta australis*	[110]

Table 4. *Cont.*

Luteolin Derivative	Stationary Phase/Column	Mobile Phase	Conditions	Detection	Analyzed Species	Ref.
Luteolin	Eclipse XDB C$_{18}$	0.025% AcOH:H$_2$O (A), 5% Ace:ACN (B)	30 °C	ESI-MS/MS	*Olea europaea*, olive oil	[111]
Luteolin 7-glucoside	Eclipse XDB C$_{18}$	1% PA:H$_2$O (A), ACN (B)	injection volume: 10 µL; flow rate: 1 mL/min; 25 °C	DAD-ESI-QTOF-MS/MS	*Verbascum ovalifolium*	[112]
Luteolin 7-rutinoside Luteolin 7-rhamnoside	Eclipse XDB C$_{18}$	ACN (A), 0.2% FA (B)	flow rate: 0.3 mL/min; 30 °C	DAD, 230, 254, 290, 334 nm; MS/MS; NMR	*Ligustrum vulgare*	[26]
Luteolin Luteolin 7,4-diglucoside Luteolin 7-rutinoside Luteolin 4-glucoside Luteolin 3-glucoside Luteolin 7-glucoside	Gemini C$_{18}$	0.5% AcOH:H$_2$O (A), ACN (B)	25 °C	ESI-QTOF-MS	*Olea europaea*	[78]
Luteolin 6-glucoside Luteolin 7-rhamnoside Luteolin 7-glucoside Luteolin 6-glucosyl-8-arabinoside	Spherisorb S5 ODS-2	5% FA:H$_2$O (A), MeOH (B)	flow rate: 1 mL/min	DAD, 280 nm	*Cymbopogon citratus*	[79]
Luteolin 2″-feruloylhexosyl-6-hexoside Luteolin 6-glucoside	Spherisorb S3 ODS-2 C$_{18}$	0.1% FA:H$_2$O (A), ACN (B)	flow rate: 0.5 mL/min; 35 °C	DAD-MS; 280, 370 nm	*Arenaria montana*	[66,67]
Luteolin Luteolin 7-rutinoside Luteolin 3-glucuronide Luteolin 3′-(2″-acetyl)-glucuronide Luteolin 6-hydroxy-7-glucoside	Kinetex 100 C$_{18}$ Superschera 100 RP C$_{18}$	1% FA:H$_2$O (A), 1% FA:ACN (B)	flow rate: 1 mL/min; 30 °C	ESI-QTOF-MS; DAD, 324 nm	*Rosmarinus officinalis*	[82]
Luteolin hexosyl-hexoside-malylester Luteolin 6-glucosyl-7-galactoside Luteolin 6-glucosyl-7-rutinoside Luteolin 6-glucosyl-7-rhamnosyl-galactoside	AquasilW C$_{18}$	TFA solution (pH 2.8) (A), ACN (B)	flow rate: 1.0 mL/min; 15 °C	UV/DAD, 340 nm	*Dianthus versicolor*	[88]
Luteolin 7-apiofuranosyl (1→2)-glucopyranoside Luteolin 7-glucopyranoside Luteolin 7-[apiofuranosyl (1→2)-(6″-malony)]-glucopyranoside	Eclipse Plus C$_{18}$	0.1% FA:MeOH (A), 0.1% FA:H$_2$O (B)	injection volume: 10 µL; isocratic mixture flow rate: 0.3 mL/min	QTOF-MS/MS	*Apium graveolens*	[80]

Table 4. *Cont.*

Luteolin Derivative	Stationary Phase/Column	Mobile Phase	Conditions	Detection	Analyzed Species	Ref.
Luteolin 7-(6-rhamnosyl)hexoside Luteolin 7-(2-rhamnosyl)hexoside Luteolin 7-(2-hexosyl[6-sulfate])hexoside Luteolin 7-hexosyl(6-sulfate)	Kinetex C_{18}	1% FA (A), ACN (B)	injection volume: 8 µL; flow rate: 0.8 mL/min	DAD-ESI-MS; 340 nm	*Phoenix dactylifera*	[94]
Luteolin Luteolin 7-glucoside	Gemini C_{18}	0.1% FA:H_2O (A), 0.1% FA:ACN (B)	0-60 min, 10–60% B; flow rate: 1 mL/min	DAD-ESI-MS/MS; 210, 270, 310 and 350 nm	*Achillea moschata*	[96]
Luteolin 7-glucoside Luteolin 7,4′-diglucoside Luteolin 6-hydroxy-7-glucoside Luteolin 4′-glucoside Luteolin 3′-glucoside	Synergy Polar RP 80Å, LiChroCART 4-4 with guard column LiChrospher 100 C_{18}	0.9% FA:0.1% AcOH:1.5% BuOH:H_2O (A), 30% MeOH:0.9% FA:0.1% AcOH:ACN (B), MeOH (C)	injection volume: 5 µL; flow rate: 1.0 mL/min; 45 °C	PAD; MS; UV; IR; NMR	*Leontopodium alpium*	[29]
Luteolin 7-diglucuronide	Eclipse Plus C_{18}	1% FA:H_2O:ACN (A), ACN (B)	injection volume: 20 µL; flow rate: 0.5 mL/min	DAD; UV-VIS, 190-450 nm; ESI-TOF-MS; ESI-IT-MS/MS	*Lippia citrodora*	[113]
Luteolin 5-rutinoside Luteolin 7-rutinoside	Luna C_{18}	0.1% FA:H_2O (A), ACN (B)	injection volume: 5 µL; flow rate: 1.0 mL/min	DAD, 200-400 nm; ESI-MS	*Nerium indicum*	[95]
Luteolin diglucoside Luteolin 7-rutinoside Luteolin 7-glucoside	Luna C_{18}	1% FA:H_2O (A), ACN (B)	injection volume: 20 µL; flow rate: 1 mL/min	PDA-ESI-MS/MS	*Taraxacum officinale*	[114]
Luteolin Luteolin 2″-Galloyl-4′-glucoside Luteolin 4′-glucoside	Gemini RP C_{18}	2% AcOH:H_2O (A), ACN (B)	—	ESI-MS; 257, 278 and 340 nm	*Pistacia atlantica*	[85]
Luteolin 6-methoxy-8-arabinosyl-7-glucoside Luteolin 8-glucoside Luteolin 8-rhamnosyl-7-rhamnoside	Symmetry Shield Waters RP_{18} Symmetry Shield RP_{18}	0.2% FA:H_2O (A), ACN (B)	flow rate: 1.2 mL/min	UV/PAD-MS; UV; 200–600 nm UV/PAD	*Saccharum officinarum*	[115]
Luteolin 8-glucoside	Kinetex PFP	0.8% AcOH:H_2O (A), ACN (B)	injection volume: 10 µL; flow rate: 0.7 mL/min; 40 °C	UV/DAD, 330 nm	*Clinacanthus nutans*	[44]

Table 4. *Cont.*

Luteolin Derivative	Stationary Phase/Column	Mobile Phase	Conditions	Detection	Analyzed Species	Ref.
Luteolin Luteolin 7-glucoside Luteolin 7-glucuronyl-3-glucoside Luteolin 6-glucoside Luteolin 6-glucosyl-2″-rhamnoside Luteolin 6-glucosyl-4′-glucoside Luteolin 3′-glucoside	Eclipse XDB C_{18}	MeOH (A), 0.2% FA:H_2O (B)	flow rate: 0.8 mL/min; 30 °C	DAD-MS/MS; UV; 254, 360 nm	*Securigera securidaca*	[75]
Luteolin hexosyl-rhamnoside Luteolin 7-rutinoside Luteolin 7-glucoside	Spherisorb S3 ODS-2 C_8	0.1% FA:H_2O (A), ACN (B)	flow rate: 0.5 mL/min; 35 °C	DAD-MS; 280, 370 nm	*Coriandrum sativum*	[77]
Luteolin Luteolin 6-glucosyl-8-arabinoside Luteolin 7-glucoside Luteolin 6-arabinosyl-8-glucoside	XDB C_{18}	1% FA:H_2O (A), ACN (B)	injection volume: 100 μL; flow rate: 4 mL/min	NMR; MS	*Casimiroa edulis*	[116,117]
Luteolin 3-glucopyranoside Luteolin 7-glucopyranoside Luteolin 7-rutinoside Luteolin methoxy-hexoside	PLRP-S 100Å	0.1% FA:H_2O (A), ACN (B)	injection volume: 100 μL; flow rate: 1.5 mL/min	MS	*Thymus alternans*	[93]
Luteolin 4′-glucopyranoside	XDB C_{18}	1% FA:H_2O (A), ACN (B),	injection volume: 100 μL; flow rate: 4 mL/min			
Luteolin 6-hydroxy-7-glucoside	Gemini C_{18}	0.1% FA:H_2O (A), ACN (B)	flow rate: 0.8 mL/min	UV-VIS, 200-400 nm; MS; DAD; NMR	*Athrixia phylicoides*	[118]
Luteolin Luteolin 7-rhamnoside	Syncronis C_{18}	0.1% AcOH:H_2O (A), ACN (B)	injection volume: 5 μL; flow rate: 0.3 mL/min	DAD-MS/MS	Honey	[86]
Luteolin Luteolin rhamnosyl hexoside Luteolin rhamnosyl dihexoside	HSS T3	0.1% FA:H_2O (A), 0.1% FA:ACN (B)	injection volume: 3.1 μL; flow rate: 0.15 mL/min	PDA-MS	*Phoenix dactylifera*	[74]
Luteolin diglucuronide Luteolin 7-glucuronide Luteolin 7-rutinoside	S3 ODS-2 C_{18}	0.1% FA:H_2O (A), ACN (B)	flow rate: 0.5 mL/min; 35 °C.	DAD-ESI-MS; 370, 330 and 280 nm	*Thymus pallescens,* *Saccocalyx satureioides,*	[119,120]
Luteolin 7-glucoside Luteolin 6-glucoside					*Ptychotis verticillata* *Coleostephus myconis*	

Table 4. *Cont.*

Luteolin Derivative	Stationary Phase/Column	Mobile Phase	Conditions	Detection	Analyzed Species	Ref.
Luteolin Luteolin rutinoside Luteolin hexoside Luteolin 6-hexoside	Kinetex C_{18}	ACN (A), 0.1% FA:H_2O (B)	injection volume: 10 µL; flow rate: 0.4 mL/min	ESI-MS	*Lathyrus pratensis* *L. aureus*	[121]
Luteolin 7-glucoside	Luna C_{18}	5% FA:H_2O (A), MeOH (B)	injection volume: 100 µL; flow rate: 1 mL/min; 35 °C	DAD, 200, 600 nm	*Thymus pulegioides*	[81]
Luteolin hexuronide		0.1% FA:H_2O (A), MeOH (B)	injection volume: 10 µL; flow rate: 0.5 mL/min; 40 °C	ESI-MS/MS		
Luteolin Luteolin hexosyl-pentoside Luteolin 6-glucoside Luteolin 8-glucoside Luteolin 7-glucoside	Eclipse Plus C_{18}	0.5% AcOH:H_2O (A), ACN (B)	injection volume: 5 µL; flow rate: 0.5 mL/min	DAD-QTOF-MS; 325–371 nm	*Ficus carica*	[87]
Luteolin (3-hydroxy-3-methylglutaroyl)-X″-deoxyhexoside Luteolin 6-hexoside Luteolin 6-rutinoside Luteolin dihexoside	PLB-2 C_{18} with ODS-2 C_{18} guard cartridge	1% FA:H_2O (A), MeOH (B)	flow rate: 0.2 mL/min; 25 °C	PDA-ESI-MS	*Urtica membranacea*	[122]
Luteolin Luteolin 6,8-dihexoside Luteolin 6-hexosyl-8-pentoside Luteolin 6-pentosyl-8-hexoside Luteolin 6-hexoside Luteolin 8-hexoside Luteolin 7-(2″-pentosyl-4″-hexosyl)hexoside Luteolin 7-(2″-pentosyl)hexoside Luteolin 7-glucoside Luteolin 7-[2″-(5‴-sinapoyl)pentosyl]hexoside Luteolin 7-(2″-pentosyl-4″-hexosyl-6″-malonyl)hexoside	Syncronis C_{18}	0.01% AcOH:H_2O (A), ACN (B)	inaction volume: 5 µL; flow rate: 0.25 mL/min	MS	*Capsicum annuum*	[76]
Luteolin 7-(2″-pentosyl-6″-malonyl)hexoside Luteolin 7-[2″-(5‴-sinapoyl)pentosyl-6″-malonyl]hexoside						

Table 4. *Cont.*

Luteolin Derivative	Stationary Phase/Column	Mobile Phase	Conditions	Detection	Analyzed Species	Ref.
Luteolin Luteolin 6,8-diglucoside Luteolin 8-glucoside Luteolin 7-rutinoside	Purospher star C_{18}	10% $FA:H_2O$ (A), ACN (B)	25 °C	DAD-ESI-MS; 320-280 nm	*Citrus aurantifolia*	[123]
Luteolin Luteolin 7-glucoside Luteolin 7-rutinoside	Zorbax SB-C_{18}	$AFNH_4:H_2O:ACN:FA$ (A), $AFNH_4:H_2O:ACN:FA$ (B)	injection volume: 5μL; flow rate: 1.0 mL/min	ESI-MS; 325 nm	*Caucalis platycarpos*	[124]
Luteolin 7-glucoside Luteolin 7-glucuronide Luteolin 7-rutinoside	Beta-Basic C_{18}	5% FA:ACN (A); 5% $FA:H_2O$ (B)	injection volume: 20 μL; flow rate: 0.9 mL/min	UV, 280 nm	*Melissa officinalis* *Mentha piperita,* *Salvia officinalis*	[54]
Luteolin Luteolin 7-glucuronide Luteolin 7-rhamnosyl-hexoside	UPLC BEH C_{18} and a Acquity UPLC BEH C_{18} VanGuardTM pre-column	0.1% $AcOH:H_2O$ (A), 0.1% AcOH:MeOH (B)	injection volume: 5 μL; flow rate: 0.5 mL/min	DAD-ESI-QTOF-MS; 370 nm	*Lactuca sativa*	[72]
Luteolin 7-rutinoside Luteolin 7-glucoside	HSS T3	0.1% $FA:H_2O$ (A), 0.1% FA:ACN (B)	injection volume: 3 μL; flow rate: 0.4 mL/min	IMS-QTOF-MS		[73]
Luteolin 7-glucuronide Luteolin hexoside Luteolin 7-rutinoside Luteolin hydroxy-glucuronide Luteolin diglucoside Luteolin 7-glucoside Luteolin 7-glucuronide Luteolin malonyl-hexoside Luteolin glucuronide	Hypersil Gold C_{18}	0.1% $FA:H_2O$ (A), ACN (B)	—	DAD-ESI-MS	*Salvia elegans,* *S. greggii,* *S. officinalis*	[125,126]
Luteolin glucoside Luteolin rutinoside			flow rate: 0.2 mL/min		*Thymus barona,* *T. pseudolanuginosus,* *T. caespititius*	[126]
Luteolin 7-rhamnosyl(1-6)galactoside	Eclipse XDB-C_{18}	0.1% $FA:H_2O$ (A), 0.1% FA:ACN (B)	flow rate: 0.5 mL/min; 25 °C	MS	*Filago germanica*	[91]
Luteolin 4′-glucuronide	C_{18}μ-Bondapak RP_{18}	$MeOH:H_2O$ (42:58 *v/v*)	flow rate: 2.0 mL/min	NMR	*Galactites elegans*	[90]
Luteolin Luteolin 3′-7-diglucoside Luteolin 7-glucoside Luteolin 6-(2-rhamnosyl)-hexoside Luteolin (pentosyl)-hexoside Luteolin hexoside	Kinetex 100 A C_{18}	1% $FA:H_2O$ (A), ACN (B)	injection volume: 5 μL; flow rate: 0.8 mL/min; 25 °C	DAD, 340 nm DAD-ESI-MS/MS; 280, 340 nm	*Allophylus africanus*	[83,127]
Luteolin 6-glucoside	ODS2 C_8	0.1% $FA:H_2O$ (A), ACN (B)	flow rate: 0.5 mL/min; 35 °C	DAD-ESI-MS; 280, 370 nm	*Achillea millefolium*	[68,69]

Table 4. *Cont.*

Luteolin Derivative	Stationary Phase/Column	Mobile Phase	Conditions	Detection	Analyzed Species	Ref.
Luteolin 6-hexosyl-8-pentoside Luteolin 2″-deoxyhexosyl-6-glucoside Luteolin 6-glucoside Luteolin 6-pentosyl-8-pentoside Luteolin-7-rhamnoside Luteolin 7-glucoside Luteolin 2″-deoxyhexosyl-pentoside Luteolin 6-pentoside Luteolin 2″-deoxyosyl-6-(6-deoxy-pento-hexosuloside	S3 ODS-2 C_{18}	0.1% FA:H_2O (A), ACN (B)	flow rate: 0.5 mL/min; 35 °C	DAD-MS; 280, 370 nm	*Cymbopogon citratus*	[70,71]
Luteolin Luteolin 8-glucoside Luteolin 4′-glucoside Luteolin 7-glucoside Luteolin 6-glucoside	C_{18}	0.005 % FA:H_2O (A), MeOH (B)	injection volume: 10 µL; flow rate: 0.5 mL/min; 30 °C	MS-Orbitrap	*Chrysanthemum trifurcatum*	[92]
Luteolin 6-glucoside Luteolin 8-glucoside Luteolin 6-deoxyhexosyl-8-pentoside Luteolin 6-fucoside Luteolin 8-deoxyhexoside Luteolin 6,8-diglucoside	HSS T3	0.1% FA:H_2O (A), 0.1% FA:ACN (B)	injection volume: 3.1 µL; flow rate: 0.15 mL/min	PDA-MS	*Passiflora edulis*	[128]
Luteolin 6,8-diglucoside Luteolin 8-glucosyl-7-glucoside isomer Luteolin 6-glucosyl-7-glucoside isomer	Kinetex C_{18}	0.05% FA:H_2O (A), 0.05% FA:ACN (B)	flow rate: 0.4 mL/min; 40 °C	MS/MS	*Eragrostis tef*	[84]
Luteolin 8-glucosyl-7-rhamnoside Luteolin 8-glucoside Luteolin 7-glucoside Luteolin 7-rhamnoside Luteolin 7-(6″-syringly)glucosyl-6-glucoside Luteolin 7-(2″-syringyl)arabinosyl-6-glucoside Luteolin 8-(6″-diacetyl)glucoside						
Luteolin Luteolin 6-glucoside	Nucleosil 100-3.5 C_{18}	FA:H_2O (A), FA:ACN (B)	injection volume: 20 µL	MS	*Arum hygrophilum*	[129]

Table 4. *Cont.*

Luteolin Derivative	Stationary Phase/Column	Mobile Phase	Conditions	Detection	Analyzed Species	Ref.
Luteolin 6-[6″-glocosy-caffeoyl-glucopyranosyl (″→2)-glucopyranoside Luteolin 6-glucopyranoside	Eclipse Plus C_{18}	0.2% FA:H_2O (A), 0.2% FA:ACN (B)	30 °C	PDA-ESI-MS/MS UV, 350 nm	*Triticum aestivum*	[130]
Luteolin 7-glucoside Luteolin 6-glucoside Luteolin 7-(2″-*p*-coumaroyl)-rhamnoside	Syncronis C_{18}	0.1% FA:H_2O (A), ACN (B)	injection volume: 5 μL; flow rate: 0.25 mL/min	MS/MS	*Iris pumila, I. variegata, I. humilis*	[131]
Luteolin Luteolin hexoside	Luna Omega Polar C_{18} with Polar C_{18} Security Guard cartridge Intersil ODS	0.1% FA:H_2O (A), ACN (B)	flow rate: 0.4 mL/min	ESI-MS/MS; 320, 350 nm	*Parentucellia latifolia*	[132,133]
Luteolin Luteolin glucoside Luteolin glucuronide Luteolin pentosyl-glucoside		ACN:H_2O:FA (10:89:1 *v/v/v*) (A), ACN:H_2O:FA (89:10:1 *v/v/v*) (B)	injection volume: 10 μL; flow rate: 0.5 mL/min; 40 °C	DAD-MS/MS; 360 nm	*Verbascum eskisehirensis*	[134]
Luteolin 4′-glucoside	Kinetex RP C_{18}	PA:H_2O pH 3 (A), PA:ACN pH 3 (B)	injection volume: 1 μL; flow rate: 1.2 mL/min; 35 °C	UV 300 nm	*Matricaria recutita*	[135]
Luteolin 7-*O*-glucuronide	C_{18}	0.02% TFA:H_2O (A), MeOH:ACN (3:7 *v/v*) (B)	injection volume: 2.5 μL; flow rate: 0.9 mL/min; 45 °C	DAD-ESI-MS 240, 254, 325 nm	*Lippia alba*	[136]
Luteolin glucoside Luteolin 7-*O*-β-glucoside Luteolin glucuronide Luteolin malonylglucoside Luteolin	YMC-Triart C_{18}	0.1% FA:H_2O (A), 0.1% FA:ACN B)	injection volume: 10 μL; flow rate: 0.8 mL/min	DAD-ESI-MA 265, 280, 330, and 360 nm	*Chrysanthemum morifolium*	[137]
Luteolin 7-diglucuronide Luteolin 7-*O*-glucuronide	Eclipse XDB C_{18}	0.03% PA:H_2O (A), solvent A:ACN (1:9 *v/v*) (B)	injection volume: 20 μL; flow rate: 0.8 mL/min; 25 °C	DAD, 210, 250, 320, 350 and 370 nm	*Thymus pannonicus*	[138]

Abbreviations: Ace, acetone; AcOH, acetic acid; ACN, acetonitrile; AFNH$_4$, ammonium formate; BuOH, butanol; DAD, diode array detector; DAD-ESI-MS, diode array detector combined with electrospray ionization and mass spectrometry; DAD-ESI-MS/MS, diode array detector combined with electrospray ionization and tandem mass spectrometry; DAD-ESI-QTOF-MS, diode array detector combined with electrospray ionization and quadrupole – time of flight mass spectrometry; DAD-MS, diode array detector combined with mass spectrometry; DAD-MS/MS, diode array detector combined with tandem mass spectrometry; DAD-QTOF-MS, diode array detector combined with quadrupole – time of flight mass spectrometry; ESI-IT-MS/MS, electrospray ionization combined with ion trap and tandem mass spectrometry; ESI-MS, electrospray ionization combined with mass spectrometry; ESI-MS/MS, electrospray ionization combined with tandem mass spectrometry; ESI-QTOF-MS, electrospray ionization combined with quadrupole – time of flight mass spectrometry; FA, formic acid; IMS-QTOF-MS, ion-mobility spectrometry combined with quadrupole – time of flight mass spectrometry; MeOH, methanol; MS/MS, tandem mass spectrometry; PA, phosphoric acid; PAD, pulsed amperometric detection; PDA-ESI-MS/MS, pulsed amperometric detection combined with electrospray ionization and tandem mass spectrometry; PDA-ESI-MS, pulsed amperometric detection combined with electrospray ionization and mass spectrometry; PDA-MS, pulsed amperometric detection combined with mass spectrometry; TFA, trifluoroacetic acid; UV/PAD-MS, UV/pulsed amperometric detection combined with mass spectrometry.

2.3. Liquid Chromatography in the Analysis of Luteolin Derivatives

LC combined with tandem mass spectrometry (LC-MS/MS) (Table 5) is especially useful in the analysis of multicomponent mixtures, such as herbal extracts because it does not require a large amount of the sample or previous separation. To further reduce the influence of other factors on the analysis, more advanced techniques, involving the combination of more than one detection method, e.g., LC combined with NMR and MS (LC-NMR-MS), are increasingly used [139].

Research on ethanol extract from *Lophatherum gracile* stems and leaves was performed using LC coupled with MS/MS. Gradient elution was performed using 0.3% FA and MeOH. Analysis of the species revealed the presence of, among others, luteolin 7-O-β-D-glucoside, and luteolin 6-C-glucoside [140].

Another raw material containing luteolin and its derivatives is the cocoa seed (*Theobroma cacao*). In the analysis of the H_2O:MeOH extract, the following mobile phase was used: H_2O:0.1% FA and ACN:0.1% FA. The elution was carried out in a linear gradient, whereas LC combined with electrospray ionization and tandem mass spectrometry (LC-ESI-MS/MS) coupling allowed the identification of the tested compounds [141].

Lin and Harnly performed a water-methanol analysis of the flower extract of *Chrysanthemum morifolium* and distinguished many compounds, including numerous derivatives of luteolin. In this case, the mobile phase was a mixture of 0.1% FA:H_2O and 0.1% FA:ACN, in varying proportions. The qualitative determination of the analyzed substances was based on a comparison of retention times as well as mass and UV/Vis spectra [142].

Table 5. Liquid chromatography in the analysis of luteolin derivatives.

Luteolin Derivative	Stationary Phase	Mobile Phase	Conditions	Detection	Analyzed Species	Ref.
Luteolin	Zorbax SB C_{18} column with Security-Guard C_{18}	MeOH (A), 0.5% AcOH:H_2O (B)	flow rate: 1.0 mL/min; injection volume: 20 µL	MS/MS	*Abri herba, A. mollis*	[143]
	Inertsil ODS-3	0.1% FA:H_2O (A), ACN (B)	flow rate: 0.5 mL/min; injection volume: 10 µL	MS/MS	*Castanea mollissima*	[144]
	RP C_{18}	0.1% FA:H_2O (A), 0.1% FA:ACN (B)	injection volume: 5 µL; flow rate: 0.3 mL/min; 40 °C	ESI-MS/MS	*Centaurea cyanus*	[145]
Luteolin 7-glucoside	Eclipse XDB C_{18}	1% FA:H_2O (A), MeOH (B)	injection volume: 5 mL; flow rate: 0.6 mL/min; 45 °C	MS/MS	*Plantago atrata, P. coronopus, P. holosteum, P. lanceolata, P. reniformis, P. schwarzenbergiana*	[146]
Luteolin	Zorbax Plus C_{18}	0.1% FA:H_2O (A), 0.1% FA:ACN (B)	injection volume: 1µL; flow rate: 0.4 mL/min	MS	*Matricaria recutita, Achillea millefolium, Thymus vulgaris, Salvia officinalis*	[30]
	Eclipse XDB C_{18}	0.05% FA:H_2O (A), MeOH (B)	flow rate: 1 mL/min	MS/MS	*Vitis vinifera*	[147]
Luteolin 7-apiosyl-glucoside Luteolin 7-glucoside Luteolin 7-malonyl-apiosyl-glucoside Luteolin 7-6′-malonyl-apiosyl-glucoside Luteolin 7-6′-malonyl-glucoside	Symmetry C_{18}	0.1% FA:H_2O (A), 0.1% FA:ACN (B)	flow rate: 1.0 mL/min; 25 °C	DAD-ESI-MS; DAD, 350, 310, 270 nm; UV, 190-650 nm	*Apium graveolens*	[148]
Luteolin 3′-glucoside Luteolin 6-glucoside Luteolin 6,8-dihexoside Luteolin 7-rutinoside	Zorbax SB C_{18}	0.1% FA:H_2O (A); 0.1% FA:ACN (B)	injection volume: 5 µL; flow rate: 0.3 mL/min; 25 °C	MS/MS; TQMS; ESI	*Phlomis persica* *Ph. eliptica*	[27]
Luteolin 7-rutinoside Luteolin 7-glucopyranoside	XTerra MS C_{18}	ACN (A), 0.05% AcOH:H_2O (B)	injection volume: 20 µL; flow rate: 1.0 mL/min	PDA-MS; NMR	*Sechium edule*	[149]

Table 5. *Cont.*

Luteolin Derivative	Stationary Phase	Mobile Phase	Conditions	Detection	Analyzed Species	Ref.
Luteolin Luteolin glucuronide Luteolin glucoside	Octadecyl silica gel ODS	H_2O:MeOH:FA (89:10:1 *v/v/v*) (A), MeOH:H_2O:FA (89:10:1 *v/v/v*) (B)	flow rate: 1 mL/min; 40 °C	MS/MS	*Nepeta cilicica*	[150]
Luteolin Luteolin dihexoside Luteolin 7-dihexoside Luteolin 7-rutinoside Luteolin glucoside Luteolin 7-acetylglucoside	Capcell Park C_{18}	0.5% FA:H_2O (A), 0.5% FA:ACN (B)	flow rate: 0.5 mL/min; 25 °C	MS/MS	*Humulus japonicus*	[151]
Luteolin 6-glucoside Luteolin glucoside	Intersil ODS	ACN:H_2O:FA (10:89:1 *v/v/v*) (A), ACN:H_2O:FA (89:10:1 *v/v/v*) (B)	flow rate: 0.7 mL/min; 40 °C	MS/MS	*Achillea sivasica*	[152]
Luteolin methoxy-2″-pentosyl-6-hexoside Luteolin 2″-rhamnosyl-6″-hexosyl-glucoside Luteolin 7-glucosyl-8-glucoside Luteolin 6-glucoside Luteolin 6-hexosyl-8-pentoside	ODS2 C_8	0.1% FA:H_2O (A), ACN (B)	flow rate: 0.5 mL/min; 35 °C	DAD-ESI-MS; 280, 370 nm	*Geranium molle*	[68,70,153]
Luteolin acetylhexoside Luteolin 6-glucoside Luteolin 3-glucuronide Luteolin glucuronide	Spherisorb S3 ODS2 C_{18}	0.1% FA:H_2O (A), ACN (B)	flow rate: 0.5 mL/min; 35 °C	DAD-ESI-MS/MS	*Achillea millefolium* *Coleostephus myconis* *Rosmarinus officinalis*	[120,154] [120] [155]
Luteolin hexosyl-pentoside Luteolin 7-glucoside Luteolin 7-hexoside Luteolin 6-hexoside Luteolin 8-glucoside	Zorbax RP C_{18}	0.1% FA:H_2O (A), 0.1% FA:ACN (B)	injection volume: 5 µL; flow rate: 0.4 mL/min; 35 °C	QTOF	*Ficus carica*	[156]
Luteolin 6-hexosyl-8-acetyl-hexoside Luteolin 8-glucosyl-7,3′-dimethoxyl-2″-*O*-glycoside Luteolin 8-glucosyl-7,3′-dimethoxyl-6-desoxyhexoside	ODS2	H_2O:ACN:FA	injection volume: 20 µL; flow rate: 0.8 mL/min	MS/MS	*Oxalis pes-caprae*	[157]

Table 5. *Cont.*

Luteolin Derivative	Stationary Phase	Mobile Phase	Conditions	Detection	Analyzed Species	Ref.
Luteolin 6-glucoside Luteolin 8-glucoside	Hydro RP	H_2O (A), MeOH (B), 5% AcOH:MeOH (C)	flow rate: 1 mL/min	MS/MS	*Passiflora morifolia*	[158]
Luteolin Luteolin 6-glucoside Luteolin 8-glucoside Luteolin 7-glucoside	Luna C_{18}	0.1% FA:H_2O (A), 0.1% FA:ACN (B)	flow rate: 0.4 mL/min	DAD, 280, 320, 365 nm; MS; MS/MS	*Theobroma cacao*	[141]
Luteolin 6-glucosie Luteolin 7-glucoside	XBridge C_{18}	0.3% FA:H_2O (A), MeOH (B)	injection volume: 1 µL; flow rate: 1 mL/min; 40 °C	MS/MS; ESI-MS	*Lophatherum gracile*	[140]
Luteolin Luteolin glucuronyl-hexoside Luteolin 7-pentosyl-hexoside Luteolin 7-rutinoside Luteolin 7-glucoside Luteolin 7-glucuronide Luteolin glucoside Luteolin 7-malonyl-6″-glucoside Luteolin 7-acetyl-6″-glucoside Luteolin 7-dihexoside	Symmetry C_{18}	0.1% FA:H_2O (A), 0.1% FA:ACN (B)	flow rate: 1.0 mL/min; 25 °C	DAD-ESI-MS; 350, 310, 270 and 520 nm	*Chrysanthemum morifolium*	[142]
Luteolin Luteolin7-glucuronide	Acquity BEH C_{18}	0.1% FA:H_2O (A), 0.1% FA:ACN (B)	flow rate: 0.3 mL/ min; 40 °C	QTOF-MS/MS	*Ageratum conyoides*	[159]
Luteolin hexoside	Zorbax SB C_{18}	0.5% FA:H_2O (A), ACN (B)	injection volume: 5 µL; flow rate: 0.4 mL/min	MS/MS	*Matricaria recutita*	[160]
Luteolin hexoside Luteolin hexosyl-deoxy-hexose	Hypersil gold C_{18}	1% FA:H_2O (A), 0.1% FA:ACN (B)	flow rate: 1 mL/min	MS/MS	*Cecropia obtusa*	[161]
Luteolin dihexoside Luteolin pentosyl-hexoside Luteolin 7-glucoside Luteolin malonyl-hexoside	Spherisorb S3 ODS-2 C_{18}	0.1% FA:H_2O (A), ACN (B)	flow rate: 0.5 mL/min; 35 °C.	DAD-ESI-MS/MS	*Cotula cinerea*	[120,162]
Luteolin Luteolin 8-glucoside Luteolin 5-glucopyranoside	Zorbax SB C_{18}	0.1% FA:AFNH$_4$ (A), 0.1% FA:ACN (B)	injection volume: 1 mL; flow rate: 0.2 mL/min	ESI-MS/MS	*Ocimum sanctum*	[163]

Abbreviations: AcOH, acetic acid; ACN, acetonitrile; AFNH$_4$, ammonium formate; DAD-ESI-MS, diode array detector combined with electrospray ionization and mass spectrometry; DAD, diode array detector; DAD-ESI-MS/MS, diode array detector combined with electrospray ionization and tandem mass spectrometry; ESI, electrospray ionization; ESI-MS, electrospray ionization combined with mass spectrometry; FA, formic acid; MeOH, methanol; MS/MS, tandem mass spectrometry; PDA-MS, pulsed amperometric detection combined with mass spectrometry; QTOF, quadrupole time-of-flight; TQMS, triple quadrupole mass spectrometer.

2.4. Gas Chromatography in the Analysis of Luteolin and Its Derivatives

Gas chromatography (GC) is a technique used to analyze volatile as well as non-volatile compounds after their derivatization. GC is characterized by chromatographic distribution of either a gas mobile phase on a solid adsorbent (gas-solid chromatography) or a liquid on an inert support (gas-liquid chromatography). GC can be hyphenated with various detection techniques such as GC combined with mass spectrometry (GC-MS), GC combined with tandem mass spectrometry (GC-MS/MS) or GC combined with time of flight mass spectrometry (GC-TOF-MS), thus greatly increasing the versatility, sensitivity and accuracy of the method [164,165]. The analyzed substances should be thermally stable, and their boiling (or sublimation) temperature should not exceed 350-400°C. To achieve this, non-volatile substances are often derivatived. Polar functional groups are transformed into their less polar counterparts, thus increasing the volatility of the prepared compounds. The most common examples include substitution with a trimethylsilyl (TM) group, organic radicals or compounds such as trimethylchlorosilane (TMCS), hexamethyldisilazane (HMDS) or *N,O*-bis(trimethylsilyl)-trifluoroacetamide (BSTFA). Such an approach greatly increases the number of possible analytes [2,17]. For example, phenolic groups of flavonoids are often transformed into their less polar trimethylsilyl counterparts allowing for rapid and effective separation of complex mixtures [166].

In adsorption GC, a gas that is chemically inert to the stationary phase as well as the components being analyzed is used as the mobile phase. Most often hydrogen, nitrogen or argon is used. Helium is being used less often due to its higher cost than other gases and the implementation of the principles of chemical safety. The mobile phase must be properly selected for compatibility with the detector used. However, the carrier gas itself does not have a significant influence on the separation effects of the analyzed mixtures [17]. In the process of separation, the method of application of the sample to the chromatography column is very important. The sample should always have as small of a volume and the shortest dosing time as possible. This ensures better separation and narrower bands [17,164].

Gas chromatography uses open-ended columns, i.e., capillary columns and packed columns. Open-ended (OT-open tubular) columns are characterized by much higher efficiency than packed columns; therefore, they are chosen much more often [164]. Capillary columns are particularly useful in the separation of substances with significantly different boiling points. Ideally, the column should have a similar polarity to the analyzed components. However, due to the higher efficiency and greater durability of stationary phases with low polarity, so capillary columns are recommended for chromatographic analyses [17,164].

Another factor that influences the efficiency of the separation of the analyzed sample components, as well as the time of analysis, is temperature. The separation temperature should be selected depending on the stationary phase used and the boiling point of the analytes [17,164].

The analysis of luteolin derivatives has been carried out in accordance with the general procedure used for analysis of other flavonoids (Table 6) [165]. MS was used for detection of volatile derivatives of luteolin, while the use of a flame ionization detector (FID) was described only in two papers [167,168]. Most commonly, helium was used as a carrier gas (e.g., [169–171]), but the use of nitrogen was also recorded [167,168]. The derivatization of luteolin, which is necessary for chromatographic separation, was mostly achieved with BSTFA/TCMS [170–172]. For the analysis of lipophilic luteolin derivatives, such as 7,3′,4′-trimethyl-luteolin in *Arnica alpina*, no derivatization was necessary [168].

2.5. Counter-Current Chromatography in the Analysis of Luteolin Derivatives

Counter-current chromatography (CCC) is a variation of liquid chromatography in which both the stationary and mobile phases are liquid. The separation of the constituents of the mixture is carried out in a system of immiscible liquids that are in equilibrium with each other. The method is simple and rapid, offering the possibility of introducing the raw sample to the column without need for previous clean-up [173]. Counter-current chromatography is mainly used for purification of natural compounds, while its use as an analytical technique is far less common. In CCC, various dividing techniques can be used, thus distinguishing centrifugal partition chromatography (CPC) and rapid- or high-speed CCC (HSCCC), often referred to as hydrodynamic chromatography [174,175]. High-speed CCC is particularly often encountered in studies involving the separation of flavonoids [176]. Table 7 presents the conditions for the separation of mixtures containing luteolin and its derivatives using rapid counter-current chromatography. Separation of luteolin derivatives from mixtures of different phytochemicals is usually carried out with mixtures of EtOAc with one of alcohol (BuOH, EtOH or AcOH) and H_2O. Upon separation, structure determination by NMR is often necessary [173,177]. However, instead of NMR, high resolution mass spectrometry (HRMS) can also be employed for structure confirmation, as evidenced by the use of this technique for differentiation of various luteolin derivatives in *Lippia origanoides* [178].

Table 6. Gas chromatography in the analysis of luteolin and its derivatives.

Luteolin Derivative	Column	Derivatization	Conditions	Detection	Analyzed Species	Ref.
Luteolin	HP-5-MS	1% TMCS:BSTFA	carrier gas: He; injector temperature: 250 °C A: (column temperature) 5 min, 170°C; 3 °C/min, 170–255 °C; 1 min, 255 °C; 2 °C/min, 255–310 °C; flow rate: 0.5 mL/min; analysis time: 70 min; B: (column temperature) 5 min, 160°C; 3°C/min, 160–188°C; 1 min, 188 °C; 15 °C/min, 188–241 °C; 1 min, 241 °C; 2 °C/min, 241–282 °C; 5 °C/min, 282–310 °C; 5 min, 310 °C; flow rate: 1.0 mL/min; analysis time: 50 min	APCI-TOF-MS	Fruits of various olives species (*Olea* L.)	[170]
	Capillary column Supelco SPBM-5	two- and three-phase transfer catalysis (PTC), methyl iodide	carrier gas: He; injector temperature: 260 °C; detector temperature: 280 °C; furnace temperature: 5 min, 50 °C; 5 °C/min, 50–150 °C; 10 °C/min, 150–210 °C; analysis time: 45 min	MS	*Mentha spicata, Hypericum perforatum*	[179]
	non-polar RSL 200 BP	0.2 M trimethylaniline hydroxide (TMAH):H_2O-free MeOH Py:HMDS:TMCS	carrier gas: N_2; linear speed of the carrier 17.5 cm/s; 0–2 min, 280 °C and 235 °C; 1 °C/min, 280–290 °C; flow rate: 30 mL/min	FID	different samples form AFRC Institute of Plant Science Research and John Innes Institute, Norwich, U.K.	[167]
	BPX5	Py:BSTFA:TMCS) (50:50:1 *v/v/v*)	carrier gas: He; injector temperature: 310 °C; 1 min, 100 °C; 30 °C/min, 100–210 °C; 2 °C/min, 210–240 °C; 4 °C/min, 240–270 °C; 5 °C/min, 270–310 °C; 5 min, 310 °C; flow rate: 1.5 mL/min	QMS	Propolis, *Chrysanthemum* sp, *Theobroma cacao* (bitter chocolate)	[171]
	Capillary column Low-bleed CP-Sil 8 CB-MS	TMCS (100 µL), BSTFA (200 µL) HMDS:TMCS:Py (3:1:9, *v/v/v*)	carrier gas: He; 2 °C/min, 70–135 °C; 10 min, 135 °C; 4 °C/min, 135–220 °C; 10 min, 220 °C; 3.5 °C/min, 220–270 °C; 20 min, 270 °C; injector temperature: 280 °C; detector temperature: 290 °C; flow rate: 1.9 mL/min	MS	*Teucrium polium*	[169]
	Quartz capillary column	Py:BSTFA (1:1 *v/v*)	carrier gas: He;injector temperature 220 °C; detector temperature: 270 °C; 2.3 °C/min 200–270 °C; 30 min, 270 °C	MS	*Aspalathus linearis*	[172]

Table 6. *Cont.*

Luteolin Derivative	Column	Derivatization	Conditions	Detection	Analyzed Species	Ref.
Luteolin 7,3′,4′-trimethyl	Capillary column Permabond OV-1	—	carrier gas: N_2, He; injector temperature: 300 °C; column temperature: 270 °C (isothermal); flow rate: 1.3 mL/min	FID	*Arnica alpina*	[168]
	Capillary column OV-1	—		MS		

Abbreviations: Py, pyridine, BSTFA, N,O-bis(trimethylsilyl)trifluoroacetamide; TMS, trimethylsily; HMDS, hexamethyldisilazane; TMCS, trimethylchlorosilane; APCI-TOF-MS, atmospheric pressure chemical ionization combined with quadrupole – time of flight mass spectrometry; FID, flame ionization detector; QMS, quadrupole mass spectrometry.

Table 7. Counter-current chromatography, as a preparative technique in the separation of luteolin derivatives.

Luteolin Derivative	Solvent System	Conditions	Detection	Analyzed Species	Ref.
Luteolin 6-glucoside	EtOAc:BuOH:H_2O (2:1:3 *v/v/v*)	rotation speed: 800 rpm; flow rate (lower phase): 2.4 mL/min	UV-VIS, 254 nm; NMR; MS	*Patrinia villosa*	[173]
Luteolin 7-glucoside	EtOAc:EtOH:AcOH:H_2O (4:1:0.25:5 *v/v/v/v*)	rotation speed: 800 rpm; flow rate (lower phase): 1.5 mL/min	UV, 254 nm; NMR; MS	*Paeonia suffruticosa*	[177]
Luteolin 6,8-dihexoside Luteolin 8-glucoside Luteolin 6-glucoside Luteolin 7-glucoside	Hx:EtOH:H_2O (4:3:1 *v/v/v*)	rotation speed: 850 rpm; flow rate (lower phase): 2 mL/min	TLC; HPLC-UV-HRMS	*Lippia origanoides*	[178]

Abbreviations: AcOH, acetic acid; BuOH, butanol; EtOAc, ethyl acetate; EtOH, ethanol; HRMS, high resolution mass spectrometry; Hx, hexane, rpm, revolutions per minute.

3. Conclusions

The presented comparison of chromatographic methods currently used to determine luteolin and its derivatives provides a systematic summary of the available knowledge. Without a doubt, chromatographic analysis may be successfully employed as an efficient method for both qualitative assessment (fingerprinting) and quantitative determination of luteolin derivatives. In tedious and time-consuming determinations of multi-ingredient plant extracts, including those that contain the most prevalent luteolin derivatives, combinations of large-scale chromatographic techniques with serially aligned detection modalities such as LC-MS/MS or LC/NMR/MS were found to be particularly useful. Such beneficial coupling of chromatography with other analytic techniques expands analytical capabilities while additionally improving the accuracy, sensitivity, and precision of assays. Despite the dominant position of LC in the analysis of natural compounds and the dynamic development of novel chromatographic methods, the TLC/HPTLC has not lost its important place in the phytochemical analysis of luteolin derivatives. The technique is relatively simple and inexpensive while facilitating rapid qualitative and quantitative analysis of test compounds. In addition, it facilitates large quantities of diluted samples being deposited in the stationary phase, allowing for a wider choice of mobile phase carriers. The technique is subject to continuous improvements and its range of applicability is expanding.

Effective chromatographic analysis in the determination of luteolin derivatives requires appropriately selected chromatographic separation conditions. The appropriate choice of stationary phase sorbent may significantly improve test conditions. Most TLC analyses of luteolin derivatives are carried out in normal phase systems featuring a polar (hydrophilic) sorbent phase. In HPLC, separation conditions can be chosen more arbitrarily as the technique allows for the use of stationary phases of varying polarity which is particularly advantageous in the analysis of compounds that are either insoluble or poorly soluble in water. In order to additionally improve the sensitivity of HPLC analyses, one should focus on the column parameters responsible for appropriate separation. Quite often, the fairly successful analyses are used as starting points for further modifications, including the development of preparative-scale analyses.

Of all chromatographic techniques, GC is the least applicable in the analysis of plant extracts, including those that contain luteolin derivatives. Despite its high sensitivity and efficiency when coupled with various detection techniques (MS, MS/MS, TOF-MS), the chromatographic separation process involves high temperatures and derivatization of analytes. Due to this important aspect, GC-MS is less frequently used in the analysis of polyphenolic compounds.

The complexity of plant matrices is unquestionably a significant problem in the chromatographic analysis of plant extracts. It has a negative impact on the efficacy of analyses and prevents complete identification of all components of the plant extract. However, chromatography remains the primary and most effective analytical technique available in the current state of the art for the determination of compounds of natural origin.

Author Contributions: Conceptualization, M.T. and M.Z.-K.; literature investigation, A.M.J.; writing—original draft preparation, A.M.J., M.T., M.Z.-K.; writing—review and editing, A.M.J., M.T.; supervision, M.T., M.Z.-K.

Funding: The work was funded by the project № POWR.03.02.00-00-I051/16 from European Union funds, PO WER 2014-2020, grant №. 05/IMSD/G/2019.

Conflicts of Interest: All authors state that they are free of any conflicts of interest related to this paper.

References

1. López-Lázaro, M. Distribution and biological activities of the flavonoid luteolin. *Mini Rev. Med. Chem.* **2009**, *9*, 31–59. [CrossRef] [PubMed]
2. Szultka, M.; Buszewski, B.; Papaj, K.; Szeja, W.; Rusin, A. Determination of flavonoids and their metabolites by chromatographic techniques. *Trends Anal. Chem.* **2013**, *47*, 47–67. [CrossRef]

3. Nabavi, S.F.; Braidy, N.; Gortzi, O.; Sobarzo-Sanchez, E.; Daglia, M.; Skalicka-Woźniak, K.; Nabavi, S.M. Luteolin as an anti-inflammatory and neuroprotective agent: A brief review. *Brain Res. Bull.* **2015**, *119*, 1–11. [CrossRef] [PubMed]

4. Bravo, L. Polyphenols: Chemistry, dietary sources, metabolism, and nutritional significance. *Nutr. Rev.* **1998**, *56*, 317–333. [CrossRef]

5. Lin, Y.; Shi, R.; Wang, X.; Shen, H.-M. Luteolin, a flavonoid with potential for cancer prevention and therapy. *Curr. Cancer Drug Targets* **2008**, *8*, 634–646. [CrossRef]

6. Aziz, N.; Kim, M.-Y.; Cho, J.Y. Anti-inflammatory effects of luteolin: A review of in vitro, in vivo, and in silico studies. *J. Ethnopharmacol.* **2018**, *225*, 342–358. [CrossRef]

7. Imran, M.; Rauf, A.; Abu-Izneid, T.; Nadeem, M.; Shariati, M.A.; Khan, I.A.; Imran, A.; Orhan, I.E.; Rizwan, M.; Atif, M.; et al. Luteolin, a flavonoid, as an anticancer agent: A review. *Biomed. Pharmacother.* **2019**, *112*, 108612. [CrossRef]

8. Narayana, K.R.; Reddy, M.S.; Chaluvadi, M.R.; Krishna, D.R. Bioflavonoids classification, pharmacological, biochemical effects and therapeutic potential. *Indian J. Pharmacol.* **2001**, *33*, 2–16.

9. Harborne, J.B.; Baxter, H. *The Handbook of Natural Flavonoids*; Harborne, J.B., Baxter, H., Eds.; John Wiley & Sons, Inc.: Chichester, UK, 1999.

10. Lam, K.Y.; Ling, A.P.K.; Koh, R.Y.; Wong, Y.P.; Say, Y.H. A review on medicinal properties of orientin. *Adv. Pharmacol. Sci.* **2016**, *2016*, 4104595. [CrossRef]

11. Zheng, H.; Zhang, M.; Luo, H.; Li, H. Isoorientin alleviates UVB-induced skin injury by regulating mitochondrial ROS and cellular autophagy. *Biochem. Biophys. Res. Commun.* **2019**, *514*, 1133–1139. [CrossRef]

12. Wu, Q.-Y.; Wong, Z.C.-F.; Wang, C.; Fung, A.H.-Y.; Wong, E.O.-Y.; Chan, G.K.-L.; Dong, T.T.-X.; Chen, Y.; Tsim, K.W.-K. Isoorientin derived from *Gentiana veitchiorum* Hemsl. flowers inhibit melanogenesis by down-regulating MITF-induced tyrosinase expression. *Phytomedicine* **2019**, *57*, 129–136. [CrossRef] [PubMed]

13. Fan, X.; Lv, H.; Wang, L.; Deng, X.; Ci, X. Isoorientin ameliorates APAP-induced hepatotoxicity via activation Nrf2 antioxidative pathway: The involvement of AMPK/Akt/GSK3β. *Front. Pharmacol.* **2018**, *9*, 1334. [CrossRef] [PubMed]

14. Anilkumar, K.; Reddy, G.V.; Azad, R.; Yarla, N.S.; Dharmapuri, G.; Srivastava, A.; Kamal, M.A.; Pallu, R. Evaluation of anti-inflammatory properties of isoorientin isolated from tubers of *Pueraria tuberosa*. *Oxid. Med. Cell. Longev.* **2017**, *2017*, 5498054. [CrossRef]

15. Jo, B.-G.; Park, N.-J.; Jegal, J.; Choi, S.; Lee, S.W.; Yi, L.W.; Kim, S.-N.; Yang, M.H. *Stellera chamaejasme* and its main compound luteolin 7-O-glucoside alleviates skin lesions in oxazolone- and 2,4-dinitrochlorobenzene-stimulated murine models of atopic dermatitis. *Planta Med.* **2018**, *85*, 583–590. [CrossRef] [PubMed]

16. Chen, S.; Yang, B.; Xu, Y.; Rong, Y.; Qiu, Y. Protection of luteolin-7-O-glucoside against apoptosis induced by hypoxia/reoxygenation through the MAPK pathways in H9c2 cells. *Mol. Med. Rep.* **2018**, *17*, 7156–7162. [CrossRef]

17. McNair, H.M.; Miller, J.M. *Basic Gas Chromatography*, 2nd ed.; John Wiley & Sons, Inc.: Hoboken, NJ, USA, 2009.

18. Monteiro, M.L.G.; Marsico, E.; Lázaro, C.; Conte Júnior, C. Thin-layer chromatography applied to foods of animal origin: A tutorial review. *J. Anal. Chem.* **2016**, *71*, 459–470. [CrossRef]

19. Kazakevich, Y.; LoBrutto, R. HPLC theory and practice. In *HPLC for Pharmaceutical Scientists*; Kazakevich, Y., LoBrutto, R., Eds.; John Wiley & Sons, Inc.: Hoboken, NJ, USA, 2007; pp. 3–13.

20. Medić-Šarić, M.; Jasprica, I.; Mornar, A.; Maleš, Ž. Application of TLC in the isolation and analysis of flavonoids. In *Thin Layer Chromatography in Phytochemistry*; Waksmundzka-Hajnos, M., Shrema, J., Kowalska, T., Eds.; CRC Press, Taylor & Francis Group: Boca Raton, FL, USA, 2008; pp. 405–423.

21. Fuchs, B.; Süβ, R.; Teuber, K.; Eibisch, M.; Schiller, J. Lipid analysis by thin-layer chromatography—A review of the current state. *J. Chromatogr. A* **2011**, *1218*, 2754–2774. [CrossRef]

22. Poole, C.F. Milestones, core concepts, and contrasts. In *Instrumental Thin-Layer Chromatography*; Poole, C.F., Ed.; Elsevier Inc.: Amsterdam, The Netherlands, 2015; pp. 1–3.

23. Santiago, M.; Strobel, S. Thin layer chromatography. *Methods Enzymol.* **2013**, *533*, 303–324.

24. Waksmundzka-Hajnos, M.; Hawrył, M.A.; Cieśla, Ł. Analysis of plant material. In *Instrumental Thin-Layer Chromatography*; Poole, C.F., Ed.; Elsevier Inc.: Amsterdam, The Netherlands, 2015; pp. 505–508.

25. Phadungrakwittaya, R. Identification of apigenin and luteolin in *Artemisia annua* L. for the quality control. *Siriraj Med. J.* **2019**, *71*, 240–245. [CrossRef]

26. Mučaji, P.; Nagy, M.; Liptaj, T.; Prónayová, N.; Švajdlenka, E. Separation of a mixture of luteolin-7-rutinoside and luteolin-7-neohesperidoside isolated from *Ligustrum vulgare* L. *JPC J. Planar Chromatogr.* **2009**, *22*, 301–304. [CrossRef]

27. Aghakhani, F.; Kharazian, N.; Lori Gooini, Z. Flavonoid constituents of *Phlomis* (Lamiaceae) species using liquid chromatography mass spectrometry. *Phytochem. Anal.* **2018**, *29*, 180–195. [CrossRef] [PubMed]

28. Kozyra, M.; Głowniak, K.; Boguszewska, M. The analysis of flavonoids in the flowering herbs of *Carduus acanthoides* L. *Curr. Issues Pharm. Med. Sci.* **2013**, *26*, 10–15.

29. Schwaiger, S.; Seger, C.; Wiesbauer, B.; Schneider, P.; Ellmerer, E.P.; Sturm, S.; Stuppner, H. Development of an HPLC-PAD-MS assay for the identification and quantification of major phenolic edelweiss (*Leontopodium alpium* Cass.) constituents. *Phytochem. Anal.* **2006**, *17*, 291–298. [CrossRef] [PubMed]

30. Jesionek, W.; Majer-Dziedzic, B.; Choma, I. Separation, identification, and investigation of antioxidant ability of plant extract components using TLC, LC–MS, and TLC–DPPH. *J. Liq. Chromatogr. Relat. Technol.* **2015**, *38*, 1147–1153. [CrossRef]

31. Amirkhanova, A.; Ustenova, G.; Krauze, M.; Poblocka, L.; Shynykul, Z. Thin-layer chromatography analysis of extract *Oxytropis glabra* lam. dc. In Proceedings of the 18th International Multidisciplinary Scientific Geoconference (SGEM 2018), Albena, Bulgaria, 30 June–9 July 2018; pp. 775–782.

32. Nile, S.; Park, S.W. HPTLC densitometry method for simultaneous determination of flavonoids in selected medicinal plants. *Front. Life Sci.* **2015**, *8*, 97–103. [CrossRef]

33. Kukkar, M.; Kukkar, R.; Saluja, A. Validation of HPTLC method for the analysis of luteolin in *Cardiospermum halicacabum* Linn. *Int. J. Green Pharm.* **2014**, *8*, 252–256. [CrossRef]

34. Hassanein, H.; Said-Al Ahl, H.; MM, A. Antioxidant polyphenolic constituents of *Satureja montana* L. Growing in Egypt. *Int. J. Pharm. Pharm. Sci.* **2014**, *6*, 578–581.

35. Satpathy, S.; Patra, A.; Ahirwar, B. Development and validation of a novel high-performance thin-layer chromatography method for the simultaneous determination of apigenin and luteolin in *Hygrophila spinosa* T. Anders. *JPC J. Planar Chromatogr. Mod. TLC* **2018**, *31*, 437–443. [CrossRef]

36. Shawky, E.; Abou El Kheir, R.M. Rapid discrimination of different *Apiaceae* species based on HPTLC fingerprints and targeted flavonoids determination using multivariate image analysis. *Phytochem. Anal.* **2018**, *29*, 452–462. [CrossRef]

37. Swar, G.; Shailajan, S.; Menon, S. Activity based evaluation of a traditional Ayurvedic medicinal plant: *Saraca asoca* (Roxb.) de Wilde flowers as estrogenic agents using ovariectomized rat model. *J. Ethnopharmacol.* **2017**, *195*, 324–333. [CrossRef]

38. Patel, N.; Patel, K.; Patel, K.; Tejal, G. Validated HPTLC method for quantification of luteolin and apigenin in *Premna mucronata* Roxb., Verbenaceae. *Adv. Pharmacol. Sci.* **2015**, *2015*, 1–7. [CrossRef] [PubMed]

39. Gupta, N.; Lobo, R.; Kumar, N.; Bhagat, J.K.; Mathew, J.E. Identity-based High-performance thin layer chromatography fingerprinting profile and tumor inhibitory potential of *Anisochilus carnosus* (L.f.) wall against ehrlich ascites carcinoma. *Pharmacogn. Mag.* **2015**, *11*, 474–480.

40. Bros, I.; Soran, M.-L.; Nașcu-Briciu, R.; Codruta, C. HPTLC quantification of some flavonoids in extracts of *Satureja hortensis* L. obtained by use of different techniques. *JPC J. Planar Chromatogr.* **2009**, *22*, 25–28. [CrossRef]

41. Ristivojevic, P.; Dimkic, I.; Trifkovic, J.; Beric, T.; Vovk, I.; Milojkovic-Opsenica, D.; Stankovic, S. Antimicrobial activity of Serbian propolis evaluated by means of MIC, HPTLC, bioautography and chemometrics. *PLoS ONE* **2016**, *11*, e0157097. [CrossRef] [PubMed]

42. Shaikh, F.; Sancheti, J.; Sathaye, S. Phytochemical and pharmacological investigations of *Eclipta alba* (Linn.) Hassak leaves for antiepileptic activity. *Int. J. Pharm. Pharm. Sci.* **2012**, *4*, 319–323.

43. Marginǎ, D.; Olaru, O.T.; Ilie, M.; Gradinaru, D.; GuTu, C.; Voicu, S.; Dinischiotu, A.; Spandidos, D.A.; Tsatsakis, A.M. Assessment of the potential health benefits of certain total extracts from *Vitis vinifera*, *Aesculus hyppocastanum* and *Curcuma longa*. *Exp. Ther. Med.* **2015**, *10*, 1681–1688. [CrossRef]

44. Chelyn, J.L.; Omar, M.H.; Mohd Yousof, N.S.A.; Ranggasamy, R.; Wasiman, M.I.; Ismail, Z. Analysis of flavone C-glycosides in the leaves of *Clinacanthus nutans* (Burm. f.) Lindau by HPTLC and HPLC-UV/DAD. *Sci. World J.* **2014**, *2014*, 1–6. [CrossRef]

45. Pereira, C.A.M.; Yariwake, J.H.; Lancas, F.M.; Wauters, J.-N.; Tits, M.; Angenot, L. A HPTLC densitometric determination of flavonoids from *Passiflora alata*, *P. edulis*, *P. incarnata* and *P. caerulea* and comparison with HPLC method. *Phytochem. Anal.* **2004**, *15*, 241–248. [CrossRef]

46. Wang, J.; Tang, F.; Yue, Y.; Guo, X.; Yao, X. Development and validation of an HPTLC method for simultaneous quantitation of isoorientin, isovitexin, orientin, and vitexin in bamboo-leaf flavonoids. *J. AOAC Int.* **2010**, *93*, 1376–1383.

47. Cook, R.; Hennell, J.R.; Lee, S.; Khoo, C.S.; Carles, M.C.; Higgins, V.J.; Govindaraghavan, S.; Sucher, N.J. The *Saccharomyces cerevisiae* transcriptome as a mirror of phytochemical variation in complex extracts of *Equisetum arvense* from America, China, Europe and India. *BMC Genomics* **2013**, *14*, 445. [CrossRef]

48. Srinivas, P.; Reddy, S.R. Screening for antibacterial principle and activity of *Aerva javanica* (Burm.f) Juss. ex Schult. *Asian Pac. J. Trop. Biomed.* **2012**, *2*, S838–S845. [CrossRef]

49. Guerrini, A.; Bruni, R.; Maietti, S.; Poli, F.; Rossi, D.; Paganetto, G.; Muzzoli, M.; Scalvenzi, L.; Sacchetti, G. Ecuadorian stingless bee (Meliponinae) honey: A chemical and functional profile of an ancient health product. *Food Chem.* **2009**, *114*, 1413–1420. [CrossRef]

50. Bazylko, A.; Tomczyk, M.; Flazińska, A.; Lęgas, A. Chemical fingerprint of *Potentilla* species by using HPTLC method. *JPC J. Planar Chromatogr.* **2011**, *24*, 441–444. [CrossRef]

51. Celep, E.; Akyüz, S.; İnan, Y.; Yesilada, E. Assessment of potential bioavailability of major phenolic compounds in *Lavandula stoechas* L. ssp. *stoechas*. *Ind. Crops Prod.* **2018**, *118*, 111–117. [CrossRef]

52. Bilušić Vundać, V.; Maleš, Ž.; Plazibat, M.; Golja, P.; Cetina-Čižmek, B. HPTLC determination of flavonoids and phenolic acids in some Croatian *Stachys* taxa. *JPC J. Planar Chromatogr.* **2005**, *18*, 269–273. [CrossRef]

53. Fecka, I.; Kowalczyk, A.; Cisowski, W. Optimization of the separation of flavonoid glycosides and rosmarinic acid from *Mentha piperita* on HPTLC plates. *JPC J. Planar Chromatogr.* **2004**, *17*, 22–25. [CrossRef]

54. Fecka, I.; Turek, S. Determination of water-soluble polyphenolic compounds in commercial herbal teas from Lamiaceae: Peppermint, melissa, and sage. *J. Agric. Food Chem.* **2007**, *55*, 10908–10917. [CrossRef]

55. Fecka, I.; Turek, S. Determination of polyphenolic compounds in commercial herbal drugs and spices from Lamiaceae: Thyme, wild thyme and sweet marjoram by chromatographic techniques. *Food Chem.* **2008**, *108*, 1039–1053. [CrossRef]

56. Fecka, I.; Raj, D.; Krauze-Baranowska, M. Quantitative determination of four water-soluble compounds in herbal drugs from Lamiaceae using different chromatographic techniques. *Chromatographia* **2007**, *66*, 87–93. [CrossRef]

57. Jug, U.; Glavnik, V.; Kranjc, E.; Vovk, I. HPTLC–densitometric and HPTLC–MS methods for analysis of flavonoids. *J. Liq. Chromatogr. Relat. Technol.* **2018**, *41*, 329–341. [CrossRef]

58. Lebot, V.; Lawac, F.; Michalet, S.; Legendre, L. Characterization of taro [*Colocasia esculenta* (L.) Schott] germplasm for improved flavonoid composition and content. *Plant. Genet. Resour.* **2017**, *15*, 260–268. [CrossRef]

59. Orsini, F.; Vovk, I.; Glavnik, V.; Jug, U.; Corradini, D. HPTLC, HPTLC-MS/MS and HPTLC-DPPH methods for analyses of flavonoids and their antioxidant activity in *Cyclanthera pedata* leaves, fruits and dietary supplement. *J. Liq. Chromatogr. Relat. Technol.* **2019**, *42*, 290–301. [CrossRef]

60. Nile, S.H.; Park, S.W. HPTLC analysis, antioxidant and antigout activity of Indian plants. *Iran. J. Pharm. Res.* **2014**, *13*, 531–539. [PubMed]

61. Chisvert, A.; Benedé, J.L.; Salvador, A. Current trends on the determination of organic UV filters in environmental water samples based on microextraction techniques—A review. *Anal. Chim. Acta* **2018**, *1034*, 22–38. [CrossRef]

62. Dias, I.H.K.; Wilson, S.R.; Roberg-Larsen, H. Chromatography of oxysterols. *Biochimie* **2018**, *153*, 3–12. [CrossRef]

63. Pascual-Maté, A.; Osés, S.M.; Fernández-Muiño, M.A.; Sancho, M.T. Analysis of polyphenols in honey: Extraction, separation and quantification procedures. *Sep. Purif. Rev.* **2018**, *47*, 142–158. [CrossRef]

64. Kazakevich, Y. HPLC theory. In *HPLC for Pharmaceutical Scientists*; Kazakevich, Y., LoBrutto, R., Eds.; John Wiley & Sons, Inc.: Hoboken, NJ, USA, 2007; pp. 40–70.

65. Kazakevich, Y.; LoBrutto, R. Stationary phases. In *HPLC for Pharmaceutical Scientists*; Kazakevich, Y., LoBrutto, R., Eds.; John Wiley & Sons, Inc.: Hoboken, NJ, USA, 2007; pp. 77–90.

66. Pereira, E.; Barros, L.; Calhelha, R.C.; Dueñas, M.; Carvalho, A.M.; Santos-Buelga, C.; Ferreira, I.C.F.R. Bioactivity and phytochemical characterization of *Arenaria montana* L. *Food Funct.* **2014**, *5*, 1848–1855. [CrossRef]

67. Barros, L.; Dueñas, M.; Carvalho, A.M.; Ferreira, I.C.F.R.; Santos-Buelga, C. Characterization of phenolic compounds in flowers of wild medicinal plants from Northeastern Portugal. *Food Chem. Toxicol.* **2012**, *50*, 1576–1582. [CrossRef]

68. Dias, M.I.; Barros, L.; Dueñas, M.; Pereira, E.; Carvalho, A.M.; Alves, R.C.; Oliveira, M.B.P.P.; Santos-Buelga, C.; Ferreira, I.C.F.R. Chemical composition of wild and commercial *Achillea millefolium* L. and bioactivity of the methanolic extract, infusion and decoction. *Food Chem.* **2013**, *141*, 4152–4160. [CrossRef]

69. Rodrigues, S.; Calhelha, R.C.; Barreira, J.C.M.; Dueñas, M.; Carvalho, A.M.; Abreu, R.M.V.; Santos-Buelga, C.; Ferreira, I.C.F.R. *Crataegus monogyna* buds and fruits phenolic extracts: Growth inhibitory activity on human tumor cell lines and chemical characterization by HPLC–DAD–ESI/MS. *Food Res. Int.* **2012**, *49*, 516–523. [CrossRef]

70. Roriz, C.L.; Barros, L.; Carvalho, A.M.; Santos-Buelga, C.; Ferreira, I.C.F.R. *Pterospartum tridentatum*, *Gomphrena globosa* and *Cymbopogon citratus*: A phytochemical study focused on antioxidant compounds. *Food Res. Int.* **2014**, *62*, 684–693. [CrossRef]

71. Barros, L.; Pereira, E.; Calhelha, R.C.; Dueñas, M.; Carvalho, A.M.; Santos-Buelga, C.; Ferreira, I.C.F.R. Bioactivity and chemical characterization in hydrophilic and lipophilic compounds of *Chenopodium ambrosioides* L. *J. Funct. Foods* **2013**, *5*, 1732–1740. [CrossRef]

72. Viacava, G.E.; Roura, S.I.; López-Márquez, D.M.; Berrueta, L.A.; Gallo, B.; Alonso-Salces, R.M. Polyphenolic profile of butterhead lettuce cultivar by ultrahigh performance liquid chromatography coupled online to UV–visible spectrophotometry and quadrupole time-of-flight mass spectrometry. *Food Chem.* **2018**, *260*, 239–273. [CrossRef] [PubMed]

73. Yang, X.; Cui, X.; Zhao, L.; Guo, D.; Feng, L.; Wei, S.; Zhao, C.; Huang, D. Exogenous glycine nitrogen enhances accumulation of glycosylated flavonoids and antioxidant activity in lettuce (*Lactuca sativa* L.). *Front. Plant. Sci.* **2017**, *8*, 2098. [CrossRef]

74. Farag, M.A.; Handoussa, H.; Fekry, M.I.; Wessjohann, L.A. Metabolite profiling in 18 Saudi date palm fruit cultivars and their antioxidant potential *via* UPLC-qTOF-MS and multivariate data analyses. *Food Funct.* **2016**, *7*, 1077–1086. [CrossRef]

75. Ibrahim, R.M.; El-Halawany, A.M.; Saleh, D.O.; El Naggar, E.M.B.; El-Shabrawy, A.E.-R.O.; El-Hawary, S.S. HPLC-DAD-MS/MS profiling of phenolics from *Securigera securidaca* flowers and its anti-hyperglycemic and anti-hyperlipidemic activities. *Rev. Bras. Farmacogn.* **2015**, *25*, 134–141. [CrossRef]

76. Mudrić, S.Ž.; Gašić, U.M.; Dramićanin, A.M.; Ćirić, I.Ž.; Milojković-Opsenica, D.M.; Popović-Đorđević, J.B.; Momirović, N.M.; Tešić, Ž.L. The polyphenolics and carbohydrates as indicators of botanical and geographical origin of Serbian autochthonous clones of red spice paprika. *Food Chem.* **2017**, *217*, 705–715. [CrossRef]

77. Barros, L.; Dueñas, M.; Dias, M.I.; Sousa, M.J.; Santos-Buelga, C.; Ferreira, I.C.F.R. Phenolic profiles of in vivo and in vitro grown *Coriandrum sativum* L. *Food Chem.* **2012**, *132*, 841–848. [CrossRef]

78. Quirantes-Pine, R.; Lozano-Sanchez, J.; Herrero, M.; Ibanez, E.; Segura-Carretero, A.; Fernandez-Gutierrez, A. HPLC-ESI-QTOF-MS as a powerful analytical tool for characterising phenolic compounds in olive-leaf extracts. *Phytochem. Anal.* **2013**, *24*, 213–223. [CrossRef]

79. Costa, G.; Nunes, F.; Vitorino, C.; Sousa, J.J.; Figueiredo, I.V.; Batista, M.T. Validation of a RP-HPLC method for quantitation of phenolic compounds in three different extracts from *Cymbopogon citratus*. *Res. J. Med. Plant.* **2015**, *9*, 331–339.

80. Liu, G.; Linwu, Z.; Song, D.; Lu, C.; Xu, X. Isolation, purification, and identification of the main phenolic compounds from leaves of celery (*Apium graveolens* L. var. dulce Mill./Pers.). *J. Sep. Sci.* **2016**, *40*, 472–479. [CrossRef] [PubMed]

81. Taghouti, M.; Martins-Gomes, C.; Schäfer, J.; Félix, L.M.; Santos, J.A.; Bunzel, M.; Nunes, F.M.; Silva, A.M. *Thymus pulegioides* L. as a rich source of antioxidant, anti-proliferative and neuroprotective phenolic compounds. *Food Funct.* **2018**, *9*, 3617–3629. [CrossRef] [PubMed]

82. Achour, M.; Mateos, R.; Ben Fredj, M.; Mtiraoui, A.; Bravo, L.; Saguem, S. A comprehensive characterisation of rosemary tea obtained from *Rosmarinus officinalis* L. collected in a sub-humid area of Tunisia. *Phytochem. Anal.* **2018**, *29*, 87–100. [CrossRef] [PubMed]

83. Ferreres, F.; Gomes, N.G.M.; Valentão, P.; Pereira, D.M.; Gil-Izquierdo, A.; Araújo, L.; Silva, T.C.; Andrade, P.B. Leaves and stem bark from *Allophylus africanus* P. Beauv.: An approach to anti-inflammatory properties and characterization of their flavonoid profile. *Food Chem. Toxicol.* **2018**, *118*, 430–438. [CrossRef]

84. Ravisankar, S.; Abegaz, K.; Awika, J.M. Structural profile of soluble and bound phenolic compounds in teff (*Eragrostis tef*) reveals abundance of distinctly different flavones in white and brown varieties. *Food Chem.* **2018**, *263*, 265–274. [CrossRef]

85. Khallouki, F.; Breuer, A.; Merieme, E.; Ulrich, C.M.; Owen, R.W. Characterization and quantitation of the polyphenolic compounds detected in methanol extracts of *Pistacia atlantica* Desf. fruits from the Guelmim region of Morocco. *J. Pharm. Biomed. Anal.* **2017**, *134*, 310–318. [CrossRef]

86. Vasić, V.; Gašić, U.; Stanković, D.; Lušić, D.; Vukić-Lušić, D.; Milojković-Opsenica, D.; Tešić, Ž.; Trifković, J. Towards better quality criteria of European honeydew honey: Phenolic profile and antioxidant capacity. *Food Chem.* **2019**, *274*, 629–641. [CrossRef]

87. Ammar, S.; Contreras, M.d.M.; Belguith-Hadrich, O.; Bouaziz, M.; Segura-Carretero, A. New insights into the qualitative phenolic profile of *Ficus carica* L. fruits and leaves from Tunisia using ultra-high-performance liquid chromatography coupled to quadrupole-time-of-flight mass spectrometry and their antioxidant activity. *RSC Adv.* **2015**, *5*, 20035–20050. [CrossRef]

88. Obmann, A.; Purevsuren, S.; Zehl, M.; Kletter, C.; Reznicek, G.; Narantuya, S.; Glasl, S. HPLC determination of flavonoid glycosides in Mongolian *Dianthus versicolor* Fisch. (Caryophyllaceae) compared with quantification by UV spectrophotometry. *Phytochem. Anal.* **2012**, *23*, 254–259. [CrossRef]

89. Chen, H.-J.; Inbaraj, B.S.; Chen, B.-H. Determination of phenolic acids and flavonoids in *Taraxacum formosanum* Kitam by liquid chromatography-tandem mass spectrometry coupled with a post-column derivatization technique. *Int. J. Mol. Sci.* **2012**, *13*, 260–285. [CrossRef]

90. Tebboub, O.; Cotugno, R.; Oke-Altuntas, F.; Bouheroum, M.; Demirtas, İ.; D'Ambola, M.; Malafronte, N.; Vassallo, A. Antioxidant potential of herbal preparations and components from *Galactites elegans* (All.) Nyman ex soldano. *Evid. Based. Complement. Alternat. Med.* **2018**, *2018*, 9294358. [CrossRef] [PubMed]

91. Saleem, H.; Htar, T.T.; Naidu, R.; Nawawi, N.S.; Ahmad, I.; Ashraf, M.; Ahemad, N. Biological, chemical and toxicological perspectives on aerial and roots of *Filago germanica* (L.) huds: Functional approaches for novel phyto-pharmaceuticals. *Food Chem. Toxicol.* **2019**, *123*, 363–373. [CrossRef] [PubMed]

92. Tahri, W.; Chatti, A.; Romero-González, R.; López-Gutiérrez, N.; Frenich, A.G.; Landoulsi, A. Phenolic profiling of the aerial part of *Chrysanthemum trifurcatum* using ultra high-performance liquid chromatography coupled to Orbitrap high resolution mass spectrometry. *Anal. Methods* **2016**, *8*, 3517–3527. [CrossRef]

93. Dall'Acqua, S.; Peron, G.; Ferrari, S.; Gandin, V.; Bramucci, M.; Quassinti, L.; Mártonfi, P.; Maggi, F. Phytochemical investigations and antiproliferative secondary metabolites from *Thymus alternans* growing in Slovakia. *Pharm. Biol.* **2017**, *55*, 1162–1170. [CrossRef]

94. Abu-Reidah, I.M.; Gil-Izquierdo, Á.; Medina, S.; Ferreres, F. Phenolic composition profiling of different edible parts and by-products of date palm (*Phoenix dactylifera* L.) by using HPLC-DAD-ESI/MSn. *Food Res. Int.* **2017**, *100*, 494–500. [CrossRef]

95. Vinayagam, A.; Sudha, P.N. Separation and identification of phenolic acid and flavonoids from *Nerium indicum* flowers. *Indian J. Pharm. Sci.* **2015**, *77*, 91–95. [CrossRef]

96. Vitalini, S.; Madeo, M.; Tava, A.; Iriti, M.; Vallone, L.; Avato, P.; Cocuzza, C.E.; Simonetti, P.; Argentieri, M.P. Chemical profile, antioxidant and antibacterial activities of *Achillea moschata* Wulfen, an endemic species from the Alps. *Molecules* **2016**, *21*, 830. [CrossRef]

97. Liu, H.; Zhang, M.; Guo, Y.; Qiu, H. Solid-phase extraction of flavonoids in honey samples using carbamate-embedded triacontyl-modified silica sorbent. *Food Chem.* **2016**, *204*, 56–61. [CrossRef]

98. Li, L.; Jiang, H.; Wu, H.; Zeng, S. Simultaneous determination of luteolin and apigenin in dog plasma by RP-HPLC. *J. Pharm. Biomed. Anal.* **2005**, *37*, 615–620. [CrossRef]

99. Klangprapun, S.; Buranrat, B.; Caichompoo, W.; Nualkaew, S. Pharmacognostical and physicochemical studies of *Enhalus acoroides* (L.F.) royle (rhizome). *Pharmacogn. J.* **2018**, *10*, s89–s94. [CrossRef]

100. Song, T.; Liu, L. A strategy for quality control of the fruits of *Perilla frutescens* (L.) Britt based on antioxidant activity and fingerprint analysis. *Anal. Methods* **2016**, *8*, 295–302. [CrossRef]

101. da Silva, J.B.; dos Temponi, V.S.; Gasparetto, C.M.; Fabri, R.L.; de Aragão, D.M.O.; de Pinto, N.C.C.; Ribeiro, A.; Scio, E.; Del-Vechio-Vieira, G.; de Sousa, O.V.; et al. *Vernonia condensata* Baker (Asteraceae): A promising source of antioxidants. *Oxid. Med. Cell. Longev.* **2013**, *2013*, 698018. [CrossRef] [PubMed]

102. Hasan, H.T.; Kadhim, E.J. Phytochemical investigation of leaves and seeds of *Corchorus olitorius* L. Cultivated in Iraq. *Asian J. Pharm. Clin. Res.* **2018**, *11*, 408–417. [CrossRef]

103. Yin, L.; Han, H.; Zheng, X.; Wang, G.; Li, Y.; Wang, W. Flavonoids analysis and antioxidant, antimicrobial, and anti-inflammatory activities of crude and purified extracts from *Veronicastrum latifolium*. *Ind. Crops Prod.* **2019**, *137*, 652–661. [CrossRef]

104. Amanpour, A.; Kelebek, H.; Selli, S. LC-DAD-ESI-MS/MS-based phenolic profiling and antioxidant activity in Turkish cv. Nizip Yaglik olive oils from different maturity olives. *J. Mass Spectrom.* **2019**, *54*, 227–238. [CrossRef]

105. Ivanov, I.; Vrancheva, R.; Petkova, N.; Tumbarski, Y.; Dincheva, I.; Badjakov, I. Phytochemical compounds of anise hyssop (*Agastache foeniculum*) and antibacterial, antioxidant, and acetylcholinesterase inhibitory properties of its essential oil. *J. Appl. Pharm. Sci.* **2019**, *9*, 72–78.

106. Ren, G.; Xue, P.; Sun, X.; Zhao, G. Determination of the volatile and polyphenol constituents and the antimicrobial, antioxidant, and tyrosinase inhibitory activities of the bioactive compounds from the by-product of *Rosa rugosa* Thunb. var. plena Regal tea. *BMC Complement. Altern. Med.* **2018**, *18*, 307. [CrossRef]

107. Tomczyk, M.; Gudej, J. Quantitative analysis of flavonoids in the flowers and leaves of *Ficaria verna* Huds. *Z. Naturforsch. C.* **2003**, *58*, 762–764. [CrossRef]

108. Félix-Silva, J.; Gomes, J.A.S.; Fernandes, J.M.; Moura, A.K.C.; Menezes, Y.A.S.; Santos, E.C.G.; Tambourgi, D.V.; Silva-Junior, A.A.; Zucolotto, S.M.; Fernandes-Pedrosa, M.F. Comparison of two *Jatropha* species (Euphorbiaceae) used popularly to treat snakebites in Northeastern Brazil: Chemical profile, inhibitory activity against *Bothrops erythromelas* venom and antibacterial activity. *J. Ethnopharmacol.* **2018**, *213*, 12–20. [CrossRef]

109. Sánchez-Roque, Y.; Ayora-Talavera, G.; Rincón-Rosales, R.; Gutiérrez-Miceli, F.; Meza Gordillo, R.; Winkler, R.; Gamboa-Becerra, R.; Ayora, T.; Ruíz-Valdiviezo, V. The flavonoid fraction from *Rhoeo discolor* leaves acts antiviral against influenza a virus. *Rec. Nat. Prod.* **2017**, *11*, 532–546. [CrossRef]

110. Reheman, A.; Aisa, H.A.; Ma, Q.L.; Nijat, D.; Abdulla, R. Quality evaluation of the traditional medicine majun mupakhi ELA via chromatographic fingerprinting coupled with UHPLC-DAD-Quadrupole-Orbitrap-MS and the antioxidant activity in vitro. *Evid. Based. Complement. Alternat. Med.* **2018**, *2018*, 1035809. [CrossRef] [PubMed]

111. Moreno-González, R.; Juan, M.E.; Planas, J.M. Table olive polyphenols: A simultaneous determination by liquid chromatography–mass spectrometry. *J. Chromatogr. A* **2019**, 460434. [CrossRef] [PubMed]

112. Luca, S.V.; Miron, A.; Aprotosoaie, A.C.; Mihai, C.-T.; Vochita, G.; Gherghel, D.; Ciocarlan, N.; Skalicka-Wozniak, K. HPLC-DAD-ESI-Q-TOF-MS/MS profiling of *Verbascum ovalifolium* Donn ex Sims and evaluation of its antioxidant and cytogenotoxic activities. *Phytochem. Anal.* **2019**, *30*, 34–45. [CrossRef] [PubMed]

113. Quirantes-Piné, R.; Funes, L.; Micol, V.; Segura-Carretero, A.; Fernández-Gutiérrez, A. High-performance liquid chromatography with diode array detection coupled to electrospray time-of-flight and ion-trap tandem mass spectrometry to identify phenolic compounds from a *Lemon verbena* extract. *J. Chromatogr. A* **2009**, *1216*, 5391–5397. [CrossRef] [PubMed]

114. Flores-Ocelotl, M.R.; Rosas-Murrieta, N.H.; Moreno, D.A.; Vallejo-Ruiz, V.; Reyes-Leyva, J.; Dominguez, F.; Santos-López, G. *Taraxacum officinale* and *Urtica dioica* extracts inhibit dengue virus serotype 2 replication in vitro. *BMC Complement. Altern. Med.* **2018**, *18*, 95. [CrossRef] [PubMed]

115. Colombo, R.; Harumi Yariwake, J.; Queiroz, E.; Ndjoko, K.; Hostettmann, K. LC-MS/MS analysis of sugarcane extracts and differentiation of monosaccharides moieties of flavone C-glycosides. *J. Liq. Chromatogr. Relat. Technol.* **2013**, *36*, 239–258. [CrossRef]

116. Elkady, W.M.; Ibrahim, E.A.; Gonaid, M.H.; El Baz, F.K. Chemical profile and biological activity of *Casimiroa edulis* non-edible fruit's parts. *Adv. Pharm. Bull.* **2017**, *7*, 655–660. [CrossRef]

117. Mattila, P.; Astola, J.; Kumpulainen, J. Determination of flavonoids in plant material by HPLC with diode-array and electro-array detections. *J. Agric. Food Chem.* **2000**, *48*, 5834–5841. [CrossRef]

118. de Beer, D.; Joubert, E.; Malherbe, C.J.; Jacobus Brand, D. Use of countercurrent chromatography during isolation of 6-hydroxyluteolin-7-O-β-glucoside, a major antioxidant of *Athrixia phylicoides*. *J. Chromatogr. A* **2011**, *1218*, 6179–6186. [CrossRef]

119. Ziani, B.E.C.; Barros, L.; Boumehira, A.Z.; Bachari, K.; Heleno, S.A.; Alves, M.J.; Ferreira, I.C.F.R. Profiling polyphenol composition by HPLC-DAD-ESI/MSn and the antibacterial activity of infusion preparations obtained from four medicinal plants. *Food Funct.* **2018**, *9*, 149–159. [CrossRef]

120. Bessada, S.M.F.; Barreira, J.C.M.; Barros, L.; Ferreira, I.C.F.R.; Oliveira, M.B.P.P. Phenolic profile and antioxidant activity of *Coleostephus myconis* (L.) Rchb.f.: An underexploited and highly disseminated species. *Ind. Crops Prod.* **2016**, *89*, 45–51. [CrossRef]

121. Llorent-Martínez, E.J.; Ortega-Barrales, P.; Zengin, G.; Uysal, S.; Ceylan, R.; Guler, G.O.; Mocan, A.; Aktumsek, A. *Lathyrus aureus* and *Lathyrus pratensis*: Characterization of phytochemical profiles by liquid chromatography-mass spectrometry, and evaluation of their enzyme inhibitory and antioxidant activities. *RSC Adv.* **2016**, *6*, 88996–89006. [CrossRef]

122. Carvalho, A.R.; Costa, G.; Figueirinha, A.; Liberal, J.; Prior, J.A.V.; Lopes, M.C.; Cruz, M.T.; Batista, M.T. *Urtica* spp.: Phenolic composition, safety, antioxidant and anti-inflammatory activities. *Food Res. Int.* **2017**, *99*, 485–494. [CrossRef] [PubMed]

123. Brito, A.; Ramirez, J.E.; Areche, C.; Sepúlveda, B.; Simirgiotis, M.J. HPLC-UV-MS profiles of phenolic compounds and antioxidant activity of fruits from three citrus species consumed in Northern Chile. *Molecules* **2014**, *19*, 17400–17421. [CrossRef]

124. Plazonić, A.; Bucar, F.; Maleš, Ž.; Mornar, A.; Nigović, B.; Kujundžić, N. Identification and quantification of flavonoids and phenolic acids in burr parsley (*Caucalis platycarpos* L.), using high-performance liquid chromatography with diode array detection and electrospray ionization mass spectrometry. *Molecules* **2009**, *14*, 2466–2490. [CrossRef]

125. Pereira, O.R.; Catarino, M.D.; Afonso, A.F.; Silva, A.M.S.; Cardoso, S.M. *Salvia elegans, Salvia greggii* and *Salvia officinalis* decoctions: Antioxidant activities and inhibition of carbohydrate and lipid metabolic enzymes. *Molecules* **2018**, *23*, 3169. [CrossRef]

126. Afonso, A.F.; Pereira, O.R.; Neto, R.T.; Silva, A.M.S.; Cardoso, S.M. Health-promoting effects of *Thymus herba-barona, Thymus pseudolanuginosus*, and *Thymus caespititius* decoctions. *Int. J. Mol. Sci.* **2017**, *18*, 1879. [CrossRef]

127. Ferreres, F.; Grosso, C.; Gil-Izquierdo, A.; Fernandes, A.; Valentão, P.; Andrade, P. Comparing the phenolic profile of *Pilocarpus pennatifolius* Lem. by HPLC–DAD–ESI/MSn with respect to authentication and enzyme inhibition potential. *Ind. Crops Prod.* **2015**, *77*, 391–401. [CrossRef]

128. Otify, A.; George, C.; Elsayed, A.; Farag, M.A. Mechanistic evidence of *Passiflora edulis* (Passifloraceae) anxiolytic activity in relation to its metabolite fingerprint as revealed *via* LC-MS and chemometrics. *Food Funct.* **2015**, *6*, 3807–3817. [CrossRef]

129. Afifi, F.U.; Kasabri, V.; Litescu, S.; Abaza, I.F.; Tawaha, K. Phytochemical and biological evaluations of *Arum hygrophilum* Boiss. (Araceae). *Pharmacogn. Mag.* **2017**, *13*, 275–280. [CrossRef]

130. Kowalska, I.; Pecio, L.; Ciesla, L.; Oleszek, W.; Stochmal, A. Isolation, chemical characterization, and free radical scavenging activity of phenolics from *Triticum aestivum* L. aerial parts. *J. Agric. Food Chem.* **2014**, *62*, 11200–11208. [CrossRef]

131. Kostic, A.Z.; Gasic, U.M.; Pesic, M.B.; Stanojevic, S.P.; Barac, M.B.; Macukanovic-Jocic, M.P.; Avramov, S.N.; Tesic, Z.L. Phytochemical analysis and total antioxidant capacity of rhizome, above-ground vegetative parts and flower of three *Iris* species. *Chem. Biodivers.* **2019**, *16*, e1800565. [CrossRef] [PubMed]

132. Llorent-Martínez, E.J.; Córdova, M.L.F.; Zengin, G.; Bahadori, M.B.; Aumeeruddy, M.Z.; Rengasamy, K.R.R.; Mahomoodally, M.F. Parentucellia latifolia subsp. *latifolia*: A potential source for loganin iridoids by HPLC-ESI-MSn technique. *J. Pharm. Biomed. Anal.* **2019**, *165*, 374–380.

133. Llorent-Martínez, E.J.; Zengin, G.; Lobine, D.; Molina-García, L.; Mollica, A.; Mahomoodally, M.F. Phytochemical characterization, in vitro and in silico approaches for three *Hypericum* species. *New J. Chem.* **2018**, *42*, 5204–5214. [CrossRef]

134. Öztürk, G.; Ağalar, H.G.; Yildiz, G.; Göger, F.; Kirimer, N. Biological activities and luteolin derivatives of *Verbascum eskisehirensis* Karavel., Ocak Ekici. *J. Res. Pharm.* **2019**, *23*, 532–542. [CrossRef]

135. Bączek, K.B.; Wiśniewska, M.; Przybył, J.L.; Kosakowska, O.; Węglarz, Z. Arbuscular mycorrhizal fungi in chamomile (*Matricaria recutita* L.) organic cultivation. *Ind. Crops Prod.* **2019**, *140*, 111562. [CrossRef]

136. Gomes, A.F.; Almeida, M.P.; Leite, M.F.; Schwaiger, S.; Stuppner, H.; Halabalaki, M.; Amaral, J.G.; David, J.M. Seasonal variation in the chemical composition of two chemotypes of *Lippia alba*. *Food Chem.* **2019**, *273*, 186–193. [CrossRef]

137. Han, A.-R.; Nam, B.; Kim, B.-R.; Lee, K.-C.; Song, B.-S.; Kim, S.H.; Kim, J.-B.; Jin, C.H. Phytochemical composition and antioxidant activities of two different color chrysanthemum flower teas. *Molecules* **2019**, *24*, 329. [CrossRef]

138. Arsenijević, J.; Drobac, M.; Šoštarić, I.; Ražić, S.; Milenković, M.; Couladis, M.; Maksimović, Z. Bioactivity of herbal tea of Hungarian thyme based on the composition of volatiles and polyphenolics. *Ind. Crops Prod.* **2016**, *89*, 14–20. [CrossRef]

139. Cheng, S.; Shiea, J. Advanced spectroscopic detectors for identification and quantification: Mass spectrometry. In *Instrumental Thin-Layer Chromatography*; Poole, C.F., Ed.; Elsevier Inc.: Amsterdam, The Netherlands, 2015; p. 251.

140. Shao, Y.; Wu, Q.; Wen, H.; Chai, C.; Shan, C.; Yue, W.; Yan, S.; Xu, H. Determination of flavones in *Lophatherum gracile* by liquid chromatography tandem mass spectrometry. *Instrum. Sci. Technol.* **2014**, *42*, 173–183. [CrossRef]

141. Sánchez-Rabaneda, F.; Jauregui, O.; Casals, I.; Andres-Lacueva, C.; Izquierdo-Pulido, M.; Lamuela-Raventós, R.M. Liquid chromatographic/electrospray ionization tandem mass spectrometric study of the phenolic composition of cocoa (*Theobroma cacao*). *J. Mass Spectrom.* **2003**, *38*, 35–42. [CrossRef] [PubMed]

142. Lin, L.-Z.; Harnly, J.M. Identification of the phenolic components of chrysanthemum flower (*Chrysanthemum morifolium* Ramat). *Food Chem.* **2010**, *120*, 319–326. [CrossRef]

143. Liu, R.; Yan, W.; Han, Q.; Lv, T.; Wang, X.; Liu, X.; Fan, X.; Meng, C.; Wang, C. Simultaneous detection of four flavonoids and two alkaloids in rat plasma by LC-MS/MS and its application to a comparative study of the pharmacokinetics between *Abri Herba* and *Abri mollis Herba* extract after oral administration. *J. Sep. Sci.* **2019**, *42*, 1341–1350. [CrossRef] [PubMed]

144. Ye, Y.; Mo, S.; Feng, W.; Ye, X.; Shu, X.; Long, Y.; Guan, Y.; Huang, J.; Wang, J. The ethanol extract of *Involcucrum castaneae* ameliorated ovalbumin-induced airway inflammation and smooth muscle thickening in guinea pigs. *J. Ethnopharmacol.* **2019**, *230*, 9–19. [CrossRef] [PubMed]

145. Escher, G.B.; Santos, J.S.; Rosso, N.D.; Marques, M.B.; Azevedo, L.; do Carmo, M.A.V.; Daguer, H.; Molognoni, L.; Prado-Silva, L.D.; Sant'Ana, A.S.; et al. Chemical study, antioxidant, anti-hypertensive, and cytotoxic/cytoprotective activities of *Centaurea cyanus* L. petals aqueous extract. *Food Chem. Toxicol.* **2018**, *118*, 439–453. [CrossRef]

146. Janković, T.; Zdunić, G.; Beara, I.; Balog, K.; Pljevljakušić, D.; Stešević, D.; Šavikin, K. Comparative study of some polyphenols in *Plantago* species. *Biochem. Syst. Ecol.* **2012**, *42*, 69–74. [CrossRef]

147. Pintać, D.; Četojević-Simin, D.; Berežni, S.; Orčić, D.; Mimica-Dukić, N.; Lesjak, M. Investigation of the chemical composition and biological activity of edible grapevine (*Vitis vinifera* L.) leaf varieties. *Food Chem.* **2019**, *286*, 686–695. [CrossRef]

148. Lin, L.-Z.; Lu, S.; Harnly, J.M. Detection and quantification of glycosylated flavonoid malonates in celery, Chinese celery, and celery seed by LC-DAD-ESI/MS. *J. Agric. Food Chem.* **2007**, *55*, 1321–1326. [CrossRef]

149. Siciliano, T.; De Tommasi, N.; Morelli, I.; Braca, A. Study of flavonoids of *Sechium edule* (Jacq) Swartz (Cucurbitaceae) different edible organs by liquid chromatography photodiode array mass spectrometry. *J. Agric. Food Chem.* **2004**, *52*, 6510–6515. [CrossRef]

150. İşcan, G.; Göger, F.; Demirci, B.; Köse, Y.B. Chemical composition and biological activity of *Nepeta cilicica*. *Bangladesh J. Pharmacol.* **2017**, *12*, 204–209. [CrossRef]

151. Choi, J.Y.; Desta, K.T.; Lee, S.J.; Kim, Y.-H.; Shin, S.C.; Kim, G.-S.; Lee, S.J.; Shim, J.-H.; Hacimüftüoğlu, A.; Abd El-Aty, A.M. LC-MS/MS profiling of polyphenol-enriched leaf, stem and root extracts of Korean *Humulus japonicus* Siebold & Zucc and determination of their antioxidant effects. *Biomed. Chromatogr.* **2018**, *32*, e4171. [PubMed]

152. Haliloglu, Y.; Ozek, T.; Tekin, M.; Goger, F.; Baser, K.H.C.; Ozek, G. Phytochemicals, antioxidant, and antityrosinase activities of *Achillea sivasica* Çelik and Akpulat. *Int. J. Food Prop.* **2017**, *20*, S693–S706. [CrossRef]

153. Graça, V.C.; Dias, M.I.; Barros, L.; Calhelha, R.C.; Santos, P.F.; Ferreira, I.C.F.R. Fractionation of the more active extracts of *Geranium molle* L.: A relationship between their phenolic profile and biological activity. *Food Funct.* **2018**, *9*, 2032–2042. [CrossRef] [PubMed]

154. Pereira, J.M.; Peixoto, V.; Teixeira, A.; Sousa, D.; Barros, L.; Ferreira, I.C.F.R.; Vasconcelos, M.H. *Achillea millefolium* L. hydroethanolic extract inhibits growth of human tumor cell lines by interfering with cell cycle and inducing apoptosis. *Food Chem. Toxicol.* **2018**, *118*, 635–644. [CrossRef]

155. Gonçalves, G.A.; Corrêa, R.C.G.; Barros, L.; Dias, M.I.; Calhelha, R.C.; Correa, V.G.; Bracht, A.; Peralta, R.M.; Ferreira, I.C.F.R. Effects of in vitro gastrointestinal digestion and colonic fermentation on a rosemary (*Rosmarinus officinalis* L) extract rich in rosmarinic acid. *Food Chem.* **2019**, *271*, 393–400. [CrossRef]

156. Soltana, H.; De Rosso, M.; Lazreg, H.; Vedova, A.D.; Hammami, M.; Flamini, R. LC-QTOF characterization of non-anthocyanic flavonoids in four Tunisian fig varieties. *J. Mass Spectrom.* **2018**, *53*, 817–823. [CrossRef]

157. Gaspar, M.C.; Fonseca, D.A.; Antunes, M.J.; Frigerio, C.; Gomes, N.G.M.; Vieira, M.; Santos, A.E.; Cruz, M.T.; Cotrim, M.D.; Campos, M.G. Polyphenolic characterisation and bioactivity of an *Oxalis pes-caprae* L. leaf extract. *Nat. Prod. Res.* **2018**, *32*, 732–738. [CrossRef]

158. Kite, G.C.; Porter, E.A.; Denison, F.C.; Grayer, R.J.; Veitch, N.C.; Butler, I.; Simmonds, M.S.J. Data-directed scan sequence for the general assignment of C-glycosylflavone O-glycosides in plant extracts by liquid chromatography-ion trap mass spectrometry. *J. Chromatogr. A* **2006**, *1104*, 123–131. [CrossRef]

159. Paulina Tambunan, A.; Bahtiar, A.; Tjandrawinata, R. Influence of extraction parameters on the yield, phytochemical, TLC-densitometric quantification of quercetin, and LC-MS profile, and how to standardize different batches for long term from *Ageratum conyoides* L. leaves. *Pharmacogn. J.* **2017**, *9*, 767–774. [CrossRef]

160. Mincsovics, E.; Ott, P.; Alberti, A.; Böszörményi, A.; Héthelyi, E.B.; Szoke, É.; Kery, A.; Lemberkovics, E.; Móricz, Á. In-situ clean-up and OPLC fractionation of chamomile flower extract to search active components by bioautography. *JPC J. Planar Chromatogr.* **2013**, *26*, 172–179. [CrossRef]

161. de Alves, G.A.D.; Oliveira de Souza, R.; Ghislain Rogez, H.L.; Masaki, H.; Fonseca, M.J.V. *Cecropia obtusa* extract and chlorogenic acid exhibit anti aging effect in human fibroblasts and keratinocytes cells exposed to UV radiation. *PLoS ONE* **2019**, *14*, e0216501. [CrossRef] [PubMed]

162. Ghouti, D.; Rached, W.; Abdallah, M.; Pires, T.C.S.P.; Calhelha, R.C.; Alves, M.J.; Abderrahmane, L.H.; Barros, L.; Ferreira, I.C.F.R. Phenolic profile and in vitro bioactive potential of Saharan *Juniperus phoenicea* L. and *Cotula cinerea* (Del) growing in Algeria. *Food Funct.* **2018**, *9*, 4664–4672. [CrossRef] [PubMed]

163. Venuprasad, M.P.; Kandikattu, H.K.; Razack, S.; Amruta, N.; Khanum, F. Chemical composition of *Ocimum sanctum* by LC-ESI-MS/MS analysis and its protective effects against smoke induced lung and neuronal tissue damage in rats. *Biomed. Pharmacother.* **2017**, *91*, 1–12. [CrossRef] [PubMed]

164. Majchrzak, T.; Wojnowski, W.; Lubinska-Szczygeł, M.; Różańska, A.; Namieśnik, J.; Dymerski, T. PTR-MS and GC-MS as complementary techniques for analysis of volatiles: A tutorial review. *Anal. Chim. Acta* **2018**, *1035*, 1–13. [CrossRef]

165. Nolvachai, Y.; Marriott, P.J. GC for flavonoids analysis: Past, current, and prospective trends. *J. Sep. Sci.* **2013**, *36*, 20–36. [CrossRef]

166. Zhang, K.; Zuo, Y. GC-MS determination of flavonoids and phenolic and benzoic acids in human plasma after consumption of cranberry juice. *J. Agric. Food Chem.* **2004**, *52*, 222–227. [CrossRef]

167. Creaser, C.S.; Koupai-Abyazani, M.R.; Stephenson, G.R. Capillary column gas chromatography of methyl and trimethylsilyl derivatives of some naturally occurring flavonoid aglycones and other phenolics. *J. Chromatogr. A* **1989**, *478*, 415–421. [CrossRef]

168. Schmidt, T.J.; Merfort, I.; Matthiesen, U. Resolution of complex mixtures of flavonoid aglycones by analysis of gas chromatographic-mass spectrometric data. *J. Chromatogr. A* **1993**, *634*, 350–355. [CrossRef]

169. Proestos, C.; Sereli, D.; Komaitis, M. Determination of phenolic compounds in aromatic plants by RP-HPLC and GC-MS. *Food Chem.* **2006**, *95*, 44–52. [CrossRef]

170. García-Villalba, R.; Pacchiarotta, T.; Carrasco-Pancorbo, A.; Segura-Carretero, A.; Fernández-Gutiérrez, A.; Deelder, A.M.; Mayboroda, O.A. Gas chromatography–atmospheric pressure chemical ionization-time of flight mass spectrometry for profiling of phenolic compounds in extra virgin olive oil. *J. Chromatogr. A* **2011**, *1218*, 959–971. [CrossRef]

171. Gao, X.; Williams, S.J.; Woodman, O.L.; Marriott, P.J. Comprehensive two-dimensional gas chromatography, retention indices and time-of-flight mass spectra of flavonoids and chalcones. *J. Chromatogr. A* **2010**, *1217*, 8317–8326. [CrossRef] [PubMed]

172. Krafczyk, N.; Glomb, M.A. Characterization of phenolic compounds in rooibos tea. *J. Agric. Food Chem.* **2008**, *56*, 3368–3376. [CrossRef] [PubMed]

173. Peng, J.; Fan, G.; Hong, Z.; Chai, Y.; Wu, Y. Preparative separation of isovitexin and isoorientin from *Patrinia villosa* Juss by high-speed counter-current chromatography. *J. Chromatogr. A* **2005**, *1074*, 111–115. [CrossRef] [PubMed]

174. Waksmundzka-Hajnos, M.; Sherma, J.; Kowalska, T. Overview of the field of TLC in phytochemistry and the structure of the book. In *Thin Layer Chromatography in Phytochemistry*; Waksmundzka-Hajnos, M., Sherma, J., Kowalska, T., Eds.; CRC Press, Taylor & Francis Group: Boca Raton, FL, USA, 2008; p. 5.

175. Skalicka-Woźniak, K.; Mroczek, T.; Kozioł, E. High-performance countercurrent chromatography separation of *Peucedanum cervaria* fruit extract for the isolation of rare coumarin derivatives. *J. Sep. Sci.* **2015**, *38*, 179–186. [CrossRef] [PubMed]

176. Sieniawska, E.; Skalicka-Woźniak, K. Isolation of chlorogenic acid from *Mutellina purpurea* L. herb using high-performance counter-current chromatography. *Nat. Prod. Res.* **2014**, *28*, 1936–1939. [CrossRef] [PubMed]

177. Wang, X.; Cheng, C.; Sun, Q.; Li, F.; Liu, J.; Zheng, C. Isolation and purification of four flavonoid constituents from the flowers of *Paeonia suffruticosa* by high-speed counter-current chromatography. *J. Chromatogr. A* **2005**, *1075*, 127–131. [CrossRef]

178. Leitão, S.G.; Leitão, G.G.; Vicco, D.K.T.; Pereira, J.P.B.; de Morais Simão, G.; Oliveira, D.R.; Celano, R.; Campone, L.; Piccinelli, A.L.; Rastrelli, L. Counter-current chromatography with off-line detection by ultra high performance liquid chromatography/high resolution mass spectrometry in the study of the phenolic profile of *Lippia origanoides*. *J. Chromatogr. A* **2017**, *1520*, 83–90. [CrossRef]

179. Fiamegos, Y.C.; Nanos, C.G.; Vervoort, J.; Stalikas, C.D. Analytical procedure for the in-vial derivatization—extraction of phenolic acids and flavonoids in methanolic and aqueous plant extracts followed by gas chromatography with mass-selective detection. *J. Chromatogr. A* **2004**, *1041*, 11–18. [CrossRef]

Review

On the Neuroprotective Effects of Naringenin: Pharmacological Targets, Signaling Pathways, Molecular Mechanisms, and Clinical Perspective

Zeinab Nouri [1], Sajad Fakhri [2], Fardous F. El-Senduny [3], Nima Sanadgol [4,5], Ghada E. Abd-ElGhani [6], Mohammad Hosein Farzaei [2,*] and Jen-Tsung Chen [7,*]

[1] Student's Research Committee, Faculty of Pharmacy, Kermanshah University of Medical Sciences, Kermanshah 6714415153, Iran; zeinab7641@yahoo.com

[2] Pharmaceutical Sciences Research Center, Health Institute, Kermanshah University of Medical Sciences, Kermanshah 6734667149, Iran; pharmacy.sajad@yahoo.com

[3] Biochemistry division, Chemistry Department, Faculty of Science, Mansoura University, Mansoura 35516, Egypt; biobotany@gmail.com

[4] Department of Biology, Faculty of Sciences, University of Zabol, Zabol 7383198616, Iran; Sanadgol.n@gmail.com

[5] Department of Physics and Chemistry, School of Pharmaceutical Sciences of Ribeirao Preto, University of Sao Paulo, Ribeirao Preto 14040-903, Brazil

[6] Department of Chemistry, Faculty of Science, University of Mansoura, Mansoura 35516, Egypt; dr_ghada_emad@yahoo.com

[7] Department of Life Sciences, National University of Kaohsiung, Kaohsiung 811, Taiwan

* Correspondence: mh.farzaei@gmail.com (M.H.F.); jentsung@nuk.edu.tw (J.-T.C.); Tel.: +98-918-3393351 (M.H.F.)

Received: 22 September 2019; Accepted: 31 October 2019; Published: 3 November 2019

Abstract: As a group of progressive, chronic, and disabling disorders, neurodegenerative diseases (NDs) affect millions of people worldwide, and are on the rise. NDs are known as the gradual loss of neurons; however, their pathophysiological mechanisms have not been precisely revealed. Due to the complex pathophysiological mechanisms behind the neurodegeneration, investigating effective and multi-target treatments has remained a clinical challenge. Besides, appropriate neuroprotective agents are still lacking, which raises the need for new therapeutic agents. In recent years, several reports have introduced naturally-derived compounds as promising alternative treatments for NDs. Among natural entities, flavonoids are multi-target alternatives affecting different pathogenesis mechanisms in neurodegeneration. Naringenin is a natural flavonoid possessing neuroprotective activities. Increasing evidence has attained special attention on the variety of therapeutic targets along with complex signaling pathways for naringenin, which suggest its possible therapeutic applications in several NDs. Here, in this review, the neuroprotective effects of naringenin, as well as its related pharmacological targets, signaling pathways, molecular mechanisms, and clinical perspective, are described. Moreover, the need to develop novel naringenin delivery systems is also discussed to solve its widespread pharmacokinetic limitation.

Keywords: neurodegenerative diseases; naringenin; pharmacological targets; signaling pathways; molecular mechanisms; drug delivery systems

---

1. Introduction

Neurodegenerative diseases (NDs) encompass several disorders, including Alzheimer's disease (AD), Parkinson's disease (PD), amyotrophic lateral sclerosis (ALS), Huntington's disease (HD), multiple sclerosis (MS) and, spinal cord injury (SCI) [1]. They are considered disabling conditions that affect millions of people worldwide [2]. Although the exact etiology of these chronic diseases is entirely unknown, the critical nerve degeneration-related pathologies including decompensated neuronal structure/function changes [3], disruption in the synthesis of neurotransmitters [4], as well as abnormal protein accumulation have been clarified [5]. From the pathophysiological point of view, the ubiquitin-proteasome system and autophagy pathway, play a pivotal role in degradation and elimination of malformed proteins and damaged organelles. Disruption of these two crucial intracellular proteolytic pathways is involved in the pathogenesis of NDs [6,7]. In addition to these pathologies, mitochondrial dysfunction, oxidative insults, and neuroinflammation in the microglial cells are proposed to be implicated in neurodegeneration [8]. Microglial cells are monocyte-like immune cells which are considered as the guardian of the nervous system. There are two phenotypes of microglial cells depending on the activation state, either classical M1 or alternative M2 phenotype [9]. The activated M1 microglial cells express toll-like receptors (TLRs) as well as other pro-inflammatory cytokines such as interferon γ (INF-γ), interleukin 1β (IL-1β) and tumor necrosis factor α (TNF-α) promote the axonal degeneration and apoptotic death of neuronal cells [9]. On the other hand, M2 microglial cells express anti-inflammatory mediators such as IL-4, IL-10, IL-13, transforming growth factor β (TGF-β) and neurotrophic factors leading to neurogenesis, angiogenesis, and oligodendrogenesis (neuronal cells) and remyelination (formation of new myelin sheath around demyelinated axons) [10,11]. Therefore, the balance between M1/M2 keeps homeostasis of the immune response [9]. Naringenin could promote a phenotypic shift of microglial polarization from M1 to M2 through down-regulation of M1 markers including TNF-α and IL-1β as well as overexpression of arginase-1 (Arg-1) and IL-10 which are markers of M2 [12].

Due to the complex pathophysiological mechanisms behind the neurodegeneration, neuroprotective therapies of NDs have remained a clinical challenge. It raises the need to investigate novel multi-target agents. Natural products extracted from medicinal plants or fruits showed promising activities in treating different types of NDs through targeting multiple signaling pathways. Therefore, there is a priority in research (either in vitro or in vivo) to discover new naturally-derived compounds to target the pathways involved in NDs. Among natural entities, flavonoids are multi-target agents affecting different signaling pathways involved in neurodegeneration. Naringenin is a natural flavonoid with neuroprotective effects. Citrus fruits are rich with naringenin and its precursor, naringin [13]. In recent years, growing interest has been focused on the potential therapeutic activities of naringenin in neurological disorders. In spite of the promising effects of naringenin to manage NDs, there is a great challenge posed by poor bioavailability and slight accessibility to the brain. Here in this review, the role of naringenin in treating different neuronal disorders is described. The nanostructured formulations of naringenin which are being investigated to solve its widespread pharmacokinetic limitation are also discussed.

2. Chemistry of Naringenin and Its Sources

Naringenin is one of the most critical naturally-occurring flavonoids (I). The basic structure of flavonoids (I) consists of three rings (A, B, and C). Naringenin (II) has a molecular formula $C_{15}H_{12}O_5$ and is chemically named as 2,3-dihydro-5,7-dihydroxy-2-(4-hydroxyphenyl) 4*H*-1-benzopyran-4-one. Its molecular weight is 272.26 g·mol^{-1}, and the melting point is 251 °C. This molecule is insoluble in water and soluble in organic solvents as alcohol [14]. Naringenin may be found in two forms of aglycone or its glycosidic form (naringenin-7-*O*-glucoside) [15] (Figure 1). Naringin (III) can also be seen as naringin (naringenin-7-rhamnoglucoside) [16] and narirutin (naringenin-7-*O*-rutinoside) [17]. Naringenin glycosides depending on their sugar moiety, attach via a glycosidic linkage at C7 to the flavonoid, and are cleaved by specific enzymes, then naringenin (aglycone) would be released [18].

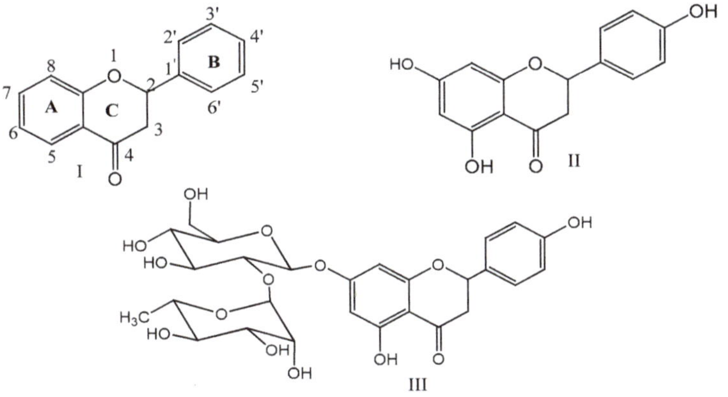

Figure 1. Structures of flavonoids (I); naringenin (II); and naringin (III).

The biosynthesis of naringenin's basic structure, as a flavonoid, occurs via the combination of shikimic acid and acylpolymalonate metabolic pathways. A starting compound is phenylpropane, a cinnamic acid derivative derived from shikimic acid, in which three acetate residues are incorporated followed by ring closure. The chalcone structure is intermediate to the flavone structures, which might be hydroxylated and reduced at different positions [19–21] (Figure 2).

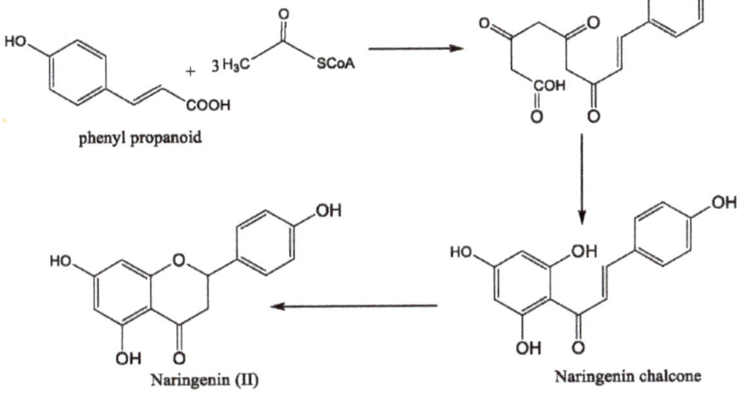

Figure 2. Biosynthesis of naringenin.

Naringenin has various pharmacological properties depending on the arrangement of related functional groups in its structure. The hydroxyl groups (OH) have high reactivity against reactive oxygen species (ROS) and reactive nitrogen species. In ring (A) 5,7-*m*-dihydroxy arrangement serves to stabilize the structure after donating electrons to free radicals. The association between 5-OH and 4-oxo substituents contributes to the chelation of compounds such as heavy metals [22,23].

Naringenin has a single chiral center at carbon 2 (C2), resulting as (2*S*)- and (2*R*)-enantiomers forms which are found in citrus fruits. It has been reported to be resistant to enantiomerization over pH 9–11 [24]. Separation of the enantiomers has been explored for over 20 years [24], primarily via high-performance liquid chromatography (HPLC) on polysaccharide-derived chiral stationary phases [25–28]. There is an evidence to suggest stereospecific pharmacokinetic and pharmacodynamic profiles, which has been proposed to be an explanation for the wide variety in naringenin's reported bioactivity [29]. Flavonols, flavones, flavanones, isoflavones, flavanols, and anthocyanidins are the most abundant flavonoids found in plants [30,31], and are potent scavengers of free radicals. As promising flavonoids, naringenin and naringin are potent antioxidants [32]; however, naringin is less potent because the sugar moiety in the former causes steric hindrance of the scavenging group. The naringenin-7-glucoside form seems less bioavailable than the aglycone form [33]. Naringenin and its glycoside has been discovered, mainly from a variety of vegetables, nuts and fruits, including grapefruit [34], bergamot [35], sour orange [36], tart cherries [37], tomatoes [38,39], cocoa [40], Greek oregano [41], water mint [42], drynaria [43], beans [44] and beverages such as coffee, tea, and red wine, as well as in medical herbs.

3. Naringenin in Neurodegenerative Diseases

3.1. Naringenin and Alzheimer's Disease

As the most common form of NDs, AD is characterized by gradual memory decline and cognitive deficits affecting all aspects of person's ability to do daily activities [2,45,46]. AD is accompanied by aggregation of amyloid plaques as well as neurofibrillary tangles, which are composed of amyloid-beta (Aβ) and hyperphosphorylated tau, respectively [47]. Activation of phosphokinase, glycogen synthase kinase 3β (GSK-3β), participates in the hyper-phosphorylation of tau [48]. Additionally, disrupted insulin signaling and brain insulin resistance are common in the pathogenesis of AD [49], associated with the inhibition of phosphatidylinositol-3 kinase (PI3K)/AKT pathway [50].

Currently, there is no effective treatment for AD as a neurodegenerative condition with multiple pathophysiological mechanisms. Therefore, discovering a multi-target agent is necessary to counteract AD. In this line, the anti-inflammatory, antioxidant, anti-apoptotic and neuroprotective effects [51] of naringenin make it a promising option for treating AD. Administration of naringenin interestingly improved spatial learning and memory in a rat model of AD through regulating the PI3K/AKT/GSK-3β pathway and reducing Tau hyper-phosphorylation. Additionally, naringenin improved brain insulin signaling as well as peroxisome proliferator-activated receptor gamma (PPAR-γ) [52]. Activation of PPAR-γ (belonging to the nuclear receptor superfamily) participates in insulin sensitization and increases metabolism [53].

An in vitro study in PC12 cells has shown that naringenin plays a vital role in the attenuation of apoptosis and neurotoxicity in Aβ-stimulated AD. The intracellular mechanisms responsible for anti-apoptotic activity as well as neuroprotective effects of naringenin is connected with the inhibition of caspase3, activation of PI3K/AKT, and modulation of GSK-3β signaling pathways [54].

In an Aβ-induced mouse model of AD, oral administration of naringenin resulted in the amelioration of memory deficit. Furthermore, naringenin effectively rescued the cells from apoptosis and inhibited lipid peroxidation (LPO) by decreasing hippocampal malondialdehyde (MDA) content. Besides, no significant effect was observed in behavior tasks in the co-administration of naringenin and fulvestrant as an estrogen receptor (ER) antagonist vs. naringenin alone. Therefore, neuroprotective role of naringenin may only be due to its estrogen-like activity and via the ER pathway [55].

Administration of Drynaria rhizome extract in mice alleviated memory impairment and showed considerable neuroprotective effects. Naringenin, naringenin-7-*O*-glucuronide, and naringenin-4′-*O*-glucuronide were shown to be responsible for the neuroprotective effects of Drynaria rhizome extract. However, while the three mentioned compounds could cross the BBB, just naringenin and naringenin-7-*O*-glucuronide significantly recovered Aβ-stimulated axonal atrophy in cultured cortical neurons and decreased collapsin response mediator protein-2 (CRMP2) hyperphosphorylation [56]. CRMP2 is a CNS protein that is involved in axonal and neuronal growth. The hyperphosphorylated CRMP2, which is implicated in neurofibrillary tangles, plays a pivotal role in the pathogenesis of AD-related disease. The C-terminal tail and H19 helix of CRMP-2 are more susceptible to phosphorylation by kinase enzymes such as GSK3β and cyclin depended kinase5 (CDK5) [57,58]. Therefore, these regions have been considered as an exciting drug target for AD. According to a molecular docking study, naringenin-7-*O*-glucuronide revealed the tighter binding, more affinity, and stability vs. naringenin in the active site of C-terminal tail of CRMP-2. Therefore, the presence of the glucuronyl group may play a key role in the neuroprotective effects of naringenin. Residues of Thr509, Thr514, Ser518, and Ser522 are phosphorylation sites located in the C-terminal tail of CRMP-2. Before the binding of naringenin and naringenin-7-*O*-glucuronide to the C-terminal tail, these residues have more flexibility, activity, and are more susceptible to phosphorylation. But these effects reversed after binding of naringenin and naringenin-7-*O*-glucuronide [59]. On this regard, another molecular docking study demonstrated that the occupation of CRMP2 by naringenin restricts the access of the phosphorylated enzyme to the terminal H19 helix especially in Tyr479 residue which is the susceptible residue for phosphorylation [60].

From another mechanistic point of view, acetylcholinesterase (AChE) has a crucial role in the regulation of the cholinergic system by hydrolyzing acetylcholine. An increase in AChE activity caused NDs associated with cholinergic impairment as observed in AD [61]. Methanolic extract of *Citrus junos* remarkably ameliorated AChE activity in cellular models. Naringenin is the major phytobioactive isolated from *C. junos* extract. Treating with naringenin reversed scopolamine, a muscarinic antagonist, stimulated amnesia in mice. Besides, a significant increase in spontaneous alteration behavior was observed in the mice pretreated with naringenin. The authors speculated that AChE inhibitory activity of the naringenin is responsible for promising effects of memory recovery [62].

In addition to AChE, amyloid precursor protein (APP) cleaving enzyme 1 (BACE1), and butyrylcholinesterase (BChE) are considered as potential key enzymes in the pathogenesis of AD [63,64]. As Lee et al. reported, naringenin exerted its anti-AD effects through a decrease in the activity of BACE1, AChE, and BChE in vitro [65].

In this line, Orhan et al., reported that 8-prenyl naringenin, a naringenin derivative, exhibited inhibitory effect against BChE with IC50 value of 86.58 ± 3.74 µM. According to their docking studies, 8-prenyl naringenin-BChE complex demonstrated a negative binding energy of −8.86 kcal/mol. In case of 8-prenylnaringenin, residues of BChE, including Ser198, Gly117, and His438 were responsible for forming hydrogen bonds as well as π-π stacking interaction [66].

Altogether, these findings suggest that naringenin may be considered as a multi-target and promising neuroprotective compound toward alleviating neurodegeneration observed in AD through different mechanisms.

3.2. Naringenin and Parkinson's Disease

PD is a progressive neurodegenerative disorder recognized by the depletion of dopaminergic neurons in the substantia nigra pars compacta (SNpc), triggering broad loss of dopamine in the striatum. PD is characterized by activated microglia and intracytoplasmic eosinophilic proteinaceous inclusions known as Lewy bodies which are made up of alpha-synuclein self-aggregation in substantia nigra (SN) neurons [67]. Oxidative insults, dopamine depletion, and neuroinflammation play a critical role in the induction and progression of this type of NDs [68,69]. Due to the potential dopamine enhancing, potent antioxidant as well as anti-inflammatory effects of naringenin, it could be used as a beneficial agent in the treatment of PD [70]. Regarding the role of naringenin in PD, Lou and colleagues showed that administration of naringenin (70 mg/kg bwt, p.o.) in mice protected them against 6-hydroxydopamine (6-OHDA)-induced nigrostriatal dopaminergic neurodegeneration and oxidative damage through activation of the nuclear factor E2-related factor 2/antioxidant response element (Nrf2/ARE) and its downstream target genes including heme oxygenase-1 (HO-1), and glutathione cysteine ligase regulatory subunit. In addition, naringenin blocked apoptotic pathway through the inhibition of the phosphorylation of c-Jun *N* terminal kinase (JNK)/p38 and abrogating caspase 3 [71].

From another point of view, astroglia cells play a supportive role in neuronal survival by producing different types of neurotrophic growth factors [72]. Recently, Wang et al. demonstrated that naringenin (50 µM, five days) improved the release of brain-derived neurotrophic factor (BDNF) and glial cell line-derived neurotrophic factor (GDNF) as astroglial neurotrophic factors from dopaminergic neurons (in neuron-glia culture) through the up-regulation of Nrf2 [73].

Another study by Zbarskt et al. investigated the use of naringenin in the amelioration of unilateral 6-OHDA infusion in adult Sprague–Dawley male rats. Naringenin treatment protected the tyrosine hydroxylase (TH)-positive cells from damage and increased the level of dopamine and its metabolites [74].

Genetic alterations in α-synuclein, parkin, leucine-rich repeat kinase 2 (LRRK2), PTEN-induced putative kinase 1 (PINK1) and DJ-1 have also been correlated with the development of PD. The genetic modification of related genes leads to the accumulation of toxic substances and oxidative stress-inducing striatal dopaminergic terminals degeneration and formation of the aggregates containing ubiquitin and α-synuclein [75]. Angeline et al. investigated the ameliorative effects of naringenin (10 mg/kg/for 10 days) in rotenone-treated rats. Their study showed the ability of naringenin to improve the motor skills and body weight via the overexpression of protective proteins including parkin, PARK 7 protein (DJ1), TH and C terminus Hsp70 interacting protein (CHIP) and reduction of the level of caspase and ubiquitin [76].

Aggregation and deposition of α-synuclein, the incidence of oxidative stress, as well as neuroinflammation, has been linked to the dopaminergic neurons loss [77]. Naringenin (100 mg/kg bwt, p.o.) potentially ameliorated 1-methyl-4-phenyl-1,2,3,6-tetrahydropyridine (MPTP)-induced dopaminergic degeneration. This effect was attributed to a decrease in α-synuclein, a restoration of the reduced dopamine and its metabolites, 3,4-dihydroxyphenylacetic acid (DOPAC) and homovanillic acid (HVA), a decrease in proinflammatory cytokines (TNFα and IL1β), and an increase in superoxide dismutase (SOD) [78]. A recent mechanistic in vitro study in human neuroblastoma cells (SH-SY5Y) showed that naringenin attenuated 1-methyl-4-phenylpyridinium (MPP+)-induced dopaminergic degeneration via decreasing ROS level, abolishing α-synuclein aggregation, as well as reducing neuroinflammation by decreasing the level of TNF-α and nuclear factor-κB (NF-κB). Moreover, naringenin reduced the level of apoptosis-induced by MPP^+ through the downregulation of Bax and overexpressed Bcl-2. Additionally, naringenin pretreatment increased the transporter for dopamine and the rate-limiting enzyme in dopamine synthesis TH [79]. Besides, naringenin interestingly abrogated MPTP-induced PD by inhibiting LPO and incrementing of catalase (CAT) and glutathione reductase (GR) and also improve locomotion efficiency. Additionally, pretreatment with naringenin in C57BL/6J mice down-expressed inducible nitric oxide synthase (iNOS) and alleviated cytoplasmic vacuolation and nuclear pigmentation in the mid-brain region [80].

Considering the attained results, naringenin could be a hopeful agent to combat PD through the modulation of several pathological pathways as well as the activation of protecting processes.

3.3. Naringenin and Neuroinflammation

Neuroinflammation is defined as an extreme dysregulation of the inflammatory response in the central nervous system (CNS) [81]. Neuroinflammation is mediated by the activation of microglia cells in CNS [82,83]. Overactivation of microglia mediated-infection or injury caused by innate immunity and release of various pro-inflammatory cytokines contribute to neuronal cell damage as well as the progression of NDs [84]. Naringenin targets several inflammatory signals involved in the neuroinflammation, including mitogen-activated protein kinase (MAPK), suppressor of cytokine signaling 3 (SOCS-3), NF-κB and signal transducer and activator of transcription-1 (STAT-1) [12,85–87]. As NFκB is known as an essential transcription factor that regulates various pro-inflammatory gene expression, naringenin induced a significant decrease in the phosphorylation and nuclear translocation of P65, a subunit of NFκB [87]. In other words, naringenin prevented the phosphorylation and degradation of IκB-α and decreased the expression of NFκB, confirming the potential anti-neuroinflammatory activity of naringenin through the suppression of NFκB [88]. From another point of view, naringenin significantly decreased neuroinflammation by reducing the phosphorylated JNK, extracellular-signal-regulated kinase (ERK1/2) [12], p38MAPK, Akt [12], and STAT-1 [87], pathways that play a principle role in the pathogenesis of neuroinflammation.

In the other study, naringenin ameliorated the levels of NO and prostaglandin E2, which indicated the inhibition of inflammation-associated enzymes, iNOS and cyclooxygenase-2 (COX-2), respectively [86], in a concentration-dependent manner, thereby decreasing neuroinflammation triggered by lipopolysaccharide (LPS) in BV-2 microglia cells. Besides, significant upregulation of the SOCS3 was observed in the cells treated with naringenin compared to small interfering RNA which was used as downregulated SOCS3 [89,90]. Administration of naringenin significantly increased the phosphorylation of AMP-activated protein kinase α (AMPKα) and protein kinase C δ (PKCδ) at Thr172 and Thr505 respectively [91,92]. Besides, in vivo results revealed that naringenin significantly recovered motor impairment and normalized morphology of activated microglia cells. Taken together, coordination between SOCS3 and AMPKα/PKCδ is of great importance in the molecular mechanism of naringenin [85].

As reported by Santa-Cecília et al., naringenin decreased IL-1β and TNF-α, as pro-inflammatory cytokines, and also reduced the expression of chemokine monocyte chemoattractant protein-1 (MCP-1) just at the gene transcription level [87].

Hence, naringenin combats against neuroinflammatory mediators and pathways, being a promising nutraceutical agent for the treatment of inflammatory related disorders.

3.4. Naringenin and Multiple Sclerosis

MS is an inflammatory and autoimmune disease in the CNS, which is characterized by progressive loss of neuronal myelin and disruption in neuronal function [93,94]. The first stage of MS progression is a self-reaction of T-cells against myelin antigens, which cause axonal demyelination, subsequently leading to clinical disability [85]. An imbalance between Th1 and Th2 secretes various pro-inflammatory and anti-inflammatory cytokines, respectively, accompanied by free radical production, which plays a critical role in the pathogenesis of MS [95,96].

As a multi-target agent, naringenin ameliorates the clinical manifestation of MS due to its ability to alleviate inflammatory signaling pathways, to increment antioxidant performance and to regulate the autoimmune response. Wang et al. revealed that naringenin diminished the infiltration of immune cells (T-cells, B-cells, neutrophils, macrophages), which is implicated in the CNS demyelination. Naringenin also downregulated CD4(+) T-cell differentiation Th1, Th9, Th17, as triggers of inflammatory processes, and reversed regulatory T cell (Treg) depletion, as modulator of inflammatory response, in the mice immunized with myelin oligodendrocyte glycoprotein (MOG)35–55 peptide. Naringenin blocked Th1, Th9, Th17 related transcription factors, including T-bet, PU.1, and RORγt, respectively. It also decreased the level of pro-inflammatory cytokines such as TNF-α and IL-1 [97].

Another mechanistic study by Wang et al. revealed that naringenin remarkably decreased CD4$^+$ T-cell proliferation vs. CD8$^+$ T-cell. Besides, Th1 cell response (IFNγ production) was more influenced by naringenin compared to Th2 cell response (IL4 production). Furthermore, naringenin could abrogate the expression of STAT3, which is the specific upstream signaling transducer RORγt and also inhibited STAT1 as well as STAT4 which are considered as corresponding signal transducers t-bet [98]. Besides, naringenin blocked IL-6-stimulated inhibition of Treg. There is also a direct link between secretion of TGF-β and generation of Treg cells as well as inhibitory effect of Treg cells on autoimmune responses [99]. TGF-β itself augmented the expression of intracellular protein Smad2/3, leading to the upregulation of T-reg transcription factor forkhead box P3 (Foxp3) [100]. Contradictory, slight mitigation of Smad2/3 expression as well as diminished Foxp3 levels were observed in cells treated with naringenin. Therefore, naringenin can regulate the development of CD4$^+$ T cell lineages by impacting their respective modulatory signals [101].

Niu et al. explored the immunosuppressive effect of naringenin and its possible mechanisms on mouse T cells. Treatment with naringenin significantly blocked the proliferation of mouse T cells either induced by anti-CD3/CD28 or autoantigen MOG35-55. Besides, naringenin significantly decreased Th1 and Th17 cells-mediated cytokines (IFN-γ and IL-17A respectively) as well as pro-inflammatory cytokines, including TNF-α, and IL-6. Naringenin not only arrested T cells at G0/G1 phase in a dose depending manner but also enhanced the level of P27 which is an inhibitor of CDK [102]. Increasing the proportion of CDK4/6 is necessary for the phosphorylation and inactivation of retinoblastoma protein, which is considered as a negative regulator of cell cycle, subsequently leading to promote cell cycling from G0/G1 to S phase transition. Naringenin suppressed the phosphorylation of retinoblastoma protein, thereby modulating the immune system in MS [103]. Treatment with naringenin also decreased the level of IL-2, IL-2R α-chain (CD25) and also inhibited STAT5 phosphorylation in IL-2-restimulated activated T cells [104].

In addition, the overexpression of adhesive molecules in vascular endothelial of CNS and their interaction with related ligands on the surface of T cells, permit the T cells to attach the endothelium cells and to develop experimental autoimmune encephalomyelitis (EAE) [105]. Naringenin effectively attenuated vascular cell adhesion molecule-1 (VCAM-1) as well as its related ligand VLA4 and also downregulated CXCL10 [97,106].

From another mechanistic point of view, as the epidermal growth factor receptor ErbB4 plays a decisive role in the proliferation and survival of oligodendrocytes [107], its upregulation is involved in the remyelination and development of nerve cells [108]. In this line, the tyrosine kinase domain of ErbB4 and extracellular region of ErbB4 receptor was selected as the target receptor for molecular docking study. Joshi et al., reported naringenin as a promising candidates for activating ErbB4 receptor [109].

Therefore, the results asserted that naringenin might be a promising agent with a prosperous future in the treatment of MS, through affecting multiple signaling pathways.

3.5. Naringenin and Cognitive Deficit

Cognitive decline is considered as one of the most important hallmarks of various neurological diseases. Oxidative stress, inflammation, disturbance in insulin signaling, and mitochondrial dysfunction play a crucial role in the progression of cognitive deficit and dementia [52,110].

Intracerebroventricular (ICV) injection of streptozocin (STZ) causes the brain glucose depletion, and the impairment of energy metabolism, accompanied by mitigation of cholinergic neurotransmitter [111], could be related to diabetes mellitus-associated dementia [112]. Naringenin displays antioxidant, antihyperglycemic, as well as cholinergic effects, and seems to be a promising agent in the treatment of diabetes-related memory dysfunction [113,114]. According to Rahigude et al. reports, chronic administration of naringenin (50 mg/kg bwt/day, p.o.) for 58 days, significantly improved memory dysfunction, decreased hyperglycemia, mitigated oxidative stress as well as AChE activity in rats with type-2 diabetes [115]. In another study, naringenin potentially improved learning and memory dysfunction, as well as cognitive deficits caused by ICV-STZ in rats [116].

In addition, Zaki and colleagues illustrated the role of AChE activity and oxidative stress as two major causes of dementia [117]. Animals receiving naringenin revealed a decline in 4-hydroxynonenal (4-HNE), MDA, thiobarbituric acid-reactive substances (TBARS) [118], hydrogen peroxide (H_2O_2), protein carbonyls and an increase in glutathione (GSH) [118], glutathione peroxidase (GPx), GR, GST, SOD, CAT and Na^+/K^+ ATPase activity as well as upregulation of choline acetyltransferase [119,120]. Hippocampal levels of Nrf2, SOD, CAT, and GSH significantly augmented in a group treated with naringenin at a dose of 100 mg/kg. In contrast, the hippocampal level of MDA and AChE activity were diminished [110]. Naringenin also ameliorated the brain level of 5HT [120] and norepinephrine (NE); however, no effects were observed on the levels of dopamine and GABA content [118]. Therefore, the neuroprotective effect of naringenin against scopolamine-induced dementia maybe associated with its auspicious cholinergic and antioxidant activity as well as change in brain neurotransmitter content.

Neuroapoptosis and neuroinflammation play a key role in the induction and progression of memory and cognitive deficit, as isoflurane (an anesthetic) does [121]. Pre-treatment with naringenin (100 mg/kg bwt/day, p.o.) significantly attenuated neuroapoptosis and cognitive dysfunction in rats, through the downregulation of Bad, caspase-3, Bax and up-expression of Bcl-2, Bcl-xL, a decline in TUNEL and the activation of PI3K/Akt/GSK-3β pathway [122]. Besides, naringenin significantly down expressed phosphatase and tensin homolog (PTEN), as a negative modulator of PI3K/Akt/GSK-3β signaling pathway, in a dose depending manner [123].

To combat against neuroinflammation in cognitive dysfunctions, naringenin decreased the NF-κB-related inflammatory cascades by decreasing TNF-α, IL-6, IL-1β. Naringenin pretreatment also prevented cognitive deficit stimulated by isoflurane through the suppression of the NF-κB signaling pathway [123]. As reported by Khajevand-Khazaei et al., naringenin alleviated LPS-induced neuroinflammation through the downregulation of iNOS, TNF-α, COX2, NF-κB, and TLR4 [110]. TLR4 is a main upstream signaling protein that exerts a pivotal role in activation of NF-κB and subsequently upregulation of NF-κB signaling cascade including inflammatory cytokines and/or mediators (TNF-α, COX2, iNOS) [124]. From another point of view, due to the crucial role of glial fibrillary acidic protein (GFAP) in neuroinflammation, and presence of a direct link between overexpression of GFAP and hyperactivity of astrocytes [125], attenuating GFAP is of great importance. The abatement of GFAP by naringenin represents its alleviation effect on astrogliosis [110].

Recently, Sarubbo and colleagues demonstrated that naringenin (20 mg/kg bwt/day, i.p.) augmented the age-induced monoaminergic neurotransmitters depletion (noradrenaline, serotonin, and dopamine), decreased NF-κB levels, and increased tryptophan hydroxylase and SIRT1 (belonging to the histone deacetylase class III family) levels in the hippocampus. Moreover, SIRT1 contributed to a decline in inflammatory responses by suppressing NF-κB signaling pathway, thereby improving brain plasticity and memory [126,127]. It has also been reported that treatment with naringenin (50 mg/kg bwt/day, i.p.) exhibited a positive effect on cognitive function in young male rats [120].

In other studies, the ability of methyl mercury (MeHg) has been reported in the induction of severe mitochondrial dysfunction, oxidative insult, and neuronal damage, which ultimately coupled with memory and cognitive deficit [128–130]. Pretreatment with naringenin reversed MeHg-induced pyramidal cell damage, mitigated oxidative stress condition (confirmed by a significant increase in GSH, and GST accompanied with a marked decrease in MDA and protein carbonyl level) and ameliorated mitochondrial dysfunction (as indicated by preservation activities of mitochondrial complex I–IV and abrogation of lesions /10kb) [131].

Overall, the findings indicated that naringenin combats oxidative stress, neuroinflammation, and neuroapoptosis to overcome cognitive dysfunction thereby demonstrating the therapeutic potential.

3.6. Naringenin and Neurotoxicity

Excessive exposure to the neurotoxicant agents including neurotransmitter glutamate, metals (iron, sodium tungstate, aluminum) as well as, hypoxia participates in brain and spinal injury as well as widespread neurobehavioral changes [132–135]. Hypobaric hypoxia is involved in neuronal loss and behavioral dysfunction by inducing excessive formation of ROS. Hypoxia plays a causative role in triggering apoptotic pathway and upregulation of hypoxia-inducible factor 1α (HIF 1α) and its main target protein, vascular endothelial growth factor (VEGF) [123]. Enhanced free radicals and depleted antioxidants levels occur following neurotoxicity and is subsequently linked to the development of NDs [136]. Due to free radical scavenging and anti-inflammatory properties, naringenin is an attractive candidate for the management of neurotoxicity. Pre-administration of naringenin attenuated behavioral impairment in hypoxia exposed mice. Besides, naringenin alleviated hypoxia-induced oxidative stress and apoptosis by down-expression of HIF1α, VEGF as well as blocking caspase-3 and ubiquitin. Moreover, naringenin elevated the levels of parkin and chip, as ubiquitin E3 ligases or accumulated proteins in NDs [137]. In addition, naringenin improved survival of mouse neuroblastoma cells following carbaryl-induced toxicity, reduced oxidative stress (by mitigating ROS), and suppressed apoptosis (by inhibiting Bax, caspase-3 and upregulating Bcl-2). It also restored a depletion of mitochondrial membrane potential induced by carbaryl [138].

Moreover, naringenin exhibited neuroprotective effects against iron-induced neurotoxicity and anxiety-like behavioral deficit by attenuation of ROS formation, enhancement of endogenous antioxidant capacity (GSH, CAT, SOD), upregulation of AChE activity and augmentation of ectonucleotidase enzymes (such as adenosine triphosphate diphosphohydrolase and 50-nucleotide enzyme) which are responsible for regulation of extracellular ATP and adenosine concentrations in the synaptic cleft. Naringenin significantly recovered the decreased level of mitochondrial complex I–V enzymes activities and mitochondrial membrane potential, thereby reducing neurotoxicity [139].

In another similar study, co-administration of naringenin and iron potentially abolished oxidative insult markers, including MDA, NO, ROS formation, and protein carbonyls content. Naringenin also significantly enhanced enzymatic/non-enzymatic antioxidant activities, accompanied by an increase in AchE as well as Na^+/K^+ ATPase activity in the cerebral cortex of iron-exposed rats. In addition, co-treatment of naringenin and iron ameliorated apoptosis damage corroborated by decrease in DNA fragmentation induced by endogenous endonucleases. Moreover, naringenin ameliorated the morphological changes (necrotic cells accompanied with pyknotic nucleus and vacuolated spaces) induced by iron [140].

As reported by Sachdeva and colleagues, co-administration of naringenin and N-acetylcysteine (NAC) for three months significantly restored biogenic amines (dopamine, 5-HT, and NE) in a rat model of sodium tungstate-induced neurological alterations. In their study, co-administration of naringenin and NAC also reduced the oxidative stress through increasing GSH and suppressing TBARS in the brain [141].

The neuroprotective effects of naringenin against glutamate and related down-stream molecular mechanisms were examined. Naringenin potentially inhibited excitotoxicity and abolished apoptosis by reducing caspase 3 and calpain1 ($Ca2^+$-dependent protease), known as caspase-independent pathway [142,143]. Phosphorylated Akt and ERK were significantly upregulated in primary culture of mouse hippocampal neurons treated with naringenin, thereby favoring neuronal survival and showing a vital role in the regulation of apoptosis [144].

From another mechanistic view, brain fatty acid-binding protein 7 (FABP7) plays a critical role in the remyelination process [145]. Naringenin provided a neuroprotective activity against oseltamivir-induced neurotoxicity via improvement of neurophysiological factors (increase in total antioxidant capacity, Ca ATPase, and FABP7) and suppression of oxidative stress through inhibiting total oxidant capacity, total nitrite oxide, as well as cytochrome P450 activation [146].

It could be concluded that naringenin can be considered as a good alternative to chemical medicines against neurotoxicants agents.

3.7. Naringenin and Other Neurodegenerative Diseases

Reports have revealed an auspicious role for naringenin to improve other NDs, including ischemic stroke, brain injury, HD, ALS, and SCI.

Ischemic stroke is defined as an obstruction of arteries in particular middle cerebral artery, which is responsible for supplying blood to the brain. Immediately after an ischemic stroke attack, decompensated neuronal death and damage occur [147]. Several factors including oxidative stress, apoptosis, edema-mediated by ionic imbalance as well as inflammation are implicated in the pathogenesis of ischemic stroke [148–150]. Naringenin, with neuroprotective properties, is an excellent therapeutic candidate for ischemic brain injury [151]. Bai et al. revealed that pre-treatment with naringenin (100 mg/kg), significantly alleviated neurological impairment, decreased the infarct size and attenuated brain water content in a rat model of permanent middle cerebral artery occlusion (pMCAO)-induced cerebral ischemia. In their study, naringenin remarkably abolished the expression of nucleotide-binding oligomerization domain (NOD2), receptor interacting protein-2 (RIP2), NF-κB, and matrix metallopeptidase 9 (MMP-9). Whereas the expression of claudin-5, protein responsible for tight junction, was elevated [152]. Belonging to the cytosolic NOD-like receptor family, NOD2 is implicated in cerebral ischemic injury by upregulation of downstream proteins, including RIP2 and NF-κB, which are considered as pivotal contributors to the initiation of the pro-inflammatory cascade [153].

A further mechanistic study by Raza et al. demonstrated that naringenin enhanced the depleted antioxidant markers, SOD and GSH. NF-κB-mediated pro-inflammatory factors including COX-2, iNOS, IL-1β as well as TNF-α were significantly alleviated in the naringenin pre-supplementation group [154]. The expression of GFAP and microglia (Iba-1) were noticeably mitigated in naringenin pre-treated group as compared to MCAO group [155]. There is a close link between enhancing expression of glial fibrillary acidic protein (GFAP) and cerebral ischemia [154].

In another study, a novel synthesized co-drug combination of naringenin and lipoic acid, named "VANL-100" was evaluated in both in vitro and in vivo models of oxidative stress. VANL-100 resulted in a significant neuroprotection against hypoxia (100-fold more potent than naringenin alone). This effect might be due to a substantial increase in intracellular antioxidant capacity by VANL-100. In an in vivo model of ischemia-reperfusion injury following transient occlusion of the middle cerebral artery (tMCAO), pre-administration of either naringenin or conjugated formulation significantly abrogated infarct size. The potency of the novel formulation was 10,000-fold compared to naringenin alone [156].

Due to the importance of oxidative stress in the occurrence of brain injury, combating oxidative mediators are very important. Nrf2, which is a prominent modulator of the oxidative stress system, plays a pivotal role in the preservation of the mitochondrial function through elevated transcription of ARE-dependent antioxidant genes [157,158]. Mitochondria dysfunction causes deadly results for neurons due to the strong dependence of neuronal survival and function on produced energy by mitochondria [159]. Discovering new antioxidant agents that regulate the Nrf2/ARE pathway may be an innovative strategy for combating oxidative stress-associated disease. Wang et al. showed that hypoxia/re-oxygenation cells treated with a high dose of naringenin (80 μM) suppressed oxidative stress by attenuating ROS generation and MDA content, as well as increasing endogenous enzymatic/non-enzymatic antioxidants such as SOD and GSH activities. Besides, naringenin treatment effectively upregulated Nrf2 downstream genes including HO-1 as well as NAD(P)H Quinone dehydrogenase1(NQO1) [160]. These findings revealed that the Nrf2/ARE pathway displayed protective effects against neuronal dysfunction.

Additionally, naringenin blocked apoptosis by increasing cell viability, accompanying with downexpression of cleaved caspase-3 protein as well as Bax protein and upregulation of Bcl-2 protein. Naringenin exerted protective effects against mitochondrial dysfunction, a primary endogenous ROS-rich source, via increasing the levels of adenylates (AMP, ADP, ATP), adenine nucleotide transporter (ANT), as a exchanger of ADP/ATP in the mitochondrial membrane, and also improvement of mitochondrial membrane potential (Δψm) [161].

A mechanistic study by Wang et al. evaluated the neuroprotective effects of naringenin in vitro on primary cultured rat cortical neurons exposed to oxygen-glucose deprivation/reoxygenation (OGD/R) and also in vivo in rats subjected to MCAO and reperfusion (MCAO/R) injury. They reported that naringenin diminished the apoptosis-related protein Bcl2, and apoptosis-related gene including Kelch-like ECH-associated protein 1, HO-1, and NQO1. It significantly reversed the translocation of Nrf2 from cytoplasm to nucleus in the primary culture of rat hippocampal neurons exposed to OGD/R. The overexpression of Nrf2 induced cell proliferation and abrogated apoptosis [162]. Naringenin and the overexpression of Nrf2 remarkably recovered OGD/R-induced mitochondrial dysfunction by increasing mitochondrial membrane potential. Besides, MCAO/R rats treated with naringenin demonstrated lower brain water content and also attenuation of the neurological impairment. Additionally, naringenin inhibited neuronal apoptosis in vivo by alleviating the number of TUNEL-positive cells [163].

As another neurodegenerative disorder with neuropathy complications, diabetes-mediated retinopathy is characterized by vascular damage accompanied by neuronal degeneration, which caused visual dimming and blindness [164]. Overproduction of free radicals has a central role in retinal cell injury [165]. Naringenin with antioxidant, antidiabetic, and anti-apoptosis effects, is a beneficial therapeutic option for combating diabetes mediated-retinal neurodegeneration [166]. Naringenin supplementation decreased the level of TBARS, as a marker of oxidative insult to lipid, protein and DNA peroxidation [167] and improved GSH level in the STZ-induced diabetic rats. Ameliorated apoptosis in the retina of diabetic rats treated with naringenin is suggested to be mediated by suppression of caspase-3, pro-apoptotic Bax as well as overregulation of anti-apoptotic Bcl-2 protein. Naringenin also restored the diminished level of BDNF, a class of neurotrophin protein, affected nerve survival and proliferation following binding with its selective receptor (tropomyosin-related kinase B (TrkB)) [168], and also amplified synaptophysin which is necessary for releasing neurotransmitter and synaptic integrity [169]. Downregulation of TrkB exacerbated apoptosis and developed neuronal injury [170]. Based on these results, it can be concluded that naringenin might be a promising candidate for prevention of diabetic retinopathy.

As other NDs, polyglutamine (polyQ) diseases are defined as an unfolding or a misfolded in the particular proteins, then leading to the aggregation of abnormal proteins in neuronal cells [171]. The endoplasmic reticulum signaling pathway has a critical role in the induction of polyQ-related disease [172]. Several studies have reported that an enhanced level of glucose-regulated protein (GRP78), a chaperon located in the endoplasmic reticulum, causes alteration in apoptosis and polyQ-related disease, such as HD [173]. HD is characterized by the accumulation of expanded polyQ tract (consists of >40 repeat of glutamine unit in huntingtin proteins) [174]. *Glycyrrhizae radix* elevated the activity of GRP78 promoter in vitro. Naringenin is the main active ingredient present in *G. radix* and disrupted the aggregation of enhanced green fluorescent protein with a pathological-length polyQ tract (EGFP-polyQ97) by overexpression of GRP78 [174,175]. Therefore, naringenin can be effective in the prevention of neurodegenerative disorders with polyQ expansion.

As a fatal neurodegenerative condition, ALS is characterized by paralysis and death within three to five years from the day of appearance of symptoms due to the impairment of respiratory systems [176,177]. The loss of bulbar and limb function is the main feature of ALS. A recent statistics study showed that around 10% of patients inherited the disease, while 90% of patients have no family history of ALS [178]. The investigation of the genetic alterations in ALS patients revealed that 20% of the familial ALS (FALS) is due to the mutations in cytosolic Cu/Zn-SOD leading to the accumulation of free radicals. The accumulation of free radicals was found to damage the homeostasis of mitochondria, the transport of axons and dysfunction in glutamate transporter [179]. Considering the critical role of naringenin in combating oxidative stress and glutamate-induce neurotoxicity, naringenin could be a promising agent in ALS patients to improve their survival rate.

As another neurodegenerative condition, SCI is a devastating and debilitating condition that resulted in neurologic deterioration accompanied by sensory-motor deficit [143]. Inflammation, apoptosis, and oxidative stress contribute to the secondary damage of SCI [180]. Pharmacological interventions that mitigate the secondary phase process may provide the therapeutic efficacy for SCI [181]. Shi et al. indicated that the administration of naringenin ameliorated SCI by down-regulating mir-223, and decreasing tissue myeloperoxidase, as a marker of neutrophil infiltration, as well as abrogating inflammatory cytokines [182].

Altogether, naringenin has shown a bright future in the treatment of NDs. Figure 3 demonstrates the molecular mechanisms underlying the neuroprotective effect of naringenin for combating NDs (Figure 3).

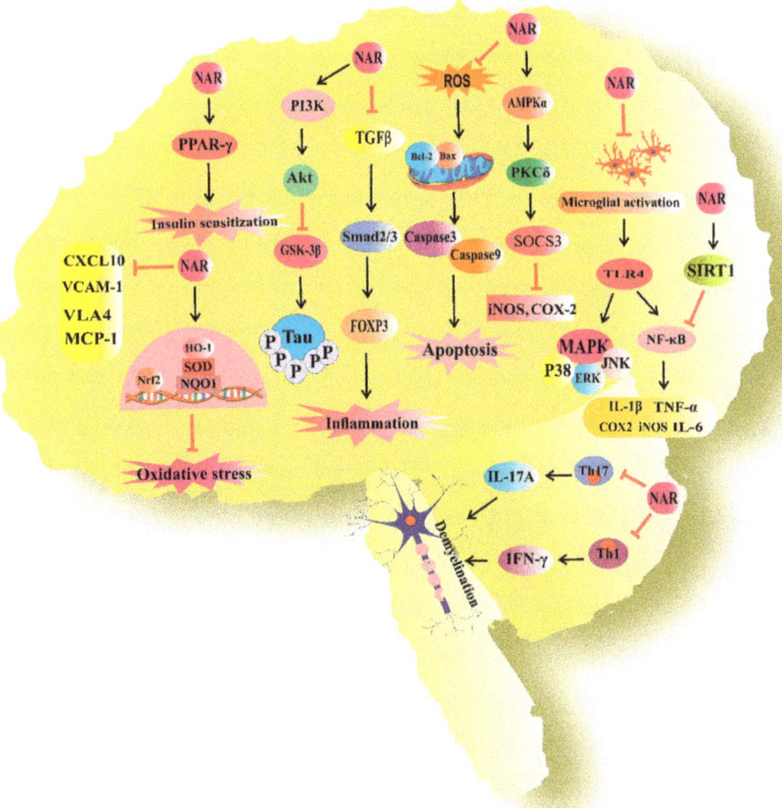

Figure 3. Neuroprotective mechanisms of naringenin. NAR, naringenin; PPAR-γ, peroxisome proliferator-activated receptor gamma; CXCL10, C-X-C motif chemokine ligand 10; VLA4, very late antigen 4; VCAM-1, vascular cell adhesion molecule-1; MCP-1, monocyte chemoattractant protein-1; Nrf2, nuclear factor E2-related factor 2; HO-1, hemoxygenase-1; SOD, superoxide dismutase; NQO-1, NAD(P)H quinone dehydrogenase1; PI3K/AKT, phosphoinositide 3-kinase/AKT; GSK3-β, glycogen synthase kinase3-β; TGF-β, transforming growth factor-β; Foxp3, forkhead box P3; Treg, regulatory T cell; Th1, T helper1; iNOS, inducible nitric oxide synthase; COX-2, cyclooxygenase-2; SOCS-3, suppressor of cytokine signaling 3; AMPKα, (AMP)-activated protein kinase α; PKCδ, protein kinase Cδ; JNK, c-Jun *N* terminal kinase; ERK, extracellular-signal-regulated kinase; MAPK, mitogen-activated protein kinase; TNF-α, tumor necrosis factor α; INF-γ, interferon γ; IL-1β, interleukin 1β; NF-κB, nuclear factor-κB; TLR4, toll-like receptor 4; ROS, reactive oxygen species.

4. Nanostructured Formulations of Naringenin for Management of Neurodegenerative Diseases.

Naringenin is a well-known antioxidant and anti-inflammatory flavanone with wide ranges of beneficial effects in the management of NDs but there are several challenges for the accessibility and accumulation of naringenin in the brain. Poor ability of naringenin to cross biological membranes, its poor water solubility, and high first-pass effect, as well as related gastrointestinal degradation, cause low bioavailability and limit its clinical applications [183].

One of the promising ways to overcome these challenges is to involve nanostructured formulations of naringenin. Several nanoformulations have been developed for naringenin to increase its bioavailability, solubility, and decrease drug degradation. The most investigated nanoformulations are based on biodegradable and biocompatible polymers that encapsulate naringenin in nanostructures including solid lipid nanoparticles, nanocapsules, nanomicelles, nanoliposomes, and nanoemulsions [14]. Md et al. investigated the potential neuroprotective effect of naringenin-based nanoemulsion against Aβ-induced AD. SHSY5Y cells pretreated with either naringenin or nanoemulsions showed attenuated phosphorylation of tau, decreased β amyloid-stimulated ROS production and abolished BACE [184]. BACE is an enzyme that breaks the APP and ultimately leads to aggregation of Aβ and induction of amyloidogenic pathway [185]. While pretreatment with naringenin was not able to hinder the production of Aβ, pretreatment with naringenin-nanoemulsion alleviated the level of tau [184], highlighting the importance of using nanoformulations in drug delivery systems.

To achieve an efficient management of PD in animal models, intranasal delivery of vitamin E-loaded naringenin nanoemulsion was successfully used. Intranasal administration of nanoemulsion of naringenin accompanied by levodopa played an essential role in the amelioration of behavioral parameters, increment of GSH and SOD, as well as decrement of MDA level in 6-OHDA-induced PD in rats. Synergism between the antioxidant effects of naringenin and vitamin E, lipophilicity, nanoscale size of nanoemulsion as well as intranasal administration of nanoemulsion which provided high concentration of naringenin in brain through olfactory pathway caused better efficacy of nanoemulsion compared to naringenin alone [186].

In another study, treatment of human mesenchymal stem cells with naringenin encapsulated into gelatin-coated polycaprolactone nanoparticle conferred neuroprotective effects on OGD-stimulated ischemic stroke [187]. The nanoformulation showed a sustained release manner and contributed to blood-brain barrier integrity as well as cellular morphology improvement. The nanoparticles significantly abrogated NF-κB activation as a trigger of inflammatory cascade. Besides, remarkable decrease in pro-inflammatory factors such as TNF-α, IFN-γ, and IL-1β along with a reduction in inducible enzymes, including COX2 and iNOS was observed in cells treated with the nanoparticle vs. OGD group [188]. Thus, the nano-formulation may be a suitable candidate to be evaluated in an animal model of ischemic stroke.

Overall, several findings suggested that nanostructured formulations of naringenin were able to overcome limitations affecting the beneficial effect of naringenin in NDs. All of the current data focused on the safety and efficacy of naringenin loaded nanoparticles in vitro and/or in vivo models of NDs. Future preclinical and clinical trials for mentioned nanostructured formulations of naringenin should be carried out to corroborate the therapeutic effect and promote health in the patients with NDs.

5. Conclusions

In spite of exhaustive research, no effective treatment has been investigated for NDs. The complex pathophysiological mechanisms of neurodegeneration, along with the lack of safe and efficient therapies for NDs, raise the need to find novel multi-target treatments. Nowadays, with a priority in research, natural products, as well as their active compounds, have attained special attention in the treatment of NDs. Among natural products, naringenin is a multi-target flavonoid, possessing promising neuroprotective effects, through targeting multiple therapeutic targets and signaling pathways (Table 1). Although, numerous in vitro and in vivo studies have elucidated the ability of naringenin to alleviate several types of NDs, the lack of clinical trials regarding therapeutic potential of naringenin is a critical limitation. There exists a dire need to conduct well-designed clinical trials in patients with NDs. On the other hand, the poor bioavailability and slight brain accessibility of naringenin, has resulted in a lack of efficacy in clinical trials. To solve these drawbacks, nanostructured formulations of naringenin have also been developed, and naringenin based nanoparticles exhibited biocompatibility and biodegradability characteristics as well as high bioactivity. Further studies and engineered methods are necessary to provide surface modification of nanoformulations of naringenin to optimize drug delivery to CNS.

Table 1. Neuropharmacological mechanisms of naringenin against different type of neurodegenerative diseases.

Type of Diseases	Method	Model	Neuropharmacological Mechanisms and Outcome	References
Neuroinflammation	LPS Induced neuroinflammation	In vitro: BV2 microgelia cells	↓iNOS ↓COX-2 ↑SOCS3 ↑ AMPKα ↑ PKCδ	[85]
	Induced by LPS	In vitro: BV2 microgelia cells	↓JNK ↓ERK ↓p38 ↓MAPK ↓TNF-α ↓IL-1β ↑ Arg-1↑IL-10	[12]
	Induced by LPS	In vitro: BV2 microgelia cells	↓TNF-α ↓IL-6 ↓IL-1β ↓MCP-1 ↓NfκB ↓MAPK ↓Akt ↓iNOS ↓ COX-2	[86]
	Induced by LPS/ IFN-γ	In vitro: microglia	↓p38 MAPK ↓ERK1/2 ↓STAT1 ↓iNOS ↓ TNF-α	[87]
AD	Aβ25-35-induced AD	In vitro: PC12 cells	↓apoptosis, ↓caspase3, ↑ PI3K/AKT, ↑ER	[54]
	Aβ25-35-induced AD	In vivo: Wistar rats	↓MDA ↓apoptosis, ↑ER, ↑spatial memory and cognition	[55]
	ICV-STZ- induced AD	In vivo: Sprague–Dawley rats In vitro: cultured cortical neurons	↓Tau hyper-phosphorylation, ↑ PI3K/AKT ↓GSK3-β ↑ PPAR- γ ↑insulin signaling	[52]
	Induced by Aβ1-42 and Aβ25-35	In vivo: 5XFAD Mice	↑spatial learning and memory ↓amyloid plaques, ↓Tau hyper-phosphorylation	[56]
Amnesia	Induced by scopolamine	In vivo: ICR mice	↓AchE activity ↑spontaneous alteration behavior	[62]
Ischemic stroke	pMCAO- induced cerebral ischemic	In vivo: Sprague–Dawley rats	↓infarct size, ↓brain water content, ↓ NOD2, RIP2, NF-κB, MMP-9, ↑ claudin-5	[152]
	Induced by hypoxia	In vitro: neurons isolated from the brain of Sprague–Dawley rats	↓ROS, MDA, ↑SOD, GSH↓caspase-3, Bax, ↑ Bcl-2, ↑AMP, ADP, ATP, ANT↑ Nrf2, HO-1, NQO1	[160]
	MCAO/R-induced ischemic stroke	In vivo: Sprague–Dawley rats	↓brain water content, ↓TUNEL-positive cells	[163]
	Induced by MCAO/R	In vivo: Wistar rats	↓infarct size, neurological deficits, brain water, ↑ motor, and somatosensory function ↑SOD, GSH, MPO, TBARS, ↓COX-2, iNOS, ↓IL-1β, TNF-α ↓ NF-κB	[155]
Diabetic retinopathy	STZ-induced diabetic retinopathy	In vivo: Wistar albino rats	↓ TBARS, ↑GSH, ↓caspase-3, Bax, ↑Bcl-2 ↑ BDNF, TrkB, synaptophysin,	[169]
Polyglutamine diseases	-	In vitro: mouse C3H10T1/2 cells, COS-7 cells, and HeLa-tetQ97 Cells	↑GRP78	[175]
EAE	(MOG)35-55-induced EAE	In vivo: C57BL/6 mice	↓ Th1, Th9, Th17, ↑ Treg, ↓T-bet, PU.1, and RORγt,	[97]
	Induced by anti-CD3/CD28	In vivo: C57BL/6 mice	↓IFNγ, ↓STAT1, STAT3, STAT4, ↓IL-6, ↑gp-130, ↓Foxp3	[101]
	Induced by anti-CD3/CD28 and (MOG)35-55	In vitro: mouse T cells	↓T cells proliferation, ↓ IFN-γ, IL-17A ↓ TNF-α, IL-6, block T cells at G0/G1 phase ↑ P27, ↓ retinoblastoma protein phosphorylation, ↓IL-2, CD25 ↓ STAT5	[104]
PD	Induced by 6-OHDA	In vitro: Human neuroblastoma SH-SY5Y cells. In vivo: C57BL/6 mice	↑Nrf2/ARE ↑HO-1, ↓ROS ↑GSH ↓ JNK and p38	[71]
	MPTP-induced PD	In vivo: C57BL/6J mice	↓α-synuclein ↑dopamine transporter ↑DOPAC ↑HVA ↑TH ↓TNFα & IL1β ↑SOD	[78]
	Rotenone-induced PD	In vivo: Wistar rats	↓ubiquitin and caspase3 improvement of motor skills ↑parkin ↑CHIP ↑PARK 7 protein ↑TH	[76]
	MPTP-induced PD	In vivo: C57BL/6J mice	↑GRx & CAT ↓LPO& iNOS ↓ nuclear pigmentation and cytoplasmic vacuolation	[80]
	-	In vitro: primary rat midbrain neuron-glia co-cultures	↑ BDNF, GDNF↑ Nrf2 ↑Dopaminergic neurons survival	[73]
	6-OHDA-induced PD	In vivo: Sprague-Dawley rats	↑DOPAC, ↑HVA, ↑Dopamine↑TH	[74]
	Induced by MPP⁺	In vitro: Human neuroblastoma SH-SY5Y cells	↓ ROS ↓NF-κB ↓TNF-α ↓Bax ↑Bcl-2	[79]

Table 1. *Cont.*

Type of Diseases	Method	Model	Neuropharmacological Mechanisms and Outcome	References
Neurotoxicity	Induced by sodium tungstate	In vivo: Wistar rat	↑GSH ↓ROS ↓ TBARS ↑Dopamine	[141]
	Induced by glutamate	In vitro: primary culture of mouse hippocampal neurons	↑ Erk1/2 & Akt phosphorylation ↓calpain-1 & caspase-3	[144]
	Induced by hypobaric hypoxia	In vivo: Swiss albino mice	↓HIF1a ↓VEGF ↓caspase-3 ↓ ubiquitin ↑CHIP ↑ parkin	[137]
	iron-induced neurotoxicity	In vivo: Wistar rat	↑ SOD, CAT ↓ROS ↑ AChE ↓MDA ↑Na$^+$/K$^+$ ATPase	[139]
	Induced by oseltamivir	In vivo: Wistar rat	↑FABP7 ↑Ca ATPase, ↑TAC ↓TOC ↓TNO ↓ cytochrome P450	[146]
	Induced by iron	In vivo: Wistar rat	↓ROS ↑GSH, CAT, SOD ↑AchE ↑ectonucleotidase enzymes ↑mitochondrial complex I–V enzymes ↑ mitochondrial membrane potential	[140]
	Induced by carbaryl	In vitro: mouse neuroblastoma cells	↓ROS ↓Bax, caspase-3 ↑Bcl-2 ↑mitochondrial membrane potential	[138]
Cognitive deficit	Induced by ICV-STZ	In vivo: Wistar rats	↑ Learning and memory performance ↑learning & memory	[116]
	Induced by ICV-STZ	In vivo: Wistar rats	↓TBARS, MDA, 4-HNE, H2O2, protein carbonyl, ↑GSH, SOD, CAT ↑Na$^+$/K$^+$ ATPase activity	[119]
	Induced by scopolamine	In vivo: albino Wistar rats	↓AChE ↑GSH ↓TBARS ↓TNFα ↓5HT, NE ↑spontaneous alternation performance & conditioned avoidance response	[118]
	Induced by isoflurane	In vivo: Sprague–Dawley rats	↓Bad, caspase-3, Bax ↑ Bcl-2, Bcl-xL ↓TUNEL ↑ PI3K/Akt ↓PTEN ↓NF-κB, TNF-α, IL -6, IL-1β, Improvement of cognitive dysfunction	[123]
	Induced by LPS	In vivo: albino Wistar rats	↓TLR4, NF-κB, TNF-α, COX2 and iNOS ↑Nrf2, SOD, CAT, and GSH ↓MDA and AChE ↓GFAP ↑ spatial recognition memory, discrimination ratio & retention and recall capability	[110]
	Age-induced cognitive deficit	In vivo: Sprague–Dawley rats	↑ SIRT1 ↓ NF-κB ↑serotonin, noradrenaline, dopamine, TH	[126]
	Induced by MeHg	In vivo: Swiss Albino mice	↑ mitochondrial complex I- IV activities, ↓lesions /10kb ↑GSH, GST ↓MDA & protein carbonyl ↑spatial and recognition memory	[131]
	-	In vivo: young adult male Albino Wistar rats	↓AChE, ↑ 5HT	[120]
	Induced by type 2 diabetes mellitus	In vivo: Young Sprague–Dawley rats	↓AChE, ↓hyperglycemia ↑memory performance	[115]

A further area of research on the safety and efficacy of naringenin and its nanoformolations in human, as well as related novel pathological signaling pathways of NDs in the human, will show new avenues in the prevention, management, and treatment of several diseases.

Author Contributions: Conceptualization, M.H.F. and Z.N.; software, Z.N.; writing—original draft preparation, Z.N., S.F., N.S., F.F.E.-S., G.E.A.-E., and J.-T.C.; review and editing, Z.N., S.F., M.H.F.; revising, Z.N., S.F., and M.H.F.

Funding: This research received no external funding.

Conflicts of Interest: The authors declared no conflicts of interest.

Abbreviations

iNOS	inducible nitric oxide synthase
COX-2	cyclooxygenase
SOCS-3	suppressor of cytokine signaling 3
AMPKα	(AMP)-activated protein kinase α
PKC	protein kinase C
JNK	c-Jun N terminal kinase
ERK	extracellular-signal-regulated kinase
MAPK	mitogen-activated protein kinase
TNF-α	tumor necrosis factor α
INF-γ	interferon γ
IL-1β	interleukin 1β
MCP-1	monocyte chemoattractant protein-1
STAT-1	signal transducer and activator of transcription-1
PI3K/Akt	phosphatidylinositol-3 kinase/Akt
ER	estrogen receptor
ROS	reactive oxygen species
MDA	malondialdehyde
BACE	β-Site amyloid precursor protein (APP) cleaving enzyme
APP	amyloid precursor protein
GSK-3β	glycogen synthase kinase 3β
PPAR-γ	peroxisome proliferator-activated receptor gamma
AchE	acetylcholinesterase
RIP2	receptor interacting protein-2
NF-κB	nuclear factor-κB
MMP-9	matrix metallopeptidase 9
SOD	superoxide dismutase
GSH	glutathione
Nrf-2/ARE	nuclear factor E2-related factor 2/ antioxidant response element
HO-1	hemoxygenase-1
NQO-1	NAD(P)H quinone dehydrogenase1
TBARS	thiobarbituric acid-reactive substances
BDNF	brain-derived neurotrophic factor
TrkB	tropomyosin-related kinase B
GRP78	glucose-regulated protein
Foxp3	forkhead box P3
DOPAC	dihydroxyphenylacetic acid
HVA	homovanillic acid
TH	tyrosine hydroxylase
CHIP	C terminus Hsp70 interacting protein
LPO	lipid peroxidation
GDNF	glial cell line-derived neurotrophic factor
HIF1a	hypoxia inducible factor1a
VEGF	vascular endothelial growth factor

5-HT	5-hydroxytryptamine
GFAP	glial fibrillary acidic protein
6-OHDA	6-hydroxydopamine
ICV-STZ	intracerebroventricular-streptozotocine
AD	Alzheimer's disease
PD	Parkinson's disease
LPS	lipopolysaccharide
MeHg	methyl mercury
MPTP	1-methyl-4-phenyl-1,2,3,6-tetrahydropyridine
MPP$^+$	1-methyl-4-phenylpyridinium
pMCAO	permanent middle cerebral artery occlusion
tMCAO	transient middle cerebral artery occlusion
EAE	experimental autoimmune encephalomyelitis

References

1. Hardy, J.; Gwinn-Hardy, K. Genetic classification of primary neurodegenerative disease. *Science* **1998**, *282*, 1075–1079. [CrossRef] [PubMed]
2. Thrall, J.H. Prevalence and costs of chronic disease in a health care system structured for treatment of acute illness. *Radiology* **2005**, *235*, 9–12. [CrossRef]
3. Sweeney, M.D.; Sagare, A.P.; Zlokovic, B.V. Blood–brain barrier breakdown in Alzheimer disease and other neurodegenerative disorders. *Nat. Rev. Neurol.* **2018**, *14*, 133–150. [CrossRef] [PubMed]
4. Singh, A.K.; Kashyap, M.P.; Tripathi, V.K.; Singh, S.; Garg, G.; Rizvi, S.I. Neuroprotection through rapamycin-induced activation of autophagy and PI3K/Akt1/mTOR/CREB signaling against amyloid-β-induced oxidative stress, synaptic/neurotransmission dysfunction, and neurodegeneration in adult rats. *Mol. Neurobiol.* **2017**, *54*, 5815–5828. [CrossRef] [PubMed]
5. Giráldez-Pérez, R.M.; Antolín-Vallespín, M.; Muñoz, M.D.; Sánchez-Capelo, A. Models of α-synuclein aggregation in Parkinson's disease. *Acta Neuropathol. Commun.* **2014**, *2*, 176. [CrossRef]
6. Nah, J.; Yuan, J.; Jung, Y.-K. Autophagy in neurodegenerative diseases: From mechanism to therapeutic approach. *Mol. Cells* **2015**, *38*, 381–389. [CrossRef]
7. Guo, F.; Liu, X.; Cai, H.; Le, W. Autophagy in neurodegenerative diseases: Pathogenesis and therapy. *Brain Pathol.* **2018**, *28*, 3–13. [CrossRef]
8. Vallee, A.; Lecarpentier, Y.; Guillevin, R.; Vallee, J.N. Effects of cannabidiol interactions with Wnt/beta-catenin pathway and PPARgamma on oxidative stress and neuroinflammation in Alzheimer's disease. *Acta Biochim. Biophys. Sin.* **2017**, *49*, 853–866. [CrossRef]
9. Tang, Y.; Le, W. Differential roles of M1 and M2 microglia in neurodegenerative diseases. *Mol. Neurobiol.* **2016**, *53*, 1181–1194. [CrossRef]
10. Cherry, J.D.; Olschowka, J.A.; O'Banion, M.K. Arginase 1+ microglia reduce Aβ plaque deposition during IL-1β-dependent neuroinflammation. *J. Neuroinflammation* **2015**, *12*, 203. [CrossRef]
11. Hu, X.; Leak, R.K.; Shi, Y.; Suenaga, J.; Gao, Y.; Zheng, P.; Chen, J. Microglial and macrophage polarization-new prospects for brain repair. *Nat. Rev. Neurol.* **2015**, *11*, 56–64. [CrossRef] [PubMed]
12. Zhang, B.; Wei, Y.Z.; Wang, G.Q.; Li, D.D.; Shi, J.S.; Zhang, F. Targeting MAPK Pathways by Naringenin Modulates Microglia M1/M2 Polarization in Lipopolysaccharide-Stimulated Cultures. *Front. Cell. Neurosci.* **2018**, *12*, 531. [CrossRef] [PubMed]
13. Youdim, K.A.; Dobbie, M.S.; Kuhnle, G.; Proteggente, A.R.; Abbott, N.J.; Rice-Evans, C. Interaction between flavonoids and the blood-brain barrier: In vitro studies. *J. Neurochem.* **2003**, *85*, 180–192. [CrossRef] [PubMed]
14. Zobeiri, M.; Belwal, T.; Parvizi, F.; Naseri, R.; Farzaei, M.H.; Nabavi, S.F.; Sureda, A.; Nabavi, S.M. Naringenin and its nano-formulations for fatty liver: Cellular modes of action and clinical perspective. *Curr. Pharm. Biotechnol.* **2018**, *19*, 196–205. [CrossRef]
15. Hernández-Aquino, E.; Muriel, P. Naringenin and the liver. In *Liver Pathophysiology*; Elsevier: Amsterdam, The Netherlands, 2017; pp. 633–651.
16. Yen, F.-L.; Wu, T.-H.; Lin, L.-T.; Cham, T.-M.; Lin, C.-C. Naringenin-loaded nanoparticles improve the physicochemical properties and the hepatoprotective effects of naringenin in orally-administered rats with CCl 4-induced acute liver failure. *Pharm. Res.* **2009**, *26*, 893–902. [CrossRef]

17. Jiménez-Moreno, N.; Cimminelli, M.J.; Volpe, F.; Ansó, R.; Esparza, I.; Mármol, I.; Rodríguez-Yoldi, M.J.; Ancín-Azpilicueta, C. Phenolic Composition of Artichoke Waste and Its Antioxidant Capacity on Differentiated Caco-2 Cells. *Nutrients* **2019**, *11*, 1723. [CrossRef]

18. Hernández-Aquino, E.; Muriel, P. Beneficial effects of naringenin in liver diseases: Molecular mechanisms. *World J. Gastroenterol.* **2018**, *24*, 1679–1707. [CrossRef]

19. Croft, K.D. The Chemistry and Biological Effects of Flavonoids and Phenolic Acids[a]. *Ann. N. Y. Acad. Sci.* **1998**, *854*, 435–442. [CrossRef]

20. Patel, K.; Singh, G.K.; Patel, D.K. A review on pharmacological and analytical aspects of naringenin. *Chin. J. Integr. Med.* **2018**, *24*, 551–560. [CrossRef]

21. Verma, A.K.; Pratap, R. Chemistry of biologically important flavones. *Tetrahedron* **2012**, *68*, 8523–8538. [CrossRef]

22. Heim, K.E.; Tagliaferro, A.R.; Bobilya, D.J. Flavonoid antioxidants: Chemistry, metabolism and structure-activity relationships. *J. Nutr. Biochem.* **2002**, *13*, 572–584. [CrossRef]

23. Rice-Evans, C.A.; Miller, N.J.; Paganga, G. Structure-antioxidant activity relationships of flavonoids and phenolic acids. *Free Radic. Biol. Med.* **1996**, *20*, 933–956. [CrossRef]

24. Yáñez, J.A.; Andrews, P.K.; Davies, N.M. Methods of analysis and separation of chiral flavonoids. *J. Chromatogr. B* **2007**, *848*, 159–181. [CrossRef] [PubMed]

25. Yáñez, J.A.; Remsberg, C.M.; Miranda, N.D.; Vega-Villa, K.R.; Andrews, P.K.; Davies, N.M. Pharmacokinetics of selected chiral flavonoids: Hesperetin, naringenin and eriodictyol in rats and their content in fruit juices. *Biopharm. Drug Dispos.* **2008**, *29*, 63–82. [CrossRef]

26. Krause, M.; Galensa, R. Analysis of enantiomeric flavanones in plant extracts by high-performance liquid chromatography on a cellulose triacetate based chiral stationary phase. *Chromatographia* **1991**, *32*, 69–72. [CrossRef]

27. Wistuba, D.; Trapp, O.; Gel-Moreto, N.; Galensa, R.; Schurig, V. Stereoisomeric Separation of Flavanones and Flavanone-7- O -glycosides by Capillary Electrophoresis and Determination of Interconversion Barriers. *Anal. Chem.* **2006**, *78*, 3424–3433. [CrossRef]

28. Krause, M.; Galensa, R. High-performance liquid chromatography of diastereomeric flavanone glycosides in Citrus on a β-cyclodextrin-bonded stationary phase (Cyclobond I). *J. Chromatogr. A* **1991**, *588*, 41–45. [CrossRef]

29. Gaggeri, R.; Rossi, D.; Collina, S.; Mannucci, B.; Baierl, M.; Juza, M. Quick development of an analytical enantioselective high performance liquid chromatography separation and preparative scale-up for the flavonoid Naringenin. *J. Chromatogr. A* **2011**, *1218*, 5414–5422. [CrossRef]

30. Peterson, J.; Dwyer, J. Flavonoids: Dietary occurrence and biochemical activity. *Nutr. Res.* **1998**, *18*, 1995–2018. [CrossRef]

31. Ross, J.A.; Kasum, C.M. Dietary flavonoids: Bioavailability, metabolic effects, and safety. *Annu. Rev. Nutr.* **2002**, *22*, 19–34. [CrossRef]

32. Renugadevi, J.; Prabu, S.M. Naringenin protects against cadmium-induced oxidative renal dysfunction in rats. *Toxicology* **2009**, *256*, 128–134. [CrossRef] [PubMed]

33. Jung, U.J.; Kim, H.J.; Lee, J.S.; Lee, M.K.; Kim, H.O.; Park, E.J.; Kim, H.K.; Jeong, T.S.; Choi, M.S. Naringin supplementation lowers plasma lipids and enhances erythrocyte antioxidant enzyme activities in hypercholesterolemic subjects. *Clin. Nutr.* **2003**, *22*, 561–568. [CrossRef]

34. Ho, P.C.; Saville, D.J.; Coville, P.F.; Wanwimolruk, S. Content of CYP3A4 inhibitors, naringin, naringenin and bergapten in grapefruit and grapefruit juice products. *Pharm. Acta Helv.* **2000**, *74*, 379–385. [CrossRef]

35. Gattuso, G.; Barreca, D.; Gargiulli, C.; Leuzzi, U.; Caristi, C. Flavonoid Composition of Citrus Juices. *Molecules* **2007**, *12*. [CrossRef] [PubMed]

36. Gel-Moreto, N.; Streich, R.; Galensa, R. Chiral separation of diastereomeric flavanone-7-O-glycosides in citrus by capillary electrophoresis. *Electrophoresis* **2003**, *24*, 2716–2722. [CrossRef] [PubMed]

37. Wang, H.; Nair, M.; Strasburg, G.M.; Booren, A.; Ian Gray, J. Antioxidant Polyphenols from Tart Cherries (Prunus cerasus). *J. Agric. Food Chem.* **1999**, *47*, 840–844. [CrossRef]

38. Llorach, R.; Martínez-Sánchez, A.; Tomás-Barberán, F.; Gil, M.; Ferreres, F. Characterisation of polyphenols and antioxidant properties of five lettuce varieties and escarole. *Food Chem.* **2008**, *108*, 1028–1038. [CrossRef]

39. Vallverdú-Queralt, A.; Odriozola-Serrano, I.; Oms-Oliu, G.; Lamuela-Raventós, R.M.; Elez-Martínez, P.; Martín-Belloso, O. Changes in the Polyphenol Profile of Tomato Juices Processed by Pulsed Electric Fields. *J. Agric. Food Chem.* **2012**, *60*, 9667–9672. [CrossRef]

40. Sánchez-Rabaneda, F.; Jáuregui, O.; Casals, I.; Andrés-Lacueva, C.; Izquierdo-Pulido, M.; Lamuela-Raventós, R.M. Liquid chromatographic/electrospray ionization tandem mass spectrometric study of the phenolic composition of cocoa (Theobroma cacao). *J. Mass Spectrom.* **2003**, *38*, 35–42. [CrossRef]

41. Exarchou, V.; Godejohann, M.; van Beek, T.A.; Gerothanassis, I.P.; Vervoort, J. LC-UV-solid-phase extraction-NMR-MS combined with a cryogenic flow probe and its application to the identification of compounds present in Greek oregano. *Anal. Chem.* **2003**, *75*, 6288–6294. [CrossRef]

42. Olsen, H.T.; Stafford, G.I.; Van Staden, J.; Christensen, S.B.; Jäger, A.K. Isolation of the MAO-inhibitor naringenin from *Mentha aquatica* L. *J. Ethnopharmacol.* **2008**, *117*, 500–502. [CrossRef]

43. Sung, Y.-Y.; Kim, D.-S.; Yang, W.-K.; Nho, K.J.; Seo, H.S.; Kim, Y.S.; Kim, H.K. Inhibitory effects of Drynaria fortunei extract on house dust mite antigen-induced atopic dermatitis in NC/Nga mice. *J. Ethnopharmacol.* **2012**, *144*, 94–100. [CrossRef] [PubMed]

44. Hungria, M.; Johnston, A.; Phillips, D.A. Effects of flavonoids released naturally from bean (Phaseolus vulgaris) on nodD-regulated gene transcription in Rhizobium leguminosarum bv. phaseoli. *Mol Plant. Microbe Interact.* **1992**, *5*, 199–203. [CrossRef] [PubMed]

45. Feng, G.; Wang, W.; Qian, Y.; Jin, H.J. Anti-Aβ antibodies induced by Aβ-HBc virus-like particles prevent Aβ aggregation and protect PC12 cells against toxicity of Aβ1–40. *J. Neurosci. Methods* **2013**, *218*, 48–54. [CrossRef] [PubMed]

46. Pedersen, W.A.; McMillan, P.J.; Kulstad, J.J.; Leverenz, J.B.; Craft, S.; Haynatzki, G.R. Rosiglitazone attenuates learning and memory deficits in Tg2576 Alzheimer mice. *Exp. Neurol.* **2006**, *199*, 265–273. [CrossRef]

47. Rajmohan, R.; Reddy, P.H. Amyloid-Beta and Phosphorylated Tau Accumulations Cause Abnormalities at Synapses of Alzheimer's disease Neurons. *J. Alzheimer's Dis. JAD* **2017**, *57*, 975–999. [CrossRef]

48. Ballard, C.; Gauthier, S.; Corbett, A.; Brayne, C.; Aarsland, D.; Jones, E. Alzheimer's disease. *Lancet* **2011**, *377*, 1019–1031. [CrossRef]

49. Arnold, S.E.; Arvanitakis, Z.; Macauley-Rambach, S.L.; Koenig, A.M.; Wang, H.-Y.; Ahima, R.S.; Craft, S.; Gandy, S.; Buettner, C.; Stoeckel, L.E. Brain insulin resistance in type 2 diabetes and Alzheimer disease: Concepts and conundrums. *Nat. Rev. Neurol.* **2018**, *14*, 168. [CrossRef]

50. De la Monte, S.M.; Tong, M.; Daiello, L.A.; Ott, B.R. Early-Stage Alzheimer's Disease Is Associated with Simultaneous Systemic and Central Nervous System Dysregulation of Insulin-Linked Metabolic Pathways. *J. Alzheimers Dis.* **2019**, *68*, 657–668. [CrossRef]

51. Fakhri, S.; Abbaszadeh, F.; Dargahi, L.; Jorjani, M. Astaxanthin: A Mechanistic Review on its Biological Activities and Health benefits. *Pharmacol. Res.* **2018**, *136*, 1–20. [CrossRef]

52. Yang, W.; Ma, J.; Liu, Z.; Lu, Y.; Hu, B.; Yu, H. Effect of naringenin on brain insulin signaling and cognitive functions in ICV-STZ induced dementia model of rats. *Neurol. Sci. Off. J. Ital. Neurol. Soc. Ital. Soc. Clin. Neurophysiol.* **2014**, *35*, 741–751. [CrossRef] [PubMed]

53. Variya, B.C.; Bakrania, A.K.; Patel, S.S. Antidiabetic potential of gallic acid from Emblica officinalis: Improved glucose transporters and insulin sensitivity through PPAR-γ and Akt signaling. *Phytomedicine* **2019**, 152906. [CrossRef] [PubMed]

54. Zhang, N.; Hu, Z.; Zhang, Z.; Liu, G.; Wang, Y.; Ren, Y.; Wu, X.; Geng, F. Protective Role Of Naringenin Against Aβ25-35-Caused Damage via ER and PI3K/Akt-Mediated Pathways. *Cell. Mol. Neurobiol.* **2018**, *38*, 549–557. [CrossRef] [PubMed]

55. Ghofrani, S.; Joghataei, M.-T.; Mohseni, S.; Baluchnejadmojarad, T.; Bagheri, M.; Khamse, S.; Roghani, M. Naringenin improves learning and memory in an Alzheimer's disease rat model: Insights into the underlying mechanisms. *Eur. J. Pharmacol.* **2015**, *764*, 195–201. [CrossRef] [PubMed]

56. Yang, Z.; Kuboyama, T.; Tohda, C. A Systematic Strategy for Discovering a Therapeutic Drug for Alzheimer's Disease and Its Target Molecule. *Front. Pharmacol.* **2017**, *8*, 340. [CrossRef] [PubMed]

57. Sumi, T.; Imasaki, T.; Aoki, M.; Sakai, N.; Nitta, E.; Shirouzu, M.; Nitta, R. Structural Insights into the Altering Function of CRMP2 by Phosphorylation. *Cell Struct. Funct.* **2018**, *43*, 15–23. [CrossRef]

58. Wilson, S.M.; Ki Yeon, S.; Yang, X.F.; Park, K.D.; Khanna, R. Differential regulation of collapsin response mediator protein 2 (CRMP2) phosphorylation by GSK3ss and CDK5 following traumatic brain injury. *Front. Cell. Neurosci.* **2014**, *8*, 135. [CrossRef]

59. Lawal, M.F.; Olotu, F.A.; Agoni, C.; Soliman, M.E. Exploring the C-Terminal Tail Dynamics: Structural and Molecular Perspectives into the Therapeutic Activities of Novel CRMP-2 Inhibitors, Naringenin and Naringenin-7-O-glucuronide, in the Treatment of Alzheimer's Disease. *Chem. Biodivers.* **2018**, *15*, e1800437. [CrossRef]

60. Lawal, M.; Olotu, F.A.; Soliman, M.E.S. Across the blood-brain barrier: Neurotherapeutic screening and characterization of naringenin as a novel CRMP-2 inhibitor in the treatment of Alzheimer's disease using bioinformatics and computational tools. *Comput. Biol. Med.* **2018**, *98*, 168–177. [CrossRef]

61. Zambrano, P.; Suwalsky, M.; Jemiola-Rzeminska, M.; Strzalka, K.; Sepúlveda, B.; Gallardo, M.J.; Aguilare, L.F. The acetylcholinesterase (AChE) inhibitor and anti-Alzheimer drug donepezil interacts with human erythrocytes. *Biochim. Biophys. Acta Biomembr.* **2019**, *1861*, 1078–1085. [CrossRef]

62. Heo, H.J.; Kim, M.J.; Lee, J.M.; Choi, S.J.; Cho, H.Y.; Hong, B.; Kim, H.K.; Kim, E.; Shin, D.H. Naringenin from Citrus junos has an inhibitory effect on acetylcholinesterase and a mitigating effect on amnesia. *Dement. Geriatr. Cogn. Disord.* **2004**, *17*, 151–157. [CrossRef] [PubMed]

63. Tamagno, E.; Bardini, P.; Obbili, A.; Vitali, A.; Borghi, R.; Zaccheo, D.; Pronzato, M.A.; Danni, O.; Smith, M.A.; Perry, G. Oxidative stress increases expression and activity of BACE in NT2 neurons. *Neurobiol. Dis.* **2002**, *10*, 279–288. [CrossRef] [PubMed]

64. Ciro, A.; Park, J.; Burkhard, G.; Yan, N.; Geula, C. Biochemical differentiation of cholinesterases from normal and Alzheimer's disease cortex. *Curr. Alzheimer Res.* **2012**, *9*, 138–143. [CrossRef] [PubMed]

65. Lee, S.; Youn, K.; Lim, G.; Lee, J.; Jun, M. In Silico Docking and In Vitro Approaches towards BACE1 and Cholinesterases Inhibitory Effect of Citrus Flavanones. *Molecules* **2018**, *23*. [CrossRef] [PubMed]

66. Orhan, I.E.; Jedrejek, D.; Senol, F.S.; Salmas, R.E.; Durdagi, S.; Kowalska, I.; Pecio, L.; Oleszek, W. Molecular modeling and in vitro approaches towards cholinesterase inhibitory effect of some natural xanthohumol, naringenin, and acyl phloroglucinol derivatives. *Phytomedicine: Int. J. Phytother. Phytopharm.* **2018**, *42*, 25–33. [CrossRef]

67. Ouchi, Y.; Yoshikawa, E.; Sekine, Y.; Futatsubashi, M.; Kanno, T.; Ogusu, T.; Torizuka, T. Microglial activation and dopamine terminal loss in early Parkinson's disease. *Ann. Neurol.* **2005**, *57*, 168–175. [CrossRef]

68. Zella, M.A.S.; Metzdorf, J.; Ostendorf, F.; Maass, F.; Muhlack, S.; Gold, R.; Haghikia, A.; Tönges, L. Novel immunotherapeutic approaches to target alpha-synuclein and related neuroinflammation in Parkinson's disease. *Cells* **2019**, *8*, 105. [CrossRef]

69. Jayaraj, R.L.; Beiram, R.; Azimullah, S.; Meeran, M.F.N.; Ojha, S.K.; Adem, A.; Jalal, F.Y. Lycopodium Attenuates Loss of Dopaminergic Neurons by Suppressing Oxidative Stress and Neuroinflammation in a Rat Model of Parkinson's Disease. *Molecules* **2019**, *24*, 2182. [CrossRef]

70. Li, J.; Long, X.; Hu, J.; Bi, J.; Zhou, T.; Guo, X.; Han, C.; Huang, J.; Wang, T.; Xiong, N. Multiple pathways for natural product treatment of Parkinson's disease: A mini review. *Phytomedicine.* **2019**, *60*, 152954. [CrossRef]

71. Lou, H.; Jing, X.; Wei, X.; Shi, H.; Ren, D.; Zhang, X. Naringenin protects against 6-OHDA-induced neurotoxicity via activation of the Nrf2/ARE signaling pathway. *Neuropharmacology* **2014**, *79*, 380–388. [CrossRef]

72. Pascua Maestro, R.; González, E.; Lillo, C.; Ganfornina, M.D.; Falcon-Perez, J.M.; Sanchez, D. Extracellular vesicles secreted by astroglial cells transport Apolipoprotein D to neurons and mediate neuronal survival upon oxidative stress. *Front. Cell. Neurosci.* **2018**, *12*, 526. [CrossRef] [PubMed]

73. Wang, G.-Q.; Zhang, B.; He, X.-M.; Li, D.-D.; Shi, J.-S.; Zhang, F. Naringenin targets on astroglial Nrf2 to support dopaminergic neurons. *Pharmacol. Res.* **2019**, *139*, 452–459. [CrossRef] [PubMed]

74. Zbarsky, V.; Datla, K.P.; Parkar, S.; Rai, D.K.; Aruoma, O.I.; Dexter, D.T. Neuroprotective properties of the natural phenolic antioxidants curcumin and naringenin but not quercetin and fisetin in a 6-OHDA model of Parkinson's disease. *Free Radic. Res.* **2005**, *39*, 1119–1125. [CrossRef] [PubMed]

75. Haavik, J.; Toska, K. Tyrosine hydroxylase and Parkinson's disease. *Mol Neurobiol* **1998**, *16*, 285–309. [CrossRef] [PubMed]

76. Angeline, M.S.; Sarkar, A.; Anand, K.; Ambasta, R.K.; Kumar, P. Sesamol and naringenin reverse the effect of rotenone-induced PD rat model. *Neuroscience* **2013**, *254*, 379–394. [CrossRef]

77. Wegrzynowicz, M.; Bar-On, D.; Anichtchik, O.; Iovino, M.; Xia, J.; Ryazanov, S.; Leonov, A.; Giese, A.; Dalley, J.W.; Griesinger, C. Depopulation of dense α-synuclein aggregates is associated with rescue of dopamine neuron dysfunction and death in a new Parkinson's disease model. *Acta Neuropathol.* **2019**, 1–21. [CrossRef]

78. Mani, S.; Sekar, S.; Barathidasan, R.; Manivasagam, T.; Thenmozhi, A.J.; Sevanan, M.; Chidambaram, S.B.; Essa, M.M.; Guillemin, G.J.; Sakharkar, M.K. Naringenin decreases α-synuclein expression and neuroinflammation in MPTP-induced Parkinson's disease model in mice. *Neurotox. Res.* **2018**, *33*, 656–670. [CrossRef]

79. Mani, S.; Sekar, S.; Chidambaram, S.B.; Sevanan, M. Naringenin protects against 1-methyl-4-phenylpyridinium-induced neuroinflammation and resulting reactive oxygen species production in SH-SY5Y cell line: An in Vitro model of parkinson's disease. *Pharmacogn. Mag.* **2018**, *14*, 458–464.

80. Sugumar, M.; Sevanan, M.; Sekar, S. Neuroprotective effect of naringenin against MPTP-induced oxidative stress. *Int. J. Neurosci.* **2019**, *129*, 534–539. [CrossRef]

81. Stephenson, J.; Nutma, E.; van der Valk, P.; Amor, S. Inflammation in CNS neurodegenerative diseases. *Immunology* **2018**, *154*, 204–219. [CrossRef]

82. Tejera, D.; Heneka, M.T. Microglia in Neurodegenerative Disorders. *Methods Mol. Biol.* **2019**, *2034*, 57–67. [PubMed]

83. Qin, H.; Roberts, K.L.; Niyongere, S.A.; Cong, Y.; Elson, C.O.; Benveniste, E.N. Molecular mechanism of lipopolysaccharide-induced SOCS-3 gene expression in macrophages and microglia. *J. Immunol.* **2007**, *179*, 5966–5976. [CrossRef] [PubMed]

84. Voet, S.; Mc Guire, C.; Hagemeyer, N.; Martens, A.; Schroeder, A.; Wieghofer, P.; Daems, C.; Staszewski, O.; Vande Walle, L.; Jordao, M.J.C.; et al. A20 critically controls microglia activation and inhibits inflammasome-dependent neuroinflammation. *Nat. Commun.* **2018**, *9*, 2036. [CrossRef] [PubMed]

85. Wu, L.H.; Lin, C.; Lin, H.Y.; Liu, Y.S.; Wu, C.Y.; Tsai, C.F.; Chang, P.C.; Yeh, W.L.; Lu, D.Y. Naringenin Suppresses Neuroinflammatory Responses Through Inducing Suppressor of Cytokine Signaling 3 Expression. *Mol. Neurobiol.* **2016**, *53*, 1080–1091. [CrossRef] [PubMed]

86. Park, H.Y.; Kim, G.Y.; Choi, Y.H. Naringenin attenuates the release of pro-inflammatory mediators from lipopolysaccharide-stimulated BV2 microglia by inactivating nuclear factor-kappaB and inhibiting mitogen-activated protein kinases. *Int. J. Mol. Med.* **2012**, *30*, 204–210.

87. Vafeiadou, K.; Vauzour, D.; Lee, H.Y.; Rodriguez-Mateos, A.; Williams, R.J.; Spencer, J.P. The citrus flavanone naringenin inhibits inflammatory signalling in glial cells and protects against neuroinflammatory injury. *Arch. Biochem. Biophys.* **2009**, *484*, 100–109. [CrossRef]

88. Santa-Cecília, F.V.; Socias, B.; Ouidja, M.O.; Sepulveda-Diaz, J.E.; Acuna, L.; Silva, R.L.; Michel, P.P.; Del-Bel, E.; Cunha, T.M.; Raisman-Vozari, R. Doxycycline suppresses microglial activation by inhibiting the p38 MAPK and NF-kB signaling pathways. *Neurotox. Res.* **2016**, *29*, 447–459. [CrossRef]

89. Kim, G.; Ouzounova, M.; Quraishi, A.A.; Davis, A.; Tawakkol, N.; Clouthier, S.G.; Malik, F.; Paulson, A.K.; D'Angelo, R.C.; Korkaya, S.J.O. SOCS3-mediated regulation of inflammatory cytokines in PTEN and p53 inactivated triple negative breast cancer model. *Oncogene* **2015**, *34*, 671–680. [CrossRef]

90. Zheng, Y.; Hou, X.; Yang, S. Lidocaine Potentiates SOCS3 to Attenuate Inflammation in Microglia and Suppress Neuropathic Pain. *Cell. Mol. Neurobiol.* **2019**, *39*, 1081–1092. [CrossRef]

91. Bair, A.M.; Thippegowda, P.B.; Freichel, M.; Cheng, N.; Richard, D.Y.; Vogel, S.M.; Yu, Y.; Flockerzi, V.; Malik, A.B.; Tiruppathi, C.J. Ca^{2+} entry via TRPC channels is necessary for thrombin-induced NF-κB activation in endothelial cells through AMP-activated protein kinase and protein kinase Cδ. *J. Biol. Chem.* **2009**, *284*, 563–574. [CrossRef]

92. Gautam, S.; Ishrat, N.; Yadav, P.; Singh, R.; Narender, T.; Srivastava, A.K. 4-Hydroxyisoleucine attenuates the inflammation-mediated insulin resistance by the activation of AMPK and suppression of SOCS-3 coimmunoprecipitation with both the IR-β subunit as well as IRS-1. *Mol. Cell. Biochem.* **2016**, *414*, 95–104. [CrossRef] [PubMed]

93. Hemmer, B.; Kerschensteiner, M.; Korn, T. Role of the innate and adaptive immune responses in the course of multiple sclerosis. *Lancet. Neurol.* **2015**, *14*, 406–419. [CrossRef]

94. Lee, M.J.; Jang, M.; Choi, J.; Chang, B.S.; Kim, D.Y.; Kim, S.H.; Kwak, Y.S.; Oh, S.; Lee, J.H.; Chang, B.J.; et al. Korean Red Ginseng and Ginsenoside-Rb1/-Rg1 Alleviate Experimental Autoimmune Encephalomyelitis by Suppressing Th1 and Th17 Cells and Upregulating Regulatory T Cells. *Mol Neurobiol* **2016**, *53*, 1977–2002. [CrossRef] [PubMed]

95. Ahmad, S.F.; Zoheir, K.M.; Abdel-Hamied, H.E.; Ashour, A.E.; Bakheet, S.A.; Attia, S.M.; Abd-Allah, A.R. Amelioration of autoimmune arthritis by naringin through modulation of T regulatory cells and Th1/Th2 cytokines. *Cell. Immunol.* **2014**, *287*, 112–120. [CrossRef] [PubMed]

96. Sanchez-Lopez, A.L.; Ortiz, G.G.; Pacheco-Moises, F.P.; Mireles-Ramirez, M.A.; Bitzer-Quintero, O.K.; Delgado-Lara, D.L.C.; Ramirez-Jirano, L.J.; Velazquez-Brizuela, I.E. Efficacy of Melatonin on Serum Pro-inflammatory Cytokines and Oxidative Stress Markers in Relapsing Remitting Multiple Sclerosis. *Arch. Med Res.* **2018**, *49*, 391–398. [CrossRef] [PubMed]

97. Wang, J.; Qi, Y.; Niu, X.; Tang, H.; Meydani, S.N.; Wu, D. Dietary naringenin supplementation attenuates experimental autoimmune encephalomyelitis by modulating autoimmune inflammatory responses in mice. *J. Nutr. Biochem.* **2018**, *54*, 130–139. [CrossRef] [PubMed]

98. Xie, L.; Gong, W.; Chen, J.; Xie, H.-W.; Wang, M.; Yin, X.-P.; Wu, W. The flavonoid kurarinone inhibits clinical progression of EAE through inhibiting Th1 and Th17 cell differentiation and proliferation. *Int. Immunopharmacol.* **2018**, *62*, 227–236. [CrossRef]

99. De Araújo Farias, V.; Carrillo-Gálvez, A.B.; Martín, F.; Anderson, P. TGF-β and mesenchymal stromal cells in regenerative medicine, autoimmunity and cancer. *Cytokine Growth Factor Rev.* **2018**, *43*, 25–37. [CrossRef]

100. Xiao, S.; Jin, H.; Korn, T.; Liu, S.M.; Oukka, M.; Lim, B.; Kuchroo, V.K. Retinoic acid increases Foxp3+ regulatory T cells and inhibits development of Th17 cells by enhancing TGF-beta-driven Smad3 signaling and inhibiting IL-6 and IL-23 receptor expression. *J. Immunol.* **2008**, *181*, 2277–2284. [CrossRef]

101. Wang, J.; Niu, X.; Wu, C.; Wu, D. Naringenin Modifies the Development of Lineage-Specific Effector CD4(+) T Cells. *Front. Immunol.* **2018**, *9*, 2267. [CrossRef]

102. Zhan, Z.; Song, L.; Zhang, W.; Gu, H.; Cheng, H.; Zhang, Y.; Yang, Y.; Ji, G.; Feng, H.; Cheng, T. Absence of cyclin-dependent kinase inhibitor p27 or p18 increases efficiency of iPSC generation without induction of iPSC genomic instability. *Cell Death Dis.* **2019**, *10*. [CrossRef]

103. Lee, Y.; Lahens, N.F.; Zhang, S.; Bedont, J.; Field, J.M.; Sehgal, A. G1/S cell cycle regulators mediate effects of circadian dysregulation on tumor growth and provide targets for timed anticancer treatment. *PLoS Biol.* **2019**, *17*, e3000228. [CrossRef]

104. Niu, X.; Wu, C.; Li, M.; Zhao, Q.; Meydani, S.N.; Wang, J.; Wu, D.J.T. Naringenin is an inhibitor of T cell effector functions. *J. Nutr. Biochem.* **2018**, *58*, 71–79. [CrossRef]

105. Chaudhary, P.; Marracci, G.H.; Bourdette, D.N. Lipoic acid inhibits expression of ICAM-1 and VCAM-1 by CNS endothelial cells and T cell migration into the spinal cord in experimental autoimmune encephalomyelitis. *J. Neuroimmunol.* **2006**, *175*, 87–96. [CrossRef] [PubMed]

106. Proost, P.; Struyf, S.; Van Damme, J.; Fiten, P.; Ugarte-Berzal, E.; Opdenakker, G. Chemokine isoforms and processing in inflammation and immunity. *J. Autoimmun.* **2017**, *85*, 45–57. [CrossRef] [PubMed]

107. Galvez-Contreras, A.Y.; Quinones-Hinojosa, A.; Gonzalez-Perez, O. The role of EGFR and ErbB family related proteins in the oligodendrocyte specification in germinal niches of the adult mammalian brain. *Front. Cell. Neurosci.* **2013**, *7*, 258. [CrossRef]

108. Aguirre, A.; Dupree, J.L.; Mangin, J.M.; Gallo, V. A functional role for EGFR signaling in myelination and remyelination. *Nat. Neurosci.* **2007**, *10*, 990–1002. [CrossRef]

109. Joshi, M.; Singh, S.; Patel, S.; Shah, D.; Krishnakumar, A. Identification of small molecule activators for ErbB 4 receptor to enhance oligodendrocyte regeneration by in silico approach. *Comput. Toxicol.* **2018**, *8*, 13–20. [CrossRef]

110. Khajevand-Khazaei, M.-R.; Ziaee, P.; Motevalizadeh, S.-A.; Rohani, M.; Afshin-Majd, S.; Baluchnejadmojarad, T.; Roghani, M.J.E. Naringenin ameliorates learning and memory impairment following systemic lipopolysaccharide challenge in the rat. *Eur. J. Pharmacol.* **2018**, *826*, 114–122. [CrossRef] [PubMed]

111. Lannert, H.; Hoyer, S. Intracerebroventricular administration of streptozotocin causes long-term diminutions in learning and memory abilities and in cerebral energy metabolism in adult rats. *Behav. Neurosci.* **1998**, *112*, 1199. [CrossRef]

112. Pathan, A.R.; Viswanad, B.; Sonkusare, S.K.; Ramarao, P. Chronic administration of pioglitazone attenuates intracerebroventricular streptozotocin induced-memory impairment in rats. *Life Sci.* **2006**, *79*, 2209–2216. [CrossRef]

113. Ren, B.; Qin, W.; Wu, F.; Wang, S.; Pan, C.; Wang, L.; Zeng, B.; Ma, S.; Liang, J. Apigenin and naringenin regulate glucose and lipid metabolism, and ameliorate vascular dysfunction in type 2 diabetic rats. *Eur. J. Pharmacol.* **2016**, *773*, 13–23. [CrossRef] [PubMed]

114. Hong, Y.; Yin, Y.; Tan, Y.; Hong, K.; Zhou, H. The Flavanone, Naringenin, Modifies Antioxidant and Steroidogenic Enzyme Activity in a Rat Model of Letrozole-Induced Polycystic Ovary Syndrome. *Med Sci. Monit. Int. Med J. Exp. Clin. Res.* **2019**, *25*, 395–401. [CrossRef] [PubMed]

115. Rahigude, A.; Bhutada, P.; Kaulaskar, S.; Aswar, M.; Otari, K.J.N. Participation of antioxidant and cholinergic system in protective effect of naringenin against type-2 diabetes-induced memory dysfunction in rats. *Neuroscience* **2012**, *226*, 62–72. [CrossRef] [PubMed]

116. Baluchnejadmojarad, T.; Roghani, M.J.P. Effect of naringenin on intracerebroventricular streptozotocin-induced cognitive deficits in rat: A behavioral analysis. *Pharmacology* **2006**, *78*, 193–197. [CrossRef] [PubMed]

117. Upadhyay, P.; Sadhu, A.; Singh, P.K.; Agrawal, A.; Ilango, K.; Purohit, S.; Dubey, G.P.J.B. Revalidation of the neuroprotective effects of a United States patented polyherbal formulation on scopolamine induced learning and memory impairment in rats. *Biomed. Pharmacother.* **2018**, *97*, 1046–1052. [CrossRef]

118. Zaki, H.F.; Abd-El-Fattah, M.A.; Attia, A.S. Naringenin protects against scopolamine-induced dementia in rats. *Bull. Fac. Pharm. Cairo Univ.* **2014**, *52*, 15–25. [CrossRef]

119. Khan, M.B.; Khan, M.M.; Khan, A.; Ahmed, M.E.; Ishrat, T.; Tabassum, R.; Vaibhav, K.; Ahmad, A.; Islam, F. Naringenin ameliorates Alzheimer's disease (AD)-type neurodegeneration with cognitive impairment (AD-TNDCI) caused by the intracerebroventricular-streptozotocin in rat model. *Neurochem. Int.* **2012**, *61*, 1081–1093. [CrossRef]

120. Liaquat, L.; Batool, Z.; Sadir, S.; Rafiq, S.; Shahzad, S.; Perveen, T.; Haider, S. Naringenin-induced enhanced antioxidant defence system meliorates cholinergic neurotransmission and consolidates memory in male rats. *Life Sci.* **2018**, *194*, 213–223. [CrossRef]

121. Li, Y.; Zeng, M.; Chen, W.; Liu, C.; Wang, F.; Han, X.; Zuo, Z.; Peng, S. Dexmedetomidine reduces isoflurane-induced neuroapoptosis partly by preserving PI3K/Akt pathway in the hippocampus of neonatal rats. *PLoS ONE* **2014**, *9*, e93639. [CrossRef]

122. Liu, P.; Cheng, H.; Roberts, T.M.; Zhao, J.J. Targeting the phosphoinositide 3-kinase pathway in cancer. *Nat. Rev. Drug Discov.* **2009**, *8*, 627–644. [CrossRef] [PubMed]

123. Hua, F.-Z.; Ying, J.; Zhang, J.; Wang, X.-F.; Hu, Y.-H.; Liang, Y.-P.; Liu, Q.; Xu, G.-H. Naringenin pre-treatment inhibits neuroapoptosis and ameliorates cognitive impairment in rats exposed to isoflurane anesthesia by regulating the PI3/Akt/PTEN signalling pathway and suppressing NF-κB-mediated inflammation. *Int. J. Mol. Med.* **2016**, *38*, 1271–1280. [CrossRef] [PubMed]

124. Dou, W.; Zhang, J.; Sun, A.; Zhang, E.; Ding, L.; Mukherjee, S.; Wei, X.; Chou, G.; Wang, Z.-T.; Mani, S.J.B. Protective effect of naringenin against experimental colitis via suppression of Toll-like receptor 4/NF-κB signalling. *Br. J. Nutr.* **2013**, *110*, 599–608. [CrossRef] [PubMed]

125. Ahshin-Majd, S.; Zamani, S.; Kiamari, T.; Kiasalari, Z.; Baluchnejadmojarad, T.; Roghani, M.J.P. Carnosine ameliorates cognitive deficits in streptozotocin-induced diabetic rats: Possible involved mechanisms. *Peptides* **2016**, *86*, 102–111. [CrossRef] [PubMed]

126. Sarubbo, F.; Ramis, M.; Kienzer, C.; Aparicio, S.; Esteban, S.; Miralles, A.; Moranta, D.J. Chronic silymarin, quercetin and naringenin treatments increase monoamines synthesis and hippocampal Sirt1 levels improving cognition in aged rats. *J. Neuroimmune Pharmacol.* **2018**, *13*, 24–38. [CrossRef]

127. Gao, J.; Wang, W.-Y.; Mao, Y.-W.; Gräff, J.; Guan, J.-S.; Pan, L.; Mak, G.; Kim, D.; Su, S.C.; Tsai, L.-H. A novel pathway regulates memory and plasticity via SIRT1 and miR-134. *Nature* **2010**, *466*, 1105–1109. [CrossRef]

128. Farina, M.; Aschner, M.; Rocha, J.B. Oxidative stress in MeHg-induced neurotoxicity. *Toxicol. Appl. Pharmacol.* **2011**, *256*, 405–417. [CrossRef]

129. Mailloux, R.J.; Yumvihoze, E.; Chan, H.M. Superoxide produced in the matrix of mitochondria enhances methylmercury toxicity in human neuroblastoma cells. *Toxicol. Appl. Pharmacol.* **2015**, *289*, 371–380. [CrossRef]

130. Sumathi, T.; Christinal, J. Neuroprotective effect of Portulaca oleraceae ethanolic extract ameliorates methylmercury induced cognitive dysfunction and oxidative stress in cerebellum and cortex of rat brain. *Biol. Trace Elem. Res.* **2016**, *172*, 155–165. [CrossRef]

131. Manu, K.A.; Christina, H.; Das, S.; Mumbrekar, K.D.; Rao, B.S. Neuroprotective Role of Naringenin against Methylmercury Induced Cognitive Impairment and Mitochondrial Damage in a Mouse Model. *Environ. Toxicol. Pharmacol.* **2019**, 103224. [CrossRef]

132. Fakhri, S.; Dargahi, L.; Abbaszadeh, F.; Jorjani, M. Astaxanthin attenuates neuroinflammation contributed to the neuropathic pain and motor dysfunction following compression spinal cord injury. *Brain Res. Bull.* **2018**, *143*, 217–224. [CrossRef] [PubMed]

133. Dusek, P.; Roos, P.M.; Litwin, T.; Schneider, S.A.; Flaten, T.P.; Aaseth, J. The neurotoxicity of iron, copper and manganese in Parkinson's and Wilson's diseases. *J. Trace Elem. Med. Biol.* **2015**, *31*, 193–203. [CrossRef] [PubMed]

134. Kumar, R.; Jain, V.; Kushwah, N.; Dheer, A.; Mishra, K.P.; Prasad, D.; Singh, S.B. Role of DNA Methylation in Hypobaric Hypoxia-Induced Neurodegeneration and Spatial Memory Impairment. *Ann. Neurosci.* **2018**, *25*, 191–200. [CrossRef]

135. Cheraghi, G.; Hajiabedi, E.; Niaghi, B.; Nazari, F.; Naserzadeh, P.; Hosseini, M.J. High doses of sodium tungstate can promote mitochondrial dysfunction and oxidative stress in isolated mitochondria. *J. Biochem. Mol. Toxicol.* **2019**, *33*, e22266. [CrossRef] [PubMed]

136. Yao, J.; Peng, S.; Xu, J.; Fang, J. Reversing ROS-mediated neurotoxicity by chlorogenic acid involves its direct antioxidant activity and activation of Nrf2-ARE signaling pathway. *Biofactors* **2019**. [CrossRef] [PubMed]

137. Sarkar, A.; Angeline, M.S.; Anand, K.; Ambasta, R.K.; Kumar, P. Naringenin and quercetin reverse the effect of hypobaric hypoxia and elicit neuroprotective response in the murine model. *Brain Res.* **2012**, *1481*, 59–70. [CrossRef]

138. Muthaiah, V.P.K.; Venkitasamy, L.; Michael, F.M.; Chandrasekar, K.; Venkatachalam, S.J. Neuroprotective role of naringenin on carbaryl induced neurotoxicity in mouse neuroblastoma cells. *J. Pharmacol. Pharmacother.* **2013**, *4*, 192–197.

139. Chtourou, Y.; Fetoui, H.; Gdoura, R. Protective effects of naringenin on iron-overload-induced cerebral cortex neurotoxicity correlated with oxidative stress. *Biol. Trace Elem. Res.* **2014**, *158*, 376–383. [CrossRef]

140. Chtourou, Y.; Slima, A.B.; Gdoura, R.; Fetoui, H. Naringenin mitigates iron-induced anxiety-like behavioral impairment, mitochondrial dysfunctions, ectonucleotidases and acetylcholinesterase alteration activities in rat hippocampus. *Neurochem. Res.* **2015**, *40*, 1563–1575. [CrossRef]

141. Sachdeva, S.; Pant, S.C.; Kushwaha, P.; Bhargava, R.; Flora, S.J.J.F.; Toxicology, C. Sodium tungstate induced neurological alterations in rat brain regions and their response to antioxidants. *Food Chem. Toxicol.* **2015**, *82*, 64–71. [CrossRef]

142. Xue, L.; Murray, J.H.; Tolkovsky, A.M.J. The Ras/phosphatidylinositol 3-kinase and Ras/ERK pathways function as independent survival modules each of which inhibits a distinct apoptotic signaling pathway in sympathetic neurons. *J. Biol. Chem.* **2000**, *275*, 8817–8824. [CrossRef] [PubMed]

143. Fakhri, S.; Dargahi, L.; Abbaszadeh, F.; Jorjani, M. Effects of astaxanthin on sensory-motor function in a compression model of spinal cord injury: Involvement of ERK and AKT signalling pathway. *Eur. J. Pain* **2019**, *23*, 750–764. [CrossRef] [PubMed]

144. Xu, X.-H.; Ma, C.-M.; Han, Y.-Z.; Li, Y.; Liu, C.; Duan, Z.-H.; Wang, H.-L.; Liu, D.-Q.; Liu, R.-H. Protective effect of naringenin on glutamate-induced neurotoxicity in cultured hippocampal cells. *Arch. Biol. Sci.* **2015**, *67*, 639–646. [CrossRef]

145. Kipp, M.; Clarner, T.; Gingele, S.; Pott, F.; Amor, S.; Van Der Valk, P.; Beyer, C. Brain lipid binding protein (FABP7) as modulator of astrocyte function. *Physiol. Res.* **2011**, *60*, S49.

146. Hegazy, H.G.; Ali, E.H.; Sabry, H.A.J.T.; Zoology, A. The neuroprotective action of naringenin on oseltamivir (Tamiflu) treated male rats. *J. Basic. Appl. Zool.* **2016**, *77*, 83–90. [CrossRef]

147. Rikhtegar, R.; Yousefi, M.; Dolati, S.; Kasmaei, H.D.; Charsouei, S.; Nouri, M.; Shakouri, S.K. Stem cell-based cell therapy for neuroprotection in stroke: A review. *J. Cell. Biochem.* **2019**, *120*, 8849–8862. [CrossRef]

148. Zhang, F.; Yan, C.; Wei, C.; Yao, Y.; Ma, X.; Gong, Z.; Liu, S.; Zang, D.; Chen, J.; Shi, F.-D. Vinpocetine inhibits NF-κB-dependent inflammation in acute ischemic stroke patients. *Transl. Stroke Res.* **2018**, *9*, 174–184. [CrossRef]

149. Nakano, T.; Nishigami, C.; Irie, K.; Shigemori, Y.; Sano, K.; Yamashita, Y.; Myose, T.; Tominaga, K.; Matsuo, K.; Nakamura, Y.J.J.; et al. Goreisan prevents brain edema after cerebral ischemic stroke by inhibiting aquaporin 4 upregulation in mice. *J. Stroke Cerebrovasc. Dis.* **2018**, *27*, 758–763. [CrossRef]

150. Dai, Y.; Zhang, H.; Zhang, J.; Yan, M. Isoquercetin attenuates oxidative stress and neuronal apoptosis after ischemia/reperfusion injury via Nrf2-mediated inhibition of the NOX4/ROS/NF-κB pathway. *Chem. Biol. Interact.* **2018**, *284*, 32–40. [CrossRef]

151. Hwang, S.-L.; Shih, P.-H.; Yen, G.-C. Neuroprotective effects of citrus flavonoids. *J. Agric. Food Chem.* **2012**, *60*, 877–885. [CrossRef]

152. Bai, X.; Zhang, X.; Chen, L.; Zhang, J.; Zhang, L.; Zhao, X.; Zhao, T.; Zhao, Y. Protective effect of naringenin in experimental ischemic stroke: Down-regulated NOD2, RIP2, NF-kappaB, MMP-9 and up-regulated claudin-5 expression. *Neurochem. Res.* **2014**, *39*, 1405–1415. [CrossRef] [PubMed]

153. Park, J.H.; Kim, Y.G.; McDonald, C.; Kanneganti, T.D.; Hasegawa, M.; Body-Malapel, M.; Inohara, N.; Nunez, G. RICK/RIP2 mediates innate immune responses induced through Nod1 and Nod2 but not TLRs. *J. Immunol.* **2007**, *178*, 2380–2386. [CrossRef] [PubMed]

154. Puspitasari, V.; Gunawan, P.Y.; Wiradarma, H.D.; Hartoyo, V.J.O.A.M. Glial Fibrillary Acidic Protein Serum Level as a Predictor of Clinical Outcome in Ischemic Stroke. *J. Med. Sci.* **2019**, *7*, 1471–1474. [CrossRef] [PubMed]

155. Raza, S.; Khan, M.; Ahmad, A.; Ashafaq, M.; Islam, F.; Wagner, A.; Safhi, M. Neuroprotective effect of naringenin is mediated through suppression of NF-κB signaling pathway in experimental stroke. *Neuroscience* **2013**, *230*, 157–171. [CrossRef]

156. Saleh, T.M.; Saleh, M.C.; Connell, B.J.; Song, Y.H. A co-drug conjugate of naringenin and lipoic acid mediates neuroprotection in a rat model of oxidative stress. *Clin. Exp. Pharmacol. Physiol.* **2017**, *44*, 1008–1016. [CrossRef]

157. Zhou, C.; Luo, D.; Xia, W.; Gu, C.; Lahm, T.; Xu, X.; Qiu, Q.; Zhang, Z. Nuclear factor (erythroid-derived 2)-like 2 (Nrf2) contributes to the neuroprotective effects of histone deacetylase inhibitors in retinal ischemia-reperfusion injury. *Neuroscience* **2019**, *418*, 25–36. [CrossRef]

158. Dinkova-Kostova, A.T.; Abramov, A.Y. The emerging role of Nrf2 in mitochondrial function. *Free Radic. Biol. Med.* **2015**, *88*, 179–188. [CrossRef]

159. Rose, J.; Brian, C.; Woods, J.; Pappa, A.; Panayiotidis, M.I.; Powers, R.; Franco, R. Mitochondrial dysfunction in glial cells: Implications for neuronal homeostasis and survival. *Toxicology* **2017**, *391*, 109–115. [CrossRef]

160. Wang, K.; Chen, Z.; Huang, L.; Meng, B.; Zhou, X.; Wen, X.; Ren, D. Naringenin reduces oxidative stress and improves mitochondrial dysfunction via activation of the Nrf2/ARE signaling pathway in neurons. *Int. J. Mol. Med.* **2017**, *40*, 1582–1590. [CrossRef]

161. Tang, G.; Zhang, C.; Ju, Z.; Zheng, S.; Wen, Z.; Xu, S.; Chen, Y.; Ma, Z. The mitochondrial membrane protein FgLetm1 regulates mitochondrial integrity, production of endogenous reactive oxygen species and mycotoxin biosynthesis in Fusarium graminearum. *Mol. Plant Pathol.* **2018**, *19*, 1595–1611. [CrossRef]

162. Wang, S.; Zhang, Y.; Song, Q.; Fang, Z.; Chen, Z.; Zhang, Y.; Zhang, L.; Zhang, L.; Niu, N.; Ma, S. Mitochondrial dysfunction causes oxidative stress and Tapetal apoptosis in chemical hybridization reagent-induced male sterility in wheat. *Front. Plant Sci.* **2018**, *8*, 2217. [CrossRef] [PubMed]

163. Wang, K.; Chen, Z.; Huang, J.; Huang, L.; Luo, N.; Liang, X.; Liang, M.; Xie, W. Naringenin prevents ischaemic stroke damage via anti-apoptotic and anti-oxidant effects. *Clin. Exp. Pharmacol. Physiol.* **2017**, *44*, 862–871. [CrossRef] [PubMed]

164. Pavlou, S.; Augustine, J.; Cunning, R.; Harkin, K.; Stitt, A.W.; Xu, H.; Chen, M.J.I. Attenuating Diabetic Vascular and Neuronal Defects by Targeting P2rx7. *Int. J. Mol. Sci.* **2019**, *20*, 2101. [CrossRef] [PubMed]

165. Ola, M.S.; Alhomida, A.S.; LaNoue, K.F. Gabapentin Attenuates Oxidative Stress and Apoptosis in the Diabetic Rat Retina. *Neurotox. Res.* **2019**, *36*, 81–90. [CrossRef] [PubMed]

166. Nyane, N.A.; Tlaila, T.B.; Malefane, T.G.; Ndwandwe, D.E.; Owira, P.M.O. Metformin-like antidiabetic, cardio-protective and non-glycemic effects of naringenin: Molecular and pharmacological insights. *Eur. J. Pharmacol.* **2017**, *803*, 103–111. [CrossRef]

167. Kiokias, S.; Proestos, C.; Oreopoulou, V. Effect of natural food antioxidants against LDL and DNA oxidative changes. *Antioxidants* **2018**, *7*, 133. [CrossRef]

168. Cui, X.; Fu, Z.; Wang, M.; Nan, X.; Zhang, B. Pitavastatin treatment induces neuroprotection through the BDNF-TrkB signalling pathway in cultured cerebral neurons after oxygen-glucose deprivation. *Neurol. Res.* **2018**, *40*, 391–397. [CrossRef]

169. Al-Dosari, D.I.; Ahmed, M.M.; Al-Rejaie, S.S.; Alhomida, A.S.; Ola, M.S. Flavonoid naringenin attenuates oxidative stress, apoptosis and improves neurotrophic effects in the diabetic rat retina. *Nutrients* **2017**, *9*. [CrossRef]

170. Tang, J.; Hu, Q.; Chen, Y.; Liu, F.; Zheng, Y.; Tang, J.; Zhang, J.; Zhang, J.H. Neuroprotective role of an N-acetyl serotonin derivative via activation of tropomyosin-related kinase receptor B after subarachnoid hemorrhage in a rat model. *Neurobiol. Dis.* **2015**, *78*, 126–133. [CrossRef]

171. Reijonen, S.; Putkonen, N.; Norremolle, A.; Lindholm, D.; Korhonen, L. Inhibition of endoplasmic reticulum stress counteracts neuronal cell death and protein aggregation caused by N-terminal mutant huntingtin proteins. *Exp. Cell Res.* **2008**, *314*, 950–960. [CrossRef]

172. Jiang, Y.; Chadwick, S.R.; Lajoie, P. Endoplasmic reticulum stress: The cause and solution to Huntington's disease? *Brain Res.* **2016**, *1648*, 650–657. [CrossRef] [PubMed]

173. Jiang, Y.; Lv, H.; Liao, M.; Xu, X.; Huang, S.; Tan, H.; Peng, T.; Zhang, Y.; Li, H. GRP78 counteracts cell death and protein aggregation caused by mutant huntingtin proteins. *Neurosci. Lett.* **2012**, *516*, 182–187. [CrossRef] [PubMed]

174. Kubota, H.; Kitamura, A.; Nagata, K. Analyzing the aggregation of polyglutamine-expansion proteins and its modulation by molecular chaperones. *Methods (San DiegoCalif.)* **2011**, *53*, 267–274. [CrossRef] [PubMed]

175. Yamagishi, N.; Yamamoto, Y.; Noda, C.; Hatayama, T. Naringenin inhibits the aggregation of expanded polyglutamine tract-containing protein through the induction of endoplasmic reticulum chaperone GRP78. *Biol. Pharm. Bull.* **2012**, *35*, 1836–1840. [CrossRef]

176. Hardiman, O.; van den Berg, L.H.; Kiernan, M.C. Clinical diagnosis and management of amyotrophic lateral sclerosis. *Nat. Rev. Neurol.* **2011**, *7*, 639. [CrossRef]

177. Bruijn, L.I.; Miller, T.M.; Cleveland, D.W. Unraveling the mechanisms involved in motor neuron degeneration in ALS. *Annu. Rev. Neurosci.* **2004**, *27*, 723–749. [CrossRef]

178. Kim, S.H.; Shi, Y.; Hanson, K.A.; Williams, L.M.; Sakasai, R.; Bowler, M.J.; Tibbetts, R.S. Potentiation of amyotrophic lateral sclerosis (ALS)-associated TDP-43 aggregation by the proteasome-targeting factor, ubiquilin 1. *J. Biol. Chem.* **2009**, *284*, 8083–8092. [CrossRef]

179. Rosen, D.R.; Siddique, T.; Patterson, D.; Figlewicz, D.A.; Sapp, P.; Hentati, A.; Donaldson, D.; Goto, J.; O'Regan, J.P.; Deng, H.-X. Mutations in Cu/Zn superoxide dismutase gene are associated with familial amyotrophic lateral sclerosis. *Nature* **1993**, *362*, 59. [CrossRef]

180. Xia, P.; Gao, X.; Duan, L.; Zhang, W.; Sun, Y.-F. Mulberrin (Mul) reduces spinal cord injury (SCI)-induced apoptosis, inflammation and oxidative stress in rats via miroRNA-337 by targeting Nrf-2. *Biomed. Pharmacother.* **2018**, *107*, 1480–1487. [CrossRef]

181. Cao, Z.; Chen, L.; Liu, Y.; Peng, T. Oxysophoridine rescues spinal cord injury via anti-inflammatory, anti-oxidative stress and anti-apoptosis effects. *Mol. Med. Rep.* **2018**, *17*, 2523–2528. [CrossRef]

182. Shi, L.-B.; Tang, P.-F.; Zhang, W.; Zhao, Y.-P.; Zhang, L.-C.; Zhang, H. Naringenin inhibits spinal cord injury-induced activation of neutrophils through miR-223. *Gene* **2016**, *592*, 128–133. [CrossRef] [PubMed]

183. Joshi, R.; Kulkarni, Y.A.; Wairkar, S. Pharmacokinetic, pharmacodynamic and formulations aspects of Naringenin: An update. *Life Sci.* **2018**. [CrossRef] [PubMed]

184. Md, S.; Gan, S.Y.; Haw, Y.H.; Ho, C.L.; Wong, S.; Choudhury, H. In vitro neuroprotective effects of naringenin nanoemulsion against β-amyloid toxicity through the regulation of amyloidogenesis and tau phosphorylation. *Int. J. Biol. Macromol.* **2018**, *118*, 1211–1219. [CrossRef] [PubMed]

185. Sun, J.; Roy, S. The physical approximation of APP and BACE-1: A key event in alzheimer's disease pathogenesis. *Dev. Neurobiol.* **2018**, *78*, 340–347. [CrossRef] [PubMed]

186. Gaba, B.; Khan, T.; Haider, M.F.; Alam, T.; Baboota, S.; Parvez, S.; Ali, J. Vitamin E Loaded Naringenin Nanoemulsion via Intranasal Delivery for the Management of Oxidative Stress in a 6-OHDA Parkinson's Disease Model. *Biomed. Res. Int.* **2019**, *2019*. [CrossRef]

187. Drenscko, M.; Loverde, S.M. Molecular dynamics simulations of the interaction of phospholipid bilayers with polycaprolactone. *Mol. Simul.* **2019**, *45*, 859–867. [CrossRef]

188. Ahmad, A.; Fauzia, E.; Kumar, M.; Mishra, R.K.; Kumar, A.; Khan, M.A.; Raza, S.S.; Khan, R. Gelatin-Coated Polycaprolactone Nanoparticle-Mediated Naringenin Delivery Rescue Human Mesenchymal Stem Cells from Oxygen Glucose Deprivation-Induced Inflammatory Stress. *ACS Biomater. Sci. Eng.* **2018**, *5*, 683–695. [CrossRef]

Review

The Exploration of Natural Compounds for Anti-Diabetes from Distinctive Species *Garcinia linii* with Comprehensive Review of the Garcinia Family

Ting-Hsu Chen [1], May-Jywan Tsai [2], Yaw-Syan Fu [3,4] and Ching-Feng Weng [3,4,*]

[1] Department of Life Science and Institute of Biotechnology, National Dong Hwa University, Hualien 97401, Taiwan; 410613031@gms.ndhu.edu.tw
[2] Neural Regeneration Laboratory, Department of Neurosurgery, Neurological Institute, Taipei Veterans General Hospital, Taipei 11221, Taiwan; mjtsai2@vghtpe.gov.tw
[3] Department of Biomedical Science and Environmental Biology, Kaohsiung Medical University, Kaohsiung city 80708, Taiwan; m805004@kmu.edu.tw
[4] Institute of Respiratory Disease, Department of Basic Medical Science, Xiamen Medical College, Xiamen 361023, China
* Correspondence: cfweng-cfweng@hotmail.com; Tel.: +886-3-890-3609; Fax: +886-3-890-0163

Received: 17 September 2019; Accepted: 21 October 2019; Published: 23 October 2019

Abstract: Approximately 400 Garcinia species are distributed around the world. Previous studies have reported the extracts from bark, seed, fruits, peels, leaves, and stems of *Garcinia mangostana*, *G. xanthochymus*, and *G. cambogia* that were used to treat adipogenesis, inflammation, obesity, cancer, cardiovascular diseases, and diabetes. Moreover, the hypoglycemic effects and underlined actions of different species such as *G. kola*, *G. pedunculata*, and *G. prainiana* have been elucidated. However, the anti-hyperglycemia of *G. linii* remains to be verified in this aspect. In this article, the published literature was collected and reviewed based on the medicinal characteristics of the species Garcinia, particularly in diabetic care to deliberate the known constituents from Garcinia and further focus on and isolate new compounds of *G. linii* (Taiwan distinctive species) on various hypoglycemic targets including α-amylase, α-glucosidase, 5′-adenosine monophosphate-activated protein kinase (AMPK), insulin receptor kinase, peroxisome proliferator-activated receptor gamma (PPARγ), and dipeptidyl peptidase-4 (DPP-4) via the molecular docking approach with Gold program to explore the potential candidates for anti-diabetic treatments. Accordingly, benzopyrans and triterpenes are postulated to be the active components in *G. linii* for mediating blood glucose. To further validate the potency of those active components, *in vitro* enzymatic and cellular function assays with *in vivo* animal efficacy experiments need to be performed in the near future.

Keywords: *Garcinia linii*; hypoglycemia; benzopyran; triterpene; bioflavonoid; phenolic; *in silico*

---

1. Introduction

1.1. Impact of Diabetes

Diabetes mellitus (DM) is a global health issue due to its high risk factors, e.g., obesity, physical inactivity, ageing, bad eating habits, genetic predisposition, hypertension, and hyperlipidemia [1]. It is worth noting that DM is a metabolic disease and more than 400 million people suffered from diabetes in 2014 [2]. Interestingly, the adult diabetic population of 2014 has risen from 4.7% to 8.5% worldwide in contrast to the population in 1980 [2]; the morbidity and mortality were 4.95% and 4.00%, respectively [3]. In the future, the diabetic population will increase to 642 million by around 2045 and this population will continue to grow [3].

1.2. Therapy Agent of Diabetes

Nowadays, many clinical medicines such as α-glucosidase and α-amylase inhibitors such as Acarbose are applied to delay the metabolism of carbohydrates and control post-meal blood glucose for diabetes patients [4]. Metformin and 5-Amino-4-Imidazolecarboxamide Riboside (AICAR) are activators that increase the glucose absorption of skeletal muscles and inhibit gluconeogenesis of the liver [5,6]. The major anti-diabetes functions of Sitagliptin increase concentrations of incretin. Incretin can further potentiate the pancreas to produce insulin and inhibit the production of glucagon to decrease blood glucose levels [7]. Rosiglitazone can reduce the resistance of insulin absorption in liver cells, skeletal muscle cells, and fatty tissues [8]. Additionally, GW-9662 is an agonist that promotes peroxisome proliferator-activated receptors gamma (PPARγ) expression, which could stimulate fat metabolism, and trigger insulin pathways to regulate blood glucose and has anti-inflammatory properties [9].

1.3. Distribution of Garcinia Plants and Recent Discovery of Anti-Diabetic Agents with Garcinia Plants

Approximately 400 Garcinia species are distributed around the world including Bangladesh, China, India, Indonesia, Taiwan, Thailand, tropical Asia, Southern Africa, and Western Polynesia. Usually, the Garcinia plants are shrubs or trees. Most of their edible fruits are used in agricultural societies or the fruits and seeds are used to produce oils and dyes, and to treat various diseases, e.g., abdominal pain, food allergies, arthritis, diarrhea, dysentery, and wound infections as past research has shown [10–16]. Most of the trunks from the Garcinia plants such as *Garcinia subelliptica* have been used as building materials to prevent destruction by typhoons in ancient Japan, which is in contrast to why *G. subelliptica* were planted as alley trees, in gardens, and as decorative plants [17,18]. Moreover, East South Asian peoples usually eat the fruit of *G. mangostana*, *G. xanthochymus*, and *G. cambogia* for calories or nutrition and use the extracts of *G. xanthochymus* and *G. cambogia* in curry powder to increase the sour flavor in India. Interestingly, the extract of *G. cambogia* is also used as an antiseptic for preserving food freshness [18,19].

The accumulated literature showed that metabolic syndromes gradually became public health problems such as obesity, hyperglycemia, hyperlipidemia, and hypertension and lead to cardiovascular diseases (i.e., atherosclerosis, stroke, or peripheral artery disease) or diabetes [20]. Hereafter, the medicinal plants or herbs have important characteristics that resist the threat of these diseases because the adverse side effects of natural compounds isolated from medicinal plants are reduced and often less severe than those from clinical drugs [21]. Interestingly, previous studies have reported that the extracts from the bark, seeds, fruits, peels, leaves, and stems of *G. mangostana*, *G. xanthochymus*, and *G. cambogia* are used to treat adipogenesis, inflammation, obesity, cancer, cardiovascular disease, and diabetes. Furthermore, these extracts could also trigger the myotubes and skeletal cells to absorb glucose and to balance blood glucose levels [20,22–26]. Remarkably, numerous studies have indicated that the Garcinia species, e.g., *G. cambogia* [10], *G. xanthochymus* [11], *G. kola* [12], *G. mangostana* [13,14], *G. pedunculata* [15], and *G. prainiana* [16] contain plenty of biflavonoids and phenolic compounds. These compounds have been found to inhibit the enzymatic activity of α-amylase and α-glucosidase for propagating an anti-diabetic effect [27]. Concurrently, the variability of biological actions includes anti-diabetic agents that are dependent on the constituents of plants that grow locality, differently used parts (root, leaves, flowers), or seasonal harvests that exhibit various compositions or ratios for each component, metabolite, or derivative including different delivery systems such as powers, pills, or teas, among others. Generally, batch to batch control or a standard operating system via index chemicals or fingerprints is crucial for the evaluation of the efficacy of crude extracts, the yields in isolation, and the purification of active ingredients. In this article, the published articles were collected and reviewed. We have addressed the isolated compounds for all Garcinia species by molecular structure and briefly described their targeted biomolecules for anti-diabetic function according to chemical category via PubChem such as benzophenone (one compound isolated from *G. mangostana*), biflavonoids (one compound isolated from *G. mangostana* and seven compounds isolated from *G. kola*), xanthones (forty-four compounds isolated from *G. mangostana*, thirty-two compounds isolated from *G. xanthochymus*, and nine compounds isolated from *G. hanburyi*), and procyanidin (one compound isolated from *G. mangostana*). The medicinal characteristics of the Garcinia species for the care of DM (Table 1) are listed to deliberate on the potential constituents of certain Garcinia species. Interestingly, *G. linii* is one endemic evergreen tree only distributed in outlying islands—Langyu land and Green island of Taiwan. To the best of our knowledge, active constituents including 15 xanthones, 6 biphenyls, 2 benzopyran, and 13 known compounds isolated from the root of *G. linii* have been reported with anti-tubercular activity and cytotoxicity [28,29]. However, the medicinal values of *G. linii* extract as an anti-diabetic agent remain to be explored. Moreover, this review article simultaneously offers insights into dissecting the molecular mechanism of isolated compounds such as the three new xanthones (linixanthones A–C), five new biphenyls (garcibiphenyls A–E), and two new benzopyran (garcibenzopyran and (*S*)-3-hydroxygarcibenzopyran) of the *G. linii* root from Taiwan [28,29], combined with the nine known xanthones (10-*O*-Methylmacluraxanthone; 1,5-Dihydroxyxanthone; 1,6-Dihydroxy-3,5,7-trimethoxyxanthone; 1,6-Dihydroxy-5-methoxyxanthone; 1,6-dihydroxy-5,7-dimethoxyxanthone; 1,6-Dihydroxy-7-methoxyxanthone; 1,7-Dihydroxyxanthone; 1,7-Dihydroxy-3-methoxyxanthone; 5-Hydroxy-1-methoxyxanthone) from the Garcinia family on various hypoglycemic targets including α-amylase, α-glucosidase, 5′-adenosine monophosphate-activated protein kinase (AMPK), insulin receptor kinase (IRK), PPARγ, and dipeptidyl peptidase-4 (DPP4) via the molecular docking approach with Gold program (Cambridge Crystallographic Data Centre, Cambridge, UK).

1.4. α-Amylase and α-Glucosidase

To regulate the postprandial blood glucose level, diabetic patients took carbohydrate hydrolase inhibitors such as α-glucosidase and α-amylase to avoid hyperglycemia. α-amylase and α-glucosidase are the key enzymes to hydrolyze carbohydrates and help glucose ingestion [30]. Therefore, diabetic patients have to control their blood glucose by using clinical drugs such as *Precose*® (Acarbose) and *Glyset*® (Miglitol) [4] or other anti-diabetic natural compounds [27] to prolong hydrolysis of carbohydrates against hyperglycemia. In cumulative studies, a few crude extracts from the Garcinia species, e.g., *G. cambogia*, *G. xanthochymus*, *G. kola*, and *G. mangostana* [10–14], containing biflavonoids, polyphenols, and xanthones, also inhibit the enzyme activity of α-amylase and α-glucosidase. Therefore, the extracts are able to help diabetic patients to control their blood glucose levels by the inhibition of carbohydrate hydrolysis. Our docking results (Figure 1) showed that benzopyrans and triterpenes had a higher binding affinity with α-amylase and α-glucosidase than with biflavonoid and phenolic compounds. Additionally, α-tocopherolquinone (a kind of benzopyrans) and squalene (a kind of triterpenes) had a high binding affinity with α-amylase and α-glucosidase to prolong carbohydrate hydrolyzation, reduce the absorption of glucose and mediate the blood glucose level.

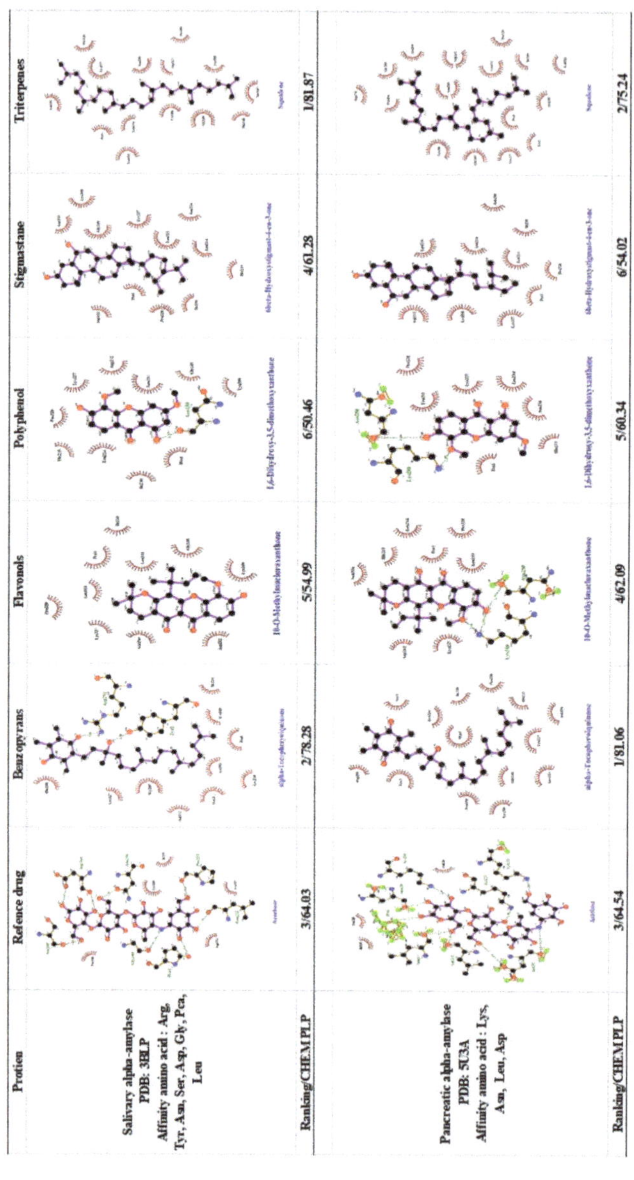

Figure 1. *Cont.*

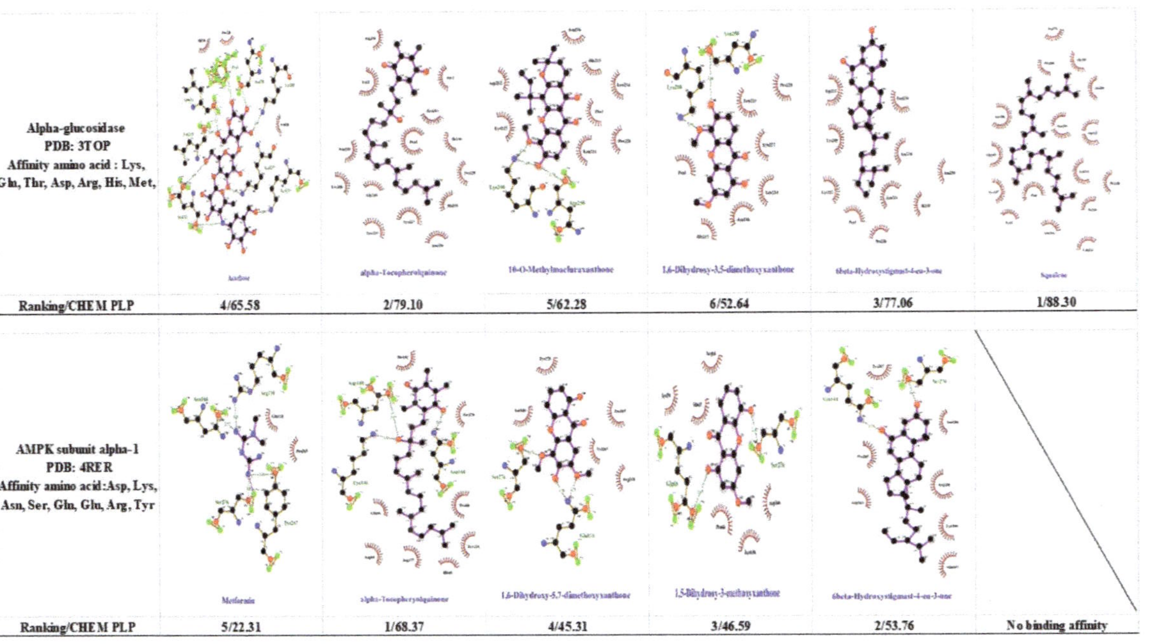

Alpha-glucosidase PDB: 3TOP Affinity amino acid : Lys, Gln, Thr, Asp, Arg, His, Met,	Isoflone	alpha-Tocopherolquinone	16-O-Methylmacluraxanthone	1,6-Dihydroxy-3,5-dimethoxyxanthone	6beta-Hydroxystigmast-4-en-3-one	Squalene
Ranking/CHEM PLP	4/65.58	2/79.10	5/62.28	6/52.64	3/77.06	1/88.30
AMPK subunit alpha-1 PDB: 4RER Affinity amino acid:Asp, Lys, Asn, Ser, Gln, Glu, Arg, Tyr	Metformin	alpha-Tocopherylquinone	1,6-Dihydroxy-5,7-dimethoxyxanthone	1,5-Dihydroxy-3-methoxyxanthone	6beta-Hydroxystigmast-4-en-3-one	No binding affinity
Ranking/CHEM PLP	5/22.31	1/68.37	4/45.31	3/46.59	2/53.76	No binding affinity

Figure 1. *Cont.*

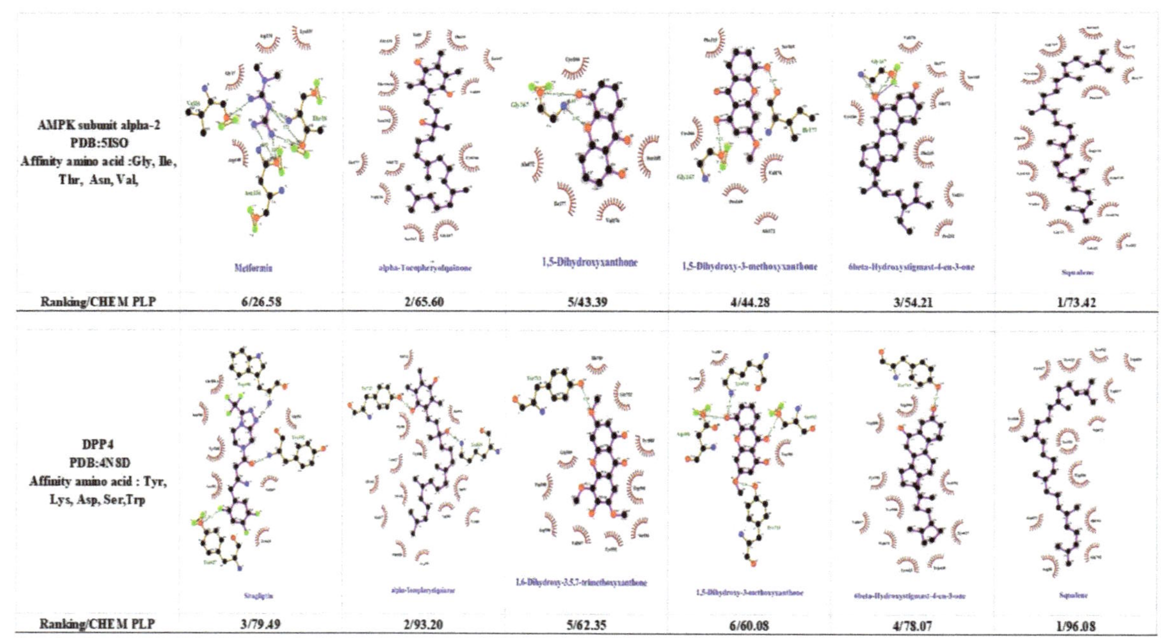

Figure 1. *Cont.*

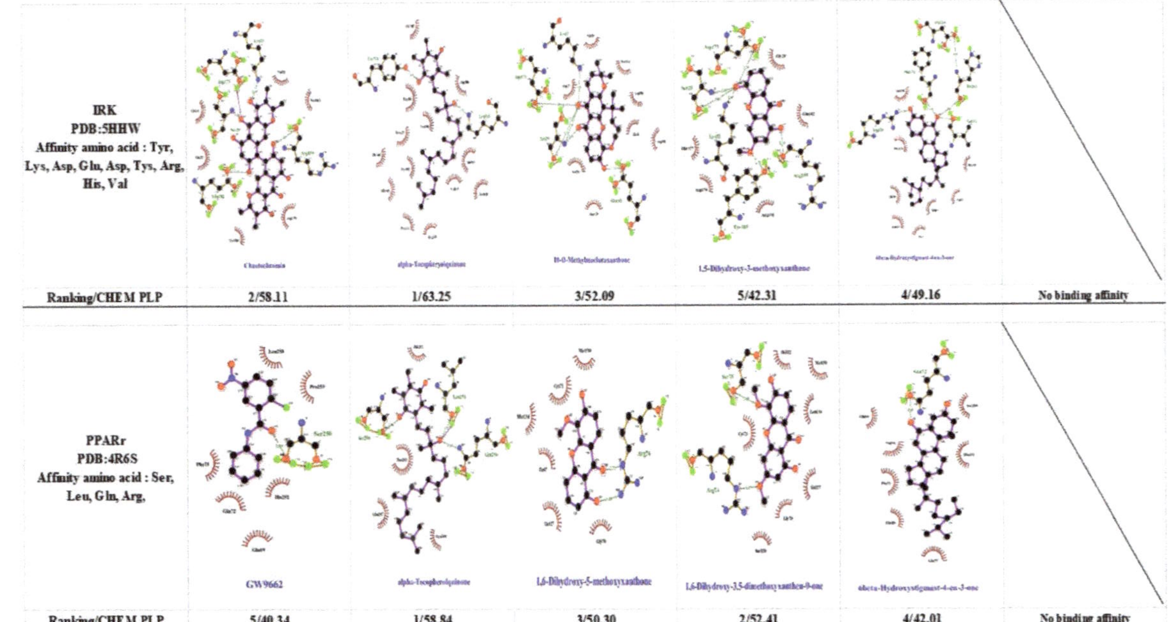

Figure 1. The binding affinity of benzopyrans, triterpenes, stigmastane, biflavonoid, and phenolic on α-amylase, α-glucosidase, AMPK, insulin receptor kinase, PPARγ, and DPP4. Molecular docking was performed by Gold program. Ranking/ChemPLP score presents the order of score value. The model setup was genetic algorithms (GA) run 10 times, a GA search efficiency 200%, removal of water and hydrogen, and ChemPLP scoring. ChemPLP used hydrogen bonding and multiple linear potentials to model Van der Waals and repulsive terms. α-Tocopherolquinone (a kind of benzopyrans) and squalene (a kind of triterpenes) had a higher binding affinity than the reference drug, Acarbose with α-amylase and α-glucosidase prolonging the carbohydrates hydrolyzed to reduce the absorption of glucose and regulate blood glucose levels. Interestingly, α-tocopherolquinone also had a higher binding affinity than reference drugs (Metformin, Chaetochromin, and GW9662) with AMPK1, AMPK2, PPARγ, and IRK templates, respectively; and binding signals would stimulate insulin secretion in contrast to Squalene, which only had a binding affinity with AMPK1. However, α-tocopherolquinone and Squalene still had a stronger binding affinity than Sitagliptin (reference drug) with DDP4 template that could prevent incretins from being digested by DDP4 and promote skeletal cells' uptake of glucose from the blood.

1.5. 5′-Adenosine Monophosphate-Activated Protein Kinase (AMPK)

The 5′-adenosine monophosphate-activated protein kinase (AMPK) is composed of α subunits, regulatory β subunits, and r subunits and is a sensor of cellular energy level. Cellular energy levels were changed by the ratio of AMP:ATP and ADP:ATP that influenced cellular growth and survival [31]. Previous research indicated that AMPK was activated and the rate of ATP-generating processes would increase while the rate of ATP-consuming processes decreased [32]. This mechanism has revealed that AMPK could restore energy homeostasis through an anabolic pathway to consume ATP or catabolic pathways for ATP production [33]. Hence, some clinical/reference drugs such as Metformin and Phenformin could assist peripheral tissues or skeletal muscles to uptake or utilize glucose and even increase insulin sensitivity [5,34]. To avoid adverse effects (diarrhea, nausea, ketonemia, etc.) from clinical drugs such as Metformin and Phenformin [5,35], some natural products such as curcumin [36], rutin, quercetin [37], and catechin [38] were applied to battle or ameliorate diabetes. Notably, *G. xanthochymus* in South East Asia, Africa, Australia, Thailand, and China [39] showed that it was folk medicine used for treating several diseases including diabetes. In previous studies, there were three major compounds identified: 12b-hydroxy-des-d-garcigerrin, 1,2,5,6-tretrahydroxy-4-(1,1-dimethyl-2-propenyl)-7-(3-methyl-2-butenyl)   xanthone,   and 1,5,6-trihydroxy-7,8-di(3-methyl-2-butenyl)-6′,6′-dimethylpyrano (2′,3′:3,4) xanthone that were isolated in the extract of *G. xanthochymus*. These compounds had a significant effect on the promotion of glucose uptake in skeletal cells when compared with Metformin [25]. Our docking results showed that α-tocopherolquinone, 6β-Hydroxystigmast-4-en-3-one, 1,6-dihydroxy-5,7-dimethoxyxanthone, 1,5-Dihydroxyxanthone, 1,5-Dihydroxy-3-methoxyxanthone, and Squalene were isolated from *G. linii* (Figure 1) and had a higher binding affinity with AMPK α1 than Metformin. Interestingly, there is a significant effect that increases glucose uptake in skeletal cells when compared with Metformin. Alternatively, *G. linii* alone or in combination with Metformin can be more prospective to alleviate side-effects or elevate applicable time (e.g., cumulative effect) by reducing Metformin dosage for clinical use.

1.6. Peroxisome Proliferator-Activated Receptor Gamma (PPARγ)

The peroxisome proliferator-activated receptor (PPAR) is a nuclear receptor superfamily and has three isotypes α, δ, and γ that can regulate lipid metabolism, inflammation, and insulin sensitivity as well as insulin production and secretion for treating diabetes [40–42]. PPARγ could mediate lipid mobilization, glucose metabolism, inflammatory response, and adipokines production and secretion [41,43]. Henceforth, cumulative studies emerged and showed PPARγ ligands that could promote triglyceride storage in fat that was implicated in insulin resistance and control adipocyte-secreted hormones [41]. In clinical treatments, Rosiglitazone is an agonist of PPARγ that could ameliorate the memory of Alzheimer patients and even increase insulin sensitivity for diabetes [44]. In traditional therapy, thiazolidinedione (TZD) was usually used to treat diabetes patients but TZD promotes triglyceride storage that causes adverse effects such as headache, muscle soreness, obesity, edema, etc. [45]. Previously, the extract of *G. cambogia* contained (−)-hydroxycitric acid (HCA), which was found to be an active ingredient used to treat obesity and obesity-related diseases, e.g., diabetes, atherosclerosis, etc. [46]. The results (Figure 1) showed that α-tocopherolquinone, 6β-Hydroxystigmast-4-en-3-one, 1,6-Dihydroxy-3,5-dimethoxyxanthen-9-one, and 1,6-Dihydroxy-5-methoxyxanthone stimulated insulin sensitivity, and in virtual screening via the binding affinity of GW9662 (reference drug), which is lower than those of compounds isolated from *G. linii*.

1.7. Dipeptidyl-Peptidase 4 (DPP-4) and Glucagon-Like Peptide 1 (GLP-1)

The dipeptidyl peptidase-4 (DPP-4) could hydrolyze glucagon-like peptide 1 (GLP-1) or gastric inhibitory polypeptide (GIP) and lead to negative effects on the concentration of incretins (GLP-1 and

GIP), insulin secretion, and glucose tolerance due to DPP4 gene expression [47]. Consequently, some diabetes patients may take a DPP4 inhibitor such as Sitagliptin to increase insulin secretion for diabetes therapy and ameliorate the therapeutic effect of GLP-1 [48,49]. The GLP-1 was treated with DPP4 inhibitors against diabetes from 2005 to 2007 and still had adverse effects such as rhinopharyngitis and upper respiratory tract infections [50]. Therefore, some cumulative studies indicated that natural compounds, e.g., rutin, curcumin, antroquinonol, quercetin, and 16-hydroxy-cleroda-3, 13-dien-15, 16-olide (HCD), could inhibit DPP4 activity, such as the inhibitory efficacy of curcumin and quercetin, better than Sitagliptin [36,37,51]. A previous report showed that the extract of the *G. cambogia* fruit, which contains hydroxycitric acid (HCA), could decrease the serum insulin levels and prolong intestinal tracts to absorb glucose as well as to potentially change incretins (GLP-1, GIP) secretions [10,52]. Taken altogether, the extract of *G. cambogia* could regulate blood glucose levels, treat metabolic syndromes, and lead to weight loss. To increase insulin sensitivity, our docking results showed that benzopyrans, triterpenes, stigmastane, and biflavonoids were found to act as insulin receptor agonists and promoted glucose uptake in skeletal cells from blood. Hereafter, incretins are degraded by DPP4 and lead to pancreatic β cells to decrease secretions of insulin (Figure 1). Of note, the reference drug, Sitagliptin, plays a major role in inhibiting the activation of DPP4. Obviously, our data indicated that α-tocopherolquinone and squalene had stronger binding affinity with DPP4 as an inhibitor than with Sitagliptin to prevent incretin (GLP-1) degraded by DPP4.

1.8. Insulin Receptor Kinase (IRK)

α-subunits of insulin receptors receive signal insulin, which triggers tyrosine kinase of β-subunits (Insulin receptor kinase, IRK) to form intracellular auto-phosphorylation at Tyr1158, Tyr1162, and Tyr1163 [53]. Once the insulin receptors are activated, they promote PI3K to phosphorylate PIP2; and, further, PIP3 leads the PDK1/2 activation. When AKT was phosphorylated by receiving the signal, the downstream AS160 would prompt glucose transporter 4 (GLUT4) translocation and uptake glucose into the cells [54]. Previously, some natural compounds have been demonstrated such as (+)-antroquinonol isolated from *Antrodia cinnamomea* [55], rutin (a kind of flavonoid) isolated from *Toona sinensis* Roem [53], and the phenolics isolated from coffee silverskins and husks [56] that result in lowered glucose levels. All of these compounds could enhance the activation of IRK to promote the skeletal tissues to absorb glucose and, consequently, ameliorate insulin resistance by reducing blood glucose levels in the diabetic patients. Therefore, in this study, we collated research from the literature by the application of the Garcinia species for various anti-diabetes treatments. Previous literature revealed that *G. xanthochymus*, *G. kola*, *G. mangostana*, *G. pedunculata*, and *G. prainiana* contained natural compounds, e.g., biflavonoids, xanthone, HCA, and depsidone, which could augment IRK activity and regulate the blood glucose levels for diabetic patients [10,15,16,25,57]. Accordingly, our docking data revealed that only α-tocopherolquinone had a higher binding affinity with IRK than a reference drug (Chaetochromin), suggesting that α-tocopherolquinone acts as an anti-hyperglycemic compound to heighten IRK activity (Figure 1).

Table 1. Summary of the Garcinia species on specific targets of anti-diabetes with basic findings.

Species.	Molecular Targets	Basic Findings
G. cambogia	α-Glucosidase, PPARγ, DPP4	Small intestinal exposure to HCA resulted in a modest reduction in glycemia of healthy individuals [20]. Mixture (GE containing HCA as an active ingredient, PE, anti-adipogenic activity) reduced the expression of adipogenesis-related factors C/EBP-α, PPARγ, and FAS [46]. Insulin resistance did not develop in HCA-SX-supplemented rats via lowered fasting plasma insulin and glucose [58,59].
G. xanthochymus	α-Amylase, α-Glucosidase, AMPK, IRK	Activated PI3K/PKB/Akt signaling pathway and AMPK signaling pathway, resulting in the translocation of GLUT4 in L6 myotubes without affecting the expression of GLUT4 [18]. Identification of α-amylase inhibitor from *G. xanthochymus* [21].
G. kola	α-Amylase, IRK	KV offered significant anti-diabetic relief via reduction of FBG, α-amylase and HbA1c [22].
G. mangostana	α-Amylase, IRK	MVR from *G. mangostana* fruit pericarp had an α-amylase inhibitor and enhanced insulin sensitivity [23,24] GME significantly reduced the blood glucose level in normoglycemic rats and STZ-induced diabetic rats [57].
G. pedunculata	IRK	Elevated insulin levels of rats [25].
G. prainiana	IRK	Increased insulin sensitivity of 3T3-L1 adipocytes [26].

G. cambogia extract (GE); (−)-hydroxycitric acid (HCA); *Pear pomace* extract (PE), CCAAT-enhancer binding protein alpha (C/EBP-α); Peroxisome proliferator-activated receptor gamma (PPARγ); fatty acid synthase (FAS); insulin receptor kinase (IRK); phosphatidylinositol-3 kinase (PI3K)/the serine/threonine kinase protein kinase B (PKB/Akt); AMP-activated protein kinase (AMPK); fasting blood glucose (FBG); kolaviron (KV); mangosteen vinegar rind (MVR); Super CitriMax hydroxycitric acid (HCA-SX), a novel calcium/potassium salt; *G. mangostana* pericarp ethanolic extract (GME); glycated hemoglobin (HbA1c); Streptozotocin (STZ).

2. Conclusions and Future Remarks

In ancient societies, the Garcinia species were used as a daily supply, e.g., building material, food additives, fruit juice, jam, and dye. However, natural compounds that are isolated from the bark, seeds, fruits, peels, leaves, and stems of some Garcinia species such as *G. kola*, *G. pedunculata*, *G. prainiana*, *G. mangostana*, *G. xanthochymus*, and *G. cambogia* have been reported to have a variety of medicinal values. These compounds are applied to treat adipogenesis, inflammation, obesity, cancer, cardiovascular diseases, and diabetes. Predominantly, the isolated natural compounds of *G. linii* in this study are employed to do molecule docking with α-amylase, α-glucosidase, AMPK, IRK, PPARγ, and DPP4, respectively. Of note, our docking data revealed that the ChemPLP scores for Benzopyrans, Flavonols, Polyphenol, Stigmastane, and Triterpenes isolated from *G. linii* had a higher AMPK affinity when compared with Metformin; and, alternatively, *Garcinia linii* alone or in combination with Metformin can have a greater potential to alleviate side-effects or elevate applicable time (e.g., cumulative effect) by reducing Metformin dosage. These results demonstrated that benzopyrans and triterpenes had a stronger binding affinity with anti-diabetic target molecules as a template than reference drugs, e.g., Acarbose with α-Amylase and α-Glucosidase, Metformin with AMPK, Sitagliptin with DPP4, Chaetochromin with IRK, and GW9662 with PPARγ. According to this evidence, benzopyrans and triterpenes are suggested to be the active components in *G. linii* for mediating blood glucose. To further validate the potency of these active components, compounds purified and subsequently the enzyme activity test, an *in vitro* cellular function assay and an *in vivo* animal efficacy experiment need to be conducted to investigate their potential role in anti-diabetes and anti-hyperglycemia in the future.

Author Contributions: Conceptualization, Y.-S.F. and C.-F.W.; Methodology, T.-H.C. and C.-F.W.; Software, T.-H.C. and C.-F.W.; Validation, Y.-S.F. and C.-F.W.; Writing and editing,—original draft preparation, T.-H.C. and M.-J.T.; writing—review M.-J.T. and C.-F.W.; Funding acquisition, C.-F.W.

Funding: This research was funded by Ministry of Science and Technology, Taiwan [grant number 107-2320-B-259-003] for C.F. Weng.

Acknowledgments: We sincerely thanks Max K Leong (Department of Chemistry, NDHU, Hualien, Taiwan) for his help for this work accomplishment.

Conflicts of Interest: The authors declare no conflict of interest.

References

1. Hay, S.I.; Abajobir, A.A.; Abate, K.H.; Abbafati, C.; Abbas, K.M.; Abd-Allah, F.; Abdulkader, R.S.; Abdulle, A.M.; Abebo, T.A.; Abera, S.F.; et al. Global, regional, and national disability-adjusted life-years (DALYs) for 333 diseases and injuries and healthy life expectancy (HALE) for 195 countries and territories, 1990–2016: A systematic analysis for the Global Burden of Disease Study 2016. *Lancet* **2017**, *390*, 1260–1344. [CrossRef]
2. World Health Organization. *Global Report on Diabetes*; WHO: Geneva, Switzerland, 2016.
3. Atlas, D.; International Diabetes Federation. *IDF Diabetes Atlas*, 7th ed.; International Diabetes Federation: Brussels, Belgium, 2015.
4. Van de Laar, F.A. Alpha-glucosidase inhibitors in the early treatment of type 2 diabetes. *Vasc. Health Risk Manag.* **2008**, *4*, 1189–1195. [CrossRef] [PubMed]
5. Thomas, I.; Gregg, B. Metformin; a review of its history and future: From lilac to longevity. *Pediatr. Diabetes* **2017**, *18*, 10–16. [CrossRef] [PubMed]
6. Mendler, M.; Kopf, S.; Groener, J.B.; Riedinger, C.; Fleming, T.H.; Nawroth, P.P.; Okun, J.G. Urine levels of 5-aminoimidazole-4-carboxamide riboside (AICAR) in patients with type 2 diabetes. *Acta Diabetol.* **2018**, *55*, 585–592. [CrossRef]
7. Scott, L.J. Sitagliptin: A Review in Type 2 Diabetes. *Drugs* **2017**, *77*, 209–224. [CrossRef]
8. Abou Daya, K.; Daya, H.A.; Eddine, M.N.; Nahhas, G.; Nuwayri-Salti, N. Effects of rosiglitazone (PPAR gamma agonist) on the myocardium in non-hypertensive diabetic rats (PPAR gamma). *J. Diabetes* **2015**, *7*, 85–94. [CrossRef]
9. Mahajan, U.B.; Chandrayan, G.; Patil, C.R.; Arya, D.S.; Suchal, K.; Agrawal, Y.O.; Ojha, S.; Goyal, S.N. The Protective Effect of Apigenin on Myocardial Injury in Diabetic Rats mediating Activation of the PPAR-gamma Pathway. *Int. J. Mol. Sci.* **2017**, *18*, 756. [CrossRef]
10. Thazhath, S.S.; Wu, T.; Bound, M.J.; Checklin, H.L.; Standfield, S.; Jones, K.L.; Horowitz, M.; Rayner, C.K. Effects of intraduodenal hydroxycitrate on glucose absorption, incretin release, and glycemia in response to intraduodenal glucose infusion in health and type 2 diabetes: A randomised controlled trial. *Nutrition* **2016**, *32*, 553–559. [CrossRef]
11. Li, Y.; Chen, Y.; Xiao, C.; Chen, D.; Xiao, Y.; Mei, Z. Rapid screening and identification of alpha-amylase inhibitors from Garcinia xanthochymus using enzyme-immobilized magnetic nanoparticles coupled with HPLC and MS. *J. Chromatogr. B Analyt. Technol. Biomed. Life Sci.* **2014**, *960*, 166–173. [CrossRef]
12. Adaramoye, O.A. Antidiabetic effect of kolaviron, a biflavonoid complex isolated from Garcinia kola seeds, in Wistar rats. *Afr. Health Sci.* **2012**, *12*, 498–506. [CrossRef]
13. Karim, N.; Rahman, A.; Chanudom, L.; Thongsom, M.; Tangpong, J. Mangosteen Vinegar Rind from Garcinia mangostana Prevents High-Fat Diet and Streptozotocin-Induced Type II Diabetes Nephropathy and Apoptosis. *J. Food Sci.* **2019**, *84*, 1208–1215. [CrossRef] [PubMed]
14. Loo, A.E.; Huang, D. Assay-guided fractionation study of alpha-amylase inhibitors from Garcinia mangostana pericarp. *J. Agric. Food Chem.* **2007**, *55*, 9805–9810. [PubMed]
15. Ali, M.Y.; Paul, S.; Tanvir, E.M.; Hossen, M.S.; Rumpa, N.N.; Saha, M.; Bhoumik, N.C.; Islam, M.A.; Hossain, M.S.; Alam, N.; et al. Antihyperglycemic, Antidiabetic, and Antioxidant Effects of Garcinia pedunculata in Rats. *Evid.-Based Complement. Altern. Med.* **2017**, *2017*, 2979760. [CrossRef] [PubMed]
16. Susanti, D.; Amiroudine, M.Z.; Rezali, M.F.; Taher, M. Friedelin and lanosterol from Garcinia prainiana stimulated glucose uptake and adipocytes differentiation in 3T3-L1 adipocytes. *Nat. Prod. Res.* **2013**, *27*, 417–424. [CrossRef] [PubMed]
17. Inoue, T.; Kainuma, M.; Baba, K.; Oshiro, N.; Kimura, N.; Chan, E.W. Garcinia subelliptica Merr. (Fukugi): A multipurpose coastal tree with promising medicinal properties. *J. Intercult. Ethnopharmacol.* **2017**, *6*, 121–127.
18. Mohamed, G.A.; Al-Abd, A.M.; El-Halawany, A.M.; Abdallah, H.M.; Ibrahim, S.R.M. New xanthones and cytotoxic constituents from Garcinia mangostana fruit hulls against human hepatocellular, breast, and colorectal cancer cell lines. *J. Ethnopharmacol.* **2017**, *198*, 302–312. [CrossRef]
19. Semwal, R.B.; Semwal, D.K.; Vermaak, I.; Viljoen, A. A comprehensive scientific overview of Garcinia cambogia. *Fitoterapia* **2015**, *102*, 134–148. [CrossRef]

20. Tousian Shandiz, H.; Razavi, B.M.; Hosseinzadeh, H. Review of Garcinia mangostana and its Xanthones in Metabolic Syndrome and Related Complications. *Phytother. Res.* **2017**, *31*, 1173–1182. [CrossRef]

21. Alam, F.; Islam, M.A.; Kamal, M.A.; Gan, S.H. Updates on Managing Type 2 Diabetes Mellitus with Natural Products: Towards Antidiabetic Drug Development. *Curr. Med. Chem.* **2018**, *25*, 5395–5431. [CrossRef]

22. Liu, Q.Y.; Wang, Y.T.; Lin, L.G. New insights into the anti-obesity activity of xanthones from Garcinia mangostana. *Food Funct.* **2015**, *6*, 383–393. [CrossRef]

23. Nguyen, C.N.; Trinh, B.T.D.; Tran, T.B.; Nguyen, L.T.; Jager, A.K.; Nguyen, L.D. Anti-diabetic xanthones from the bark of Garcinia xanthochymus. *Bioorg. Med. Chem. Lett.* **2017**, *27*, 3301–3304. [CrossRef] [PubMed]

24. Widowati, W.; Laksmitawati, D.R.; Wargasetia, T.L.; Afifah, E.; Amalia, A.; Arinta, Y.; Rizal, R.; Suciati, T. Mangosteen peel extract (Garcinia mangostana L.) as protective agent in glucose-induced mesangial cell as in vitro model of diabetic glomerulosclerosis. *Iran. J. Basic Med. Sci.* **2018**, *21*, 972–977. [PubMed]

25. Li, Y.; Zhao, P.; Chen, Y.; Fu, Y.; Shi, K.; Liu, L.; Liu, H.; Xiong, M.; Liu, Q.H.; Yang, G.; et al. Depsidone and xanthones from Garcinia xanthochymus with hypoglycemic activity and the mechanism of promoting glucose uptake in L6 myotubes. *Bioorg. Med. Chem.* **2017**, *25*, 6605–6613. [CrossRef] [PubMed]

26. Maia-Landim, A.; Ramirez, J.M.; Lancho, C.; Poblador, M.S.; Lancho, J.L. Long-term effects of Garcinia cambogia/Glucomannan on weight loss in people with obesity, PLIN4, FTO and Trp64Arg polymorphisms. *BMC Complement. Altern. Med.* **2018**, *18*, 26. [CrossRef]

27. Liu, S.; Li, D.; Huang, B.; Chen, Y.; Lu, X.; Wang, Y. Inhibition of pancreatic lipase, alpha-glucosidase, alpha-amylase, and hypolipidemic effects of the total flavonoids from Nelumbo nucifera leaves. *J. Ethnopharmacol.* **2013**, *149*, 263–269. [CrossRef]

28. Chen, J.J.; Chen, I.S.; Duh, C.Y. Cytotoxic xanthones and biphenyls from the root of Garcinia linii. *Planta Med.* **2004**, *70*, 1195–1200. [CrossRef]

29. Chen, J.J.; Peng, C.F.; Huang, H.Y.; Chen, I.S. Benzopyrans, biphenyls and xanthones from the root of Garcinia linii and their activity against Mycobacterium tuberculosis. *Planta Med.* **2006**, *72*, 473–477. [CrossRef]

30. Pantidos, N.; Boath, A.; Lund, V.; Conner, S.; McDougall, G.J. Phenolic-rich extracts from the edible seaweed, ascophyllum nodosum, inhibit α-amylase and α-glucosidase: Potential anti-hyperglycemic effects. *J. Funct. Foods* **2014**, *10*, 201–209. [CrossRef]

31. Hardie, D.G.; Schaffer, B.E.; Brunet, A. AMPK: An Energy-Sensing Pathway with Multiple Inputs and Outputs. *Trends Cell Biol.* **2016**, *26*, 190–201. [CrossRef]

32. Olivier, S.; Foretz, M.; Viollet, B. Promise and challenges for direct small molecule AMPK activators. *Biochem. Pharmacol.* **2018**, *153*, 147–158. [CrossRef]

33. Scott, J.W.; Ling, N.; Issa, S.M.; Dite, T.A.; O'Brien, M.T.; Chen, Z.P.; Galic, S.; Langendorf, C.G.; Steinberg, G.R.; Kemp, B.E.; et al. Small molecule drug A-769662 and AMP synergistically activate naive AMPK independent of upstream kinase signaling. *Chem. Biol.* **2014**, *21*, 619–627. [CrossRef] [PubMed]

34. Oh-Hashi, K.; Irie, N.; Sakai, T.; Okuda, K.; Nagasawa, H.; Hirata, Y.; Kiuchi, K. Elucidation of a novel phenformin derivative on glucose-deprived stress responses in HT-29 cells. *Mol. Cell. Biochem.* **2016**, *419*, 29–40. [CrossRef] [PubMed]

35. Dujic, T.; Causevic, A.; Bego, T.; Malenica, M.; Velija-Asimi, Z.; Pearson, E.R.; Semiz, S. Organic cation transporter 1 variants and gastrointestinal side effects of metformin in patients with Type 2 diabetes. *Diabet. Med.* **2016**, *33*, 511–514. [CrossRef] [PubMed]

36. Seo, K.I.; Choi, M.S.; Jung, U.J.; Kim, H.J.; Yeo, J.; Jeon, S.M.; Lee, M.K. Effect of curcumin supplementation on blood glucose, plasma insulin, and glucose homeostasis related enzyme activities in diabetic db/db mice. *Mol. Nutr. Food Res.* **2008**, *52*, 995–1004. [CrossRef] [PubMed]

37. Nazeri, S.; Farhangi, M.; Modarres, S. The effect of different dietary inclusion levels of rutin (a flavonoid) on some liver enzyme activities and oxidative stress indices in rainbow trout, Oncorhynchus mykiss (Walbaum) exposed to Oxytetracycline. *Aquac. Res.* **2017**, *48*, 4356–4362. [CrossRef]

38. Meena, K.P.; Vijayakumar, M.R.; Dwibedy, P.S. Catechin-loaded Eudragit microparticles for the management of diabetes: Formulation, characterization and in vivo evaluation of antidiabetic efficacy. *J. Microencapsul.* **2017**, *34*, 342–350. [CrossRef] [PubMed]

39. Che Hassan, N.K.N.; Taher, M.; Susanti, D. Phytochemical constituents and pharmacological properties of Garcinia xanthochymus—A review. *Biomed. Pharmacother.* **2018**, *106*, 1378–1389. [CrossRef]

40. Wang, Q.; Imam, M.U.; Yida, Z.; Wang, F. Peroxisome Proliferator-Activated Receptor Gamma (PPARgamma) as a Target for Concurrent Management of Diabetes and Obesity-Related Cancer. *Curr. Pharm. Des.* **2017**, *23*, 3677–3688. [CrossRef]

41. Janani, C.; Kumari, B.D.R. PPAR gamma gene—A review. *Diabetes Metab. Syndr.* **2015**, *9*, 46–50. [CrossRef]

42. Polvani, S.; Tarocchi, M.; Tempesti, S.; Bencini, L.; Galli, A. Peroxisome proliferator activated receptors at the crossroad of obesity, diabetes, and pancreatic cancer. *World J. Gastroenterol.* **2016**, *22*, 2441–2459. [CrossRef]

43. Siersbaek, R.; Nielsen, R.; Mandrup, S. PPARgamma in adipocyte differentiation and metabolism—novel insights from genome-wide studies. *FEBS Lett.* **2010**, *584*, 3242–3249. [CrossRef] [PubMed]

44. Sebastiao, I.; Candeias, E.; Santos, M.S.; de Oliveira, C.R.; Moreira, P.I.; Duarte, A.I. Insulin as a Bridge between Type 2 Diabetes and Alzheimer Disease - How Anti-Diabetics Could be a Solution for Dementia. *Front. Endocrinol.* **2014**, *5*, 110. [CrossRef] [PubMed]

45. Elaidy, S.M.; Hussain, M.A.; El-Kherbetawy, M.K. Time-dependent therapeutic roles of nitazoxanide on high-fat diet/streptozotocin-induced diabetes in rats: Effects on hepatic peroxisome proliferator-activated receptor-gamma receptors. *Can. J. Physiol. Pharmacol.* **2018**, *96*, 485–497. [CrossRef] [PubMed]

46. Sharma, K.; Kang, S.; Gong, D.; Oh, S.H.; Park, E.Y.; Oak, M.H.; Yi, E. Combination of Garcinia cambogia Extract and Pear Pomace Extract Additively Suppresses Adipogenesis and Enhances Lipolysis in 3T3-L1 Cells. *Pharmacogn. Mag.* **2018**, *14*, 220–226.

47. Bohm, A.; Wagner, R.; Machicao, F.; Holst, J.J.; Gallwitz, B.; Stefan, N.; Fritsche, A.; Haring, H.U.; Staiger, H. DPP4 gene variation affects GLP-1 secretion, insulin secretion, and glucose tolerance in humans with high body adiposity. *PLoS ONE* **2017**, *12*, e0181880. [CrossRef]

48. Nahon, K.J.; Doornink, F.; Straat, M.E.; Botani, K.; Martinez-Tellez, B.; Abreu-Vieira, G.; van Klinken, J.B.; Voortman, G.J.; Friesema, E.C.H.; Ruiz, J.R.; et al. Effect of sitagliptin on energy metabolism and brown adipose tissue in overweight individuals with prediabetes: A randomised placebo-controlled trial. *Diabetologia* **2018**, *61*, 2386–2397. [CrossRef]

49. Araujo, F.; Shrestha, N.; Gomes, M.J.; Herranz-Blanco, B.; Liu, D.; Hirvonen, J.J.; Granja, P.L.; Santos, H.A.; Sarmento, B. In vivo dual-delivery of glucagon like peptide-1 (GLP-1) and dipeptidyl peptidase-4 (DPP4) inhibitor through composites prepared by microfluidics for diabetes therapy. *Nanoscale* **2016**, *8*, 10706–10713. [CrossRef]

50. Andel, M.; Skrha, P.; Kraml, P.; Potockova, J.; Hoffmanova, I.; Silhova, E.; Fontana, J.; Richterova, A.; Gadiredi, M.; Busek, P.; et al. Annual monitoring of side effects of administering sitagliptin in patients with type 2 diabetes mellitus. *Vnitr. Lék.* **2016**, *62*, 455–461.

51. Huang, P.K.; Lin, S.X.; Tsai, M.J.; Leong, M.K.; Lin, S.R.; Kankala, R.K.; Lee, C.H.; Weng, C.F. Encapsulation of 16-Hydroxycleroda-3,13-Dine-16,15-Olide in Mesoporous Silica Nanoparticles as a Natural Dipeptidyl Peptidase-4 Inhibitor Potentiated Hypoglycemia in Diabetic Mice. *Nanomaterials* **2017**, *7*, 112. [CrossRef]

52. Hayamizu, K.; Hirakawa, H.; Oikawa, D.; Nakanishi, T.; Takagi, T.; Tachibana, T.; Furuse, M. *Effect of* Garcinia cambogia extract on serum leptin and insulin in mice. *Fitoterapia* **2003**, *74*, 267–273. [CrossRef]

53. Hsu, C.-Y.; Shih, H.-Y.; Chia, Y.-C.; Lee, C.-H.; Ashida, H.; Lai, Y.-K.; Weng, C.-F. Rutin potentiates insulin receptor kinase to enhance insulin-dependent glucose transporter 4 translocation. *Mol. Nutr. Food Res.* **2014**, *58*, 1168–1176. [CrossRef] [PubMed]

54. Świderska, E.; Strycharz, J.; Wróblewski, A.; Szemraj, J.; Drzewoski, J.; Śliwińska, A. Role of PI3K/AKT Pathway in Insulin-Mediated Glucose Uptake. In *Glucose Transport [Working Title]*; 2018; Available online: https://www.intechopen.com/online-first/role-of-pi3k-akt-pathway-in-insulin-mediated-glucose-uptake (accessed on 23 October 2019). [CrossRef]

55. Hsu, C.Y.; Sulake, R.S.; Huang, P.K.; Shih, H.Y.; Sie, H.W.; Lai, Y.K.; Chen, C.; Weng, C.F. Synthetic (+)-antroquinonol exhibits dual actions against insulin resistance by triggering AMP kinase and inhibiting dipeptidyl peptidase IV activities. *Br. J. Pharmacol.* **2015**, *172*, 38–49. [CrossRef] [PubMed]

56. Rebollo-Hernanz, M.; Zhang, Q.; Aguilera, Y.; Martin-Cabrejas, M.A.; de Mejia, E.G. Phenolic compounds from coffee by-products modulate adipogenesis-related inflammation, mitochondrial dysfunction, and insulin resistance in adipocytes, via insulin/PI3K/AKT signaling pathways. *Food Chem. Toxicol.* **2019**, *132*, 110672. [CrossRef] [PubMed]

57. Taher, M.; Zakaria, T.M.F.S.T.; Susanti, D.; Zakaria, Z.A. Hypoglycaemic activity of ethanolic extract of Garcinia mangostana Linn. in normoglycaemic and streptozotocin-induced diabetic rats. *BMC Complement. Altern. Med.* **2016**, *16*, 135.

58. Asghar, M.; Monjok, E.; Kouamou, G.; Ohia, S.E.; Bagchi, D.; Lokhandwala, M.F. Super CitriMax (HCA-SX) attenuates increases in oxidative stress, inflammation, insulin resistance, and body weight in developing obese Zucker rats. *Mol. Cell. Biochem.* **2007**, *304*, 93–99. [CrossRef]

59. Talpur, N.; Echard, B.W.; Yasmin, T.; Bagchi, D.; Preuss, H.G. Effects of niacin-bound chromium, Maitake mushroom fraction SX and (-)-hydroxycitric acid on the metabolic syndrome in aged diabetic Zucker fatty rats. *Mol. Cell. Biochem.* **2003**, *252*, 369–377. [CrossRef]

 *biomolecules*

Article

Variation in Chemical Composition and Biological Activities of *Flos Chrysanthemi indici* Essential Oil under Different Extraction Methods

Chang-Liang Jing [1,†], **Rui-Huan Huang** [1,†], **Yan Su** [2], **Yi-Qiang Li** [1,*] and **Cheng-Sheng Zhang** [1,*]

[1] Ocean Agriculture Research Center, Tobacco Research Institute of Chinese Academy of Agricultural Sciences, Qingdao 266101, China; jingchangliang@caas.cn (C.-L.J.); 18779368572@163.com (R.-H.H.)

[2] Chongqing key laboratory of traditional Chinese medicine resources, Chongqing Academy of Chinese Materia Medica, Chongqing 400065, China; cqsuyan@126.com

[*] Correspondence: liyiqiang@caas.cn (Y.-Q.L.); zhangchengsheng@caas.cn (C.-S.Z.); Tel.: +86-532-8870-2115 (Y.-Q.L.)

[†] These authors contributed equally to this work.

Received: 15 August 2019; Accepted: 19 September 2019; Published: 21 September 2019

Abstract: *Flos Chrysanthemi indici*, an important medicinal and aromatic plant in China, is considered to have many different preservative and pharmacological properties. Considering the capability of essential oils (EOs), the present study is conducted to compare different extraction methods in order to improve yield and biological activities. Hydro-distillation (HD), steam-distillation (SD), solvent-free microwave extraction (SFME), and supercritical fluid extraction (SFE) are employed to prepare EOs from *Flos Chrysanthemi indici*. A total of 71 compounds are assigned by gas chromatography/mass spectrometry (GC–MS) in comparison with retention indices. These include 32 (HD), 16 (SD), 31 (SFME) and 38 (SFE) compounds. Major constituents of EOs differ according to the extraction methods were heptenol, tricosane, camphor, borneol, and eucalyptol. EOs extracted by SFME exhibit higher antioxidant activity. All EOs show varying degrees of antimicrobial activity, with minimum inhibitory concentration (MIC) ranging from 0.0625 to 0.125 mg/mL and SFME and SFE prove to be efficient extraction methods. EOs alter the hyphal morphology of *Alternaria alternata*, with visible bumps forming on the mycelium. Overall, these results indicate that the extraction method can significantly influence the composition and biological activity of EOs and SFME and SFE are outstanding methods to extract EOs with high yield and antimicrobial activity.

Keywords: *Flos Chrysanthemi indici*; essential oil; extraction method; chemical composition; biological activity

---

1. Introduction

Flos Chrysanthemi indici derived from *Chrysanthemum indicum* L. is listed in the Chinese Pharmacopoeia (2005) for medical use and distributed across most parts of China. Studies show that *Flos Chrysanthemi indici* has anti-inflammatory and antibacterial activities and is effective in neutralizing infectious diseases [1,2]. Cineole, borneol, and bornyl acetate are the principal constituents of essential oils (EOs) contributing to their medicinal properties [3].

EOs can be obtained from different plants by many methods including hydro-distillation (HD), steam-distillation SD, solvent-free microwave extraction SFME, and supercritical fluid extraction SFE [4–6]. Previous studies showed that extraction methods could affect the composition of EOs and further result in different bioactivities [7]. Currently, HD is the principal approach for the extraction of essential oil from *Flos Chrysanthemi indici*. This approach suffers some disadvantages such as time-consumption, the requirement of extensive energy consumption, and low efficiency. However,

SFME and SFE have emerged as efficient alternative methods among the others for essential oil separation from plant materials [8,9].

Diseases that are caused by microorganisms lead to large postharvest losses. Fruits and vegetables can be attacked by microorganisms during the production and postharvest stages, leading to significant economic losses for the food industry [10]. In the case of rice, fruits, and vegetables, the most common diseases were caused by *Xanthomonas oryzae* pv. *oryzae*, *Ralstonia solanacearum*, *Pseudomonas syringae* pv. *lachrymans*, *Acidovorax citrulli*, *Rhizoctonia solani*, and *Erwinia carotovora* [11,12]. Chemical agents are typically used to inhibit the growth of contaminating microorganisms. However, due to growing concerns over the safety of foods containing chemical additives, natural antibacterial products for fruit and vegetable preservation are attracting increasing attention. EOs are naturally synthesized in aromatic plants as secondary metabolites [13] and exhibit potent suppressive activity against bacteria, molds, and yeasts, and have therefore been used in food preservation and natural remedies [14–16]. EOs have the added advantages of being volatile, eco-friendly, and biodegradable [17,18], which makes them acceptable to consumers [19].

To the best of our knowledge, application of SFME and SFE to obtain essential oil from *Flos Chrysanthemi indici* has not been previously reported. There is also no reported research to compare the chemical composition and biological activities, including the antimicrobial and antioxidant properties of essential oils, obtained from different extraction methods. To address these issues, the present study investigates the yield, chemical composition, and biological activities of EOs extracted from *Flos Chrysanthemi indici* by four different methods, including HD, SD, SFME and SFE, analyzed by gas chromatography/mass spectrometry (GC–MS). In order to identify the most effective extraction method in terms of their activities, we investigated by 2,2-diphenyl-1-picryl-hydrazyl-hydrate (DPPH), 2,2'-Azinobis-(3-ethylbenzthiazoline-6-sulphonate) (ABTS) assays for antioxidant activities and minimum inhibition concentration for antimicrobial activities. Moreover, the results of the present study might be of help to finding high quality essential oil with antioxidant and antimicrobial activities.

2. Materials and Methods

2.1. Plant Material and Chemicals

Flos Chrysanthemi indici were identified by Prof. Chengsheng Zhang [20] at the sampling sites. Fresh flowers were collected from coastal shoals in Shandong province of China in October 2017. The materials were dried at room temperature and mechanically grounded in a laboratory mill to a homogenous powder and then sieved through a 0.45 mm sifter. All chemicals and solvents used in this study were of analytical grade.

2.2. Extraction Method

2.2.1. Hydro-Distillation (HD)

Flos Chrysanthemi indici powder was weighted (100 g) and placed in a round bottom flask with 1000 mL distilled water as extraction solvent; the material-water mixture was refluxed for about 240 min (until no more extracts was obtained), during which the extract was collected in the side arm of the Clevenger-type apparatus. The product was dried with anhydrous sodium sulfate and stored at 4 °C until analysis. The process was performed three times. Yield percentage was calculated using the volume (mL) of EO per 100 g of dried *Flos Chrysanthemi indici* powder.

2.2.2. Steam-Distillation Method (SD)

Glassware and operating conditions were the same as those used in HD. The vapor produced by the steam generator passed through the EO-rich plant material before condensing in a receiving Clevenger-type apparatus [21]. The product was dried over anhydrous sulfate and stored at 4 °C until analysis. The extraction was performed three times.

2.2.3. Solvent-Free Microwave Extraction (SFME)

Solvent-free microwave extraction was carried out using Uwave-200, Sineo Microwave (Chemistry Technology Co., LTD, Shanghai, China). This instrument was equipped with an infrared temperature sensor, an electromagnetic stirrer, a time controller and a circulating water-cooling system. The extraction procedure followed the method previously described [22] with some modifications. Briefly, 100 g of *Flos Chrysanthemi indici* were placed in a 1 L flask and soaked for 30 min. The flask was placed inside the microwave oven and a condenser was fitted to the top (outside the oven) to collect extracted EO. The vapor produced by microwaves passed through the materials before condensing in a receiving Clevenger-type apparatus. The microwave oven was operated at 800 W for a period of 30 min, during which EO was collected from the top of the Clevenger apparatus. The product was dried over anhydrous sodium sulfate, weighed, and stored in dark brown vials at 4 °C until use. The extraction was performed three times.

2.2.4. Supercritical Fluid Extraction (SFE)

Supercritical fluid extractions were performed on a laboratory scale supercritical fluid extraction system (Waters, SFE-500M1-2-C10, USA). All extractions were performed with supercritical CO_2 in dynamic mode. The ground sample (100 mg) was loaded into 50 mL high pressure vessel connected in series and the CO_2 was pressurized using a high pressure pump and then charged into the vessel at the rate of 20 g/min to maintain the desired pressure of 20 MPa during the cycle. The supercritical CO_2 containing the extract was then passed through a temperature-controlled micrometer valve and was expanded to ambient pressure. The volume of EOs obtained were measured and filled up to 10 mL with hexane and stored in dark brown vials at 4 °C until use. The sample was extracted in triplicate.

2.3. Gas Chromatography-Mass Spectrometry (GC–MS) Analysis

The gas chromatography-mass spectrometry (GC–MS) analysis of the EOs was performed using an Agilent GC–MSD system as previously described [23]. The system consisted of an Agilent Technologies 7890A gas chromatograph (Santa Clara, CA, USA) equipped with a Mars 6100 ion trap mass detector (USA). The separation was done on a DB-5MS capillary column (30 m × 0.25 mm id × 0.25 μm, Bellefonte, PA, USA). The vector gas was helium and the flow rate was 1.0 mL/min, injection volume was 1 μL, 1:20 split ratio; injection temperature was 250 °C, oven temperature was held at 40 °C for 5 min and then increased to 250 °C by 20 °C/min and held at 250 °C for 5 min; 70 eV ionization energy; 30–300 μscan range; quadrupole temperature at 150 °C, ion source temperature at 230 °C. The identity of each component was assigned by comparison of their retention time and mass spectrum with that of a standard from the NIST08 database provided with the GC–MS system software. The relative contents of the components were calculated by comparing its GC peak area to the total areas that are summed from all detected peaks.

2.4. Antioxidant Capacity Evaluation

2.4.1. DPPH Radical-Scavenging Activity Assay

The DPPH radical scavenging activity was measured based on the method described previously [24] with some modifications. In brief, 50 μL of EOs at different concentrations were mixed with 2 mL of DPPH (60 μM) methanol solution. The EOs were mixed well with the solution. After 30 min sealed incubation at room temperature in darkness, absorbance was recorded at 517 nm. The DPPH antioxidant activity was calculated using the following formula: DPPH scavenging activity (%) = [1 − (A_{sample} − $A_{sample\ blank}$)/$A_{control}$] × 100, where A_{sample} represents absorbance of the ethanol solution of DPPH with tested samples; A_{sample} blank represents absorbance of the extracts without the ethanol solution of DPPH and $A_{control}$ was prepared without sample (which was replaced by distilled water) and the sample concentration providing 50% inhibition (IC_{50}) was calculated by plotting inhibition percentages against concentration of the sample. The results were reported as the average of three replicates.

2.4.2. ABTS Radical Scavenging Activity

The ABTS radical scavenging activity of extracts was determined using the method described previously [25] with some modifications. The ABTS+ solution was prepared from the reaction of equal volumes of 2.45 mM potassium persulfate and 7 mM ABTS in a dark place at ambient temperature for 16 h. The ABTS+ solution was adjusted to the absorbance of 0.700 ± 0.02 at 734 nm with ethanol. An aliquot of 0.5 mL EOs (2 mg/mL) was mixed with 1.5 mL ABTS+ solution and sealed and incubated at room temperature for 6 min, the absorbance was measured at 734 nm. The EOs were mixed well with the solution. The ABTS scavenging activity of extracts was calculated as follows: ABTS scavenging activity (%) = [($A_{control}$ − A_{sample})/$A_{control}$] × 100, where $A_{control}$ represented absorbance without sample and A_{sample} represented absorbance with sample. The results were reported as the average of three replicates.

2.5. Evaluation of Antibacterial and Antifungal Activities

Ralstonia solanacearum (tobacco bacterial wilt) were obtained from the Tobacco Research Institute (Qingdao, China). *Acidovorax citrulli* (bacterial fruit spot), *Pseudomonas syringae* pv. *lachrymans* (angular leaf spots in cucumbers), *Erwinia carotovora* (Chinese cabbage soft rot), *Xanthomonas oryzae* pv. *oryzae* and *Rhizoctonia solani* were provided by the Plant Protection Research Institute of CAAS.

The antibacterial activity of EOs obtained by different method against six plant pathogenic bacteria including *Xanthomonas oryzae* pv. *oryzae*, *Ralstonia solanacearum*, *Pseudomonas syringae* pv. *lachrymans*, *Acidovorax citrulli*, *Rhizoctonia solani*, and *Erwinia carotovora* as tested by disk diffusion method. *Ralstonia solanacearum* (tobacco bacterial wilt) were obtained from the Tobacco Research Institute. *Acidovorax citrulli* (bacterial fruit spot), *Pseudomonas syringae* pv. *Lachrymans* (angular leaf spots in cucumbers), *Erwinia carotovora* (Chinese cabbage soft rot), *Xanthomonas oryzae* pv. *Oryzae* and *Rhizoctonia solani* were provided by the Plant Protection Research Institute of CAAS.

Briefly, the six plant bacteria were cultured in Luria–Bertani (LB) broth at 37 °C overnight and then adjusted to a concentration of 10^6 CFU/mL. One hundred microliters of bacterial suspensions were spread on petri dishes containing 10 mL LB broth with agar and the petri dishes were divided into four parts. Sterile filter paper disks (6 mm in diameter) impregnated with 10 μL of EO, dimethylsulfoxide (DMSO) (negative control), and aqueous streptomycin sulfate solutions (10 mg/mL) (positive control) were placed on the cultured plates. The treated petri dishes were sealed and incubated at 37 °C for 24 h.

Antifungal activity of EOs was tested using the method described previously [26] with some modifications. A mycelial agar disk (diameter = 7 mm) isolated from a 7-day culture was placed at the center of a petri dish containing 15 mL Potato-Dextrose-Agar (PDA) medium. EOs were dissolved in DMSO to a final concentration of 1.0 mg/mL, and 15 μL of EOs solutions were added to each of the symmetrically holes (diameter = 6 mm) on the dishes and then incubated at 28 °C for five days. Dishes with added DMSO served as a negative control and streptomycin sulfate solution (10 mg/mL) as a positive control. Antifungal activity was evaluated by observing the inhibition zones after five days.

2.6. Determination of Minimum Inhibitory Concentration (MIC)

The minimal inhibitory concentration (MIC) was determined by the microdilution method as described previously [27]. Subsequent serial dilutions were performed on sterile 96-well microplates. EO was first diluted to the highest concentration of 0.5 mg/mL and serially diluted (two fold dilutions) with DMSO to a final concentration of 0.03125 mg/mL. The bacterial suspension was adjusted to 10^6 CFU/mL using a 0.5 McFarland turbidity standard. Finally, 100 μL of standard bacterial suspension were added to each well containing different concentrations of EO, and the plates were incubated at 37 °C for 24 h. The MIC of the EO was determined as the lowest concentration that visibly inhibited bacterial growth.

The MIC of EO against a plant pathogenic fungal strain was determined as previously described [28] with some modifications. The strain was grown in potato dextrose water (PDW) cultures at 28 °C for 72–120 h until the plate was covered. Fungal samples were harvested using a 5 mm sterilized puncher along the edge of the colonies and added to the center of plates containing different concentrations of EOs. The plates were cultured on a rotary shaker at 150 rpm for 20 s and incubated at 28 °C for 48 h. Each treatment was prepared in triplicate. The MIC values were determined as the lowest concentration required to inhibit fungal growth.

2.7. Scanning Electron Microscopy (SEM)

To evaluate morphological changes in fungi treated with EO, SEM analysis was performed according to previously described methods [23] with some modifications. About five 10 mm mycelium segments were excised from the cultured plates with EOs and washed 3 times with normal saline. The segments were fixed with 2.5% (*v/v*) glutaraldehyde (4 °C, 2 h) and washed with 0.1 M PBS (pH 7.0). Then the samples were dehydrated in a graded series of ethanol (30%, 50%, 70%, and 100%) for 30 min each, dried in liquid CO_2, and viewed with a scanning electron microscope (Hitachi S3400N, Hitachi Science Systems, Ltd., Ibaraki, Japan) operated at 20 kV at a magnification of 100,000×.

2.8. Data Analysis

Statistical analysis was performed using Statistical Analysis System version 9.2 software (SAS Institute, Cary, NC, USA.). All analyses were performed in triplicate and the data were reported as means ± standard deviation (SD) of three samples. $p < 0.05$ was considered statistically significant.

3. Results and Discussion

3.1. Effect of Extraction Method on EO Yield

The results illustrated in Figure 1 showed that *Flos Chrysanthemi indici* EO yields were affected by extraction method. It was reported that SFME and SFE are more eco-friendly and efficient technologies for extracting EOs than HD [29,30]. In this study, the highest yield was obtained by SFE (0.87 ± 0.11% *v/w*, dry weight basis) and the lowest by HD (0.42 ± 0.05%, *v/w*). This is consistent with a previous report that SFE was the optimal process for obtaining a high yield of EOs from clove buds [31]. Supercritical fluid possesses gas-like properties of diffusion, surface tension, liquid-like density, viscosity, and solvation power. Thus, the higher yield may be due to the favorable transports properties of fluids near their critical points allow deeper penetration into matrix and more efficient and faster extraction [8,32].

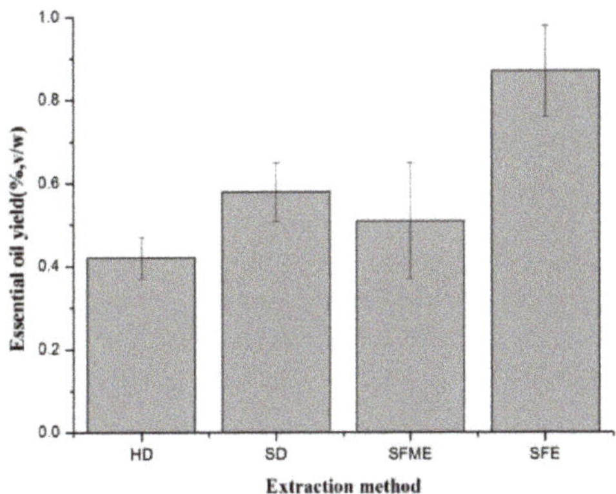

Figure 1. Changes in essential oil yields (%, *v/w*) of *Flos Chrysanthemi indici* under different extraction methods.

3.2. Effect of Extraction method on EOs Composition

In total, 60 compounds were identified by GC–MS analysis which were obtained by four extraction methods, including eight alcohols and acid compounds, eight esters, five aldehydes and ketones, 19 aliphatic hydrocarbons, 11 monoterpenes and sesquiterpenes and nine other compounds. The categorization of assigned compounds in various extracts is shown in Figure 2. It can be observed that the EOs obtained by SFE has a higher content of alcohols and esters and the EOs obtained by HD contains higher terpenes.

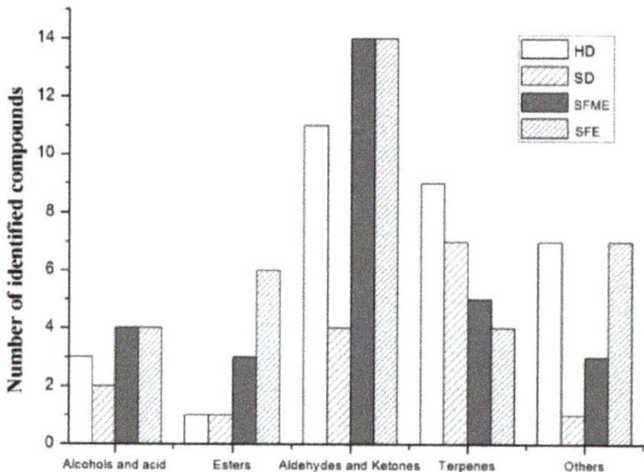

Figure 2. Numbers of the classes of compounds in extracts obtained by different extraction methods.

The main differences among the four extraction methods are composition of individual groups. Comparative composition of the EOs obtained from different extraction method are shown in Table 1, 35 compounds were identified in SFE extracts, and 31 compounds in HD extracts, 15 compounds in SD extracts, 29 compounds in SFME extracts, respectively. Only five compounds were identified in all different EOs (Bornyl acetate, 1,7,7-Trimethylbicyclo[2.2.1]hept-5-en-2-one, Eucalyptol, Camphor and isobenzofuranone). The main compounds detected of individual extraction method were heptenol, borneol, eucalyptol, camphor, and caryophyllene oxide. Other detected compounds were present in concentrations less than 2%.

From the extraction yield and the total numbers of assigned compounds, SFE was a more effective method. Nevertheless, borneol and caryophyllene oxide were not detected in the SFE extract. If we consider the main compounds content, SFME was a more effective technique and contains the highest amount of heptenol, borneol, caryophyllene oxide, and trans-β-Farnesene. The SFME and SFE produced essential oils with higher quantities of valuable oxygenated compounds. This may be due to the reduced heating time required, which partially prevented decomposition oxygenated compounds. Comparatively, it could be noticed that camphor (relative content 39.71 ± 2.89%) and eucalyptol (relative content 11.38 ± 2.47%) were found in the highest quantities in the SD extract and the two compounds are regarded as parameters for assessing the quality of *Flos Chrysanthemi indici* [33]. Camphor has various biological effects including antimicrobial, insecticidal, and antiviral activities [34]. Camphor was also reported in Chuzhou *chrysanthemum* and *Chrysanthemum boreale Makino* essential oil and showed the highest component at vegetative stage in *Chrysanthemum boreale Makino* [35,36].

Table 1. Volatile compounds identified in extracts of *Flos Chrysanthemi indici* prepared by different extraction methods.

No.	Name of Compound	Molecular Weight	CAS Number	Extraction Method [a]							
				HD		SD		SFME		SFE	
				RT [b]	%A [c]	RT	%A	RT	%A	RT	%A
1	Camphene	136.125	000079-92-5	9.642	0.42 ± 0.05	n.i	n.i	n.i	n.i	n.i	n.i
2	Eucalyptol	154.136	000470-82-6	11.388	3.59 ± 1.64	11.418	11.38 ± 2.47	11.38	0.69 ± 0.04	11.38	0.95 ± 0.96
3	1,4-Cyclohexadiene, 1-methyl-4-(1-methylethyl)-	136.125	000099-85-4	11.862	0.22 ± 0.05	11.862	0.28 ± 0.29	n.i	n.i	n.i	n.i
4	Methyl ethyl cyclopentene	110.11	019780-56-4	n.i	n.i	n.i	n.i	12.75	0.66 ± 0.21	12.746	0.66 ± 0.78
5	(Z,Z)-3,5-Octadiene	110.11	007348-80-3	12.754	1.45 ± 2.28	12.767	2.40 ± 0.47	n.i	n.i	n.i	n.i
6	lsobenzofuranone	182.094	054346-06-4	12.923	1.32 ± 0.47	12.936	1.97 ± 2.25	12.919	0.86 ± 0.98	12.923	0.57 ± 0.41
7	(E)-2-Hexen-4-yn-1-ol	96.058	053497-80-6	13.718	2.34 ± 1.12	n.i	n.i	n.i	n.i	n.i	n.i
8	1,7-Octadiene, 3,6-dimethylene	134.11	003382-59-0	n.i	n.i	n.i	n.i	13.722	2.99 ± 0.74	n.i	n.i
9	4-Ethyl-2-hexynal	124.089	071932-97-3	n.i	n.i	13.743	2.30 ± 0.87	n.i	n.i	n.i	n.i
10	endo-Borneol	154.136	000507-70-0	n.i	n.i	n.i	n.i	n.i	n.i	13.866	1.55 ± 0.06
11	Borneol	154.136	010385-78-1	13.909	7.07 ± 1.17	13.925	5.22 ± 2.74	13.913	8.01 ± 2.44	n.i	n.i
12	4-methyl-1-(1-methylethyl)-3-Cyclohexen-1-ol	154.136	000562-74-3	n.i	n.i	14.018	1.20 ± 0.24	n.i	n.i	n.i	n.i
13	1-butenylidene-Cyclohexane	136.125	036144-40-8	14.234	1.05 ± 0.25	n.i	n.i	14.234	0.96 ± 0.47	n.i	n.i
14	heptenol	194.131	1000195-66-0	14.746	17.99±1.25	n.i	n.i	14.746	16.66±2.31	14.716	6.93± 0.09
15	tert-butyl-Benzene	134.11	000098-06-6	n.i	n.i	15.13	24.12±2.36	14.746	0.15 ± 0.47	n.i	n.i
16	Bicyclo[3.1.1]hept-2-en-6-ol, 2,7,7-trimethyl-, acetate, [1S-(1. 1α,5α,6β)]-	194.131	050764-55-1	n.i	n.i	n.i	n.i	15.122	1.71 ± 1.21	n.i	n.i
17	Bornyl acetate	196.146	000076-49-3	15.545	2.06 ± 0.04	15.558	3.15 ± 2.14	15.545	1.33 ± 0.05	15.537	1.51 ± 0.23
18	Cosmene	134.11	000460-01-5	16.074	0.28 ± 0.06	16.078	0.26 ± 0.12	n.i	n.i	n.i	n.i
19	Caryophyllene	204.188	000087-44-5	n.i	n.i	17.473	0.43 ± 0.25	n.i	n.i	17.469	0.31 ± 0.01
20	Isocaryophyllene	204.188	000118-65-0	n.i	n.i	n.i	n.i	n.i	n.i	17.765	0.68 ± 0.99
21	a-Gurjunene	204.188	000489-40-7	18.137	0.27 ± 0.01	n.i	n.i	18.137	0.28 ± 0.04	n.i	n.i
22	Curcumene	202.172	000644-30-4	18.179	0.58 ± 0.11	n.i	n.i	18.184	0.84 ± 0.31	18.175	0.33 ± 0.11
23	Naphthalene	204.188	000473-13-2	18.37	1.28 ± 0.25	n.i	n.i	18.441	3.26 ± 1.02	18.366	0.21 ± 0.14
24	Spathulenol	204.188	025246-27-9	n.i	n.i	n.i	n.i	18.37	1.12 ± 0.08	n.i	n.i
25	1,1,7-Trimethyl-4-methylenedecahydro-1H-cyclopropa[e]azulene	204.188	025246-27-9	18.37	0.53 ± 0.25	n.i	n.i	18.37	1.49 ± 0.79	n.i	n.i
26	trans-β-Farnesene	204.188	018794-84-8	18.251	2.66 ± 0.17	18.725	0.33 ± 0.45	20.213	5.07 ± 2.14	n.i	n.i
27	(Z)-3-Undecen-1-yne	150.141	074744-32-4	n.i	n.i	n.i	n.i	n.i	n.i	19.528	1.23 ± 0.28
28	Caryophyllene oxide	220.183	001139-30-6	19.541	4.23 ± 1.11	19.537	1.09 ± 0.52	19.554	7.33 ± 1.28	n.i	n.i
29	(E,Z)-α-Famesene	204.188	1000293-03-2	n.i	n.i	n.i	n.i	20.213	1.0 ± 0.04	n.i	n.i
30	1R,3Z,9S-4,11,11-Trimethyl-8-methylenebicyclo[7.2.0] undec-3-ene	204.188	1000140-07-3	20.416	1.35 ± 0.28	n.i	n.i	n.i	n.i	n.i	n.i
31	cis-Z-α-Bisabolene epoxide	220.183	1000131-71-2	20.501	0.49 ± 0.74	n.i	n.i	n.i	n.i	20.818	0.39 ± 0.08
32	trans-α-Bergamotene	204.188	013474-59-4	20.97	3.76 ± 0.45	n.i	n.i	n.i	n.i	n.i	n.i
33	2,6-dimethyl-6-(4-methyl-3-pentenyl)-Bicyclo[3.1.1] hept-2-ene	204.188	017699-05-7	n.i	n.i	n.i	n.i	20.974	4.98 ± 2.47	n.i	n.i
34	1,3-Bis-(2-cyclopropyl, 2-methylcyclopropyl)-but-2-en-1-one	258.198	1000222-08-6	n.i	n.i	n.i	n.i	n.i	n.i	20.975	1.31± 0.04

Table 1. *Cont.*

No.	Name of Compound	Molecular Weight	CAS Number	Extraction Method [a]							
				HD		SD		SFME		SFE	
				RT [b]	%A [c]	RT	%A	RT	%A	RT	%A
35	1,7,7-Trimethylbicyclo [2.2.1] hept-5-en-2-one	150.104	022516-10-5	21.288	5.89 ± 0.87	21.266	0.67 ± 0.12	21.296	8.18 ± 2.14	21.262	3.90 ± 1.25
36	Camphor	150.104	022516-10-5	21.288	23.53±5.21	21.266	39.71±2.89	21.296	13.51±1.57	21.262	7.14 ± 2.36
37	Naphthalenone	218.167	091416-23-8	21.381	1.59 ± 0.21	n.i	n.i		1.34 ± 0.51	n.i	n.i
38	6,6-Dimethyl-2-vinylidenebicyclo [3.1.1] heptane	148.125	039021-75-5	n.i	n.i	n.i	n.i	21.389	2.31 ± 0.58	n.i	n.i
39	5-benzyloxy-Pent-1-yne	174.104	057618-47-0	21.609	1.27 ± 0.28	n.i	n.i	n.i	n.i	n.i	n.i
40	octahydro-1,4,9,9-tetramethyl-1*H*-3a, 7-Methanoazulene	206.203	025491-20-7	21.93	0.34 ± 0.14	n.i	n.i	21.93	1.05 ± 0.43	21.93	0.22 ± 0.08
41	1,5-Heptadiyne	92.063	000764-56-7	n.i	n.i	n.i	n.i	n.i	n.i	22.07	1.06 ± 0.41
42	Phenylethyne	102.047	000536-74-3	n.i	n.i	n.i	n.i	n.i	n.i	23.152	0.25 ± 0.09
43	2,3-dihydroxy-1*H*-Cyclopenta[b] quinoxalin-1-one	214.038	023774-23-4	23.512	1.34 ± 0.96	n.i	n.i	23.512	4.76 ± 1.12	23.512	7.63 ± 1.25
44	2,3,4,5-tetramethyl-Tricyclo [3.2.1.02,7] oct-3-ene	162.141	062338-44-7	n.i	n.i	n.i	n.i	n.i	n.i	23.668	7.62 ± 1.47
45	Octacosane	394.454	000630-02-4	n.i	n.i	n.i	n.i	n.i	n.i	24.789	1.81 ± 0.21
46	Heneicosane	296.344	000629-94-7	24.793	0.81 ± 0.22	n.i	n.i	24.793	1.10 ± 0.17	n.i	n.i
47	Linoleic acid	294.256	002566-97-4	n.i	n.i	n.i	n.i	24.729	± 0.03	25.106	2.22 ± 0.09
48	7,10,13-Hexadecatrienoic acid, methyl ester	264.209	056554-30-4	n.i	n.i	n.i	n.i	n.i	n.i	25.161	3.45 ± 1.12
49	Cyclododecyne	164.157	001129-90-4	25.313	0.56 ± 0.12	n.i	n.i	n.i	n.i	25.313	1.75 ± 0.25
50	Tricosane	324.376	000638-67-5	26.569	0.96 ± 0.08	n.i	n.i	26.573	1.25 ± 0.21	26.573	4.03 ± 1.25
51	1-(2,2-dimethyl-1-phenylethynylcyclopropyl)-Ethanol	214.136	1000268-53-7	n.i	n.i	n.i	n.i	n.i	n.i	26.641	5.31± 0.14
52	Acetic acid, 9,9-dioxo-9-thiabicyclo [3.3.1] non-6-en-2-yl ester	230.061	1000185-79-6	n.i	n.i	n.i	n.i	n.i	n.i	27.005	2.86 ± 0.89
53	Pentacosane	352.407	000629-99-2	28.206	0.34 ± 0.08	n.i	n.i	28.205	0.45 ± 0.22	28.206	4.56 ± 1.12
54	2-Hexenedioic acid, 2-methoxy-, dimethyl ester	202.084	056114-71-7	n.i	n.i	n.i	n.i	n.i	n.i	28.903	1.06 ± 0.95
55	ʟ-Cysteine, *N*,*S*-bis (cyclohexylcarbonyl)-, methyl ester	355.182	1000282-52-4	n.i	n.i	n.i	n.i	n.i	n.i	29.254	1.26 ± 0.02
56	Hentriacontane	436.501	000630-04-6	n.i	n.i	n.i	n.i	n.i	n.i	29.719	1.10 ± 0.29
57	Eicosane	282.329	000112-95-8	n.i	n.i	n.i	n.i	n.i	n.i	31.132	0.73 ± 0.28
58	4-Trimethyl-3-hydroxymethyl-5a-(3-methyl-but-2-enyl)cyclhexene	222.198	000144-10-5	32.185	1.79 ± 0.74	n.i	n.i	n.i	n.i	n.i	n.i
59	Ursodeoxycholic acid	392.293	000128-13-2	n.i	n.i	n.i	n.i	n.i	n.i	35.504	0.55± 0.25
60	Chola-5,22-dien-3-ol, (3. β.,22Z)-	342.292	057597-14-5	n.i	n.i	n.i	n.i	n.i	n.i	35.656	1.82 ± 0.87

[a] Relative content of individual compounds are expressed as average relative percent peak area after three replicated ($n = 3$), [b] RT-Retention times; [c]%A—Peak areas, n.i. = not identified.

3.3. Bioactivity of EOs

3.3.1. Antioxidant Activities

Antioxidant activity is a critical indicator to reveal the bioactivity of natural products. It is necessary to use different antioxidant assays to estimate antioxidant activity of a product. In this study, we employed two assays (DPPH radical scavenging assay and ABTS scavenging assay) to determine the antioxidant activity of the four EOs. Ascorbic acid was used as positive controls. The results are given in Table 2.

Table 2. Antioxidant activities of essential oils (EOs) extracted by different methods.

Samples	IC_{50} (mg/mL)	
	DPPH	**ABTS**
Ascorbic acid	0.08 ± 0.01^a	0.12 ± 0.02 [a]
HD	0.53 ± 0.03 [c]	1.14 ± 0.08 [c]
SD	0.58 ± 0.05 [c]	0.84 ± 0.07 [b]
SFME	0.43 ± 0.01 [b]	0.75 ± 0.03 [b]
SFE	0.42 ± 0.03 [b]	1.05 ± 0.11 [c]

[a–c] Means in the same row followed by different letters are significantly different ($p < 0.05$). Each value is represented in terms of mean ($n = 3$) ± SD (Standard deviation).

Regardless of the extraction method and antioxidant assays, all studied EOs possessed significantly weaker free radical scavenging activity than the synthetic antioxidant ascorbic acid. The radical scavenging ability of examined EOs was higher in the DPPH assay than in the ABTS assay. EOs extracted by HD and SD possessed weaker free radical scavenging power than that by SFE and SFME by DPPH assay and EOs extracted by SD and SFME possessed higher antioxidant than that by HD and SFE. These differences may be explained by the complex constituents which are considered to be effective antioxidant compounds. A number of reports are available in the literature that the antioxidant activities of essential oils may act synergistically and the activity may be higher than a single compound [37].

3.3.2. Antibacterial Activity

The antibacterial activities of the obtained EOs are presented in Table 3. This is the first report describing the antibacterial activities of EOs extracted from *Flos Chrysanthemi indici* by different methods. We evaluated the antibacterial activity of EOs against six plant pathogenic bacteria including *X. oryzae*, *R. solanacearum*, *P. syringae*, *A. avenae*, *R. solani, and E. carotovora* and found that the growth of all six strains was inhibited in the presence of EOs (Figure 3).

Table 3. Minimal inhibitory concentrations (MIC) of essential oils extracted from *Flos Chrysanthemi indici* by different methods.

Method	MIC (μg/mL)					
	X. oryzae	*R. solanacearum*	*P. syringae*	*A. avenae*	*R. solani*	*E. carotovora*
Positive control	0.1	25.0	20.0	2.50	–	1.95
HD	62.5	125.0	62.5	62.5	125.0	125.0
SD	125.0	125.0	62.5	125.0	125.0	125.0
SFME	62.5	125.0	62.5	62.5	125.0	62.5
SFE	62.5	62.5	62.5	125	125.0	62.5

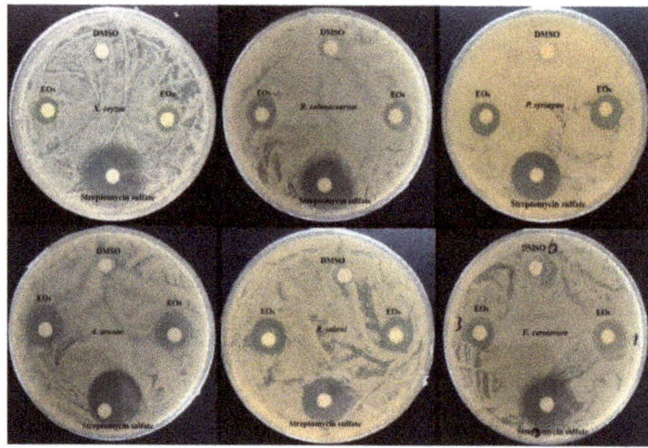

Figure 3. Antibacterial activity of EOs extracted from *Flos Chrysanthemi Indici* evaluated by disk diffusion method.

We compared the antibacterial activities of EOs obtained by different extraction methods based on the MIC against the six bacterial strains. All of the EOs showed varying degrees of antibacterial activity, with MIC ranging from 0.0625–0.125 mg/mL (Table 3). EOs extracted by the four extraction methods had the same MIC against *P. syringae* and *R. solani*. EOs extracted by SFE showed the lowest MIC against *R. solanacearum* (MIC of 62.5 μg/mL), whereas those extracted by HD and SFME showed the lowest MIC against *A. avenae* (MIC of 62.5 μg/mL). They were all higher than the positive control, with MIC 17.15 and 1.953 μg/L, respectively. Compared with other EOs, Mohamed et al. reported that EOs from *Cupressus sempervirens*, *Lantana camara* and *Corymbia citriodora* showed antibacterial effects against *Ralstonia solanacearum* from potato with MIC ranged from 8–5000 μg/mL [38]. Sabir et al. reported that EOs from local plants showed antibacterial against *Pseudomonas syringae* with MIC ranged from 31.25–500 μg/mL [39]. These observations could be explained by differences in the chemical composition of obtained EOs. All EOs obtained by the four methods were rich in active camphor (39.71–7.14%), which could account for the antibacterial activity of *Flos Chrysanthemi indici* EOs.

3.3.3. Antifungal Activity

Several EOs have been shown to be effective in the control of postharvest fungal pathogens, including *A. alternate*. We also tested the antifungal activity of EOs from *Flos Chrysanthemi indici* against *A. alternate* and found that fungal growth was significantly inhibited by SFME obtained EOs with a MIC of 0.625 mg/mL. In prior reports, Castro et al. [40] who tested six kinds of EOs against *A. alternata* in dragon fruit, found that the MIC value ranged from 250–1000 μg/mL. The MIC of EOs of *Flos Chrysanthemi indici* in this study were generally lower than some of the EOs from other plants, such as *Cymbopogon flexuosus*, *Eucalyptus globulus* and *Rosmarinus officinalis* [40].

To investigate the mechanism associated with the antifungal activity of the EO, we examined the cell membrane structure of spores treated with the MIC by SEM (Figure 4C,D). In the control group, the mycelium retained a normal morphology with intact cell wall and membrane (Figure 4C). However, the mycelium of *A. alternata* treated with 0.625 mg/mL EO showed significant morphological changes, including bumps on the surface (Figure 4D). This differs from the reported effects of other EOs on this strain [23].

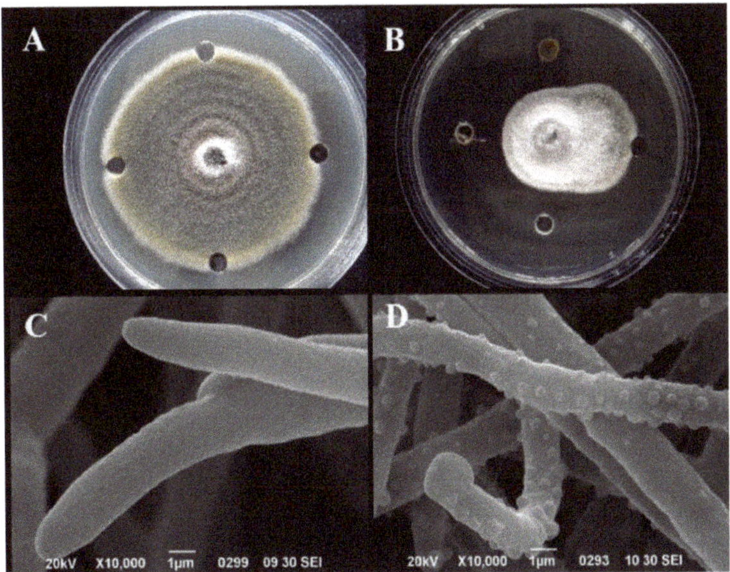

Figure 4. Antifungal bioassay against *A. alternata* and scanning electron micrographs (10,000×) of the morphology of *A. alternata* with and without EO treatment. (**A**) Control group; (**B**) *A. alternate* treated with EOs; (**C**) mycelium of untreated culture (control); (**D**) mycelium of *A. alternata* treated with EOs at MIC.

4. Conclusions

This study presents results of the analysis of EOs obtained from *Flos Chrysanthemi indici* by using various extraction methods. We found that EOs yield, chemical composition, and bioactivities varied according to the extraction method. The extraction yield of SFE was higher than other methods. The SFME and SFE method produced EOs with higher quantities of valuable oxygenated while HD method produced higher quantities terpenes. Bioassay indicated that EOs contain compounds with antioxidant and antimicrobial activities against several pathogenic strains and varied according to different extraction methods. Nonetheless, it was detected that EOs of *Flos Chrysanthemi indici* extracted by any one of the four methods inhibited the growth of these plant pathogens and SFME and SFE proved to be higher in bioactivity. These results can provide essential information for the application of EOs obtained from different extraction methods in food, beverage and pharmaceutical industries.

Author Contributions: Writing—review and editing, C.-L.J., data curation, R.-H.H., methodology, Y.S., Visualization, C.-S.Z., funding acquisition, Y.-Q.L.

Funding: The work was supported financially by the Agricultural Science and Technology Innovation Program (ASTIP-TRIC09).

Conflicts of Interest: We declare that there are no conflicts of interest. We will store the samples for first three years after publication.

References

1. Cheng, W.; Li, J.T.; Hu, C. Anti-inflammatory and immunomodulatory activities of the extracts from the inflorescence of *Chrysanthemum indicum* Linne. *J. Ethnopharmacol.* **2005**, *101*, 334. [CrossRef] [PubMed]
2. Doyeon, L.; Goya, C.; Taesook, Y.; Myeongsook, C.; Byungkil, C.; Hokyoung, K. Anti-inflammatory activity of *Chrysanthemum indicum* extract in acute and chronic cutaneous inflammation. *J. Ethnopharmacol.* **2009**, *123*, 149–154.

3. Wu, Q.; Deng, C.H.; Shen, S.; Song, G.X.; Hu, Y.M.; Fu, D.; Chen, J.K.; Zhang, X.M. Solid-phase microextraction followed by gas chromatography-mass spectrometry analysis of the volatile components of *Flos Chrysanthemi indici* in different growing areas. *Chromatographia* **2004**, *59*, 763–767. [CrossRef]

4. Hashemi-Moghaddam, H.; Mohammadhosseini, M.; Basiri, M. Optimization of microwave assisted hydrodistillationon chemical compositions of the essential oils from the aerial parts of *thymus pubescens* and comparison with conventional hydrodistllation. *J. Essent. Oil Bear Plants* **2015**, *18*, 884–893. [CrossRef]

5. Shrigod, N.M.; Swami Hulle, N.R.; Prasad, R.V. Supercritical fluid extraction of essential oil from mint leaves (*Mentha spicata*): Process optimization and its quality evaluation. *J. Food Process Eng.* **2017**, *40*, 1–9. [CrossRef]

6. Asl, R.M.Z.; Niakousari, M.; Gahruie, H.H.; Saharkhiz, M.J.; Khaneghah, A.M. Study of two-stage ohmic hydro-extraction of essential oil from *Artemisia aucheri* Boiss.: Antioxidant and antimicrobial characteristics. *Food Res. Int.* **2018**, *107*, 462–469.

7. Khalili, G.; Mazloomifar, A.; Larijani, K.; Tehrani, M.S.; Azar, P.A. Solvent-free microwave extraction of essential oils from *Thymus vulgaris* L. and *Melissa officinalis* L. *Ind. Crops Prod.* **2018**, *119*, 214–217. [CrossRef]

8. Bajer, T.; Surmova, S.; Eisner, A.; Ventura, K.; Bajerova, P. Use of simultaneous distillation-extraction, supercritical fluid extraction and solid-phase microextraction for characterisation of the volatile profile of *Dipteryx odorata* (Aubl.) Willd. *Ind. Crops Prod.* **2018**, *119*, 313–321. [CrossRef]

9. Khalvandi, M.; Amerian, M.; Pirdashti, H.; Keramati, S.; Hosseini, J. Essential oil of peppermint in symbiotic relationship with *Piriformospora indica* and methyl jasmonate application under saline condition. *Ind. Crops Prod.* **2019**, *127*, 195–202. [CrossRef]

10. Barth, M.; Hankinson, T.R.; Zhuang, H.; Breidt, F. Microbiological spoilage of fruits and vegetables. *Comp. Micro. Foods Bever.* **2009**, 135–183.

11. Mansfield, J.; Genin, S.; Magori, S.; Citovsky, V.; Sriariyanum, M.; Ronald, P.; Toth, I.A.N. Top 10 plant pathogenic bacteria in molecular plant pathology. *Mol. Plant Pathol.* **2012**, *13*, 614–629. [CrossRef] [PubMed]

12. Tian, Y.L.; Zhao, Y.Q.; Zhou, J.J.; Sun, T.; Luo, X.; Kurowski, C.; Hu, B.S.; Walcott, R.R. Prevalence of *Acidovorax citrulli* in commercial cucurbit seedlots during 2010–2018 in China. *Plant Dis.* **2019**. [CrossRef]

13. El Asbahani, A.; Miladi, K.; Badri, W.; Sala, M.; Aït Addi, E.H.; Casabianca, H.; El, M.A.; Hartmann, D.; Jilale, A.; Renaud, F.N. Essential oils: From extraction to encapsulation. *Int. J. Pharm.* **2015**, *483*, 220–243. [CrossRef] [PubMed]

14. Baczek, K.B.; Kosakowska, O.; Przybyl, J.; Pioro-Jabrucka, E.; Costa, R.; Mondello, L.; Gniewosz, M.; Synowiec, A.; Weglarz, Z. Antibacterial and antioxidant activity of essential oils and extracts from costmary (*Tanacetum balsamita* L.) and tansy (*Tanacetum vulgare* L.). *Ind. Crops Prod.* **2017**, *102*, 154–163. [CrossRef]

15. Hu, Y.; Zhang, J.; Kong, W.; Zhao, G.; Yang, M. Mechanisms of antifungal and anti-aflatoxigenic properties of essential oil derived from turmeric (*Curcuma longa* L.) on *Aspergillus flavus*. *Food Chem.* **2017**, *220*, 1–8. [CrossRef] [PubMed]

16. Tao, N.; Jia, L.; Zhou, H. Anti-fungal activity of *Citrus reticulata Blanco* essential oil against *Penicillium italicum* and *Penicillium digitatum*. *Food Chem.* **2014**, *153*, 265–271. [CrossRef] [PubMed]

17. Araújo, F.M.; Dantas, M.C.S.M.; Silva, L.S.E.; Aona, L.Y.S.; Tavares, I.F.; Souza-Neta, L.C.D. Antibacterial activity and chemical composition of the essential oil of *Croton heliotropiifolius* Kunth from Amargosa, Bahia, Brazil. *Ind. Crops Prod.* **2017**, *105*, 203–206. [CrossRef]

18. Tzortzakis, N.G.; Economakis, C.D. Antifungal activity of lemongrass (*Cympopogon citratus* L.) essential oil against key postharvest pathogens. *Innov. Food Sci. Emerg.* **2007**, *8*, 253–258. [CrossRef]

19. Ribes, S.; Fuentes, A.; Talens, P.; Barat, J.M. Use of oil-in-water emulsions to control fungal deterioration of strawberry jams. *Food Chem.* **2016**, *211*, 92–99. [CrossRef]

20. Zhang, C.S.; Xin, H.; Zou, P. *Plants at Beach in Shandong*; Agricultural Science and Technology Press: Beijing, China, 2017.

21. Sandrine, P.; Christian, G.; Giancarlo, C.; Farid, C. A comparison of essential oils obtained from lavandin via different extraction processes: Ultrasound, microwave, turbohydrodistillation, steam and hydrodistillation. *J. Chromatogr. A* **2013**, *1305*, 41–47.

22. Memarzadeh, S.M.; Pirbalouti, A.G.; Adibnejad, M. Chemical composition and yield of essential oils from Bakhtiari savory (*Satureja bachtiarica* Bunge.) under different extraction methods. *Ind. Crops Prod.* **2015**, *76*, 809–816. [CrossRef]

23. Jing, C.L.; Zhao, J.; Han, X.B.; Huang, R.H.; Cai, D.S.; Zhang, C.S. Essential oil of *Syringa oblata* Lindl. as a potential biocontrol agent against tobacco brown spot caused by *Alternaria alternata*. *Crop Prot.* **2018**, *104*, 41–46. [CrossRef]

24. Babahmad, R.A.; Aghraz, A.; Boutafda, A.; Papazoglou, E.G.; Tarantilis, P.A.; Kanakis, C.; Hafidi, M.; Ouhdouch, Y.; Outzourhit, A.; Ouhammou, A. Chemical composition of essential oil of *Jatropha curcas* L. leaves and its antioxidant and antimicrobial activities. *Ind. Crops Prod.* **2018**, *121*, 405–410. [CrossRef]

25. Clain, E.; Baranauskiene, R.; Kraujalis, P.; Sipailiene, A.; Mazdzieriene, R.; Kazernaviciute, R.; Kalamouni, C.E.; Venskutonis, P.R. Biorefining of *Cymbopogon nardus* from Reunion Island into essential oil and antioxidant fractions by conventional and high pressure extraction methods. *Ind. Crops Prod.* **2018**, *126*, 158–167. [CrossRef]

26. Yangui, I.; Zouaoui Boutiti, M.; Boussaid, M.; Messaoud, C. Essential oils of *Myrtaceae* species growing wild in Tunisia: Chemical variability and antifungal activity against *Biscogniauxia mediterranea*, the causative agent of charcoal canker. *Chem. Biodivers.* **2017**, *14*, e1700058. [CrossRef] [PubMed]

27. Gokhan, A. Composition and antibacterial effect on fodd borne pathogens of *Hibiscus surrattensis* L. calyces essential oil. *Ind. Crops Prod.* **2019**, *137*, 285–289.

28. Jing, C.; Gou, J.; Han, X.; Wu, Q.; Zhang, C. In vitro and in vivo activities of eugenol against tobacco black shank caused by *Phytophthora nicotianae*. *Pestic. Biochem. Phys.* **2017**, *142*, 148–154. [CrossRef] [PubMed]

29. Golmakani, M.T.; Rezaei, K. Comparison of microwave-assisted hydrodistillation with the traditional hydrodistillation method in the extraction of essential oils from *Thymus vulgaris* L. *Food Chem.* **2008**, *109*, 925–930. [CrossRef]

30. Uquiche, E.; Cirano, N.; Millao, S. Supercritical fluid extraction of essential oil from *Leptocarpha rivularis* using CO_2. *Ind. Crops Prod.* **2015**, *77*, 307–314. [CrossRef]

31. Guan, W.; Li, S.; Yan, R.; Tang, S.; Quan, C. Comparison of essential oils of clove buds extracted with supercritical carbon dioxide and other three traditional extraction methods. *Food Chem.* **2007**, *101*, 1558–1564. [CrossRef]

32. Gupta, A.; Naraniwal, M.; Kothari, V. Modern extraction methods for preparation of bioactive plant extracts. *Int. J. Appl. Nat. Sci.* **2012**, *1*, 8–26.

33. Shen, S.; Sha, Y.; Deng, C.; Zhang, X.; Fu, D.; Chen, J. Quality assessment of *Flos Chrysanthemi indici* from different growing areas in China by solid-phase microextraction-gas chromatography-mass spectrometry. *J. Chromatogr. A* **2004**, *1047*, 281–287. [CrossRef]

34. Chen, W.Y.; Vermaak, I.; Viljoen, A. Camphor-A fumigant during the black death and a coveted fragrant wood in ancient Egypt and Babylon—A review. *Molecules* **2013**, *18*, 5434–5454. [CrossRef] [PubMed]

35. Kim, D.Y.; Won, K.J.; Hwang, D.I.; Park, S.M.; Kim, B.; Lee, H.M. Chemical composition, antioxidant and anti-melanogenic activities of essential oils from *Chrysanthemum boreale* Makino at Different Harvesting Stages. *Chem. Biodivers.* **2018**, *15*, e1700506. [CrossRef] [PubMed]

36. Cui, H.; Bai, M.; Sun, Y.; Abdel-Samie, M.A.S.; Lin, L. Anbacterial activity and mechanism of Chuzhou chrysanthemum essential oil. *J. Funct. Foods* **2018**, *48*, 159–166. [CrossRef]

37. Miguel, M.G. Antioxidant and anti-inflammatory activities of essential oils: A short review. *Molecules* **2010**, *15*, 9252–9287. [CrossRef] [PubMed]

38. Abeer, A.M.; Said, I.B.; Hosny, A.Y.; Nader, A.A.; Mohamed, Z.M.S.; Ofelia, M.; Alberto, B. Antibacterial activity of three essential oils and some monoterpenes against *Ralstonia solanacearum* phylotypeIIisolated from potato. *Microb. Pathog.* **2019**, *135*, 103604.

39. Sabir, A.; EI-Khalfi, B.; Errachidi, F.; Chemsi, I.; Serrano, A.; Soukri, A. Evaluation of the potential of some essential oils in biological control against phytopathogenic agent *Pseudomonas syringae* pv. Tomato DC3000 responsible for the tomatoes speck. *J. Plant Pathol. Microbiol.* **2017**, *8*, 1000420.

40. Castro, J.C.; Endo, E.H.; Souza, M.R.D.; Zanqueta, E.B.; Polonio, J.C.; Pamphile, J.A.; Ueda-Nakamura, T.; Nakamura, C.V.; Filho, B.P.D.; Filho, B.A.D.A. Bioactivity of essential oils in the control of *Alternaria alternata* in dragon fruit (*Hylocereus undatus* Haw.). *Ind. Crops Prod.* **2017**, *97*, 101–109. [CrossRef]

 *biomolecules*

Article

Polysaccharide Fraction Extracted from Endophytic Fungus *Trichoderma atroviride* D16 Has an Influence on the Proteomics Profile of the *Salvia miltiorrhiza* Hairy Roots

Wei Peng [1,2,†], Qian-liang Ming [3,†], Xin Zhai [1], Qing Zhang [2], Khalid Rahman [4], Si-jia Wu [1], Lu-ping Qin [5,*] and Ting Han [1,*]

[1] Department of Pharmacognosy, School of Pharmacy, Second Military Medical University, 325 Guohe Road, Shanghai 200433, China
[2] School of Pharmacy, Chengdu University of Traditional Chinese Medicine, No.1166 Liutai Avenue, Chengdu 611137, China
[3] Department of Pharmacognosy, School of Pharmacy, Army Medical University, 30 Gaotanyan Street, Chongqing 400038, China
[4] Faculty of Science, School of Pharmacy and Biomolecular Sciences, Liverpool John Moores University, Byrom Street, Liverpool L3 3AF, UK
[5] School of Pharmacy, Zhejiang Chinese Medical University, Hangzhou 310053, China
* Correspondence: lpqin@zcmu.edu.cn (L.-p.Q.); hanting@smmu.edu.cn or than927@163.com (T.H.); Tel.: +86-21-81871306 (T.H.)
† Authors contributed equally to this manuscript.

Received: 29 July 2019; Accepted: 20 August 2019; Published: 26 August 2019

Abstract: *Trichoderma atroviride* develops a symbiont relationship with *Salvia miltiorrhiza* and this association involves a number of signaling pathways and proteomic responses between both partners. In our previous study, we have reported that polysaccharide fraction (PSF) of *T. atroviride* could promote tanshinones accumulation in *S. miltiorrhiza* hairy roots. Consequently, the present data elucidates the broad proteomics changes under treatment of PSF. Furthermore, we reported several previously undescribed and unexpected responses, containing gene expression patterns consistent with biochemical stresses and metabolic patterns inside the host. In summary, the PSF-induced tanshinones accumulation in *S. miltiorrhiza* hairy roots may be closely related to Ca^{2+} triggering, peroxide reaction, protein phosphorylation, and jasmonic acid (JA) signal transduction, leading to an increase in leucine-rich repeat (LRR) protein synthesis. This results in the changes in basic metabolic flux of sugars, amino acids, and protein synthesis, along with signal defense reactions. The results reported here increase our understanding of the interaction between *T. atroviride* and *S. miltiorrhiza* and specifically confirm the proteomic responses underlying the activities of PSF.

Keywords: polysaccharide fraction; *Trichoderma atroviride*; *Salvia miltiorrhiza*; proteomics; tanshinones

---

1. Introduction

Salvia miltiorrhiza is a Chinese traditional medical herb widely used for preventing and treating disorders of liver, vascular, menstrual, and blood circulation systems. Diterpenoid pigments are the main bioactive constituents of *S. miltiorrhiza* roots, and exert anti-inflammatory properties via significant inhibition of production of NO, IL-1β, and TNF-α [1,2]. In particular, tanshinones, the predominant active constituents in the roots of *S. miltiorrhiza*, possess various promising bioactivities, such as anti-inflammation, anticoagulation, and liver protection, etc. [2,3]. However, the current production of tanshinones is not meeting the current medicinal market needs. Hairy root can yield a great number of secondary metabolites due to its fast growth and biochemical stability. As a typical material in plant science research, *S. miltiorrhiza* hairy roots were investigated for increased production of diterpenoid constituents for pharmaceutical usages.

Endophytic fungi, as non-harm microbes, can form long-term beneficial relationships with the host plant and could also produce some active natural secondary metabolites mainly including small molecules, which can be explained by the hypothesis of balance antagonism. Endophytic fungi is distinctive from normal chemical elicitors with the same effects on secondary metabolites accumulation as biotic inducer as reported [4–6]. Endophytic fungi can break through the first defense line of plants' microbe-associated molecule patterns (MAMPs) [7], and the cell wall is the primary structure of plants that comes in contact with its microbes. Chitin has been identified and studied as a well-known elicitor in pathogens or beneficial fungi; and great importance is also attached to β-glucan in microbe–plant relationship research. As previously reported, polysaccharides are another type of elicitors in the cell walls of fungus and bacterium to induce the resistance of plants, as well as promoting the accumulation of secondary metabolites [7,8]. Consequently, chitins, β-glutan, and polysaccharides are the components of the fungal cell wall [9]; although Chen et al. [10] have implicated that mannose can be listed as a new elicitor which is different from other known elicitors, thus broadening our knowledge and perception of other components of the fungi cell wall as elicitors [11]. In our previous study, we found that the *Trichoderma atroviride*, an endopyhtic fungus isolated from *S. miltiorrhiza*, was able to produce tanshinone I and tanshinone IIA. In addition, we also reported that the extracts of mycelium (EM) and the polysaccharide fraction (PSF) of *T. atroviride* have more efficient influences than live fungus, especially the PSF. Therefore, PSF may be the key active constituent as elicitor of *T. atroviride* in eliciting tanshinones accumulation [6].

Though the balance antagonism hypothesis can explain the interaction mechanism between endophytic fungi and host plant, the molecular signaling transduction process of plants in response to endophytic fungi is still unclear [12]. Due to post-translational events and modifications, proteomics is more specific compared to mRNA when investigating signaling and metabolic processes in plants [13,14]. Currently, the isobaric tags for relative and absolute quantitation (iTRAQ) have developed into significant technology which can analyze, quantify, and compare protein levels linked with liquid chromatography-quadrupole mass spectrometry (LC-MS/MS), providing greater efficiency and accuracy when compared with gel-based techniques. Some reports of novel differently-expressed types of proteins in the interaction between host plant and endophytic fungi do exist [15,16]. For example, the endophytic fungus *Gilmaniella* sp. AL12 can induce *Atractylodes lancea* to yield volatile oils through G protein-mediated signal transduction and the mannose-binding lectin pathway according to proteome analysis [17]. The aim of the present study was to investigate the influences of PSF from *T. atroviride* D16 (D16 PSF) on the proteomics profile of the *S. miltiorrhiza* hairy roots and its possible mechanisms based on iTRAQ strategy.

2. Materials and Methods

2.1. The Culture and Treatment of Hairy Root

Simple hairy roots were separated into different 250 mL conical flasks equipped with 100 mL half-strength B5 liquid medium at 25 °C, 135 rpm in the dark. Every Erlenmeyer flask was loaded with 0.1 g hairy root and was cultured for 3 weeks and the liquid medium was changed weekly. The blank and the treatment group were cultured in half B5 liquid medium, except that the treatment group also contained 60 mg/L PSF. The hairy root samples were taken out for further experiments on the third and sixth days after treatment commenced.

The samples were handled with liquid nitrogen to remove the moisture and then deposited in −80 °C refrigerator. The blank group samples taken on day 3 were numbered A1 and A2, while treatment group samples were numbered B1 and B2. Similarly, the blank group samples taken on day 6 were numbered C1 and C2, while treatment group samples were numbered D1 and D2. However, hairy root samples were disposed by the same method on the twelfth day after treatment for RNA extraction.

2.2. The Preparation of PSF

The hyphae of *T. atroviride* D16 was transferred into the newly-made Potato Dextrose Agar (PDA) medium for 4 days and then moved into 250 mL conical flasks containing 100 mL half B5 liquid medium and cultured for 3 days. Finally, the cultured D16 was added into 5 L conical flask for amplified culturing between 7 to 10 days. Subsequently, the mycelia were collected with vacuum suction filtration, washed three times with distilled water and homogenized by placing in a blender for 10 min, followed by addition of three volumes of distilled water equivalent to the wet weight of mycelia. The water solution of mycelia was then heated to a temperature of 121 °C under high pressure for 60 min and was filtered as fungal elicitor. The filtrate was decompressed and concentrated to an appropriate volume at 75 °C and mixed with 4 times volume of 95% ethanol for 2 days at room temperature. After lyophilization, the precipitate was further subjected to deproteinization with Sevag reagent (chloroform:n-butanol, 4:1, *v/v*), and small molecule impurities were removed as well as proteins less than 2000 kDa. Thereafter, the solution was freeze-dried under vacuum into powder as PSF.

2.3. Monosaccharides Composition of PSF

A total of 5.00 mg D16 PSF was hydrolyzed with trifluoroacetic acid (TFA, 2 M) for 2 h at 120 °C hermetically, the TFA was then evaporated under vacuum, and the residue was subsequently washed three times with methanol. The residue was then mixed with 30 mg $NaBH_4$ and 1mL H_2O and left overnight for reduction followed by the addition of methanol and acetic acid at a ratio of 5:1, the solvent was evaporated under reduced pressure, followed by the addition and evaporation of methanol three times. The sample was then left to dry at 105 °C for 10 min followed by acetylation of the sample by the addition of 3 mL acetic anhydride (AC_2O), for 60 min at 105 °C, after which the reaction was terminated by the addition of 1 mL water followed by vortex mixing. The acetylation products were extracted with 2 mL chloroform, which was then dried by the addition of anhydrous sodium sulfate prior to GC-MS analysis. The standard products (including rhamnose, arabinose, xylose, fructose, mannose, glucose, and galactose, which were purchased from the National Institute for Food and Drug Control, Beijing, China) were subjected to the same procedure as the sample without hydrolyzation.

GC-MS analysis conditions: The size of TR-5MS (Thermo) chromatographic column was 60 m × 0.25 mm × 0.25 mm; initial conditions: The programmed temperature was 140 to 198 °C at 2 °C/min and kept for 4 min. The temperature was then increased to 214 °C at 4 °C/min and 217 °C at 1 °C/min for 4 min. Finally, the temperature was raised to 250 °C at 3 °C/min for 5 min; injection port temperature was 250 °C; the carrier gas was He; the flow rate was 1 mL/min; ion source 250 °C; $m/z = 40$-500; EI = 70 ev.

2.4. HPLC Analyses

The hairy roots were dried at 50 °C in an oven until a constant weight was obtained and then grounded into powder form. The powder was then extracted with chromatographic pure methanol for 60 min under sonication. Next, the methanol extract was applied to the high-performance liquid chromatography (HPLC) system-Agilent-1100-for analysis. The Agilent-1100 was carried out with a H_2O (containing 0.5% HCOOH) (A)/acetonitrile (B) gradient by a ZORBAX SB-C_{18} chromatographic column (250 × 4.6 mm, 5 μm) at 30 °C [6]. The reference standards of dihydrotanshinone I (DT-I), tanshinone I (T-I), cryptotanshinone (CT), and tanshinone IIA (T-IIA) were purchased from the Chengdu Mansite Pharmaceutical Co. Ltd. (Chengdu, China). In addition, dihydrotanshinone I (0.0025, 0.0050, 0.0100, 0.0200, and 0.0400 mg/mL), tanshinone I (0.002, 0.004, 0.006, 0.008, and 0.010 mg/mL), cryptotanshinone (0.001, 0.005, 0.010, 0.025, 0.050, and 0.100 mg/mL), and tanshinone IIA (0.0005, 0.001, 0.002, 0.004, 0.006, and 0.008 mg/mL) were used to prepare the standards curves.

The methanol extract of hairy roots was analyzed and the peaks identified and contrasted in comparison with the available standards.

2.5. RNA Isolation and Real-Time Quantitative PCR Analysis

TRIzol reagent (Invitrogen Co., Carlsbad, CA, USA) was used to extract the total RNA of hairy root and 1 μL RNA was diluted to 100 μL with RNase ddH$_2$O. According to the RNA concentration ratio of 260 and 280 nm UV absorbance value, OD260/280 should be between 1.9 to 2.1 and the total RNA concentration was calculated according to the formula total RNA (μg/mL) = OD260 × 100 (dilution multiple) × 40 μg/mL. The reverse transcription reaction was conducted at 37 °C for 15 min, 85 °C for 5 s termination reaction at ABI 9700 PCR. Finally, 90 μL RNase-free ddH$_2$O was added to 100 μL for storing at −20 °C for Real-time PCR. Primers with the following sequences, LRRK F (5′-TGTGGTAGCTTTGTGGGGTT-3′) and LRRK R (5′-CAGACCGGAGATTGAGTCCG-3′), LRRK-PEPR2 F (5′-TGTGGTAGCTTTGTGGGGTTT-3′) and LRRK-PEPR2 R (5′-GCCAGACCGGAGATTGAGTC-3′), Nucleoporin F (5′-GTCAAAACCTGCAACCACCT-3′) and Nucleoporin R (5′-AGAATGCTGGAGAAATGCCG-3′), probable cation-transporting ATPase F (5′-TGTCCCCATGAATTAGAACTGGT-3′) and probable cation-transporting ATPase R (5′-TGGCGACTTTTGCAGTCAAC-3′), primers identifying the Smactin gene, Smactin F (5′-ATGATAACTCGACGGATCGC-3′) and Smactin R (5′-CTTGGATGTGGTAGCCGTTT-3′), were used as a reference gene to normalize cDNA loading. The real-time PCR amplification was performed in a 384-well plate Roche LC480 thermocycler (Roche Diagnostic Basel, Switzerland) with Super Real PreMix kit (TIANGEN, Beijing, China). Each reaction contained a mixture of 1 μL of diluted cDNA, 0.2 μL of forward primer (10 μM), 0.2 μL of reverse primer (10 μM), 5 μL of SYBR Green PCR Master Mix (TIANGEN, Beijing, China), and 3.6 μL of RNase-free H$_2$O. The reaction mixture was incubated for 15 min at 95 °C, and for 40 cycles of 10 s at 95 °C and 30 s at 60 °C. The relative gene expression was quantified using the comparative CT method.

2.6. Protein Extraction of Salvia Miltorrhiza Hariy Root

A total of 0.5000 g of *S. miltiorrhiza* hairy root sample was grounded into powder in liquid nitrogen and transferred into 10% trichloroacetic acid and precooling acetone solution containing 65 mM Dithiothreitol (DTT). Followed precipitation at −20 °C for 6 h, the sample was centrifuged at 10,000 rpm, at 4 °C for 45 min, and the sediment was re-suspended by the addition of PH standard reagent (STD buffer) according to the volume ratio of 10:1. The sample was then vortexed and blended and placed in a boiling water bath for 5 min and then subjected to ultrasound 10 times (ultrasound for 10 s, interval for 15 s). Finally, the sample was heated in a water bath for 5 min and subjected to centrifuging for 15 min at 10,000 rpm. The supernatant containing the total protein extract of *S. miltiorrhiza* hairy root was then stored at −80 °C until assayed for protein content by the bicinchonininc acid (BCA) and Bradford method.

2.7. The Mass Spectrum Analysis of Protein

The protein samples were enzymolyzed and the peptides solution was qualified by measuring at OD280. The sample were then analyzed by LC-MS preliminarily and performed on AKTA Purifier 100 (GE Healthcare, Chicago, IL, USA) for SCX classification via strong cation chromatographic column. Finally, the samples were separated by the Easy nLC liquid phase system of the nanoliter velocity. Q-Exactive mass spectrometer (Thermo Finnigan, Silicon Valley, CA, USA) was used to analyze the mass spectra of the sample solution after capillary HPLC separation. The analysis time was 120 min, testing method was positive ion mode, and the mother ion scanning ranged from 300 to 1800 *m/z*. The level of mass spectrum resolution was 70,000 (*m/z* 200), AGC target was 3e6, first level of Maximum IT was 10 ms, number of scan ranges was 1, and dynamic exclusion was 40 s. The ion-charge ratio of peptide and the peptide fragments was collected according to the following methods: Full scan acquiring 10 secondary debris each map; secondary excitation types: heated capillary dissociation

(HCD); separate window: 2 *m/z*; secondary mass spectrum resolution: 17,500 (200) *m/z*; microscans: 1; secondary maximum IT: 60 ms; standard collision energy: 30 eV; under fill the wire: 0.1%.

2.8. Data Analysis

The raw data of mass spectrometry was analyzed with software Mascot 2.2 and the identified proteins were blasted with UniProt database for the Gene Ontology (GO) classification annotation. The peak strength value of the reported ion was analyzed quantitatively with Proteome Discoverer1.4 software and the differentially expressed genes of Fc (fold change, Fc (B/A) = B/A, Fc (D/C) = D/C) were screened according to the standard of Fc ≥ 1.50 or Fc ≤ 0.66. The differentially expressed proteins identified in the Arabidopsis database were compared to the database (UniProt, SWISSPROT and TREMBL) by using the software BLASTP (2.2.23+) and the comparison results were extracted. These comments were corresponded to the GO number by the gene association file (gene association. Goa uniprot) and the number of the corresponding small items were counted in the three large catalogs of the biological process, cellular component, and molecular function. According to the bi-directional best hits (BBH) analysis, the corresponding number was mapped to KEGG pathway maps and the total protein of Arabidopsis was calculated by using the hyper-geometric distribution of *p*-value. The false discovery rate (FDR) was then used to correct it and z-score was calculated at the same time. The pathway of significant enrichment was the pathway of FDR is less than 0.05 and z-score > 0. INTACT and String was used to construct differentially expressed protein interaction networks: (a) Searching INTACT to build network (http://www.ebi.ac.uk/intact/); (b) using the software String to build the network (http://string.embl.de).

All experiments, including both control and different treatments of hairy root cultures, HPLC analysis, and semi-quantitative real-time PCR, were performed in triplicate. The results are presented as their mean values and standard deviations (SD). The error bars in the figures represent the standard deviation in biological triplicates. The statistical significance of the differences in root growth and the accumulation of phenolic acids and tanshinones were analyzed by one-way analysis of variance (ANOVA) with SPSS software (version 18.0, Chicago, IL, USA), the heatmap was drawn using Excel software (version: 2010, Microsoft Co., Redmond, WA, USA), and the term significant was used to denote the differences for which *p*-value is < 0.05. The statistical significance of differences in gene transcripts was analyzed by one-sample *t*-test.

3. Results

3.1. Monosacharide Composition and Its Effects on Accumulation of Tanshinones of the D16 PSF

D16 PSF was extracted from *T. atroviride* D16 and it had greatly induced the accumulation of tanshinones in *S. miltiorrhiza* hairy roots, as reported in the previous study [6]. To further determine the effects of hydrolytic constitutes on *S. miltiorrhiza* hairy root, the D16 PSF was hydrolyzed to determine its monosaccharides compositions, and the results revealed that D16 PSF was composited by mannose, glucose, and galactose at a ratio of 5:10:2 (Figure 1). The standard samples of mannose, glucose, and galactose were made into the solution with the concentration of 60 mg/L, respectively, and then put into the *S. miltiorrhiza* hairy root culture medium in triplet for 14 days. The results showed the content of tanshinones in *S. miltiorrhiza* hairy root after 14 days' culture varied in the presence of PSF and the monosaccharides (Table 1). Compared with the control group, PSF (60 mg/L) stimulated the tanshinones accumulation at the greatest extent. Different components of PSF had various effects on promoting synthesis of tanshinones; however, the singular result of each D16 PSF constitute could not catch up with the influence of PSF on the accumulation of tanshinones. Under the treatment of PSF (60 mg/L), the content of dihydrotanshinone I was similar to the results at the same concentration of mannose and galactose respectively, while the equally-increased effects on tanshinone I were among the different treatments of PSF (60 mg/L), glucose, and mannose. Under the treatment of PSF (60 mg/L), the content of cryptotanshinones was 3.08-fold higher than the control group, while less increased effects occurred followed by the treatment of glucose. Galactose made the cryptotanshinones content decreased, while mannose had no effect on the accumulation of cryptotanshinones.

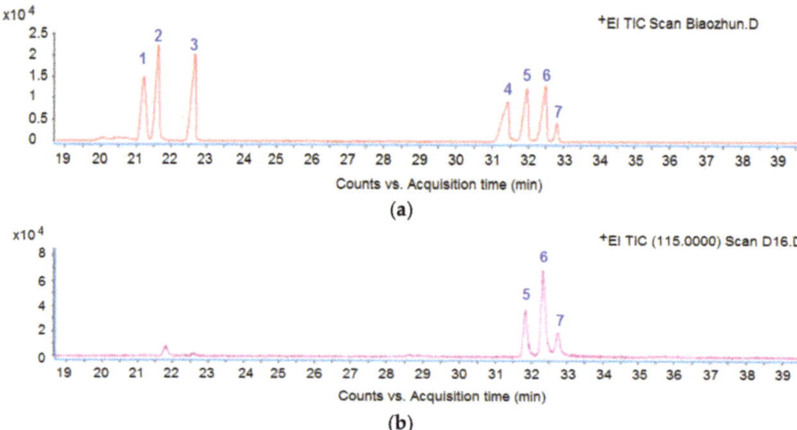

Figure 1. GC-MS analysis of D16 PSF monosaccharides components. (**a**) GC-MS chromatogram of monosaccharide standards; (**b**) GC-MS chromatogram of the D16 PSF sample. The numbers 1 to 7 represented the rhamnose, arabinose, xylose, fructose, mannose, glucose, and galactose, respectively. D16 PSF means polysaccharide fraction from *T. atroviride* D16.

Table 1. The effects of D16 PSF and its monosaccharide constitutes on tanshinones accumulation on the fourteenth day.

Content (mg/g dw)	Control	60 mg/L PSF	60 mg/L Glucose	60 mg/L Mannose	60 mg/L Galactose
Dihydrotanshinone I	0.3548 ± 0.0084	1.0380 ± 0.0455 ***	0.4271 ± 0.0380	1.1641 ± 0.0133 ***	1.0559 ± 0.0367 ***
Cryptotanshinone	0.6421 ± 0.0211	1.9798 ± 0.0708 ***	1.2007 ± 0.0198 ***	0.6799 ± 0.0138	0.4209 ± 0.0099 ***
Tanshinone I	7.1378 ± 0.1455	11.6405 ± 0.1581 ***	9.3103 ± 0.1214 ***	10.7130 ± 0.1163 ***	7.4991 ± 0.1744
Tanshinone IIA	0.2321 ± 0.0022	0.4713 ± 0.0186 ***	0.3630 ± 0.0054 ***	0.2811 ± 0.0049 **	0.2639 ± 0.0056 *

D16 PSF means polysaccharide fraction from *T. atroviride* D16. Data were expressed as mean ± SD ($n = 3$), * $p < 0.01$, ** $p < 0.01$, *** $p < 0.001$, vs. control.

3.2. Differential Proteomics of S. miltiorrhiza in Response to D16 PSF

The differences of protein content in *S. miltiorrhiza* hairy roots treated with D16 PSF were studied by proteomics technique of iTRAQ. Based on the database of *Arabidopsis thaliana*, 3685 feature peptides (unique peptide) belonged to 1476 proteins, which were identified and analyzed for GO classification annotation. As shown in Figure 2, most of proteins can be annotated into the three ontologies of biological process, cellular component, and molecular function. In the biological process annotation, the number of proteins in the metabolic process (1162) and response to stimulus (584) was the largest, and 187 proteins were not annotated. In cellular components annotation, positioning to the cytoplasm (1036), the cell membrane (602), and the cytosol (517) account for the greatest number of proteins, while 211 proteins had not been annotated. In the molecular function annotation, a large number of proteins were divided into catalytic activity (894), nucleotide binding (507), protein binding (417), as well as metal ion binding (315) under the entry and 146 proteins had not been annotated.

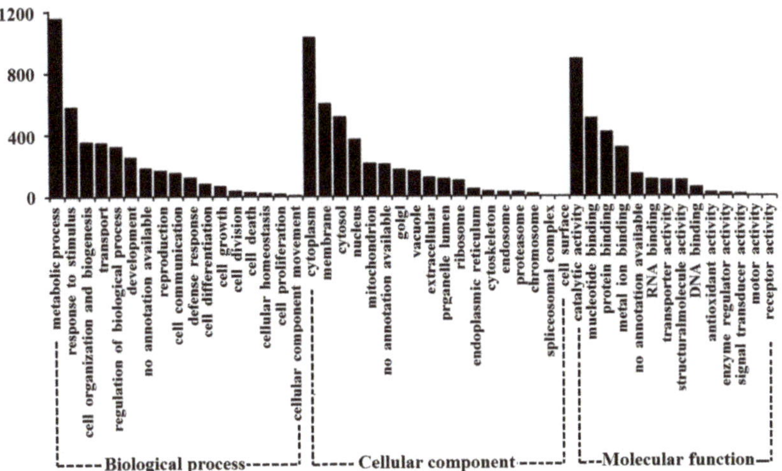

Figure 2. Gene ontology classification of the proteins identified base on *Arabidopsis thaliana* database.

Among the 1476 proteins, 89 differentially expressed proteins were picked out for quantitative analysis (Figure 3). The classification of 89 different proteins was identified and analyzed in the *Arabidopsis thaliana* database. In addition, the number of proteins was corresponded to each category, as shown in Figure 4. The 89 proteins corresponded to 39 KEGG pathways, including amino acid metabolism, sugar metabolism, energy metabolism, and secondary metabolites synthesis. The first 10 KEGG pathways were annotated as shown in Table 2. The 89 different proteins were analyzed by adopting the INTACT and String interaction networks and the results are shown in Figure 5.

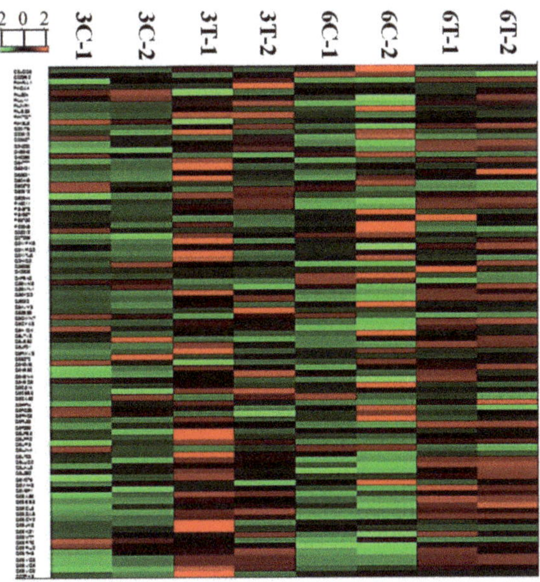

Figure 3. Heatmap of statistically-changed proteins identified based on *Arabidopsis thaliana* database. High- and low-expression are shown in green and red; C and T indicate control and PSF-treated, respectively.

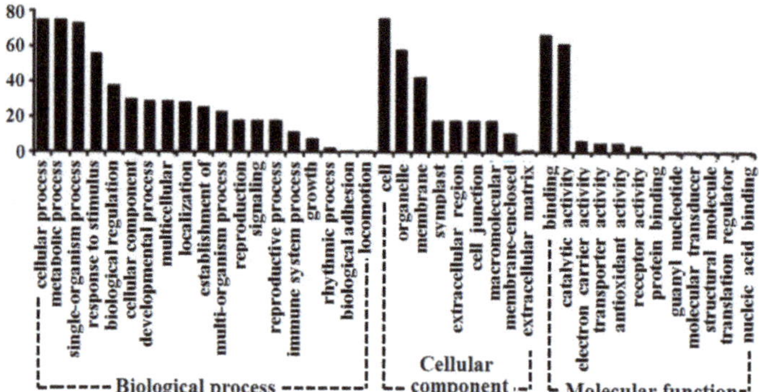

Figure 4. Gene ontology classification of the differential proteins identified based on *Arabidopsis thaliana* database.

Table 2. The 10 KEGG pathways annotated by major differential proteins.

Pathway	Pathway Name	Protein Num.
ath00900	Terpenoid backbone biosynthesis	5
ath00010	Glycolysis/gluconeogenesis	3
ath00460	Cyanoamino acid metabolism	3
ath00940	Phenylpropanoid biosynthesis	3
ath03040	Spliceosome	3
ath00061	Fatty acid biosynthesis	2
ath00480	Glutathione metabolism	2
ath00590	Arachidonic acid metabolism	2
ath00630	Glyoxylate and dicarboxylate metabolism	2
ath03013	RNA transport	2

KEGG means Kyoto Encyclopedia of Genes and Genomes.

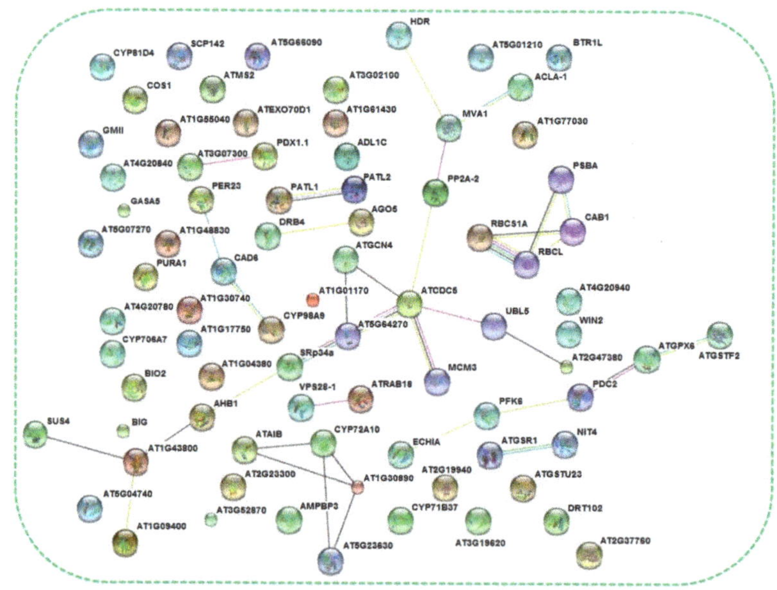

Figure 5. String network of the differential proteins identified based on *Arabidopsis thaliana* database.

3.3. Quantitative Analysis of the 89 Differential Proteins

For the quantitative analysis, 89 differential proteins were mainly involved in signal transduction, redox, amino acid synthesis and metabolism, protein synthesis and metabolism, the biosynthesis of terpenoids, carbohydrate synthesis, metabolism and transport, and other physiological and biochemical process (Table 3, Figure 6).

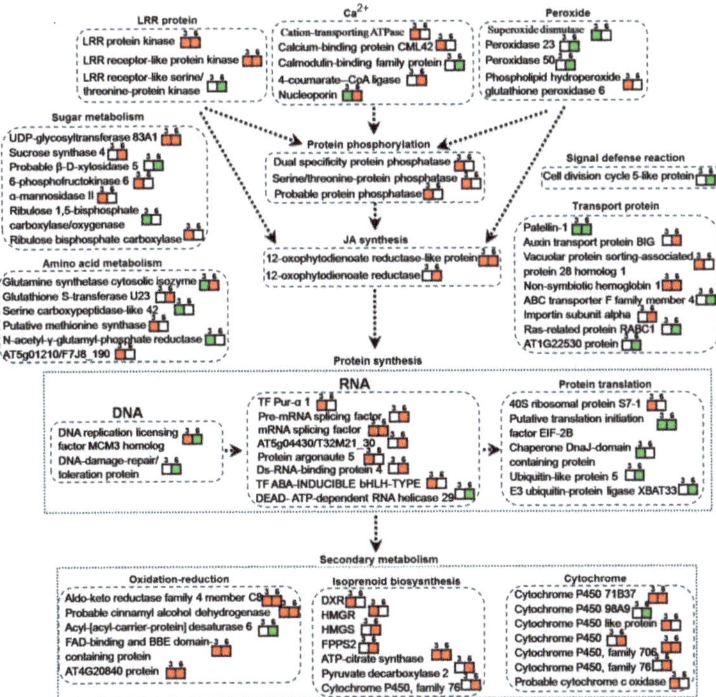

Figure 6. The action network map of 89 differential proteins according to the *Arabidopsis thaliana* database. The number in the box indicates the number of PSF-treated days, red indicated significant increase and green indicated significant reduction.

The 89 differential proteins in signal transduction involves the leucine-rich repeat (LRR) protein, Ca^{2+}-related protein, peroxidase-related protein, protein phosphorylation protein, and the synthesis of jasmonic acid (JA). We found that two LRR-related proteins significantly increased leucine-rich repeat protein kinase family protein and leucine-rich repeat receptor-like protein kinase PEPR2 under the action of *T. atroviride* D16PSF on the third day and sixth day in this study, which revealed that the LRR protein plays an important role in promoting *S. miltiorrhiza* hairy root growth and tanshinone synthesis under the action of *T. atroviride* D16 PSF. The relative expression of LRR and LRR-PEPR on the fourteenth day of *S. miltiorrhiza* after culturing in the presence of D16 PSF was also detected. The results revealed that under the sustained action of D16 PSF, a trend showing an increase in LRR and LRR-PEPR is observed (Figure 7). In our study, we also detected that Ca^{2+}-related protein expression changed significantly. The cation-transporting ATPase and calcium-binding protein increased significantly at day three and calmodulin-binding protein (AT3g52870/F8J2_40) decreased significantly on the sixth day and the relative expression of cation-transporting ATPase still increased greatly on the fourteenth day (Figure 7). 4-coumarate-CoA ligase-like 10 increased significantly responding to Ca^{2+} on the sixth day and nucleoporin related to the level of intracellular Ca^{2+} decreased on the third day, but increased on the sixth day significantly [18]. These significant changes of Ca^{2+}-related proteins showed calcium plays a very important role under the action of *T. atroviride* D16 PSF in *S. miltiorrhiza* hairy root. At the same time, we also found superoxide dismutase, peroxidase, and peroxidase 50 associated with peroxide was lowered significantly. 12-oxophytodienoate reductase is a very important key enzyme in the jasmonic acid (JA) biosynthetic pathway, and two proteins, 12-oxophytodienoate reductase-like protein and 12-oxophytodienoate reductase (At1g04380), related with jasmonic acid were found to be significantly higher in this study [19]. Therefore, the effect of *T. atroviride* D16 PSF on *S. miltiorrhiza* hairy root may also be related to peroxide, protein phosphorylation, and jasmonic acid.

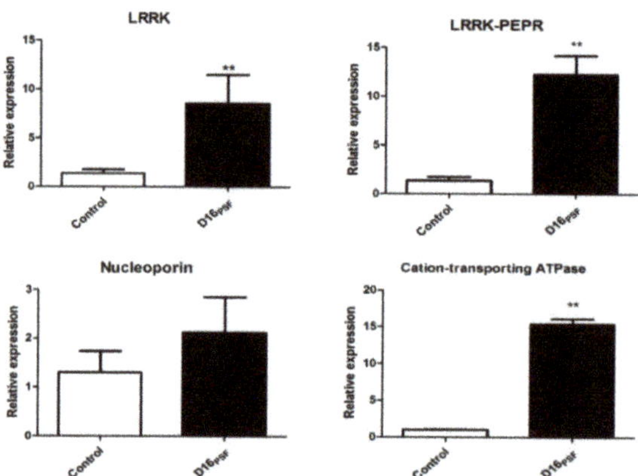

Figure 7. Relative expression of leucine-rich repeat protein kinase (LRRK), leucine-rich repeat protein kinase PEPR (LRR-PEPR), cation-transporting ATPase, and nucleoporin after D16 PSF treatment on the fourteenth day. Values are presented as means ± SD ($n = 3$). * $p < 0.05$; ** $p < 0.01$ versus the control culture.

The 89 differential proteins are mainly involved at the levels of DNA, RNA, and protein translation in protein synthesis. At the level of DNA, the protein–DNA replication licensing factor MCM3 appeared significantly decreased after the first significant rise. DNA and protein damage repair/toleration protein expression was significantly lower on the sixth day. At the level of RNA, transcription factor ABA-inducible bHLH TYPE and transcription factor Pur-alpha 1 was raised significantly on the third day; the pre-mRNA splicing factor SF2-like protein and nuclear protein was also significantly increased. At the level of protein translation, 40 S ribosomal protein was raised significantly on the third day and the rest of the proteins related to the protein degradation were significantly lower such as chaperone DNAJ domain containing protein, the ubiquitin-like 5 and E3 ubiquitin protein–protein ligase XBAT33. These changes showed *T. atroviride* D16 PSF played a positive regulatory role in protein synthesis of *S. miltiorrhiza* hairy root at the protein expression level.

In terms of the differential proteins in secondary metabolism, they were mainly involved in the cytochrome P450 enzymes, redox enzyme, and enzymes related to the terpenoid biosynthetic pathway, and most of these proteins increased significantly under the influence of *T. atroviride* D16 PSF. At the same time, significant changes have taken place in some metabolic activities, like synthesis and metabolism of sugar for providing energy, the amino acid synthesis, and metabolism to provide raw materials for protein synthesis and transporters which have transportation function.

With the deep analysis of these differential expression proteins, it can be presumed that D16 PSF regulates the synthesis and metabolism of saccharides and amino acids, the transcription of genes and the translation of protein through the signal transduction pathways involved in leucine-rich repeat proteins, calcium, peroxides, protein phosphorylation, and jasmonic acid (JA) synthesis. This then results in changes of protein expression involved in secondary metabolism and thus induces metabolic profile changes in *S. miltiorrhiza* hairy roots.

Table 3. The differential proteins of S. miltiorrhiza hairy roots after PSF treatment on third day and sixth day.

Accession	Signal Transduction Description	B/A	D/C	Accession	Signal Transduction Description	B/A	D/C
	Leucine repeated cell proteins and receptors			O49289	Putative DEAD-box ATP-dependent RNA helicase 29	–	0.61
O22178	Leucine-rich repeat protein kinase family protein	1.54	2.09	Q9SCL3	PRE-MRNA SPLICING FACTOR SF2-like protein	–	1.63
Q9FZ59	Leucine-rich repeat receptor-like protein kinase PEPR2	1.70	3.38	Q56YD2	Nuclear protein-like	1.77	1.83
C0LGQ9	Probable LRR receptor-like serine/threonine-protein kinase At4g20940	–	0.65	Q9LZ82	AT5g04430/T32M21_30	–	3.06
	Calcium ion and its related proteins			Q9SJK3	Protein argonaute 5	1.69	–
Q9SVG9	Calcium-binding protein CML42	1.72	–		Protein level		
Q9LT02	Probable cation-transporting ATPase	1.53	–	Q9C514	40S ribosomal protein S7-1	1.71	0.30
Q9SMT7	4-coumarate-CoA ligase-like 10	–	1.50	Q9SRT5	Putative translation initiation factor EIF-2B beta subunit, 3′ partial (Fragment)	0.57	0.64
Q8L7V5	AT3g52870/F8J2_40	–	0.51	F4K8L9	Chaperone DnaJ-domain containing protein	0.51	–
F4IGA4	Nucleoporin, Nup133/Nup155-like protein	0.41	12.75	Q9FGZ9	Ubiquitin-like protein 5	–	0.64
	Protein phosphorylase			Q4FE45	E3 ubiquitin-protein ligase XBAT33	–	0.49
F4K7Q7	Dual specificity protein phosphatase family protein	2.04	–		Terpene biosynthesis		
Q07098	Serine/threonine-protein phosphatase PP2A-2 catalytic subunit	1.50	–	Q9SGY2	ATP-citrate synthase alpha chain protein 1	1.65	2.07
Q8RXV3	Probable protein phosphatase 2C 59	1.75	–	Q9FFT4	Pyruvate decarboxylase 2	–	1.72
	Jasmonic acid synthesis			Q9XFS9	1-deoxy-D-xylulose 5-phosphate reductoisomerase, chloroplastic	1.54	–
Q8GYA3	Putative 12-oxophytodienoate reductase-like protein 1	1.64	2.96	Q94B35	4-hydroxy-3-methylbut-2-enyl diphosphate reductase, chloroplastic	1.54	–
Q593I2	At1g04380 (Fragment)	–	2.08	P54873	Hydroxymethylglutaryl-CoA synthase	–	1.60
	Peroxidase			F4JNF1	Farnesyl diphosphate synthase 2	–	1.50
F4J504	Superoxide dismutase	0.64	–	F4HRA1	Cytochrome P450, family 76, subfamily C, polypeptide 5	–	2.24
O80912	Peroxidase 23	–	0.35		Carbohydrate synthesis and metabolism		
F4JS33	Peroxidase 50	–	0.49	Q9SGA8	UDP-glycosyltransferase 83A1	3.47	2.41
O48646	Probable phospholipid hydroperoxide glutathione peroxidase 6	1.52	–	Q9LXL5	Sucrose synthase 4	–	1.59
	Signal transduction defense			Q9LJN4	Probable beta-D-xylosidase 5	–	0.56
P92948	Cell division cycle 5-like protein	–	0.60	Q9M076	6-phosphofructokinase 6	1.85	–
	Oxidation and reduction			Q9LFR0	Alpha-mannosidase II	1.71	–
	Oxidoreductase			Q85B88	Ribulose 1,5-bisphosphate carboxylase/oxygenase large chain (Fragment)	2.71	–

Table 3. *Cont.*

Accession	Signal Transduction Description	B/A	D/C	Accession	Signal Transduction Description	B/A	D/C
O80944	Aldo-keto reductase family 4 member C8	2.65	1.56	Q42306	Ribulose bisphosphate carboxylase small chain (Fragment)	–	12.36
O65621	Probable cinnamyl alcohol dehydrogenase 6	1.85	2.69		Transport protein		
Q84VY3	Acyl-[acyl-carrier-protein] desaturase 6, chloroplastic	–	0.66	Q56WK6	Patellin-1	0.65	0.57
Q9SA89	FAD-binding and BBE domain-containing protein	2.25	1.67	Q9SRU2	Auxin transport protein BIG	–	1.71
Q9SVG3	AT4G20840 protein	2.68	2.61	O65421	Vacuolar protein sorting-associated protein 28 homolog 1	2.40	–
	Cytochrome			O24520	Non-symbiotic hemoglobin 1	1.92	5.16
Q9LIP3	Cytochrome P450 71B37	1.73	1.65	Q9M1H3	ABC transporter F family member 4	–	0.62
Q9CA60	Cytochrome P450 98A9	–	0.58	F4JL11	Importin subunit alpha	–	1.87
Q0WTJ5	Cytochrome P450 like protein (Fragment)	1.81	–	O23657	Ras-related protein RABC1	–	0.59
Q9LUD0	Cytochrome P450	–	1.51	C0Z3B2	AT1G22530 protein	–	0.60
Q9STH8	Cytochrome P450, family 706, subfamily A, polypeptide 7	1.52	2.09		Other functional protein		
F4HRA1	Cytochrome P450, family 76, subfamily C, polypeptide 5	–	2.24	P54967	Biotin synthase	–	0.38
O22912	Probable cytochrome c oxidase subunit 5C-1	1.72	–	O80575	6,7-dimethyl-8-ribityllumazine synthase, chloroplastic	–	0.64
	Synthesis and metabolism of amino acids			P46011	Bifunctional nitrilase/nitrile hydratase NIT4	–	1.57
Q56WN1	Glutamine synthetase cytosolic isozyme 1-1	0.52	1.67	Q9C8S5	Chlorophyll A-B-binding protein 2, 5′ partial; 1-750 (Fragment)	1.64	–
Q9M9F1	Glutathione S-transferase U23	–	1.63	Q8LF21	Dynamin-related protein 1C	–	1.5
Q9FH05	Serine carboxypeptidase-like 42	0.66	–	Q0WRQ2	Enoyl-CoA hydratase like protein	–	1.52
Q94BN4	Putative methionine synthase	1.63	–	Q38939	GASA5	–	1.78
Q93Z70	Probable N-acetyl-gamma-glutamyl-phosphate reductase, chloroplastic	0.6	–	Q949M6	Putative uncharacterized protein At1g55040	0.48	–
Q9LFB5	AT5g01210/F7J8_190	1.73	–	Q0WPK8	Putative uncharacterized protein At1g72470	1.77	–
	Protein synthesis and degradation				Unknown protein		
	DNA level			Q94BQ9	Integral membrane HRF1-like protein	–	1.64
Q9FL33	DNA replication licensing factor MCM3 homolog	1.90	0.53	P83755	Photosystem Q(B) protein	–	1.97
Q05212	DNA-damage-repair/toleration protein DRT102	–	0.64	Q8GWN7	Putative uncharacterized protein At5g66090/K2A18_17	0.66	0.53
	RNAlevel			Q8LB85	Putative uncharacterized protein	0.58	–
Q8H1D4	Double-stranded RNA-binding protein 4	–	2.28	Q9SBE3	T14P8.11 (Fragment)	3.91	5.23
Q9ZPY8	Transcription factor ABA-INDUCIBLE bHLH-TYPE	1.75	–	Q2HIQ2	At1g01170	–	0.65
Q9SKZ1	Transcription factor Pur-alpha 1	1.91	–				

Note: B/A and D/C are 3 and 6 days, respectively, dealing with the ratio of the amount of protein group and the blank group, 1.5 or higher to increase protein, 0.66 or less for cut protein, "–" indicates no difference. GASA5, GA-Stimulated in Arabidopsis; LRR, Leucine-rich repeat.

4. Discussion

It is known that protein is the carrier of gene function and the executor of the life activities. The high-throughput and large-scale proteomics to study time changes of the expression of all the proteins in the cell or tissue can reveal the physiological and biochemical processes in detail [20,21]. At present, proteomics research mainly is based on the separation of the gel and the liquid phase. The methods based on gel separation include two-dimensional electrophoresis proteomics research and fluorescent difference gel electrophoresis (DIGE) proteomics research, mainly through the gel electrophoresis separation of mixed protein samples. After getting various single protein points, we can screen differential proteins according to the dyed color and identify proteins through the MALDI-MS finally. Methods based on the liquid phase include label-free proteomics research, marked iTRAQ/SILAC proteomics research, and SRM/MRM proteomics research, mainly by means of liquid phase to separate mixed protein samples and through the mass spectrometer connected directly with liquid phase for qualitative and quantitative analysis [22]. Proteomics technology has been widely used in plant growth and development, plant adaptation mechanism to biological and abiotic stress, and the interaction mechanism of microbes [23,24]. iTRAQ (isobaric tags for relative and absolute quantitation) technology is a kind of proteomics technology through four or eight kinds of isotope labels specifically to mark amino groups of polypeptides, and then to identify and separate proteins by LC-MS and analyze relative and absolute content of different protein samples according to the isotope intensity [25,26]. This experiment adopted the iTRAQ technology to tag eight *S. miltiorrhiza* hairy root samples of control group and *T. atroviride* D16 PSF on the third and sixth day.

Leucine-rich repeat (LRR) protein kinase family proteins are the largest known type of transmembrane receptor protein kinases in plants, which have the function of regulation in plant growth and development, hormone signal transduction, and biological and non-biological stress response [27,28]. LRR proteins play a critical role in the process of *Piriformospora indica* promoting growth of *Arabidopsis thaliana* [29,30]. As studied, Pep1 is a 23-amino acid peptide that enhances resistance to a root pathogen, *Pythium irregulare*. Pep1 and its homologs (Pep2 to Pep7) are endogenous amplifiers of innate immunity of *Arabidopsis thaliana* that induce the transcription of defense-related genes and bind to PEPR1, a plasma membrane leucine-rich repeat (LRR) receptor kinase [31]. Our present results revealed that *T. atroviride* D16 PSF could gradually enhance the LRR protein kinase family protein and LRR receptor-like protein kinase PEPR2 in hairy roots of *S. miltiorrhiza*. The LRR receptor-like protein kinase PEPR2 might strengthen the ability of resisting *T. atroviride* D16 PSF, whereas the expression of probable LRR receptor-like serine/threonine-protein kinase At4g20940 decreased. CML42 are calcium-binding proteins that are thought to function as plant signal transduction elements, and were up-regulated and induced by *Spodoptera littoralis* in *Arabidopsis thaliana* followed by Ca^{2+} and phytohormone elevation [32,33]. In addition, CML42 enhanced the content of aliphatic glucosinolate and hyperactivated transcript accumulation of the jasmonic acid (JA)-responsive genes through the negative regulation of the jasmonate receptor Coronatine Insensitive1 (COI 1). As proteomics analysis revealed, this was expressed 1.7-fold higher in *S. miltiorrhiza* hairy root on the third day, which indicated *T. atroviride* D16 PSF may regulate Ca^{2+} and JA elevation. It has been reported that CML42 was not only involved in abiotic stress responses and insect herbivory defense, but also related to trichomes branching and endophytic fungi stimulation as a Ca^{2+} sensor, indicating CML42 is an important calcium-binding protein in plant growth and defense processes. Nucleoporin is a component of the nuclear pore complex, which has strongly attracted the attention of its involvement in hormonal and pathogen/symbiotic signaling [34,35]. Ca^{2+} signal transduction is important in the interaction between microorganisms and plants [36,37]. In the present study, as a key nucleoprotein, the Nup133/Nup155-like protein had a sharp rise on the third day as well as a substantial decrease in *S. miltiorrhiza* hairy root under the treatment of *T. atroviride* D16 PSF. The present results revealed that Nup133 are essential for mRNA export, and the Nup133/Nup155-like protein may regulate the symbiotic signaling transduction and *S. miltiorrhiza* response via mRNA export.

Cation-transporting ATPase is an enzyme protein widely found in biological membranes. Its structure and function is complex and plays an important regulatory role in the biological activities of cells [38]. PDE1 is an encoded P-type ATPase, which is required for the maintenance of phospholipid asymmetry in biological membranes, as membrane asymmetry may play a critical role in the development of infection hyphae by phytopathogenic fungi. The cation-transporting ATPase was up-regulated and induced by *T. atroviride* D16 PSF on the third day, and may provide a chance for interacting with *S. miltiorrhiza* hairy root. The 4-coumarate-Co-A ligase (4CL)-like proteins belong to the adenosine monophosphate (AMP)-binding domain-containing proteins family and widely exist in various plant species [39]. AMP-binding domain-containing 4CLs are critical enzymes in the phenylpropanoid metabolism pathway and drive the carbon flow from primary metabolism to different branches of secondary metabolism in plants [40]. Along with the tanshinones accumulation, the 4-coumarate-Co-A ligase (4CL)-like 10 protein were raised on the sixth day. In support, the 4-coumarate-Co-A ligase (4CL) like protein was up-regulated by *Magnaporthe oryzae* infection, which may be a defense-related AMP-binding protein (AMPBP) that is involved in the regulation of the defense response through salicylic acid (SA) and/or jasmonic acid (JA)/ ethylene (ET) signaling pathways. Phosphoprotein phosphatases (PPP) are present in all eukaryotic organisms, which is an ancient and important regulatory enzyme. The protein phosphatase 2Cs (PP2Cs) from various organisms have been implicated to act as negative modulators of protein kinase pathways involved in diverse environmental stress responses and developmental processes [41]. The Ser/Thr-specific phosphatases are metal-dependent enzymes divided into two major families: The PPP family, which includes protein phosphatases 1, 2A, and 2B (PP1, PP2A, PP2B/calcineurin); and the PPM family, which includes PP2C, which is a highly ABA-induced protein in guard cells once found in Arabidopsis and induced by endophyte fungi *T. atroviride* D16 PSF as well in this study. In addition, dual specificity protein phosphatases were significantly higher on the third day, referring to serine/threonine protein phosphatase and probable protein phosphatase related to protein phosphorylation [42]. *T. atroviride* D16 PSF induced high expression of protein phosphatase 2C and serine/threonine-protein phosphatase PP2A-2 through ABA signaling in *S. miltiorrhiza* hairy root. 12-oxophytodienoate reductase-like protein 1 is an enzyme in the jasmonic acid (JA) biosynthesis pathway [43], and the expression of 12-oxophytodienoate reductase-like protein 1 was gradually enhanced under the treatment of *T. atroviride* D16 PSF involved in the JA biosynthetic pathway. Superoxide dismutases (SODs) are involved in plant adaptive responses to biotic and abiotic stresses although the upstream signaling process which modulates their expression is not clearly understood [44]. Surprisingly, the level of SOD and peroxides were decreased at the treatment of *T. atroviride* D16 PSF, while plant overexpressing antioxidant enzymes had higher tolerance to external stress [45]. Phospholipid hydroperoxide glutathione peroxidase (PHGPX) is the principal enzymatic defense against oxidative destruction of biomembranes [46], and its major function appears to be the scavenging of phospholipid hydroperoxides, thereby protecting cell membranes from peroxidative damage. Gene expression analysis has shown an increase in the levels of PHGPX mRNA in several plant species undergoing different biotic and abiotic stresses, such as pathogen infections, high salt concentrations, exposure to heavy metals, mechanical stimulation, aluminum toxicity, seed germination, salt and osmotic stress, oxidative stress, and chilling stress [47]. PHGPXs may play dual roles as a redox transducer in addition to acting as a H_2O_2 scavenger under stress, thus PHGPX proteins may have different functions in plant cells, with some isoforms functioning in the signal transduction pathway, while others are involved in catalyzing the reduction of harmful products formed by ROS. The effect of various signaling molecules (salicylic acid, JA, ABA, ethylene) and certain protein phosphatase inhibitors (cantharidin and endothall) on the expression of the PHGPX gene in rice seedling leaves has demonstrated an up-regulation of the mRNA levels. These data suggest the role of this enzyme in the induction of defense mechanisms in plant cells subjected to oxidative damage, as a result of exposure to various environmental stresses, is by reducing the phospholipid hydroperoxides formed in the biomembranes.

5. Conclusions

Collectively, the analysis indicated that 89 differentially abundant proteins were involved mainly in protein synthesis, protein folding and degradation, biotic stress defense, photosynthesis, RNA process, signal transduction, and other functions. When induced by *T. atroviride* D16 PSF, *S. miltiorrhiza* hairy roots generally respond through elucidating the synthesis of tanshinones. These results provided valuable information for *T. atroviride* D16 PSF inducing tanshinones accumulation of *S. miltiorrhiza*. Characterization of PSF and proteomics provided an important bioinformatic resource for investigating mechanisms in inducing tanshinones accumulation.

Author Contributions: Conceptualization, L.-p.Q. and T.H.; methodology, W.P., Q.-l.M. and X.Z.; software, X.Z. and Q.Z.; validation, L.-p.Q. and T.H.; formal analysis, Q.-l.M., X.Z. and Q.Z.; investigation, W.P., Q.-l.M., X.Z. and S.-J.W.; resources, T.H.; data curation, Q.-l.M. and X.Z.; writing—original draft preparation, W.P., X.Z. and Q.Z.; writing—review and editing, K.R., T.H. and L.-p.Q.; visualization, Q.-l.M. and X.Z.; supervision, T.H. and L.-p.Q.; project administration, T.H.; funding acquisition, T.H. and L.-p.Q.

Funding: This research was funded by the National Natural Science Foundation of China, grant number 81872953 and the Shanghai Pujiang Program, grant number 18PJD061.

Conflicts of Interest: The authors declare no conflicts of interest.

References

1. Wang, L.; Ma, R.; Liu, C.; Liu, H.; Zhu, R.; Guo, S.; Tang, M.; Li, Y.; Niu, J.; Fu, M.; et al. *Salvia miltiorrhiza*: A potential red light to the development of cardiovascular diseases. *Curr. Pharm. Des.* **2017**, *23*, 1077–1097. [CrossRef] [PubMed]

2. Ren, J.; Fu, L.; Nile, S.H.; Zhang, J.; Kai, G. *Salvia miltiorrhiza* in treating cardiovascular diseases: A review on its pharmacological and clinical applications. *Front. Pharmacol.* **2019**, *10*, 753. [CrossRef] [PubMed]

3. Jiang, Z.; Gao, W.; Huang, L. Tanshinones, critical pharmacological components in *Salvia miltiorrhiza*. *Front. Pharmacol.* **2019**, *10*, 202. [CrossRef] [PubMed]

4. Zhang, B.; Zheng, L.P.; Wang, J.W. Nitric oxide elicitation for secondary metabolite production in cultured plant cells. *Appl. Microbiol. Biotechnol.* **2012**, *93*, 455–466. [CrossRef] [PubMed]

5. Wang, H.; Yang, X.; Guo, L.; Zeng, H.; Qiu, D. PeBL1, a novel protein elicitor from *Brevibacillus laterosporus* strain A60, activates defense responses and systemic resistance in *Nicotiana benthamiana*. *Appl. Environ. Microbiol.* **2015**, *81*, 2706–2716. [CrossRef] [PubMed]

6. Ming, Q.; Su, C.; Zheng, C.; Jia, M.; Zhang, Q.; Zhang, H.; Rahman, K.; Han, T.; Qin, L. Elicitors from the endophytic fungus *Trichoderma atroviride* promote *Salvia miltiorrhiza* hairy root growth and tanshinone biosynthesis. *J. Exp. Bot.* **2013**, *64*, 5687–5694. [CrossRef] [PubMed]

7. Bohlmann, H.; Vignutelli, A.; Hilpert, B.; Miersch, O.; Wasternack, C.; Apel, K. Wounding and chemicals induce expression of the *Arabidopsis thaliana* gene Thi2.1, encoding a fungal defense thionin, via the octadecanoid pathway. *FEBS Lett.* **1998**, *437*, 281. [CrossRef]

8. Smeekens, S. Sugar- induced signal transduction in plants. *Ann. Rev. Plant Physiol. Plant Mol. Biol.* **2000**, *51*, 49–81. [CrossRef]

9. Wang, H.; Zhang, X.; Dong, P.; Luo, Y.; Cheng, F. Extraction of polysaccharides from *Saccharomyces cerevisiae* and its immune enhancement activity. *Int. J. Pharmacol.* **2013**, *9*, 288–296.

10. Chen, F.; Ren, C.G.; Zhou, T.; Wei, Y.J.; Dai, C.C. A novel exopolysaccharide elicitor from endophytic fungus *Gilmaniella* sp. AL12 on volatile oils accumulation in *Atractylodes lancea*. *Sci. Rep.* **2016**, *6*, 120–125. [CrossRef]

11. Escribano, J.; Rubio, A.; Alvarez-Ortí, M.; Molina, A.; Fernández, J.A. Purification and characterization of a mannan-binding lectin specifically expressed in corms of saffron plant (*Crocus sativus* L.). *J. Agric. Food Chem.* **2000**, *48*, 451–457. [CrossRef] [PubMed]

12. Schulz, B.; Rommert, A.K.; Dammann, U.; Aust, H.J.; Strack, D. The endophyte-host interaction: A balanced antagonism? *Mycol. Res.* **1999**, *103*, 1275–1283. [CrossRef]

13. Zhang, H.Y.; Lei, G.; Zhou, H.W.; He, C.; Liao, J.L.; Huang, Y.J. Quantitative iTRAQ-based proteomic analysis of rice grains to assess high night temperature stress. *Proteomics* **2017**, *2*, 160–165. [CrossRef] [PubMed]

14. Zhang, Z.; Zhou, H.; Yu, Q.; Li, Y.; Mendoza-Cózatl, D.G.; Qiu, B.; Liu, P.; Chen, Q. Quantitative proteomics investigation of leaves from two *Sedum alfredii* (Crassulaceae) populations that differ in cadmium accumulation. *Proteomics* **2017**, *17*, 210–216.

15. Kaul, S.; Sharma, T.; Dhar, M.K. "Omics" tools for better understanding the plant endophyte interactions. *Front. Plant Sci.* **2016**, *7*, 955. [CrossRef] [PubMed]

16. Alidrus, A.; Carpentier, S.C.; Ahmad, M.T.; Panis, B.; Mohamed, Z. Elucidation of the compatible interaction between banana and *Meloidogyne incognitavia* high-throughput proteome profiling. *PLoS ONE* **2017**, *12*, e0178438.

17. Gadjeva, M.; Thiel, S.; Jensenius, J.C. The mannan-binding-lectin pathway of the innate immune response. *Curr. Opin. Immunol.* **2001**, *13*, 74–78. [CrossRef]

18. Liu, H.; Guo, Z.; Gu, F.; Ke, S.; Sun, D.; Dong, S.; Liu, W.; Huang, M.; Xiao, W.; Yang, G. 4-Coumarate-CoA ligase-Like gene OsAAE3 negatively mediates the rice blast resistance, floret development and lignin biosynthesis. *Front. Plant Sci.* **2016**, *7*, 13–21. [CrossRef]

19. Wasternack, C. Jasmonates: An update on biosynthesis, signal transduction and action in plant stress response, growth and development. *Ann. Bot.* **2007**, *100*, 681–697. [CrossRef]

20. Song, C.; Zeng, F.; Wu, F.; Ma, W.; Zhang, G. Proteomic analysis of nitrogen stress-responsive proteins in two rice cultivars differing in N utilization efficiency. *J. Integr. Omics* **2011**, *1*, 12–19.

21. He, C.Y.; Zhang, G.Y.; Zhang, J.G.; Duan, A.G.; Luo, H.M. Physiological, biochemical, and proteome profiling reveals key pathways underlying the drought stress responses of *Hippophae rhamnoides*. *Proteomics* **2016**, *16*, 2688–2697. [CrossRef] [PubMed]

22. Ruan, S.L.; Ma, H.S.; Wan, S.H.; Xin, Y.; Qian, L.H.; Tong, J.X.; Wang, J. Advances in plant proteomics–I. Key techniques of proteome. *Yi Chuan* **2006**, *28*, 1472–1486. [CrossRef] [PubMed]

23. Jiang, Q.; Li, X.; Niu, F.; Sun, X.; Hu, Z.; Zhang, H. iTRAQ-based quantitative proteomic analysis of wheat roots in response to salt stress. *Proteomics* **2017**, *17*, 160–165. [CrossRef] [PubMed]

24. Yang, C.; Xu, L.; Zhang, N.; Islam, F.; Song, W.; Hu, L.; Liu, D.; Xie, X.; Zhou, W. iTRAQ-based proteomics of sunflower cultivars differing in resistance to parasitic weed *Orobanche cumana*. *Proteomics* **2017**, *1*, 120–127. [CrossRef] [PubMed]

25. Kambiranda, D.; Katam, R.; Basha, S.M.; Siebert, S. iTRAQ-based quantitative proteomics of developing and ripening muscadine grape berry. *J. Proteome Res.* **2014**, *13*, 555–569. [CrossRef] [PubMed]

26. Zheng, B.B.; Fang, Y.N.; Pan, Z.Y.; Sun, L.; Deng, X.X.; Grosser, J.W.; Guo, W.W. iTRAQ-based quantitative proteomics analysis revealed alterations of carbohydrate metabolism pathways and mitochondrial proteins in a male sterile cybrid pummelo. *J. Proteome Res.* **2014**, *13*, 299–318. [CrossRef]

27. Keerthisinghe, S.; Nadeau, J.A.; Lucas, J.R.; Nakagawa, T.; Sack, F.D. The Arabidopsis leucine-rich repeat receptor-like kinase MUSTACHES enforces stomatal bilateral symmetry in Arabidopsis. *Plant J. Cell Mol. Biol.* **2015**, *81*, 684–694. [CrossRef]

28. Zhou, F.; Yong, G.; Qiu, L.J. Genome-wide identification and evolutionary analysis of leucine-rich repeat receptor-like protein kinase genes in soybean. *BMC Plant Biol.* **2016**, *16*, 58–64. [CrossRef]

29. Shahollari, B.; Vadassery, J.; Varma, A.; Oelmüller, R. A leucine-rich repeat protein is required for growth promotion and enhanced seed production mediated by the endophytic fungus *Piriformospora indica* in *Arabidopsis thaliana*. *Plant J.* **2007**, *50*, 1–13. [CrossRef]

30. Park, S.J.; Moon, J.C.; Yong, C.P.; Kim, J.H.; Dong, S.K.; Jang, C.S. Molecular dissection of the response of a rice leucine-rich repeat receptor-like kinase (LRR-RLK) gene to abiotic stresses. *J. Plant Physiol.* **2014**, *171*, 1645–1649. [CrossRef]

31. Yamaguchi, Y.; Huffaker, A.; Bryan, A.C.; Tax, F.E.; Ryan, C.A. PEPR2 is a second receptor for the Pep1 and Pep2 peptides and contributes to defense responses in Arabidopsis. *Plant Cell* **2010**, *22*, 508–522. [CrossRef] [PubMed]

32. Stephanie, D.; David, C.; Polly, L.; Steven, P.S.; Snedden, W.A. The calmodulin-related calcium sensor CML42 plays a role in trichome branching. *J. Biol. Chem.* **2009**, *284*, 31647–31657.

33. Vadassery, J.; Reichelt, M.; Hause, B.; Gershenzon, J.; Boland, W.; Mithöfer, A. CML42-mediated calcium signaling coordinates responses to spodoptera herbivory and abiotic stresses in Arabidopsis. *Plant physiol.* **2012**, *159*, 1159–1175. [CrossRef] [PubMed]

34. Vasu, S.; Shah, S.; Orjalo, A.; Park, M.; Fischer, W.; Forbes, D. Novel vertebrate nucleoporins Nup133 and Nup160 play a role in mRNA export. *J. Cell Biol.* **2001**, *155*, 339–341. [CrossRef] [PubMed]

35. Martin, B.; Thoe, F.A.; Doan-Trung, L.; Hervé, S.; Isabelle, G.; Isabelle, C. Reduced expression of AtNUP62 nucleoporin gene affects auxin response inArabidopsis. *BMC Plant Biol.* **2016**, *16*, 2–10.

36. Wenping, H.; Yuan, Z.; Jie, S.; Lijun, Z.; Zhezhi, W. De novo transcriptome sequencing in *Salvia miltiorrhiza* to identify genes involved in the biosynthesis of active ingredients. *Genomics* **2011**, *98*, 272–279. [CrossRef]

37. Snedden, W.A.; Fromm, H. Calmodulin as a versatile calcium signal transducer in plants. *New Phytol.* **2001**, *151*, 35–66. [CrossRef]

38. Balhadère, P.V.; Talbot, N.J. PDE1 encodes a P-type ATPase involved in appressorium-mediated plant infection by the rice blast fungus *Magnaporthe grisea*. *Plant Cell* **2001**, *13*, 1987–1991. [CrossRef]

39. Zhang, X.C.; Yu, X.; Zhang, H.J.; Song, F.M. Molecular characterization of a defense-related AMP-binding protein gene, OsBIABP1, from rice. *Biomed Biotechnol.* **2009**, *10*, 731–739. [CrossRef]

40. Soltani, B.M.; Ehlting, J.; Douglas, C.J. Genetic analysis and epigenetic silencing of At4CL1 and At4CL2 expression in transgenic Arabidopsis. *Biotechnol. J.* **2006**, *1*, 1124–1136. [CrossRef]

41. Xue, T.; Wang, D.; Zhang, S.; Ehlting, J.; Ni, F.; Jakab, S.; Zheng, C.; Zhong, Y. Genome-wide and expression analysis of protein phosphatase 2C in rice and Arabidopsis. *BMC Genom.* **2008**, *9*, 550–561. [CrossRef] [PubMed]

42. Cohen, P. The structure and regulation of protein phosphatases. *Ann. Rev. Biochem.* **1989**, *58*, 453–459. [CrossRef] [PubMed]

43. Sanders, P.M.; Lee, P.Y.; Biesgen, C.; Boone, J.D.; Beals, T.P.; Weiler, E.W.; Goldberg, R.B. The arabidopsis DELAYED DEHISCENCE1 gene encodes an enzyme in the jasmonic acid synthesis pathway. *Plant Cell* **2000**, *12*, 1041–1047. [CrossRef] [PubMed]

44. Xing, Y.; Chen, W.; Jia, W.; Zhang, J. Mitogen-activated protein kinase kinase 5 (MKK5)-mediated signalling cascade regulates expression of iron superoxide dismutase gene in Arabidopsis under salinity stress. *J. Exp. Bot.* **2015**, *66*, 5971–5978. [CrossRef] [PubMed]

45. Novo-Uzal, E.; Gutiérrez, J.; Martínez-Cortés, T.; Pomar, F. Molecular cloning of two novel peroxidases and their response to salt stress and salicylic acid in the living fossil Ginkgo biloba. *Ann. Bot.* **2014**, *114*, 923–929. [CrossRef]

46. Chen, S.; Vaghchhipawala, Z.; Li, W.; Asard, H.; Dickman, M.B. Tomato phospholipid hydroperoxide glutathione peroxidase inhibits cell death induced by Bax and oxidative stresses in yeast and plants. *Plant Physiol.* **2004**, *135*, 1630–1641. [CrossRef] [PubMed]

47. Jain, P.; Bhatla, S.C. Signaling role of phospholipid hydroperoxide glutathione peroxidase (PHGPX) accompanying sensing of NaCl stress in etiolated sunflower seedling cotyledons. *Plant Signal. Behav.* **2014**, *9*, 977746–977747. [CrossRef]

 *biomolecules*

Article

The Methodological Trends of Traditional Herbal Medicine Employing Network Pharmacology

Won-Yung Lee [1], Choong-Yeol Lee [1], Youn-Sub Kim [2,*] and Chang-Eop Kim [1,*]

[1]   Department of Physiology, College of Korean Medicine, Gachon University, Seongnam 13120, Korea
[2]   Department of Anatomy-Pointology, College of Korean Medicine, Gachon University, Seongnam 13120, Korea
*   Correspondence: ysk5708@hanmail.net (Y.-S.K.); eopchang@gachon.ac.kr (C.-E.K.);
    Tel.: +82-31-750-5416 (Y.-S.K.); +82-31-750-5416 (C.-E.K.)

Received: 18 June 2019; Accepted: 6 August 2019; Published: 13 August 2019

Abstract: Natural products, including traditional herbal medicine (THM), are known to exert their therapeutic effects by acting on multiple targets, so researchers have employed network pharmacology methods to decipher the potential mechanisms of THM. To conduct THM-network pharmacology (THM-NP) studies, researchers have employed different tools and databases for constructing and analyzing herb–compound–target networks. In this study, we attempted to capture the methodological trends in THM-NP research. We identified the tools and databases employed to conduct THM-NP studies and visualized their combinatorial patterns. We also constructed co-author and affiliation networks to further understand how the methodologies are employed among researchers. The results showed that the number of THM-NP studies and employed databases/tools have been dramatically increased in the last decade, and there are characteristic patterns in combining methods of each analysis step in THM-NP studies. Overall, the Traditional Chinese Medicine Systems Pharmacology Database and Analysis Platform (TCMSP) was the most frequently employed network pharmacology database in THM-NP studies. Among the processes involved in THM-NP research, the methodology for constructing a compound–target network has shown the greatest change over time. In summary, our analysis describes comprehensive methodological trends and current ideas in research design for network pharmacology researchers.

Keywords: network pharmacology; traditional herbal medicine; methodological trend

---

1. Introduction

Traditional herbal medicine (THM) has maintained the health of Asian people for thousands of years and built a unique medical system based on empirically accumulated knowledge. Billions of people around the world are taking THM daily, and the drug development field considers THM to be a source of inspiration [1,2]. The research indicates that THM's therapeutic effects may involve various biomolecules [3]. However, due to the complexity of THM and the limitations of experimental applications, specific mechanisms of action have been fully elucidated for only a few THMs [4]. This is a major obstacle to THM's modernization and wider application to modern healthcare.

Network pharmacology has emerged as a promising approach to accelerate drug development and elucidate the mechanisms of action of multiple target components [5]. It understands disease as a perturbation of interconnected complex biological networks and identifies the mechanisms of drug action by network topology [6,7]. The conceptual elements of network pharmacology were derived from systems biology, which can address both the connectivity and the interdependence of individual components [8]. The core idea of network pharmacology is well suited for analyzing the multi-targeted agents, so network pharmacology methods may be appropriate for identifying the complex mechanisms of THM.

In the last decade, researchers have employed network pharmacology methods to elucidate the potential targets and toxicity of THMs [9]. THM-network pharmacology (THM-NP) studies are conducted by constructing an herb–compound–target (H-C-T) network by integrating the herbal constituent data and drug-target interactions (DTIs) information. Then, the target network is analyzed to interpret related biological functions, pathways, and diseases (Figure 1). Since there are no gold standard methods for THM-NP studies yet, researchers have developed and applied various tools and databases for each step.

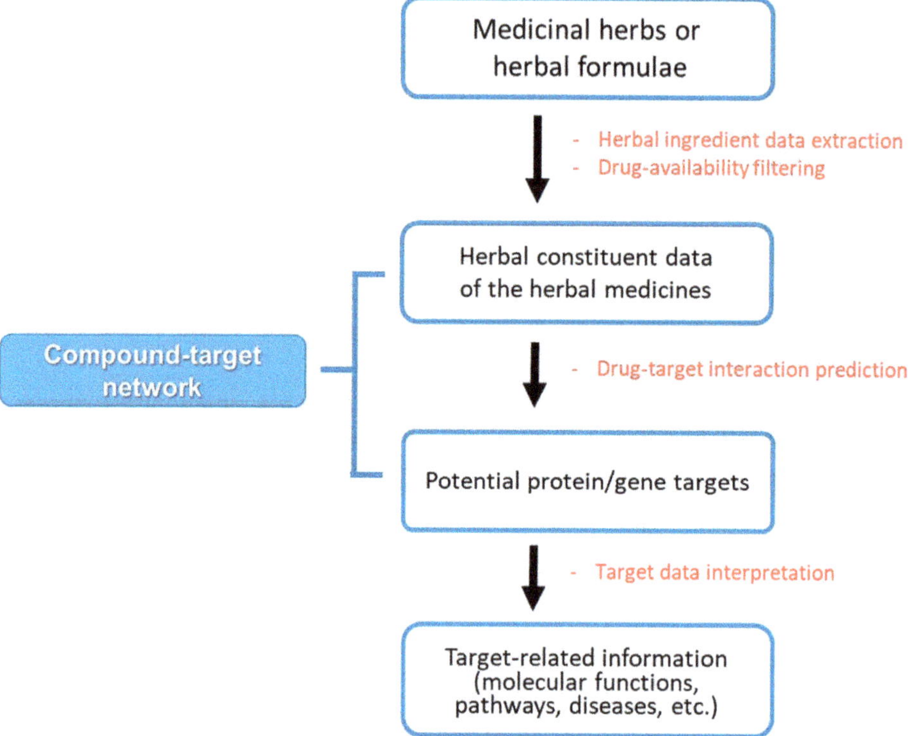

Figure 1. The general framework of network pharmacology analysis of herbal medicine.

Several studies have described THM-NP researches by summarizing network pharmacology databases for THM and illustrating several representative applications [10–15]. Although these studies contributed to a better understanding of THM-NP studies, they were limited in providing quantitative information on the frequencies, variations, and combinatorial patterns of the employed methods in THM-NP studies. In this study, we systematically attempted to capture the methodological trends of the THM-NP research field. We identified the tools and databases employed to conduct THM-NP studies and visualized their frequency, variation, and combinatorial pattern in a step-by-step manner. The THM-NP studies were identified by searching PubMed and then preprocessed. We also constructed and analyzed the co-author and affiliation networks to identify how the diverse methods for THM-NP studies are employed and shared among researchers in the field. We believe that analyzing the methodological trends will provide a comprehensive understanding and valuable insights into the THM-NP research field.

2. Materials and Methods

2.1. Search Strategy

The literature search was performed in PubMed (https://www.ncbi.nlm.nih.gov/pubmed/) from January 2000 to December 2018. The search language was restricted to English. The search terms used were NP-related terms ("network pharmacology" OR "network analysis" OR "system-level" OR "systems-level" OR "systems pharmacology" OR "systems biology" OR "bioinformatics") in [title] AND THM-related terms ("oriental medicine" OR "traditional medicine" OR "traditional Asian medicine" OR "Chinese medicine" OR "Kampo medicine" OR "Korean medicine") in [title/abstract]. The search range of THM-related terms was extended to [title/abstract] since the titles of THM studies generally contain only the name of herbs or herbal formulae that are difficult to search.

2.2. Inclusion Criteria

We considered a THM-NP study as the original article that analyzed a THM's mode of action through the construction of a compound-target network. Full-text articles from the literature search were checked to determine their eligibility. There was no restriction regarding in vivo, in vitro, and in silico studies. THM was considered as (1) extract(s) from a single herb; (2) preparation(s) containing multiple herbs; (3) proprietary herbal product(s); and (4) molecule(s) derived from a single herb.

2.3. Study Selection and Data Extraction

Two authors (W.Y. Lee and C.E. Kim) independently examined titles, abstracts, and journals to select eligible THM-NP studies. When articles were duplicated, only the most recent information was included. Then the full text of potentially relevant studies was retrieved. Two authors (W.Y. Lee and C.E. Kim) independently examined the full-text records to determine which studies met the inclusion criteria. Disagreements about the study selection were resolved by rechecking whether the studies met our criteria for inclusion.

Authors extracted the following data from the included THM-NP studies: authors, affiliations, publication years, tools, and databases. Synonyms for tools and databases were merged and counted under a single keyword. DTpre and SysDT [16] were considered to be the same method as Traditional Chinese Medicine Systems Pharmacology Database and Analysis Platform (TCMSP, http://lsp.nwu.edu.cn/tcmsp.php) [17] since these methods were originally developed and implemented in TCMSP.

2.4. Categorizing Drug-Target Interaction Methods

To capture the trends in the methods for constructing compound-target networks, we categorized DTI methods into four groups by their hypothesis and which information was used: the chemogenomic approach, docking simulation approach, ligand-based approach, and others [18–20]. The chemogenomic approach predicts potential compound–target pairs similar to validated compound-target pairs. This method is based on the assumption that a compound–target pair is likely to interact with high similarity to a validated compound-target interaction in terms of chemo–physical properties [21]. The docking simulation approach predicts the binding conformation of small-molecule ligands to the appropriate binding site of the target using 3D structural information on the compounds and protein targets [22]. The key hypothesis of this approach is that compounds with a high binding affinity at the binding site are likely to interact with the target [23]. The ligand-based approach predicts interactions by comparing a new ligand to known proteins' ligands based on the hypothesis that similar molecules usually bind to similar proteins [24]. DTI methods that do not belong to the above categories were assigned to the category "others", such as data mining techniques, high-throughput screening, and databases that integrate drug-target interaction information from heterogeneous sources.

2.5. Construction of the Co-Author Network and Affiliation Network

The author network and affiliation network were constructed to identify the methodological characteristics of corresponding authors and affiliations. The nodes in each network represent authors or affiliations, and the edges represent co-occurrences of authors or affiliations in THM-NP studies. The frequencies of employed DTI and drug availability methods were counted for each corresponding author or affiliation. These methodologies were mapped to the author network and the affiliation network. Cytoscape 3.7.1 (http://www.cytoscape.org/) was used to visualize the networks [25].

3. Results

3.1. Description of the Search

We initially found 233 potentially relevant articles from PubMed. The search was conducted using combined keywords consisting of THM-related terms and NP-related terms. Another 15 potentially relevant articles were included by searching references in other THM-NP studies or review articles. Titles, abstracts, and journal names were screened, and 167 studies were considered potentially eligible for inclusion. Of these, 20 articles were excluded after screening the full texts. Finally, a total of 147 THM-NP studies were included in our study (Figure 2). The included THM-NP studies are listed in Supplementary Table S1.

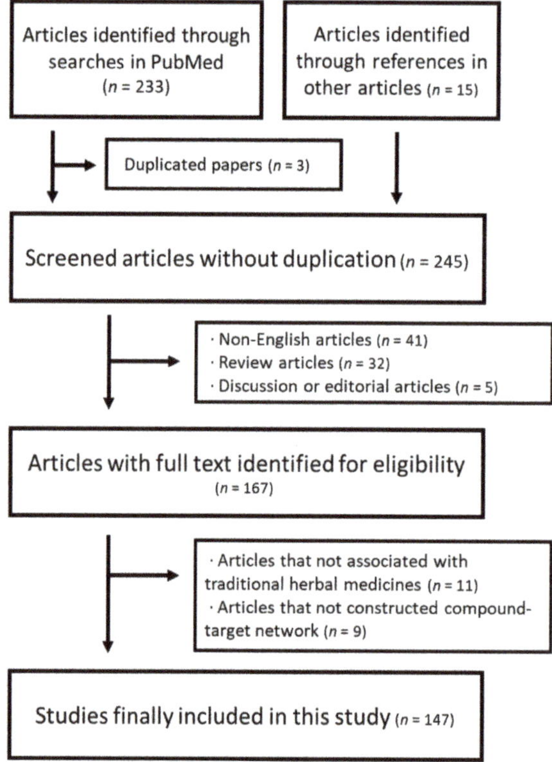

Figure 2. The flowchart of the study selection process.

We identified the number of published THM-NP studies over time, along with the affiliations involved in the studies. THM-NP studies that met our criteria for inclusion have appeared since 2011. In 2011, only three affiliations published two THM-NP studies, but in 2018, the number of affiliations and studies increased to 83 and 52, respectively (Figure 3). The increased number of papers and of affiliations involved indicates that the THM-NP research fields have continuously grown.

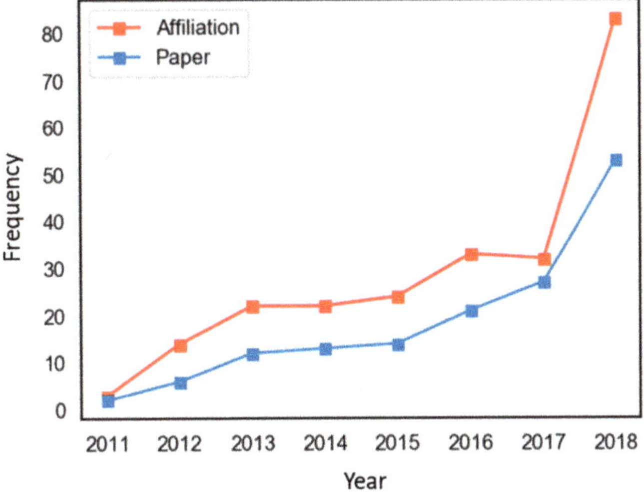

Figure 3. Annual publication trends of traditional herbal medicine-network pharmacology (THM-NP) studies.

3.2. Methodological Trends in Constructing the Herb-Compound Network

We next investigated the trends in employed methods in THM-NP studies. Commonly used databases and tools are described in Table 1 (see Supplementary Table S2 for complete lists). The construction of the herb-compound network is the first step of a THM-NP study. Among the databases for herbal medicines, TCMSP was most commonly used to construct an herb-compound network. Additionally, some THM-NP researchers used their own experimental results (e.g., Ultra Performance Liquid Chromatography (UPLC) or High-performance liquid chromatography) to identify ingredients of the herbal medicines in their studies (Figure 4A).

Table 1. The public databases related to traditional herbal medicine-network pharmacology (THM-NP) studies.

Name	Providing Information			Description	Website	PMID (Reference)
	H-C	C-T	TI			
TCMSP	○	○	○	A system of pharmacology platforms that provide information about ingredients, ADME-related properties, targets, and diseases of herbal medicines.	http://lsp.nwu.edu.cn/tcmsp.php	24735618 [26]
TCMID	○	○	○	An integrative database which stores the information of herbs, herbal compounds, targets, and their related information from different resources and through text-mining method	http://www.megabionet.org/tcmid/	23203875 [17]
TCM Databasetaiwan	○			A database that includes the information of molecular properties and substructures, TCM ingredients with their 2D and 3D structures.	http://tcm.cmu.edu.tw/	21253603 [27]
PharmMapper		○		A web server for potential drug target identification by reversed pharmacophore matching the query compound against an in-house pharmacophore model database	http://lilab.ecust.edu.cn/pharmmapper/	20430828 [28]

Table 1. *Cont.*

Name	Providing Information			Description	Website	PMID (Reference)
	H-C	C-T	TI			
STITCH		○		A database that integrates disparate data sources of interactions between proteins and small molecules	http://stitch.embl.de/	18084021 [29]
TTD		○	○	A database that provides information about the therapeutic targets in the literature, targeted disease condition, and the corresponding drugs/ligands directed at each of these targets.	http://xin.cz3.nus.edu.sg/group/ttd/ttd.asp	11752352 [30]
SEA		○		A computational tool that relates proteins and chemicals based on the set-wise chemical similarity among their ligands.	http://sea.bkslab.org/	17287757 [31]
HIT		○	○	A comprehensive and fully curated database for herbal ingredients with protein target information	http://lifecenter.sgst.cn/hit/	21097881 [32]
Drugbank		○	○	A unique bioinformatics and cheminformatics resource that combines detailed drug data with comprehensive drug target information	https://www.drugbank.ca/	16381955 [33]
KEGG			○	A database resource for understanding high-level functions and utilities of the biological system from molecular-level information	https://www.genome.jp/kegg/	9847135 [34]
Gene ontology			○	The world's largest source of information on the functions of genes	http://geneontology.org/	18792943 [35]
OMIM			○	A comprehensive and authoritative compendium of human genes and genetic phenotypes	https://www.omim.org/	11752252 [36]
PharmGkb			○	A database for the aggregation, curation, integration, and dissemination of knowledge regarding the impact of human genetic variation on drug response	https://www.pharmgkb.org/	11752281 [37]
Genecards			○	A searchable and integrated database of human genes that provides concise genomic related information, on all known and predicted human genes.	https://www.genecards.org/	12424129 [38]

H-C, herb-compound network construction; C-T, compound-target network construction; TI, target interpretation. TCMSP, Traditional Chinese Medicine Systems Pharmacology Database; TCMID, Traditional Chinese Medicine Integrated Database; STITCH, Search Tool for Interactions of Chemicals; TTD, Therapeutic Target Database; SEA, Similarity Ensemble Approach; HIT, Herb Ingredients' Targets; KEGG, Kyoto Encyclopedia of Genes and Genomes database; OMIM, Online Mendelian Inheritance in Man; PharmGkb, The Pharmacogenetics Knowledge Base.

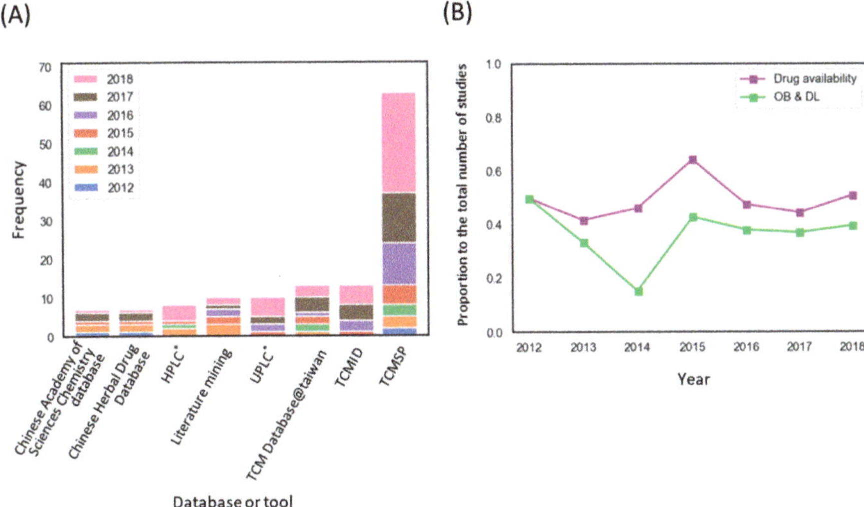

Figure 4. Methodological trends in the construction of the herb–compound network. (**A**) Frequency of databases and tools used to identify constituents of herbal medicine. * Analytical techniques to identify the ingredients of herbal medicines; Ultra Performance Liquid Chromatography (UPLC), High-performance liquid chromatography (HPLC) (**B**) The application rate of the drug availability method by year. Note that Obioavail (OB) and drug-likeness (DL) are among the most commonly used drug availability assessment methods in THM-NP studies. THM-NP studies in 2011 were excluded from the visualization due to the low frequency (*n* = 2). TCMSP, Traditional Chinese Medicine Systems Pharmacology Database; TCMID, Traditional Chinese Medicine Integrated Database.

Because information on the absorption, distribution, metabolism, and excretion (ADME) properties of herbal medicines in humans are lacking, researchers have employed evaluation methods or machine learning tools to predict those properties. We counted the number of THM-NP studies that evaluated the drug availability of herbal ingredients. We found that approximately half of (72/147, 49.0%) THM-NP studies evaluated the drug availability of herbal ingredients, and the majority of the studies (54/72, 75.0%) employed Obioavail and drug-likeness in combination (Figure 4B). Obioavail is an in silico model that predicts the fraction of an administered dose of a drug that reaches the systemic circulation unchanged [39]. Drug-likeness measures the structural similarity between herbal ingredients and the drugs in the Drugbank database (http://www.drugbank.ca/) using the Tanimoto coefficient [40]. They are applied to screen ADME-favorable compounds and pharmacologically suitable compounds in herbal medicines, respectively.

3.3. Methodological Trends for Constructing Compound-Target Networks

We next attempted to determine the frequency of each DTI method for constructing compound–target (C-T) networks (Note that some of the THM-NP studies combined several methods to identify DTIs. Therefore, the total frequency of the DTI method is greater than the total number of THM-NP studies). The results showed that TCMSP (47/222, 21.1%) and molecular docking (44/222, 19.8%) were the most frequently used. In addition, experimental methods, such as microarrays, were also applied (Figure 5A). It is noteworthy that DTI methods of TCMSP have existed for less than 10 years since its development but have been used most frequently in THM-NP studies [16]. More than one-third of THM-NP studies (54/147, 36.7%) combined several DTI methods for constructing C-T networks, and most of them included TCMSP (e.g., TCMSP-molecular docking and TCMSP-STITCH) (Figure 5B).

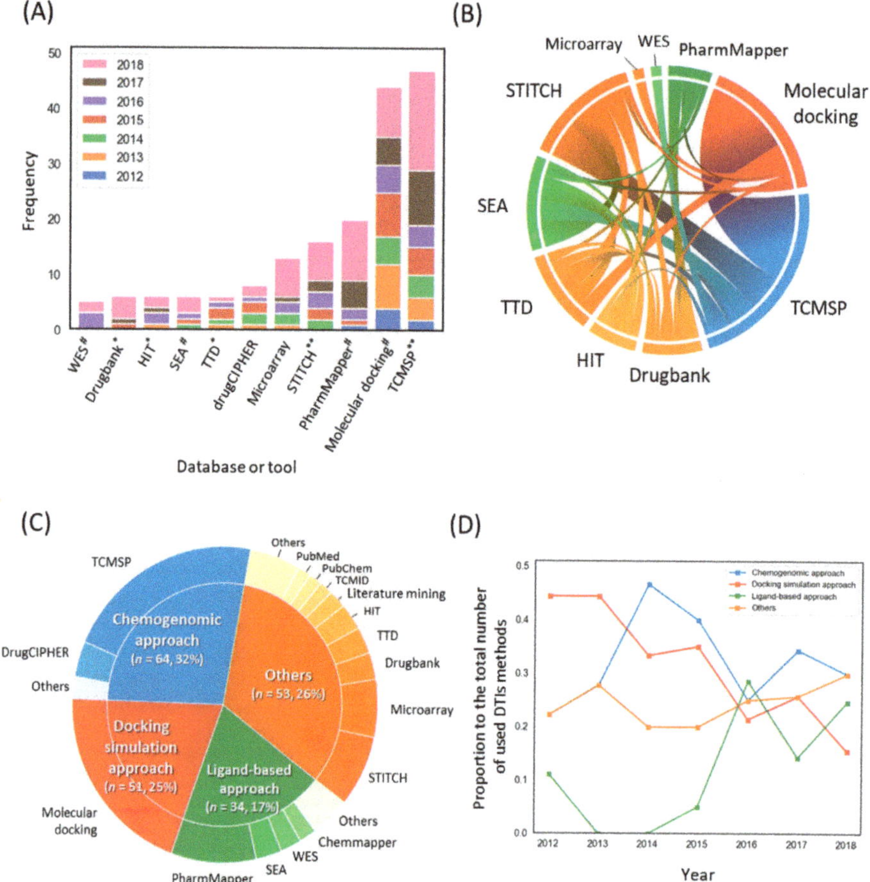

Figure 5. Methodological trends in the construction of the compound–target network. (**A**) The frequencies of databases and tools used to identify targets of herbal ingredients. * databases that provide validated drug-target interactions (DTIs); ** databases that provide validated databases that provide both validated and predicted DTIs; # Computational tools to predict DTIs. (**B**) A co-occurrence pattern of the DTI method. (**C**) Categories of DTI methods and their composition. The outer circle and inner circle represent the DTI methods and their categories, respectively. (**D**) The annual rate of the groups of DTI methods. Note that THM-NP studies in 2011 were excluded from the visualization due to their low frequency ($n = 2$). TCMSP, Traditional Chinese Medicine Systems Pharmacology Database; TTD, Therapeutic Target Database; SEA, Similarity Ensemble Approach; HIT, Herb Ingredients' Targets; WES, Weighted Ensemble Similarity.

We categorized the methods into four groups: the chemogenomic approach, docking simulation approach, ligand-based approach, and others (Figure 5C, see Materials and Methods for details). To identify trends in DTI methods, we counted the frequency of each DTI group each year (Figure 5D). In the early stage, approximately half of THM-NP studies (4/9, 44.4% and 8/18, 44.4% in 2012 and 2013, respectively) employed molecular docking simulation, but the proportion of molecular docking simulations decreased gradually and was the lowest (12/77, 15.9%) in 2018.

3.4. Methodological Trends for Target Interpretation

We identified the frequency of biomedical databases employed to analyze the biological processes, pathways, and diseases from the targets of herbal medicines (Figure 6). Most THM-NP studies (94/96, 98.0%) employed single databases to analyze biological functions and pathways, such as Gene Ontology (GO, http://geneontology.org/) [35] for biological processes or the Kyoto Encyclopedia of Genes and Genomes database (KEGG, https://www.genome.jp/kegg/) [41] for pathways. On the other hand, more than half of the studies (44/76, 57.9%) integrated several databases to analyze target-related diseases, such as the Therapeutic Target Database (TTD, http://bidd.nus.edu.sg/group/cjttd/) [30], Online Mendelian Inheritance in Man (OMIM, https://www.omim.org/) [36], and Drugbank [42].

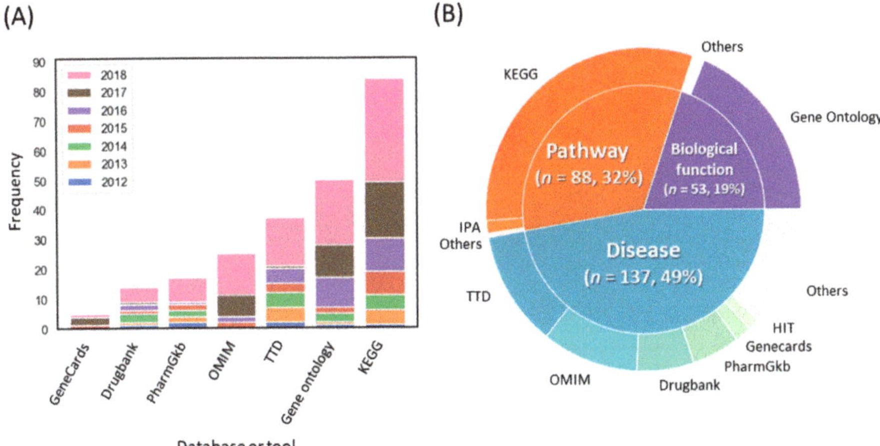

Figure 6. Trends in the biomedical databases used for target interpretation. (**A**) The frequencies of databases used for target interpretation. (**B**) Categories of databases and their composition. The outer circle and inner circle represent the databases used for target interpretation and their categories, respectively. Note that THM-NP studies in 2011 were excluded from the visualization due to their low frequency (*n* = 2). KEGG, Kyoto Encyclopedia of Genes and Genomes; TTD, Therapeutic Target Database; OMIM, Online Mendelian Inheritance in Man; IPA, Ingenuity Pathway Analysis.

3.5. Combinatorial Patterns in Methodologies of THM-NP Studies

We identified the combinatorial patterns of each step in THM-NP studies by a Sankey diagram-like representation (Figure 7). The Sankey diagram is a visualization tool used to depict quantitative information about flows from one set to another within a network [43]. The nodes in each layer (vertical lines) represent the methods of herb–compound (H-C) network construction, compound–target (C-T) network construction, and target interpretation, respectively. The edges (connected lines) between layers indicate that these methods are used together in the same THM-NP studies.

The Sankey diagram-like representation shows the diversity of databases and tools used in THM-NP studies and their combination patterns (Figure 7). We found that the nodes in the first layer (H-C network construction) tend to be connected to specific nodes in the second layer (C-T network construction), which indicates that the combinatorial pattern between the first and second layer is biased by the methods for H-C network construction. For example, TCMSP in the first layer is mainly connected to TCMSP and molecular docking in the second layer, and Traditional Chinese Medicine Integrated Database (TCMID), UPLC, and literature mining in the first layer are not linked to molecular docking in the second layer. On the other hand, the nodes in the second layer tended to be evenly connected to the nodes in the third layer (target interpretation), which indicates that the combinatorial pattern between the second layer and third layer are relatively independent of the methods for C-T network construction.

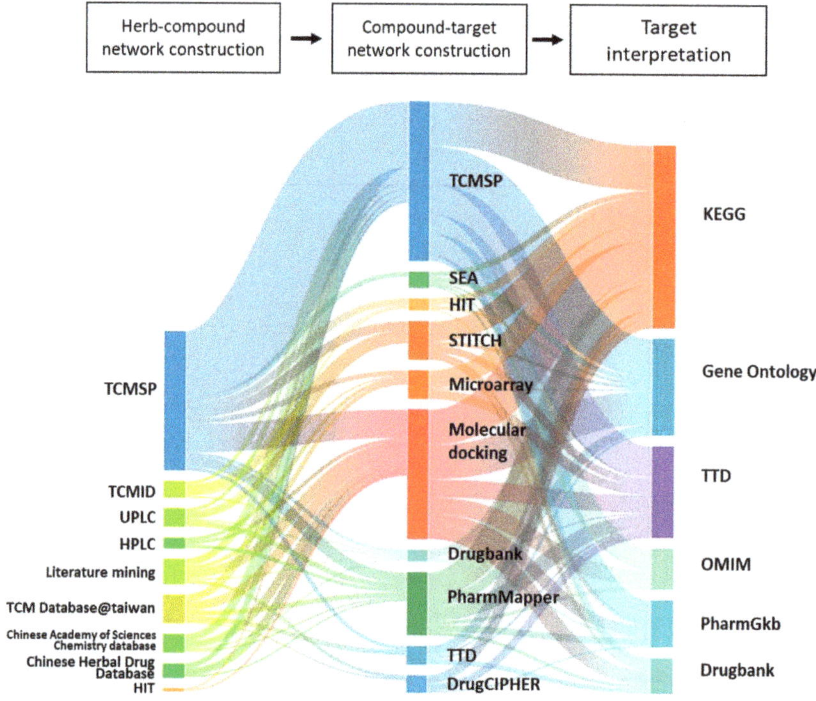

Figure 7. The combinatorial pattern of tools and databases employed in THM-NP studies. Each layer represents the process of network pharmacology analysis of herbal medicines, and the components of each layer represent the employed tools and databases. A connecting line between components indicates that the connected tools and databases are used together in the same THM-NP studies. The thickness of each connecting line indicates the frequency with which the two methodologies are used together in the THM-NP studies. TCMSP, Traditional Chinese Medicine Systems Pharmacology Database; TCMID, Traditional Chinese Medicine Integrated Database; HIT, Herb Ingredients' Targets; KEGG, Kyoto Encyclopedia of Genes and Genomes; TTD, Therapeutic Target Database; OMIM, Online Mendelian Inheritance in Man.

3.6. Co-Author Network and Affiliation Network

To further understand how the methodologies of THM-NP studies are employed among researchers, we constructed a co-author network and an affiliation network that were mapped with drug availability and DTI methods. The nodes in each network denote the author and affiliation, and the edges indicate that two of them appear on the same paper. The methods of DTI and drug-availability used by the corresponding author and affiliation are represented by the pie chart and the outline, respectively.

In the author network, Yonghua Wang ($n = 18$) and Shao Li ($n = 8$) appeared most frequently as the corresponding author (Figure 8). More than a third of corresponding authors (52/147) combined DTI methods, such as the chemogenomic approach, docking simulation approach, and ligand-based approach. Approximately half of the corresponding authors (69/147) employed evaluation tools to screen for compounds with favorable pharmacokinetic properties, and most of them (50/69) used Obioavail and drug-likeness in combination.

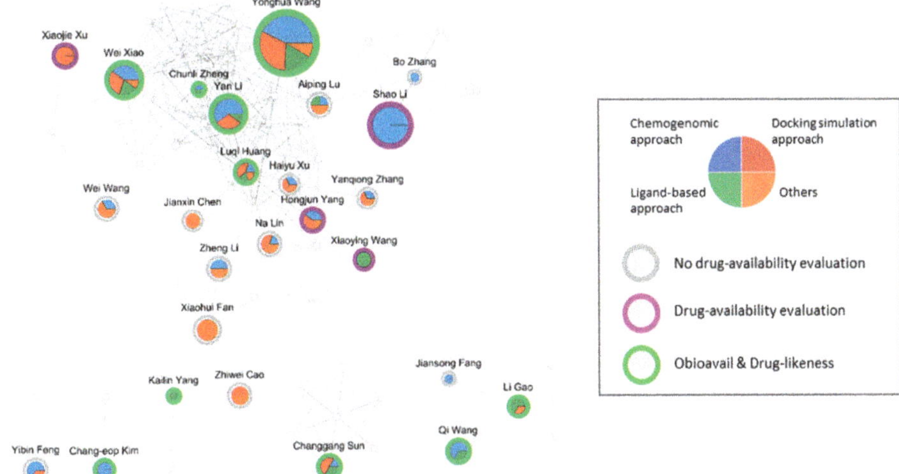

Figure 8. The co-author network of THM-NP studies. Circles represent corresponding authors, and squares represent non-corresponding authors. The size of the circles and squares reflect the number of occurrences in the THM-NP studies. Nodes that appeared fewer than three times were removed. The box to the right of the network represents the index for the pie chart and the outline of the circle.

We also constructed and visualized the affiliation network (Supplementary Figure S1). Northwest A&F University (n = 22) and China Academy of Chinese Medical Sciences (n = 17) appeared most frequently. Most affiliations combined various DTI methods (68/143) and employed drug-availability methods (86/143).

4. Discussion

In this study, we successfully identified the complex methodological trends of THM-NP research fields by analyzing the frequency of the employed methods in THM-NP studies over time and visualizing the combinatorial patterns between them. Our results showed that the number of THM-NP studies and employed databases/tools have been dramatically increased in the last decade. We also found characteristic patterns exist in combining methods of each analysis step in THM-NP studies. Finally, we showed how the diverse methods for THM-NP studies are employed and shared among researchers in the field by analyzing the co-authorship and affiliation networks.

Among the network pharmacology databases, TCMSP was the most frequently employed database for constructing herb-compound-target networks. This database was developed in 2014 and has been predominantly employed among THM-NP studies [17]. TCMSP provides a network pharmacological analysis of 499 medicinal herbs registered in the Chinese pharmacopeia along with information on ADME properties, such as bioavailability, drug-likeness, and P450, in a one-step manner. Recently, other network pharmacology databases, such as BATMAN-TCM (http://bionet.ncpsb.org/batman-tcm) and TCM-Mesh (http://mesh.tcm.microbioinformatics.org/), were developed [44,45]. They are expected to further facilitate THM-NP research fields by providing network pharmacological analysis for more than 5000 medicinal herbs.

Among the processes used in THM-NP research, the methodology for constructing a C-T network has shown the greatest change over time. In the early stages of THM-NP research, DTI methods for identifying targets of herbal ingredients relied on molecular docking simulations, which require high computational resources (Figure 5D). With the advancement of DTI prediction methods, several methodologies have been applied to THM-NP research fields that can efficiently identify the multiple targets of multiple ingredients in herbal medicines. First, the development and application of machine learning techniques and network-based methods enabled large-scale prediction of the targets of herbal medicines in terms of efficient computational costs [17,44,46,47]. Second, increased computational power made it possible to comprehensively explore potential targets of the compounds using the pharmacophore model [48]. Last, the development of databases that integrate disparate data sources provides comprehensive and high-quality information on DTIs [49]. Furthermore, recently developed DTI prediction models based on deep learning showed higher performance than other state-of-the-art models [50,51]. Such innovation in the machine learning field is expected to facilitate the development of the THM-NP research field.

To conduct THM-NP research, various tools and databases are combined in each phase of a study (Figure 7). We found that the methods for H-C network construction tend to be linked with specific methods for C-T network construction. This result indicates that there might be a preferred combinatorial pattern when choosing the methods for constructing H-C-T network. On the other hand, the combinatorial patterns between the methods for C-T network construction and target interpretation are relatively independent when compared to the previous step. This result indicates that the databases used for target interpretation tend to be chosen for the purpose of the study, while each method was preferred by different researchers in the previous steps. Further studies are needed to evaluate the reliability of network pharmacological analysis by evaluating the consistency between predicted results according to the methodologies of THM-NP studies.

There are some limitations to our study that should be noted. First, we identified potentially relevant articles using combined keywords consisting of THM-related terms and NP-related terms. Although we carefully selected search terms, we cannot guarantee that our search strategy can fully identify THM-NP studies. Second, we found potentially relevant articles only in PubMed. It is one of the largest electronic database in the world. However, there are other databases which may include other potential THM-NP studies, such as Embase, China Knowledge Resource Integrated Database (CNKI), Research Information Sharing Service (RISS), and Japan Science Technology Information Aggregator (J-stage). Last, we limited the search range of our study to English literature, which might have introduced some bias. In spite of these limitations, our results will help to improve the understanding of the methodological trends of the THM-NP research fields.

5. Conclusions

In conclusion, we investigated the methodological trends in THM-NP studies. Our results provide researchers with the current status of which methodologies are used in THM-NP studies and how they are applied.

Supplementary Materials: The following are available online at http://www.mdpi.com/2218-273X/9/8/362/s1. Supplementary Figure S1. The affiliation network of THM-NP studies; Supplementary Table S1. Selected THM-NP studies; Supplementary Table S2. Providing information and their types of methods employed in THM-NP studies.

Author Contributions: Conceptualization, Y.-S.K. and C.-E.K.; Data curation, W.-Y.L.; Formal analysis, W.-Y.L. and C.-Y.L.; Visualization, W.-Y.L.; Writing—original draft, W.-Y.L.; Writing—review and editing, C.-Y.L., Y.-S.K., and C.-E.K.

Funding: This work was supported by the National Research Foundation of Korea (NRF) grant funded by the Korea government (MSIP; Ministry of Science, ICT & Future Planning) (No. 2017-0129). This work was also supported by the Gachon University research fund of 2019 (GCU-2019-0025).

Conflicts of Interest: The authors declare no conflict of interest. The funders had no role in the design of the study; in the collection, analyses, or interpretation of data; in the writing of the manuscript; or in the decision to publish the results.

References

1. Kong, D.X.; Li, X.J.; Zhang, H.Y. Where is the hope for drug discovery? Let history tell the future. *Drug Discov. Today* **2009**, *14*, 115–119. [CrossRef] [PubMed]
2. Verpoorte, R.; Crommelin, D.; Danhof, M.; Gilissen, L.J.; Schuitmaker, H.; van der Greef, J.; Witkamp, R.F. Commentary: "A systems view on the future of medicine: Inspiration from Chinese medicine?". *J. Ethnopharmacol.* **2009**, *121*, 479–481. [CrossRef] [PubMed]
3. Zheng, J.; Wu, M.; Wang, H.; Li, S.; Wang, X.; Li, Y.; Wang, D.; Li, S. Network Pharmacology to Unveil the Biological Basis of Health-Strengthening Herbal Medicine in Cancer Treatment. *Cancers (Basel)* **2018**, *10*, 461. [CrossRef] [PubMed]
4. Ma, X.H.; Zheng, C.J.; Han, L.Y.; Xie, B.; Jia, J.; Cao, Z.W.; Li, Y.X.; Chen, Y.Z. Synergistic therapeutic actions of herbal ingredients and their mechanisms from molecular interaction and network perspectives. *Drug Discov. Today* **2009**, *14*, 579–588. [CrossRef] [PubMed]
5. Berg, E.L. Systems biology in drug discovery and development. *Drug Discov. Today* **2014**, *19*, 113–125. [CrossRef] [PubMed]
6. Barabási, A.-L.; Gulbahce, N.; Loscalzo, J. Network medicine: A network-based approach to human disease. *Nat. Rev. Genet.* **2011**, *12*, 56–68. [CrossRef] [PubMed]
7. Hopkins, A.L. Network pharmacology: The next paradigm in drug discovery. *Nat. Chem. Biol.* **2008**, *4*, 682–690. [CrossRef]
8. Van der Greef, J. Perspective: All systems go. *Nature* **2011**, *480*, S87. [CrossRef]
9. Li, S.; Fan, T.-P.; Jia, W.; Lu, A.; Zhang, W. Network Pharmacology in Traditional Chinese Medicine. *Evid. Based Complement. Altern. Med.* **2014**, *2014*, 138460. [CrossRef]
10. Li, S.; Zhang, B. Traditional Chinese medicine network pharmacology: Theory, methodology and application. *Chin. J. Nat. Med.* **2013**, *11*, 110–120. [CrossRef]
11. Zhang, R.; Zhu, X.; Bai, H.; Ning, K. Network Pharmacology Databases for Traditional Chinese Medicine: Review and Assessment. *Front. Pharmacol.* **2019**, *10*, 123. [CrossRef] [PubMed]
12. Yang, M.; Chen, J.-L.; Xu, L.-W.; Ji, G. Navigating Traditional Chinese Medicine Network Pharmacology and Computational Tools. *Evid. Based Complement. Altern. Med.* **2013**, *2013*, 731969. [CrossRef] [PubMed]
13. Kibble, M.; Saarinen, N.; Tang, J.; Wennerberg, K.; Mäkelä, S.; Aittokallio, T. Network pharmacology applications to map the unexplored target space and therapeutic potential of natural products. *Nat. Prod. Rep.* **2015**, *32*, 1249–1266. [CrossRef] [PubMed]
14. Hao, D.C.; Xiao, P.G. Network pharmacology: A rosetta stone for traditional chinese medicine. *Drug Dev. Res.* **2014**, *75*, 299–312. [CrossRef] [PubMed]
15. Liu, Y.; Ai, N.; Keys, A.; Fan, X.; Chen, M. Network Pharmacology for Traditional Chinese Medicine Research: Methodologies and Applications. *Chin. Herb. Med.* **2015**, *7*, 18–26. [CrossRef]
16. Yu, H.; Chen, J.; Xu, X.; Li, Y.; Zhao, H.; Fang, Y.; Li, X.; Zhou, W.; Wang, W.; Wang, Y. A systematic prediction of multiple drug-target interactions from chemical, genomic, and pharmacological data. *PLoS ONE* **2012**, *7*, e37608. [CrossRef] [PubMed]
17. Ru, J.; Li, P.; Wang, J.; Zhou, W.; Li, B.; Huang, C.; Li, P.; Guo, Z.; Tao, W.; Yang, Y.; et al. TCMSP: A database of systems pharmacology for drug discovery from herbal medicines. *J. Cheminform.* **2014**, *6*, 13. [CrossRef]
18. Katsila, T.; Spyroulias, G.A.; Patrinos, G.P.; Matsoukas, M.T. Computational approaches in target identification and drug discovery. *Comput. Struct. Biotechnol. J.* **2016**, *14*, 177–184. [CrossRef]
19. Macalino, S.J.Y.; Gosu, V.; Hong, S.; Choi, S. Role of computer-aided drug design in modern drug discovery. *Arch. Pharm. Res.* **2015**, *38*, 1686–1701. [CrossRef]
20. Yang, S.Y. Pharmacophore modeling and applications in drug discovery: Challenges and recent advances. *Drug Discov. Today* **2010**, *15*, 444–450. [CrossRef]
21. Ding, H.; Takigawa, I.; Mamitsuka, H.; Zhu, S. Similarity-basedmachine learning methods for predicting drug-target interactions: A brief review. *Brief. Bioinform.* **2013**, *15*, 734–747. [CrossRef] [PubMed]
22. Lengauer, T.; Rarey, M. Computational methods for biomolecular docking. *Curr. Opin. Struct. Biol.* **1996**, *6*, 402–406. [CrossRef]
23. Cherfils, J.; Duquerroy, S.; Janin, J. Protein-protein recognition analyzed by docking simulation. *Proteins Struct. Funct. Bioinform.* **1991**, *11*, 271–280. [CrossRef] [PubMed]

24. Vidal, D.; Garcia-Serna, R.; Mestres, J. Ligand-Based Approaches to In Silico Pharmacology. *Methods Mol. Biol.* **2010**, *672*, 489–502.

25. Shannon, P.; Markiel, A.; Ozier, O.; Baliga, N.S.; Wang, J.T.; Ramage, D.; Amin, N.; Schwikowski, B.; Ideker, T. Cytoscape: A software environment for integrated models of biomolecular interaction networks. *Genome Res.* **2003**, *13*, 2498–2504. [CrossRef]

26. Xue, R.; Fang, Z.; Zhang, M.; Yi, Z.; Wen, C.; Shi, T. TCMID: Traditional Chinese medicine integrative database for herb molecular mechanism analysis. *Nucleic Acids Res.* **2013**, *41*, 1089–1095. [CrossRef]

27. Chen, C.Y.C. TCM Database@Taiwan: The world's largest traditional Chinese medicine database for drug screening In Silico. *PLoS ONE* **2011**, *6*, e15939. [CrossRef]

28. Liu, X.; Ouyang, S.; Yu, B.; Liu, Y.; Huang, K.; Gong, J.; Zheng, S.; Li, Z.; Li, H.; Jiang, H. PharmMapper server: A web server for potential drug target identification using pharmacophore mapping approach. *Nucleic Acids Res.* **2010**, *38*, 5–7. [CrossRef]

29. Kuhn, M.; von Mering, C.; Campillos, M.; Jensen, L.J.; Bork, P. STITCH: Interaction networks of chemicals and proteins. *Nucleic Acids Res.* **2008**, *36*, 684–688. [CrossRef]

30. Chen, X.; Ji, Z.L.; Chen, Y.Z. TTD: Therapeutic Target Database. *Nucleic Acids Res.* **2002**, *30*, 412–415. [CrossRef]

31. Keiser, M.J.; Roth, B.L.; Armbruster, B.N.; Ernsberger, P.; Irwin, J.J.; Shoichet, B.K. Relating protein pharmacology by ligand chemistry. *Nat. Biotechnol.* **2007**, *25*, 197–206. [CrossRef]

32. Ye, H.; Ye, L.; Kang, H.; Zhang, D.; Tao, L.; Tang, K.; Liu, X.; Zhu, R.; Liu, Q.; Chen, Y.Z.; et al. HIT: Linking herbal active ingredients to targets. *Nucleic Acids Res.* **2011**, *39*, 1055–1059. [CrossRef]

33. Wishart, D.S. DrugBank: A comprehensive resource for in silico drug discovery and exploration. *Nucleic Acids Res.* **2005**, *34*, D668–D672. [CrossRef]

34. Ogata, H.; Goto, S.; Sato, K.; Fujibuchi, W.; Bono, H.; Kanehisa, M. KEGG: Kyoto encyclopedia of genes and genomes. *Nucleic Acids Res.* **1999**, *28*, 27–30. [CrossRef]

35. Harris, M.A.; Clark, J.; Ireland, A.; Lomax, J.; Ashburner, M.; Foulger, R.; Eilbeck, K.; Lewis, S.; Marshall, B.; Mungall, C.; et al. The Gene Ontology (GO) database and informatics resource. *Nucleic Acids Res.* **2004**, *32*, D258–D261.

36. Hamosh, A.; Scott, A.F.; Amberger, J.S.; Bocchini, C.A.; McKusick, V.A. Online Mendelian Inheritance in Man (OMIM), a knowledgebase of human genes and genetic disorders. *Nucleic Acids Res.* **2005**, *33*, D514–D517. [CrossRef]

37. Hewett, M.; Oliver, D.E.; Rubin, D.L.; Easton, K.L.; Stuart, J.M.; Altman, R.B.; Klein, T.E. PharmGKB: The pharmacogenetics knowledge base. *Nucleic Acids Res.* **2002**, *30*, 163–165. [CrossRef]

38. Safran, M.; Solomon, I.; Shmueli, O.; Lapidot, M.; Shen-Orr, S.; Adato, A.; Ben-Dor, U.; Esterman, N.; Rosen, N.; Peter, I.; et al. GeneCards^{TM} 2002: Towards a complete, object-oriented, human gene compendium. *Bioinformatics* **2002**, *18*, 1542–1543. [CrossRef]

39. Wang, X.; Xu, X.; Tao, W.; Li, Y.; Wang, Y.; Yang, L. A Systems Biology Approach to Uncovering Pharmacological Synergy in Herbal Medicines with Applications to Cardiovascular Disease. *Evid. Based Complement. Altern. Med.* **2012**, *2012*, 519031. [CrossRef]

40. Vistoli, G.; Pedretti, A.; Testa, B. Assessing drug-likeness—What are we missing? *Drug Discov. Today* **2008**, *13*, 285–294. [CrossRef]

41. Kanehisa, M.; Furumichi, M.; Tanabe, M.; Sato, Y.; Morishima, K. KEGG: New perspectives on genomes, pathways, diseases and drugs. *Nucleic Acids Res.* **2017**, *45*, D353–D361. [CrossRef]

42. Wishart, D.S.; Feunang, Y.D.; Guo, A.C.; Lo, E.J.; Marcu, A.; Grant, J.R.; Sajed, T.; Johnson, D.; Li, C.; Sayeeda, Z.; et al. DrugBank 5.0: A major update to the DrugBank database for 2018. *Nucleic Acids Res.* **2018**, *46*, D1074–D1082. [CrossRef]

43. Soundararajan, K.; Ho, H.K.; Su, B. Sankey diagram framework for energy and exergy flows. *Appl. Energy* **2014**, *136*, 1035–1042. [CrossRef]

44. Liu, Z.; Guo, F.; Wang, Y.; Li, C.; Zhang, X.; Li, H.; Diao, L.; Gu, J.; Wang, W.; Li, D.; et al. BATMAN-TCM: A Bioinformatics Analysis Tool for Molecular mechANism of Traditional Chinese Medicine. *Sci. Rep.* **2016**, *6*, 21146. [CrossRef]

45. Zhang, R.Z.; Yu, S.J.; Bai, H.; Ning, K. TCM-Mesh: The database and analytical system for network pharmacology analysis for TCM preparations. *Sci. Rep.* **2017**, *7*, 2821. [CrossRef]

46. Zhao, S.; Li, S. Network-based relating pharmacological and genomic spaces for drug target identification. *PLoS ONE* **2010**, *5*, e11764. [CrossRef]

47. Wu, Z.; Lu, W.; Wu, D.; Luo, A.; Bian, H.; Li, J.; Li, W.; Liu, G.; Huang, J.; Cheng, F.; et al. In silico prediction of chemical mechanism of action via an improved network-based inference method. *Br. J. Pharmacol.* **2016**, *173*, 3372–3385. [CrossRef]

48. Wang, X.; Shen, Y.; Wang, S.; Li, S.; Zhang, W.; Liu, X.; Lai, L.; Pei, J.; Li, H. PharmMapper 2017 update: A web server for potential drug target identification with a comprehensive target pharmacophore database. *Nucleic Acids Res.* **2017**, *45*, W356–W360. [CrossRef]

49. Szklarczyk, D.; Santos, A.; Von Mering, C.; Jensen, L.J.; Bork, P.; Kuhn, M. STITCH 5: Augmenting protein-chemical interaction networks with tissue and affinity data. *Nucleic Acids Res.* **2016**, *44*, D380–D384. [CrossRef]

50. Bahi, M.; Batouche, M. Deep semi-supervised learning for DTI prediction using large datasets and H2O-spark platform. In Proceedings of the 2018 International Conference on Intelligent Systems and Computer Vision (ISCV), Fez, Morocco, 2–4 April 2018; pp. 1–7.

51. Wen, M.; Zhang, Z.; Niu, S.; Sha, H.; Yang, R.; Yun, Y.; Lu, H. Deep-Learning-Based Drug-Target Interaction Prediction. *J. Proteome Res.* **2017**, *16*, 1401–1409. [CrossRef]

MDPI
St. Alban-Anlage 66
4052 Basel
Switzerland
Tel. +41 61 683 77 34
Fax +41 61 302 89 18
www.mdpi.com

Biomolecules Editorial Office
E-mail: biomolecules@mdpi.com
www.mdpi.com/journal/biomolecules